4th International Symposium on LAPAN-IPB Satellite for Food Security and Environmental Monitoring (LISAT-FSEM 2017)

IOP Conference Series: Earth and Environmental Science
Volume 149

Bogor, Indonesia
9 - 11 October 2017

Editors:

Syartinilia
Prita Ayu Permatasari
Arif Kurnia Wijayanto
Vina Nurul Husna
Sri Malahayati Yusuf
Irza Arnita Nur
Luisa Febrina Amalo

ISBN: 978-1-5108-6310-1
ISSN: 1755-1307

Printed from e-media with permission by:

Curran Associates, Inc.
57 Morehouse Lane
Red Hook, NY 12571

Some format issues inherent in the e-media version may also appear in this print version.

Printed by Curran Associates, Inc. (2018)

For permission requests, please contact the Institute of Physics
at the address below.

Institute of Physics
Dirac House, Temple Back
Bristol BS1 6BE UK

Phone: 44 1 17 929 7481
Fax: 44 1 17 920 0979

techtracking@iop.org

Additional copies of this publication are available from:

Curran Associates, Inc.
57 Morehouse Lane
Red Hook, NY 12571 USA
Phone: 845-758-0400
Fax: 845-758-2633
Email: curran@proceedings.com
Web: www.proceedings.com

Table of contents

Volume 149

The 4th International Symposium on LAPAN-IPB Satellite for Food Security and Environmental Monitoring

9–11 October 2017, Bogor, Indonesia

Accepted papers received: 1 May 2018
Published online: 16 May 2018

Preface

Papers

Agriculture

Marine and Fisheries

Satellite Technology and Data Processing

IOP Conf. Series: Earth and Environmental Science **149** (2018) 011001 doi:10.1088/1755-1315/149/1/011001

Foreword

Bogor Agricultural University (IPB) and National Institute of Aeronautics and Space of Indonesia (LAPAN) have committed to work together in developing microsatellite namely LAPAN-A3/ LAPAN-IPB Satellite (LISAT) in order to support the food security and environmental monitoring program especially in Indonesia. This satellite has been launched on June 22, 2016 aboard the India-owned Polar Satellite Launch Vehicle (PSLV-C34) in Sriharikota, India. Moreover, the series of international symposium is a part of the agenda that will support the LISAT satellite programs and it is also intended to support the research related to agriculture, forestry, marine and fisheries, and climate change.

The 4th International Symposium on LAPAN-IPB Satellite for Food Security and Environmental Monitoring 2017 (LISAT-FSEM 2017), to which this issue of IOP Conference Series: Earth and Environmental Science (EES) is dedicated, was held on October 9-11, 2017 at the IPB International Convention Center, Bogor, Indonesia. This symposium is jointly organized by the Bogor Agricultural University (IPB), Center for Environmental Research, Research and Community Services Institution, IPB (PPLH-LPPM, IPB), Directorate of Research and Innovation, IPB (DRI-IPB) and the National Institute of Aeronautics and Space of Indonesia (LAPAN), Indonesia. The 4th LISAT International Symposium theme emphasized on application of satellite remote sensing data for supporting the food security and environmental monitoring especially on the field of agriculture, marine and fisheries, environmental management, sattelite technology and data processing.

This symposium was being participated by around 100 participants that come from various government institutions such as: Indonesian Institute of Science (LIPI), Geospatial Information Agency (BIG), PT Hatfield Indonesia, and some local government's agencies; from diverse universities such as: Bogor Agricultural University, Gadjah Mada Univeriesty, Bandung Institute of Technology (ITB), Udayana University, University of Indonesia, University of Trunojoyo Madura, Riau University, Sriwijaya University, Surya University, Padjajaran University, Hokkaido University, Borneo University, University of Maryland, The University of Tokyo, Saga Univeristy, and University Malaysia Sabah.

The committee has received 170 research titles for both category of oral and poster presentation, which come from various scientists and students from 5 countries and 33 institutions. These research papers have been reviewed by the scientific committee of LISAT-FSEM 2017. Regarding the peer review result, 73 papers have been selected to be published in this issue. Indeed, the result from this symposium will provide useful information from satellite technology in order to support the government's programs for food security and environmental monitoring. Moreover, this symposium will strengthen the collaboration between multi-institutions and develop a better cooperation among the different research groups, either for basic research or technology developments.

We wish to thank the authors who have contributed to yield a high scientific standard to this issue. We are also grateful to IOP Publishing, especially Stephanie Gill (IOP Conference Commissioning Editor), for allowing us to share some of the symposium results with a large audience of colleagues through the IOP Conference Series: Earth and Environmental Science. We would like to thank IPB World Class University Program for giving us opportunity to get the additional funding for our

IOP Conf. Series: Earth and Environmental Science **149** (2018) 011001 doi:10.1088/1755-1315/149/1/011001

international symposium. This symposium was supported also by sponsorship partners such as Geospatial Information Agency (BIG) and PT Leica Geosystem Indonesia.

Syartinilia
Editor-in-Chief
Proceeding LISAT-FSEM 2017

IOP Conf. Series: Earth and Environmental Science **149** (2018) 011001 doi:10.1088/1755-1315/149/1/011001

Editorial Committee

Scientific Committee

Assistant Editors

IOP Conf. Series: Earth and Environmental Science **149** (2018) 011001 doi:10.1088/1755-1315/149/1/011001

Sponsorship Partners

BADAN INFORMASI
GEOSPASIAL

Leica
Geosystems

IOP Conf. Series: Earth and Environmental Science **149** (2018) 011001 doi:10.1088/1755-1315/149/1/011001

IOP Conf. Series: Earth and Environmental Science **149** (2018) 011001 doi:10.1088/1755-1315/149/1/011001

Peer review statement

All papers published in this volume of *IOP Conference Series: Earth and Environmental Science* have been peer reviewed through processes administered by the proceedings Editors. Reviews were conducted by expert referees to the professional and scientific standards expected of a proceedings journal published by IOP Publishing.

IOP Conf. Series: Earth and Environmental Science **149** (2018) 012001 doi:10.1088/1755-1315/149/1/012001

Selection of vegetation indices for mapping the sugarcane condition around the oil and gas field of North West Java Basin, Indonesia

Tri Muji Susantoro[1,2,3], Ketut Wikantika[2,3,4], Asep Saepuloh[2,4], Agus Handoyo Harsolumakso[3]
[1]Research and Development Center for Oil and Gas Technology"LEMIGAS" Indonesia
[2]Center for Remote Sensing, Bandung Institute of Technology (ITB), Indonesia
[3]Faculty of Earth Sciences and Technology, Bandung Institute of Technology (ITB), Indonesia
[4]ForMIND Institute (Indonesian Young Researcher Forum)

E-mail: trimuji_s@yahoo.com

Abstract: Selection of vegetation indices in plant mapping is needed to provide the best information of plant conditions. The methods used in this research are the standard deviation and the linear regression. This research tried to determine the vegetation indices used for mapping the sugarcane conditions around oil and gas fields. The data used in this study is Landsat 8 OLI/TIRS. The standard deviation analysis on the 23 vegetation indices with 27 samples has resulted in the six highest standard deviations of vegetation indices, termed as GRVI, SR, NLI, SIPI, GEMI and LAI. The standard deviation values are 0.47; 0.43; 0.30; 0.17; 0.16 and 0.13. Regression correlation analysis on the 23 vegetation indices with 280 samples has resulted in the six vegetation indices, termed as NDVI, ENDVI, GDVI, VARI, LAI and SIPI. This was performed based on regression correlation with the lowest value R^2 than 0,8. The combined analysis of the standard deviation and the regression correlation has obtained the five vegetation indices, termed as NDVI, ENDVI, GDVI, LAI and SIPI. The results of the analysis of both methods show that a combination of two methods needs to be done to produce a good analysis of sugarcane conditions. It has been clarified through field surveys and showed good results for the prediction of microseepages.

1. Introduction

The vegetation indices is an important algorithm used to extract the information of vegetation condition [1]. Many vegetation indices have been published but only small of them have the important biophysical fundamental for plant monitoring and have been tested systematically [2]. Vegetation indices are employed to enhance the vegetation conditions and they represent a single value for converting the reflectance spectrum for measuring vegetation characteristics [2]. Vegetation characteristics, in general, can be distinguished into three categories: structure, biochemistry and plant physiology or stress condition [2].

For the past two decades, the normalized difference vegetation index (NDVI) is used as the important algorithm to analyze and map the spatial distribution of vegetation. Generally, NDVI is calculating based on the ratio of the near-infrared reflectance and red reflectance [4]. NDVI is the popular index that corresponds to the photosynthetic rate and the capability to absorb energy in the vascular plants [5]. NDVI also is related to the plant cover linearly [6] and has a significant positive relationship to the Landscape function analysis (LFA) [7].

Similar to NDVI, many vegetation indices were developed to analyze the vegetation characteristics successfully. These indices include EVI which was designed to optimize the vegetation signal in regions of leaf area index [8], GARI is demonstrated to sense the concentration of chlorophyll, to measure the rate of photosynthesis and to monitor plant stress [9], GNDVI is calculating based on the ratio of near-infrared and green spectrum. This index was found more sensitive to the chlorophyll concentration in wide range variation and enabled precise assessment of pigment concentration [10], GRVI tested showed significant linear relationships with biomass [11]. The purpose of this study is to select the related vegetation indices for mapping the sugarcane conditions around the oil and gas field. The results of the vegetation indices selection are expected to identify vegetation anomalous due to the impact of hydrocarbon seepages.

2. Data and Methods

2.1 Study sites

The study site is located at Jatitujuh sugarcane plantation in Indramayu and Majalengka Regency. Jatitujuh sugarcane plantation area is the largest sugar producer in West Java Province. Jatitujuh plantation has been operating since 1975 based on Agriculture Minister Decree No. 795/Mentan/VI/1975 regarding Principle License of Establishment of Jatitujuh Sugar Company [12]. This area also covers active oil and gas fields operated by Pertamina. Several wells have been drilled since 1979 [13]. Hence we choose this site for our study. The study site is located approximately ±32 km from Indramayu and Majalengka [14]. The study site can be seen in figure 1.

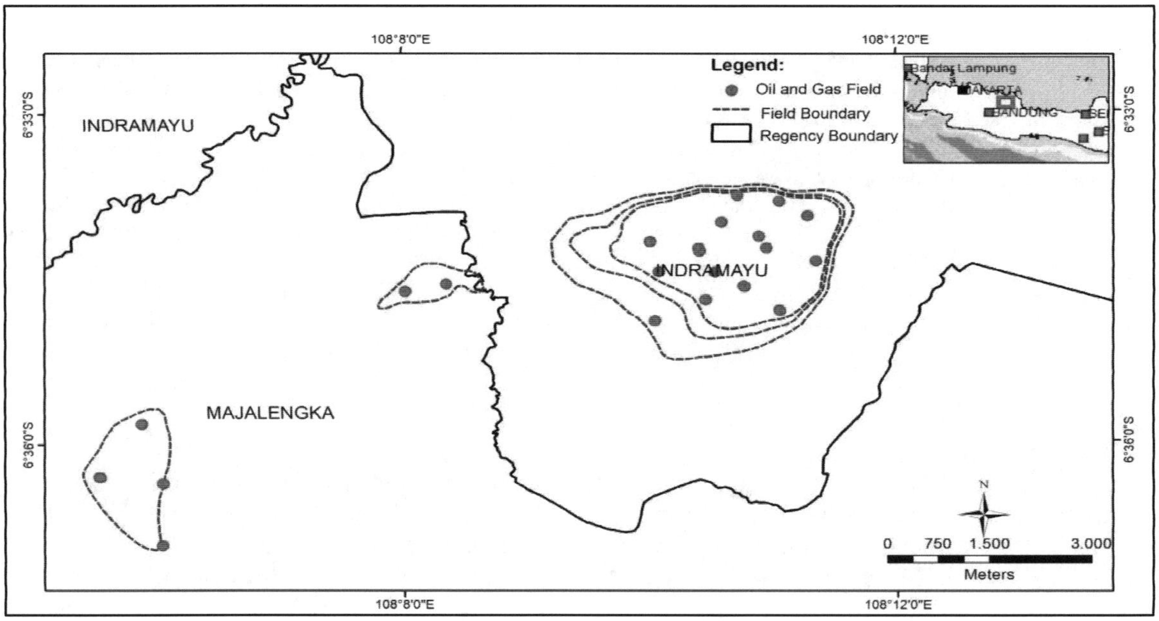

Figure 1. Study site map overlay with oil and gas wells in red dot and boundary of oil and gas field in brown lines.

2.2 Data and Processing

To choose and select the vegetation indices for mapping of sugarcane conditions around oil and gas field, we employed Landsat 8 with an acquisition date on September 25, 2015, and path/row 121/165.

Landsat 8 started recording of the earth surface changes after Landsat 5 discontinued and Landsat 7 has the SLC problem [15]. Comparing with Landsat 7 ETM+, Landsat 8 has higher near-infrared band values in the vegetation land cover and lower values in non-vegetation cover. Landsat 8 also has lower values for shortwave infrared (2.11-2.29) and had higher values in shortwave infrared (1.57-1.65) in all land cover types [17].

Radiometric calibration is a multi-step process to convert 8-bit digital numbers to satellite reflectance [18]. The atmospheric radiometric correction is conducted by using the FLAASH. This correction is based on the MODTRAN4 model. It is one of the most accurate preprocessing for remote sensing data in atmospheric radiometric correction [19]. The geometric correction is conducted using an image to image correction. The process of registration converts each pixel of Landsat 8 into a new coordinate system. Landsat 8 is corrected and registered using Universal Transverse Mercator (UTM) projection Zone 49 South. The process of registration is following the cubic convolution technique. The root means squares (RMS) error of this registration is 0.345 pixel.

2.3 Analysis

For the purpose of this research, 23 vegetation indices were evaluated. The spectral band combination and vegetation indices used in the analysis are in table 1. These vegetation indices have been used extensively in the vegetation properties studies. The vegetation indices were evaluated using the standard deviation and the linear regression to select the vegetation indices. The analysis to select the vegetation indices was conducted to map sugarcane conditions in three stages. In the first stage, the vegetation indices were selected using standard deviation. The standard deviation is a measure of spreading a set of data from its mean. In this research, we used 27 samples that had similar vegetation features near the oil and gas field (figure 2). The high values indicated that the data points are spread out over a wider range of values. Its mean there was an anomaly in the vegetation.

Table 1. The spectral band combination and vegetation indices used in the analysis.

Vegetation Indices	Formulas	Ref.
Enhanced Vegetation Index	EVI = 2.5 x (NIR-Red)/(NIR + (6 x Red)-(7.5 x Blue) + 1)	[8]
Green Difference Vegetation Index	GDVI = NIR - Green	[11]
Green Ratio Vegetation Index	GRVI = NIR/Green	[11]
Atmospherically Resistant Vegetation Index	ARVI = (NIR - (Red – γ(Blue-Red)))/(NIR + (Red – γ(Blue-Red)))	[20]
Difference Vegetation Index	DVI = NIR - Red	[21]
Global Environmental Monitoring Index	$\text{GEMI} = \text{eta x } (1 - 0.25 \text{ x eta}) + \frac{(\text{Red} - 0.125)}{(1 - \text{Red})}$ $\text{eta} = \frac{2(\text{NIR}^2 - \text{Red}^2) + 1.5 \text{ x NIR} + 0.5 \text{ x Red}}{\text{NIR} + \text{Red} + 0.5}$	[22]
Green Normalized Difference Vegetation Index	GNDVI = (NIR- Green)/NIR+Green)	[10]
Green Atmospherically Resistant Index	GARI = (NIR-[Green-γ(Blue-Red)])/(NIR+[Green-γ(Blue-Red)])	[10]
Green Vegetation Index	GVI = (-0.2848xTM_1) + (-0.2435xTM_2) + (-0.5436xTM_3) + (0.7243xTM_4) + (0.0840xTM_5) + (-0.18xTM_7)	[23]
Infrared Percentage Vegetation Index	IPVI = NIR/(NIR+Red)	[24]

Table 1. *Continued*

Vegetation Indices	Formulas	Ref.
Leaf Area Index	LAI = 3.618 x EVI – 0.118	[25]
Modified Non Linear Index	MNLI = ((NIR^2-Red) x(1+L))/(NIR^2+Red+L)	[26]
Modified Simple Ratio	MSR = ((NIR/Red)-1)/(√(NIR/Red)+1)	[27]
Non Linear Index	NLI = (NIR^2 – Red)/ (NIR^2 + Red)	[28]
Normalized Difference Vegetation Index	NDVI = (NIR-Red)/(NIR + Red)	[29]
Optimized Soil Adjusted Vegetation Index	OSAVI = (NIR-Red)/(NIR+Red+0.16)	[30]
Renormalized Difference Vegetation Index	RDVI = (NIR-Red)/√(NIR+Red)	[31]
Soil Adjusted Vegetation Index	SAVI = (1.5 x (NIR-Red))/(NIR+Red+0.5)	[32]
Simple Ratio	SR = NIR/Red	[33]
Visible Atmospherically Resistant Index	VARI = (Green – Red)/(Green + Red - Blue)	[34]
Transformed Difference Vegetation Index	TDVI = √(0.5 + (NIR-Red)/(NIR+Red))	[35]
Structurally Independent Pigment Index	SIPI = (NIR–Blue)/(NIR–Red)	[36]
Enhanced Normalized Difference Vegetation Index	ENDVI = ((NIR+Green)-(2xBlue))/((NIR+Green)+(2xBlue))	[37]

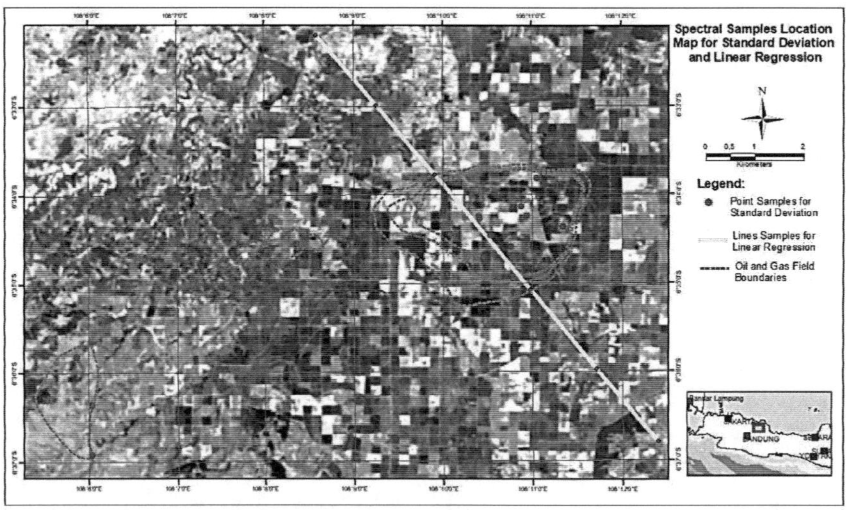

Figure 2. Samples of vegetation indices values for standard deviation analysis in red dot and linear regression analysis in yellow lines overlaid with Landsat 8 OLI/TIRS 432 RGB.

In the second stage, the method employed to select the vegetation indices is the linear regression. In this research, we used 280 samples to create the linear regression. The samples on the image of vegetation indices are taken out continuously cross the oil and gas field with the south-north direction (figure 2). The correlation analysis between vegetation indices is done by the coefficient of determination (R^2) method on linear regression results of each vegetation indices. The coefficient of determination (R^2) is a statistical calculation showing of how well the regression line approaches the real data points. The limit of R^2 in this research is 0.8. The correlation results between vegetation indices are expected to obtain the indices used for this research.

In the third stage, the final analysis is done by combining the standard deviation analysis results and the linear regression correlation to choose the vegetation indices employed. This analysis is done based on the role of each index to map the effect of oil and gas field on sugarcane conditions. The result of its combination is expected to obtain the vegetation indices that can map the sugarcane plantation conditions due to the influence of oil and gas fields.

3. Results and Discussion

The results of standard deviation analysis of 23 vegetation indices algorithm showed the values range from 0.03 to 0.47 (figure 3). The standard deviation is a description of the spread of data and how widely it spreads from the mean. Smaller standard deviations show the data clustered near the mean. A larger one indicates that the data are scattered. This indicates that the vegetation indices can differentiate vegetation conditions in more detail. So it can illustrate the health of vegetation and the changes that occur in leaf structure. Based on these results, the vegetation indices used for the next analysis are GRVI, SR, NLI, LAI, SIPI and GEMI. The standard deviation values are 0.47, 0.43, 0.30, 0.18, 0.17 and 0.16. The vegetation indices are used with large standard deviation values as to map anomalous sugarcane conditions that are influenced by the oil and gas seepages.

NO	VEGETATION INDICES	STANDARD DEVIATION
1	GRVI	0,47
2	SR	0,43
3	NLI	0,30
4	LAI	0,18
5	SIPI	0,17
6	GEMI	0,16
7	OSAVI	0,12
8	ENDVI	0,10
9	ARVI	0,10
10	TDVI	0,09
11	NDVI	0,09
12	GARI	0,09
13	GNDVI	0,09
14	SAVI	0,08
15	RDVI	0,07
16	MSR	0,07
17	GDVI	0,07
18	DVI	0,06
19	EVI	0,06
20	MNLI	0,05
21	IPVI	0,05
22	VARI	0,04
23	GVI	0,04

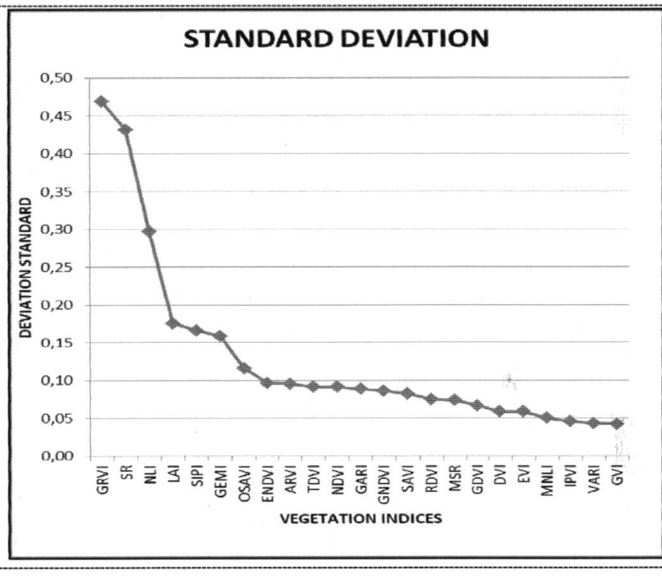

Figure 3. Standard deviation values of 23 vegetation indices.

The impact of oil and gas below the earth has resulted anomalous in the surface conditions. This phenomenon occurs due to the hydrocarbon seepages reaches the surface. This has an impact on the health conditions of vegetation, changes in clay mineral content, iron oxide, carbon delta and soil gas [38]. The condition occurs because of oil and gas seepage either in the form of microseepage and macroseepage [39]. Vegetation anomalies that occur are usually by vegetation stress characterized by slow growth and reduced chlorophyll [40]; low vegetation density and reduced oxygen in the soil [42]; the leaves tend to be yellowish [43]. In the internal leaf, there will be an increase in carotenoids [44].

The regression correlation results on 23 vegetation indices with the coefficient of determination limit (R^2) = 0.8 obtained NDVI, GDVI, ENDVI, VARI, LAI, SIPI. The results of correlation based on the coefficient of determination are as follows:

IOP Conf. Series: Earth and Environmental Science **149** (2018) 012001 doi:10.1088/1755-1315/149/1/012001

1. NDVI correlates with TDVI (R^2 = 1), SAVI (R^2 = 0.8619), RDVI (R^2 = 0.8829), MSR (R^2 = 0.9976), MNLI (R^2 = 0.8497), IPVI (R^2 = 1), GNDVI (R^2 = 0.8002), GARI (R^2 = 0.9467), EVI (R^2 = 0.8715), ARVI (R^2 = 0.9585), GRVI (R^2 = 0.8308), SR (R^2 = 0.9244), LAI (R^2 = 0.8715) and OSAVI (R^2 = 0.9518).
2. GDVI correlates with GVI (R^2 = 0.8564), DVI (R^2 = 0.9202), GEMI (R^2 = 0.8475) and, NLI (R^2 = 0.9384).
3. ENDVI correlated with SAVI (R^2 = 0.8516), RDVI (R^2 = 0.8394), GVI (R^2 = 0.8348), GNDVI (R^2 = 0.8244), DVI (R^2 = 0.8226), GRVI (R^2 = 0.8074), NLI (R^2 = 0.8526) and OSAVI (R^2 = 0.8114).
4. LAI correlated with TDVI (R^2 = 0.8715), SAVI ((R^2 = 0.9909), RDVI (R^2 = 0.9925), NDVI (R^2 = 0.8715), MSR (R^2 = 0.8836), MNLI (R^2 = 0.8903), IPVI (R^2 = 0.0,8715), GVI (R^2 = 0.0,9634), GNDVI (R^2 = 0.8108), EVI (R^2 = 1), DVI (R^2 = 0.9441), GRVI (R^2 = 0.9076), SR (R^2 = 0.8840), NLI (R^2 = 0.9057) and OSAVI (R^2 = 0.9719).
5. VARI and SIPI are not correlated well with other vegetation indices on the R^2 = 0.8.

The results of selecting the vegetation indices based on the standard deviation and the linear regression obtained the 10 vegetation indices, namely GRVI, SR, NLI, SIPI, GEMI, LAI, NDVI, GDVI, ENDVI dan VARI. Further analysis is needed to minimize the number of vegetation indices used for mapping. The combination of the regression analysis and the Standard deviation showed that GRVI correlated with NDVI, ENDVI and LAI; SR correlated with NDVI and LAI; NLI correlated with GDVI, ENDVI and LAI; and GEMI correlated with GDVI. Based on these correlations, GRVI, SR, NLI and GEMI can be represented with NDVI, GDVI, LAI and ENDVI. Based on the final analysis result, the mapping of sugarcane vegetation condition can be done with 6 vegetation indices, that is NDVI, ENDVI, GDVI, VARI, LAI and SIPI. Although LAI correlates well with NDVI, it is still used in this study. LAI is required for analyzing the leaf area index. It is important to know the influence of hydrocarbon microseepages in the leaf cover.

Visual analysis showed that NDVI, ENDVI and GDVI were used to assess vegetation anomalies suspected to be in the south to the east of the study site. The anomaly is described as lower index values around the oil and gas field area. It is confirmed using LAI and SIPI results. The LAI results provide the information that at the location of the anomalous vegetation results still have vegetation with the rare condition. The SIPI results indicate that the location of the suspected vegetation anomaly has a higher SIPI value compared to its vicinity. These SIPI values indicate that the carotenoids in sugarcane vegetation increase, while the chlorophyll of leaves is reduced as a sign of stress vegetation [45]. VARI is less able to provide a significant picture of anomalies in sugarcane crops. A low standard deviation score (0.04) is thought to cause the result of VARI value in relatively uniform, so the anomalous does not appear well.

Survey results in March 2017 provide the confidence in the existence of microseepages in the south to the east of the oil and gas field. In this site, the sugarcane plant is not growing well. This indication is clearly visible with the small number of sugarcane clumps, ie between 2 - 7 trees and the height of the trees are less than 2 meters including leaves. This is a very different to the normal vegetation, where the clumps of sugarcane range from 11 to 20 trees. Surveys in August 2017 showed that sugarcane crops died at the site of the anomalous vegetation. NDVI, ENDVI, GDVI, LAI, SIPI and VARI images showed in figure 4.

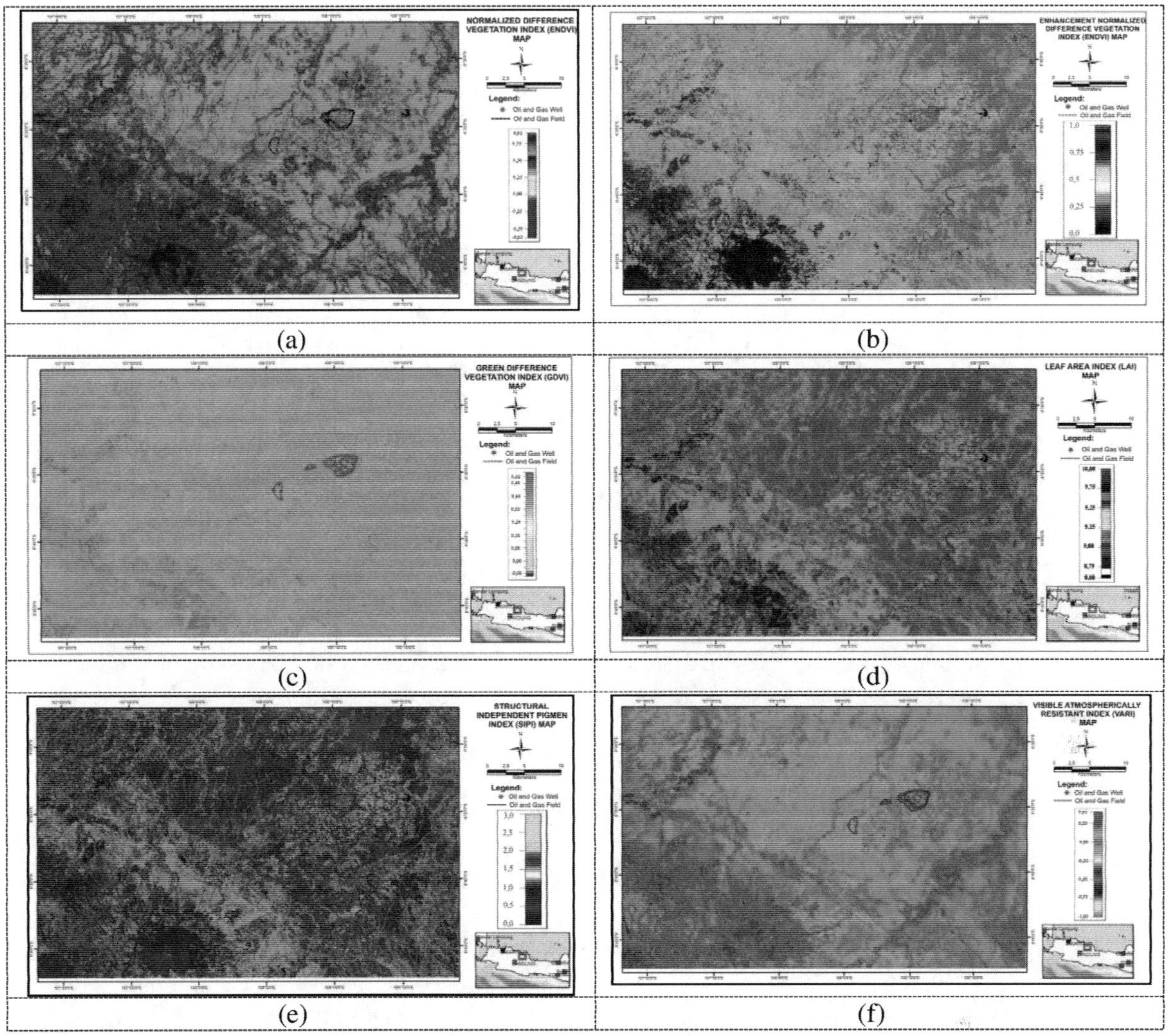

Figure 4. Maps of vegetation indices were extracted from Landsat 8 as the selection results for mapping sugarcane condition around of oil and gas field, (a) NDVI; (b) ENDVI; (c) GDVI; (d) LAI; (e) SIPI and (f) VARI.

4. Conclusion

The selection of vegetation indices to map the sugarcane conditions around the oil and gas field can be done with the combination of the standard deviation and the linear regression methods with the value of the coefficient of determination of (R^2) > 0.8. The combination of both obtained six vegetation indices for mapping the sugarcane vegetation conditions around the oil and gas field, namely NDVI, ENDVI, GDVI, LAI and SIPI. The NDVI, ENDVI and GDVI provide an overview of vegetation conditions around the oil and gas fields. The vegetation anomalies in the oil and gas fields are characterized by the low value of the three vegetation indices. The LAI describes the vegetation cover conditions while the SIPI describes the stress vegetation conditions with increasing index values as an illustration of increased carotenoids and decreased chlorophyll.

Acknowledgements

We would like to express our thanks to Head of Remote Sensing and GIS Group, Head of Exploration Division in the Research and Development Center for Oil and Gas Technology "LEMIGAS", Center for Remote Sensing and Faculty of Earth Sciences and Technology which has facilitated this research. We would like to thank the LISAT Symposium organizers who have given the opportunity to present the results of this research.

5. References

[1] Salas E A L and Henebry G M 2014 A new approach for the analysis of hyperspectral data: Theory and sensitivity analysis of the moment distance method *Remote sensing* **6** pp 20-41

[2] Pettorelli N 2013 *The Normalized Difference Vegetation Index* First Edition (United Kingdom: Oxford University Press) p 193

[3] Roberts D A, Roth K L and Perroy R L 2011 Hyperspectral Vegetation Indices Chapter 14, eds Thenkabail P S, Lyon J G and Huete A 2011 *Hyperspectral Remote Sensing of Vegetation* p 782 (United States: CRC Press Taylor & Francis Group)

[4] Brantley S T, Zinnert J C and Young D R 2011 Application of hyperspectral vegetation indices to detect variations in high leaf area index temperate shrub thicket canopies *Remote Sensing of Environment* **115** pp 514-523

[5] Tucker C J, Townshend J R and Goff T E 1985 African land-cover classification using satellite data *Science* **227** pp 369–375

[6] Paruelo J, Garbulsky M F, Guerschman J P and Jobbagy E G 2004 Two decades of normalized difference vegetation index changes in South America: identifying the imprint of global change *Int. J. Remote Sens.* **25** pp1–14

[7] Gomez M G and Maestre F T 2011 Remote sensing data predict indicators of soil functioning in semi arid steppes, Central Spain. Short Communication *Ecological Indicators* **11** pp 1476-1481

[8] Huete A, Didan K, Miura T and Ferreira L G 2002 Overview of the radiometric and biophysical performance of the MODIS vegetation indices *Remote Sensing of Environment* **83** pp 195–213

[9] Gitelson A A, Kaufman Y and Merzylak M 1996 Use of a green channel in remote sensing of global vegetation from EOS-MODIS *Remote Sensing of Environment* **58** pp 289-298

[10] Gitelson A A and Merzlyak M N 1998 Remote sensing of chloropyll concentration in higher plant leaves *Adv. Space Res.* **22** No. 5 pp 689-692

[11] Sripada R P, Heinigerb R W, Whitec J G and Meijer A D 2006 Aerial color infrared photography for determining early in-season nitrogen requirements in Corn *Agronomy Journal* **98** pp 968-977

[12] Rohman T L 2007 Decision Support for Product Diversification of Sugarcane in PT PG. Rajawali II PG. Jatitujuh Unit Majalengka-West Java *Thesis Undegraduate Program* Faculty of Agriculture Technology, Bogor Agriculture University

[13] Pertamina 1994 *Exploitation Wells Drilling Proposal Year 1995-1996* (Jakarta, Geological Production Pertamina Unit EP III)

[14] Lohjayanti A 2007 Performance and Decision Support System of Sugarcane Process Control in PT. Rajawali II Unit Pabrik Gula Jatitujuh Majalengka *Thesis Undegraduate Program* Faculty of Agriculture Technology, Bogor Agriculture University

[15] Teillet P M and Ren X 2008 Spectral band difference effects on vegetation indices derived from multiple satellite sensor data *Canadian Journal Remote Sensing* 2008 **34** pp 159–173

[16] Chen X Vogelmann J E, Chander G, Ji L, Tolk B, Huang C and Rollins M 2013 Cross-sensor comparisons between Landsat 5 TM and IRS-P6 AWiFS and disturbance detection using integrated Landsat and AWiFS time-series images *International Journal Remote Sensing* **34** pp 2432–2453

[17] Li P, Jiang L and Feng Z 2014 Cross-comparison of vegetation indices derived from Landsat-7 Enhanced Thematic Mapper Plus (ETM+) and Landsat-8 Operational Land Imager (OLI) sensor *Remote Sensing* **6** pp 310-329

IOP Conf. Series: Earth and Environmental Science **149** (2018) 012001 doi:10.1088/1755-1315/149/1/012001

[18] Schroeder T A, Cohen W B, Song C, Canty M J and Yang Z 2006 Radiometric correction of multi-temporal Landsat data for characterization of early successional forest patterns in Western Oreon *Remote Sensing of Environment* **103** pp 16-26
[19] Jiapaer G, Chen X and Bao A 2011 A comparison of methods for estimating fractional vegetation cover in Arid regions *Agric. For. Meteorol.* **151** pp 1698-1710
[20] Kaufman Y and Tanre D 1992 Atmospherically resistant vegetation index (ARVI) for EOS-MODIS *IEEE Transactions on Geoscience and Remote Sensing* **30** No 2 pp 261-270
[21] Tucker C 1979 Red and photographic infrared linear combinations for monitoring vegetation *Remote Sensing of Environment* **8** pp 127–150
[22] Pinty B and Verstraete M 1992 GEMI: a non-linear index to monitor global vegetation from satellites *Vegetation* **10** pp 15-20
[23] Kauth R and Thomas G 1979 The tasselled sap-A graphic description of the spectral-temporal development of agricultural crops as seen By Landsat *Symposium of Machine Processing of Remotely-Sensed Data: Proc. of the LARS* (Purdue University, West Lafayette, Indiana, 29 June – 1 July 1979) pp 4B41-4B51
[24] Crippen R 1990 Calculating the vegetation index faster *Remote Sensing of Environment* **34** pp 71-73
[25] Boegh E, Soegaard H, Broge N, Hasager C, Jensen N, Schelde K and Thomsen A 2002 Airborne multi-spectral data for quantifying leaf area index, nitrogen concentration and photosynthetic efficiency in agriculture *Remote Sensing of Environment* **81** no 2-3 pp 179-193
[26] Yang Z, Willis P and Mueller R 2008 Impact of band-ratio enhanced AWIFS image to crop classification accuracy *Remote Sensing Symposium: Proc. of the Pecora 17* (Denver Colorado, United States, 18-20 November 2008) pp 18-20
[27] Chen J 1996 Evaluation of vegetation indices and modified simple ratio for boreal applications *Canadian Journal of Remote Sensing* **22** pp 229-242
[28] Goel N and Qin W 1994 Influences of canopy architecture on relationships between various vegetation indices and LAI and Fpar: A computer simulation *Remote Sensing Reviews* **10** pp 309-347
[29] Rouse J, Haas R, Schell J and Deering D 1973 Monitoring vegetation systems in the great llains with ERTS *Third ERTS Symposium* NASA (United States, 10 – 14 December 1993) pp 309-317
[30] Rondeaux G, Steven M and Baret F 1996 Optimization of soil-adjusted vegetation indices *Remote Sensing of Environment* **55** pp 95-107
[31] Roujean J and Breon F 1995 Estimating PAR absorbed by vegetation from bidirectional reflectance Measurements *Remote Sensing of Environment* **51** pp 375-384
[32] Huete A 1988 A soil-adjusted vegetation index (SAVI) *Remote Sensing of Environment* **25** pp 295-309
[33] Birth G and McVey G 1968 Measuring the color of growing turf with a reflectance spectrophotometer *Agronomy Journal* **60** 640-643
[34] Gitelson A A, Strark R, Grits U, Rundquist D, Kaufman and Derry D 2002 Vegetation and soil lines in visible spectral space: A concept and technique for remote estimation of vegetation fraction *International Journal of Remote Sensing* **23** pp 2537–2562
[35] Bannari A, Asalhi H and Teillet P 2002 Transformed difference vegetation index (TDVI) for vegetation cover mapping IGARSS 2002: *Proc. of the Geoscience and Remote Sensing Symposium* (Canada, 24 – 28 June 2002) (IEEE International vol 5) pp 3053-3055
[36] Penuelas J, Baret F and Filella I 1995 Semi-empirical indices to assess carotenoids/chlorophyll-a ratio from leaf spectral reflectance *Photosynthetica* **31** pp 221-230
[37] Maxmax 2015 *Enhanced Normalized Difference Vegetation Index (ENDVI)* Available online https://www.maxmax.com/endvi.htm (accessed on 4 December 2015)

IOP Conf. Series: Earth and Environmental Science **149** (2018) 012001 doi:10.1088/1755-1315/149/1/012001

[38] Yang H 1999 Imaging spectrometry for hydrocarbon microseepage *Dissertation* (Negara: TU Delft) Master of Science in Geology ITC (Publication Number 76)

[39] Clarke R H and Cleverly R W 1991 Petroleum seepage and post-accumulation migration *Petroleum Migration (Geological Society Special Publication N 59)* ed England W A and Fleet A J (London: Geological Society of London) pp 265–271

[40] Noomen M F 2007 Hyperspectral Reflectance of Vegetation Affected by Underground Hydrocarbon *Dissertation* (The Netherlands: International Institute for Geo-information Science & Earth Observation Enschede)

[41] Lakkaraju V R, Zhou X, Apple M E, Chunningham A and Dobeck L M 2010 Studying the vegetation response to simulated leakage of sequestered CO_2 using spectral vegetation indices *Economic Informatics* **5** pp 379 -389

[42] Werff H M A V D, Noomen M F, Meijde V D and Meer F D V D 2006 Remote sensing of onshore hydrocarbon seepage: Problems and solutions *Geological Society London Special Publication* **283** pp 125-133

[43] Susantoro T M, Wikantika K, Saepuloh A and Harsolumakso A H 2017 Utilization of vegetation indices to interpret the possibility of oil and gas microseepages at ground surface *Paper Presented at the 1st UPI International Geography Seminar* (Bandung, Indonesia, 8 August 2017)

[44] Strachan I B, Pattey E and Boisvert J B 2002 Impact of nitrogen and environmental conditions on corn as detected by hyperspectral reflectance *Remote Sensing of Environment* **80** pp 213-224

[45] Blackburn G A 1998 Quantifying chlorophylls and careteniods at leaf and canopy scales: An evaluation of hyperspectral approaches *Remote Sensing Environmental* **66** pp 273-285

Modelling landscape change in paddy fields using logistic regression and GIS

E E Franjaya[1], Syartinilia[1] and Y Setiawan[2]

[1] Department of Landscape Architecture, Faculty of Agriculture, Bogor Agricultural University, Kampus IPB Darmaga, Bogor 16680, Indonesia
[2] Center for Environmental Research, Bogor Agricultural University, Kampus IPB Darmaga, Bogor 16680, Indonesia

E-mail: eduwinekofranjaya@yahoo.com

Abstract. Paddy field in karawang district, as an important agricultural land in west java, has been decreased since 1994. From previous study, paddy fields dominantly turned into built area. The changes were almost occured in the middle area of the district where roadways, industries, settlements, and commercial buildings were existed. These were estimated as driving forces. But, we still need to prove it. This study aimed to construct the paddy field probability change model, subsequently the driving forces will be obtained. GIS combined with logistic regression using environmental variables were used as main method in this study. Ten environmental variables were elevation 0–500 m, elevation>500 m, slope<8%, slope>8%, CBD, build up area, river, irrigation, toll and national roadway, and collector and local roadway. The result indicated that four variables were significantly played as driving forces (slope>8%, CBD area, build up area, and collector and local roadway). Paddy field has high, medium, and low probability to change which covered about 27.8%, 7.8%, and 64.4% area in Karawang respectively. Based on landscape ecology, the recommendation that suitable with landscape change is adaptive management.

1. Introduction

Rice paddy is one of the important agricultural commodity in Indonesia [1]. Even in the world, Indonesia is known as one of the top three rice paddy producers. This is because paddy is staple food for Indonesian people [2]. But nowadays, the paddy fields faced the alteration. From previous study, we know that paddy fields has been changed a lot in the middle area of Karawang District. The paddy mostly turned into built area. There are much thing that make this happened. From other researches, we got that the development of physical and economic growth has become the main factor for the changed [3,4]. The location of Jakarta that so close to Karawang has an influence, as we know that Jakarta is a metropolitan city. The increasing of socio-economic activities in Jakarta have led to massive land use changes [5] and urbanization [6,7]. But, for the spesific and detailed reason about driving forces that causing the alteration happened in Karawang District, we need to do a deeper analysis.

We need land cover change map from previous result, some environmental variables related to the changes, and presence and absence point. This research aimed to constructing the paddy field probability change model (1994-2015) in Karawang District-West Java. Using this probability change model, we can give recommendations so that more changing would not be happen in the future.

2. Materials and methods

2.1. Study area
This study was conducted in Karawang District-West Java, which located on latitude S5°56′ - S6°34′ and longitude E107°02′ - E107°40′.

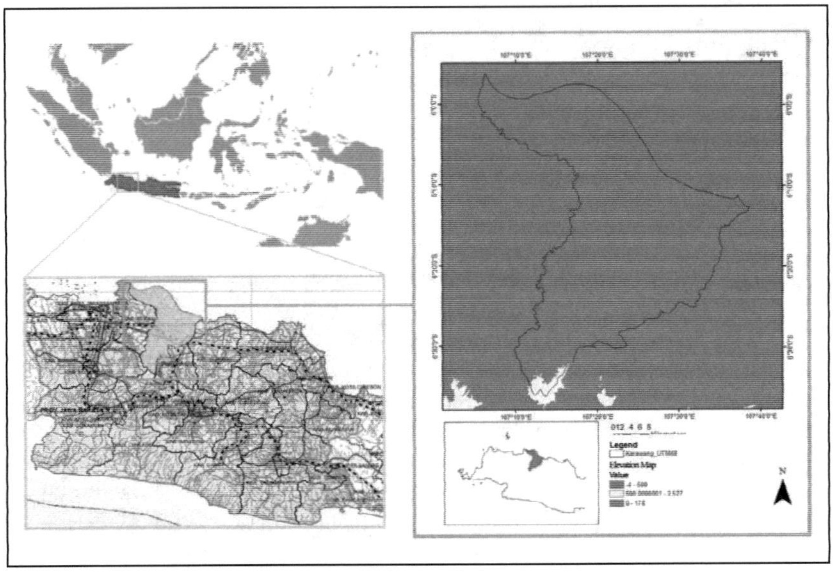

Figure 1. Study area.

2.2. Data collection
The main data used in this research is land cover change map from previous research. Besides the land cover change map, this research also need Indonesia Digital Map (RBI) and Aster GDEM map for landscape change modelling analysis. The tools that used in this research were varied. The main tools are ArcGIS software, Erdas Imagine 9.1, and SPSS program to get the value and as an input for constructing the probability change model.

IOP Conf. Series: Earth and Environmental Science **149** (2018) 012002 doi:10.1088/1755-1315/149/1/012002

Table 1. Research's material, source, and its function.

Material	Unit	Source	Tool's function
Digital Map (RBI)	Vector	Geospatial Information Agency	Landscape change modelling analysis and environmental variables
Land cover change map (1994-2015)	Raster 30x30 m pixel resolution	National Aeronautics and Space Administration	Landscape change modelling analysis and environmental variables
Aster GDEM	Raster 30x30 m pixel resolution	National Aeronautics and Space Administration	Elevation and Slope fr environmental variables

2.3. Landscape change modelling analysis
On this stage, the result of land cover analysis from previous study will be processed further to detect the driving forces using Logistic Regression (figure 2). Presence and absence points data is needed to get value from each environmental variables. The environmental variables used in this research are stated on the figure of list of environmental variables below (table 2). The data will be executed in Model Maker ArcGIS and SPSS using Logistic Regression method [8]. The result of analysis is land cover change variables that significant enough as driving forces [9]. This driving force variables will be executed to construct the probability change model. The validation is needed to evaluate the model in the end.

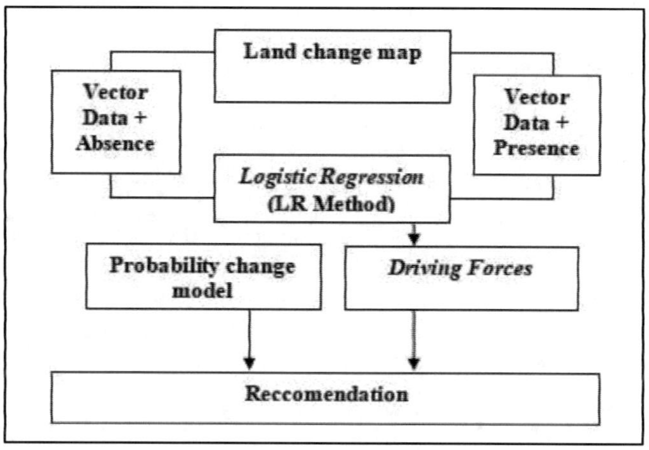

Figure 2. Study flowchart.

Table 2. List of environmental variables.

No	Environmental variables	Abbreviation	Preferences	Sources
		Elevation		
1.	Euclidean distance to Elevation 0-500 m above sea level	JTE1	Suitability level in planting paddies and geographical condition of Karawang County	Extracted from ASTER GDEM Data that become euclidean distance map
2.	Euclidean distance to Elevation >500 m above sea level	JTE2		
		Slope		
3.	Euclidean distance to Slope <8%	JTS1	Suitability level of slope on planting paddies and geographical condition of Karawang County	Extracted from ASTER GDEM Data that become euclidean distance map
4.	Euclidean distance to Slope >8%	JTS2		
		Land Cover and Infrastructures		
5.	Euclidean distance to River	JTSu	One of water sources of paddy field	Extracted from RBI Map that become euclidean distance map
6.	Euclidean distance to Main Irrigation	JTIr	Water Canal	
7.	Euclidean distance to CBD Area	JTPe	Land, social, and cultural value	
8.	Euclidean distance to Build Up Area	JTAt	Land, social, and cultural value	Extracted from land cover map that become euclidean distance map
9.	Euclidean distance to National and Arterial road	JTJ1	Accesibility, land, social, and cultural value	Extracted from RBI Map that become euclidean distance map
10.	Euclidean distance to Collector and Local road	JTJ2	Accesibility, land, social, and cultural value	

3. Results and discussions

Based on the result at logistic regression process, there are 4 environmental variables that significantly has an influence as driving factors. Those are JTS2, JTPE, JTAt, and JTJ2 (Table 3). Model is proper based on Hosmer and Lemeshow test, because the model have a value more than 5% (93 %). Nagelkerke R^2 also showed that 86.7% from paddy field land change probability model can be explained by variable on model. Overall, this result will produce logistic regression formula as stated on figure 3.

Table 3. T-test, Coefficient, Hosmer and Lemeshow, and Nagelkerke R2 Stepwise method.

Variabel	β	p-value (Sig.)	Hosmer and Lemeshow	Nagelkerke R2
JTS2	-0.0234	0.000	93%	86,7%
JTPe	-0.0015	0.000		
JTAt	-0.0152	0.000		
JTJ2	0.0015	0.003		
Constants	7.9371	0.000		

$$Pi = \frac{1}{1 + \exp[-(7{,}9371 - 0{,}0234(JTS2) - 0{,}0015(JTPe) - 0{,}0152(JTAt) + 0{,}0015(JTJ2))]}$$

Figure 3. Logistic regression formula of paddy field land change probability model.

From table 3 and figure 3, we know that Slope > 8%, CBD Area, Build Up Area, and Collector & Local road are the main factors for the alteration in Karawang District. The increasing of human population that illustrated from the development of build up area and the exictence of CBD area have affected the change in paddy field area. Along with the geographical condition and development of infastructures, they become a significant driving forces in Karawang District.

The probability change model is the development of the logistic regression formula. From 4 significant driving forces, we will get the map of probability change model (figure 4).

From the figures 4, the red colour area shows the high probability, it means that the paddy field in red area is easily to change to another land cover. It is covered 27.8 % of all area of Karawang District. The orange colour area is a medium level of change, covered just 7.8% of total area. The green colour area is the biggest area in the map, took up more than half of total area (64.4%). This area is an area that does not affected by the alteration.

The area in each probability has several significant subdistrict. In high probabilty consists of subdistrict Cikampek, Jatisari, Kotabaru, Klari, Teluk Jambe Timur, Karawang Timur, Banyusari, Lemahabang, Telaga Sari, Rengasdengklok, Tegalwaru, Batujaya, Tempuran, Jayakerta, and Pedes. Medium probability consists of subdistrict Tirtamulya, Purwasari, Telukjambe Timur, Pangkalan, Klari, Rawamerta, Tempuran, Cilamaya, Cilebar, Pedes, and Jayakerta. Low probabilty area consists of subdistrict Pakisjaya, Tirtajaya, Tirtajaya, Cilebar, Cilamaya, Rawamerta, Tirtamulya, Klari, and Ciampel.

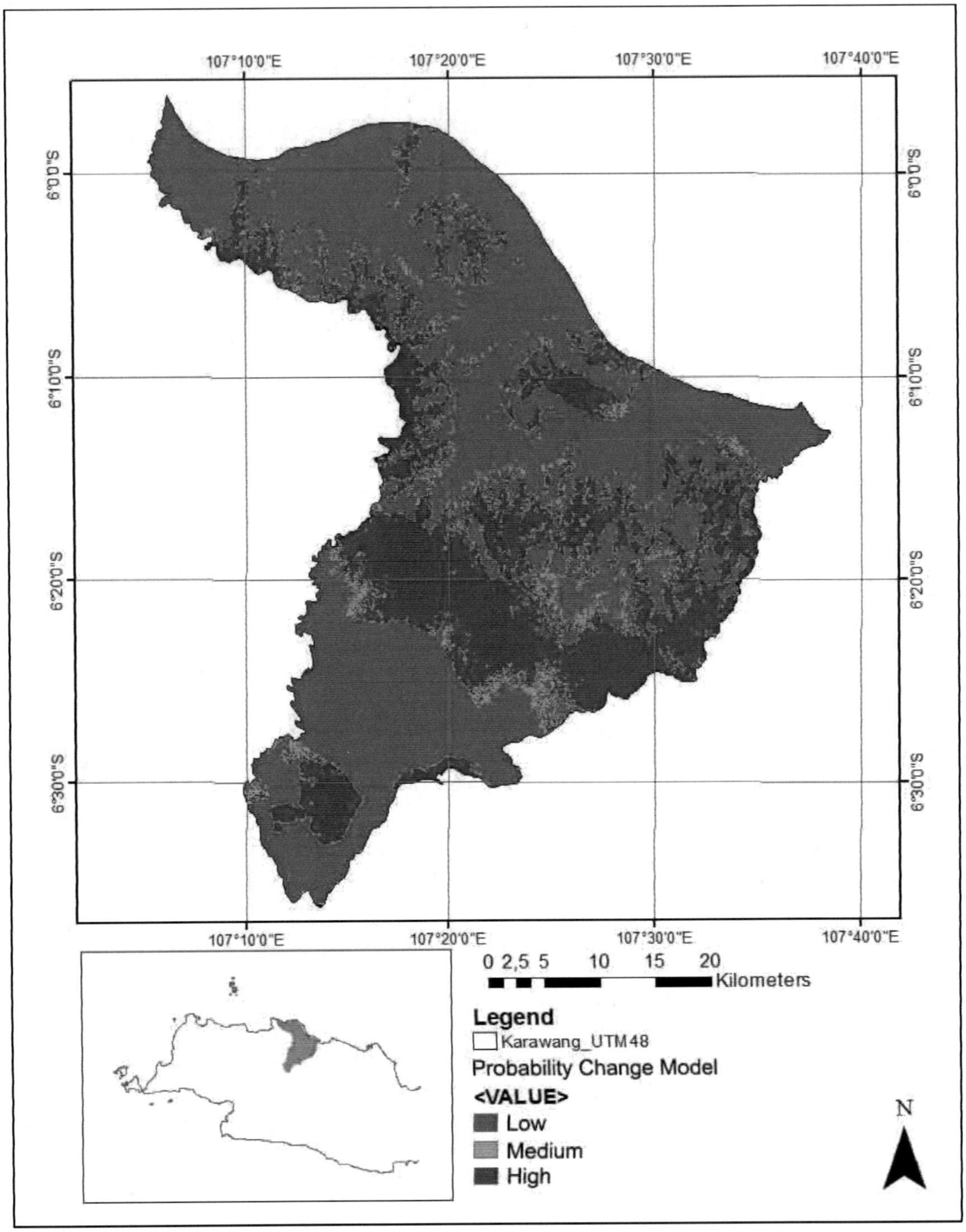

Figure 4. Probability change model

4. Recommendation

The recommendation is given based on each kind of probability. The recommendation for high probability are such collecting the data of paddy field & strategic value of Landscape Services, collecting the data of strategic food protection area: Subdistrict Karawang Timur, Teluk Jambe Timur, Cikampek, Jatisari, and Kotabaru, applying policies: free tax, seed incentive, fertilizer, etc, applying adaptive management [10]: Agroedutourism/Agrotourism/Integrated farming, applying law and rule enforcement about RTRW and agricultural area, and do supervision of the development of 4 driving forces. The recommendation for medium probability are like applying law and rule enforcement about strategic food protection area (Subdistrict Teluk Jambe and Klari) and adaptive management, agricultural elucidation and ease of access of saprotan, and do supervision of the development of RTRW and 4 driving forces. The last one, recommendation for low probability are such applying agricultural elucidation and ease of access of saprotan, and do supervision of the development of RTRW and 4 driving forces.

Acknowledgments

We gratefully acknowledged for the help that given by *Badan Informasi Geospasial* (en: Geospatial Information Agency, abbreviation: BIG), Staff of *Pusat Penelitian Lingkungan Hidup* (en: Center for Environmental Research, abbreviation: PPLH), and the government of Karawang so that this research is well done. We should also thanks to the other for the helpful suggestions.

References

[1] BPS 2013 *The table of crop area, productivity, production of paddy throughout Indonesia*. Retrieved from *http://www.bps.go.id*. on February 28th, 2014; Dated 2013 (in Indonesian)

[2] Franjaya EE, Syartinilia, Setiawan Y 2017 Monitoring of landscape change in paddy fields: Case study of Karawang District - West Java Province *IOP Conf. Ser.: Earth Environ. Sci.* **54** 012016

[3] Rodriguez MP, Butsic V, Gartner P, Macchi L, Baumann M, Pizzaro GG, Volante JN, Gasparri IN, Kuemmerle T 2018 Drivers of agricultural land-use change in the Argentine Pampas and Chaco regions*Applied Geography* **91** 111-122

[4] Kumar R, Nandy S, Agarwal R, Kushwaha SPS 2014 Forest cover dynamics analysis and prediction modeling using logistic regression model *Ecological Indicators* **45** 444-455

[5] Indraprahasta GS 2013 The potential of urban agriculture development in Jakarta *Procedia Environmental Sciences* **17** 11-19

[6] Rustiadi E, Panuju DR and Trisasongko BH 2009 *Environmental impacts of urbanization in Jabodetabek area* S Uchida *et al* Bogor *Proc. of The ICALRD-JIRCAS Workshop on Enhancement of Remote Sensing and GIS Technologies for Sustainable Utilization of Agricultural Resources in Indonesia* Indonesian Center for Agricultural Land Resources Research and Development Ministry of Agriculture p 44-55

[7] Handayani W 2013 Rural-urban transition in central java: population and economic structural changes based on cluster analysis *Land* **2** 419-436

[8] Hendayana R 2013 Application Method of Logistic Regression on analyzing the Agricultural Technology Adoption *Informatika Pertanian* **22** 1-9

[9] Xie C, Huang B, Claramunt C, Chandramouli M 2005 Spatial Logistic Regression and GIS to Model Rural-Urban Land Conversion. *In Second International Colloquium on the Behavioral Foundations of Integrated Land-use and Transportation Models:*

Frameworks, Models and Applications. 2005 July; Toronto, Canada.
[10] Liu J, Taylor WW 2004 Integrating Landscape Ecology into Natural Resource Management. Cambridge (UK): Cambridge University Press.

IOP Conf. Series: Earth and Environmental Science **149** (2018) 012003 doi:10.1088/1755-1315/149/1/012003

Impact of spatial plan on the conversion of Subak rice fields and food security, in Badung and Gianyar Regencies, Bali Province

Indayati Lanya[1], N. Netera Subadiyasa[1], and Gst. P. Ratna Adi[1]
[1]Study Program of Agroecotechnology, Faculty of Agicultural, Udayana University

E-mail: indahnet@yahoo.co.id

Abstract. Regional Spatial Plan of Bali Province 2009-2029, allocating rice fields can be converted 10% (± 10.800 ha). Over the next 20 years, the conversion of rice field is permitted 540 ha year^{-1}, the real condition in Bali is 800 ha year^{-1}.Research location in Badung and Gianyar Regencies. Visual satellite image interpretation methods, digitization of on-screen, delineation of subak rice field, field survey, superimpose analysis of Spatial Plan (SP) map with rice field map, trough toolbox-analysis tools–overlay-intersect using QGIS, Harvest Index (HI) of cropping pattern in one year. SP has a negative impact on agricultural land resources and food security. Local Regulation (SP), subak rice fields outside the agricultural area licensed to be converted, and food deficits. Regency of Badung, potential land conversion of 3,324.97 ha (34.44%) from 119 subak with paddy field area of 24,184.85 ha. There are 10 subak 100% and 8 subak > 95% can be converted; projected food deficit -115.343 tons of rice by 2020 for HI 2. In Gianyar Regency, potential land conversion 13,021.41 ha (53.51%) of 66 subak with an area of 24184.85 ha; 8 subak 100% and 8 subak with area < 5 ha can be converted; projected food deficit is about -194438 tons of rice in 2040 for HI 2.

1. Introduction

Subak rice field in Bali experienced conversion of 800 ha year^{-1} for the last 10 years. Greater than the average land inventory (540 ha year^{-1}) allocated in SP [1]. Chapter 60, paragraph (1) letter a, Local Regulation No. 16 of 2009, on the SP of Bali Province, establishes 76,337 ha or 13.5% of the total area of Bali Province [2], spread over districts/cities. In the Spatial Plan Map, Bali set 11 tourism areas that the existing condition is still a subak rice field. This is contrary to the Subak conservation effort that is designated as world cultural heritage by UNESCO, 2012 [3]. Rice fields in Bali in the form of Subak land. Subak as a traditional institution, preserved and strengthened in Regional Regulation (Perda) No. 9 of 2012 [4]. As a follow-up conservation efforts Subak, the Regional Government of Bali, the Regency has programmed through grants for institutional management and farming activities.

The grant aid is not based on the Subak area, meaning the narrow Subak region (<5 ha) and large (> 100 ha) receive the same grant. Therefore, there are Subak that do expansion which has an impact on the increasing of rice field area [5]; only based on statistical data [6]. The research of Subak land function control strategy, through zoning text and zoning map with remote sensing technology and GIS, produces Subak of the conservation area, buffer and it can be converted [7]. The results of these studies can assist local governments in drafting legislation for Academic Studies, Sustainable Food

Agricultural Land Protection, as mandated in the Act No. 41 of 2009 [8] and Government Regulation No. 1 of 2011 [9].

The regencies of Badung and Gianyar are selected as a research area, as these two districts are experiencing very rapid tourism development compared to the other 6 districts in Bali. Prior research, the highest land conversion function is located around the centers of tourism areas [6].

Based on the above description, research on SP impact on the sustainability of Subak rice field and food balance in Bali needs to be done. Given the subak as a local genius of Bali potentially conflict of interest with the needs of tourism development. The purpose of this study is to analyze the existence of Subak, such as the potential loss of number, name, and area of Subak rice field, the potential conversion of the Subak rice field in a certain year. Furthermore, it is used to calculate the potential of food supply and projection of food balance in Badung and Gianyar regency. Both districts are experiencing with the development of tourism which is very rapid in Bali Province. The hope of Bali SP map can be refined to minimize conversion of Subak rice field and food deficit in both Regencies and as well as in other regencies in Bali.

2. Material and Methods

Research location is in two Regencies, namely Badung and Gianyar Regency, Bali Province. The 2013 Quick Bird satellite image for research location and image map creation in 2015 was used. The research material consists of Quick Bird satellite images 2013, SP map for Badung and Gianyar Regency, a map of SP, and Quantum GIS (QGIS) software version 10.3. Research methods are multi-analytics, ie: (1) literature study through secondary data collection: Subak data from the Central Bureau of Statistics, map of SP, administration map, (2) visual interpretation of Quick Bird satellite images: delimitation and delineation land use, (3) the field survey to obtain information boundaries of Subak, area, production yield, and cropping pattern, (4) analysis of thematic maps using QGIS: (a) digitization of on-screen delineation of subak rice field, (b) super impose analysis of thematic map between the paddy rice field map and SP map, (c) toolbox - analysis tools – overlay - intersect, (d) identify the result of overlapping of Subak rice filed inside and outside the agricultural land in the SP map, (5) food balance analysis: projected population and food needs, projected area of rice field and food supply. Harvest index (HI) is calculated from rice cropping pattern in one year, HI 1 = rice production only one harvest in one year and HI 2 = rice production in one year twice harvest.

3. Result and Discussion

3.1. Map of the rice fields of Badung and Gianyar regencies

The visual interpretation results of the 2013 Quick Bird satellite imagery and the 2015 field survey, the digitization of on-screen border of subak rice fields, and the layout of rice field use maps for Badung and Gianyar regencies are shown in figure 1 and figure 2. Figure 1 and figure 2 shows that the rice fields spread throughout the region. Especially for Badung regency, the upstream (north) in Petang District and in the southern part (South Kuta District) is dominated by dryland farming. In contrast to Gianyar, where dry land only in the upstream (Tegallalang District).

Dryland is not discussed further because in this study the objective is to obtain subak rice data used for the analysis of land conversion and food balance in a certain year. These data include the names and extent of subak in each village used. District map of subak is used for overlay process with Plan map of Space of each district. Based on the Subak rice field map, it was obtained 119 Subak for Badung Regency and while for Gianyar Regency there are 66 Subak.

IOP Conf. Series: Earth and Environmental Science **149** (2018) 012003 doi:10.1088/1755-1315/149/1/012003

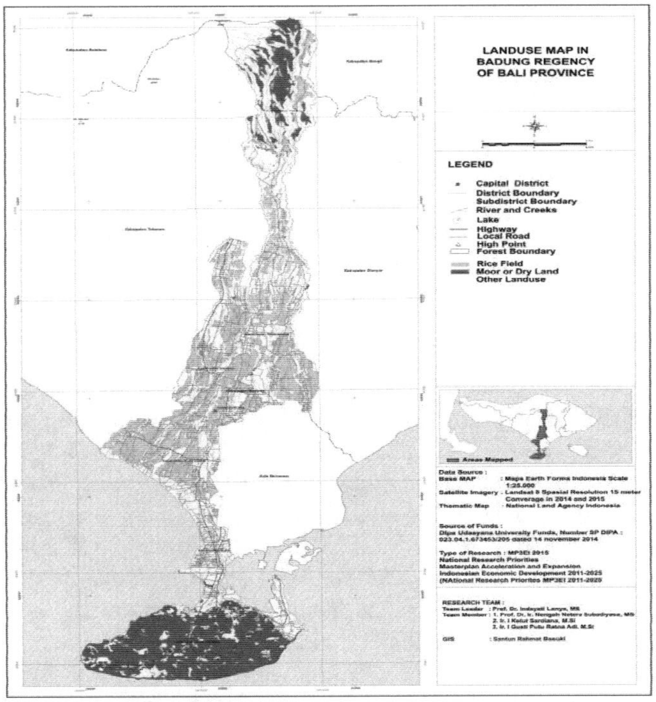

Figure 1. Land use map of Subak rice field in Badung Regency (analysis, 2016).

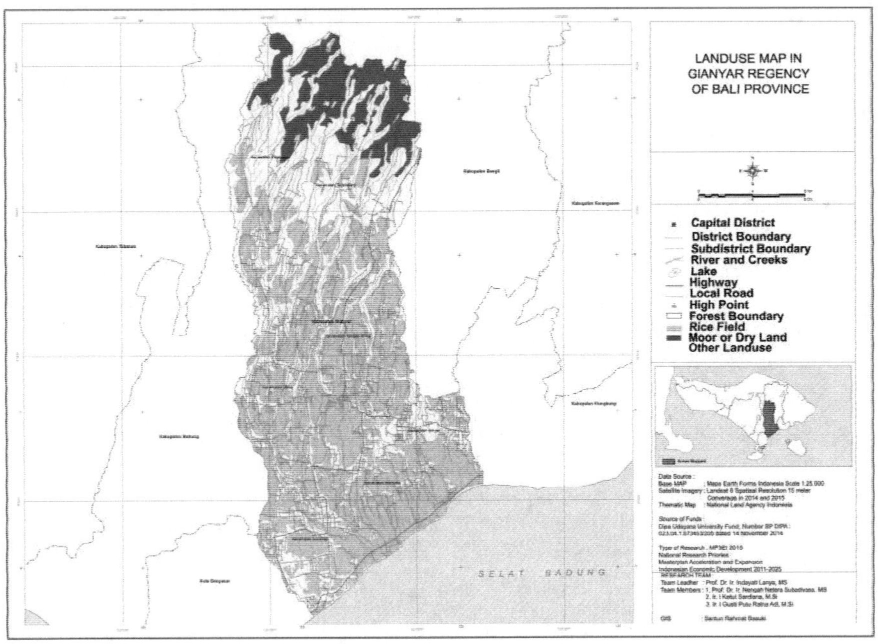

Figure 2. Land use map of Subak rice field in Gianyar Regency (analysis, 2016).

3.2. Map of Land Use of Subak Rice Field with Area of Food Crop Farming in Map of Spatial Plan of Badung and Gianyar Regency

The result of the overlap between Subak rice field map and the area of crop farming land in the spatial planning map of Badung and Gianyar regencies is shown in figure 3 and figure 4. Figure 3 and 4 show that Subak rice fields inside (green color) and outside (brown color) areas of cropland farming in Badung Regency of 3,324.97 ha (34.44%) and in Gianyar Regency 13,021.41 ha (53.51%). The brown color is Subak rice field located in the tourism area, in which outside the agricultural area is generally allocated for tourism accommodation facilities. The potential for conversion of Subak rice fields land is around Kuta paradise area, Ubud, western Gianyar area which is bordered by Denpasar City and Badung Regency, and Ida Bagus Mantra road access. This means that legally aspect of the rice field is allowed to be converted, because it is allocated for tourism accommodation, such as hotels, restaurants and supporting facilities. Examples of Subak rice data in each Subak and village for each district are presented in table 1 and table 2.

Data of the result of identification and analysis of all paddy fields (119) Subak in Badung regency got a total area of 9,653.77 ha. Spread throughout the District, except South Kuta District is dominated by dry land. The number of Subak in each district is Petang 16 subak with the area as follow of 993.84 ha, Mengwi 48 Subak with wide of 3,303.34 ha, Abiansemal 34 Subak with wide 2,918.12 ha, North Kuta 19 Subak with wide of 1,406.75 ha, and Kuta 2 Subak with wide 31.12 ha. Rice field area inside agricultural land area 6,328.80 ha (65.56%) and rice field area outside agricultural land area 3,324.97 ha (34.44%). Subak that will be lost 100% as many as 10 subak with an area of 237.7 ha and potentially lost (area <5 ha) as much as 8 Subak with an area of 286.96 ha.

The total area of Subak (66) in Gianyar Regency is 24,184.85 ha, located in seven districts. Blahbatu District 9 Subak area of 1,859 ha, Gianyar District 19 Subak area 2,102 ha, District Payangan 7 Subak area 1,411 ha, Sukawati District 12 area 2,721 ha, sub-district Tampaksiring 9 Subak area 1,160 ha, Tegallalang District 7 Subak area 1,238 ha, Ubud District 7 Subak area 1,553 ha. Rice field area inside agricultural land area is about 11,163.46 ha (46.49%) and rice field area outside/ converted the agricultural land area 13,021.41 ha (53.51%). 100% of the Subak rice field will be lost, converted to non-farm use by 8 subak, with a total area of 1,630. 69 ha. The potentially lost (area < 5 ha) amounted to 8 Subak with an area of 1,849.18 ha.

Example of rice field allocation in agricultural area in SP of Badung Regency for central tourism area in Kuta District 0.07 ha from 31.12 ha, it means that it can be converted 99.77% area, for North Kuta area of rice field 282.34 ha from 625.06 ha or 54.83% can be converted, while for Mengwi District rice field area of 99. 47 ha from 133.88 ha, may be converted 25.70%. Similarly, the allocation of rice fields in the area of agricultural land in SP Gianyar regency for central tourism area of Ubud only 34.48 ha from 1,555 ha, or 97% of rice fields may be converted, while for District Tampaksiring rice field 850.19 ha from 1161 ha or 26.77%.

Whereas the potentially lost with an area <5 ha is located in the city centers. Especially for Gianyar Regency, the potential loss of paddy field Subak is located in the border area with Badung Regency and Denpasar City by the beach, and around the main access road tourism Ida Bagus Mantera road. The main cause is the supply of land for non-agricultural development in the two districts is only a Subak rice field. Legally aspect, the transfer of rice field is allowed to be converted to non-agriculture, due to the Regional Regulation and Spatial Plan Map of the territory. In other words, the development of tourism in both districts has a negative impact on agricultural land resources in the form of rice field Subak, especially in the food supply.

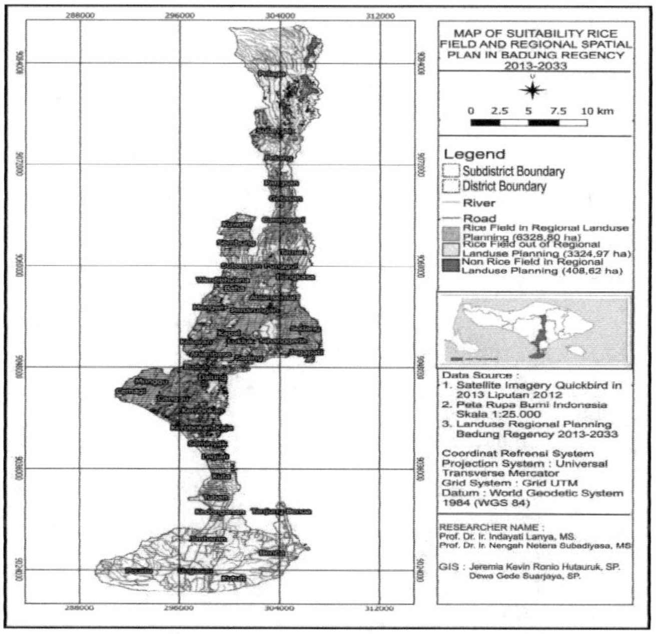

Figure 3. The result of the overlap between paddy field map and agricultural land area on the spatial map in every Subak and village area in map of SP Badung Regency 2012-2032 [10].

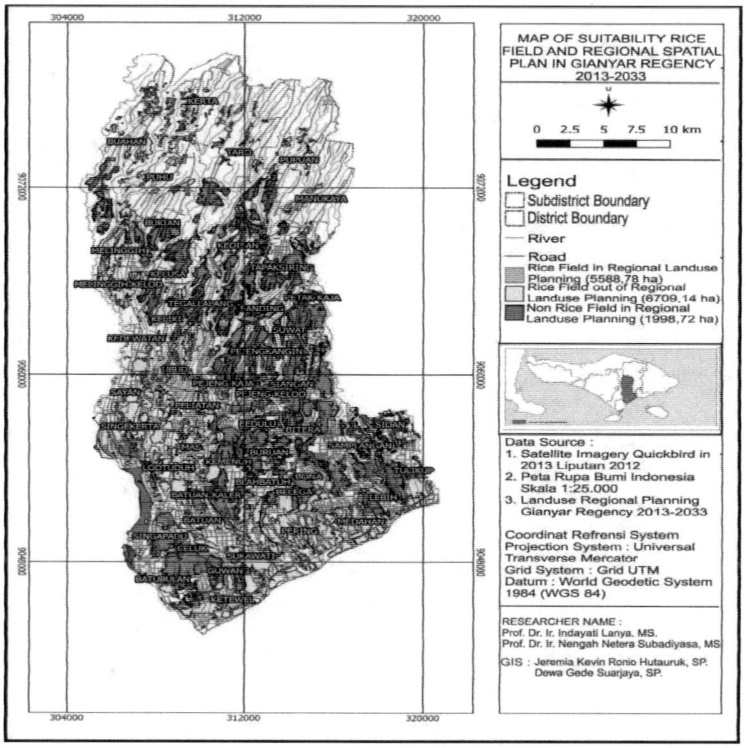

Figure 4. The result of the overlap between paddy field map and agricultural land area on the spatial map in every Subak and village area in Map of SP Gianyar Regency 2012-2032 [11].

Table 1. The Sample of Subak rice field in each village in District of Badung Regency.

No	Subak Name	Name of Village	Area of Rice Field Subak (ha)	Area of Subak Rice Field in Agricultural Area (ha)	(%)
Mengwi District: 5 samples from 48 Subak					
1	Babakan Sobangan	Sobangan	42.34	37.46	88.47
2	Balangan	Kuwum	44.94	34.10	75.88
3	Batan Asem	Mengwi Tani	3.74	1.73	46.26
4	Batan Badung	Gulingan	30.00	25.59	85.30
5	Batan Wani	Sading	12.86	0.59	4.59
	Total		133.88	99.47	74.30
North Kuta District: 10 samples from 19 Subak					
1	Banjar Sari	Tibubeneng	47.08	3.46	7.35
2	Bantan	Tibubeneng	50.75	37.70	74.29
3	Daksina	Tibubeneng	57.56	37.35	64.89
4	Dawas	Tibubeneng	16.89	1.18	6.99
5	Gaji	Dalung	60.94	28.21	46.29
6	Petitenget	Kerobokan Klod	59.82	0.00	0.00
7	Saih	Dalung	140.10	34.95	24.95
8	Tegal	Kerobokan Kaja	88.34	68.58	77.63
9	Umaalas	Canggu	16.72	0.00	0.00
10	Umadesa	Canggu	86.86	70.91	81.64
	Total		625.06	282.34	45.17
Kuta District : 2 Subak					
1	Abianbase	Kuta	19.09	0.07	0.37
2	Seminyak	Seminyak	12.03	0.00	0.00
	Total		31.12	0.07	2.25

Table 2. The Sample of Subak rice field in each village in District in Gianyar Regency.

No	Name of Village	Area of Subak Rice Field (ha)	Area of Subak Rice Field in Agricultural Area Ha	%	Area of Subak Rice Field outside Agricultural Area Ha	%
Tampaksiring District: 9 Subak						
1	Manukaya	111	30.63	27.59	58.16	52.40
2	Pejeng	99	65.45	66.2	33.41	33.80
3	Pejeng Kaja	146	113.64	77.59	32.82	22.41
4	Pejong Klod	64	41.99	65.87	21.76	34.13
5	Pejeng	193	153.01	79.38	39.75	20.62
6	Pejeng Kangin	163	152.79	93.85	10.00	6.14
7	Pejeng Kawan	22	7.00	32.36	14.63	67.64
8	Sanding	80	52.32	65.2	27.92	34.80
9	Tampaksiring	283	233.36	82.48	49.58	17.52
	Total	1161	850.19	73.23	310.81	26.77
Ubud District: 8 Subak						
1	Lohtunduh	268	3.01	1.12	264.52	98.87
2	K edewatan	72	0.00	0.00	71.64	100
3	Mas	279	3.45	1.23	270.41	96.92
4	Peliatan	121	3.66	3.02	117.37	96.98
5	Petulu	121	0.36	0.3	120.19	99.70
6	Sayan	154	0.00	0.00	153.60	100
7	Singakerta	359	8.00	2.23	350.57	97.77
8	Ubud	181	16.00	8.86	164.58	91.20
	Total	1161	850.19	73.23	310.81	26.77

3.3 Food Balances

Regional Regulation (Perda) No 16 of 2009 on Spatial Plans Bali, impact on the decrease of wetland area. Similarly, for the SP of Badung Regency in 2012 -2032 and SP of Gianyar Regency in 2013-2033. Potential decreased food security, with a deficit food balance of -115.343 and -679,984 tons of rice in 2030 and 2050 in Badung regency. For Gianyar Regency, food balance also deficit = -225,516.5 for harvest index (HI) 1 or -695227 tons of rice for HI 2 by 2030. Food deficit increased by 2050 as much as - 728799 tons of rice for HI 1 or -1016319 tons of rice for HI 2. Analysis population projection, food demand and food supply for Badung and Gianyar Regencies are shown in figure 5 and figure 6. Based on figure 5 and 6, the projected food demand is inversely proportional to the food supply in these two regions. Food demand increases in line with population growth over time, while food supply declines as the impact of uncontrolled conversion of rice field.

The permissible allocation of land in both districts is very high, affecting food security is weakening, it is necessary to revise the local regulation of SP. The revision of the SP is very urgent to be conducted in order to reduce rice field allowed to be converted. This is related to the preservation of Subak rice field as a source of sustainable food. In line with the process of control over the function of rice field into the non-agricultural land, a strategy is required by law, the Provincial Regulation on the Protection of Agricultural Land for Sustainable Food, so that the fields are not converting Subak continuously.

Another strategy to reduce food deficit is the development agricultural program of increasing paddy rice production of more than 6.5 tons ha^{-1} each harvest [4], and an increase of HI more than 2, through the five farming. The rapid increase of production can be through the addition of fertilizer in accordance with the needs of the plant, it is expected that the production will increase by 50-100% from the national average production of 4.5 tons ha^{-1}. However, besides the development, maintenance and conservation of irrigation water are very necessary to increase the HI from 1 to 2.5.

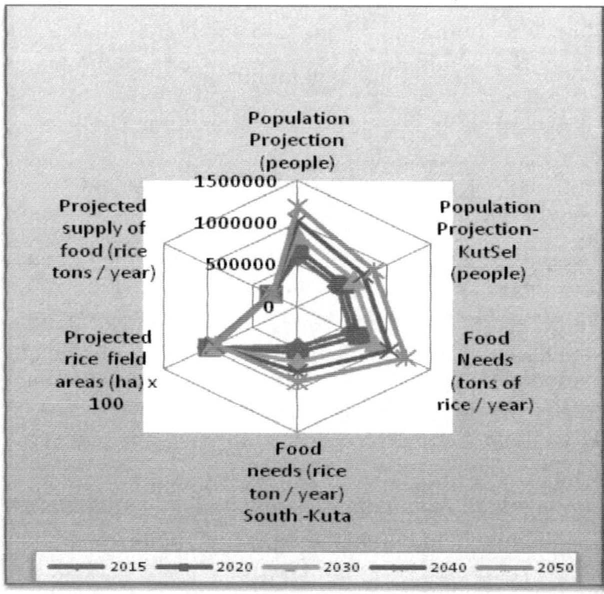

Figure 5. The Balance of food for Badung Regency in 2015-2050.

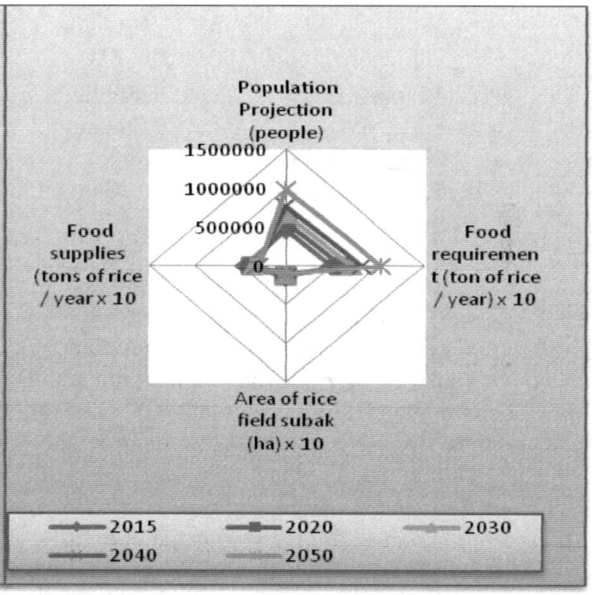

Figure 6. The Balance of food for Gianyar regency in 2015-2050.

4. Conclusion and Recomendation

4.1. Conclusion

Regional Spatial Plans have negative impacts on agricultural land resources and increasing food deficit, especially in the center of the area satisfied tourist destinations, such as District of Kuta and Ubud.

Legally aspect (Perda SP), Subak paddy field that can be converted function to tourism area and its supporting in Badung Regency is 99.77% in Kuta District, decreasing 54.83% in South Kuta District and 25.70% in Mengwi Sub-district. For Gianyar Regency, Subak rice field which is designated as the highest tourism area is 97.78% in Ubud sub-district and 26.77% in Tampaksiring District.

Food deficit in Badung and Gianyar Regencies continues to increase, in line with population growth, the uncontrolled rate of rice field conversion, low rice production, and the establishment of Subak rice fields as non-agricultural areas in SP.

4.2. Suggestion

Revision of Spatial Plans is needed to protect the Subak rice fields as cultural heritage, nature conservation, and local food security. Especially in the area of Ubud tourism center, accessibility around the road Ida Bagus Mantera, Gianyar border with Denpasar, and North Kuta district.

Acknowledgments

Thanks and appreciation go to Rector and Chairman of the Institute for Research and Community Services University of Udayana, with the approval and guidance in this study; Dean of the Faculty of Agriculture, University of Udayana, with the approval and assistance in implementation research.

5. References

[1] Lanya I N, Subadiyasa N, Sardiana K, and Ratna Gst P A 2015 Transfer fungsion control strategy Subak rice field through zoning map and zoning teks *Proc. Senastek II (National Seminar on Science and Technology)* (Badung Bali, Indonesia: Institute for Research and Community Services, Bukit Jimbaran) pp 2039-2046

[2] Bali Provencial Government 2009 Regional Regulation No. 16 In 2009 The Spatial Plans (RTRW) Bali Province from 2009 to 2027 (Bali Government, Denpasar)

[3] UNESCO 2012 Subak System as Manifestation of the Tri Hita Karana Philosophy (Rusia: St Petersburg)

[4] Bali Provincial Government 2012 Regional Regulation No. 9 of 2012 on Subak (Bali Province Gazette of 2012 number 9) Supplement to Bali Province Gazette Number 8

[5] Lanya I N 2010 Research Technology Applications in Remote Sensing and GIS Inventory and Mapping of Natural Resources and Environmental Potential of Bali Province *The final report of the research results* (Udayana University, Bali)

[6] Central Bureau of Statistics (BPS) Bali Privince 2013 *Bali in Figures* (Denpasar: BPS)

[7] Lanya I N, Subadiyasa N, Sardiana K and Ratna Gst P A 2014 Numerical clasification, Subak zoning and land transfer function rice field in Province of Bali *Remote Sensing and GIS. Proc. Int. Symposiumon LAPAN-IPB Satelite (LISAT) for Food Security and Enverimental monitoring* (IPB, 25-26 November 2014) (IOP Conference Series vol **24)** pp 47-55

[8] Republik Indonesia 2009 Act No. 41 Year 2009 about Sustainable Food Agricultural Land Protection State Gazette of the Republic of Indonesia Year 2009 Number 149 Gazette of the Republic of Indonesia Number 5068 Secretariat of State of the Republic of Indonesia. Jakarta Available online www.dpr.go.id/dokjdih/document/uu/UU 2009 41.pdf [accessed on 7 March 2014]

[9] President of the Republic of Indonesia 2011 Government Regulation No. 1 Year 2011 About Determination and Sustainable Food Agricultural Land Transfer Function State Gazette of the Republic of Indonesia Year 2011 Number 2 Gazette of the Republic of Indonesia

Number 5185 Secretariat of State of the Republic of Indonesia Available online www.hukumonline.com/pusatdata/downloadfile [accessed on 7 March 2014]
[10] Badung Regency Government 2012 Regional Regulation Number 11 In 2012 The Spatial Plans (RTRW) Badung from 2012 to 2032 Badung government, Sempidi, Badung, Bali
[11] Gianyar Government 2012 Regional Regulation Number 16 In 2012 The Spatial Plans (RTRW) Gianyar from 2013 to 2033 Badung government, Gianyar, Bali

IOP Conf. Series: Earth and Environmental Science **149** (2018) 012004 doi:10.1088/1755-1315/149/1/012004

Spatial Analysis to Determine Paddy Field Changes in Indonesia: A Case Study in Suburban Areas of Jakarta

Nadia Putri Utami[1], Tofael Ahamed[2]

[1]School of Life and Environmental Sciences, College of Agrobiological Resource Sciences, University of Tsukuba, Japan
[2]Faculty of Life and Environmental Sciences, University of Tsukuba, Japan

E-mail: dhea_tsukuba@yahoo.com

Abstract. Karawang, a suburban area of Greater Jakarta, is known as the second largest rice-producing region in West Java, Indonesia. However, expansion of urban sprawl and industrial area from Greater Jakarta have created rapid agricultural land use/cover changes, especially paddy field, in Karawang. This study analyzed the land use/cover changes of paddy field from 2000 to 2016. Landsat 4-5 TM and Landsat 8 OLI/TIRS images were acquired from USGS Earth Explorer, UTM zone 48 south. Satellite image pre-processing, ground truth data collection, supervised maximum likelihood classifications, and Post-Classification Comparison (PCC) were performed in ArcGIS 10.3®. It was observed between 2000 and 2016, urban area increased 4.46% (8530 ha) from initial area of 10,004 ha. Meanwhile paddy field decreased 3.18% (6091 ha) from initial area of 115,720 ha. The spatial analysis showed that paddy field in the fringe of urban area are more susceptible for changes.

1. Introduction

Agriculture is one of the primary sectors in Indonesia. Relying on natural and human resource potential, the government of Indonesia has made agricultural sector as one of the pillars for promoting economic development of a country. Increasing food self-sufficiency on targeted commodities like rice, corn, soybean, and sugar has been considered as priority program constructed on National Long Term Development Plan 2015-2019.

Food self-sufficiency defines as dependency on domestic production to meet consumption needs of country's population [1]. Prior to promote rice self-sufficiency, Directorate General of Food Crops [2] has made target on paddy production up to 81.9 million tons in 2019. West Java is therefore targeted to produce 13.4 million tons of paddy, equal to 16.4% of total national target production. Land protection and utilization has been considered as one of the solution to increase rice production particularly in West Java [3]. It has been a concern since rapid paddy field changes dominantly occurred around suburban areas of Greater Jakarta.

The development of Jakarta as the capital city of Indonesia has led to the emergence of megacities known as Greater Jakarta which covers Jakarta and adjoin cities. Jakarta and its expanding urban regions are the most urbanized region with 12.41% of Indonesia total population in 2016. Rapid population growth have resulted in urbanization beyond municipal boundaries (also known as suburbanzation) prior to accommodate the need of growing population [4]. The inevitable consequences from expansion of urban sprawl in suburban areas of Greater Jakarta is land use/cover changes, particularly paddy field.

This is due to the high economic development that lead to excessive migration across the cities, which significantly contributes to the population growth in suburban area of Greater Jakarta [5]. Therefore, the suburban areas have a significant contribution in food production scheme in the West Java and Greater Jakarta.

Therefore the objective of this research is understanding the pattern and the factor behind the higher rates of paddy field changes. This study used integrated satellite remote sensing and GIS approach to develop geo-spatial model and analyzed paddy field changes between 2000 and 2016 in Karawang. Karawang, a suburban area of Greater Jakarta, currently attracts considerable attention due to land use/cover changes.

2. Geographical Extent of Study Area

Karawang is located between longitude 107°02'E to 107°40'E and latitude 5°56'S to 6°34'S in the suburban areas of Greater Jakarta. Karawang consists of 30 districts and bordered by Bekasi and Bogor on the west, Subang on the east, and Java Sea on the north. Located in relatively flat terrain with 0-5 m above sea level has made Karawang becomes a potential area for irrigated paddy field.

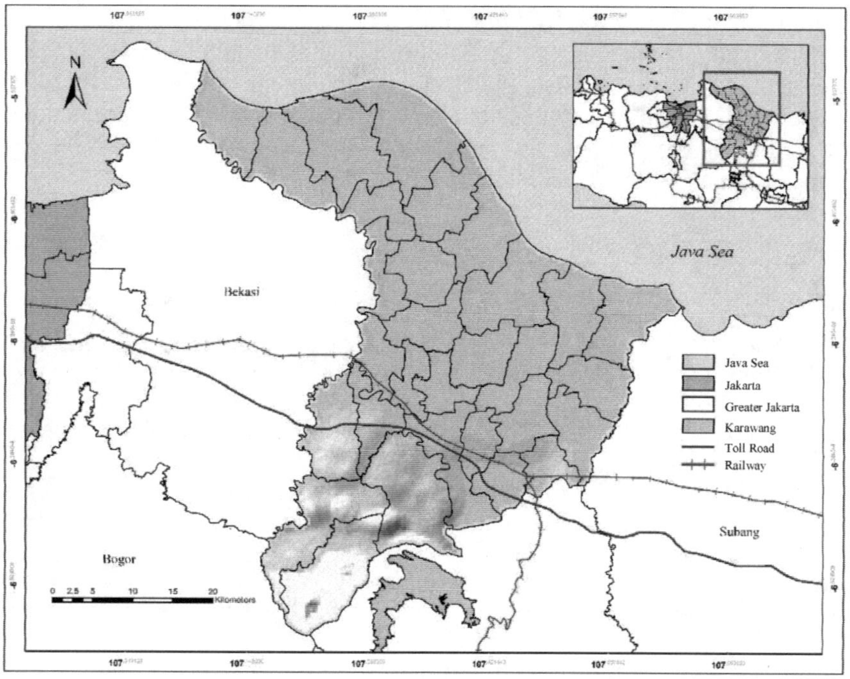

Figure 1. Land use/cover in 2000 and 2016 for Karawang areas

3. Methodology

Prior to develop spatial analysis on land use/cover changes, several methodological approaches were conducted as follows: (a) Data acquisition; (b) Ground truth data collection; (c) Image pre-processing; (d) Image classification and accuracy assessment; (e) Change detection analysis.

3.1. Data Acquisition

This study used Landsat 4-5 TM and Landsat 8 OLI/TIRS acquired from USGS Earth Explorer on September 6, 2000 and May 13, 2016 respectively. Geo-referencing was done by correctly register satellite images with reference images and projected into Universal Transverse Mercator (UTM), Datum WGS 1984 zone 48 south.

IOP Conf. Series: Earth and Environmental Science **149** (2018) 012004 doi:10.1088/1755-1315/149/1/012004

3.2. Ground truth data collection

Ground observation was conducted to identify types of land use/cover distribution in study area. The ground control points in the fields and visual observation of high resolution images on google maps were used to complete supervised classification and evaluate the accuracy of the land use/cover map. Ground control points were taken based land cover type and accessibility of the location.

3.3. Image classification and accuracy assessment

The satellite images were classified into nine classes; forest cover, mixed garden, paddy field, upland agriculture, shrub land, fallow land, aquaculture use, water body, and urban settlements (Table 2). False color composite for vegetation analysis was applied to help classification process due to contrast color between vegetation, water, and soil (Table 1). Maximum likelihood supervised classification was used to generate land use/cover from satellite images by assigning pixels of interest into defined class. All the classification methods were employed in ArcGIS 10.3®.

Multiple accuracy assessment was conducted to measure the acceptability rate of the land use/cover map. The accuracy assessment was carried out by using random sampling of reference points based on ground truth data and visual interpretation on Google Earth Pro®. The comparison data between land use/cover and reference points were served in the form of confusion matrix. The confusion matrix had been widely used to promote accuracy of the land use/cover for various analysis such as watershed analysis [6] and urban expansion analysis [7]. Series of analytical statistics such as overall accuracy, Kappa coefficient, omission error, commission error, producer's accuracy (corresponds to omission error), and user's accuracy (corresponds to commission error) were derived from the confusion matrix and generally reported to assess land use/cover accuracy.

Table 1. False color composite for vegetation analysis in Landsat 4-5 TM and Landsat 8.

Satellite	Band combination (RGB)
Landsat 4-5 TM	5,4,3 (SWIR 1, NIR, Red)
Landsat 8 OLI/TIRS	6,5,4 (SWIR 1, NIR, Red)

SWIR: Shortwave Infrared, NIR: Near-Infrared

Table 2. Description of different land use/cover classes of study area.

Class	Description
Forest Cover (FC)	Large area covered with dense trees, including mountainous area
Mixed Garden (MG)	Mixture of vegetation planted behind settlement and vegetation along the street and the river
Paddy Field (PF)	Field where paddy is grown, mostly dominated by irrigated paddy field
Upland Agriculture (UA	Cultivated area, usually planted by maize, cassava, upland rice (non-irrigated)
Shrub Land (SL)	Sparse vegetation and abandon land
Fallow Land (FL)	Area with exposed soil surface
Aquaculture Use (AU)	Covered both inshore (open sea farming) and offshore (inland fishpond) fish farming
Water Body (WB)	Part of earth surface covered with water: river and lake
Urban Settlements (US)	Area dominated by human. It covers settlement, commercial building, public infrastructure, and industrial estate

3.4. Change detection analysis

Post-classification comparison (PCC) method was used to identify the differences between two independently produced classified land use/cover images [8]. In this study, land use/cover in 2000 and 2016 were overlaid. Post-classification comparison (PCC) method was employed to determine changed of paddy field and into settlements area and unchanged of paddy field during the time period studied. For accuracy assessment, points from Ground Control Points and visual classification based on Google map are utilized. Regarding further analysis, the research focused on paddy field, therefore other land use/land cover were not analyzed.

4. Results

4.1. Land use/cover in 2000
The land use/cover image in 2000 was produced through supervised classification method using Landsat TM datasets (Figure 2). In the land use/cover image, Karawang covered about 191416 ha. The land use/cover distribution consisted of paddy field (60.45%), aquaculture use (9.09%), upland agriculture (6.78%), forest cover (6.47%), urban settlements (5.23%), mixed garden (4.11%), shrub land (4.08%), fallow land (2.47%), and water body (1.31%). Accuracy assessment of the land use/cover map was found with an overall accuracy of 84% and Kappa coefficient of 0.83 (Table 3).

4.2. Land use/cover in 2016
The land use/cover image in 2016 was produced using Landsat OLI/TIRS datasets (Figure 2). The data revealed that in 2016, the land use/cover distribution consisted of paddy field (57.27%), urban settlements (9.68%), aquaculture use (8.29%), forest cover (6.95%), shrub land (6.4%), mixed garden (5.55%), upland agriculture (4.09%), water body (1.59%), and fallow land (0.15%). Accuracy assessment of the land use/cover map was observed with an overall accuracy of 89.89% and Kappa coefficient of 0.88 (Table 4). Higher accuracy assessment of land use/cover map of 2016 was obtained due to availability of recent ground truth information.

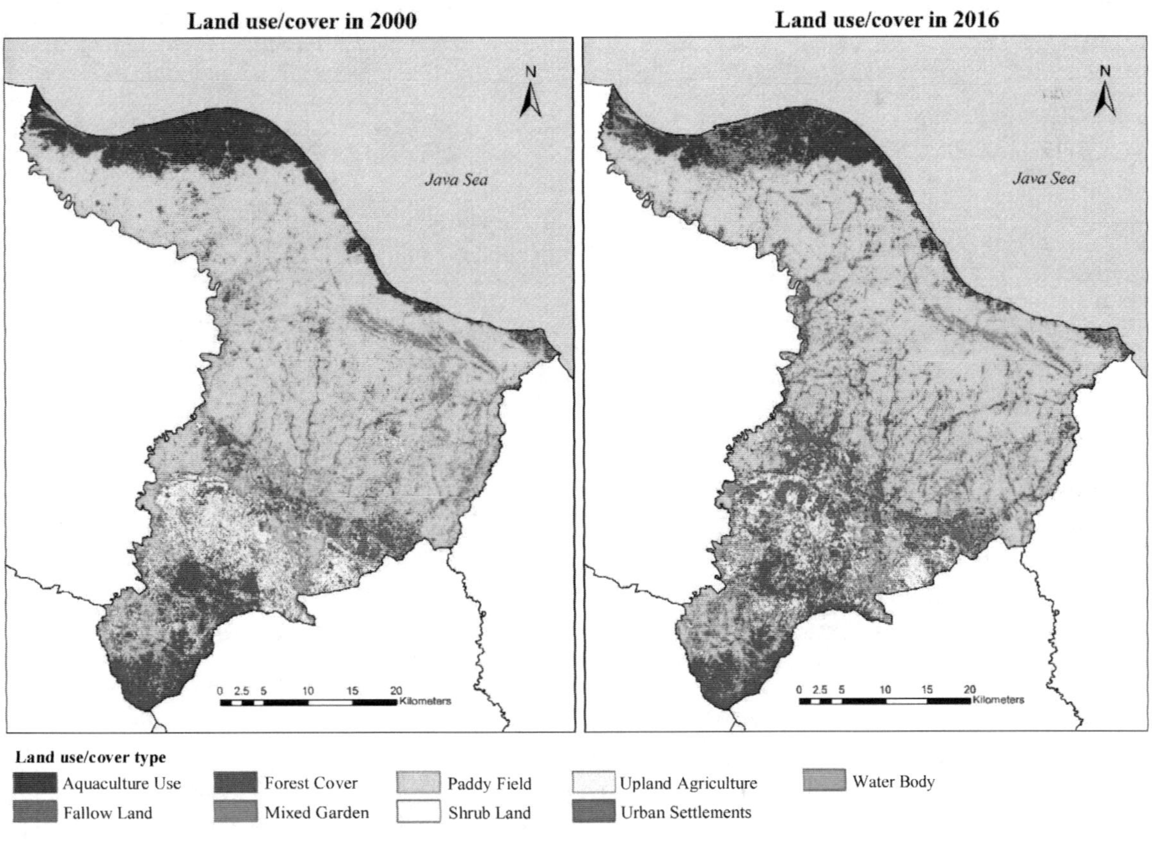

Figure 2. Land use/cover image in 2000 and 2016 for Karawang areas.

4.3. Accuracy assessment of land use/cover in 2000 and 2016
Multiple accuracy assessments of land use/cover in 2000 and 2016 were described in the form of confusion matrix. Total of 482 and 396 reference points based on random sampling of ground truth data

and visual interpretation were taken for accuracy assessments of land use/cover in 2000 and 2016, respectively. Diagonal line showed correctly classified pixels between land use/cover class and reference points. While off-diagonal line corresponded to misclassified class (commission and omission error). Sufficient accuracy of paddy field and urban area were therefore required to evaluate the paddy field land use/cover change. The accuracy assessment is listed in Table 3 and 4.

Low accuracy of land use/cover map in 2000 and 2016 were found in aquaculture use and water body. It happened due to similar surface characteristics. Misclassified pixel was known as salt and paper or pixel noise. Removing the pixel noise could improve the quality of classification.

Table 3. Accuracy assessment of land use/cover (LU) 2000.

LU	Ground truth reference points											
	FC	MG	PF	UA	SL	FL	AU	WB	US	Total	UA (%)	PA (%)
FC	40	2	0	2	0	0	0	0	0	44	90.91	86.96
MG	1	46	0	2	1	0	0	0	0	50	92.00	92.00
PF	1	0	71	0	1	1	3	3	2	82	86.59	89.87
UA	1	0	0	42	0	0	0	0	0	45	93.33	84.00
SL	0	0	1	2	31	1	0	0	1	36	86.11	77.50
FL	2	0	0	2	1	18	0	0	3	26	69.23	81.82
AU	0	0	0	0	0	0	46	13	0	59	77.97	79.31
WB	0	0	1	0	0	0	6	39	0	46	84.78	70.91
US	1	0	6	0	0	2	3	0	76	94	80.85	92.68
Total	46	48	79	50	34	22	58	55	82	482		

Overall Accuracy: 84.85%; Kappa Coefficient: 0.83

FC: Forest Cover; MG: Mixed Garden; PF: Paddy Field; UA: Upland Agriculture; SL: Shrub Land; FL: Fallow Land; AU: Aquaculture Use; WB: Water Body; U: Urban Settlements; UA: User's Accuracy; PA: Producer's Accuracy

Table 4. Accuracy assessment of land use/cover (LU) 2016.

LU	Ground truth reference points											
	FC	MG	PF	UA	SL	FL	AU	WB	US	Total	UA (%)	PA (%)
FC	45	2	0	3	0	0	0	0	0	50	90.00	100.00
MG	0	25	1	2	0	0	1	0	0	29	86.21	80.65
PF	0	0	69	0	0	1	1	3	0	74	93.24	97.18
UA	0	2	0	34	3	0	0	0	0	39	87.18	79.07
SL	0	2	0	4	27	1	0	1	0	35	77.14	90.00
FL	0	0	0	0	0	13	0	0	0	13	100.00	76.47
AU	0	0	1	0	0	0	49	7	0	57	85.96	90.74
WB	0	0	0	0	0	0	3	34	0	37	91.89	75.56
US	0	0	0	0	0	2	0	0	60	62	96.77	100.00
Total	45	31	71	43	30	17	54	45	60	396		

Overall Accuracy: 89.89%; Kappa Coefficient: 0.88

FC: Forest Cover; MG: Mixed Garden; PF: Paddy Field; UA: Upland Agriculture; SL: Shrub Land; FL: Fallow Land; AU: Aquaculture Use; WB: Water Body; U: Urban Settlements; UA: User's Accuracy; PA: Producer's Accuracy

4.4. Land use/cover changes in 2000-2016

Summary of land use/cover changes from 2000 until 2016 is described in Table 5. Within 17 years, considerable land use/cover changes occurred in paddy field and urban settlements in Karawang. Increased of urban settlements were followed by decreased of paddy field areas. In 2016, urban settlements increased 4.46% (8530.3 ha) from initial area of 10,004 ha in 2000 (Table 5). Meanwhile paddy field in 2016 reduced 3.18% (6091.02) from initial area of 115,720 ha in 2000 (Table 5). Loss of paddy field was occurred dominantly in marginal area of urban settlement. According to the [9], extensive road network, industrial decentralization, and affordable house price were the main factor that contributed the most to the rapid population growth in suburban area. In spite of the economic benefit that came from industrial decentralization, adverse effect of industrial growth has therefore threaten Karawang as the second largest rice-producing region in West Java [10].

Table 5. Land use/cover change of Karawang from 2000 until 2016.

No.	Land use/cover Type	Year				Change	
		2000		2016			
		(ha)	(%)	(ha)	(%)	(ha)	(%)
1	Forest Cover	12397.05	6.48	13305.3	6.95	908.25	0.47
2	Mixed Garden	7865.91	4.11	10639.8	5.56	2773.89	1.45
3	Paddy Field	115720.02	60.45	109629	57.27	-6091.02	3.18
4	Upland Agriculture	12981.96	6.78	7839.18	4.10	-5142.78	2.69
5	Shrub Land	7801.83	4.08	12259.3	6.40	4457.47	2.33
6	Fallow Land	4726.12	2.47	298.94	0.16	-4427.18	2.31
7	Aquaculture Use	17402.85	9.09	15864	8.29	-1538.85	0.8
8	Water Body	2515.86	1.31	3045.78	1.59	529.92	0.28
9	Urban Settlements	10004.4	5.23	18534.7	9.68	8530.3	4.46
Total		191416	100	191416	100		

During the same period, other land use/cover types in Karawang were relatively stable with moderate changes. Decreased of land use/cover were occurred in upland agriculture (5142.78 ha), fallow land (4427.18 ha), and aquaculture use (1538.85 ha). While increased of land use/cover occurred in forest cover (908.25 ha), mixed garden (2773.89 ha), shrub land (44 57.47 ha), and water body (529.92 ha) (Table 5).

Socio-economic factor had been considered as the reason of agricultural land abandonment in Karawang. According to Benayas [11], new economic opportunities offered to the rural people created rapid migration that lead to the agricultural land abandonment.

4.5. Change detection analysis

Change detection analysis of paddy field was described on the Figure 3. The figure showed changed and unchanged of paddy field during period of studies. In terms of urban expansion, paddy field was the highest land use/cover type that experienced changes to urban settlements. The data revealed that changed of paddy field into urban settlements from 2000 to 2016 (red area) was about 8273.37 ha (7.15%). Meanwhile, unchanged of paddy field from 2000 to 2016 (green area) covered about 95932.06 ha (82.9%).

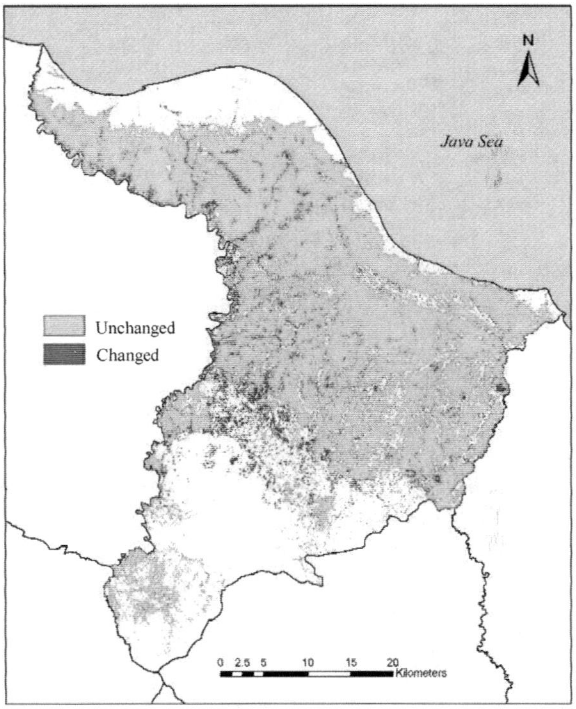

Figure 3. Change detection analysis of paddy field

5. Conclusion and recommendation

A geo-spatial model was developed to assess agricultural land use/cover changes for the suburban areas of Greater Jakarta, especially paddy field. Population pressure and industrial decentralization from greater Jakarta creates high demand on land resulted in conversion of paddy fields. Despite all the economic benefit that comes from industrial decentralization, high demand of land for industrial area, housing and infrastructure near the Greater Jakarta therefore threatened Karawang's potential as rice-producing region. Appropriate uses of land under Sustainable Agricultural Land Regulation (LP2B) should be made and implemented to control urban expansion towards high productive paddy fields.

In the further research, the land use inventory for suburban areas of major cities in Indonesia will be included using time series form with collaboration of other research contributors to provide information for decision making of land use planning. Logistics regression will also be conducted to understand the influencing factors on agricultural land use/cover changes in Karawang. The developed model could help in planning of agricultural lands in the suburban areas to increase production of rice and urban growth with strong vision of sustainability.

References

[1] Thomson A M and Metz M 1999 *Implications of economic policy for food security: a training manual* (Rome: Food and Agriculture Organization of the United Nations)
[2] Directorate General of Food Crops 2015 *Program dan Kegiatan Pembangunan Pertanian Tanaman Pangan Tahun 2015-2019* (Indonesia: Ministry of Agriculture)
[3] Deputy Minister for Food and Agriculture 2015 *Rencana Strategis Tahun 2015-2019* (Indonesia: Coordinating Ministry for Economic Affairs)
[4] Mosammam H M, Nia J T, Khani H, Teymouri A and Kazemi M 2017 Monitoring land use change and measuring urban sprawl based on its spatial forms. *The Egyptian Journal of Remote Sensing and Space Science* **20** 103-116
[5] Novianty E 2015 Balancing Local Government Capacity for a Sustainable Peri-Urban Development: The Case of Karawang Regency *Jurnal Perencanaan Wilayah dan Kota* **26** 71-85.
[6] Butt A, Shabbir R, Ahmad S S and Aziz N 2015 Land use change mapping and analysis using Remote Sensing and GIS: A case study of Simly watershed, Islamabad, Pakistan *The Egyptian Journal of Remote Sensing and Space Science* **18** 251–9
[7] Dewan A M and Yamaguchi Y 2009 Land use and land cover change in Greater Dhaka, Bangladesh: Using remote sensing to promote sustainable urbanization *Applied Geography* **29** 390–401
[8] Singh A 1989 Review Article Digital change detection techniques using remotely-sensed data *International Journal of Remote Sensing* **10** 989–1003
[9] Parés-Ramos I K, Gould W A and Aide T M 2008 Agricultural Abandonment, Suburban Growth, and Forest Expansion in Puerto Rico between 1991 and 2000 *Ecology and Society* **13**
[10] Singh R 2015 *Urban Development Challenges, Risks and Resilience in Asian Mega Cities* (Tokyo: Springer Japan)
[11] Benayas J R 2007 Abandonment of agricultural land: an overview of drivers and consequences. *CAB Reviews: Perspectives in Agriculture, Veterinary Science, Nutrition and Natural Resources* **2**

Integration of logistic regression and multicriteria land evaluation to simulation establishment of sustainable paddy field zone in Indramayu Regency, West Java Province, Indonesia

Irmadi Nahib, Jaka Suryanta, Niedyawati, Priyadi Kardono, Turmudi, Sri Lestari and Rizka Windiastuti

Geospatial Information Agency
Jl Raya Jakarta Bogor KM 46 Cibinong, Jawa Barat, 16911, Indonesia

E-mail: irmnahib@gmail.com, irmadi.nahib@big.go.id

Abstract. Ministry of Agriculture have targeted production of 1.718 million tons of dry grain harvest during period of 2016-2021 to achieve food self-sufficiency, through optimization of special commodities including paddy, soybean and corn. This research was conducted to develop a sustainable paddy field zone delineation model using logistic regression and multicriteria land evaluation in Indramayu Regency. A model was built on the characteristics of local function conversion by considering the concept of sustainable development. Spatial data overlay was constructed using available data, and then this model was built upon the occurrence of paddy field between 1998 and 2015. Equation for the model of paddy field changes obtained was: logit (paddy field conversion) = -2.3048 + 0.0032*X_1 – 0.0027*X_2 + 0.0081*X_3 + 0.0025*X_4 + 0.0026*X_5 + 0.0128*X_6 – 0.0093*X_7 + 0.0032*X_8 + 0.0071*X_9 – 0.0046*X_{10} where X_1 to X_{10} were variables that determine the occurrence of changes in paddy fields, with a result value of Relative Operating Characteristics (ROC) of 0.8262. The weakest variable in influencing the change of paddy field function was X_7 (paddy field price), while the most influential factor was X_1 (distance from river). Result of the logistic regression was used as a weight for multicriteria land evaluation, which recommended three scenarios of paddy fields protection policy: standard, protective, and permissive. The result of this modelling, the priority paddy fields for protected scenario were obtained, as well as the buffer zones for the surrounding paddy fields.

1. Introduction

The most dominant land use change in Indonesia occurs in forest, settlement and paddy field. This happens because the development of existing cities and paddy field are in the region. Paddy field function is common in many suburban areas, especially along the north coast of Java Island. Directorate General of Land and Water Management showed that in Java conversion of wetland into housing was 58.7%, while into industry, offices and shops was 21.8% [1]. National Land Agency reported that during 1994-2004 there had been a change of paddy field function in Java and Bali by 36,000 ha or about 3,600 ha/year [2]. Comparison of land use maps produced in 1990 and 2011 yielded that paddy field in Indramayu Regency in 1990 was 54,961 ha (20% from all area in the regency), while in 2011 it decreased to 13,006 ha.

In the period of twenty-one years there was a decrease of wetland area by 34,557 ha or decreased by 72.65%. The rate of depreciation of paddy fields in Indramayu Regency was 1,645 ha/year. As the impact of increasing public demand for various purposes of land use, the existence of land resources became increasingly scarce. Thus, space as the basis of economic activity must be well laid out in order to accommodate various interests.

IOP Conf. Series: Earth and Environmental Science **149** (2018) 012005 doi:10.1088/1755-1315/149/1/012005

Land use that does not consider the land suitability aspect will lead to environmental degradation so that it is against the principle of sustainable development which requires the realization of a sustainable environment so as not to harm the interests of future generations to come. For the agricultural sector land use that is not in accordance with the characteristics of the land in addition to environmental damage will also cause the decline in productivity of the plant [3].

Indramayu Regency is one of the main agricultural regions, which accounts for 43% of the total Gross Regional Domestic Product (GRDP) of Indramayu Regency. In 2011, rice production was 1,704,956.71 tons produced from 239,465 ha harvested area with paddy productivity of 7.12 tons/ha. Ministry of Agriculture have targeted food self-sufficiency by production of 1.718 million tons of dry grain harvesting during the period of 2016-2021. In order to accelerate the achievement of food self-sufficiency, it was pursued through the optimization of special commodities, including rice, soybean and corn.

Land use changes could be predicted quantitatively by incorporating physical, social, economic, and policy factors [4]. Predicted land use changes could be analyzed through a spatial-based model approach, one of which was based on Cellular Automata (CA). CA is a spatial-based model that is able to predict future conditions of local interactions between cells on a regular grid [5], in which cells represent land use. Rules were made as a consideration of the neighboring cell that became the basis of land use change. CA consists of several components: cell (pixel), state, neighborhood (neighborhood) and transition ruler/transition function.

The land use model using CA had been applied as a tool to support land use planning and policy analysis as well as exploring scenarios for future development [3]. Changes in land use, at a given location and within a certain time, could be studied as phenomena or dichotomous events. Land use change, as a binary phenomenon, consisted of only two categories: change or unchanged. Factors affecting land use change were generally a combination of continuous and categorical variables [6]. The analysis of factors influencing land use change was approached by logistic binistic regression equation (logit model). Logistic regression is a mathematical model to analyze the relationship of independent variables in the form of data continue, discrete, dichotomous, or a combination that affects one dependent variable [7]. This analysis technique has been carried out on forest conversion [8, 9], urban growth [7] and changes in wetland [10].

The purpose of this research was to know the change of land use in Indramayu Regency in period 1998-2015, and to build spatial model of change of paddy field in Indramayu Regency. The spatial model was designed by considering sustainable development aspects and characteristics of local land use conversions, which were analyzed using logistic regression. Results of this analysis were then used as values in determining the establishment of sustainable paddy fields using multicriteria evaluation.

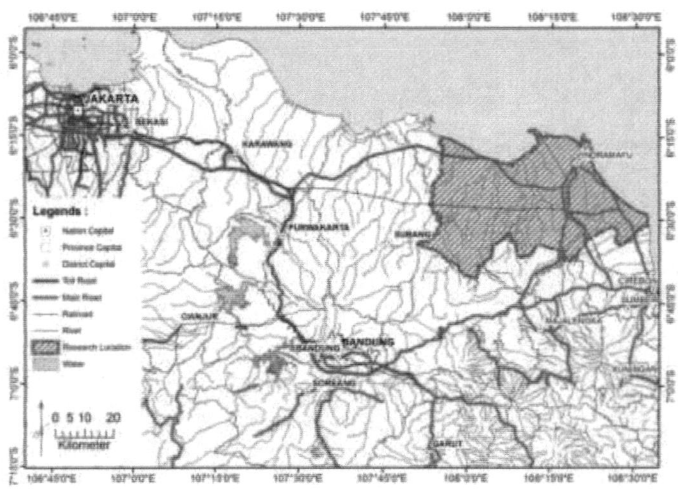

Figure 1. Location map of the research.

2. Methods

This research used ArcGIS Software, Idrisi TerrSet Software and MS Excel Software. Location map of the research can be seen in Figure 1.

2.1 Materials and tools

Data used in this study were (a) Digital Map RBI scale 1:25,000 Indramayu District year 1994 (layers: road network, river network, cities); (b) IKONOS Imagery year 2008, GeoEye Imagery year 2008, SPOT 6 Imagery year 2014 and 2015; (c) Land use and land cover map scale 1:25,000 scale year 1998, 2008 and 2015, (d) Map of the administrative area of Indramayu Regency scale 1:50,000, Development Planning Agency of Indramayu Regency year 2014, (e) Land suitability maps for paddy rice crops year 2016, (f) Map of population density of Indramayu Regency, scale 1: 50,000 year 2008. and (g) Demographic and land data from Statistics Agency year 2016. The research activity stage was presented in Figure 2

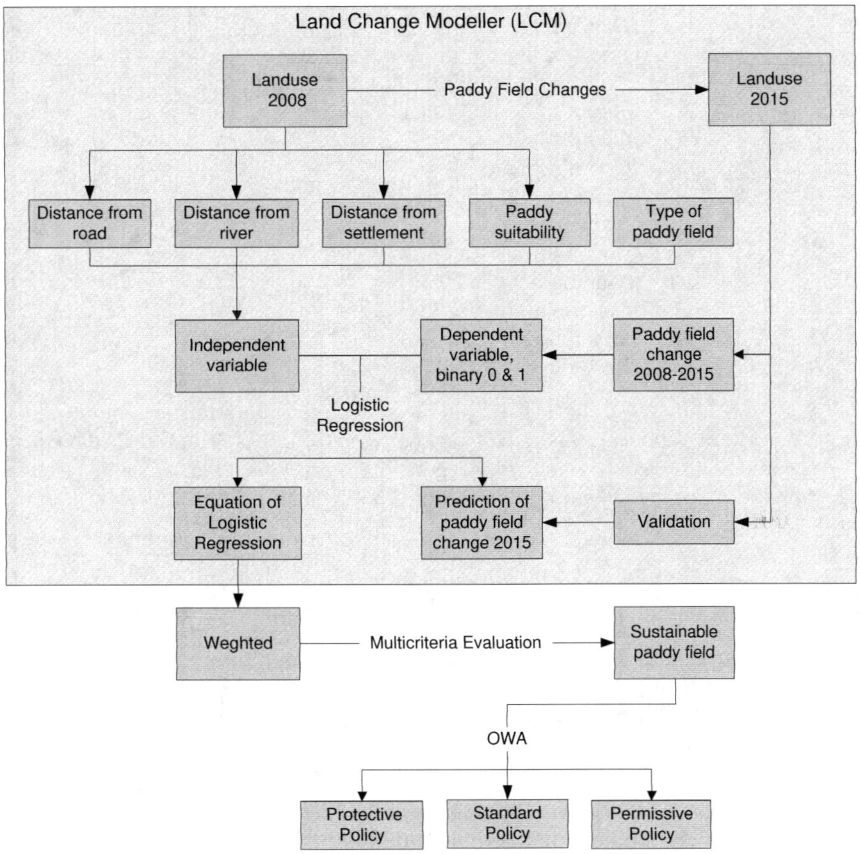

Figure 2. Research flow diagram. [11]

2.2 Analysis of land cover changes

Analysis of changes in land cover was performed by comparing the land cover maps produced by Regional Development Agency of Indramayu Regency. Determination of land cover area used spatial analysis which was done by Land Change Modeler in Indrisi TerrSet process of the maps of Indramayu Regency in 1998,

2008 and 2015. Spatial analysis of multi layer could be integrated when the data had the same format. The spatial data format used was presented in Table 1.

2.3 The Logistic Regression Model

In the modeling process land use and land cover maps for 1998, 2008 and 2015 were reclassified. The existing land cover consisted of 7 classes, namely: water bodies, forests, gardens, plantations, Irrigation paddy fields, Rain fed paddy fields and others. It then was reclassified into 3 classes, namely irrigated paddy fields, rain-fed paddy fields, and non-paddy fields. Vector-format maps were converted to raster format with cell size 20. Furthermore, using the land use map of 1998 as baseline and land use map of 2008 as the final data, analysis of land changes using TerrSet Idrisi software was performed. Analysis of land use d change was done with the Land Change Modeller (LCM) module in TerrSet Software as shown in Figure 2.

Table 1. Standard of spatial data format of Indramayu Regency.

No	Parameter	Information
1	Data type	Real
2	File type	Binary
3	Column	4,102
4	Line	2,697
5	Coordinate system	UTM Zone 49 S
6	Coordinate unit	Meter
7	Distance units	1.00
8	Minimum x	148,692.2088
9	Maximum x	230,732.2088
10	Minimum y	9,259,649.7982
11	Maximum y	9,313,589.7982
12	Resolution X	20.00
13	Resolution Y	20.00

The probability of paddy fields change was considered as a function of the explanatory variables. It was a monotonic curves linear response bounded between 0 and 1 [11] and defined by the logistic function. Changes (Y) were 0 or 1, where 0 meant no change and 1 meant there was a change. The logistic regression equation could be formulated by the form of equation 1 and equation 2 as follows:

$$p = E(Y)\frac{e^{\beta_0+\beta_1 X_1+\beta_2 X_2+\beta_3 X_3+\beta_4 X_4}}{1+e^{\beta_0+\beta_1 X_1+\beta_2 X_2+\beta_3 X_3+\beta_4 X_4}} \quad (1)$$

$$p' = \beta_0 + \beta_1 X_1 + \beta_2 X_2 + \beta_3 X_3 + \beta_4 X_4 \quad (2)$$

Where:

p = the probability of paddy fields change,
E(Y) = the expected value of the dependent variable Y,
β_0 = a constant to be estimated,
β_i = the coefficient to be estimated for each variable (X_i)

As mentioned earlier, land use change analysis was done using LCM module in TerrSet sotware. LCM is one of many tools which concern for enviromental spatial analyst. Future land cover could be predicted based on actual and previous land cover, and the module also considered some driver variables. We conducted the study in the area by cross tabulating the derived land cover map of 2008 as the land cover early input image and the 2015 land cover map as the later input image. Potential transition modelling was used to group land use transitions into a set of sub-models and then utilized to explore the potential power of the chosen explanatory variables. In the current study, the change analysis showed that there were two sets of transitions, from irrigated paddy field to non-paddy field and from rain-fed paddy field to non-paddy. Meanwhile, some independent variables used in this research were: distance from road (X_1), distance from settlement (X_2), distance from river (X_3), distance from district capital (X_4), type of paddy field (X_5), productivity of paddy fields (X_6), price of paddy field (X_7), intensity of harvesting of paddy fields (X_8), population density (X_9) and land suitability of paddy fields (X_{10})

In order to predict change, the transition was empirically modelled using Multi-Layer Perception (MLP) neural network and Logistic Regression Model (LRM). LRM was used to model and analyze the land cover change in Idrisi TerrSet. The binary presence or absence was the dependent variable for the periods 2008–2015. The predicted land cover of 2015 was validated using ROC/AUC (Relative Operating Characteristic/ Area Under Curve) module of Idrisi TerrSet. The ROC module compared suitability image by depicting the likelihood of that class occurance (the input image) and a boolean image showing where that class actually existed (the reference image) [12, 13]. The ROC curve was the true vs false positive fraction and the AUC was a measurement of overall performance.

Logistic regression was a special form of regression that was formulated to predict and explain a binary category variable [14]. Meanwhile logistic regression model required that all variables be calibrated so as to align with the change of paddy fields [8, 9]. Thus, all variables needed to be normalized between 0.1 and 0.9. Based on the land use change from paddy fields to non-paddy fields, after validating the estimation model by comparing the value with the actual changes, then the trend of land use changes during a certain period could be predicted.

2.4 Multicriteria evaluation

Multicriteria evaluation was carried out using Idrisi TerrSet. The objective was to establish sustainable paddy field area to the whole region of Indramayu. Weighting in this analysis used the results of the logistic regression equation. Multi Criteria Evaluation (MCE) with a weighted spatial analysis approach was used to establish the sustainable paddy fields. MCE consisted of two things: goals setting and criteria setting. These criteria were divided into factors and constraints. Factors consisted of sub factors which were the criteria to achieve goal. Weighted Linear Combination (WLC) equation is shown in equation 3.

$$WLC = \left(\sum_{i=1}^{n} X_{ij} \times W_{ij}\right) \times C_j \qquad (3)$$

Where:

X_{ij} = the suitability degree of j-th factor in i-th location,
W_{ij} = the weight of j-th factors/sub factors in factor location,
C_j = the constraint on the j-th factors/sub factors.

The WLC method was based on the assumption that each factor was weighted differently. The process of determining the weight was done by using the result of logistic regression analysis, which obtained the value of the variables on the change of paddy field function in Indramayu. While for variables that were categorical, namely: types of paddy field, production of paddy, paddy field price, intensity of harvest, population density and land suitability, the value of regression analysis was not used and therefore replaced with a value of one. The values were then used as weighted values in multicriteria evaluation. Furthermore the weight values were normalized by dividing the value for each variable with the sum of variable values. Weighting results and normalization were presented in Table 2.

Table 2. Weighting of independent variables in establishment of sustainable paddy field zone.

Land suitability for paddy	LR Value	Weight	Normalized
Distance from road	-0.0032	0.32	0.0418
Distance from settlement	-0.0027	0.27	0.0353
Distance from the river	+0.0081	0.81	0.1059
Distance from district	0.0025	0.25	0.0327
Types of paddy field	+0.0026	1.00	0.1307
Production of paddy	0.0128	1.00	0.1307
Paddy Field Price	-0.0093	1.00	0.1307
Intensity of harvest	0.0032	1.00	0.1307
Population density	0.0071	1.00	0.1307
Land suitability	-0.0046	1.00	0.1307
		7.65	

Where LR = logistic regression

The results of multicriteria evaluation were then tested by Ordered Weighted Average analysis (OWA), which was characterised by two sets of weights: criterion importance weights and order weights. An interactive way of choosing, modifying, and fine-tuning the decision strategy defined by the order weights was proposed [15]. Meanwhile, OWA method was a technique used to rank the criteria and manage the uncertainty interactions [16]. OWA method would produce continuous scale scenario between slices (risk adverse) with the union (risk-taking). OWA method could provide many options between two extreme sides, such as only focusing on protecting the best paddy fields and protecting almost all existing paddy fields.

Establishment of paddy field conservation using OWA method had flexibility in accomodating other land use needs for regional development. The conservation could be limited to areas having high priority or expanded to medium priority. The priority of paddy field conservation was determined by combining characteristics of the paddy field function change and its internal potential [17]. In this research OWA was performed using MCE module in TessSet software, which could give an outlook on the risks of policy scenarios taken, whether it was standard, protective or permissive, and thus could adjust trade-off and compensation between criteria. Standard policy was the result of multicritera evaluation without OWA, which could be used directly with a standard protection level. Standard policy was basically a safeguard which was equal to threat magnitude. Protective scenario was to be chosen to protect almost all existing paddy fields, whereas permissive scenario would provide government flexibility to converse in order to encourage the region development [17].

Other advantages of OWA were its ability to integrate heterogeneous datasets, such as qualitative and quantitative criteria using expert knowledge, its flexibility to choose certain criteria for different study areas or different problems, its flexibility to alter criteria priorities, and its freedom to develop various scenario models on risk level [16]. Based on result of Analytic Hierarchy Process (AHP), evaluation criteria were obtained. Weighting results and normalization were as presented in Table 3.

Table 3. Ordered weight of OWA for three different scenarios.

Ordered Weight	Policy		
	Standard	Protective	Permissive
Weight 1	0.1000	0.2000	0.0015
Weight 2	0.1000	0.1500	0.0015
Weight 3	0.1000	0.1500	0.0900
Weight 3	0.1000	0.1250	0.0900
Weight 5	0.1000	0.1000	0.0900
Weight 6	0.1000	0.0900	0.1000
Weight 7	0.1000	0.0900	0.1250
Weight 8	0.1000	0.0900	0.1500
Weight 9	0.1000	0.0015	0.1500
Weight 10	0.1000	0.0015	0.2000

Parameters used to support sustainable development included (a) physical factor; (b) economic factor and (c) social factor. The factors of multicriteria evaluation design were as shown in Figure 3.

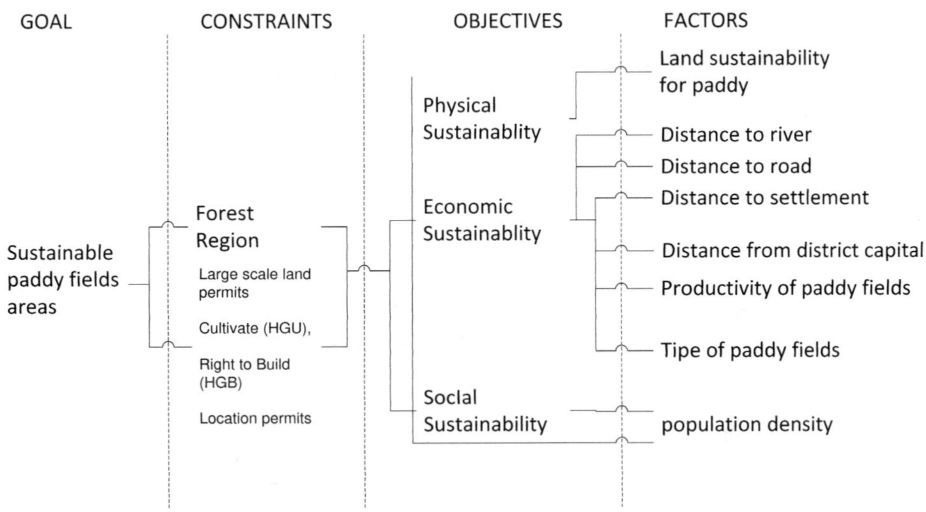

Figure 3. The design of multicriteria evaluation on sustainable paddy field establishment. [11]

In practice, logistic regression test was carried out on various combinations of independent variables in both quantity and type. The combinations of variables were implemented from 2 to 10 combinations of variables.

Results of OWA for all the three scenarios (standard, protective, and permissive) would result a value ranging from 0 to 255. The values would then be grouped into 5 classes with the same intervals. The highest value represented the paddy field that was prioritized to be protected the most due to its change function characteristics and internal potential (land suitability and productivity).

3. Results and Discussion

3.1 Determinants of Rice Function

After passing logistic regression test of various combinations on number and type of variables, then the best equation was obtained. The results in Table 4 showed that the best equation was the equation in the Set 2 combination column which was indicated by the highest pseudo R-square value of 0.1950, which was closest to the minimum standard value of 0.2.

Table 4. Results of logistic regression analysis with various variables.

No	Variabel / parameter	Set 1	Set 2	Set 3	Set 4
	Intercept	-5.2958	-2.3048	-2.1582	-4.6195
X_1	Distance from road	0.000751	+0.0032	+0.0032	
X_2	Distance from settlement	0.000753	-0.0027	-0.0027	
X_3	Distance from the river	-0.000512	+0.0081	+0.0079	0.0054
X_4	Distance from district	-0.0000112	0.0025	0.0024	
X_5	Types of paddy field	-0.34423	+0.0026	+0.0028	0.0026
X_6	Production of paddy	1.0220	0.0128	0.0127	0.0115
X_7	Paddy Field Price	0.8301	-0.0093	-0.0096	-0.0104
X_8	Intensity of harvest	0.4525	0.0032	-0.0030	0.000167
X_9	Population density		0.0071	0.0076	
X_{10}	Land suitability		-0.0046	-0.0046	
	-2logL0	72,252.8935	72,158.1203	72,514.4209	72,535.7134
	-2log(likelihood)	62,055.2679	59,963.5606	60,223.0069	61,815.0977
	Goodness of fit	386,838.6960	406,342.7648	412,298.1619	367,547.6517
	Pseudo R^2	0.1411	0.1950	0.1690	0.1478

IOP Conf. Series: Earth and Environmental Science **149** (2018) 012005 doi:10.1088/1755-1315/149/1/012005

ROC	0.7990	0.8262	0.8158	0.8032

By using ROC method, the result of validation of prediction result was 0.8262. The ROC result gave the optimum predicted model value. The results of the 2015 paddy field prediction is presented in Figure 4a, where it was clear that the threat of wetland function was based on the largest logistic regression equation along the main roads in Indramayu. This region is a path of economic growth, tourism, industrial areas, and settlement growth. The column of Set 2 in Table 4 could be arranged in the form of the equation as follow:

$$\text{Logit (paddy field conversion)} = -2.3048 + 0.0032*X_1 - 0.0027*X_2 + 0.0081*X_3 + 0.0025*X_4 + 0.0026*X_5 + 0.0128*X_6 - 0.0093*X_7 + 0.0032*X_8 + 0.0071*X_9 - 0.0046*X_{10} \quad (4)$$

Equation 5 showed that variables to the conversion of land influenced both positively and negatively. The weakest variable in influencing the change of paddy field function was the paddy field price, while the most influential factor in the change of paddy field function was distance from river and from road. The large change of paddy fields adjacent to the main road showed that economic growth was still focused around the main road lane. This relationship was useful for selection of the explanatory variables for paddy fields change in the study area. Figure 4 showed projected land cover in 2015 and the potential of paddy field conversion. This was due to the relationship between dependent variables and the explanatory variables done in this research.

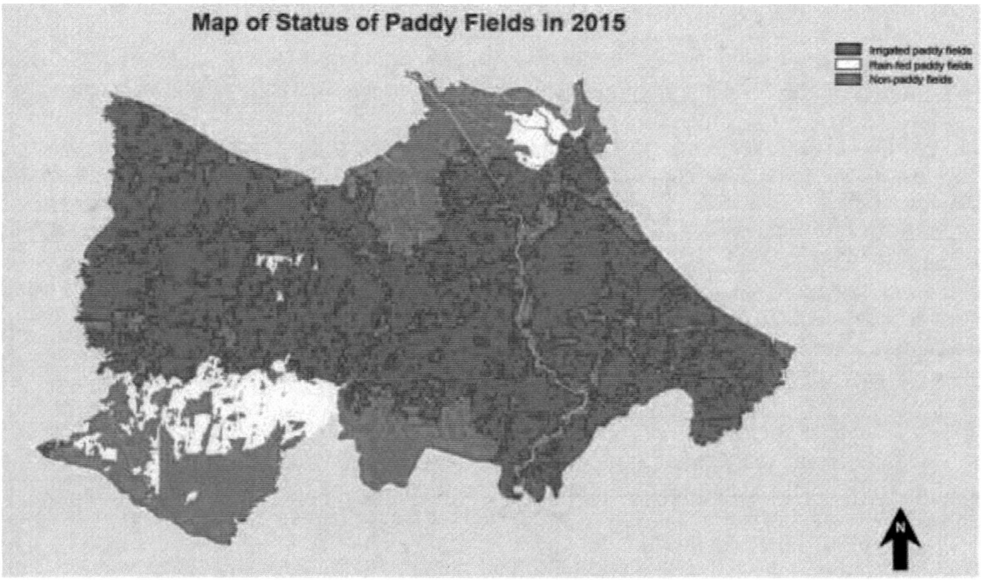

(a) Projected Land Cover 2015

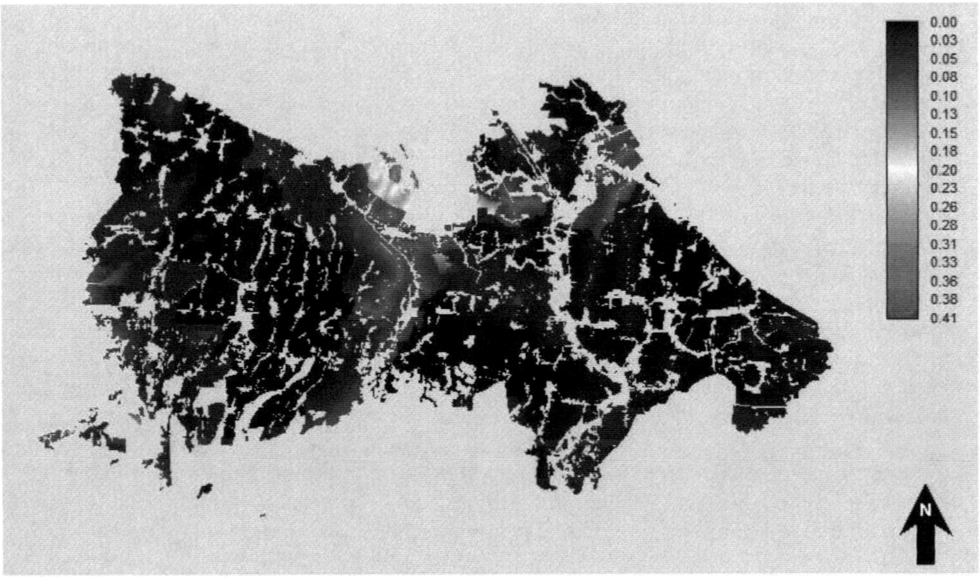

(b) Potential for Paddy Field Conversion
Figure 4. Prediction of Paddy Field in 2015.

Table 5. The paddy fields conversion data from 1998 to 2015.

No	Classes	Year			Changes	
		1998	2008	2015	1998-2008	2008-2015
1	Irrigated paddy fields	122,054.04	119,448.44	118,913.84	-2,605.60	-534.60
2	Rain-fed paddy fields	15,495.52	15,313.08	15,352.20	-182.44	39.12
3	Non-paddy fields	68,708.52	71,496.56	71,992.04	2,788.04	495.48
	Total irrigated and rain-fed rice fields conversion				-2,788.040	-495.48

From Table 5 above, irrigated paddy fields continued to decline within 1998-2015, whereas about 3,140 Ha had been converted to other uses. In periode 1998-2008, about 2.605.60 ha (260 ha per year) and periode 2008-2015 about 534.60 ha (76 ha per year) had been converted to other uses. Meanwhile, rain-fed paddy fields were not significantly decreased, which was only about 143 Ha (8 ha per year). The case of paddy field conversion was also discovered by Asmarani that showed that rice production in Indramayu district from year 2011-2015 tended to decrease with an average of 4.16% per year, probably caused by paddy field conversion [18].

The majority of paddy fields in Indramayu district were technically irrigated ones. These paddy fields were generally located in the lowlands area, in which the industrial development, settlement, and other facilities/infrastructure were more easily developed. The strategic geographical location and accessibility to the business areas such as Bandung, Bogor, and Bekasi, could be other reasons for conversion. Agricultural land usually had a lower rent price compared to residential and industrial land [19]. The calculation from Kamilah (2016) research indicated that the ratio of land rent for agricultural land to residential was 13.89 and 329.26 to industrials. No wonder, majority of land conversion in Indramayu occurred because of settlements and industrial developments.

Meanwhile other research, Changes in land use in Indramayu Regency mostly occur in paddy fields that turn into settlements. This change occurred mainly in the sub-districts of Indramayu, Patrol, Sukra, Losarang, and territories crossed by the main route of Java's northern coast. The rate of land use change between the period of 1994 - 2008 was 101.5 ha / year, whereas 2008-2015 was only 14 ha / year. Conversion of agricultural land to nonagriculture if it continues can be one of the causes of the food crisis. Conversion of this land will remove the function of rice fields as land intended to produce rice [20]

3.2 Establishment of Sustainable Paddy fields

Three policy scenarios could be arranged from sustainable paddy field protection model, namely: standard policy, protective policy and permissive policy. Standard policy was the result of multicriteria evaluation without OWA, which could be used directly with a standard protection level. Standard policy basically is a safeguard which is equal to threat magnitude. Protective scenario was chosen to protect almost all existing paddy fields, whereas permissive scenario provided government flexibility to convert in order to encourage regional development [17].

Paddy field protection priority was obtained using sustainable paddy field protection model that was demonstrated earlier. The paddy fields were mostly located on land suitability class S1, which had the most suitable land physical characteristics. The values of the results from the standard, protective and permissive scenarios were classified with range of 0-255 to 5 paddy field protection classes. The classification result and its calculation of each class can be seen in Figure 5 and Table 6. By using the three scenarios, the model had high flexibility to accommodate other land use interests of development. Paddy field protection was limited to areas with high to medium protection weight. Likewise, the regional development, should aim the paddy fields with low protection weight.

Based on the results of classification and calculation in Table 2 shown the results there were 5 categories. Category 1 meant the land area was more important to be protected than 2 to 5, and on the contrary category 5 meant that the land would change more quickly. Thus, the most priority land area to be protected for protective policy was 4.28%, for standard policy was 4.35% and for permissive policy was 2.12% as shown in Table 6.

Table 6. The area of each protection class in three different scenarios.

Category	Protective		Standart		Permissive	
	Hectare	percent	Hectare	percent	Hectare	percent
1 (200 – 255)	9,017.28	4.28	8,927.80	4.35	4,575.24	2.12
2 (151 – 200)	50,308.84	23.85	48,498.92	23.62	24,717.28	11.47
3 (101 – 150)	82,662.64	39.19	76,774.64	37.40	89,754.96	41.64
4 (51 – 100)	41,919.96	19.88	29,059.88	14.16	60,921.44	28.27
5 (0 – 50)	26,994.64	12.80	42,027.84	20.47	35,560.08	16.50
Total	210,903.36	100.00	205,289.08	100.00	215,529.00	100.00

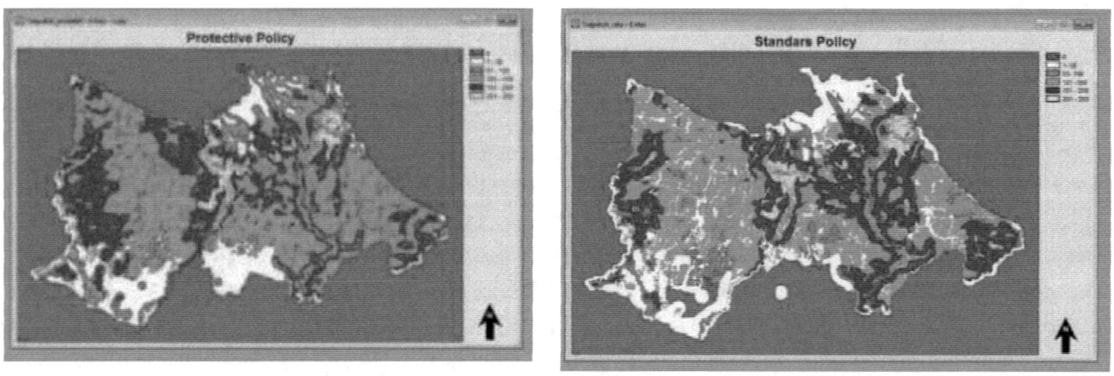

(a) Protective Policy (b) Standard Policy

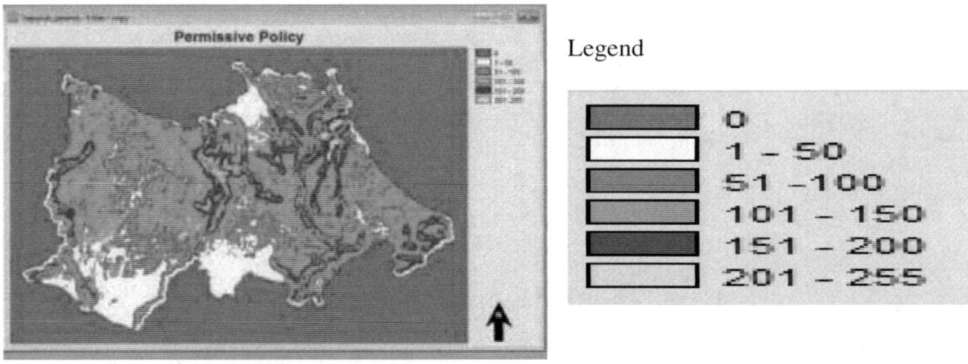

(c) Permissive Policy

Figure 5. Different policy scenarios to preserve paddy fields.

In order to accelerate the food self-sufficiency as targeted by the Ministry of Agriculture to produce 1.718 million tons of dry grain harvest within 2016-2021, optimization of special commodities such as paddy, soybean and corn should be performed. As a sample application of Table 6 above, suppose Indramayu was required to provide only 100,000 to 120,000 ha paddy fields with best to medium suitability (category 1 to 3), then permissive policy could be applied, because the sum of the areas in those category would be 119,047 ha and it would be sufficient. However, should Indramayu be required to provide 140,000 to 150,000 ha, then protective policy should be applied (sum of the areas would be 141,989 ha).

4. Conclusion

This research presented a sustainable paddy field setting model developed using logistic regression and multicriteria evaluation with respect to the characteristics of local paddy field transfers and the concept of sustainable development. Logistic regression was able to reveal the key variables in the transfer of wetland functions in a region quickly. Variable determinants of the transfer of paddy fields in Indramayu included distance from roads, settlements, and industries.

This model was built upon the occurrence of paddy field between 1998 and 2015. Equation of the paddy field changes models obtained were: logit (paddy field conversion) = $-2.3048 + 0.0032*X_1 - 0.0027*X_2 + 0.0081*X_3 + 0.0025*X_4 + 0.0026*X_5 + 0.0128*X_6 - 0.0093*X_7 + 0.0032*X_8 + 0.0071*X_9 - 0.0046*X_{10}$ with a value of Relative Operating Characteristics (ROC) of 0.8262.

Through multicriteria evaluation priority fields that must be protected and also a buffer for the fields behind it could be obtained. Sustainable paddy field protection policy could then be applied through three policy scenarios, namely standard, protective, and permissive. This would make it easier for the government to provide subsidies appropriately and efficiently while still opening investment opportunities for regional development.

The model built in this study involved spatial and temporal data that could be applied for other regions. With more complete and accurate data support, the results could be more optimal. The model was composed by clear stages so it is possible to be tested on other areas with different characteristics.

Acknowledgments

Acknowledgments are conveyed to the Indramayu Regency Government who has provided both data and assistance during field surveys. Also thanks to the Head of PPKS-BIG who has provided opportunity and supports so that the research could run well.

References

[1] Directorate General of Land and Water Management 2008 *Proc. Central and Regional Spatial Planning Coordination Meeting*

[2] BPN 2007 *ATLAS Neraca Penatagunaan Tanah Nasional* (Jakarta: Badan Pertanahan Nasional) *ATLAS National Land Use Account* (Jakarta: National Land Agency)
[3] van Vliet J, White R, and Dragicevic S 2009 *Computers, Environment and Urban Systems* **33** 35
[4] Nowar W, Baskoro D P T, and Tjahjono B 2015 *Jurnal Tata Loka* **17**
[5] Munibah K, Sitorus S R P, Rustiadi E, Gandasasmita K and Hartrisari 2010 *Jurnal Tanah dan Iklim* **32** 55
[6] Xie Y, Mei Y, Guangjin T and Xuerong X 2005 *Global Environmental Change* **15** 238
[7] Arsanjani J J, Helbich M, Kainz W and Boloorani A D 2013 *International Journal of Applied Earth Observation and Geoinformation* **21** 265
[8] Kumar R, Nandy S, Agarwal R and Kushwaha S.P.S 2014 *Ecological Indicators* **45** 444
[9] Siles N J S 2009 *Spatial Modelling and Predicton of Tropical Forest Conversion in the Isiboro Secure National Park and Indigenous Territory (TIPNIS), Bolivia* (Enschede: International Institute for Geoinformation Science and Earth Observation)
[10] Santosa S., Murtilaksono M. K, and Rachman W. N. F. 2015 *Journal of Environmental and Earth Science* **5** 79.
[11] Pontius R.G, and Schneider L.C 2001. *Agriculture, Ecosystems & Environment* **85** 239
[12] Eastman, J. R. 2014 Idrisi 18.2. The TerrSet Edition. Clark Labs. Worchester, MA**,**
[13] Reddy, C. S., Singh, S., Dadhwal, V. K., Jha, C. S., Rao, N. R., & Diwakar, P. G. 2017 *Journal of Earth System Science*, **126** 8
[14] Hair J F, Black W C, Babin B J 2010 *Multivariate Data Analysis: A Global Perspective* 7th Edition (New Jersey: Pearson Prentice Hall)
[15] Rinner, C., & Malczewski, J. 2002 *Journal of Geographical Systems* **4** 385
[16] Gorsevski, P. V., Donevska, K. R., Mitrovski, C. D., & Frizado, J. P. 2012 *Waste Management*, **32** 287
[17] Santosa, S., Rustiadi, E., Mulyanto, B., Murtilaksono, K., Widiatmaka, W., & Rachman, N F 2014 *Majalah Ilmiah Globe (Scientific Journal Of Globe),* ***16*** 181
[18] Asmarani, D P 2016. *Analisis Kemandirian Beras Di Kabupaten Indramayu Tahun Tahun 2011-2015* (Bogor : Sekolah Pascasarjana Institut Pertanian Bogor) */ Analysis of Rice Independence in Indramayu Regency Year 2011-2015.* (Bogor: Graduate School of Bogor Agricultural University)
[19] Kamilah, A 2013 *Jurnal Agribisnis dan Pengembangan Wilayah (Journal of Agribusiness and Regional Development)* **5** 1
[20] Murdaningsih, Widiatmaka , Munibah K Ambarwulan W 2017 Majalah Ilmiah Globë **19** 175

IOP Conf. Series: Earth and Environmental Science **149** (2018) 012006 doi:10.1088/1755-1315/149/1/012006

Land use/land cover and land capability data for evaluating land utilization and official land use planning in Indramayu Regency, West Java, Indonesia

W Ambarwulan[1], Widiatmaka[2] I Nahib[1]

[1]Geospatial Information Agency, Jalan Raya Jakarta-Bogor Km 46 Cibinong, 16911 Bogor, Indonesia

[2]Dept. of Soil and Land Resources, Bogor Agricultural University, Kampus IPB Darmaga, Bogor 16680, Indonesia

Email : w_ambarwulan@yahoo.com

Abstract. Land utilization in Indonesia is regulated in an official spatial land use planning (OSLUP), stipulated by government regulations. However in fact, land utilizations are often develops inconsistent with regulations. OSLUP itself is also not usually compatible with sustainable land utilizations. This study aims to evaluate current land utilizations and OSLUP in Indramayu Regency, West Java. The methodology used is the integrated analysis using land use and land cover (LU/LC) data, land capability data and spatial pattern in OSLUP. Actual LU/LC are interpreted using SPOT-6 imagery of 2014. The spatial data of land capabilities are derived from land capability classification using field data and laboratory analysis. The confrontation between these spatial data is interpreted in terms of future direction for sustainable land use planning. The results shows that Indramayu regency consists of 8 types of LU/LC. Land capability in research area range from class II to VIII. Only a small portion of the land in Indramayu has been used in accordance with land capability, but most of the land is used exceeding its land capability.

1. Introduction

Indonesia is fourth most populated country in the world after China, India and the USA with the population at more than 255 million in 2015 [1,2]. One of the problem related to this high number of population is the imbalance of the population between islands. Actually, a total of 60% of the population stay in Java Island, which have only 6.7 percent of total Indonesian territory [1]. This high density of the population in Java island give high pressure on land due to the need for settlements, industrial sector and other need for human being. As a result, land utilization is often done in the way so as it exceeds its land capability. Land utilization exceeding its land capability will in turn lead to land degradation [3,4]. The existence of cultivation on steep slopes, existence of urban slums as well as dense settlements on river banks is an example of land utilization that exceeds land capability and environmental carrying capacity [5].

In Indonesia, land utilization is actually regulated in a government regulation namely official spatial land use planning (OSLUP), which is regulated in the Act No. 26/2007. In such regulation, land allocation is given for each sector in a spatial pattern, for example land allocation for agriculture, for settlements, for

IOP Conf. Series: Earth and Environmental Science **149** (2018) 012006 doi:10.1088/1755-1315/149/1/012006

industry and others. Such regulation is intended to regulate that this limited natural resource can be used for various purposes with sustainable manner, which means minimize land degradation. This spatial pattern of OSLUP is given at any governmental level, i.e. provinces, districts and cities.

Although land has been allocated formally, factually however land utilization break often such regulation [6]. Factually also, the regulations themselves do not guarantee whether the formal allocations are correct from the point of view of land carrying capacity. This is because agencies that allocate such land, give less technical considerations that ensure sustainability. For this reason, it is necessary to review both the existing land utilizations as well as formal allocations in the spatial pattern of OSLUP in order to know whether both land utilizations and land allocations are done in a sustainable way [7, 8].

One way to evaluate existing land utilizations as well as formal land allocations in spatial pattern of OSLUP is using a reference which reflect land potency. In this case, Indonesian Ministry of Environment suggest the use of land capability as parameter control. The rationale is that both the existing land utilization as well as land allocation in OSLUP should be in accordance with land capability in order to use land in sustainable way. Such an evaluation has been done for example study in Tuban Regency, East Java [8] and in West Sulawesi [9]. With such an evaluation, it can be shown that the existing land utilization as well as OSLUP requires improvements in order to be conform with land capability, or in other word, to environmental land carrying capacity.

Based on such background, the research objectives of this paper are: (i) to delineate existing land utilizations (ii) to delineate land capability, and (iii) to evaluate the conformity between existing land utilization and spatial pattern of OSLUP with land capability.

2. Methodology

The study was done in Indramayu Regency, West Java, Indonesia (Figure 1). The regency covers an area of 209,942 ha, and is consisted of 31 districts and 309 villages. Geographically, the regency is located at 107°52'- 108°36'E and 06°15'- 06°40' S. Indramayu Regency is spread out along the north coast of Java Island and has coastline approximately 147 km. Referring to the climatologic factors, Indramayu Regency is suitable for agriculture. It has a tropical climate with daily air temperature ranges between 27°C to 34°C. The air humidity range from 70 to 80%. The average rainfall is 200.08 $mm.month^{-1}$ with an average rainy days of 3.25 days per month. Average annual rainfall was 1,428.45 mm with the number of rainy days 75 days. The minimum rainfall of 47 mm occurred in December, while the maximum rainfall of 602.4 mm occurred in February.

In term of physiography, most of the regency is dominated by lowland area with slope less than 8% reach 56% of the area. It has 14 rivers which flows to the Sea of Java. Due to its climate, soil fertility and hydrology, Indramayu regency is one of the rice production center in Indonesia. The altitude of Indramayu Regency range from 0 to 18 m asl., where the lowland area occupies the widest part, ± 90% of the area.

This research was generally divided into 4 steps, which are: (i) establishing actual land utilization by making land use/land cover (LU/LC) map, (ii) analyzing land capability, and (iii) confrontation between land capability, actual land utilization and spatial pattern of OSLUP map (Figure 2). The data applied in this research are satellite remote sensing imagery (SPOT 6), soil map, topographic map, spatial pattern of OSLUP map of Indramayu Regency as well as statistical data and climatology data of study area.

Actual land utilization was derived from remotely sensed satellite imagery SPOT 6 of 2014. SPOT 6 is one of the passive sensor satellite imagery which can be used for mapping LU/LC [10]. The SPOT 6 which was launched on 2012, is an optical imaging satellite, capable for imaging the earth with a resolution of 1.5 meter panchromatic and 6 meter multispectral (blue, green, red, near-IR). SPOT 6 are the ideal solution for national mapping 1:25.000 map series [11]. In this research, SPOT 6 was used for deriving Land Use Land Cover map. Pre-processing has been done to correct the imagery from radiometric and geographic distortion. Image classification was performed by supervised classification. Then the tentative map was validated on

September 2016. All the image processing was done by using ERDAS Imagine software. The Indonesia National Standard [12] classification system was applied in this research for making LU/LC map.

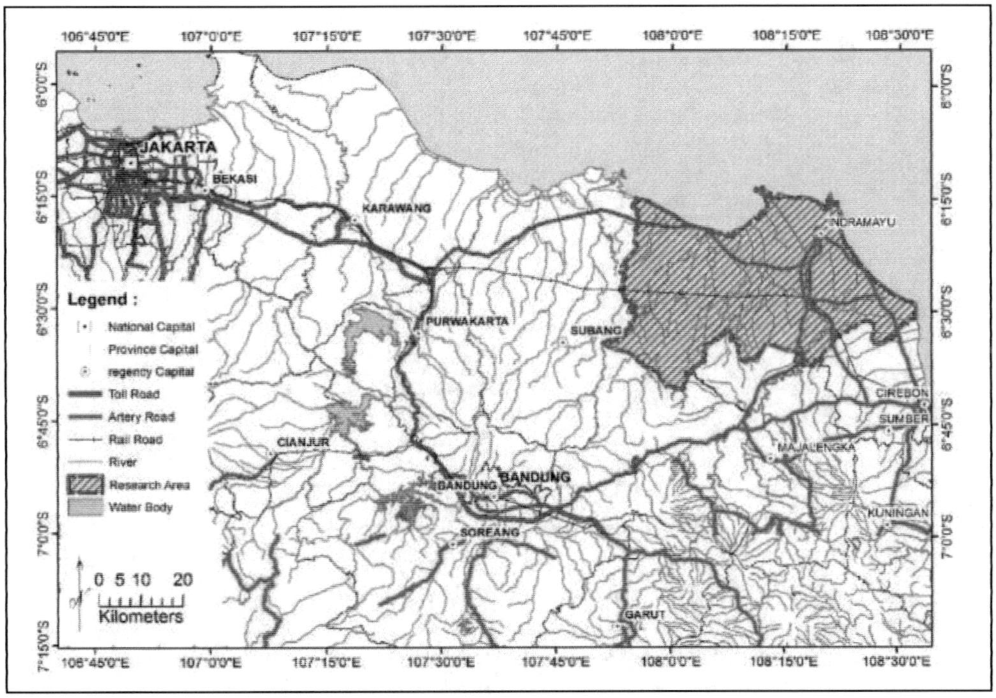

Figure 1. Situation map of the research area of Indramayu Regency, West Java Province, Indonesia.

Land capability evaluation is done, based on analysis in land mapping unit (LMU). For that reason, land mapping unit on a scale of 1: 50.000 was first established. Determination of LMU at 1: 50,000 scale is required for analysis at the regency's level. The LMU is designated using a slope class map, derived from a 1: 25,000 scale of topographic map of Indonesian Geospatial Information Agency, assisted with the soil map of 1:100,000 and field observation. In the study area there are a total of 35 LMU. At each LMU, the soil samples were taken for land capability analysis [13]. The field work for checking soil morphology and soil sampling was done on September 2016. In this field work, verification of actual land utilizations was also carried out. The soil samples were collected from each LMU from soil depth of 0-25 cm for laboratory analysis, which was conducted in the Laboratory of the Department of Soil Science and Land Resources at Bogor Agricultural University. A total of 58 soil samples were analyzed.

Land capability is an assessment of the land capacity for a general use [14]. This method assessed limiting factor of each LMU. In this research, we use matching method described by [15,13]. The criteria used are given in Table 1. LU/LC which is allowed for each land capability class was presented in Table 2. The land capability class I is very suitable for all land utilizations, while land capability class VIII is only permitted for forest conservation and natural reserve area. Land with land capability class I to IV can be considered as moderate to high class of land capability which can be used for cultivation use, while land with land capability V to VIII can be considered as moderate to low class of capability, where cultivation was not allow, they only can be used as conservation area [15].

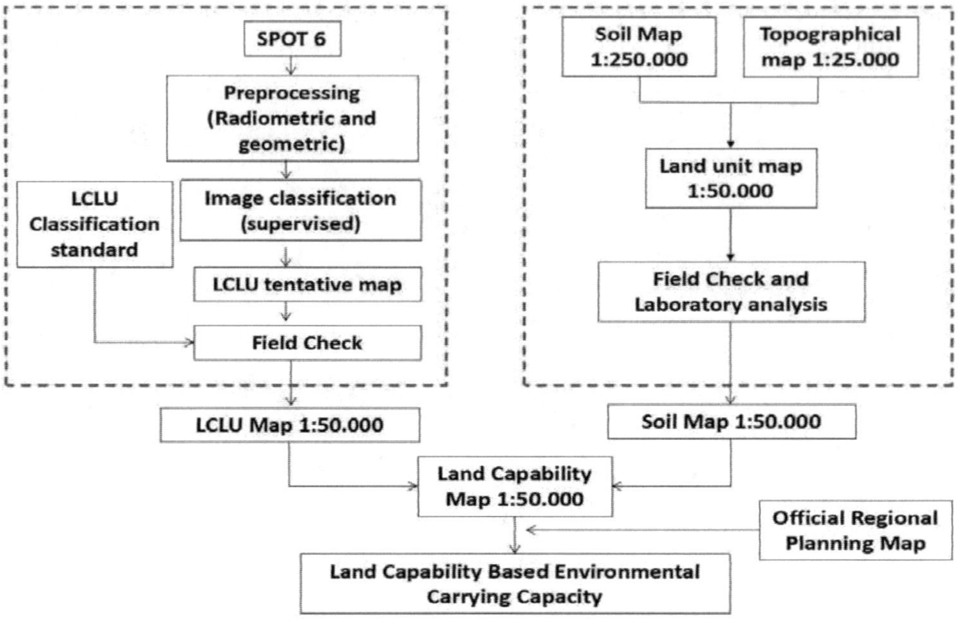

Figure 2. Flow chart of the research.

Table 1. Criteria for land capability assessment.

No.	Land Characteristics	Land Capability Class							
		I	II	III	IV	V	VI	VII	VIII
1.	Soil texture (t)[1)]								
	a. Upper layer (40 cm)	t2/t3	t1/t4	t1/t4	(*)	(*)	(*)	(*)	t5
	b. Bottom layer	t2/t4	t1/t4	t1/t4	(*)	(*)	(*)	(*)	t5
2.	Slope (%)[2)]	i_0	i1	i2	i3	(*)	l4	i5	i6
3.	Drainage[3)]	d0/d1	d2	d3	d4	(**)	(*)	(*)	(*)
4.	Effective soil depth [4)]	k_0	k_0	k1	k2	(*)	k_3	(*)	(*)
5.	Erosion[5)]	e_0	e1	e1	e2	(*)	e_3	e_4	(*)
6.	Gravel/stone[6)]	b_0	b_0	b_0	b1	b2	(*)	(*)	b3
7.	Flooding [7)]	O_0	01	02	03	04	(*)	(*)	(*)

Source: [15] (*modified*); [13]; [6]

Description:

(*) : It can have any limiting properties of the lower classes

(**) : the ground surface is always inundated

[1)] Texture : t_1: fine; t_2: rather fine; t_3:medium; t_4: rather coarse; t_5 :coarse;

[2)] Slope: I_0: (0 – 3 % ; I_1 (3 – 8 %) ; I_2 (8 – 15%); I_3 (15–30%); I_4 (30–45%); I_5 (45–65%); I_6 (> 65%);

[3)]Drainage: D_0 (good): D_1 (rather good); D_2 (rather bad); D_3 (bad); D_4 (very bad)

[4)]Effective soil depth: K_0 (deep); K_1 (moderately deep); K_2 (shallow); K_3 (very shallow);

[5)]State of Erosion: e_0 (no erosion); e_1 (light); e_2 (moderate); e_3 (heavy); e_4 (very heavy).

[6)]Gravel/Stone: b_0: nothing or little b_1 :medium; b_2 : many; b_3 :very much

[7)]Flooding O_0: none; O_1 :rarely; O_2 : sometimes; O_3 : Frequently; O_4 : very frequently

Table 2. Land utilization criteria which are permitted on each land capability class.

No.	Land Capability Class	Permitted Land Usage
1	I	All types of land use
2	II	All types of land use, except Via
3	III	All types of land use, except Via dan Ia
4	IV	La, all types of grazing and all types of forest
5	V	all types of grazing (Ig, Mg, Ltg), all types of forest
6	VI	Mg, Ltg, all types of forest
7	VII	Ltg, forest
8	VIII	Nature preserve and protected forest

Sources: [16] (*modified*), [6]
Description: Via: Very intensive agriculture; Ia: intensive agriculture; Mia: moderate intensive agriculture; La: limited agriculture; Ig: intensive grazing; Mg: moderate grazing; Ltg: limited grazing.

3. Results and Discussions

Result of LU/LC analysis are presented in Figure 3. Based on the LU/LC map, a total of 8 classes of land utilizations have been identified, which were forest, shrub, plantation, built area, rice field, dry land agriculture, ponds and water body. The rice field is the main land utilizations in the study area which is spread out in the middle of the study area. The second main land utilization is pond which is distributed in the north part of the Indramayu Regency, in the coastal area. Built area is the third land utilization and located in along the infrastructure such as road and river. The plantation and forest mostly found in the southern part of study area and in the more highly altitude.

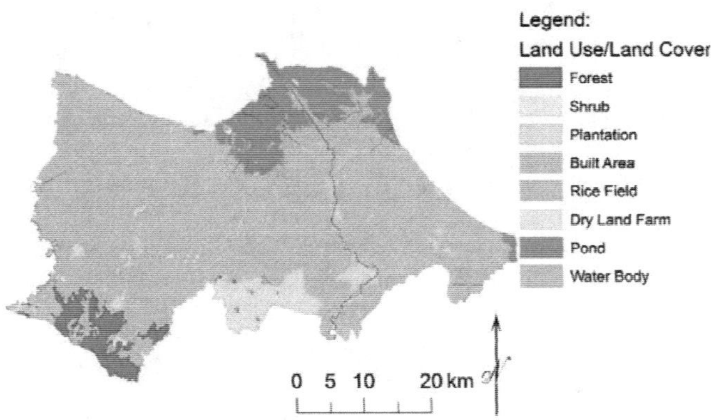

No	Land use/ Land cover	Area	
		ha	%
1	Forest	7,919.5	3.8
2	Built area	18,213.8	8.8
3	Dry Land Farm	4,099.2	2.0
4	Water body	1,452.8	0.7
5	Plantation	12,485.4	6.0
6	Rice field	144,429.0	69.4
7	Shrubs	75.0	0.0
8	Pond	19,333,9	9.3
	Total	**208,008.6**	**100.0**

Figure 3. Land use/land cover of Indramayu Regency.

Result of land capability analysis is presented in Figure 4a. The study area are composed of 4 classes of land capability, which are class II, class III, class IV and class VIII of land capability. The dominant class of land capability in the study area is land capability class III that is spread out on the mostly rice field and pond as well as settlement. The class III signify that the land that is capable for all land utilizations except for very intensive agriculture and intensive agriculture. The second main class of land capability is class II which is land that capable for many land utilization except for very intensive agriculture. This class of land capability is found in rice field area, forest and plantation of study area. Land capability class IV is found only on some part of built area. The major limiting factor in land capability classes II to IV are soil texture, drainage, effective depth, slope and erosion. There are few areas that belong to land capability class VIII. The main limiting factor on this class VIII is high erosion, located on very sloping land and rough soil texture.

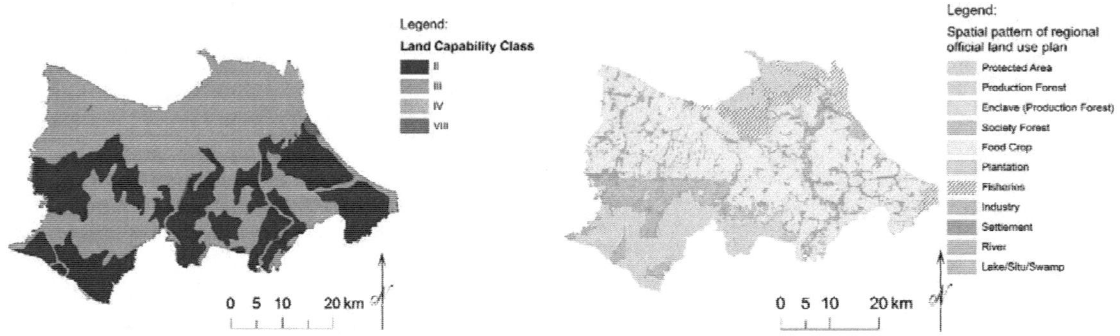

a b

Figure 4. (a) Land capability map and (b) OSLUP map of Indramayu Regency.

The result of conformity analysis between actual land utilization (LU/LC) and land capability class is given in Figure 5a. In the case of actual land utilization conform to land capability class, such land utilizations can be considered as sustainable. Contrarily, land utilizations which are more intensive than its land capability class permitted will cause land degradation. The result shows that the study area is dominated by state of conditionally conform (yellow color), at more than 90% of the area. The southern part of study area, which are dominated by forest is the area the most conform (green color) between actual land utilization and land capability. There are view area in Indramayu Regency (red color) where land utilization was not conform to its land capability. The conformity between spatial patterns of OSLUP (Figure 4b) with land capability is given in Figure 5b. Although condition is better in term of conformity, but still most of the region belong to conditionally conform to the spatial pattern, using a reference of land capability. Detailed conformity between actual land utilization and spatial pattern of OSLUP with land capability is presented in Table 3.

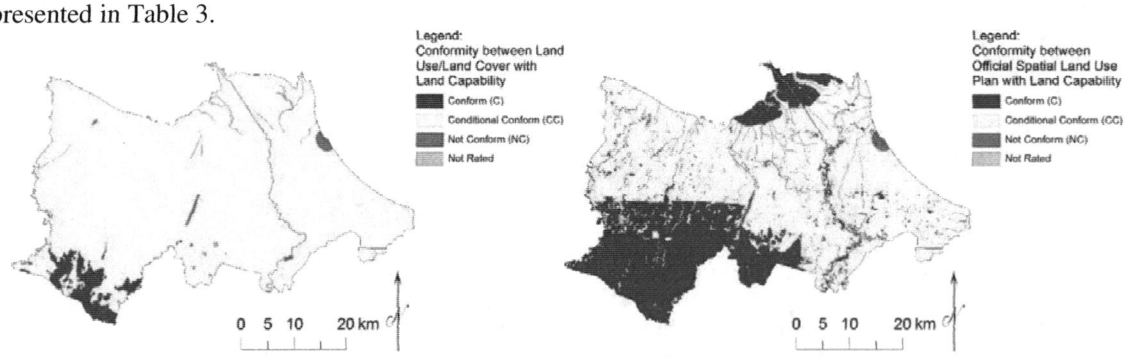

(a) (b)

Figure 5. (a) Conformity between land capability with actual land utilizations, and (b) conformity between land capability with spatial patterns of OSLUP in Indramayu Regency.

Recapitulation of calculations is presented in Table 3 and Table 4. From these tables and figures, it can be seen that on land with high to moderate land capability that can be used for agriculture (that is class I to class IV but in case of Indramayu Regency, only class II until class IV), most of its land utilization is conditionally conform to its land capability. The total area of class II to class IV is 205.311.6 ha or 98.8% of the total area. In this group, land for which its utilizations is conform to land capability is 7,463.6 ha (or only 3.6% of the high to moderate land capability). For land of class II to class IV, its use for agriculture is still possible, but certainly should be given with input to compensate the limiting factor, or in other words, conditionally conform. The land which are classified as a conditionally conform is 197.847.9 ha, or 96.4% of the area of high to moderate land capability.

Table 3. Conformity between land capability, actual land utilizations and spatial pattern in OSLUP of Indramayu Regency.

LCC[1)]	ALU[2)]	LC-LU[3)]	Area		SP[5)]	LC-SP[6)]	Area	
			ha	%[4)]			Ha	%[4)]
II	FOR, SHR	C[7)]	5,858.3	2.8	ENC, PDF, SF	C	33,860.9	16.3
	DLF, PLT, BA, RF, PND	CC[8)]	78,916.3	38.0	IDT, FSH, PLT, STL, FDC	CC	50,342.9	24.2
III	FOR, SHR	C	1,446.5	0.7	ENC, PDF, SF, PFA	C	34,446.3	16.6
	DLF, PLT, BA, RF, PND	CC	118,931.6	57.2	IDT, FSH, PLT, STL, FDC	CC	85,377.4	41.1
IV	FOR	C	158.8	0.1	PDF, SF	C	154.4	0.1
	FOR	C	0.3	0.0	PDF, SF	C	174.4	0.1
VIII	DLF, PLT, BA, RF, PND	NC	1,086.0	0.5	IDT, FSH, STL, FDC	NC	876.9	0.4
	Water Body	Not Rated	1,358.4	0.7	Water Body	Not Rated	2,523.0	1.2

Note:
1)LCC: Land Capability Class;
2)ALU: Actual Land Use; FOR: Forest; DLF: Dryland Farm; PLT: Plantation; BA: Built Area; RF: Rice Field; SHR: Shrub; PND: Pond;
3)LC-LU: Suitability between Land Capability with Actual Land Use
4)Persentage according to land unit
5)SP: Spatial Land Use Plan; PTA: Protected Area; PDF: Production Forest; ENC: Enclave (Production Forest); SCF: Society Forest; FDC: Food Crop; PLT: Plantation; FSH: Fisheries; IDT: Industry; STL: Settlement; RVR: River; LKE: Lake
6)LC-SP: Suitability between Land Capability with Spatial Land Use Plan
7)C: Conform
8)CC: Conditionally Conform
9)NC: Not Conform

In class III of land which is classified as conditionally conform, especially on poorly drained land with slow permeability, drainage improvements are required. Cropping pattern should be chosen ones that can improve soil structure for easy tillage. To prevent puddling and increase soil permeability, it is necessary to add organic matter, in addition to not till the soil in wet condition. Land class IV which is classified as conditionally conform with limiting factor of steep slope, when it is used for seasonal crops, they need a terrace bench or rotation with cover crops, fodder crops or green manure for 3 to 5 years.

Table 4. Recapitulation of conformity between land capability with current land utilization and spatial pattern in OSLUP.

No	LCC[1)]	LC-LU Area[2)] (ha)	%	LC-SP Area[3)] (ha)	%
		1. High untill Moderate Capability Land (I-IV)			
	C	7,463.6	3.6	68,461.6	33.5
	CC	197,847.9	96.4	135,720.3	66.5
	NC	0.0	0.0	0.0	0.0
	Total	205,311.6	100.0	204,181.9	100.0
		2. Rather low untill low Capability Land (V-VIII)			
	C	0.3	0.0	174.4	0.1
	CC	0.0	0.0	0.0	0.0
	NC	1,086.0	0.5	876.9	0.4
	Total	1,086.3	0.5	1,051.4	0.5
		3. All Class (I-VIII)			
	C	7,463.9	3.6	68,636.0	33.0
	CC	197,847.9	95.2	135,720.3	65.3
	NC	1,086.0	0.5	876.9	0.4
	Not Rated	1,358.4	0.7	2,523.0	1.2
	Total	207,756.3	100.0	207,756.3	100.0

Notice:
[1)]LCC : Land Capability Class
[2)]LC-LU: Conformity between Land Capability with Actual Land Use
[3)] LC-SP: Conformity between Land Capability with Spatial Land Use Plan

Lands that should be managed more carefully are land with slightly low to low land capability, such as land of class V to class VIII, in the case of Indramayu Regency, this is land class VII. The total area of land class V to VIII, or slightly low to low land capability in Indramayu Regency is 1,086.3 ha (or 0.5% of the total area). Although the extent is small, but its use is not conform to the land capability, or it is used with more than its land capability.

The results of conformity analysis between land capability with land allocation in the spatial pattern of OSLUP indicated that land with high to moderate capability (land of class I to class IV, or in the case of Indramayu Regency are class II to class IV), several part (68,461.6 ha or 33.5% of land with high to moderate capability) has been allocated for utilizations in accordance with its capability. Several part of this land, which are 135,720.3 ha (or 66.5% of land with high to moderate capability), the allocation in the spatial pattern of OSLUP exceeds the land capability based carrying capacity, although still in the state of conditionally conform. This indicates that the quality of the OSLUP is rather poor, it is shown by the high amount of the land allocated exceeding land capability in spatial pattern of OSLUP.

Overall results analysis for all Indramayu Regency show that both actual land utilization as well as land allocation in the spatial pattern of OSLUP requires to be improved. In the parts of the area where land allocations in the spatial pattern of OSLUP are still exceeding their carrying capacity, OSLUP should be revised.

4. Conclusion

Indramayu Regency has land with land capability class range from class II to class VII. Most of the land has land capability that can support agricultural land utilization (class I to class IV), but there are some small parts of the region that should not be used for cultivation (class V to class VIII). The limiting factors determining land capability in Indramayu Regency are soil texture, effective depth, drainage, slope, and flooding. Actual land utilizations in Indramayu Regency includes forest, shrub, plantation, built area, rice field, dry land agriculture, ponds and water body. The rice field is the main land utilizations in the study area followed by pond and built area. Overall, a small portion of the land in Indramayu has been used in accordance with land capability, but most of the land is used exceeding its land capability.

Spatial pattern of OSLUP in Indramayu Regency allocates most of the land by exceeding land-based carrying capacity. The results of this study indicate that adequate efforts are required for improving land utilizations in the area where actually been used beyond the land capacity. This research can also be used as input for revision of official spatial land use plan on the cycle a 5 year revision in term of allocation of land in the spatial plan in order to more in line with the land carrying capacity. In other word, the land utilization as well as land OSLUP should be improved in term of conformity and consistency with land capability.

Refferences

[1] Indonesian Statistics 2015 *Statistics of Indonesia* Indonesian Statistics Agency Jakarta: Indonesia
[2] Widiatmaka, Ambarwulan W, Setiawan Y, Walter C 2016 Assessing the suitability and availability of land for agriculture in Tuban Regency East Java Indonesia. *App. Environ. Soil Sc.* 1-13.
[3] Boix L R, Zinck J A 2008 Land-use planning in the Chaco Plain (Burruyacu´ Argentina) Part 2: Generating a consensus plan to mitigate land-use conflicts and minimize land degradation *Environ. Mgmt.*, **42** 200–209
[4] Adamu, G K, Yusuf M A, Ahmed M 2014 Soil degradation in dry lands *Acad. Res. Int.* 5 **1** 78-91
[5] Susetyo, B, Arifin H S, Machfud, Widiatmaka, Arifin N H S 2016. Model of trees appraisal using an International shading tree evaluation method at the Ciliwung Riparian area, Indonesia *Ecol. Environ. Conserv.* **21** 4 2067-2080.
[6] Widiatmaka, Ambarwulan W, Purwanto M Y J, Setiawan Y, Effendi H 2015 Land capability based environmental carrying capacity in Tuban, East Java. *J. Man. Lingk.* 22 **2** 247-259
[7] Munasinghe M 1993 Environmental Economics and Sustainable Development. World Bank Environment **3** Washington DC
[8] Feizizadeh B, Blaschke T 2012 Land suitability analysis for Tabriz County, Iran: A multi-criteria evaluation approach using GIS. *J. Environ. Planning Mgmt.* **4** 1–23
[9] Fatmawaty D, Baskoro D P T, Widiatmaka. 2015 Strategi pengembangan komoditas perkebunan berbasis daya dukung lahan di Kabupaten Majene Sulawesi Barat *Globe* 17 **1** 25-32
[10] Ambarwulan W, Widiatmaka, Suparwati T, Firmansyah I, Niendyawati 2017 Land suitability and dynamic system model for land use planning of paddy field in Indramayu Regency, Indonesia *Proc. FIG Working Week* 29 May - 2 June 2017 Helsinki Finland
[11] Spot 6 7 satellite imagery 2017. Available on http://www.intelligence-airbusds.com/en/147-spot-6-7-satellite-imagery [cited 2 September 2017]
[12] Standard National of Indonesia (SNI) 2010 Standard National of Indonesia SNI 7645:2010 Land Cover Classification
[13] Hardjowigeno S, Widiatmaka 2011 *Evaluasi Lahan dan Perencanaan Tataguna Lahan* Gadjah mada Univ. Press: Yogyakarta
[14] Fenton T E 2014 *Land capability classification. In Encyclopedia of Natural Resources: Land* Taylor and Francis: New York 299-301
[15] Arsyad S 2010 *Konservasi Tanah dan Air* IPB Press: Bogor
[16] Klingebiel A A, Montgomery PH 1961 Land capability classification. Agriculture Handbook **210** *Soil Conservation Service* U.S. Department of Agriculture Washington DC 1-3

IOP Conf. Series: Earth and Environmental Science **149** (2018) 012007 doi:10.1088/1755-1315/149/1/012007

Monitoring 2015 drought in West Java using Normalized Difference Water Index (NDWI)

Luisa Febrina Amalo[1], Ummu Ma'rufah[2], and Prita Ayu Permatasari[1]
[1]Environmental Research Centre, Bogor Agricultural University (PPLH-LPPM), Bogor, 16680, Indonesia
[2]Department of Geophysics and Meteorology, Bogor Agricultural University, Bogor, 16680, Indonesia

E-mail: luisafebrina@gmail.com

Abstract. Drought is a slow developing phenomenon that accumulates over period and affecting various sectors. It is one of natural hazards that occurs each year, particularly in Indonesia over Australian Monsoon period. During drought event, vegetation's cover can be affected by water stress. Normalized Difference Water Index (NDWI) is a method for water resource assessment and known to be strongly related to the plant water content. NDWI is produced from MODIS bands Near-infrared (NIR) and Short Wave Infrared (SWIR). This research aims to monitor drought using NDWI in West Java during El Niño 2015 and its impact on rainfall variability. The result showed rainfall was decreased significantly starting from May-June, then increased in November. According to NDWI, it also showed that mostly West Java Region affected by drought during May-November. Very strong drought occurred on September-November. On December, areal extent of drought was decreasing significantly because rainfall had increased during November. Generally, areal extent of drought in West Java was dominated by strong and moderate drought. It implied that El Niño 2015, give great impact on increasing drought and decreasing rainfall in West Java. NDWI can be detected drought occurrence as it have good correlation with rainfall spatially.

1. Introduction

Drought is a slow developing phenomenon that accumulates over period and affecting various sectors (i.e. economy, agriculture, etc.) [1]. It is one of natural hazards that occurs each year, particularly in Indonesia over Australian Monsoon period (June to August) [2]. During drought event, vegetation's cover can be effected by water stress. It can be lead to crop failure and agricultural production might be decreased. Early mitigation by using near-real time information on plant water stress can be used as early warning system to minimize the drought impact. Therefore, drought assessment can be used for drought monitoring in a quick and cost-effective way to increase adaptability of water retention. One of common methods for drought monitoring is by using satellite based indices in which has proven to be an effective way and an easy tool from previous study [3-4].

The Normalized Difference Water Index (NDWI) is a method for analyzing data for water resource assessment [5]. NDWI had a quicker response to drought conditions than NDVI [1] and it is known to be strongly related to the plant water content. NDWI is produced satellite bands Near-infrared (NIR) and Short Wave Infrared (SWIR). NIR reflectance is affected by leaf internal structure and leaf dry matter content but not by water content while SWIR reflectance reflects changes in both the vegetation water content and the spongy mesophyll structure in vegetation canopies. These two combination remove variations induced by leaf internal structure and improving the accuracy in retrieving the vegetation water content [6]. NDWI is a remote sensing based indicator sensitive to the change in the water content of leaves [7].

The occurrence of El Niño event in 2015 was impacting rainfall variability in Indonesia and triggered more severe drought to be happened [4]. West Java has potentially great contribution to food security in Indonesia. Climate in West Java is strongly influenced by monsoon (Asian and Australian monsoon) and ENSO phenomena, such as El Niño and La Niña. Analysis of rainfall variability during El Niño and La Niña has been conducted in West Java and resulted a good correlation [8]. Thus, this research aims to monitor drought using Normalized Difference Water Index (NDWI) in West Java during El Niño year 2015 and its impact on rainfall variability.

2. Methods

2.1. Study area

West Java is located at 5°50' – 7°50' S and 104°48' – 108°48' E (figure 1) with an area of 35,377.76 km^2 [9]. It covers 18 districts and 9 cities. The temperature ranges between 9°C in the Peak of Pangrango Mountain to 30°C in North Coast. The average rainfall is 2,000 mm per year but in some mountainous areas might reach 3,000 to 5,000 mm per year.

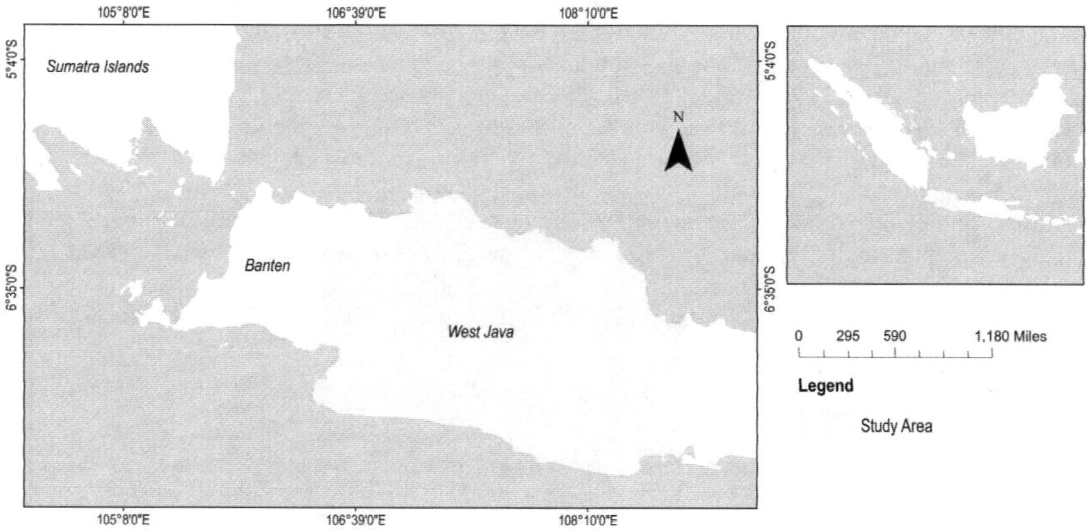

Figure 1. Geographical map of West Java Region.

2.2. Data background and processing

In this study, NDWI was calculated from NIR and SWIR band spectral of MODIS satellite data, namely MOD09A1 datasets. MOD09A1 provides MODIS band 1-7 surface reflectance 8-day at 500 m resolution [10]. These data were available and obtained from https://lpdaac.usgs.gov. NIR reflectance was retrieved from MOD09A1 surface reflectance band 2 products (841-876 nm), while SWIR from surface reflectance band 7 products (2,105-2,155 nm). The 8-day NIR and SWIR were composited into month from January to December 2015. NDWI then calculated using following equation (equation 1) and classified into eight categories (table 1) based on Gulacsi and Kovaks [11]. CHIRPS (Climate Hazards Infrared Precipitation with Stations) data were used for analyzing rainfall variability in West Java during El Niño 2015 spatially. Rainfall data were obtained from IRI/LDEO Climate Data Library.

$$NDWI = \frac{NIR-SWIR}{NIR+SWIR} \tag{1}$$

Table 1. Drought category of NDWI [11].

Category	Values
Very High Moisture Content	NDWI ≥ 0.7
High Moisture Content	0.6 ≥ NDWI > 0.7
Moderate Moisture Content	0.5 ≥ NDWI > 0.6
Low Moisture Content	0.4 ≥ NDWI > 0.5
Weak Drought	0.3 ≥ NDWI > 0.5
Moderate Drought	0.2 ≥ NDWI > 0.3
Strong Drought	0 ≥ NDWI > 0.2
Very Strong Drought	NDWI < 0

3. Result and discussion

3.1 Spatial analysis of monthly rainfall data

Rainfall in Indonesia is influenced by various climate variability, namely IOD, ENSO, Monsoon, and MJO. One of them, El Niño, has high impact on decreasing rainfall in Indonesia including West Java. The type of rainfall in West Java is monsoonal that has two season, dry and wet season. Dry season usually occurs on June, July, August (JJA), while wet season on December, January, and February.

During El Niño 2015, decreasing rainfall was significant. In northern West Java, decreasing of rainfall was starting from May, while in the southern part of West Java decreasing of rainfall was starting from June (figure 2). From June to October, the rainfall was below 60 mm/month in southern part of West Java and below 30 mm/month in northern part of West Java. Rainfall in northern part of West Java mostly lower than the other area. On November, rainfall in southern West Java increased up to 300 mm/month (figure 2). This increasing rainfall on November can caused by Monsoon Asia [12]. The difference of spatial rainfall pattern may cause different water availability. Therefore, it will give impact on difference of spatial drought pattern.

3.2 Monitoring of drought

Drought is an effect of deficit rainfall in some period of time [3]. El Niño in 2015 given specific impact on water availability in West Java Province. NDWI on this study used to identify drought based on plant water content. West Java Province is one of central rice production in Java Island, therefore monitoring and understanding drought based on plant water content is important to be carried out.

According to NDWI, almost all region in West Java was affected by drought in 2015 (figure 3). Weak to moderate drought already occurred on January in some area of West Java. From January to July, total area and severity of drought did not increase. During this period, severe drought occurred in central and northwest of West Java such as Bekasi, Bogor, Indaramayu, Subang, and Cirebon. Data from Department of Agriculture in West Java Province also shown that one of region in North West Java, namely Indramayu, experienced drought with high areal extent [13].

During August 2015, area of drought increased significantly. Almost all region experienced drought. Low moisture content to moderate drought dominance occurred in southern West Java, while in the opposite region was dominated by strong to very strong drought. During September, areal extent of very strong drought increase. It is occurred due to accumulation of the decreasing rainfall in West Java region. Very strong drought on September continued to November. On December, area of drought decreased significantly, particularly in southern area of West Java Province. It occurred because rainfall had increased during November.

Rainfall deficit and drought in West Java Province did not occur at the same time. There was a lag time between one to three months. Wang *et al.* [1] also point out that there was a lag time between deficit rainfall and drought around three month on crop production and six month on forest. The lag time occurred because soil water content did not reduce directly when rainfall gradually decreased [14].

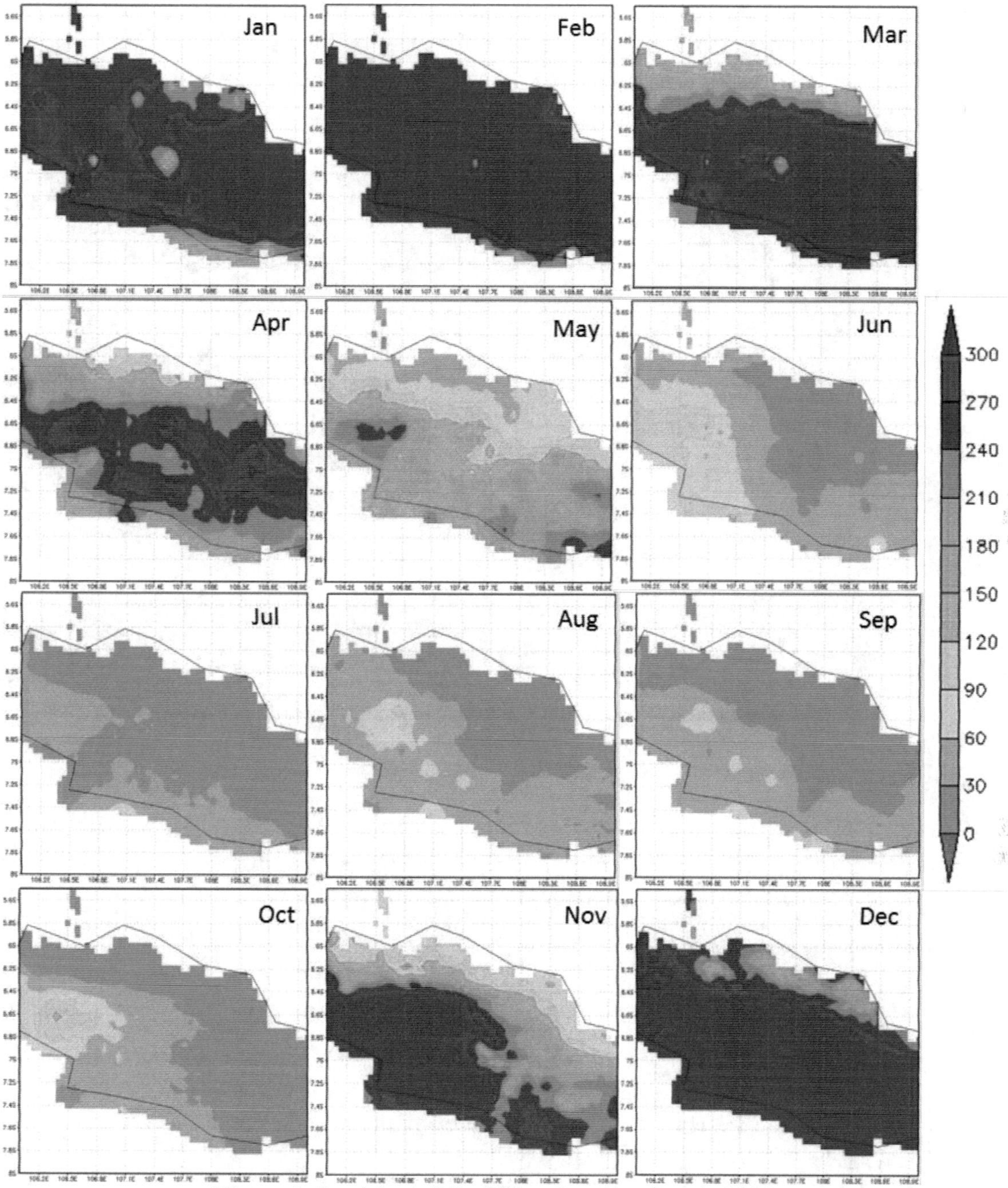

Figure 2. Spatial rainfall of West Java Province in 2015.

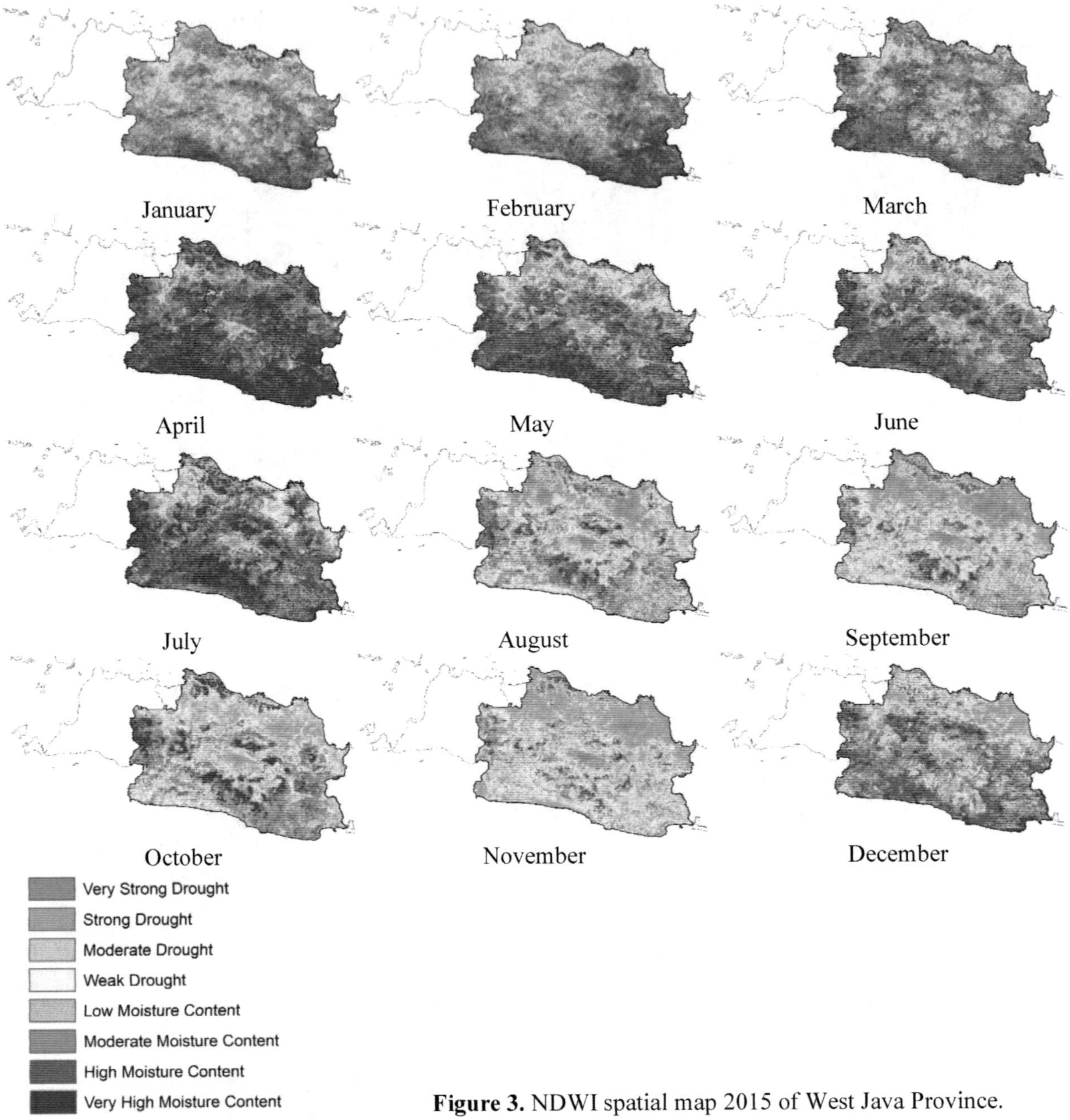

Figure 3. NDWI spatial map 2015 of West Java Province.

3.3 Areal extent of NDWI in percentage

Severity and area of drought in West Java increased during August to November 2015 (figure 3). Therefore we analyze areal extent for every level of drought during this period. On August, moderate moisture content dominantly occurred with percentage 23.41% (table 2). Strong and very strong drought cover around 17% area of West Java. During September, strong and very strong drought dominantly occurred. This condition was caused by accumulation of deficit rainfall from June to August. Severity of drought decreased during October. High moisture content was dominant in this month. During November, area of strong and very strong drought increased again around 25%. Generally, areal extent of drought in West Java was dominated by strong and moderate drought. It implied that El Niño in 2015, give great impact on increasing drought in West Java.

Table 2. Percentage of NDWI area.

NDWI/ Month	Very strong drought	Strong Drought	Moderate Drought	Weak Drought	Low Moisture Content	Moderate Moisture Content	High Moisture Content	Very High Moisture Content
	0	0.2	0.3	0.4	0.5	0.6	0.7	>0.7
Aug	2.78	14.18	12.46	13.73	11.95	23.41	13.56	2.84
Sep	4.28	22.76	13.77	18.76	18.18	14.67	6.59	1.15
Oct	0.66	4.28	21.32	0.00	29.44	0.00	31.82	12.35
Nov	1.16	24.45	21.31	19.31	16.97	11.54	4.15	1.23

4. Conclusion

The occurrence of El Niño event in 2015 was impacting rainfall variability and drought in West Java. During El Niño 2015, rainfall decreased significantly starting from May to June in all over West Java region approximately 60 mm/monthly in southern part and 30 mm/monthly in northern part of West Java. Then, rainfall increased up to 300 mm/month in November. According to NDWI as an index for drought monitoring, it also showed that mostly West Java Region affected by drought during May to November. Very strong drought occurred on September to November. On December, affected area of drought decreased significantly, particularly in southern part of West Java Province because rainfall had increased during November. Generally, affected area of drought in West Java was dominated by strong and moderate drought. It implied that El Niño in 2015, give great impact on increasing drought and decreasing rainfall in West Java. For drought monitoring, NDWI can be detected drought occurrence as it have good correlation with rainfall spatially.

References

[1] Zhang N, Hong Y, Qin Q, and Zhu L 2013 Evaluation of the visible and shortwave infrared drought index in China *Int. J. Disaster Risk Sci.* **4** 68-76

[2] Aldrian E, Susanto RD 2003 Identification of three dominant rainfall regions within Indonesia and their relationship to sea surface temperature *Int. J. Climatol* **23** 1435-1452

[3] Dalezios, N R, Blanta A., Spyropoulos N V, Tarquis A M 2014 Risk identification of agricultural drought for sustainable Agroecosystems *Nat. Hazards Earth Syst. Sci.* **14** 2435-2448.

[4] Amalo L F, Hidayat R, Sulma S 2018 Analysis of agricultural drought using vegetation health index *Journal of Agricultural Science* **40** 63-73

[5] McFeeters 1996 The use of the normalized Difference Water Index (NDWI) in the delineation of open water features *Int. J. Remote Sensing* **17** 1425-1432

[6] Ceccato P, Flasse S, Tarantola S, Jacquemond S, Gregoire J M 2001 Detection vegetation water content using reflectance in the optical domain *Remote Sensing of Environment* **77** 22-23

[7] Gao BC 1996 NDWI – A normalized difference water index for remote sensing of vegetation liquid water from space Remote Sensing of Environment **58** 257-266

[8] Nabilah F, Prasetyo Y, Sukmono A 2017 Analisis pengaruh fenomena El Niño dan La Niña terhadap curah hujan tahun 1998-2016 menggunakan indikator ONI (Oceanic Nino Index) *Jurnal Geodesi Undip* **6** 402-412

[9] Statistics Indonesia 2016 *East Java in Figures 2015* (Bandung: CV Media Kontruksi)

[10] Roger J C, Vermote E F, Ray J P 2015 MODIS surface reflectance user's guide collection 6 *MODIS Land Surface Reflectance Science Computing Facility*

[11] Gulacsi A, Kovacs F 2015 Drought monitoring with spectral indices calculated from MODIS satellite images in Hungary *Journal of Environmental Geography* **8** 11-20

[12] D'Arrigo R, Wilson R, Palmer J, Krusic P, Curtis P, Curtis A, Sakulich J, Bijaksana S, Zulaikah

S, Ngkoimani L O 2006 Monsoon drought over Java, Indonesia, during the past two centuries *Geophysical Research Letters* **33** 1-4
[13] Dinas Ketahanan Pangan dan Peternakan 2015 Keadaan bencana kekeringan di Jawa Barat sampai dengan bulan Agustus tahun 2015 http://dkpp.jabarprov.go.id/keadaan-bencana-kekeirngan-di-jawa-barat-sampai-dengan-bulan-agustus-tahun-2015/
[14] Wang H, Lin H, and Liu D 2014 Remotely sensed drought index and its responses to meteorological drought in Southwest China *Remote Sensing Letters* **5** 413-422

Measuring Environmental and Socio-economic Impact of Deforestation at Kalimantan Island

Irmadi Nahib, Soma Trenggana, Turmudi, Jaka Suryanta, Sri Lestari Munajati and Rizka Windiastuti

Geospatial Information Agency
Jl Raya Jakarta Bogor KM 46 Cibinong, Jawa Barat, 16911, Indonesia

E-mail: irmnahib@gmail.com, irmadi.nahib@big.go.id

Abstract. Indonesia's forests in the period of 2000-2009 has been deforested by about 15.158 million ha out of 103.309 milion ha. Deforestation caused carbon emissions. One method for measuring emissions from deforestation and forest degradation is GeOSIRIS model. A modeled GeOSIRIS policy used a carbon payment system to incentivize emission reductions. Data used in this study were maps of forest cover in 2005 and 2010, map of deforestation 2005-2010, carbon and agricultural price and driver variables for deforestation such as slope, elevation, logarithmic distance to the nearest road or provincial capital, or the amount of area per pixel included in a national park, or a timber plantation. The result of this study showed rate of deforestation was 1.417 million ha/5 years (observed). The REDD policy could decrease deforestation in Kalimantan Island by 0.170 million ha (16.70%), with assumption that international carbon price of US$ 10/tCO_2e. The change of emissions due to REDD was 22.29%, or reduced emissions by 245.03 million tCO_2e/5 years. Finally, Gross National Revenue from carbon payments (NPV 5 years) was US$ 2,450.34 billion, where incentivize emission reductions to sub-national entities (NPV, 5 years) was US$ 2,150.07 million and net central government surplus from carbon payments was US$ 300.26 million (NPV, 5 years).

1. Introduction

Tropical forests and other vegetated landscapes like grasslands and wooded savannahs play a major role in the global carbon sequestration process and their conservation and protection offers immense potential for reducing greenhouse gas emissions and global warming [1]. Referring to Barlow, et al [2] that clearing of primary forests also results in the destruction of unique tropical forest habitats, thus causing the loss of biodiversity.

Among tropical countries Indonesia experiences the second highest rate of deforestation. Therefore, accurate and up-to-date forest data are required to fight deforestation and forest degradation to support initiatives of climate change mitigation and biodiversity conservation policy [3]. Meanwhile, Sumargo, et al [4] explained that the largest of deforestation in Indonesia, occurred in Kalimantan and Sumatra with a percentage of 36.32% and 24.49% respectively, followed by Sulawesi 11.00%, Java 9.12%, Maluku 8.30%, Bali-Nusa Tenggara 6.62%. Papua became the smallest area contributing to deforestation of 4.15%. It could be seen that deforestation in Indonesia until 2009 was concentrated in Kalimantan and Sumatra.

Out of the 15.79 Mha of forest cover loss for Indonesia, reported 38% (6.02 Mha) happened inside primary intact or damaged forests [5]. Meanwhile, Margono et al., (2014) said that over the study period annual primary forest cover loss increased with the highest total loss happened in 2012 (0.84Mha). The number was greater than the reported forest loss of Brazil (0.46Mha; ref. 33), which was the historical leader in the tropical forest clearing. Referring to [6], Borneo Island in the period 2000-2011 has deforestation amounted to 3.040 million ha, namely deforestation in peat forests of 0.560 million (18.42%) and deforestation in mineral land (non-peatland) for 2,480 million (81.58%). Based on the period of time of deforestation, 48.5 % of deforestation occurred in the period 2006-2011, i.e. deforestation on peat forests of 0.334 million ha (59.69%) and deforestation in mineral forests of 1.144 million (46.15%).

In Indonesia deforestation is usually linked with production of timber and expansion of settlement and agricultural area. When this existing trend continues without implementing any corrective measures, it is projected to result in a reduction of forest cover by 15% between 2015 and 2030, going from approximately 880,000 km^2 to 749,941 km^2. On average, 8,300 km^2 of forest would be cleared for timber extraction or land conversion every year between 2015 and 2030. When the forest cover declines, so does the amount of carbon stored. The cumulative emissions from 2015 to 2030 due to forest loss would reach 2.5 billion tCO_2, which, assuming an average carbon price of USD 5 to USD 10 per ton (based on international average market prices), would translate in a cumulative loss of about USD 10 billion to USD 25 billion between 2015 and 2030. [7]

REDD is not directed at stopping planned conversion of forests to other economic uses, nor at stopping the use of forests for timber. REDD signifies a way to value natural resource of carbon so that it can be considered along with other regular forest assets, when making decisions about land use and forest use [8].

In the calculation and modeling for carbon emissions, there are several methods and approaches. One model is the GeOSIRIS model developed by Jonah Busch at Conservation International. The GeOSIRIS model was originally developed as OSIRIS as a transparent decision support tool for REDD+ policy makers [9].

The GeOSIRIS modeler is different from the REDD modeler found in Land Change Modeler (LCM). The REDD modeler in LCM predicts how carbon emissions and deforestation would change if a certain reference area were shielded from deforestation. Meanwhile, the GeOSIRIS modeler adopt an alternate strategy. A carbon payment system is used by a modelled GeOSIRIS policy to give incentives to emission reductions. The policy can be governed at various administrative levels, such as province or district. Rather than defending a specific section of land from deforestation, scope of work for GeOSIRIS projects would be on regional or national scale, by setting a certain price to every ton of carbon dioxide emitted ($/tCO2e). The GeOSIRIS model assumes forest users encounter a trade-off between the carbon revenue obtained by protecting the forests and the agricultural revenue obtained from deforesting the land. Given some variables such as a proposed carbon price and maps of previous deforestation, the model predicts how carbon emissions, deforestation, and agricultural and carbon revenues would change if such policy were implemented [9].

The model designs balance incentives to lower usually high deforestation emissions with incentives to keep usually low deforestation emissions. Approximations of emission reductions under REDD depend significantly on the degree to which demand for tropical agriculture in the borderline generates leakage. This emphasizes the potential importance to REDD of balancing strategies to supply agricultural needs outside the forest borderline [10].

The purposes of this study were to measure deforestation and carbon emissions in period of 2005-2010 at Kalimantan Island and to know the incentive value to be paid by the government.

2. Material and Methods

2.1 Data used

This study used data from https://clarklabs.org/download/terrset-tutorial-data/, accessed on April 4, 2017, either: (a) forest cover maps in 2005 and 2010, deforestation map 2005-2010 (see Figure 1); (b) map of potential driver variables for deforestation, consisting of maps: slope, elevation, logarithmic distance to the nearest road, distance from the provincial capital, national park map, plantation area map (see Figure 2). These data are global data with spatial resolution of 3 km x 3 km. These data include global data that can be used for monitoring a large area (such as the whole Indonesia), due to the availability of sufficient data. However, for more specific planning, medium and detail scale data are needed to obtain more accurate results.

The disadvantage of these data is that the spatial resolution is too small (where one pixel represents an area of 900 ha). Therefore, areas with less than 900 ha (one pixel) will be combined into a more dominant class. The map actually covered the entire territory of Indonesia, but for this study it was cropped to cover only Kalimantan Island.

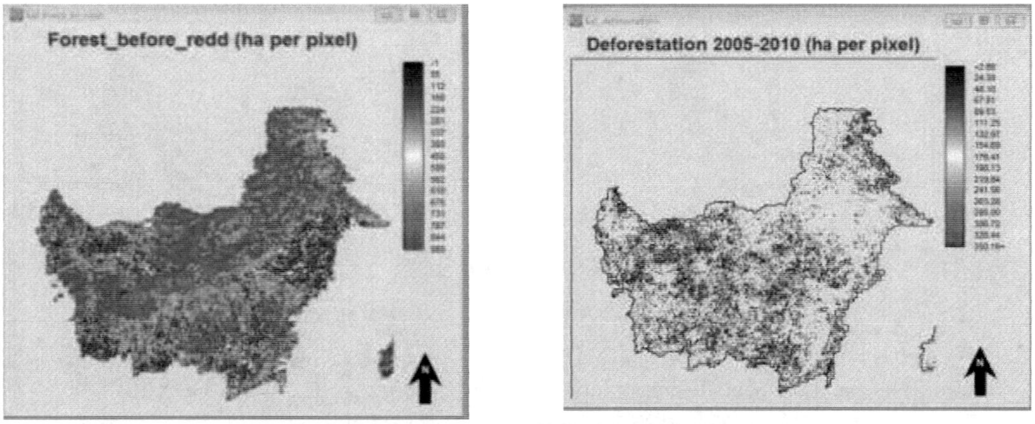

Figure 1. Forest area at Kalimantan Island.

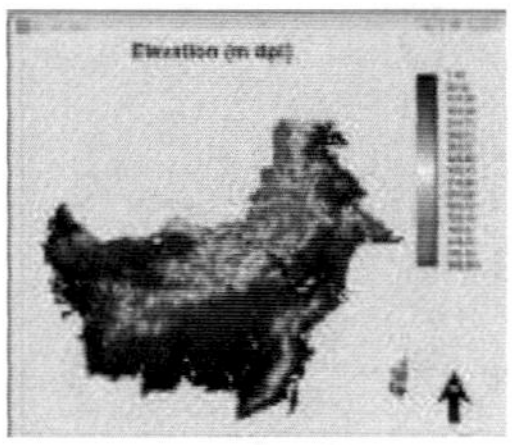

a. Elevation (m dpl)

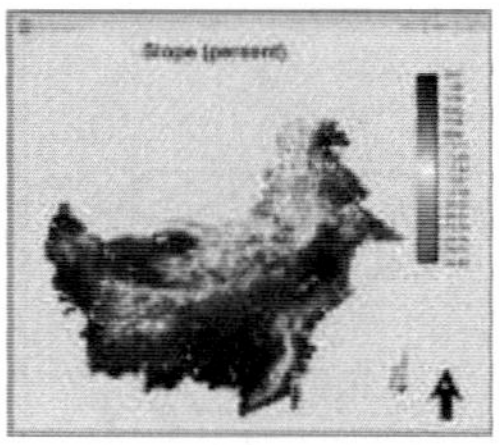

b. Slope (percent)

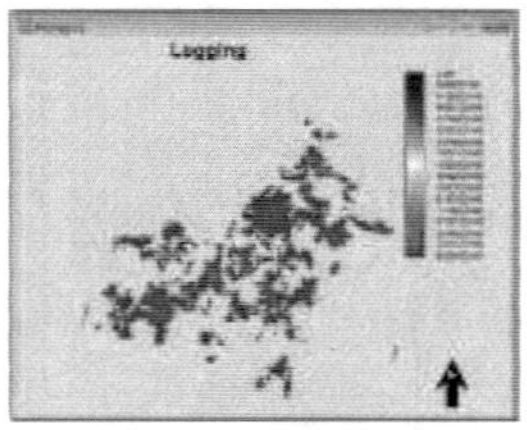

c. Logging (Ha per pixel)

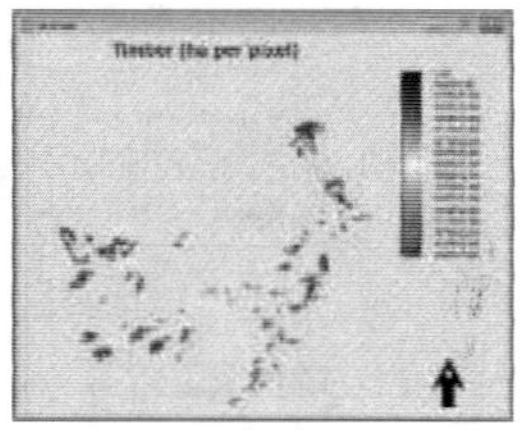

d. Timber (ha per pixel)

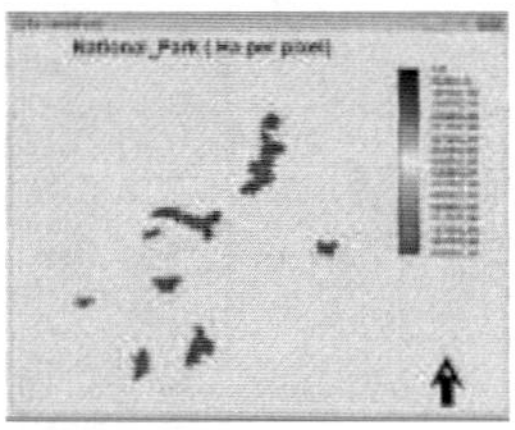

e. National Park (ha per pixel)

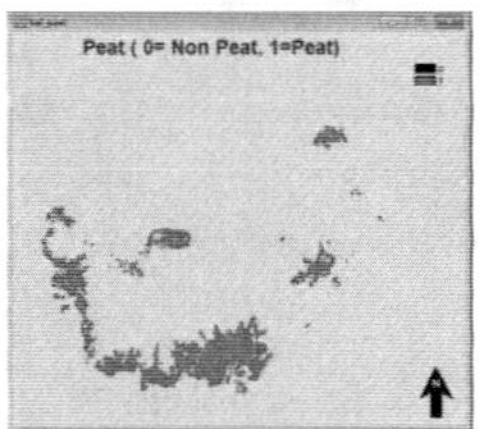

f. Peat Swamp (0= Non Peat, 1= peat)

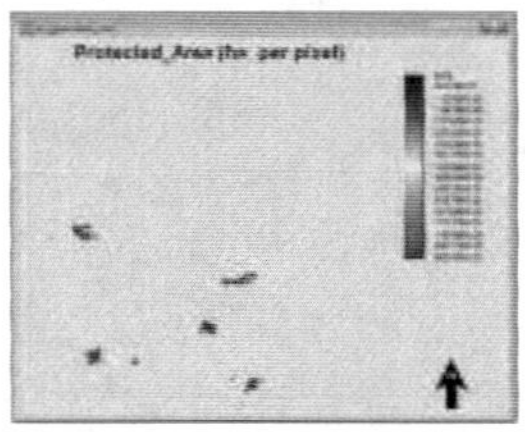

g. Protected areal (ha per pixel)

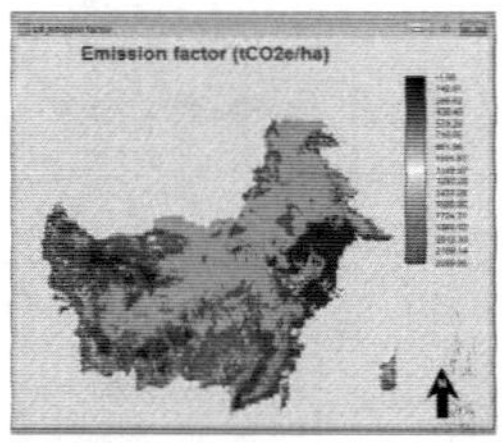

h. Emission factor (ton CO_2e per ha)

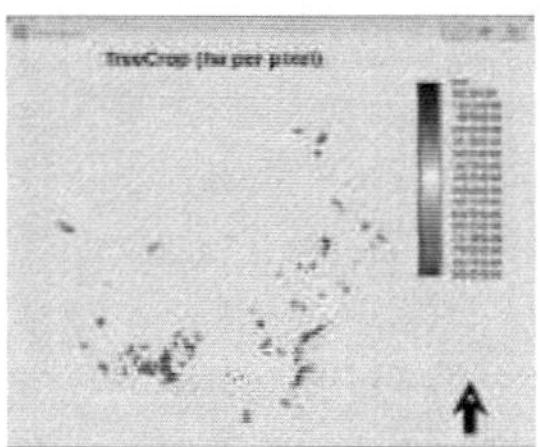

i. Tree Crops (ha per pixel)

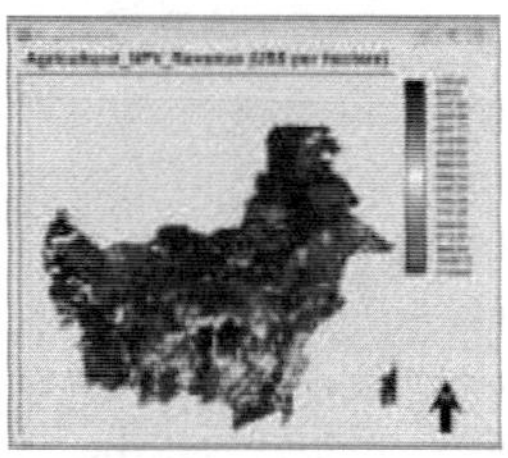

j. Agricultural Revenue (USD per hectare)

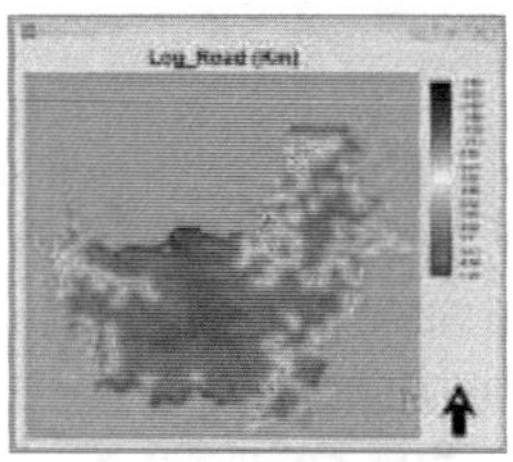

k. logarithmic distance to the nearest road (km)

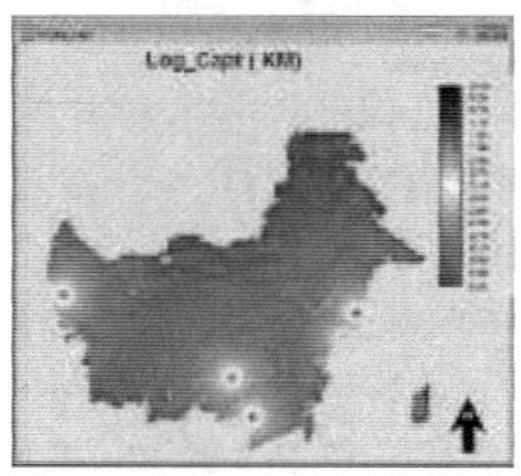

l. logarithmic distance to the nearest province capital

m. Above and Ground carbon (tons/hectare)

n. Solil carbon (tons/hectare)

Figure 2. Potential driver variables for deforestation.

The GeOSIRIS model in REDD impact calculations is based on an enhanced OSIRIS model [9]. The flow chart of the GeOSIRIS modeling stage is presented in figure 3. In general, GeOSIRIS model has two main steps: (1) regression analysis, where the regression coefficient(s) and Effective Opportunity Cost image are calculated, and (2) calculations of proportional national change in agricultural price and output images, then the summary Excel spreadsheet is generated.

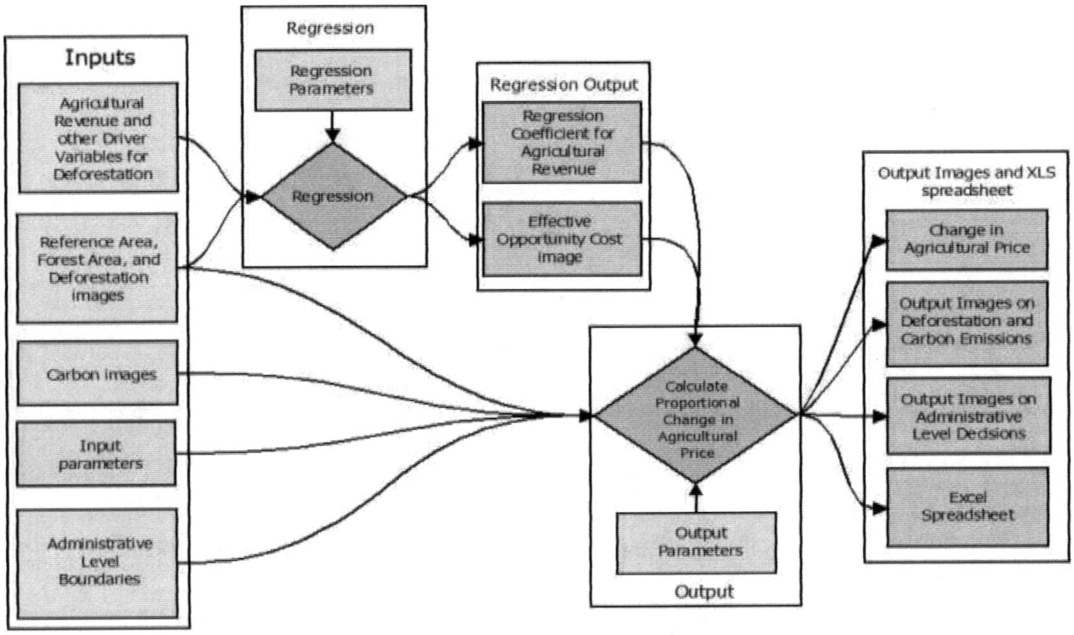

Figure 3. Flowchart stage research activities.

2.2. Regression Analysis

Stage of activity in this research refers to Eastman [9]. The regression step of the GeOSIRIS modeler calculates the correlation between deforestation and some individual variables (14 variables), including agricultural revenue. There are several options to classify this regression, where GeOSIRIS will run a separate regression for several different classes. These classes can be based on the amount of preexisting forest cover. This study is based on geographic regions, such as provinces (5 provinces) or districts (55 districts).

The regression model used in this study is Poisson regression, in which the deforestation is counted by assuming that each pixel is composed of smaller subsections which may be individually deforested [9]. The Poisson regression uses the following formula:

$$E(Y / X) = e^{\sum_{i=0}^{i=N} Bt.Xt} \quad (1)$$

where:
$(Y|X)$ = the expected count of deforestation (Y) given certain input conditions (X)
Xi =independent variable (X_0=1 for the constant term)
Bi =variable coefficients (or parameters)

The model parameters consist of external variables (economic variables) and parameters that affect the price of agricultural products. Net Present Value formula:

$$NPV = \sum_{t=1}^{T}(Bt - Ct)(1 - i)^{-t} \quad (2)$$

where :
Rt = total revenue generated in year t,
Ct = total costs in year t,
i = interest rate
T = expected lifetime (5 years)

The model parameters are economic and those affecting the price of agriculture products. The price elasticity is a measure on how sensitive the agriculture production price is to the change in deforestation. The external factors causing the increase in agricultural price (exogenous change) is a part of the final change in agricultural price as shown in figure 4.

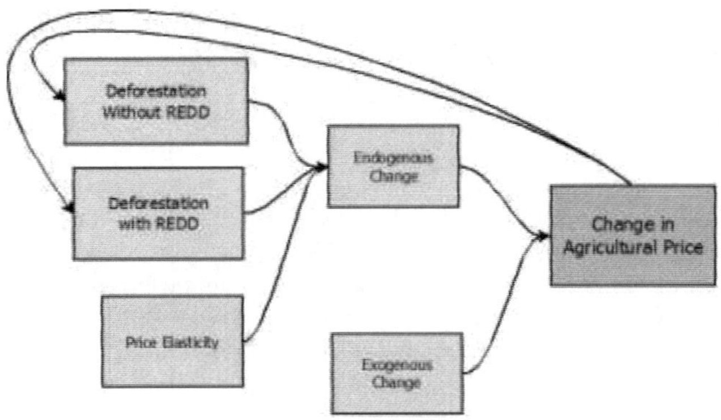

Figure 4. The exogenous increasing in agricultural price.

The GeOSIRIS model can be applied at different administrative levels, such as the district or provincial level. Image files are inputted in the administrative levels table of the input image files panel. The emission factor map is used to calculate the amount of CO_2 (in tons) that will be emitted per hectare of deforestation. There are three components for the emission factor in the GeOSIRIS model: soil carbon, above and below-ground carbon, and peat. The calculations of the emission factor for each pixel are:

$$EE = (AB+SC*fs)*3.67 \text{ where peat P=0} \quad (3)$$

$$E = AB*3.67+fp \text{ where peat } P>0 \quad (4)$$

where:

E = emission factor (tCO_2e/ha)

AB = above and ground carbon

SC = soil carbon

fs = soil carbon factor

fp = emission factor for peat soil

The final proportional change in the price of agricultural product is calculated in the output parameters panel. An iterative loop and two input parameters, which are model precision and maximum number of iterations, are used in this calculation. The price change is then calculated as the sum of endogenous change and exogenous change.

$$\text{Change in Agricultural Price} = \text{endogenous change (independent)} + \text{exogenous changes} \quad (5)$$

$$\text{endogenous Change} = \left[\frac{\text{Deforestation without REDD}}{\text{Deforestation with REDD}}\right]^{e} \quad (6)$$

where:

the exponent e = price elasticity

2.3 Calculating the Proportional Change in Agricultural Price

The GeOSIRIS model compares two consecutive values of changes in agricultural product price to see whether the value is appropriate. The model will keep on going until either the precision model or the maximum number of iterations are exceeded. The last iteration value obtained will be used for final calculation. Analysis of changes in agricultural prices, where proportional changes in agricultural prices are calculated, the image as a result of the analysis, and summary of the calculation results (in Excel worksheet) are then generated.

3. Results and discussion

3.1 Deforestation

The total forest area at Kalimantan Island in 2005 was 29.321 million ha or about 56.33% of the whole Kalimantan Island. It consists of 4.032 million ha (13.75 %) peatland forest and of 25.259 million ha (86.25 %) non-peatland forest or mineral forest. Based on forest type, it consists of primary forest and secondary forest. In the period of 2005-2010 deforestation at Kalimantan Island was 1.417 million ha (4.83%), comprising 0.350 million ha (26.48%) of peatland forests and 1.066 million ha (75.26%) of mineral forests. The rate of deforestation at peatland forest was 8.69% and at non-peatland forest was 4.21%, as presented in Table 1 and Figure 5.

Deforestation that occured at mineral forests was higher than at peatland forests because people prefer to utilize forests in mineral land first, where accessibility is easier and the existence of forests is also wider. Reduced forests in mineral land would then trigger people to take advantage of peatland forests.

The deforestation was relatively similar to the results of [4, 6]. Refer to Sumargo [4] explained that deforestation at Kalimantan Island in the period 2000-2009 amounted to 3.040 million ha, namely deforestation at peat forests of 0.560 million (18.42%) and deforestation at mineral land (non-peatland) of 2.480 million (81.58%). Meanwhile, motivated by [6] deforestation at Kalimantan Island in the period 2006-2011 amounted to 1.479 million ha (5.04%), namely deforestation at peat forests of 0.334 million ha (22.60%) and deforestation at mineral land (non-peatland) of 1.144 million ha (77.40%).

Tabel 1 Results of Deforestation Estimation Year 2005-2010 in Kalimantan Island.

No	Kalimantan Island	All land	Peatland	Non-peatland
1	Land area (ha)	52,049,200	8,175,614	43,873,588
2	Starting forest area (ha)	29,321,650	4,032,375	25,289,276
3	Deforestation without REDD (observed; ha/5 years)	1,417,493	350,562	1,066,931
4	Deforestation rate without REDD (observed; %/ years)	0.986%	0.905%	0.430%
5	Deforestation without REDD (modeled; ha/5 years)	1,020,521	299,114	721,406
6	Deforestation rate without REDD (modeled; %/ years)	0.706%	0.768%	0.289%
7	Deforestation with REDD (modeled; ha/5 years)	850,074	213,817	636,257
8	Deforestation rate with REDD (modeled; %/years)	0.587%	0.543%	0.254%
9	Reduction in deforestation (ha/5 years)	170,447	85,298	85,149
10	change in deforestation due to REDD (percent)	-16.70%	-28.52%	-11.80%

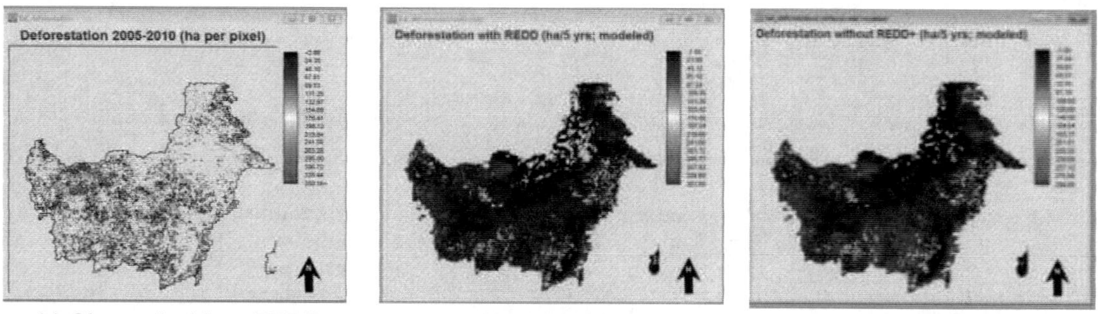

(a) Observed without REDD (b) with REDD (c) Without REDD

Figure 5. Map of deforestation at Kalimantan Islands.

The rate of deforestation in Kalimantan Island was lower than both deforestation occured in Sumatera Island and average deforestation in Indonesia over the same period. Deforestation in Sumatera Island was 1.622 million ha (8.77%), comprising of deforestation at peat forest 0.824 million (50.50%) and deforestation at mineral soils 0.808 million ha (49.50%). While average deforestation in Indonesia was 4.647 million ha (5.00%), comprising of deforestation at peat forests 1.696 million ha (13.01%) and deforestation at mineral soils 2.950 million ha (3.69%). Based on type of forest, the rate of deforestation at peat forest in Kalimantan Islands (8.69%) was lower than in Sumatera Island (13,01%) and in average of Indonesia (26,49%). Furthermore, Margono [11] declared deforestation of primary forest at Kalimantan in 2000-2012 amounted to 2.377 million ha, comprising of deforestation at wetland forest 0.897 million ha and at dryland forest 1.390 million ha. The rate of deforestation of total primary forest was 7.92%, at wetland forest was 5.25%, and at dryland forest was 5.77%. The rate of deforestation at Kalimantan Islands varied depending on the level of spatial resolution of data sources used. Margono's research used Landsat Image data, therefore he got larger amount of deforestation. This was because spatial resolution of the image was 30 m, more meticulous than the global data used in this study with spatial resolution of 3 km.

The deforestation in 2005-2010 happened as a result of government policy in the development of agricultural areas, the development of oil palm plantations and industrial plantations. This is in line with the findings of study of expansion of agricultural policy, timber extraction and infrastructure expansion [12]. The main reasons of forest cover deficit in Kalimantan were related to the expansion of worldwide markets for pulp, wood and palm oil [13,14]. While Margono [15] asserted that in the period of 2000-2010 the cause of deforestation was the expansion of agricultural areas, especially palm oil plantations, expansion of pulp and paper plantation industrial areas and industrial forest clearance.

3.2 Carbon Emissions The impact of REDD

Implementation of REDD policies, which have an impact on reducing forest degradation, also directly impact on reductions of carbon emissions. Based on the variables affecting deforestation, carbon emissions and peat swamp factors have a strong effect to deforestation as shown in figure 6.

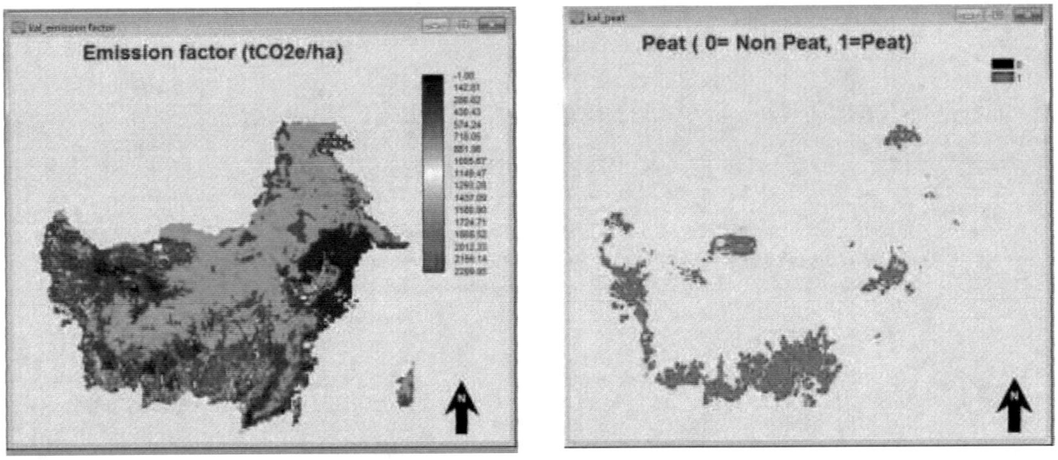

Figure 6. Carbon emission factor and peat-swamp.

Carbon emission factors in West Kalimantan Province and Central Kalimantan have relatively higher value compared to other provinces. This is related to the presence of large peat forest located in this area, while peatland is the highest contributor to emissions. The forest emissions (emitable CO_2) at Kalimantan Island was 16.792 million tCO_2e, donation from peat forest 6,798 million tCO_2e (40,48%) and from mineral forest 9.994 million tCO_2e (59,52%). The amount of carbon emissions in Kalimantan Island was 27,163 million tCO_2e. Compared to Sumatra and average of Indonesia, carbon emissions at Kalimantan was lower (22,29%), followed by average carbon emissions that occur at whole Indonesia (24,75%) and Sumatra Island (46,83%). Both islands (Kalimantan and Sumatra) contributes carbon emissions as much as 69.14%. Meanwhile, according to Alliance (2008) stated that Indonesia had various emission levels from deforestation on each island. The highest emissions came from Sumatra, which were almost 56% of all emissions, and the second was Kalimantan with 28%, thus total for both islands was 84%. Therefore, it is important to focus on these two islands in implementing emission reduction strategies. The high emissions from Sumatra and Kalimantan were caused by the high deforestation rate on both islands, reaching 77% of Indonesia's total deforestation. Deforestation in Sumatra contributed the greatest importance of the existing focus on clearance of peat forest [16].

The REDD policy targeted carbon emissions of 1.099 million ha. Meanwhile, the gross emission reduction that could be obtained was 1.099 million tCO_2e, and emissions that could be absorbed by forests was 0.854 million tCO_2e. The REDD policy was capable of reducing carbon emissions at Kalimantan Island by 0.245 million tCO_2e (47.83%). Meanwhile, the reduction of carbon emission in peat forest area was 0.80 million tCO_2e (28.30%) and in mineral soil forest area was 0.64 million tCO2e (14.02 %) as presented in Table 3 and Figure 7.

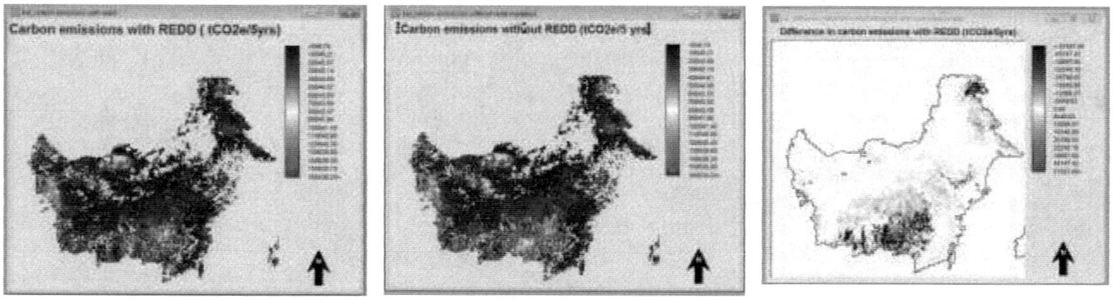

a. with REDD b. Without REDD c. Difference in Carbon Emission

Figure 7. Map of Carbon Emission at Kalimantan Island.

Table 2. Result of carbon emission expectation year 2005- 2010 in Kalimantan Island (x 1,000).

No	Kalimantan island	All land	Peatland	Non-peatland
1	Emittable CO_2 from forest carbon stock (estimated; tCO_2)	27,163.52	8,760.07	18,403.45
2	Emissions without REDD (estimated; tCO_2/5 years)	1,392.88	739.68	653.19
3	Emissions without REDD (modelled; tCO_2/5 years)	1,099.32	636.74	462.58
4	National reference level of emissions (tCO_2e/5 years)	1,099.32	-	-
5	Emissions with REDD (modelled; tCO_2/5 years)	854.29	456.54	397.75
6	Gross emission reductions (tCO_2e/5 years)	265.50	194.60	70.90
7	Gross emission increases (tCO_2e/5 years)	20.47	14.41	6.06
8	Net emission reductions (tCO2e/5 years)	245.03	180.20	64.84
9	Credited emission reductions (tCO2e/5years)	245.03	-	-
10	Change in emissions due to REDD (percent)	-22.29%	-28.30%	-14.02%

The decline in carbon emission levels at Kalimantan island was lower than the reduced emission carbon that occurred at Sumatera Islands and also at Indonesia in the same period. The carbon emission reduction at Kalimantan Island was 245 million tCO_2e (22.29%), consisting of 180 million tCO2e (28.52%) at peatland forest and 64 million tCO2e (14.22%) at mineral soil. Meanwhile, the decline in carbon emissions in Indonesia was 1.091 million tCO2e (24.75%), comprising of 858 million (32.42%) at peatland forests and a decrease in mineral soil carbon emissions of 233 million tCO2e (13.22%).

Changes in carbon emissions were proportional to the rate of deforestation that occured. The relatively smaller peatland forest area compared to the mineral forests caused the reductions deforestation rate (percentage of deforestation) in peatland forests to be greater than the rate of deforestation in mineral forests, with the same forest area.

Assuming that world carbon price was US \$ 10 / tCO2e, impact of REDD Policy at Kalimantan Island was that the gross national revenue from carbon payments (NPV, 5 years) would be \$ 2,450.34 million, with allocation for local government (provincial and district) as incentives (NPV, 5 years) was \$ 2,150.07 million (87.56%). Net government surplus originating from carbon payments was US \$ 300.28 million (NPV, 5 years).

Results of the study [17] that calculated carbon emissions in Bolivia, GeOSIRIS could also be used to evaluate how much reduction of deforestation could be achieved with the price of alternative carbon. With international CO_2 price of \$5/tCO_2, deforestation at the Kalimantan Island could be reduced by about 15% (assuming there was no transaction costs). With a price of \$10 it could be reduced by about 30% and at \$30 by around 57% (see Figure 8).

The increase in carbon prices will spur activities to protect the forests so that the forests will be better protected and deforestation will also occur. Conversely, if there is an increase in price of agricultural products, then the rate of deforestation will also increase, because more forest areas will be cultivated into agricultural areas.

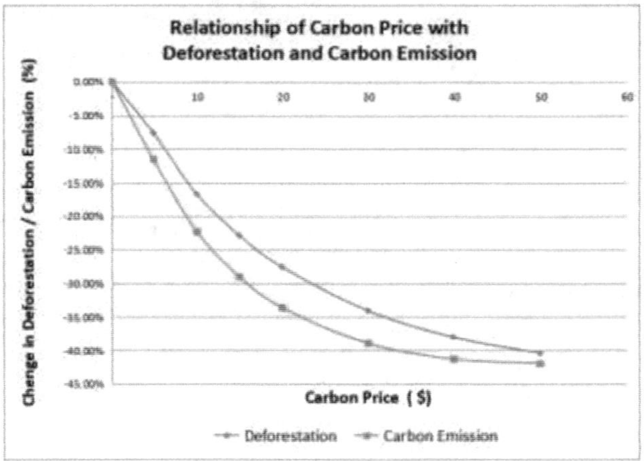

Figure 8. Relationship of carbon price with deforestation and carbon emission.

Similarly, a success in reducing deforestation is linearly related to reduction of carbon emissions. The more forests that can be protected from logging, the more economically beneficial they will be. If the REDD policy scenario will be applied at the provincial or district level, then based on the trade-off between carbon revenues and agricultural income (minus fines for carbon payments), 4 provinces (80%) and 43 districts (78%) will be involved for the success of REDD programs in Kalimantan Island.

4. Conclusion

In the period 2005-2010, deforestation at Kalimantan Island was 1.417 million ha (4.83 %). The simulation result, impact of REDD policy could reduce deforestation at Kalimantan Island by 0.170 million ha (16.70%) and also could reduce carbon emission by 245.03 million tCO_2e/5 years (22,29%), causing net government surplus originating from carbon payments by US \$ 300.28 million (NPV, 5 years).

Acknowledgment

Thanks to the Head of Center for Research, Promotion and Cooperation, and Head of Research Division of Geospatial Information Agency of Indonesia (BIG) who have facilitated this research activity. Also, a huge thanks to TerrSett that provided data.

References

[1] Bununu Y A, Ludin A N M, and Hosni N 2016 *Proc 10th SEATUC Symposium* (Tokyo: Shibaura Institute of Technology)
[2] Barlow J, Gardner T A, Araujo I S, Ávila-Pires T C, Bonaldo A B, Costa J E, and Hoogmoed M S 2007 *Proc of The National Academy of Sciences of the United States of America*
[3] FAO 2010 *Global Forest Resources Assessment 2010 Country Report Indonesia Forest Resource Assessment (FRA) 2010/095* (Rome: UNFAO).
[4] Sumargo W, Nanggara S G, Nainggolan FA, Apriani I 2011 Portrait of Indonesia's Forest *2000-2009* 1st Edition (Jakarta: Forest Watch Indonesia)

[5] Hansen M C, Potapov P V, Moore R, Hancher M, Turubanova S, Tyukavina A, and Kommareddy A 2013 *High-resolution global maps of 21st-century forest cover change (Science vol 34)*

[6] MoEF 2016 *National Forest Reference Emission Level for Deforestation and Forest Degradation: In the Context of Decision 1/CP.16 para 70 UNFCCC (Encourages developing country Parties to contribute to mitigation actions in the forest sector)* (Jakarta: Directorate General of Climate Change. The Ministry of Environment and Forestry)

[7] Bassi A, Varma K, and Toppo W 2015 *Forest ecosystem valuation study: Indonesia. United Nations Office for REDD Coordination in Indonesia (UNORCID)*

[8] Ministry of Forestry of the Republic of Indonesia 2008. COnsolidation Report Reducing Emissions From Deforestation And Forest Degradation In Indonesia.

[9] Eastman JR 2014 *Manual Terrset Manual (Chapter Eight : Geosiris).* Clark Labs, Clark University, Worcester, Massachussetts.

[10] Busch, J., Strassburg, B., Cattaneo, A., Lubowski, R., Bruner, A., Rice, R., ... & Boltz, F. 2009. Comparing climate and cost impacts of reference levels for reducing emissions from deforestation. *Environmental Research Letters*, *4*(4), 044006

[11] Margono B A, Potapov P V, Turubanova S, Stolle F, and Hansen M C 2014 *Nature Climate Change* **4** 730-735

[12] Fuller, D. O., Jessup, T. C., & Salim, A. 2004. Loss of forest cover in Kalimantan, Indonesia, since the 1997–1998 El Nino. Conservation Biology, 18(1), 249-254.

[13] Nawir A A, Murniati and Rumboko L 2007 *Forest Rehabilitation in Indonesia: Where to After More Than Three Decades?* (Bogor: CIFOR)

[14] Uryu Y, Mott C, Foead N, Yulianto K., Budiman, A, Setiabudi, F T, and Jaenicke J 2008 *Deforestation, forest degradation, biodiversity loss and CO2 emissions in Riau, Sumatra, Indonesia* (Jakarta: WWF Indonesia)

[15] Margono B A, Turubanova S, Zhuravleva I, Potapov P, Tyukavina A, Baccini A, and Hansen M C 2012 *Environmental Research Letters* **7** 034010.

[16] Andersen L E, Busch J, Curran E, Ledezma J C, Mayorga J, and Bellier M 2012 *Environmental and socio-economic consequences of forest carbon payments in Bolivia: Results of the OSIRIS model* (Bolivia: Institute for Advanced Development Studies) p 35

[17] Alliance I F C 2008 *Reducing emissions from deforestation and forest degradation in Indonesia: IFCA consolidation report* (Jakarta: Forestry Research and Development Agency, Ministry of Forestry of Republic of Indonesia)

IOP Conf. Series: Earth and Environmental Science **149** (2018) 012009 doi:10.1088/1755-1315/149/1/012009

Monthly variations of diurnal rainfall in north coast of West Java Indonesia during boreal winter periods

E Yulihastin[1,2]andTrismidianto[2]
[1]Program in Earth Sciences Department, Faculty of Earth Science and Technology, Bandung Institute of Technology, Bandung 40132, Indonesia
[2]Center of Atmospheric Science and Technology, National Institute of Aeronautics and Space, Bandung 40173, Indonesia

E-mail: erma.yulihastin@lapan.go.id

Abstract. Diurnal rainfall during the active monsoon period is usually associated with the highest convective activity that often triggers extreme rainfall. Investigating diurnal rainfall behavior in the north coast of West Java is important to recognize the behavioral trends of data leading to such extreme events in strategic West Java because the city of Jakarta is located in this region. Variability of diurnal rainfall during the period of active monsoon on December-January-February (DJF) composite during the 2000-2016 period was investigated using hourly rainfall data from Tropical Rainfall Measuring Mission (TRMM) 3B41RT dataset. Through the Empirical Mode Decomposition method was appears that the diurnal rain cycle during February has increased significantly in its amplitude and frequency. It is simultaneously shows that the indication of extreme rainfall events is related to diurnal rain divergences during February shown through phase shifts. The diurnal, semidiurnal, and terdiurnal cycles appear on the characteristics of the DJF composite rainfall data during the 2000-2016 period.The significant increases in amplitude occurred during February are the diurnal (IMF 3) and terdiurnal (IMF 1) of rainfall cycles.

1. Introduction

Research on diurnal rainfall is important to improving regional climate model output due to there are not yet high-resolution regional models which can accurately depict diurnal rainfall cycles in the tropics either in amplitude or phase [1-6]. Simulation of diurnal rainfall in the western part of Indonesia Maritime Continent using MIT Regional Climate Model (MRCM) with 12 km high resolution for 30 years (1982-2011) in DJF shows that the general model can capture diurnal rain amplitude but not able to capture the rainfall phase especially in the northern region of West Java with phase difference between 12-15 hours compared to rainfall from TRMM satellite [6].

In general, there are significant differences in diurnal rainfall patterns between land and ocean related to the phase and time of maximum rainfall. Diurnal rainfall in the mainland occurs from Late-Afternoon (LA) to Middle-Night (MN), while rainfall over the ocean occurs from MN to Early Morning (EM) [7-12]. In addition, based on maximum and phase, diurnal rainfall variability also has different patterns with respect to the frequency or number of peaks between on land and sea. Diurnal cycle patterns (one peak) and semidiurnal (two peaks) are formed on land and sea with different frequency of occurrence. Globally in the tropics, both on land and at sea, the main pattern with 45%

frequency (PC 1) and 22% (PC 2) is diurnal then followed by semidiurnal pattern with frequency 6% and 4% appearing on PC 3 and PC 4 [12].

Seasonal variability of diurnal rainfall occurredespecially over the ocean. In the Pacific Ocean, diurnal rainfall reaches themaximum during the boreal winter period of DJF while in the Indian Ocean diurnal rainfall has maximum value in the May-June-July-August(MJJA) of boreal summer period [11] while convection activity reaches a maximum in September-November. In the South China Sea, the highest convection activity occurred in October-January and slightly increased in July and August at 07:00-19:00 LT(Local Time) [13]. Similarly, in the Java Sea, convection activity reached themaximum in October-January during the period of 07:00-19:00 LT [13].

In Java Island, which belongs to areas with strong monsoon activity, diurnal rainfall occurs from LA to MN [14-15]. However, interannual variation of diurnal rain shows that interaction between ENSO and topography in southern Java Island produces more intensive if compared the El-Nino years with normal periods [16]. On the other hand, in some cases of floods occurring in the north of West Java, especially Jakarta, the heavy rainfall over land occurred from EM to noon [17-20]. Also, there is evidence that daily rainfall tends to increase for over a hundred years [21]. In addition, diurnal variations in precipitation also affect diurnal variations of the Ciliwung River that supplies the main river flows that cause Jakarta floods [22].

Currently, there has been no research that discusses diurnal rainfall in monthly variations during the rainy season of DJF in the north coast of West Java. Though such knowledge is needed to understand the characteristics of diurnal rainfall pattern in the north coast of West Java from time to time in climatologically. Therefore, this study discusses month to month variations during the DJF period of diurnal rainfall in the north coast of West Java.

2. Data and Method

The rainfall data during the DJF in the period of 2000-2016 were obtained from the TRMM Multi-satellite Precipitation Analyses Real-Time (TRMMTMPA)3B41RT dataset, which has hourly temporal resolution and 0.25° × 0.25° spatial resolution [23-24]. This data provides a calibration-based sequential scheme for combining precipitation estimates from multiple satellites. TMPA is available both after and in real time, based on calibration by the TRMM Combined Instrument and TRMM Microwave Imager precipitation products, respectively. Only the after-real-time product incorporates gauge data at present. These data could be downloaded free from https://pmm.nasa.gov/data - access/downloads.

To analyze the seasonal variability of non-stationary rainfall data (DJF) on DJF related to trends, the power spectrum, and frequency of signals data wereperformed by Empirical Mode Decomposition (EMD) method. This method decomposes the signals by performing direct extraction of signal energythatcan be regarded as an intrinsic function in the scale of space and time [25].

EMD decomposes a non-stationary signal into a set of intrinsic oscillatory modes. It component called Intrinsic Mode Functions (IMFs). By knowing the signal of each IMF, then we can calculate ofmulticomponent frequency instantly based on Hilbert transformation. In this way, the analysis of each IMF signal can be done in detail by localizing events in time and frequency scales [25]. In the current research, we composite DJF in the period of 2000-2016 and apply the EMD method to determine the characteristics of diurnal rainfall data during the period.

3. Result and Discussion

EMD analysis of DJF composite rainfall data during the period 2000-2016 have generated 10 IMFs and residuals. Based on the EMD analysis, IMF 1 represented to 3 peaks of cycles, then so-called "terdiurnal" pattern, IMF 2 showed 2 pekas of cycles or "semidiurnal", and IMF 3 described "diurnal" pattern. Residual data shows a trend line of data that has a peak cycle during January (figure 1).

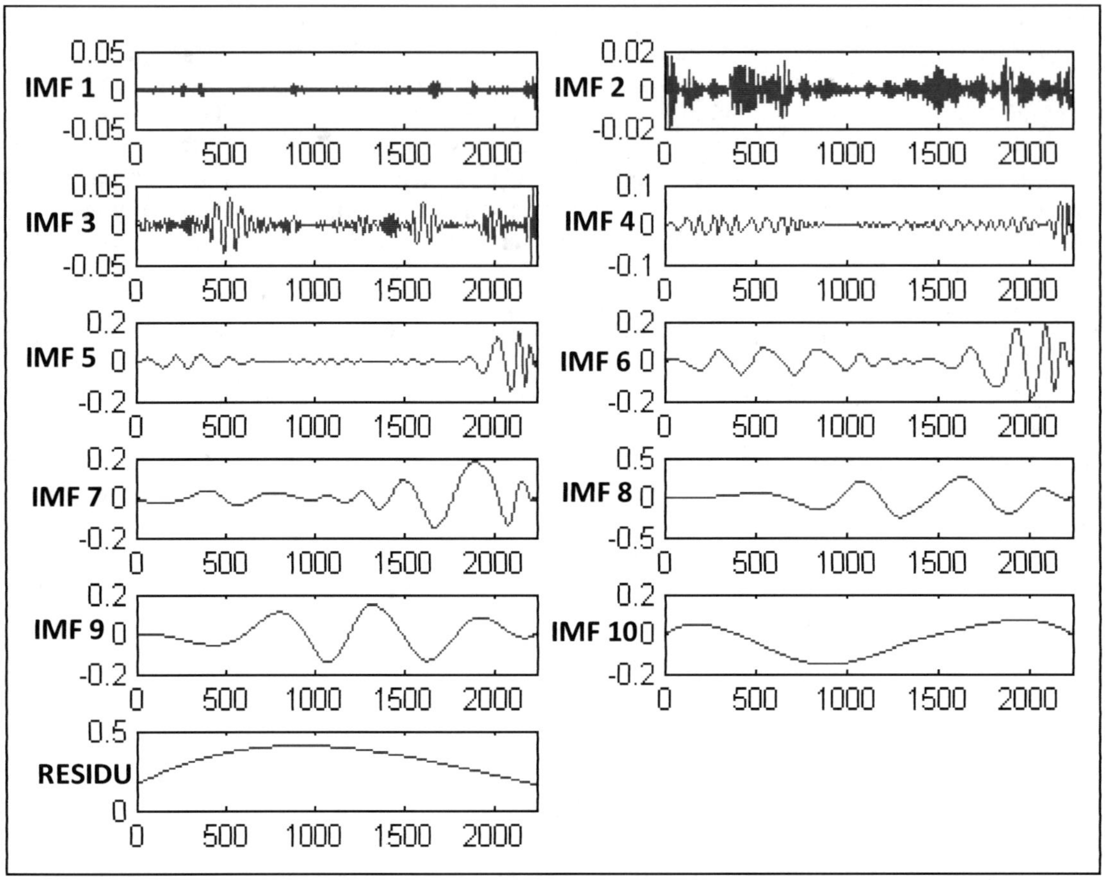

Figure 1. IMF 1-11 (a-j) based on hourly rainfall data of TRMM 3B41RT during DJF in the period of 2000-2016.

Figure 2 shows that at IMF 1, the rain cycle period is 5.5 hours or 0.3 daily or also known as the diurnal pattern ie 3 peaks in 24 hours (terdiurnal). In December and February, terdiurnal rainfall data variations were greater than in January. In addition, the soaring data variations appear at the end of February. Significant phase differences emerged in the first 1-3 hours indicating a shift in the afternoon rainy phase starting at 19:00 LT may shift to 20:00 LT until midnight at 22:00 LT. The phase shift from late afternoon to midnight or early morning corresponds to recent research results by Im and Eltahir [6] which show that the maximum diurnal rainfall phase in northern Java occurs at midnight or early morning. However, previous findings by Qian [14] show that diurnal rainfall in Java has a maximum phase in the afternoon.

Figure 3 shows that IMF 2 has a 14 hour or 0.6 daily cycle period (semidiurnal) with variations and amplitude in December and February larger than January. In addition, the signal amplitude also shows that semidiurnal cycles are larger than the terdiurnal cycles (IMF 1). The phase difference in the semidiurnal cycle is 1-5 hours faster or 7-15 hours late from the original phase. It indicated that there is a semidiurnal phase shift from the afternoon rain to the night into the night until midnight or morning until noon.The current results corresponds to earlier findings regarding the maximum phase of diurnal rainfall in northern Java occurring at midnight or early morning [6]. However, the semidiurnal rainfall pattern found in northern West Java is uncommon with the general pattern of diurnal rainfall in tropical regions that should occur over the oceans with frequencies less than 10% [28].

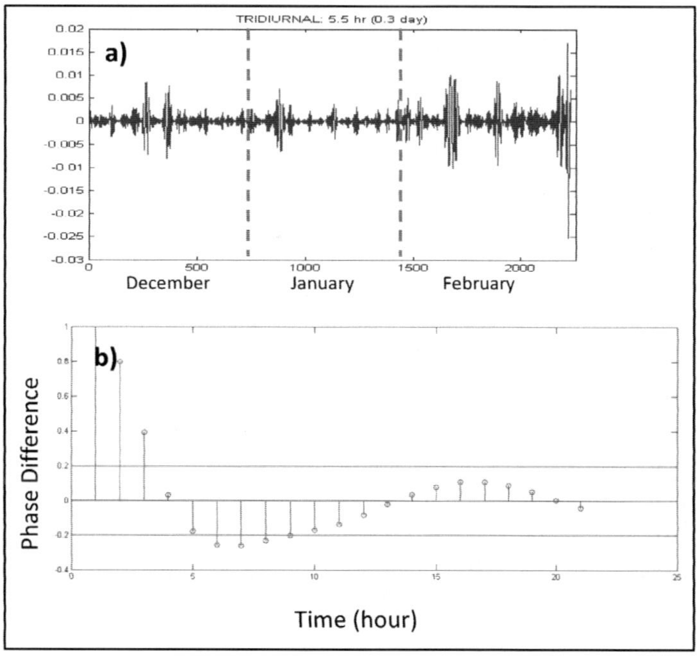

Figure 2. IMF 1 (a) and autocorrelation of hourly rainfall data of TRMM 3B41RT during DJF in the period of 2000-2016.

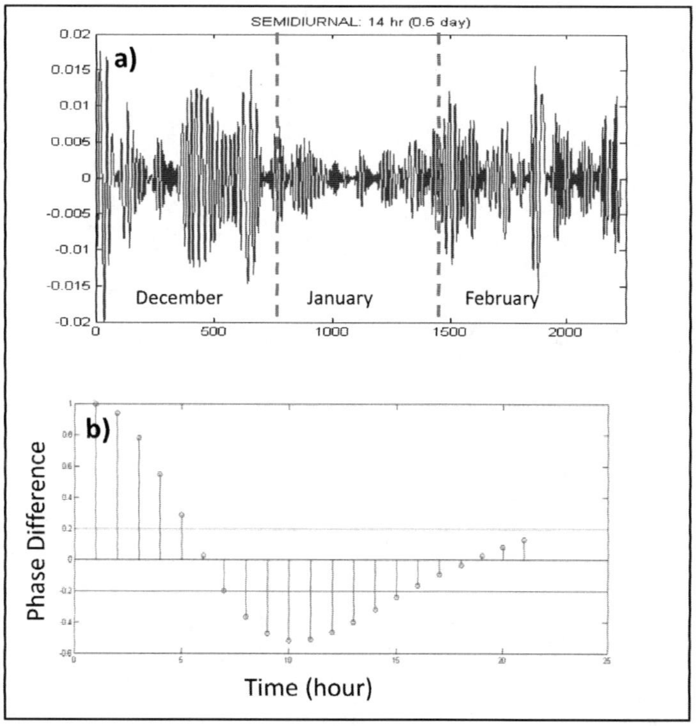

Figure 3. Same as figure 2, but for IMF 2 (semidiurnal).

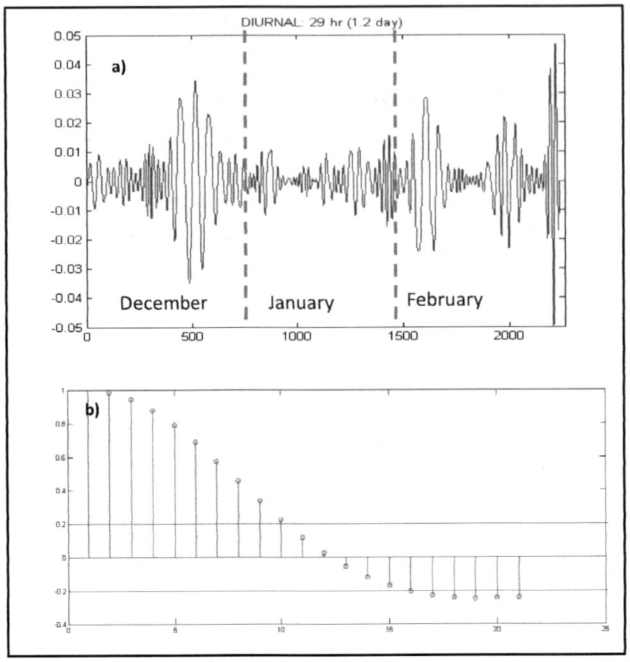

Figure 4. Same as Figure 3, but for IMF 3 (diurnal)

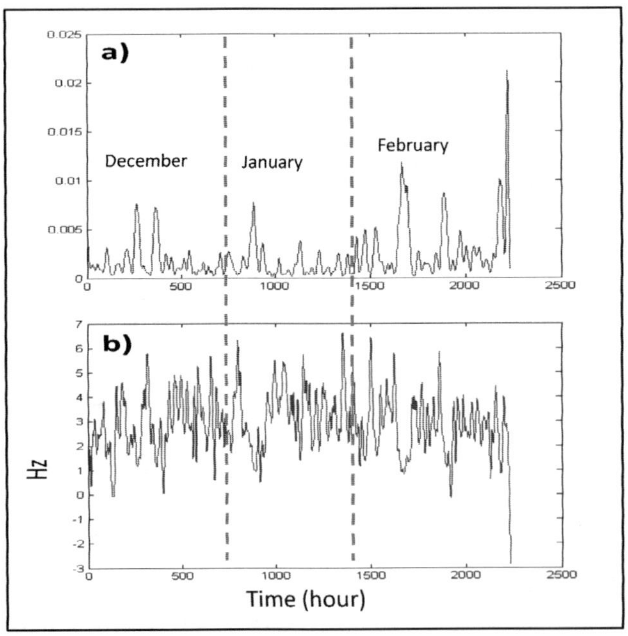

Figure 5. Absolute value or amplitude (a) and frequency (b) of IMF 1 (terdiunal).

Figure 4 shows IMF 3 which is represents a diurnal cycle with a peak of cycle during 24 hours. Regarding to figure 4, we can see that the variation and amplitude of December and February data are larger than January. In addition, there was a significant increase in data variations by the end of February. The phase difference for diurnal cycle categories can be faster between 1-10 hours than the original phase. It indicated that the phase of diurnal rainfall have shifted from Late Afternoon (LA)-Middle Night (MN) to MN-morning.

For semidiurnal patterns as shown in figure 5, the data amplitude looks large in December and February. However, the data frequency has a small value during that sharp rise. It indicated that the sharp increase in the data amplitudes corresponds to the decrease in data frequency. In contrast, the increasing the amplitude of the diurnal cycle occurs at the end of February which is also followed by a sharp rise in data frequency as shown in figure 6. It may indicate an increase in extreme rain events at the end of February for the diurnal cycle.

Figure 7 describes the category of diurnal cycles that shows the IMF 1 amplitude has increased significantly in February, especially at the end of February.However, the frequency of amplitude that experienced a significant increase is actually low or decreased when compared with data frequency in December and January. The EMD analysis was also shown the trend of changes in the data cycle pattern over time, ie IMF 5, 6, and 7 (figure 8). IMF 5-7 confirms that in the February, there was an increase in the amplitude of the cycle about 2-3 times compared with December or January. It indicated that the occurrence of extreme rain events during February that could be captured by IMF 5-7 results already added to the residu (figure 8).

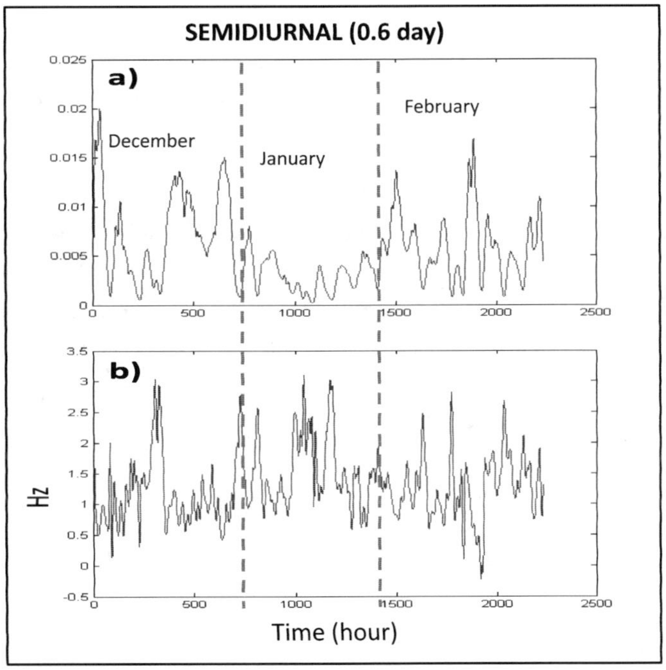

Figure 6. Same as figure 5, but for IMF 2 (semidiunal).

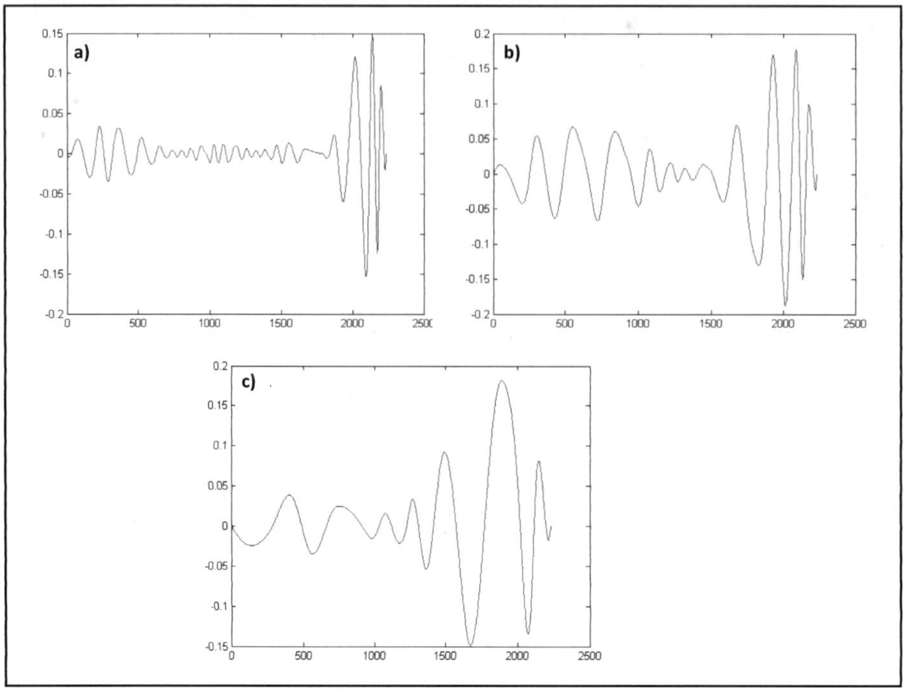

Figure 7. IMF 5-7 (a-c) based on hourly rainfall data of TRMM 3B41RT DJF 2000-2016.

The EMD method has proved that in February there was a high variation of data in amplitude and frequency of terdiurnal, semidiurnal, and diurnal rainfall cycles. It was also found that the three types of cycles had undergone phase shifts into the maximum phase of rainfall occurring in the middle of the night or early morning and this corresponds to the recent discovery of diurnal rain in northern Java [6]. Figure 9 shows an example of the diurnal rainfall mean pattern that shifted the phase from the afternoon into the early morning to February 2002 (figure 9-a), terdiurnal pattern with the main peak occurring in the early hours of February 2008 (figure 9-b), semidiurnal pattern with the main peak occurred in the early hours of February 2011 (figure 9-c), and the normal pattern of diurnal rainfall with rainfall peak occurred in the afternoon until evening as it was in February 2012 (figure 9-d).

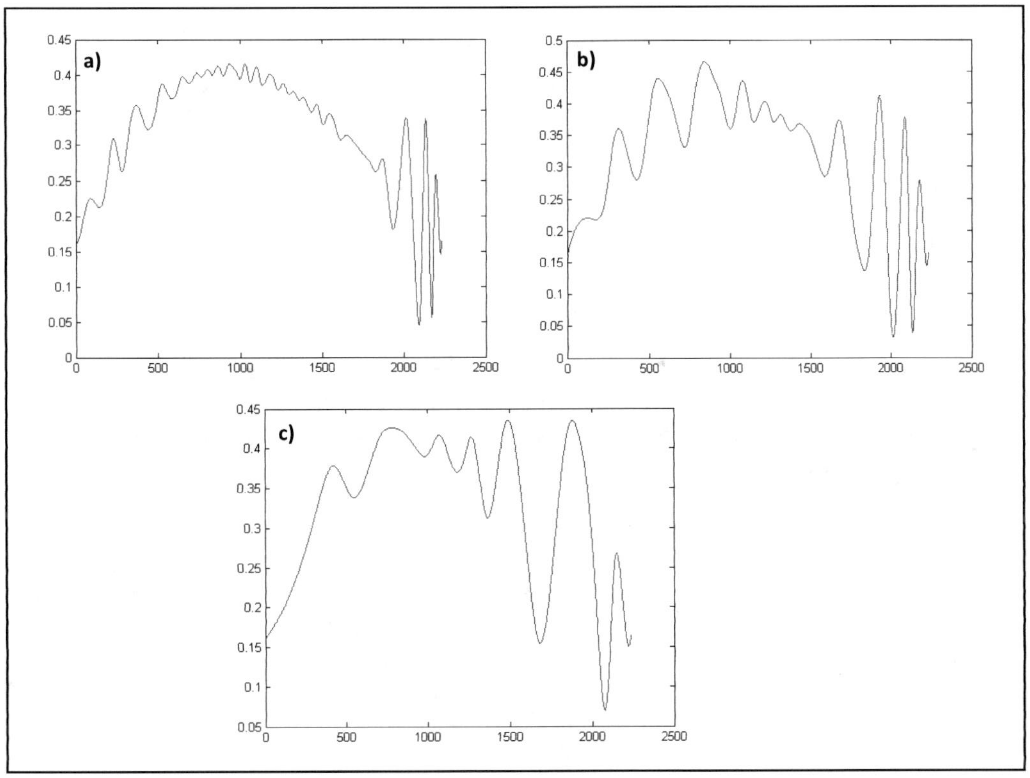

Figure 8. IMF 5-7 and itsresidu (a-c).

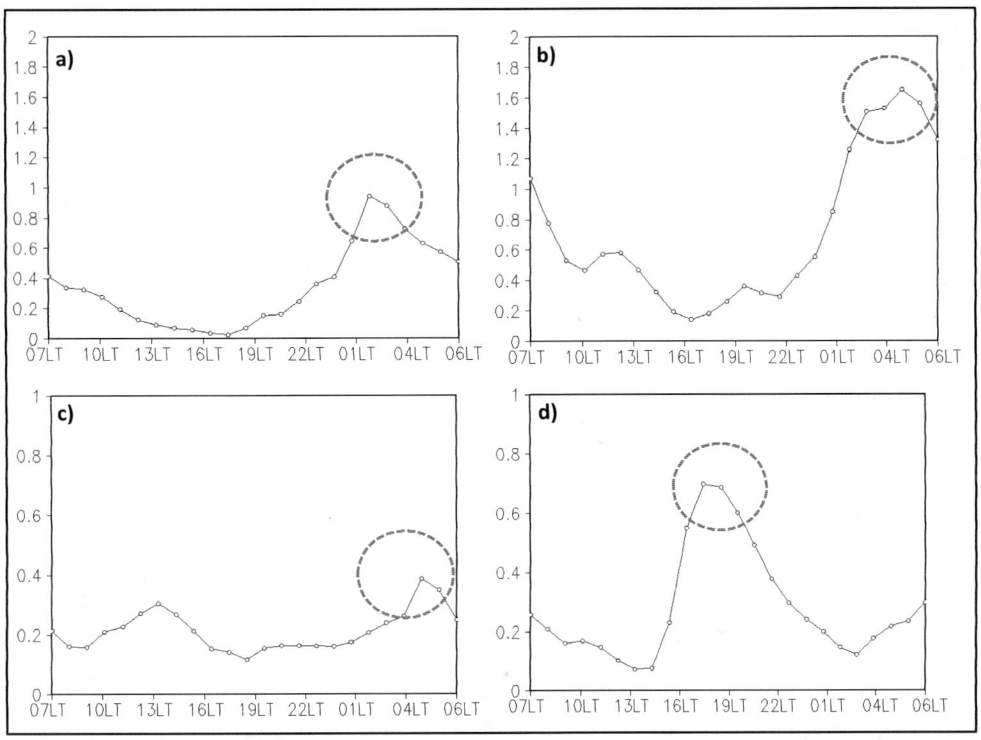

Figure 9. Diurnal rainfall cycle mean during February: 2002 (a), 2008 (b), 2011 (c), 2012 (d).

4. Conclusion

The monthly variations for the diurnal rainfall during the DJF period in the north coast of Java Island shows that there is a higher rainfall trend in January than in December and February. However, data variations in December and February were larger than in January. In addition, the diurnal rain cycle has three patterns: diurnal (one peak, IMF3), semidiurnal (two peaks, IMF 2), terdiurnal (three peaks, IMF 1). The amplitude of rain cycle with the diurnal pattern in February has increased by about 2-3 times compared to December or January. In addition, the frequency of rainfall also increased strongly in February. For semidiurnal patterns, the amplitudes of cycles with semidiurnal patterns larger in December and February, but the data frequency has minimum value during that sharp rose.For the category of terdiurnal rainfall, amplitude increased significantly followed by a decrease in frequency in February, especially at the end of February. The phase difference for the diurnal category, the maximum rainfall can experience a phase shift from afternoon rain to evening into the evening until the morning. For semidiurnal, there is a phase shift from the late afternoon to late night into the night until midnight or morning until noon. For the diurnal, there is a shift phase of the afternoon rain starting at 19:00 LT could be shifted to 20:00 LT until midnight at 22:00 LT.

Acknowledgments

The corresponding author would like to thank Indonesia Endowment Fund for Education, Ministry of Finance of the Republic of Indonesia for their scholarship in the completion of doctoral studies in Earth Sciences, Faculty of Science and Technology of Earth, Bandung Institute of Technology.

5. References

[1] Barnaby S L, Matthews A J and Lister G M S 2011 The diurnal cycle of precipitation over The Maritime Continent in a high-resolution atmospheric model *Q. J. R. Met. Soc.* **137** pp 934-947

[2] Biasutti M, Yuter S E, Burleyson C D and Sobel A H 2011 Very high resolution rainfall patterns measured by TRMM precipitation radar: Seasonal and diurnal cycles *The Int. J. of Clim. Dyn.* **39** pp 239 – 258

[3] Love B S, Matthews A J and Lister G M S 2011 The diurnal cycle of precipitation over Maritime Continent in a High-Resolution atmospheric model *Quart. J. of The Royal Met. Soc.*

[4] Evans J P and Westra S 2012 Investigating the mechanism of diurnal rainfall variability using Regional Climate Model *J. of Clim.* **25** pp 7232-7247

[5] Seo H, Subramanian A C, Miller A J and Cavanaugh N R 2014 Coupled impact of the diurnal cycle of sea surface temperature on the Madden Julian Oscillation *J. of Clim.* **27** pp 8422-8443

[6] Im E S and Eltahir E A B 2017 Simulation of the diurnal variation of rainfall over the western maritime continent using a Regional Climate Model *Clim. Dyn.* 1-16, DOI 10.1007/s00382-017-3907-3

[7] Nesbitt S W and Zipser E J 2002 The diurnal cycle of rainfall and convective intensity according to three years of TRMM measurements *J. of Clim.* **16** pp 1456-1475

[8] Sorooshian S, Gao X, Hsu K, Maddox R A, Hong Y, Gupta H V and Imam B 2002 Diurnal variability of tropical rainfall retrieved from combined GOES and TRMM satellite information *J. of Clim.* **15** pp 983-1000

[9] Mori S, Hamada J-I, Tauhid Y I and Yamanaka M D 2004 Diurnal land-sea rainfall peak migration over Sumatera Island, Indonesia Maritime Continent, bbserved by TRMM satellite and intenseve Rawinsonde soundings *Mont. Weat. Rev.* **132** pp 2021-2039

[10] Ichikawa H and Yasunari T 2005 Time-space characteristics of diurnal rainfall over Borneo and surrounding oceans as observed by TRMM-PR *J. of Clim.* **19** pp 1238-1260

[11] Yang S and Smith E A 2006 Mechanism for diurnal variabality of global tropical rainfall observed from TRMM *J. of Clim.* **19** pp 5190-5226

[12] Kikuchi K and Wang B 2007 Diurnal precipitation regimes in the global tropics *J.of Clim.* **21** pp 2680-2696
[13] Harjana T 2006 Estimation of Spatial/Temporal Variations of Rainfall over Indonesia Maritime Continent Using Satellite IR Data *Dissertation* (Japan: Kobe University)
[14] Qian J H 2008 Why precipitation is mostly concentrated over islands in the maritime continent *J. of the Atm. Sci.* **65** pp 1428–1441
[15] Sato T, Miura H, Satoh M, Takayabu Y N and Wang Y 2009 Diurnal cycle of precipitation in the tropics simulated in a global cloud-resolving model *J. of Clim.* **22** pp 4809-4826
[16] Qian J H, Robertson A W and Moron V 2010 Interaction among ENSO, the Monsoon, and diurnal cycle in rainfall variability over Java, Indonesia *J. of Atm. Sci.* **67** pp 3509-3525
[17] Yulihastin E and Trilaksono N J 2014 Evolution of heavy rainfall in Jakarta flood case 2013 based on COSMO Model *Proc. Int. Conf. On Eco.* (LIPI-UNESCO, 10-12 November 2014) (Asia Pacific Center for Ecohidrology UNESCO-Research Center for Limnology LIPI, ISBN 978-979-8163-22-7) pp 167-173
[18] Yulihastin E, Ningsih S N and Hadi T W 2016 Identification of changing in diurnal pattern of rainfall in case of flooding in Northern Coastal of West Java 2014 *Proc. Int. Sym. For Sus. Hum.* 2016 (LIPI-RISH, Indonesia, 15-16 November 2016) (Research Center for Biomaterial LIPI, ISSN 2579-5554) pp 83-90
[19] Trilaksono N J, Otsuka S, Yoden S, Saito K and Hayashi S 2011 Dependence of model simulated heavy rainfall on the horizontal resolution during the Jakarta flood event in January-February 2007 *SOLA* **7** pp 193–196
[20] Trilaksono N J, Otsuka S and Yoden S 2012 A time-lagged ensemble simulation on the modulation of precipitation over West Java in January-February 2007 *Month.Weath. Rev.* **140** pp 601–616
[21] Siswanto, Oldenborgh G J V, Schrier G V D, Jilderda R and Hurk B V D 2015 Temperature, extreme precipitation, and diurnal rainfall changes in the urbanized Jakarta City during the past 130 Years *Int. J. of Clim.* pp 1-19, DOI: 10.1002/joc.4548
[22] Sulistyowati R, Hapsari R I, Syamsudin F, Mori S, Oishi S T and Yamanaka M D 2014 Rainfall-driven diurnal variations of water level in the Ciliwung River, West Java, Indonesia *SOLA* **10** pp 141-144
[23] Huffman G J, Bolvin D T 2013 *TRMM and Other Data Precipitation Data Set Documentation* Available online ftp://meso-a.gsfc.nasa.gov/pub/trmmdocs/3B42_3B43_doc.pdf (accessed on 23 February 2018)
[24] Amanda H, Sayma R, Faisal H, Lance Y, Amvrossios C B, and Greg E 2007 Satellite-based flood modeling using TRMM-based rainfall products *Sensors* **7** pp 3416-3427
[25] Molla M D and Rahman M S 2006 Empirical mode decomposition analysis of climate change with special reference to rainfall data *Disc. Dyn. in Nat. and Soc.* **10** pp 1-17

IOP Conf. Series: Earth and Environmental Science **149** (2018) 012010 doi:10.1088/1755-1315/149/1/012010

Dynamic model of forest area on flood zone of Padang City, West Sumatra Province-Indonesia

Indang Dewata[1], Iswandi U[2]
[1]Departement of Environmental Science, State University of Padang
[2]Department of Geography, Faculty of Social Science, State University of Padang

Email : i_dewata@yahoo.com

Abstract. The flood disaster has caused many harm to human life, and the change of watershed characteristic is one of the factors causing the flood disaster. The increase of deforestation due to the increase of water causes the occurrence of flood disaster in the rainy season. The research objective was to develop a dynamic model of forest on flood hazard zone using powersim 10.1. In model development, there are three scenarios: optimistic, moderate, and pessimistic. The study shows that in Padang there are about 13 percent of high flood hazard zones. Deforestation of 4.5 percent/year is one cause that may increased the flooding intensity in Padang. There will be 14 percent of total forest area when management policy of forest absence in 2050.

1. Introduction

Indonesia is a wet tropical climate country with relatively high rainfall intensity throughout the year. The high intensity of rainfall caused flood disaster in many regions in Indonesia [1]. The factors causing the flood that the high bulk factor or the decreasing factor of forest area in the upstream river [2]. Furthermore, changes in land use to built up area will increase the rate of surface water and reduced water absorption causing an increasingly wide area of flood risk [3-5].

That there are three factors causing the occurrence of flood disasters: meteorological factors, factors of watershed characteristics, and human factors. Meteorological factors that cause the occurrence of floods that rainfall intensity and distribution of rainfall. There is great opportunity of flood disaster in high rainfall intensity area. Other than that watershed characteristics including slope, altitude, landform, and soil type contribute to the occurrence of floods. Furthermore, human behavior is an important factor for the occurrence of flood disasters. Forest encroachment and land cover change into built up area is a driving factor for flood disaster [6].

The human factor is the big role in causing flood. Increasing deforestation and conversion of forest areas into other uses as well as improper in land use are factors causing the expansion of floods [5]. Mudelsee *et al*. (2003) explains that the impact of deforestation on climate change globally accelerate with hydrological cycle. With the acceleration of the hydrologic cycle resulted in the emergence of flood disasters [7].

Mitigation is a series of efforts and or actions to reduce disaster risk [8]. Measures to safeguard and preserve forest areas are one of the non-structural forms of flood mitigation [9,10]. Therefore, prediction and simulation of flood is needed to prevent the disaster. The research objective was to develop a dynamic model of forest on flood hazard zone using powersim 10.1.

2. Research Method

This research was conducted in Padang City, West Sumatera Province of Indonesia. Geographically, the research area is located on 100^0 05' 05"-100^0 34' 09" E and 0^0 44' 00"-1^0 08' 35" S with research area of 69.496 ha. The location of the study is presented in Figure 1.

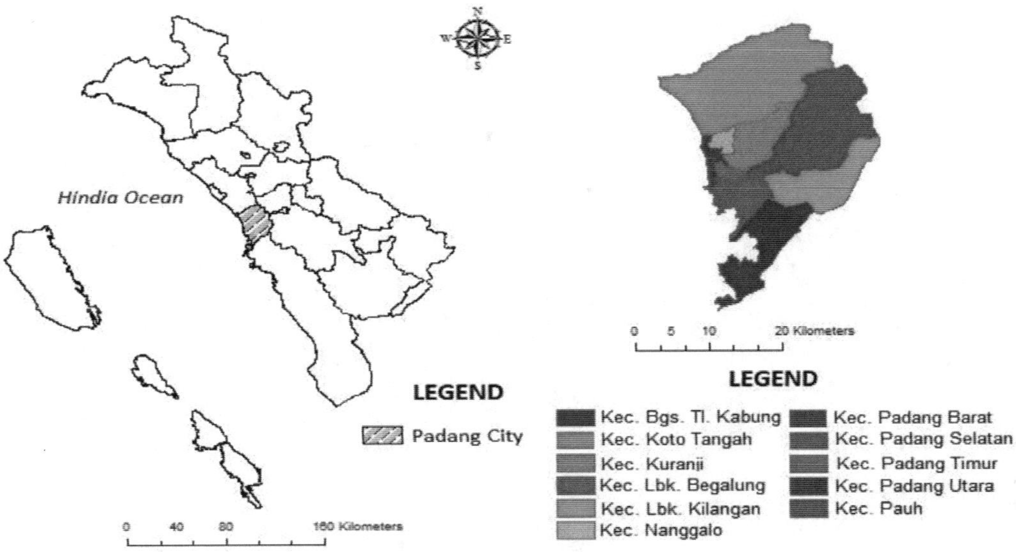

Figure 1. Location of research.

To determine the flood-prone zones, seven indicators are used: soil type, slope, elevation, rainfall, landform, land use, and flood frequency. The land use map is derived from 1: 250,000 scale. Slope maps are generated from the 1: 50,000 scale of Earth Map (RBI). The elevation maps are generated from the Digital Elevation Model (DEM) using the Shuttle Radar Topography Mission (SRTM) image. The land cover of Padang City resulted from Landsat ETM 7+ image interpretation in 2016 and soil type. The rainfall map is produced from interpolation of rainfall data from 1975-2012 period. Flood fraction map resulted from data analysis of Regional Disaster Management Agency (BPBD) of Padang City. Land use maps result from Landsat ETM 7+ image interpretation in 2016.

The indicator of determination of flood-prone zones used is presented in Table 1. For making the value of the flood prone interval class using equation 1. The highest total score is obtained at 3,661, while the lowest total score is obtained at 543. The desired interval class is three classes i.e. low vulnerability zone, moderately prone zones, and high-risk zones. There are many indicator with different weight, sub-indicator as specific parameters of indicator, ratings and score is important to built the model based on previous study (mention in source of table).

Here is equation 1.

$$I = \frac{c-b}{k} \quad .... (1)$$

Information:

I = large distance interval
c = highest score
b = lowest score
k = number of classes to be selected

IOP Conf. Series: Earth and Environmental Science **149** (2018) 012010 doi:10.1088/1755-1315/149/1/012010

Table 1. Flood zone prone indicator.

Indicator/weight	Sub. Indicator	Ratings	Score
Type of soil/10,1	Aluvial	24.7	2495
	Andosol	11.5	116.2
	Organosol	36.6	369.7
	Regosol	7.8	78.8
	Latosol	7,6	76.8
	Complexs podsolik red yellow	12.1	122.2
Slope (%)/20,6	0-9	53.5	1,102.1
	9 -16	25.9	533.5
	16-27	13.1	269.9
	>27	7.5	154.5
Form of land/15,9	Aluvial plans	8.3	132.0
	Aluvial fan	9.8	155.8
	Plans aluvial beach	9.5	151.1
	Beting gisik	12.9	205.1
	Depresion between Beting	19.6	311.6
	Vulcanic mountain complexs	3.1	49.3
	Vulcanic hills	3.0	47.7
	Rear swamp	21.2	337.1
	Karst hills	3.1	49.3
	Natural embankment	9.5	151.1
Rainfall(mm/years) /11,3	> 5000	37	418.1
	4500-5000	24.6	278.0
	4000-4500	16.2	183.1
	3500-4000	9.9	111.9
	3000-3500	7.1	80.2
	2500-3000	5.2	58.8
Elevation (mdpl)/24	0-10	38.5	924.0
	10-30	21.8	523.2
	30-50	14.4	345.6
	50-150	9.3	223.2
	150-450	6.6	158.4
	450-1000	5.3	127.2
	> 1000	4.1	98.4
Land use/10,7	Settlement	33.3	356.3
	Rice fields	21.7	232.2
	Mixed garden	8.9	95.2
	Shurbs	7.9	84.5
	Empty land	15.1	161.6
	Forest	4.8	51.4
Frequency/7,4	Always	56.9	421.1
	Often	23.7	175.4
	Rarely	11.8	87.3
	Without	7.5	55.5

Source: Iswandi (2016), Iswandi *et al.* (2016), and Iswandi *et al.* (2017)

Using the equation 1 we get the interval of 1,039. The distribution of flood hazard interval classes can be seen in Table 2. Prone class divided into three category with different interval class and and prone index. The interval class described the value of model and will be have prone index for each values.

Table 2. Interval class of flood prone.

Prone class	Interval class	Prone Index
Lowest class	543 – 1,582	Lowest prone zone
Moderate class	1,583 – 2,621	Moderate prone zone
Highest class	1.1.– 3,661	Highest prone zone

Source: Iswandi (2016)

Dynamic modeling that will be developed in the flood prone areas of upper watershed conservation area. That human behavior due to misuse of land, deforestation and development is a driving factor for floods [7,5,13]. The Landsat 1994, 2007 and 2016 image used for analysis the land use change. Based on the three images can be described the change of forest area that will be modeled in dynamic system. In dynamic modeling using software powersim 10.1. Figure 2 shows the causal loop and dynamic model structure of flood prone zones in Padang City. After all of the process and result well done, the three scenarios have to examine.

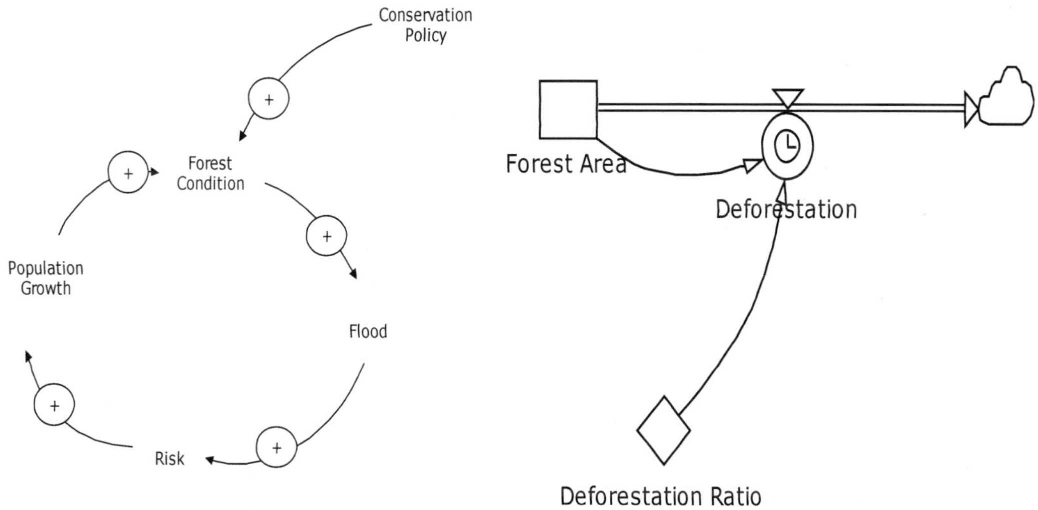

Figure 2. The relationship of causal loop (a), and model structure (b) flood prone areas.

3. Result

Padang city is the central government of West Sumatra Province. Geomorphologically the area is formed by three processes namely the fluvial origin, the process of volcanic origin, and the process of structural origin. The origin region of the fluvial process is a region with relatively flat slopes. In the research area based on the results of the zonation analysis the flood level showed that 13 percent were high vulnerable zones, 14 percent were moderately prone, and 73 percent were low-risk zones. Figure 3 presents the flood-prone zones in Padang City.

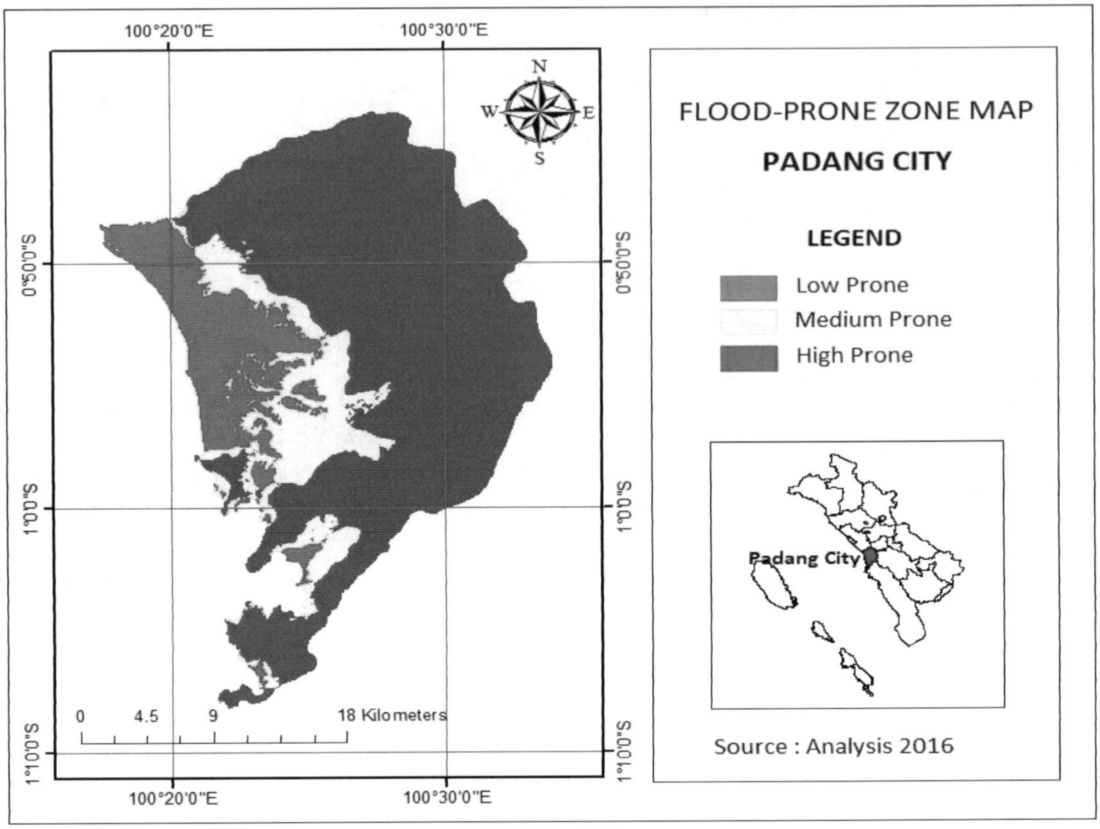

Figure 3. Flood-prone zone in Padang City.

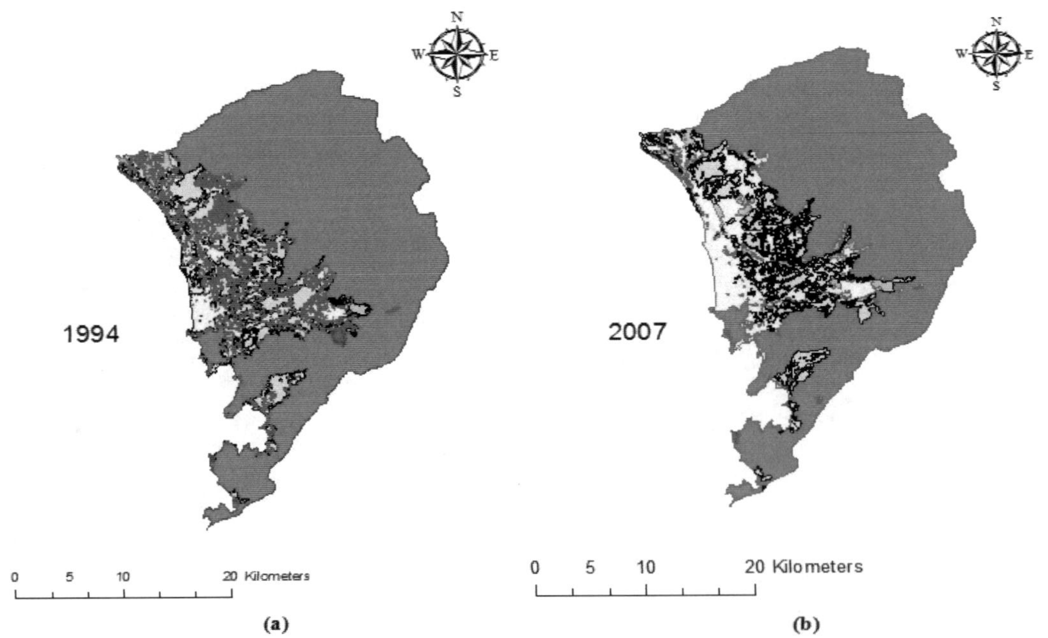

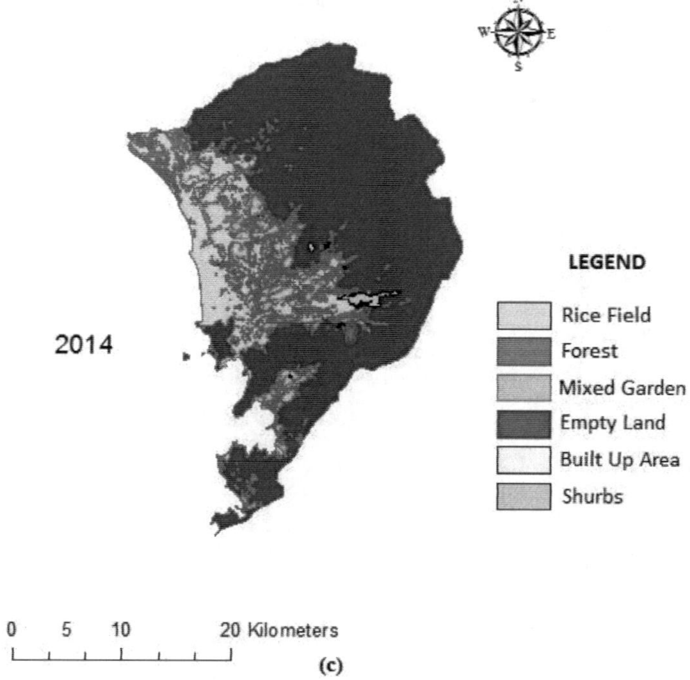

Figure 4. Land Use Change in Padang City.

Padang have four major rivers are Kandih steam, Airdingin steam, Kuranji steam, and Arau steam. The high intensity of rainfall in the upstream area caused the overflow of river water which resulted in the occurrence of flood disasters. Based on BPBD Padang City in the period 2000-2017 there was an increase and extent of flood disaster. Deforestation and conversion into bum land is considered as the causal factor of flood disaster in Padang City. That uncontrolled deforestation and land conversion into built up area will cause floods [4,5,14]. That forests have an important role in storing water during rain and preventing floods, it is necessary to maintain a minimum of 30 percent of the forest's upstream function [2].

Table 3. The dynamics of land use in Padang City.

Type of use	**Large (ha)**		
	1994	**2007**	**2016**
Settlement	7.316,8	11.287,6	16.608,0
Rice fields	7.794,8	6.823,8	6.706,0
Mixed garden	3.589,1	1.506,1	1.214,7
Bush	938,6	479,2	424,1
Forest	49.392,9	48.976,8	44.139,4
Empty land	463,9	422,4	404,0
Total	69.496	69.496	69.496

Source : Citra *landsat ETM 7+* 1994, 2007, 2016

Table 3 presents land use in Padang City in 1994, 2007 and 2016 analyzed using Landsat ETM 7+ imagery. The period of 1994-2016 in Padang City has been a reduction of forest area of 5.253.5 ha, and annually forest area is reduced by 4.5 percent. Forest areas can be presented in dynamic modeling for periods up to 2050. Figure 5 shows that if the government does not intervene in controlling deforestation, then by 2050 it is estimated that the forest area will only be 9659 ha.

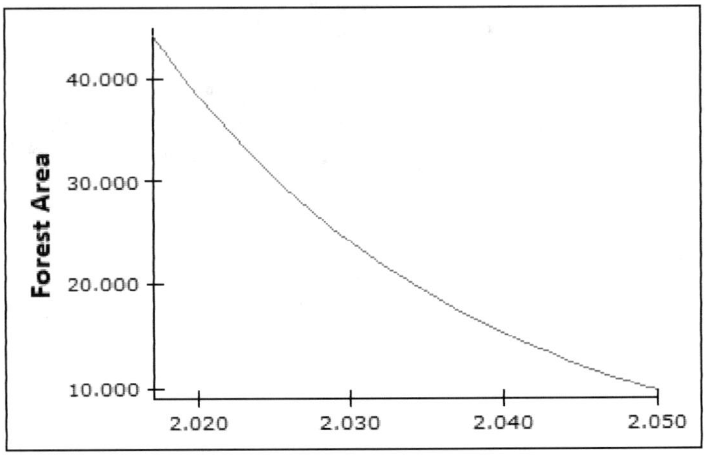

Figure 5. Forest area modeling without policy.

The importance of forest areas in the groundwater storage system in the rainy season will be impact on the disruption of the hydrological cycle [7]. That on the edge of the forest, the increase of number people affected destruction of forest area. Without strict supervision and law enforcement, the impact on deforestation by the forest community increases [13]. That to describe the future state can be explained by dynamic system modeling with three scenarios namely pessimistic, moderate, and optimistic [15,16].

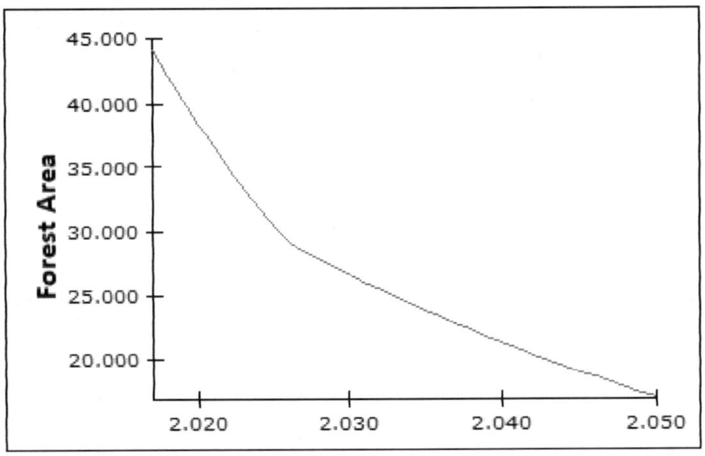

Figure 6. Pessimistic scenario of forest area modeling.

Under a pessimistic scenario that the rate of forest destruction can be reduced to half of the reality condition (2.3 percent) by 2050 forest area in the study area will remain 16,685 ha. Figure 6 presents a pessimistic scenario in reducing forest destruction.

IOP Conf. Series: Earth and Environmental Science **149** (2018) 012010 doi:10.1088/1755-1315/149/1/012010

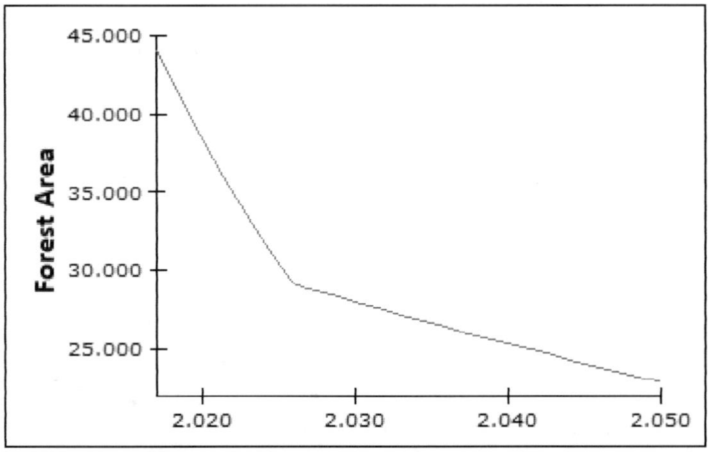

Figure 7. Moderate scenario of forest area modeling.

Moderate scenario with deforestation rate of 1 percent in 2050 forest area in Padang city will increase to 22,119 ha. Indicators of success in the moderate scenario include: implementation in law enforcement in forest encroachment, consistency in the utilization of spatial pattern, and implementation of synchronization of usage between government institutions.

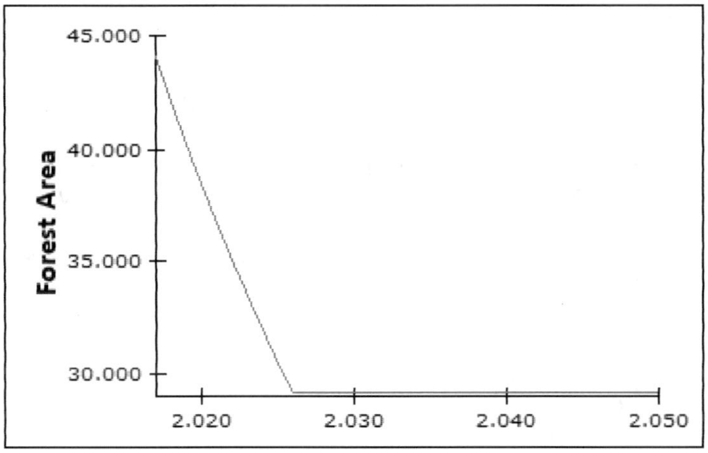

Figure 8. Optimistic scenario of forest area modeling.

Optimistic scenario with deforestation rate of 5 percent in 2050 forest area in Padang city will increase to 28,512 ha. Indicators of success in the optimistic scenario include: implementation in law enforcement in forest encroachment, consistency in the utilization of spatial pattern, and implementation of synchronization of usage between government institutions.

4. Conclusion

Padang City has about 13 percent of the area is very vulnerable (high vulnerability) to flood disaster. Approximately 50 % settlements built in the region is very prone to flooding. The model utilization success to present what will be happen to flood if there is deforestation. Deforestation is one of the factors causing flood disaster in Padang city, about 4.5 percent of forest area lost every year. In dynamic modeling is estimated in 2050 the forest area of Padang City is only 9659 ha remaining. It suggest that the regulation for sustainable environmental management should be apply well.

IOP Conf. Series: Earth and Environmental Science **149** (2018) 012010 doi:10.1088/1755-1315/149/1/012010

References

[1] Iswandi U 2016 Mitigasi Bencana Banjir pada Kawasan Permukiman di Kota Padang Provinsi Sumatera Barat. Disesertation. Graduated school Bogor Agricultural University. Bogor.
[2] Kodra H S A and Syaurkani 2004 *Bumi Makin Panas Banjir Makin Luas*. Yayasan Nuasa Cendikia. Bandung.
[3] Pribadi D, Shiddiq D and Ermyanila M 2006 Model Perubahan Tutupan Lahan dan faktor-faktor yang Mempengaruhinya *Jurnal Teknologi Lingkungan BPPT* **3** 77-91
[4] Kodoatie R 2013 *Rekayasa dan Banjir Kota* Andi:Yogyakarta
[5] Yuksek O, Kankal M and Ucuncu O 2013 Assessment of Big Flood in the Eastern Black Sea Basim of Turkey Environment Monitoring and Assesment **185** 797-814.
[6] Asdak C 1995 *Hidrologi dan Pengelolaan Daerah Aliran Sungai* UGM Press: Yogyakarta.
[7] Mudelsee M, Borngen M, Terzlaff G and Grunewald U 2003 No Upward Trends in the Occurrence of Extreme Floods in Central Europe j. Nature **425** 1-9
[8] RI [Republik Indonesia] 2007 Undang-undang No 24 Tahun 2007 tentang Penanggulangan Bencana Lembaran Negara RI Tahun 2007 66 Jakarta
[9] Iwan W Cluff L, Kimpel J and Kunreuther H 1999 Mitigation Emergen as Major Strategy for Reducing Losses Caused by Natural Disasters *Science* **28** 78-82.
[10] Bechtol V and Laurian L 2005 Restoring Straightened River for Sustainable Flood Mitigation. *Disaster Prevention and Management* **14** 6-12
[11] Iswandi U, Widiatmaka, Pramudya B and Barus B 2016 Delineation of Flood Hazard Zones by Using a Multi Criteria Evaluation Approach in Padang West Sumatera Indonesia. *Journal of Environment and Earth Science* **6** 205-2015
[12] Iswandi U, Widiatmaka, Pramudya B and Barus B 2017 Prioritas Pengembangan Kawasan Permukiman pada Zona Rawan Banjir di Kota Padang Provinsi Sumatera Barat. *Majalah Ilmiah Globe* **19** 83-94
[13] Stoica A and Iancu I 2011 Flood Vulnerability Assesment Based on Mathematical Modeling. *Technical University of Civil Enginering* **6** 32-41
[14] Popovska C, Jovanovski M, Ivanoski D and Pesevski I 2010. Storm Sewer System Analysis in Urban Areas and Flood Risk Assessment *Technical University of Civil Enginering* **3** 23-33
[15] Muhammadi, Aminullah E and Soesilo B 2001 *Analisis Sistem Dinamis* UMJ Press. Jakarta
[16] Kholil, Intan EKP and Listyarini S 2014 *Pendekatan Sistem Universitas* Terbuka Press: Jakarta

IOP Conf. Series: Earth and Environmental Science **149** (2018) 012011 doi:10.1088/1755-1315/149/1/012011

Influence of QBO on stratospheric Kelvin and Mixed Rossby gravity waves in high-top CMIP5 models

Karina Indah Solihah[1], Sandro W Lubis[2] and Sonni Setiawan[1]
[1]Department of Geophysics and Meteorology, Bogor Agricultural University (IPB), Indonesia
[2]Department of the Geophysical Sciences, University of Chicago, Chicago, USA

E-mail: karinaindahsolihah@gmail.com

Abstract. It is well established that quasi-biennial oscillation (QBO) has a substantial influence on Kelvin and mixed Rossby gravity (MRG) wave activity in the tropical lower stratosphere. In this study, we examined how QBO influences Kelvin and MRG wave activity in the lower stratosphere, based on nine high-top CMIP5 models. The results show that the Kelvin and MRG wave signals are stronger in the models with QBO, and relatively weaker in the models without QBO. The results are consistent with established theory, whereby upward-propagating Kelvin waves occurs more frequently during the easterly QBO phase, while upward-propagating MRG waves occurs during the westerly QBO phase. Without the QBO, the mean flow exhibits a near-zero easterly wind, which prevents the waves from propagating and penetrating into the stratosphere. Our analysis also shows that models with the QBO tend to have more robust signatures (in terms of amplitude and phase speed) of Kelvin and MRG waves.

1. Introduction

Equatorial planetary waves (EPWs) are atmospheric waves that are trapped close to the Equator, generated by a large-scale convective system. EPWs contribute to the western and eastern phases of quasi-biennial oscillation (QBO) [1]. QBO is a stratospheric vascillation, with a period of two years, which occurs simultaneously around the equatorial belt. Observational data have shown that Kelvin waves can transfer their momentum to the mean flow, resulting in an easterly phase of the QBO. In contrast, mixed Rossby gravity (MRG) waves can produce momentum that contributes to the westerly phase of the QBO [2-4].

The stratosphere plays a significant role in both global atmospheric circulation and surface climate. The fifth-comparison coupled model intercomparison project (CMIP5) improves the representation of the stratosphere by expanding the lid of the model up into this region. In general, most CMIP5 models include the stratosphere and QBO. This type of model can be used to analyse the effect of QBO on generating larger-amplitude waves better than models without QBO [5].

The focus of this study was to examine the characteristics of Kelvin and MRG waves in the lower stratosphere (level 50 hPa), and their interaction with QBO activity, with the main focus being on horizontal structure, based on high-top CMIP5 models. In addition, we evaluated the ability of the CMIP5 models to represent Kelvin and mixed Rossby gravity wave activity on a subset of high-top CMIP5 models.

2. Methods

The data used in this research are daily zonal wind and temperature data from nine high-top CMIP5 models, and ERA-Interim reanalysis obtained from the European Centre for Medium-Range Weather Forecasts (ECMWF), for the period January 1985 to December 1995. We focused on the equatorial latitudinal band between 10°LU and 10°LS, at a height of 50 hPa.

2.1 Space-Time Spectral Analysis (STSA)

STSA was used to study the zonal propagation of waves by decomposing tropical disturbances as a function of frequency and zonal wavenumber into symmetric and asymmetric components. This technique decomposes the time-dependent data fields (zonal winds) into wave and frequency components for eastward and westward travelling waves [6-8].

2.2 Amplitude Distribution

We also examined the amplitude variations of Kelvin and MRG waves associated with the zonal wind in the tropical lower stratosphere during different QBO phases. The amplitude was calculated as the variance of filtered fields, wherein a high value of variance indicates that the effect of the associated waves is large.

2.3 Composite Analysis

Composite analysis is a method used to determine atmospheric distribution under certain conditions [9].

3. Results

3.1 Determining the Phases of QBO

High-top CMIP5 models can represent the stratosphere and QBO vertically. Some models may indicate a good enough QBO signal, but there are also models that are unable to simulate the QBO. Kelvin wave activity is very strong in the easterly phase of QBO, while MRG wave activity is relatively stronger during the westerly phase of QBO. To determine whether the model could simulate the presence of QBO, or vice versa, we first calculated the wind speed anomaly for each model in the lower stratosphere [5]. In principle, in the equatorial lower stratosphere, EPW waves (including Kelvin and MRG waves) can propagate vertically, or grow, if there is a favorable windshear associated with the QBO.

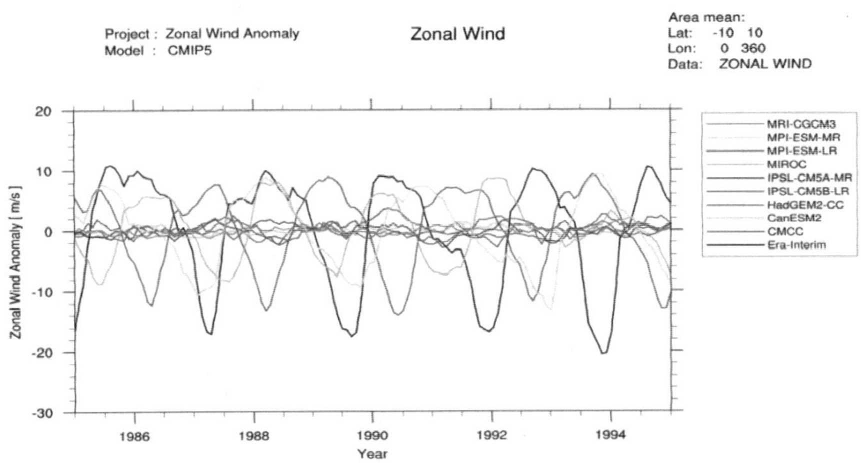

Figure 1. Time-series of monthly zonal-mean wind anomalies for 11 years (1985-1995), averaged, at 10°N-10°S (50 hPa) from ERA-Interim data and nine high-top CMIP5 models.

The time-series of the tropical stratopheric wind, from ERA-Interim data and nine high-top CMIP5 models, is shown in figure 1. In this study, the westerly and easterly phases of QBO were defined as zonal wind velocity anomalies of less than -5 m/s, or more than 5 m/s, respectively [10]. Figure 1 shows that the ERA-Interim and MPI-ESM-MR, HadGEM2-CC, and MIROC-ESM models, provided wind speeds of above 5 m/s and below -5 m/s. In contrast, the line graph, representing the models CanESM2, CMCC, MPI-ESM-LR, IPSL-CM5B-LR, IPSL-CM5A-MR, MRI-CGCM3, does not indicate the presence of QBO. Therefore, from figure 1, we can see that there are two groups of models; the first group (MPI-ESM-MR, HadGEM2-CC, MIROC-ESM) simulated QBO, but the second (CanESM2, CMCC, MPI-ESM-LR, IPSL-CM5B-LR, IPSL-CM5A-MR, MRI-CGCM3) did not.

3.2 Characteristics of Kelvin Waves

Kelvin waves are generated by large-scale convective heating oscillations in the equatorial tropospheric layer. Based on previous research [2,3,11], the Kelvin wave is the oscillation of the zonal wind component near the Equator, in the form of eastward propagating waves, having a period of between 12 and 20 days, a vertical wavelength of 10 km, and a zonal wavelength of 40,000 km [12].

Figure 2 shows the symmetric component of the STSA. The color spectrum shows the wave power, and the dispersive curve shows the relation between the frequency and the wave number. The STSA results show that the Kelvin wave moved eastward, characterized by a positive wave number and strong Kelvin wave activity, based on the ERA-Interim data and model with QBO; it is indicated by the dominance of red in the color spectrum.

The models that could not simulate QBO show weaker amplitude, dominated by green in the color spectrum. CanESM2, IPSL-CM5B-LR, IPSL-CM5A-MR, MPI-ESM-LR, and MRI-CGCM3 are the models wherein the period of propagation was longer than for the other models, but which had weaker wave strength. It is interesting to examine the spatial distribution of each wave amplitude. Figure 3 shows the variance distribution of Kelvin waves in the ERA-Interim (reference) and high-top CMIP5 models.

MPI-ESM-MR has the greatest variance, compared to the other models that could simulate QBO. Models that could not simulate QBO seem to be underestimated, since the value of its variance is smaller than in the ERA-Interim model. The higher the value of variance, the greater the wave activity. Figure 3 shows that strong Kelvin wave activity occurs at longitudes 30°-120°E, over the Indian Ocean and the Maritime Continent. The presence of cloud cover, due to the effects of the Asian monsoon, evokes Kelvin wave over the Maritime Continent. This may be a reason why the Kelvin wave signal in that region is stronger than in other regions [13].

Figure 4 shows Kelvin wave variance as a function of longitude. Almost all models that could simulate QBO tend to follow the ERA-Interim (black line) model. The variance shows the maximum value at longitude 70°, and its minimum at 227.5°. Almost all of the models overestimated the amplitude, with the MPI-ESM-MR (purple line) and MIROC-ESM (turquoise line) indicating a very distant representation of the ERA-Interim model. Based on the residual graph (figure 4b), the closer the line is to zero, the better the model in representing the Kelvin waves. The IPSL-CM5B-LR, IPSL-CM5A-MR, and CANESM2 models have residual values close to zero, while models that simulated QBO have residual values far from zero. Almost all of the models overestimated the amplitude, except for MRI-CGCM3, and IPSL-CM5A-MR. Similar results were found, based on filtered temperature data.

Furthermore, we wanted to examine the propagation of Kelvin waves, using the Hovmoeller diagram. Based on the results of composite analysis, all of the models were able to simulate the presence of eastward zonal wind propagation. Models that could simulate QBO have stronger wave signals and longer propagation phases than non-QBO models. This shows that the Kelvin wave is moving eastward and has a long period (figure 5).

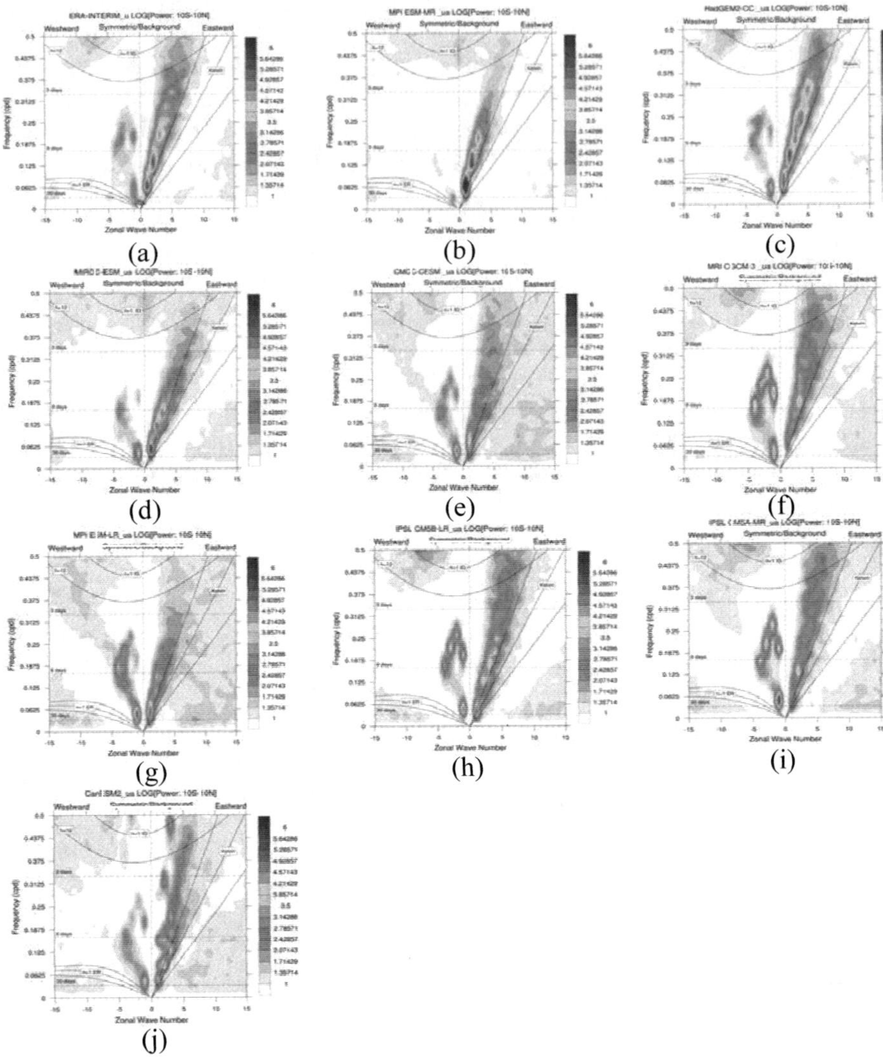

Figure 2. STSA (symmetric) daily zonal wind at 10°S-10°N (50 hPa), from (a) ERA-Interim as base, (b-d) models with QBO, and (e-j) models without QBO. The y-axis shows the frequency of the wave, while the x-axis shows the number of waves in the west and east zonal direction.

Based on the wave's magnitude, as indicated by the composite results, MPI-ESM-MR exhibits a longer propagation phase and lighter color than the ERA-Interim model. This indicates that the MPI-ESM-MR wave signal is stronger than the ERA-Interim (overestimate). In contrast, HadGEM2-CC shows weaker signals than the ERA-Interim. Overall, almost all of the models show Kelvin wave signal propagation. The models that could simulate QBO show stronger signals and longer propagation phases than models that could not demonstrate QBO.

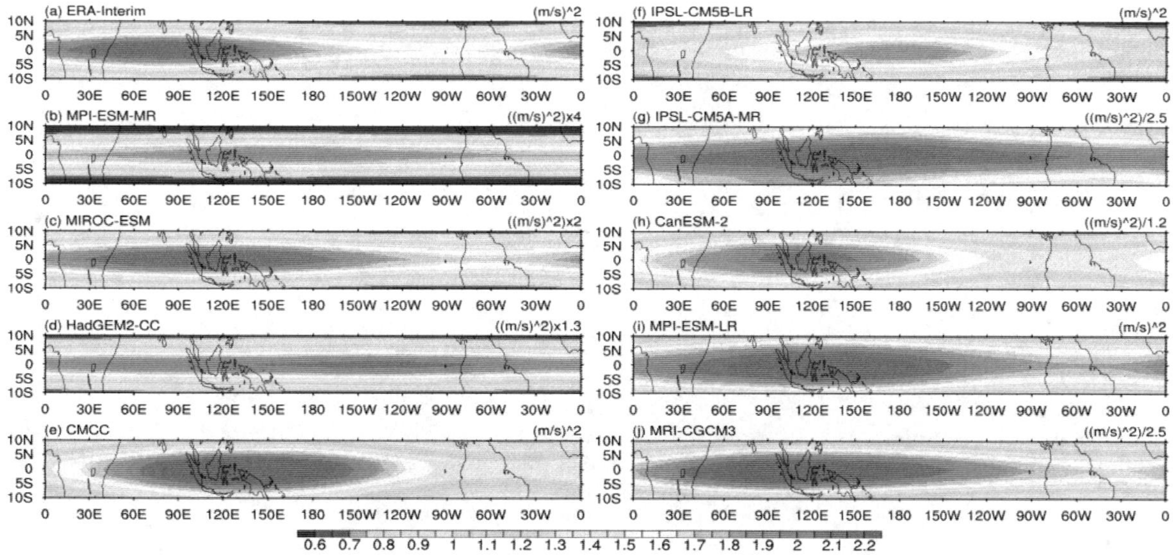

Figure 3. Spatial distribution of Kelvin wave amplitude based on filtered zonal-mean wind in (a) ERA-Interim, and (b-j) nine high-top CMIP5 models.

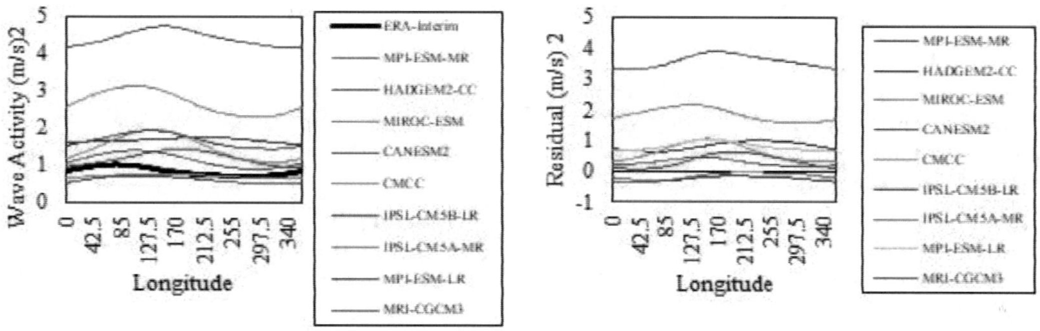

(a) (b)

Figure 4. Kelvin wave variance as a function of longitude, based on filtered zonal winds at 50 hPa (a), and the residual of Kelvin wave variance, i.e., the model minus observation (b).

3.3 Characteristics of MRG Waves

The MRG wave is a short-period wave, clearly visible in the meridional wind component, propagating westward and downward. This wave has a period of between four and five days, a zonal wavelength of 10,000 km, and a vertical wavelength of approximately 5 km [14]. MRG waves are detected by STSA (anti-symptomatic) each season, under all conditions [4].

The STSA (antisymmetric) results indicate the presence of MRG waves propagating in the lower stratosphere layer (figure 6). These also show that the models that could simulate QBO have stronger wave signals than the models that could not, characterized by lighter spectrum colors. The zonal wavenumber for the MRG wave has a negative value, indicating that the MRG wave propagates from east to west. The spectral value of the ERA-Interim dispersive curve, and the model that could simulate QBO, is larger than the spectrum value of the model that was unable to simulate QBO.

IOP Conf. Series: Earth and Environmental Science **149** (2018) 012011 doi:10.1088/1755-1315/149/1/012011

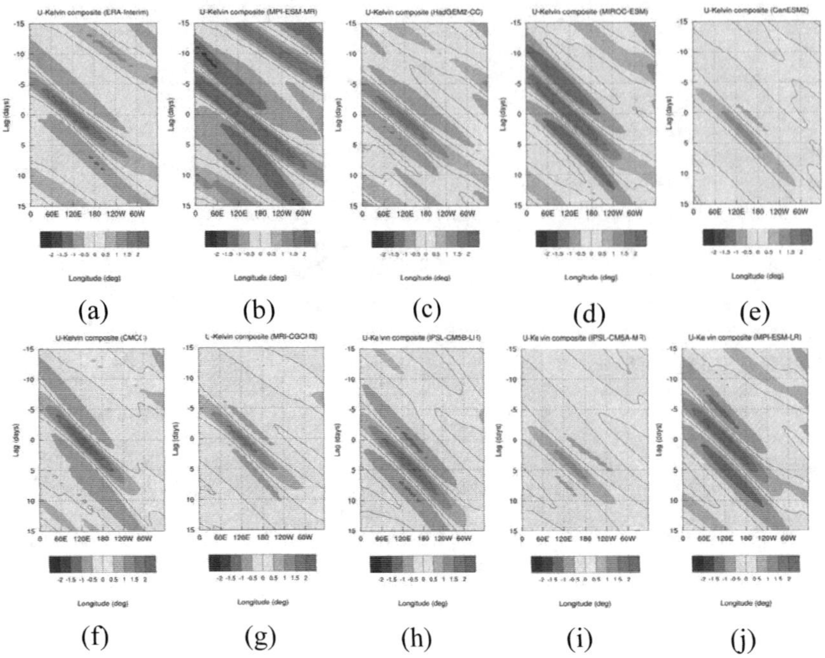

Figure 5. Composite analysis of Kelvin waves, using a filtered seasonal zonal wind (DJF) basepoint at 0°N and 120°E, at 50 hPa. (a-d) models with QBO, (e-j) models without QBO.

HadGEM2-CC is one of the models determined to be overestimated because it shows strong MRG wave activity with the highest spectrum value than ERA-Interim, which is equal to 2.78-5. Overall, not all models exhibit strong MRG wave signals, but figure 6 shows that the period of the MRG wave, as shown by the models (b-d), is longer.

Figure 7 shows the MRG wave spatial distribution based on zonal wind filter data. The variance results show that almost all models that could simulate QBO have higher variance values. Based on the results of the variance distribution of the ERA-Interim model, MRG wave propagation shows high activity in the Pacific Ocean region, characterized by a dominant red contour in this region. Based on the results of the variance values, models capable of simulating QBO have a higher variance value, compared to the reference variance value (ERA-Interim), whereas models that could not simulate QBO show lower variance values than that of the reference.

Figure 8 shows the value of the MRG wave amplitude as a function of longitude. Almost all of the models that could simulate QBO tend to follow the ERA-Interim model (black line). The variance shows the maximum value at 140° longitude, but almost all of the models show peak variance at 270° longitude. All of the models exhibit overestimation, with MPI-ESM-MR (purple line) showing a very far-reaching representation of the ERA-Interim model. Based on the residual graph (figure 8b), if the line is closer to zero, then the model represents Kelvin waves better (approaching the ERA-Interim). IPSL-CM5B-LR has residual values close to zero, while the other models have residual values far from zero.

Based on Hovmoeller analysis (figure 9), Kelvin wave signals show longer periods, compared to the MRG wave period; this is consistent with the STSA results. All of the models are able to show MRG wave signals. HadGEM2-CC, MIROC-ESM, and MPI-ESM-LR show longer MRG wave phases than the ERA-Interim model. In contrast, ERA-Interim, MPI-ESM-MR, CanESM2, CMCC-ESM, IPSL-CM5B-LR, IPSL-CM5A-MR, and MRI-CGCM3 indicate a shorter wave period. Each model shows a wave signal with different magnitudes. MPI-ESM-LR, HadGEM2-CC, and MIROC-ESM overestimate the amplitude of the propagation.

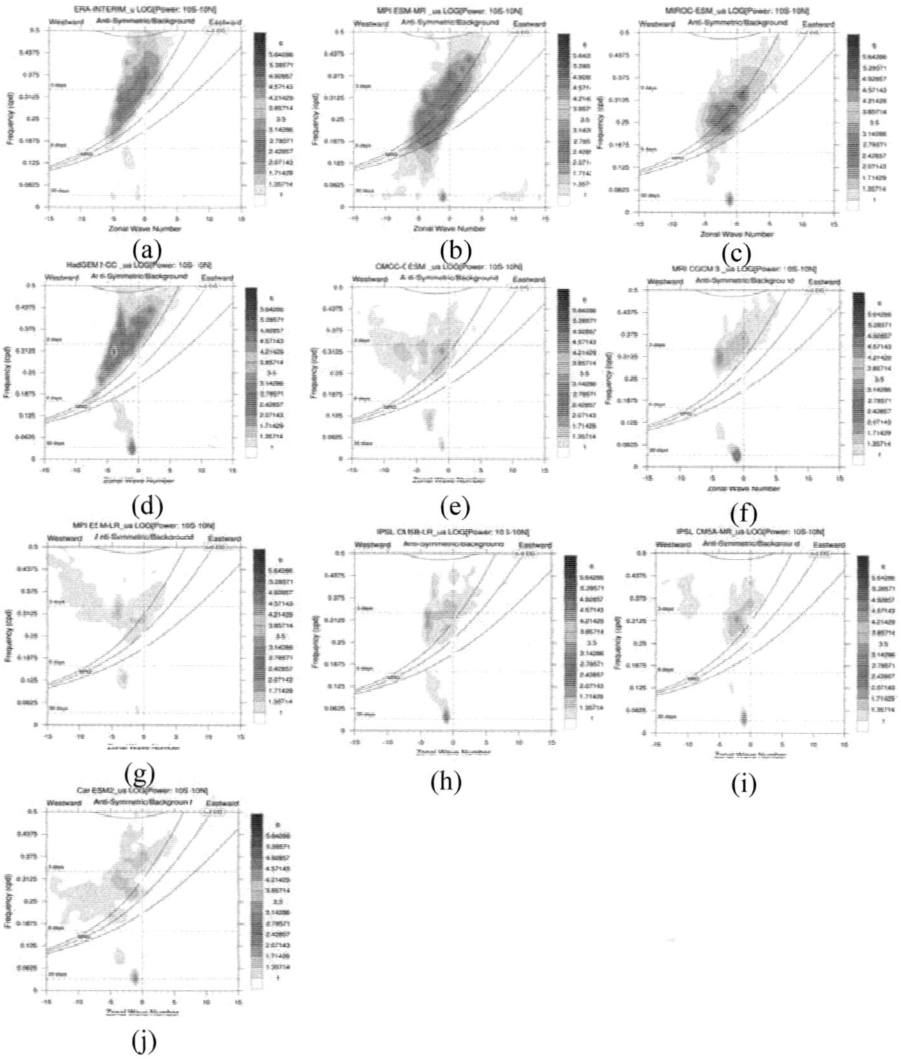

Figure 6. STSA (antisymmetric) daily zonal wind at 10°S-10°N (50 hPa), from (a) ERA-Interim as base, (b-d) models with QBO, and (e-j) models without QBO.

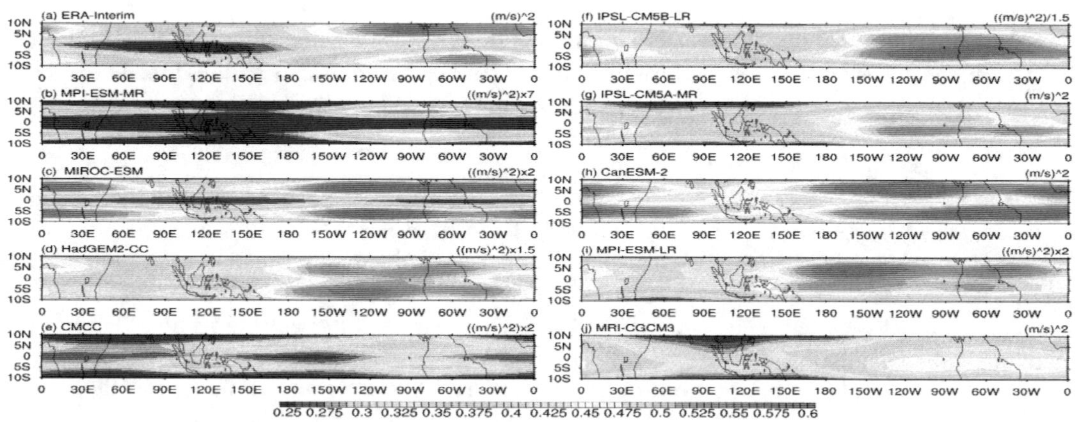

Figure 7. Spatial distribution of MRG wave amplitude, based on filtered zonal-mean wind in (a) the ERA-Interim, and (b-j) nine high-top CMIP5 models.

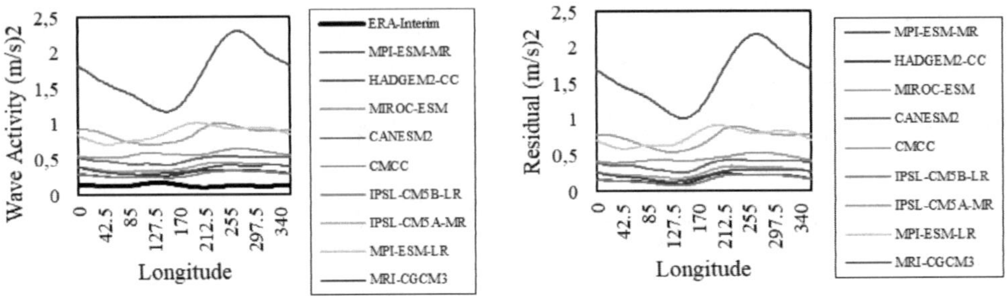

(a) (b)

Figure 8. MRG wave variance as a function of longitude, based on filtered zonal winds at 50 hPa (a), and the residual Kelvin wave variance, i.e., model minus observation (b).

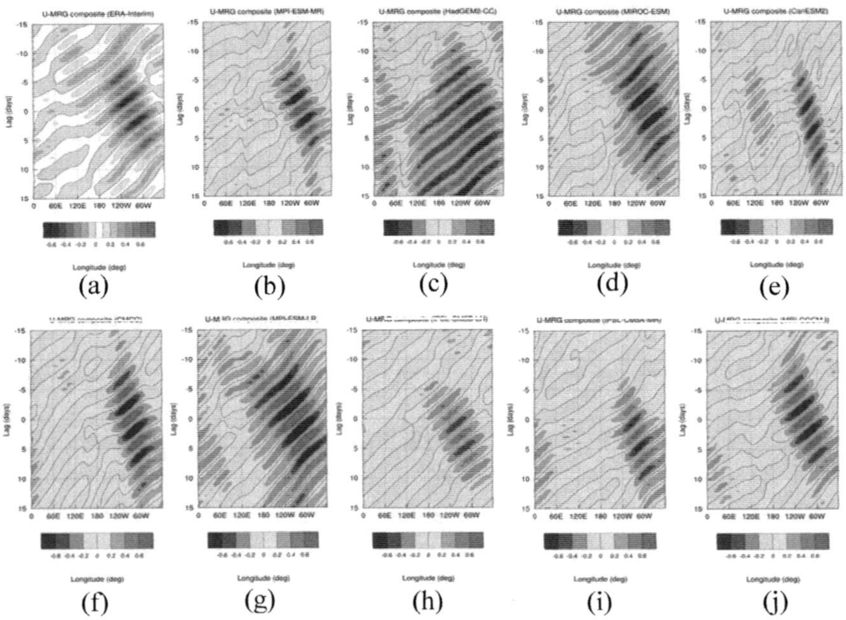

Figure 9. Composite analysis of the MRG wave, using a filtered seasonal zonal wind (DJF) basepoint at 0°N and 270°E at 50 hPa. (a-d) models with QBO, (e-j) models without QBO.

Figure 7 also illustrates that the spatial distribution of MRG waves that dominate the Pacific Ocean region. Strong MRG waves are detected for most of the year in the Pacific Ocean region. This is because the Pacific Ocean region experiences high convective activity, due to the high sea-surface temperature distribution, and its coincidence with the ITCZ region [15].

4. Conclusions

We studied the characteristics of Kelvin waves and mixed Rossby-Gravity waves in nine high-top CMIP5 models, with and without QBO. The wave properties were analysed using STSA and composite analysis. The key results of this study are:

1) Models with QBO better represent Kelvin and MRG waves, the best being the MPI-ESM-MR, HadGEM2-CC, and MIROC-ESM models.
2) Models with QBO produce larger-amplitude Kelvin and MRG waves, compared to those without QBO.

3) Consistent with established theory, upward-propagating Kelvin waves occur more frequently during the easterly QBO phase, while upward-propagating MRG waves occur during the westerly QBO phase.
4) Without QBO, mean flow exhibits a near-zero easterly wind, preventing the waves from propagating and penetrating into the stratosphere, thus weakening Kelvin and MRG wave activities in the models without QBO.

Future studies are aimed at understanding the mechanisms that control upward propagation of Kelvin and MRG waves from the troposphere to the stratosphere in CMIP5 models.

Acknowledgment
We thank the reviewers for the constructive comments and WCRP for providing CMIP5 datasets.

References

[1] Holton J R 1979 Equatorial wave means flow interaction: a numerical study of the role of latitudinal shear *Journal of the Atmospheric Science* **36** 1030-1040
[2] Baldwin M P, Gray L J, Dunkerton T J, Hamilton K, Haynes P H, Randel W J, Holton J R, Alexander M J, Hirota I, Horinouchi T, Jones D B A, *et al.* 2001 The Quasi-Biennial Oscillation *Rev. J. Geophysics.* **39** 179-229
[3] Alsepan G, Lubis S W, Setiawan S 2016 Analysis of the equatorial lower stratosphere Quasi-Biennial Oscillation (QBO) using ECMWF-Interim Reanalysis Data Set *Earth Environ. Sci.* **31** 1-14
[4] Fatullah N Z, Lubis S W and Setiawan S 2017 Characteristics of Kelvin waves and Mixed Rossby-Gravity waves in opposite QBO phases *Journal IOP science.* **54** 1-9
[5] Lott F, Denvil S, Butchart N, Cagnazzo C, Giorgetta M A, Hardiman S C, Manzini E, Krismer T, Duvel J P, Maury P, *et al.* 2014 Kelvin and Rossby-gravity wave packets in the lower stratosphere of some high-top CMIP5 models *Journal of Geophysics Res. Atmos.* **119** 2156–2173
[6] Hayashi Y 1981 Space-Time Spectral Analysis and its applications to atmospheric wave *Journal of the Meteorological Society of Japan* **60** 156-170
[7] Wheeler M, Kiladis G N 1999 Convectively coupled equatorial waves: analysis of clouds and temperature in the wavenumber-frequency domain *J. Atmos. Sci.* **56** 374-399.
[8] Lubis S W, Jacobi C 2015 The modulating influence of convectively coupled equatorial waves (CCEWs) on the variability of tropical precipitation *Int. J. Climatol.* **35** 1465-1483
[9] Rakhman S, Lubis S W, Setiawan S 2017 Impact of ENSO on seasonal variations of Kelvin Waves and Mixed Rossby-Gravity waves *Earth and Environmental Science* **54** 1-12
[10] Yamashita Y, Akiyoshi H, Takahashi M 2011 Dynamical Response in the Northern Hemisphere midlatitude and high-latitude winter to the QBO simulated by CCSR/NIES CCM *J. Geophysics* **116** 1-9
[11] Wallace J M, Kousky V E 1968 Observational evidence of Kelvin waves in the tropical stratosphere *J atmos sci.* **25** 900-907
[12] Holton J R 1972 Waves in the equatorial stratosphere generated by tropospheric heat sources *Journal of The Atmospheric Sciences* **29** 368-375
[13] Lubis S W, Setiawan S 2010 Identifikasi gelombang Kelvin atmosfer ekuatorial di Indonesia berbasis data NCEP/NCAR Reanalysis I *Jurnal Fisika* **10** 71-82
[14] Yanai M, Maruyama 1966 Stratospheric wave disturbances propagating over the equatorial Pacific *Journal of the Meteorology Society Japan* **44** 291-294
[15] Muraleedharan P M, Prasannakumar S, Mohankumar K, Sijikumar S, Sivakumar K U, Mathew T 2015 Observational evidence of Mixed Rossby Gravity waves at the central Equatorial Indian ocean *Meteorology and Atmospheric* Physics **127** 407-417

IOP Conf. Series: Earth and Environmental Science **149** (2018) 012012 doi:10.1088/1755-1315/149/1/012012

Activity of convective coupled equatorial wave in tropical Tropopause layer in reanalysis and high-top CMIP5 models

Alia Harza[1], Sandro W. Lubis[2] and Sonni Setiawan[1]
[1] Department of Geophysics and Meteorology, Bogor Agricultural University (IPB), Indonesia
[2] Department of the Geophysical Sciences, University of Chicago, Chicago, USA.

E-mail: alia.harza@gmail.com

Abstract. The activity of convectively coupled equatorial waves (CCEWs), including Kelvin waves, Mixed Rossby-Gravity (MRG), and Equatorial Rossby (ER), in the tropical tropopause layer (TTL) is investigated in the Reanalysis and nine high-top CMIP5 models using the zonal wave number-frequency spectral analysis with equatorially symmetric-antisymmetric decomposition. We found that the TTL activities in the high-top CMIP5 models show significant difference among the high-top CMIP5 models with respect to the observation. The MIROC and HadGEM2-CC models work best in simulating Kelvin wave in the TTL, while the HadGEM2-CC and MPI-ESM-LR models work best in simulating MRG waves. The ER waves in TTL are best simulated in the MRI-CGCM model. None of the models are good in simulating all waves at once. It is concluded that the broad range of wave activity found in the different CMIP5 models depend on the convective parameterization used by each model and the representation of the tropical stratosphere variability, including the QBO.

1. Introduction

Convectively coupled equatorial waves (CCEWs) are the tropical waves generated by large-scale convective system and have strong influences on tropical rainfall and circulation [1-7]. CCEWs include Kelvin waves, Mixed Rossby-Gravity (MRG), and Equatorial Rossby (ER). CCEWs have strong influence on tropical weather system and stratospheric circulation, e.g. they play an important role in transferring momentum and energy from the troposphere into the stratosphere. In the troposphere, CCEWs control the variation of the tropopause temperature and thus determine the variability of tropical tropopause layer. Since the waves are coupled with temperature, the variation of the tropical tropopause layer can be also used as a proxy of wave activity at this level.

The efforts in developing climate model have been made recently within the Coupled Model Intercomparison Project Phase 5 (CMIP5) models. CMIP5 is one of the IPCC assessment reports, which is a continuation of the previous assessment report of AR4. CMIP5 is expected to project climate change in two groups: close group (close to 2,035) and distant group (close to 2,100 to over), adding several other factors such as cloud and carbon cycle. The representation of tropical tropopause layer is varying among CMIP5 models, and the activity of CCEWs in these models is not well documented yet. In this study, we will identify the characteristics of tropical waves (Kelvin, MRG, ER, TD-type) in the tropical tropopause layer (TTL) and then, to examine the capability of high-top CMIP5 models in reproducing tropical waves in the TTL and its implication on transport processes in that region.

2. Data and Methods

2.1 Data
The research used reanalysis data from ERA-Interim and 9 models of High-Top CMIP5. The CMIP5 models used are CMCC-CESM, CanESM2-CC, HadGEM-CC, IPSL-CM5A-MR, MPS-ESM for January 1985 to December 1995. We focus on latitude 20S to 20N with a resolution of 2.5 x 2.5 ^{o}.

2.2 Zonal Wave Number Frequency Spectrum
This method is used to study zonal propagating waves and decompose the time-dependent data field (in this study using zonal winds) into the wave and frequency components for waves that travel west and east [8-10] Then a further complex FFT process is performed for the function in time against the Fourier coefficients to obtain the spectrum of the number of frequencies for each latitude. Furthermore, the spectrum of zonal wind power data is averaged over the time of the study then summed for all latitude [9].

2.3 *Spatial Distribution*
Spatial distribution used to analyze variability of equatorial waves in space, by variance and residual values. The background spectrum analysis serves to eliminate the "red noise" found in the spectrum with small frequency values [4, 9]. The bias in this analysis comes from the difference in meteorological parameter values. Disturbance is seen as a small increment associated with a red noise spectrum or in other words can be seen at the top or bottom in the output of the zonal wave number-frequency spectrum.

2.4. Taylor Diagram
Taylor diagrams make it easy to find the best models using RMS values, correlations, and variance ratios [11]. The model having a correlation value close to 1 and the RMS value closest to the observation curve illustrates the best model. The line between RMS values marks the curve of observation. The diagram will help in summarize the relative quality of each model compared with observation.

3. Results

3.1 Basic Comparison of Temperature in Tropical Tropopause Layer
The tropopause layer is a layer with a constant temperature value with respect to height but varies with time and space. The average temperature distribution of each model when compared with observation tends to be the same with different temperature values. The highest temperature value according to observation is 200° K while the model's highest temperature of 205° K is seen in CanESM2, HadGEM, IPSL-CM5B-MR and MRI-CGCM models. When viewed from the closest value, then four of the nine models are considered good. Based on spatial distribution of temperature, the models such as HadGEM, IPSL-CM5B-LR and MIROC can be considered the best models (figure 1).

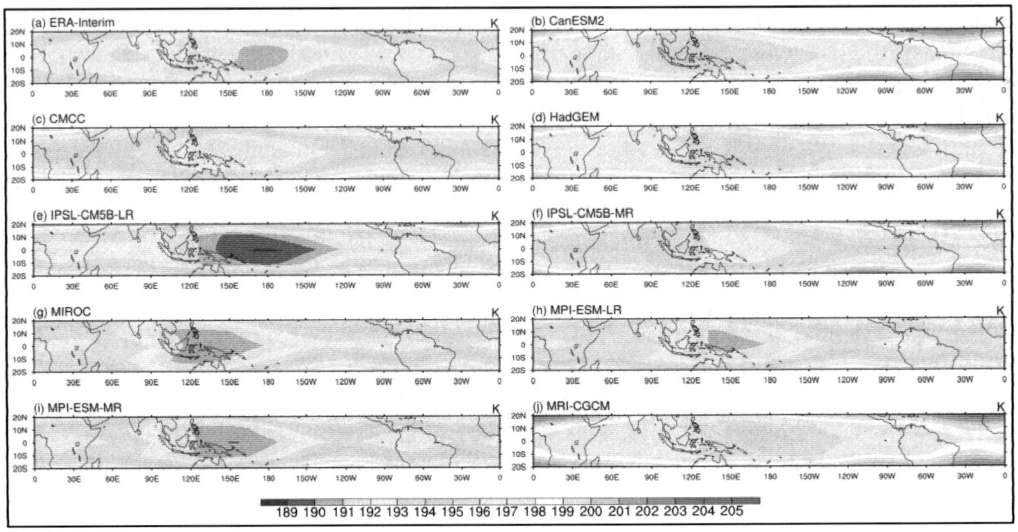

Figure 1. Temperature annual mean spatial distribution from (a) ERA-Interim and (b-j) nine *high-top* CMIP5 models.

3.2 Statistical Significance and Dispersion Relation of Equatorial Waves

The symmetric component shows the frequency and wave number spectra of the Kelvin and equatorial Rossby (ER). Kelvin wave signal are observed between wave numbers 1 to 5. The difference between Kelvin and ER waves can be seen from the shape of the dispersion curve and their phase speed. Kelvin wave has mode of $n \leq 1$ whereas ER wave has mode of $n \geq 1$, this will be related to wave propagation direction. This result is consistent with the previous observationally-based analysis of CCEWs [4, 9]. The best models that have similar signatures of Kelvin and ER wave spectra as in observation are IPSL-LR, HadGEM, MRI-CGCM, and MIROC (figure 2). In addition, the asymmetric component shows the frequency and wave number spectra of the MRG and Inertia Gravity Waves. The MRG wave has $n \geq 1$ mode, similar to ER wave. The difference between MRG and ER waves lies in its frequency and wave period. The best models that have similar signatures of MRG wave spectra as in observation are IPSL-LR, HadGEM, MPI-ESM-LR, MPI-ESM-MR, and MIROC (figure 3).

3.3 The Characteristics of Kelvin Waves

Kelvin wave amplitudes maximize in the equator from 10S to 10 N latitude. There are 5 CMIP models that well represent the distribution of Kelvin wave activity compared to the observation (figure 4). They are including the CMCC, HadGEM, MPI-ESM-MR and MIROC. Among these models, the MPI-ESM-MR model is the most representative model with respect to ERA-Interim. MIROC is the model with residual value farther than the ensemble mean model's value.

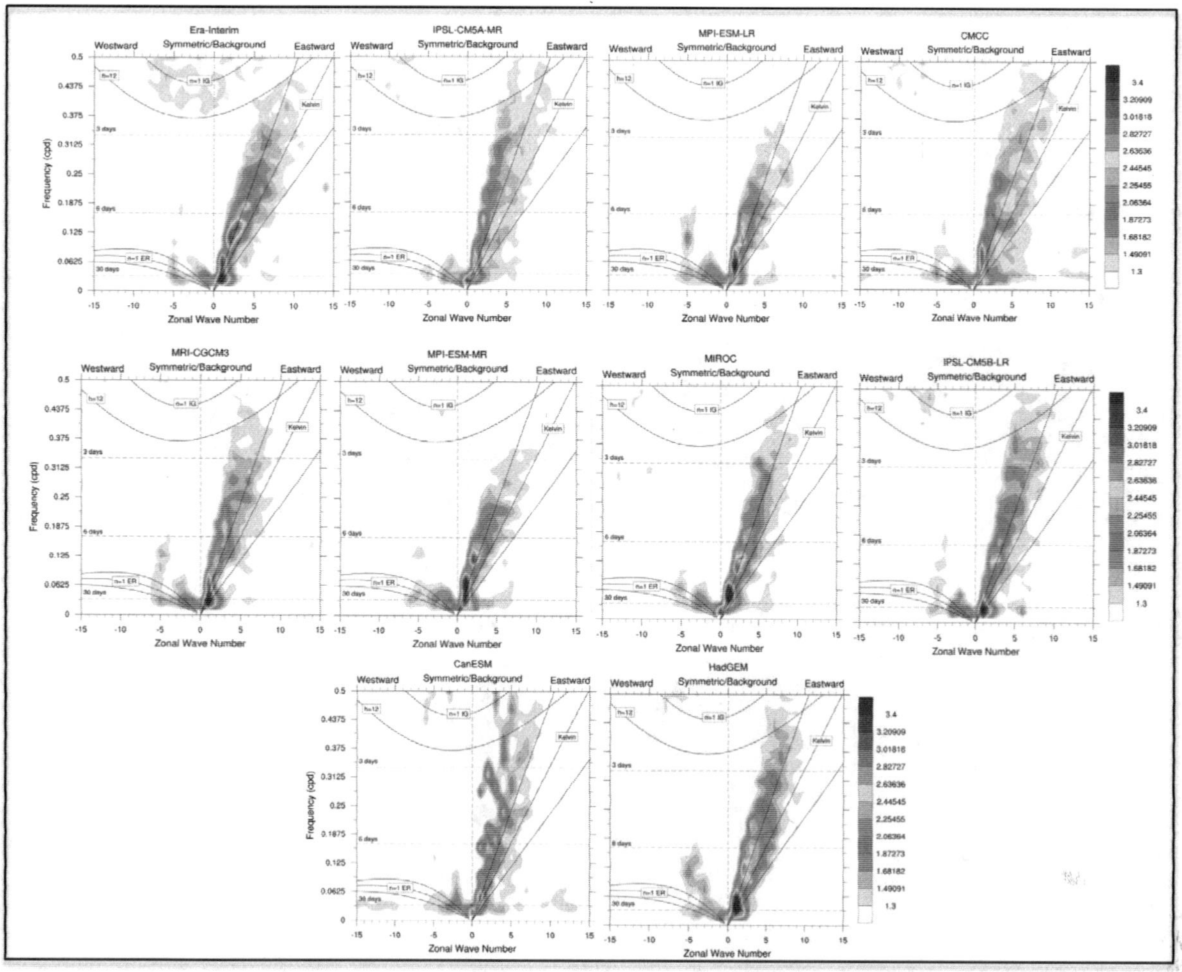

Figure 2. Space-time spectral analysis (Symmetric) of daily temperature at 20°S-20°N (1,000 hPa) from (a) ERA-Interim as base (b-d) with QBO models (e-j) without QBO models.

IOP Conf. Series: Earth and Environmental Science **149** (2018) 012012 doi:10.1088/1755-1315/149/1/012012

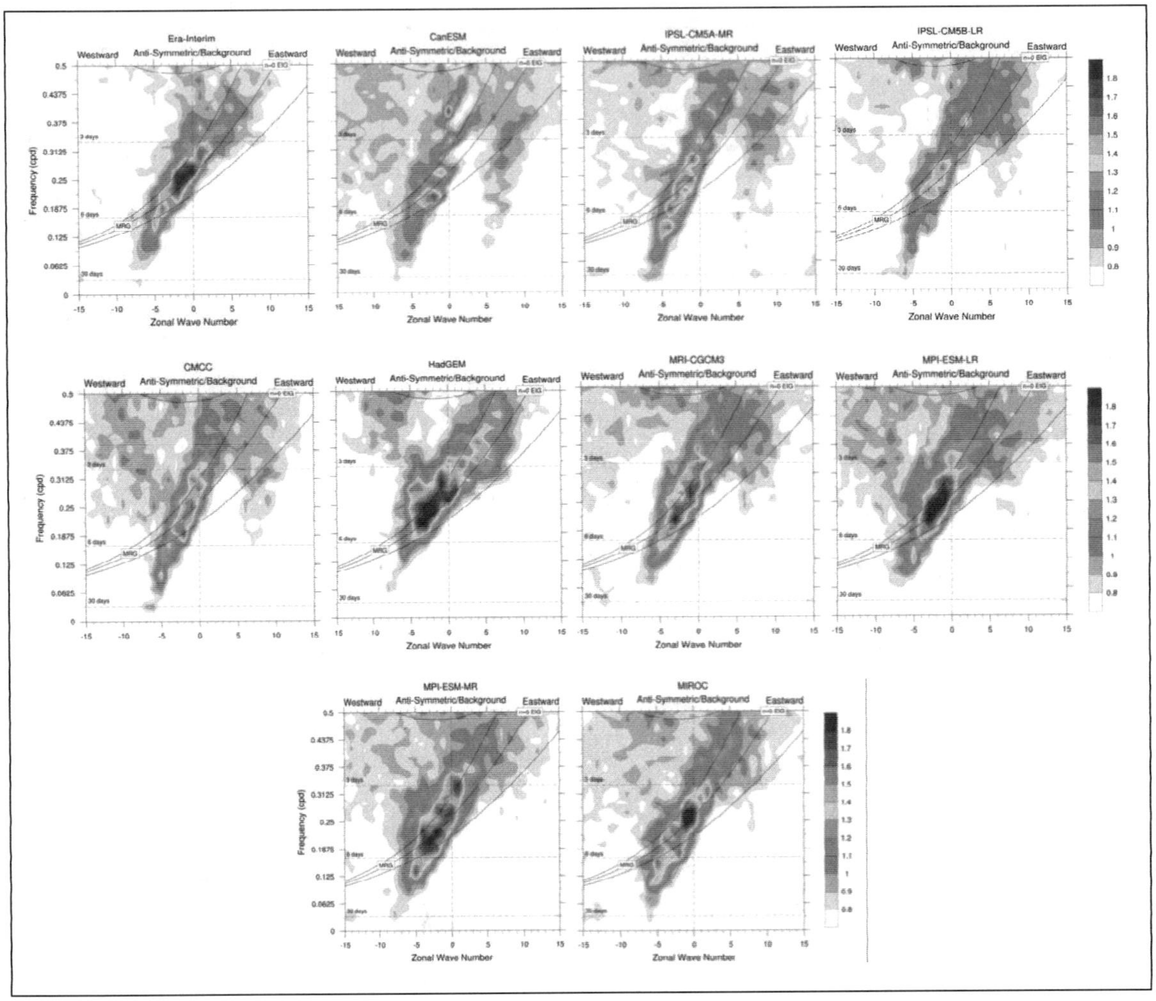

Figure 3. *Space-Time Spectral Analysis* (Antisymetric) daily temperature at 20°S-20°N (1000 hPa) from (a) ERA-Interim as base (b-d) with QBO models (e-j) without QBO models.

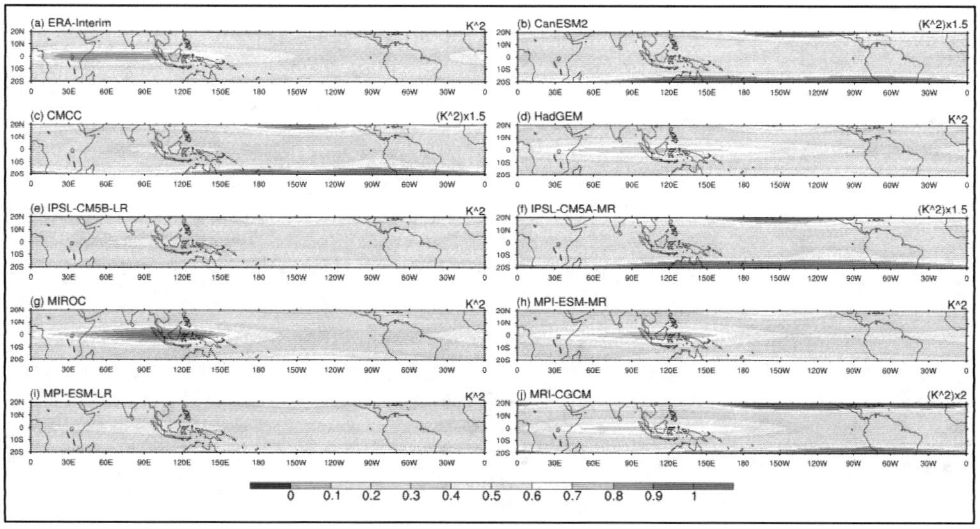

Figure 4. Spatial distributions of Kelvin wave variances using temperature filter from (a) ERA Interim and (b-j) nine *high-top* CMIP5 models.

3.4 The Characteristics of Mixed Rossby-Gravity Waves

Mixed Rossby-gravity (MRG) is a wave with zonal wavenumber of 3-5, period of 3-5 days, travels westward with center point above the equator and with maximum activity at 110-190 hPa [4, 5, 9]. It has a zonal wavelength is 10,000 km and has vertical wavelength approximately 5 km [12]. In contrast to kelvin wave that has high distribution and easily identifiable between 10S-10N latitude, the MRG wave distribution is seen at higher latitudes and spreading over the Pacific Ocean region. Figure 5 shows high MRG wave activity at 0-20 N latitudes from June to October. In particular, the MRG waves in northern Pacific Ocean have exceptionally high values (figure 5). MRG waves are most easily identified in the boreal autumn [4, 9]. In the CMIP5 models, the MPI-ESM-MR is the worst model in showing the wave activity of MRG waves that tend to overestimate the amplitudes. This model also has the farthest residual value away from the ensemble mean (not shown).

3.5 The Characteristics of Equatorial Rossby Waves

The largest ER wave activity lies in the off equatorial regions. Some models have overestimated the ER wave amplitudes compared to the ERA-Interim model. The largest ER wave activity occurs in the northeast of the Pacific Ocean during boreal summer [4, 9].

Based on the spatial distribution of ER waves, the ER wave activity in TTL from the high-top CMIP5 model is best simulated in the MRI-CGCM, IPSL-ESM-LR and CAN-ESM2 models (figure 6).

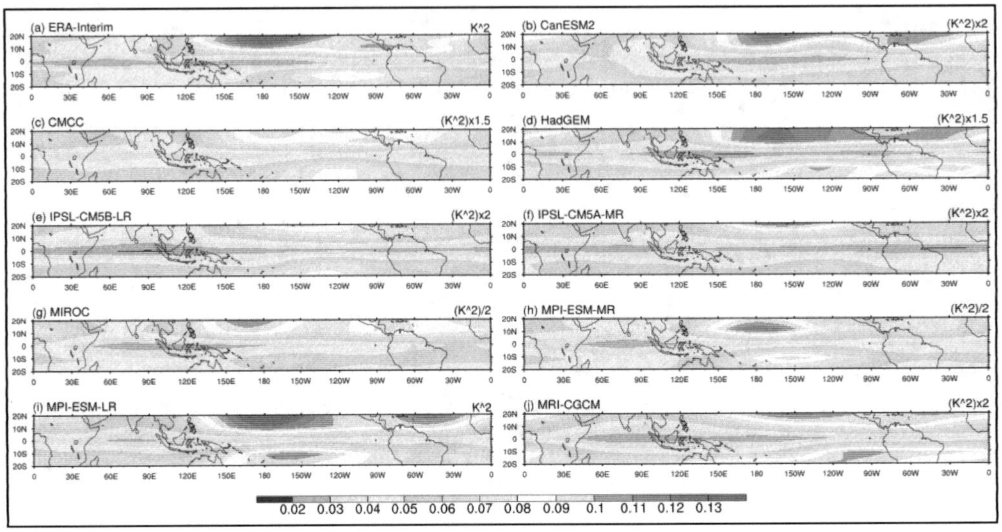

Figure 5. Spatial distributions of Mixed Rossby-Gravity wave variances using temperature filter from (a) ERA-Interim and (b-j) nine *high-top* CMIP5 models.

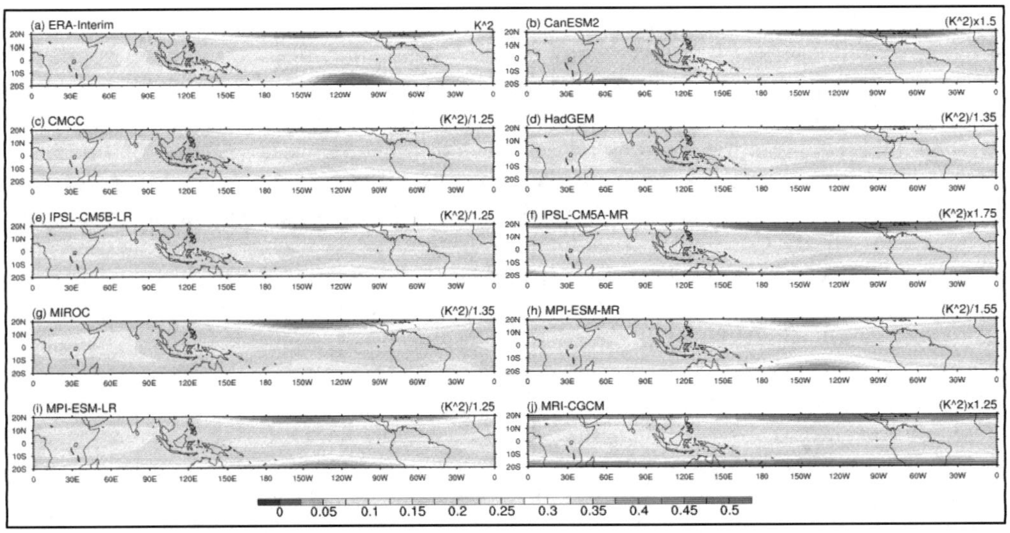

Figure 6. Spatial distributions of Equatorial Rossby wave variances using temperature filter from (a) ERA-Interim and (b-j) nine *high-top* CMIP5 models.

4 Conclusion

In this study, the characteristics of Kelvin waves, Mixed Rossby-gravity waves, and Equatorial Rossby waves in the tropical tropopause layer are analyzed using the ERA-Interim reanalysis dataset and high-top CMIP5 models from 1985-1995. We implemented the space-time spectral analysis (STSA) technique in order to understand their horizontal characteristics. The key results of this research are concluded as follow:

- TTL wave activities show significant difference among the high-top CMIP5 models with respect to the observation.

- The MIROC and HadGEM2-CC models work best in simulating Kelvin wave in the TTL, while the HadGEM2-CC and MPI-ESM-LR models work best in simulating MRG waves.
- The ER waves in TTL are best simulated in the MRI-CGCM model.
- It is concluded that the broad range of wave activity found in the different CMIP5 models rely on the convective parameterization used by each model and the representation of the tropical stratosphere variability, including the QBO and MJO.

Future studies are still required in order to understand the mechanisms causing the strong biases of the CCEWs in the CMIP5 models, and to investigate whether such biases have a significant impact on the variability (synoptic time scale) of the tropical tropopause layer.

Acknowledgment
We thank the reviewers for the constructive comments and WCRP for providing CMIP5 datasets.

References

[1] Kiladis G N, Wheeler M C, Haertel P T, Straub K H and Roundy P E 2009 Convectively coupled equatorial waves *Geophys. Rev.* **47** RG2003 pp 1-42
[2] Lubis S W 2010 Theoretical simulation of free atmospheric planetary waves on an equatorial beta-plane *J. of Theortical Compuational Studies* **48** pp 1–17
[3] Lubis S W, Setiawan S 2010 Identifikasi gelombang Kelvin atmosfer ekuatorial di Indonesia berbasis data NCEP/NCAR Reanalysis I *Jurnal Fisika.* **10** (2) pp 71-82
[4] Lubis S W, Jacobi C 2015 The modulating influence of convectively coupled equatorial waves (CCEWs) on the variability of tropical precipitation *Int. J. Climatol* **35** pp 1465-1483
[5] Fathullah N Z, Lubis S W, Setiawan S 2017 Characteristics of Kelvin waves and Mixed Rossby-Gravity waves in opposite QBO phases *Earth and Environmental Sciences* **54** pp 1-9
[6] Fathurochman I, Lubis S W and Setiawan S 2017 Impact of Madden-Julian Oscillation (MJO) on global distribution of total water vapor and column ozone *IOP Conf. Ser: Earth Environ. Sci.* **54** 012034 pp 1-9
[7] Rakhman S, Lubis S W, Setiawan S 2017 Impact of ENSO on seasonal variations of Kelvin Waves and Mixed Rossby-Gravity waves *Earth and Environmental Science* **54** pp 1-12
[8] Hayashi Y 1981 Space-Time Spectral Analysis and its applications to atmospheric wave *J. Meteorological Society of Japan* **60** (1) pp 156-170
[9] Wheeler M, Kiladis G N 1999 Convectively coupled equatorial waves: analysis of clouds and temperature in the wavenumber-frequency domain *J. Atmos. Sci.* ***56*** pp 374-399
[10] Taylor K E 2001 Summarizing multiple aspect of model performance in a single diagram *J. Geographical Research* **106** pp 7183-7192
[11] Yoshikazu H 1974 Spectral analysis of tropical disturbance appearing in a GFDL General Circulation Model *J. Atmos. Sci.* **31** pp 180-218
[12] Hendon H H, Liebmann B 1991 The structure and annual variation of antisymmetric fluctuations of tropical convection and their association with Rossby-gravity waves *J. Atmos. Sci.* **48** pp 2127–2140

IOP Conf. Series: Earth and Environmental Science **149** (2018) 012013 doi:10.1088/1755-1315/149/1/012013

Land use changes and plantation crop development in selected provinces in Sumatra and Kalimantan

S D Tarigan

Department of Soil Sciences and Natural Resources Management, Bogor Agricultural University, Bogor, Indonesia.

E-mail: sdtarigan@apps.ipb.ac.id

Abstract. Most institutions stated that biofuel will not qualify the standard of GHG emission reduction if it was produced in the plantation associated with the forest conversion. Therefore, knowing previous land use before the development of plantation is very important. In Indonesia, plantation development occurs mainly in Sumatra and Kalimantan. A number of studies had been published showing historical LUCC before plantation development. Objective of this study was to review various studies on LUCC carried out in four selected provinces, namely West Kalimantan, Central Kalimantan, East Kalimantan, and Riau. The analysis and comparison was based on the different source of historical data including online spatial data sources and various studies published in various journals. Each data source of LUCC shows significant variation on the amount of plantation developed directly from forest and other land use types. But, our review showed that the plantation areas associated with the forest cover changes far less than those claimed by several international journals. But, the debate concerning which plantation developments indirectly contributed to LUCC and which are directly will probably continue until the information on the land ownership and history of plantation development is available publicly.

1. Introduction

In Indonesia, the area of the oil palm plantation increased from 0.7 million ha in 1990 to 12 million ha in 2017 [1,2], and further millions of ha are projected to be required in 2020–2050 [3,4]. This rapid expansion of the plantation crop has been partly triggered by an increased demand for biofuel production [5]. In addition, rubber plantations, which are also prevalent in Southeast Asia [6], currently cover 3.5 million ha of land in Indonesia [1].

The development of the plantation crop as a result of the conversion of primary and secondary forests will release respectively between 270–530 and 120–190 g CO2 equivalent for each MJ produced biodiesel. In contrast, the conversion of the degraded lands results in the capture of between 23 and 85 g CO2 equivalent per MJ of biodiesel produced. According to EPA's analysis, biodiesel produced from plantation grown by conversion of primary and secondary forests do not meet the minimum 20% lifecycle GHG reduction threshold required to qualify as renewable biofuel [7].

Some international publications [8,9] claimed that more than 50% of Indonesian plantation crop were developed on the forest areas. The objective of this study was to review various data sources related to the amount of plantation developed on the forest area. We selected 4 provinces in Sumatra and Kalimantan representing areas with extensive plantation crop development, namely West Kalimantan, Central Kalimantan, East Kalimantan, and Riau. According to Ditjenbun [1], plantation area in all four provinces entail to almost 50% of total plantation in Indonesia equivalent to 5.5 Mha. Sumatra and Kalimantan are the main plantation area and they entail to 89% (10.7 Mha) of the total plantation crop area in Indonesia [1].

2. Methodology

We reviewed various published studies related to the plantation development associated with the LUCC in four provinces where intensive development of plantation has been going on including West Kalimantan, Central Kalimantan, East Kalimantan, and Riau.

In Indonesia, development of plantation crop occurred since 1970s, but rapid increase of the plantation expansion by smallholders and private companies started since 1990 [1,10]. The plantation concession maps for the study areas were obtained from different sources. The concession map is an indicative area owned by plantation company. Real planted areas are usually less than the concession area, the reason is that some of the concession area have not been planted by crops. On the other hand, some plantations were found to be developed outside the concession (9].

To obtain actual planting area inside concession map, we used 2 approaches a) cross checking with statistical data from National Plantation Office or Ditjenbun [1], and b) comparing with spatial data from Atlas-CIFOR (cifor.org/map/atlas, [10]). The spatial data from Atlas-CIFOR was based on the detail study carried out by Gaveau et al. [10]. The latest version of the Atlas-CIFOR (launched November 2017) allows user to verify the location and ownership of the concessions and planted oil palm in Kalimantan. It can tract the LUCC footprint of plantation companies (cifor.org/map/atlas). All the maps presented in the Atlas-CIFOR (cifor.org/map/atlas, [10]) underwent a rigours peer-review and published here [10].

3 Result and Discussion

3.1 Plantation development and LUCC based on various data sources

According to Koh and Wilcove [8] during the period of 1990–2005 most of the plantation expansion in Indonesia occurred by forest conversion based on FAO data. However, the accuracy of this statement has been questioned [9,11].

Carlson et al. [9] quantified plantation development across Kalimantan from 1990 to 2010 using Landsat data. Gunarso et al. [16] reported that from 1990 and 2010, only 36.5% of all oil palm plantations developed in the forest area, including both upland and swamp habitats and only 0.1% and 4% of oil palm plantations were developed on the primary forest and natural swamp forest respectively from 1990 and 2010.

Tarigan et al. [14] investigated in Merangin and Bungo Districts, Jambi province and showed that in the year from 1988 and 2013, only a small fraction of the oil palm expansion (8%) occurred by direct conversion of primary forest in the study area. Most of the oil palm expansion occurred in other land use types such as logged-over forest, shrub land and agroforest.

Among all available studies dealing with the plantation development and associated LUCC, Gaveau [10] is considered as one of the most reliable and comprehensive source to tract the LUCC footprint of plantation companies in this study. According to Gaveau [10], Kalimantan experienced extensive forest conversion before the expansion of industrial plantations. In the period of 1973-2015, the total development of the plantation (both plantation crop and pulpwood) in Kalimantan summed up to 5,6 Mha. Some 1.3 Mha and 1.2 Mha of those were developed in non-forest area and scrub respectively in the period of 1973-2015. In Kalimantan, only 11–13% of all forest conversion was related with the plantation expansion as the plantation development dominantly occurred on the lands previously cleared before 1973 [10]. Gaveau [10] further concluded that the debate concerning which plantation developments cause indirectly forest conversion and which are directly will continue until the data of the land ownership is made available to the public.

3.2 Plantation concession and planted area based on the different data sources

Our review showed that 3 Mha concession area have not yet been converted to plantation (figure 1) in West Kalimantan. This figure is similar to that showed by Atlas-CIFOR [10]. In West Kalimantan, there is data consistency between Ditjenbun [1] and Atlas-CIFOR [10] on the actual plantation area which is of ~1.4 Mha (figure 1).

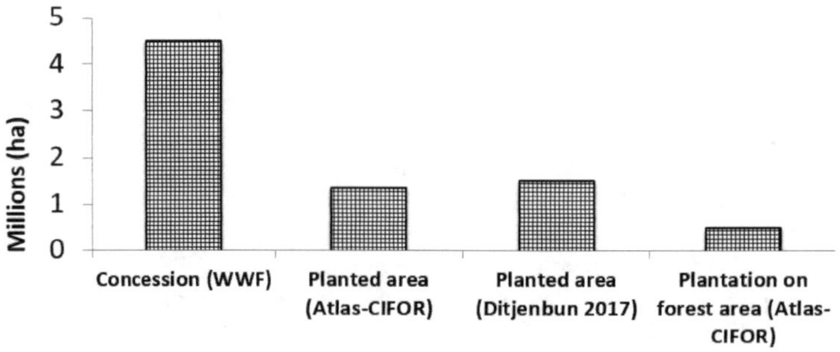

Figure 1. Concession and planted area based on the different data sources (WWF, Atlas-CIFOR, Ditjenbun) in West Kalimantan Province.

Our review in Central Kalimantan showed that 3 Mha of the concession area have not yet been converted to plantation (figure 2). Meanwhile, Atlas-CIFOR [10] showed that 1.2 Mha of the concession area have not yet been converted to plantation. In Central Kalimantan, there is data consistency between Ditjenbun [1] and Atlas-CIFOR [10] on the actual planted oil palm area which is of ~1.2 Mha.

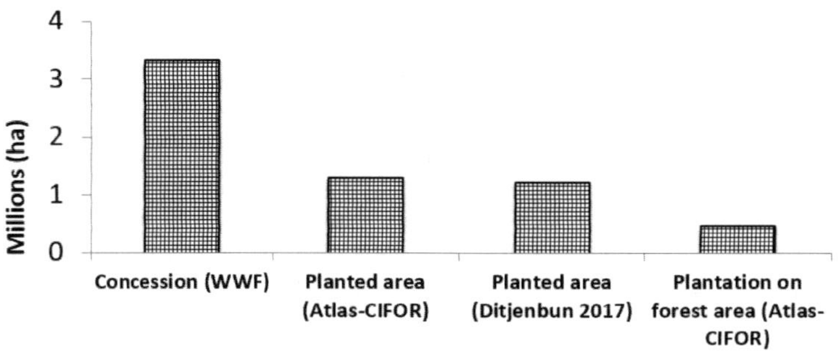

Figure 2. Concession and planted areas based on the different data sources (WWF, Atlas-CIFOR, Ditjenbun) in Central Kalimantan Province.

Our analysis showed that 2.4 Mha oil palm concession area have not yet been utilized (figure 3) in the East Kalimantan. Meanwhile, Atlas-CIFOR [10] showed that 1.2 Mha oil palm concession area have not yet been converted to plantation. In East Kalimantan, there is difference between Ditjenbun [7] and Atlas-CIFOR [10] on the actual planted oil palm area during 1973-2015 which is of 0.5 and 0.7 Mha respectively.

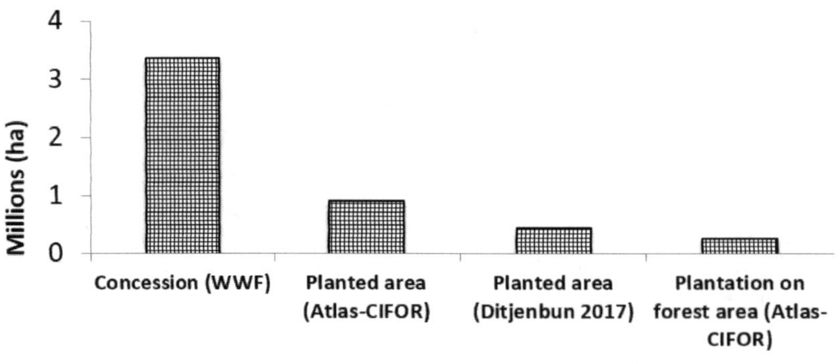

Figure 3. Concession and planted area based on the different sources (WWF, Atlas-CIFOR, Ditjenbun) in East Kalimantan Province.

Our review showed that the total amount of the oil palm concession area, which have not been planted in West, Central and East Kalimantan was 8.4 Mha. According to Austin et al. [15], there are 8.6 Mha of concession area planned for oil palm expansion in Kalimantan which have not yet been developed to plantations. Thus, our analysis and that of Austin et al. [15] showed similarity. Meanwhile

Gaveau et al [10, Atlas-CIFOR] reported that 6.6 Mha out of 10 Mha concession area planned for oil palm expansion in Kalimantan have not yet been developed to plantations. On the other hand, two studies in Kalimantan using medium-resolution LANDSAT satellite imagery found that nearly a quarter of oil-palm plantations from 1990 and 2010 were outside legal concessions, proofing the need to investigate outside concessions to identify impacts of plantation development on the forest conversion [9,15].

The Atlas-CIFOR [10] does not cover Riau province. But our review on the total area of the oil palm concession in Riau showed similarity with those showed by Ditjenbun [1] which are ~ 2.2 Mha (figure 4). According to Susanti and Burgers [12], some forest area allocated for oil palm plantation Riau province. Santosa et al. [13] investigated previous land use of oil palm in 8 big companies of oil palm plantation covering an area of 46,372.38 ha and concluded that most of the plantation was established in non-forest area.

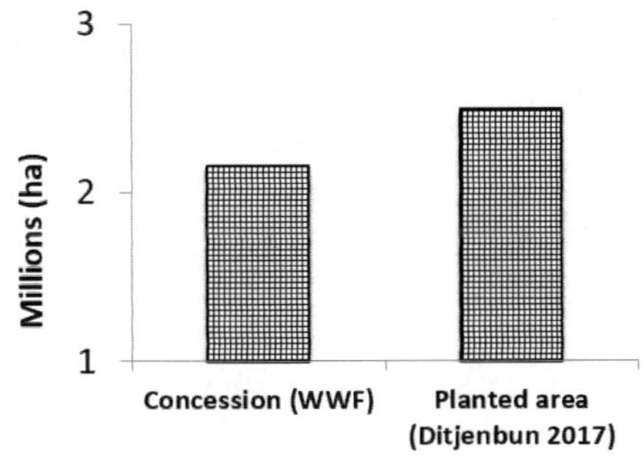

Figure 4. Concession and planted area based on the different sources in Riau Province.

3.3 Comparison of forest cover change in the study areas

Despite the largest oil palm plantation area (~2.2 Mha), Riau Province has the lowest percentage of oil palm plantation associated with direct LUCC. On the average, the proportion of oil palm plantation associated with direct LUCC in the selected provinces is less than those claimed in the international journal publication.

Concern of international communities on the tropical LUCC often focus on the GHG emission and biodiversity issues. These issues often not become of primary concern of the local communities around the LUCC sites. We need to investigate the relation of the LUCC with local environmental services issues [17, 20, 21]. These issues directly affect local communities around LUCC sites. According to Tarigan [2], flooding frequencies increased in the last decades in Jambi province along with the land use change this province. By associating LUCC with the local issues, there is incentives for local communities to implement mitigation option associated with the oil palm expansion. Simple mitigation option [18, 19] are often effective to mitigate impact of LUCC.

4. Conclusions

According to our review, the plantation development associated with direct LUCC in the four selected provinces is far less than those stated by the several international journals. The plantation activities occurred dominantly on logged-over area cleared before the plantation development.

The debate concerning which plantation developments indirectly contributed to LUCC and which are directly will probably continue until the information on the land ownership and history of plantation development is available publicly.

Acknowledgments

Authors wishing to acknowledge excellent data presentation made publicly available by Atlas-CIFOR (cifor.org/map/atlas, [10]) concerning oil palm development.

References

[1] Ditjenbun 2017 The crop estate statistic of Indonesia 2015-2017. Directorate General for Plantations, Ministry of Agriculture Indonesia

[2] Tarigan SD 2016 Land cover change and its impact on flooding frequency of Batanghari Watershed, Jambi Province, Indonesia, Procedia. Environ. Sci., 33, 386–392

[3] Wicke B, Sikkema R, Dornburg V, Faaij A 2011 Exploring land use changes and the role of palm oil production in Indonesia and Malaysia, Land. Use. Policy., 28, 193–206

[4] Afriyanti D, Kroeze C, and Saad A 2016 Indonesia palm oil production without deforestation and peat conversion by 2050, Sci. Total. Environ., 557-558, 562–570

[5] Mukherjee I, Sovacool B K 2014 Palm oil-based biofuels and sustainability in southeast Asia: a review of Indonesia, Malaysia, and Thailand, Renew. Sustainable. Energy. Rev., 37, 1–12, doi:10.1016/j.rser.2014.05.001

[6] Ziegler AD, Fox JM, Xu J 2009 The rubber juggernaut, Science., 324, 1024–1025. doi: 10.1126/science.1173833

[7] Hasan MNA, Jaramillo P, Griffin WM 2011 Life cycle GHG emissions from Malaysian oil palm bioenergy development: The impact on transportation sector's energy security. Energy Policy 39 (2011) 2615–2625

[8] Koh LP, Wilcove DS 2008 Is oil palm agriculture really destroying tropical biodiversity? Conserv. Lett., 1, 2, 60–64

[9] Carlson KM, Curran LM, Ratnasari D, Pittman AM, Soares BS, Asner GP, Trigg SN, Gaveau DA, Lawrence D, Rodrigues HO 2012 Committed carbon emissions, deforestation, and community land conversion from oil palm plantation expansion in West Kalimantan, Indonesia. Proc Nat Acad Sci USA 2012;109:7559–7564

[10] Gaveau DLA et al. 2016 Rapid conversions and avoided deforestation: examining four decades of industrial plantation expansion in Borneo. Sci. Rep. 6, 32017; doi: 10.1038/srep32017 (2016)

[11] Fitzherbert EB et al. 2008 How will oil palm expansion affect biodiversity? Trends in Ecology & Evolution 23, 538–545

[12] Susanti A, Burgers P 2012. Oil palm expansion in Riau province, Indonesia: serving people, planet, profit ? The European Report on Development by the Overseas Development Institute (ODI) in partnership with the Deutsches Institut für Entwicklungspolitik (DIE) and the European Centre for Development Policy Management (ECDPM)

[13] Santosa Y et al. 2016 Sejarah Perkembangan Status, Penggunaan Lahan, dan Keanekaragaman Hayati Kebun Kelapa Sawit Indonesia. Badan Pengelola Dana Perkebunan Kelapa Sawit and LPPM IPB, Bogor

[14] Tarigan S, Sunarti, Widyaliza S 2015 Expansion of oil palm plantations and forest cover changes in Bungo and Merangin Districts, Jambi Province, Indonesia. Elsevier Procedia Environmental Sciences 2015;5(6):199-205

[15] Austin KG, Kasibhatla PS, Urban DL, Stolle F, Vincent J 2015 Reconciling Oil Palm Expansion and Climate Change Mitigation in Kalimantan, Indonesia. PLOS ONE |DOI:10.1371/journal.pone.0127963

[16] Gunarso P, Hartoyo ME, Fahmuddin A, Killeen TJ 2013 Oil palm and land use change in Indonesia, Malaysia and Papua New Guinea. Reports from the Technical Panels of the 2nd Greenhouse Gas Working Group of the Roundtable on Sustainable Palm Oil (RSPO)

[17] Tarigan S, Wiegand K, Sunarti, Slamet B 2018 Minimum forest cover required for sustainable water flow regulation of a watershed: a case study in Jambi Province, Indonesia, Hydrol. Earth Syst. Sci., 22, 581-594, https://doi.org/10.5194/hess-22-581-2018

[18] Tarigan SD, Wiegand K, Dislich C, Slamet B, Heinonen J, Meyer K 2016 Mitigation options for improving the ecosystem function of water flow regulation in a watershed with rapid expansion of oil palm plantations, Sustainability of Water Quality and Ecology, 8, 4–13, 2016. http://dx.doi.org/10.1016/j.swaqe.2016.05.001

[19] Tarigan SD 2016 Modeling effectiveness of management practices for flood mitigation using GIS spatial analysis functions in Upper Cilliwung watershed. IOP Conf. Series: Earth and Environmental Science 31 (2016)012030 doi:10.1088/1755-1315/31/1/012030.

[20] Dislich C, Faust H, Kisel Y, Knohl A, Otten F, Meyer K, Pe'er G, Salecker J, Steinebach S, Tarigan S, Tölle M, Wiegand K 2017 A review of the ecosystem functions in oil palm plantations, using forests as a reference system, Biol. Rev. in press. doi: 10.1111/brv.12295.

[21] Merten J, Röll A, Guillaume T, Meijide A, Tarigan S, Agusta H, Dislich C, Dittrich C, Faust H, Gunawan D, Hendrayanto, Knohl A, Kuzyakov Y, Wiegand K, Hölscher D 2016 Water scarcity and oil palm expansion: social views and environmental processes. Ecology and Society, 21, 2, 5-16

IOP Conf. Series: Earth and Environmental Science **149** (2018) 012014 doi:10.1088/1755-1315/149/1/012014

Ocean heat budget analysis on sea surface temperature anomaly in western Indian Ocean during strong-weak Asian summer monsoon

Ibnu Fathrio[1,2], Atsuyoshi Manda[3], Satoshi Iizuka[4], Yasu-Masa Kodama[1], Sachinobu Ishida[1]
[1]Meteorological Laboratory, Graduate School of Science and Technology, Hirosaki University, Hirosaki, Aomori, Japan.
[2]National Institute of Aeronautics and Space of Indonesia, Bandung, West Java, Indonesia.
[3]Climate and Ecosystems Dynamics Division, Graduate School of Bioresources, Mie University, Tsu, Mie, Japan.
[4]Storm, Flood, and Landslide Division, National Research Institute for Earth Science and Disaster Resilience, Tsukuba, Ibaraki, Japan.

E-mail: ibnu.fathrio@lapan.go.id

Abstract. This study presents ocean heat budget analysis on seas surface temperature (SST) anomalies during strong-weak Asian summer monsoon (southwest monsoon). As discussed by previous studies, there was close relationship between variations of Asian summer monsoon and SST anomaly in western Indian Ocean. In this study we utilized ocean heat budget analysis to elucidate the dominant mechanism that is responsible for generating SST anomaly during weak-strong boreal summer monsoon. Our results showed ocean advection plays more important role to initate SST anomaly than the atmospheric prcess (surface heat flux). Scatterplot analysis showed that vertical advection initiated SST anomaly in western Arabian Sea and southwestern Indian Ocean, while zonal advection initiated SST anomaly in western equatorial Indian Ocean.

1. Introduction

Past studies revealed that there was close relationship between sea surface temperature (SST) in western Indian Ocean (WIO) and Asian summer monsoon (Southwest monsoon). Early numerical simulation found that colder SST anomalies in western Arabian Sea are responsible for reduced monsoon precipitation over Indian and adjoining areas [1]. This is accompanied with the increase in the moisture flux convergence and precipitation over south of the equator. Izumo et al. showed there is close relation between SST anomaly near Somali-Oman and precipitation in western Indian Continent [2]. They highlighted variations of ocean upwelling in boreal spring generates SST anomaly in Somali-Oman coast. Warming (cooling) of SST could increase (decreases) evaporation, which results in greater (less) moisture transport towards Indian continent. Izumo et al. argued that suppressed upwelling could be traced back to weak southwesterly monsoon wind generated by warm SST anomaly in the southwestern Indian Ocean [2].

Recent intensive ocean observation over Indian Ocean supported development of ocean reanalysis models by ocean data assimilation technique. It is expected that better understanding of ocean internal process over Indian Ocean can be achieved. Previous ocean heat budget studies revealed that seasonal SST variation in WIO could be explained by variation of surface heat flux and oceanic process [3]. However, the role of internal ocean process to initiate boreal summer (June, July and August) SST anomalies in WIO was not fully discussed yet. Hence, this study aims to portray the ocean advection role to initiate strong-weak summer monsoon related SST anomaly by revisiting the ocean heat budget analysis over WIO.

2. Data and Method

In this study, four ocean reanalysis were used for analysis: GECCO2, SODA3, ORAS4 and GODAS. The time period of data was chosen from 1982 to 2012 (31 years) and we limit our area of analysis to WIO in 40-80E and 30S-30N. The SST was obtained from National Oceanic and Atmospheric Administration (NOAA) Optimum Interpolation (OI) version 2 (V2) data [4]. Surface winds data and surface radiation flux data that includes latent heat flux, shortwave radiation flux, longwave radiation flux, latent heat flux and sensible heat flux are obtained from National Centers for Environmental Prediction (NCEP)-National Center for Atmospheric Research (NCAR) [5]. These data were obtained from NCEP Reanalysis data provided by the NOAA/OAR/ESRL PSD, Boulder, Colorado, USA, from their Web site at https://www.esrl.noaa.gov/psd/.

We examined the SST anomaly over the WIO in ocean reanalysis models by analyzing the heat budget in the upper-50-m which is similar to [6-8] i.e.:

$$\frac{\partial T}{\partial t} = -u\frac{\partial T}{\partial x} - v\frac{\partial T}{\partial y} - w\frac{\partial T}{\partial z} + \frac{Q}{\rho_0 C_p H} + \mathrm{Re}\,sidual \qquad (1)$$

where T and u, v, and w represent the climatological monthly means of potential temperature (mixed layer temperature) and zonal, meridional, and vertical ocean current velocities, respectively, Q represents the surface heat flux into the ocean, H is a thickness of the water column, and C_p and □$_0$ denote the specific heat of seawater at constant pressure (set to 3994 J $kg^{-1}K^{-1}$) and a reference density of seawater (set to 1025 kg m^{-3}), respectively. All the terms were averaged over the upper 50 m in this study. The use of a fixed depth of 50 m avoids complications associated with an MME such as variations in the MLD amongst the models. In heat budget analysis, the ocean subsurface temperature over 50m is considered as mixed layer temperature (MLT). In chapter 3 we used MLT as proxy for SST. Because not all models provide surface heat flux data, we used surface radiation flux from NCEP-NCAR Reanalysis 1. All data were interpolated horizontally onto a 1° × 1° uniform horizontal grid, and interpolated vertically onto a 5-m uniform vertical grid (from the ocean surface down to 100 m) and a 25-m uniform vertical grid (below 100 m), using linear interpolation.

Monsoon index introduced by Kawamura was utilized to distinguish weak and strong monsoon during boreal summe [9]. This index was defined as meridional differences in area-averaged upper troposhpheric (200-500 hPa) thickness between Tibetian Plateau region (20-40N) and Northern Indian Ocean (0-20N and 50-100E) averaged in June, July and August (JJA) was calculated to determine the summer monsoon index. The positive (negative) index was considered as strong (weak) monsoon (figure 1). In this study, we only considered strong-weak years whose value greater than 0.5 of standard deviations. Kawamura [9] showed that this index correlate well with Webster and Yang's monsoon index [10].

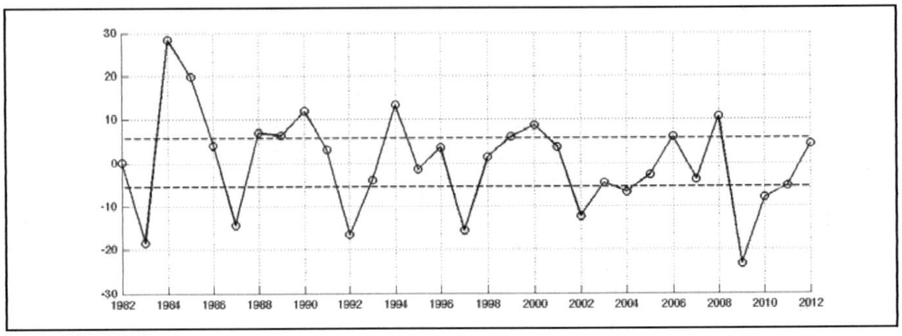

Figure 1. Time series of summer monsoon index for JJA defined by Kawamura (1998). The dashed line shows 0.5 of standard deviation value.

3. Results and Discussion

The differences between weak and strong composite of summer monsoon years in SST and surface wind are displayed in figure 2. The composite of weak monsoon years shows that warmer SST anomaly distributed over western Indian Ocean. This is accompanied by anomaly of surface northeasterly wind in the Arabian Seas and anomaly of northwesterly wind at the south of equator. Based on the SST anomaly distribution in WIO, we divided western Indian Ocean into three regions (shown by blue squares in figure 2) i.e. western equatorial Indian Ocean (WEIO, 45-60E and 10S-10N), western Arabian Sea (WAS, 50-65 °E and 13-23°N), and southwestern Indian Ocean (SWIO, 50-65°E and 5-15°S).

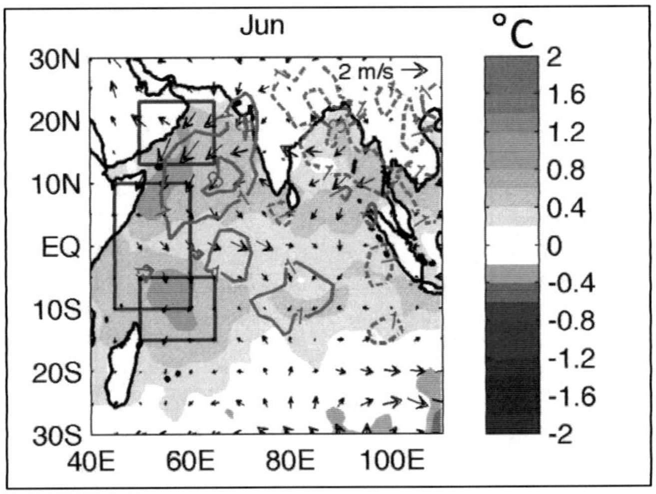

Figure 2. Composite of weak summer monsoon years minus composite of strong summer monsoon years for SST (°C, shaded), surface wind (m s^{-1}, vectors) and precipitation (mm day^{-1}, green contour with 2 mm day^{-1} interval) in February to July. Blue rectangles show region of study (see text for details)

3.1. Western equatorial Indian Ocean (WEIO)

In WEIO, MLT experiences annual variations with the warmest MLT presents in April-May (figure 3). The surface heat flux is the dominant cause of spring warming that peaks in March. In summer, cooling of MLT is caused by cooling of surface heat flux, vertical, zonal advection and residual term. Cooling in WEIO presents earlier than in WAS since the southwesterly monsoon wind was established first in WEIO. MLT recovers in autumn due to warming by meridional advection during June to August and surface heat flux in August to September. In contrast, vertical advection is against warming. It cools the MLT during boreal autumn.

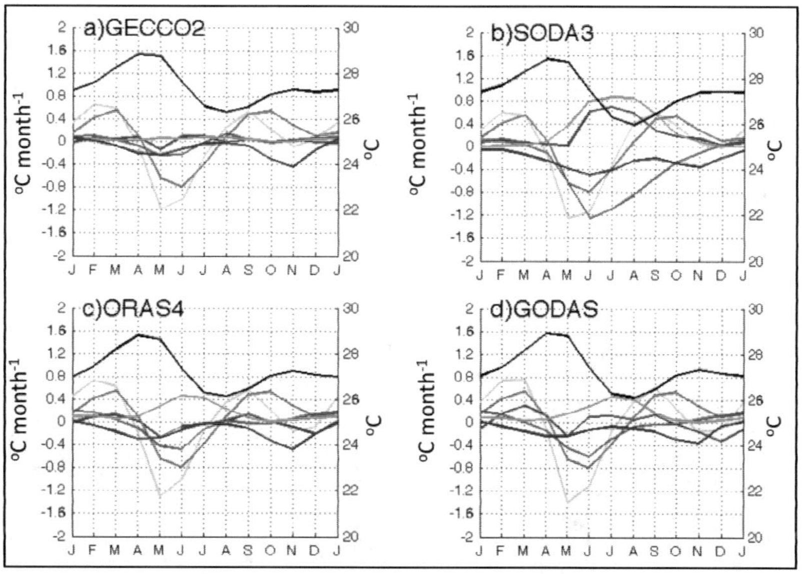

Figure 3. Area averaged monthly climatological time series of heat budget over western equatorial Indian Ocean for a) GECCO2, b) SODA3, c) ORAS4 and d) GODAS. Mixed layer temperature, mixed layer tendency, surface heat fluxes are denoted by black, yellow and red colors respectively. Vertical, zonal and meridional heat advections are denoted by blue, green and orange colors respectively. Residual term is denoted by purple color. The left y-axis is temperature tendency in °C month^{-1} and the right y-axis is mixed layer temperature in °C.

Differences between composite of weak and strong summer monsoon years are shown in figure 4. All models show that peak of positive MLT difference presents in June. This is preceded by positive MLT tendency difference in May. Scatterplot shows that MLT tendency in May and MLT monthly anomaly in June are well correlated (figure 5). It suggests that the ocean process in May could contribute to initiate the MLT monthly anomaly in June. We can clearly distinguish weak (blue dots) and strong (red dots) summer monsoon years. This implies that warm (cool) MLT anomalies in WEIO tend to occur during weak (strong) summer monsoon condition. In WEIO, the anomaly of MLT in June is well correlated with zonal advection (utm) process in May (figure 6 and table 1). This relation is shown as significant in all models. ORAS4 and GODAS showed that variation of vertical advection (wtm) and meridional advection (vtm) respectively could affect MLT monthly anomaly. Meanwhile, the zonal advection could be related to eastward transport of cold water from Omani and Somali upwelling regions [11-12]. Only GODAS that shows significant contribution of surface heat flux (hfx).

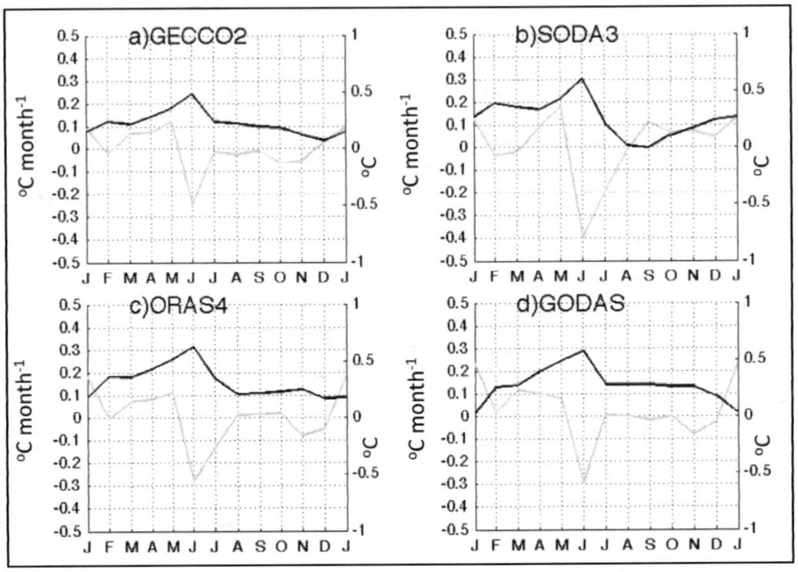

Figure 4. Time series differences for mixed layer temperature (black) and mixed layer temperature tendency (yellow) over western equatorial Indian Ocean for a) GECCO2, b) SODA3, c) ORAS4, d) GODAS. The left y-axis is temperature tendency anomaly in °C month^{-1} and the right y-axis is mixed layer temperature anomaly in °C.

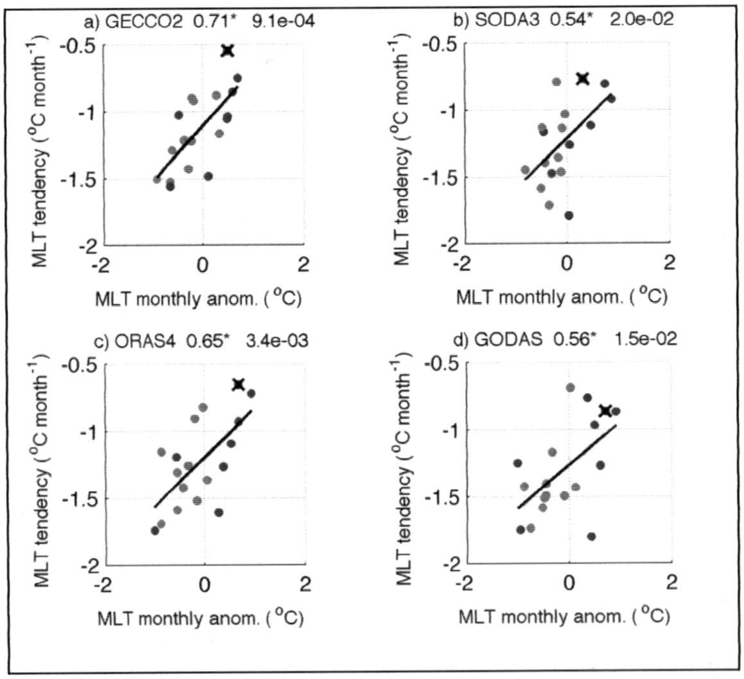

Figure 5. Scatterplot between MLT monthly anomaly in July and MLT tendency in June for western equatorial Indian Ocean. Blue and red dots indicate weak and strong summer monsoon years respectively. Cross (x) marker denotes year of 1997.

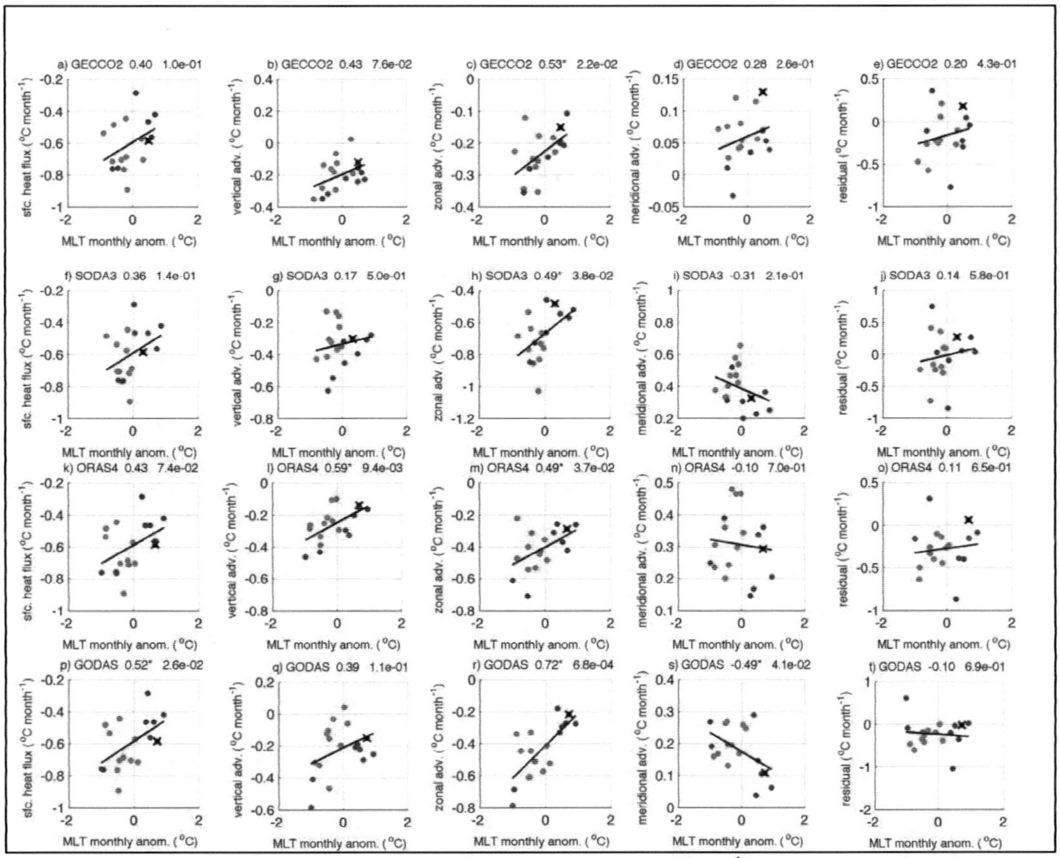

Figure 6. Scatter plot between heat budget terms (°C month^{-1}) in May and mixed layer temperature monthly anomaly (°C) in June for western equatorial Indian Ocean (45°-60°E, 10°S-10°N). The title in each panel, from left to right, indicate model's name, Pearson correlation coefficient and p-value respectively. Student t-test is used to determine p-value of correlations coefficient that is statistically significant at 5% level (P-value < 0.05) denoted by asterisk. Black solid lines indicate linear trend line. Cross (x) marker denotes year of 1997.

Table 1. Correlation between heat budget terms in May and mixed layer temperature anomaly in June for all models over western equatorial Indian Ocean. The abbreviation for dtm, hfx, utm, vtm, wtm and res stand for mixed layer tendency, surface heat flux, zonal advection, meridional advection, vertical advection, and residual term respectively. Correlation values that are significant at 95% level using student t-test are in blue color.

	GECCO2	SODA3	ORAS4	GODAS
dtm	**0.71**	**0.54**	**0.65**	**0.56**
hfx	0.40	0.36	0.43	**0.52**
utm	**0.53**	**0.49**	**0.49**	**0.72**
vtm	0.28	-0.31	-0.10	**-0.49**
wtm	0.43	0.17	**0.59**	0.39
res	0.2	0.14	0.11	-0.10

3.2 Western Arabian Sea (WAS)

In western Arabian Sea (WAS), MLT (black) experiences semiannual variations with two maximum in May-June and October-November (figure 7). The former maximum is warmer than the later one. The maximum temperature in May-June is preceded by warmest MLT tendency (yellow) in boreal spring (March-April). The main contributor for warming in boreal spring is the surface heat flux (red), which peaks in April. Penetrative of solar radiation is maximum due to clear-sky conditions and the thin mixed layer [11]. During boreal summer, the MLT tendency become negative accompanied by cooling of MLT.

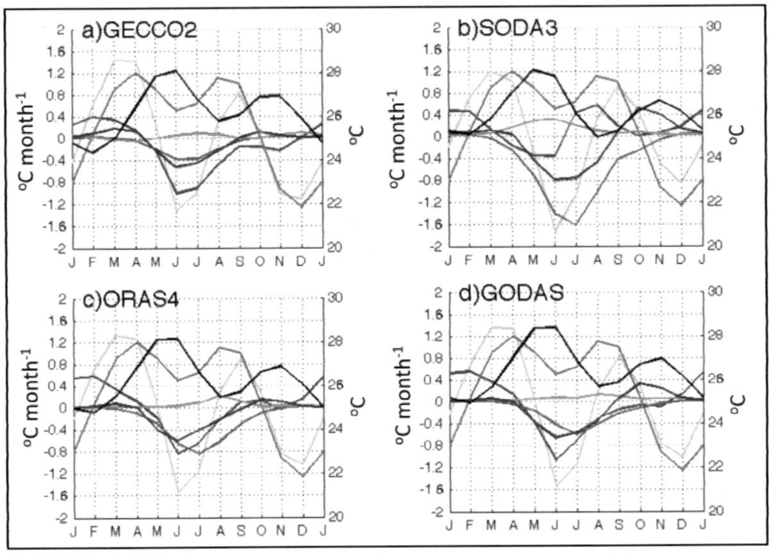

Figure 7. Similar to Figure 3 but for western Arabian Sea.

Warming of surface heat flux is suppressed in boreal summer, probably due to evaporation cooling. Summer cooling contribution also comes from vertical advection (blue) through wind-induced-upwelling, zonal advection (green) and residual term (purple) from May to September. In September-October, MLT start to warm again supported by warming of surface heat flux and suppressed cooling by vertical and meridional advection. After MLT reaches the second maxima in October-November, it slowly cools until reach the minimum point in January-February. This is probably caused by cooling of surface heat flux through evaporation induced by northeasterly monsoon wind.

Using similar approach applied in WEIO, time series difference between composite of weak and strong summer monsoon years were calculated. We found in all models that the largest positive MLT differences (black) present in July, preceded by largest positive MLT tendency differences (yellow) in May-June. To confirm this relation, scatter analysis was carried out. Figure 8 shows scatter plot between MLT in June and MLT tendency in May displays good correlation greater than 0.5 in all models. This robust relation explains that MLT monthly anomaly in July could be initiated by variation of heat budget components in May. Weak composite (blue dots) and strong composite (red dots) can be distinguished, except some outliers (blue dots) are caused due to unknown reason. This indicates that warm (cool) MLT anomalies in WAS tend to occur during weak (strong) summer monsoon years. Table 2 displays that MLT monthly anomaly is well correlated with vertical advection (wtm), surface heat flux (hfx) and zonal advection (utm). Highest correlation is shown by vertical advection in three models, except GECCO2. The second contributor is surface heat flux shown in three models, except SODA3. Zonal advection (utm) is another possible contributor process in GECCO2 and GODAS.

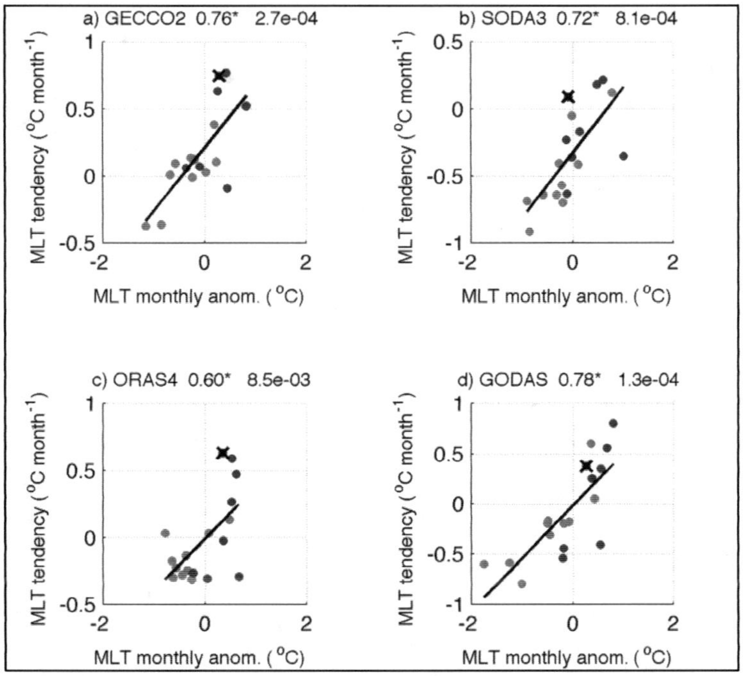

Figure 8. Scatterplot between MLT monthly anomaly in July and MLT tendency in May for western Arabian Sea. Cross (x) marker denotes year of 1997.

Table 2. Correlation between heat budget terms in May and Mixed layer temperature anomaly in July for all models over western Arabian Sea. The abbreviations and color are similar to Table 1.

	GECCO2	SODA3	ORAS4	GODAS
dtm	**0.76**	**0.72**	**0.60**	**0.78**
hfx	**0.54**	0.36	**0.58**	**0.55**
utm	**0.61**	0.24	0.27	**0.53**
vtm	0.11	0.34	0.13	0.16
wtm	0.11	**0.69**	**0.75**	**0.80**
res	0.16	0.15	0.04	0.43

3.3 Southwestern Indian Ocean

Mixed layer temperature displays annual variations that reaches maximum around March-April (figure 9). This is mainly attributed to warming by surface heat flux that occurs from October to March. In this season, radiation warming reaches maximum since the sun is located in the southern hemisphere. MLT reaches the coolest point during summer, preceded by cooling of surface heat flux contribution in boreal spring to summer. The vertical advection is against the surface heat flux. It tends to cool the MLT during boreal winter to spring. The meridional advection then contributes to warm the MLT in the late of summer. This owes to the southward Ekman heat transport associated strong meridional SST gradient and easterly wind stress associated with the Indian summer monsoon [13].

Similar approach (figure 4) was used by calculating time series difference in MLT and MLT tendency between weak and strong composite summer monsoon years in MLT and MLT tendency difference presents in March-April, preceded by positive MLT tendency difference in March. Figure 10 confirms that these two variables are well correlated. The MLT monthly anomaly in April could be initiated by atmospheric and/or oceanic process in March. Warm (cool) MLT anomalies also tend to present during weak (strong) summer monsoon years. Table 3 displays correlation between MLT monthly anomaly in April and heat budget terms in March. Note that all models show significant contribution of vertical advection (wtm) in March to initiate MLT monthly anomaly in April, while the second contributor is surface heat flux (hfx). Our results are consistent with [13] which found that vertical advection process is responsible for warm MLT monthly anomaly on interannual time scale. Other possible contributor is residual term (res) that only shown in GECCO2 and SODA3.

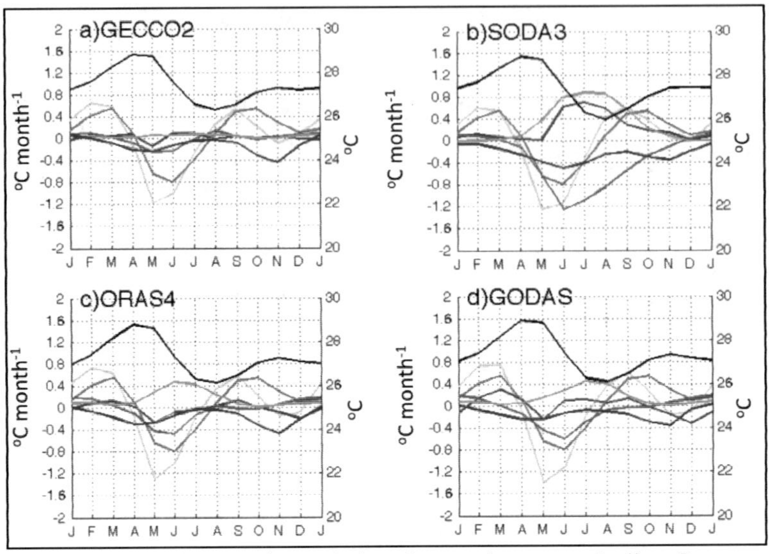

Figure 9. Similar to Figure 3, but for southwestern Indian Ocean.

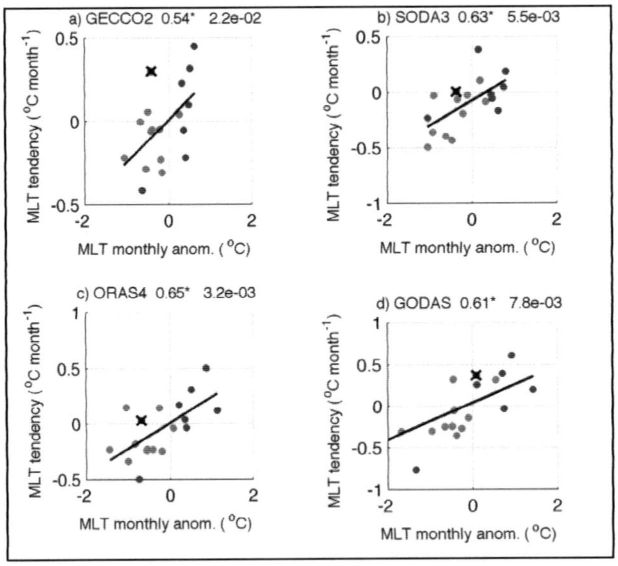

Figure 10. Scatterplot between MLT monthly anomaly in July and MLT tendency in May for southwestern Indian Ocean. Cross (x) marker denotes year of 1997.

During early development of El Nino 1997/1998 in March, it was recorded that Indian Ocean's SST also experience warming that exceed normal conditions in JJA 1997. This El Nino suppressed the monsoon circulation and induced convective activity in the east of Somalia coast [14]. Scatter analysis in Figure 5(WEIO), Figure 8(WAS) and Figure 10 (SWIO) clearly show that relatively warm MLT and MLT tendency in summer 1997 that correspond to growing El Nino. Our results display (figure 6c, h, m and r) shows that during El Nino in 1997, cooling by zonal advection is more suppressed than in other years.

Table 3. Correlation between heat budget terms in March and mixed layer temperature anomaly in April for all models over southwestern Indian Ocean. The abbreviation is similar to Table 1.

	GECCO2	SODA3	ORAS4	GODAS
dtm	**0.54**	**0.63**	**0.65**	**0.61**
hfx	**0.59**	**0.67**	**0.54**	**0.57**
utm	0.37	0.13	**-0.54**	-0.31
vtm	-0.16	0.15	0.35	-0.02
wtm	**0.67**	**0.77**	**0.80**	**0.57**
res	**-0.48**	**-0.48**	-0.46	0.01

It is interesting to note that during boreal winter zonal advection becomes less important than the vertical advection in WEIO (table 4). Vertical advection (wtm) displays significant correlation in all models. It could be understood since during winter, climatologically, the thermocline is shallower than summer. As a result, the upwelling is easily affects the MLT/SST. During winter monsoon, the possible cause of thermocline variations is not only strong northeasterly wind. Indian Ocean Dipole (IOD) or El-Nino Southern Oscillation also could modulate the thermocline variation. The strongest impact may present when the IOD and ENSO co-occurred as in 1997 case. Figure 11 shows suppressed vertical advection in boreal autumn that is mainly contribute to warming compared to zonal advection which is associated with deepening of thermocline (isothermal of 20°C) up to 30m than in normal condition. Our results also corroborate [15] which emphasizes the importance of realistic ocean process to represent realistic SST in WIO. Insufficient cooling by ocean process may contribute to warm SST bias in WIO.

Table 4. Correlation between heat budget terms averaged in September, October, November and mixed layer temperature anomaly averaged in December, January and February for all models over western equatorial Indian Ocean. The abbreviation is similar to Table 1.

	GECCO2	SODA3	ORAS4	GODAS
dtm	**0.78**	**0.48**	**0.8**	**0.60**
hfx	**0.70**	0.16	-0.04	-0.10
utm	0.13	**0.48**	0.38	0.23
vtm	**0.70**	-0.25	**0.76**	0.35
wtm	**0.48**	**0.52**	**0.69**	**0.80**
res	-0.02	-0.43	-0.02	-0.13

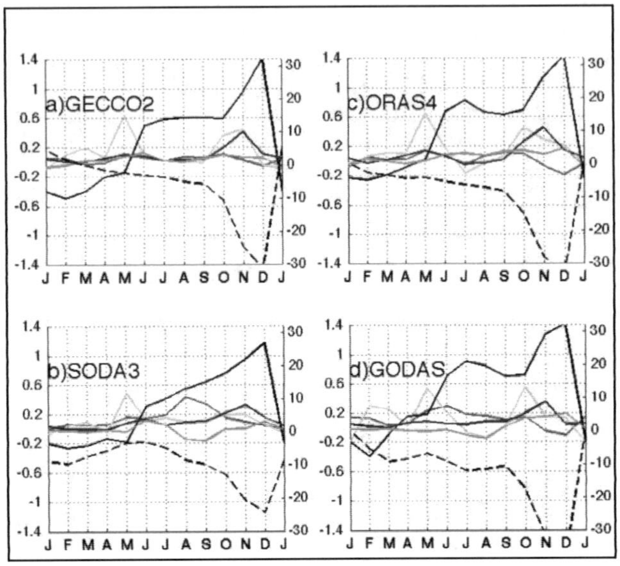

Figure 11. Time series difference of heat budget between El Nino 1997 case and monthly climatology. Black solid lines indicate difference in MLT (°C, right y-axis) and black dashed lines indicate thermocline (m, left y-axis). Solid lines in yellow color indicate difference in MLT tendency (°C month^{-1}, left y-axis). Solid lines in blue, green and orange colors indicate vertical advection, zonal advection and meridional advection respectively (°C month^{-1}, left y-axis).

4. Conclusion

Our result portrayed that ocean process plays more important role than atmospheric process (surface heat flux) to initiate the SST anomaly during extreme weak/strong summer monsoon conditions. Ocean heat budget analysis has been applied to analyze the SST anomaly during weak-strong southwest monsoon over on four ocean reanalysis models: GECCO2, SODA3, ORAS4 and GODAS. During weak summer monsoon, the SST anomalies tend to be warmer than usual, while during strong summer monsoon the SST anomalies tend to be colder than normal condition. During weak summer monsoon, at first the SST anomaly presents in SWIO intiated by vertical advection. Then the SST anomaly propagates northwards initiated by zonal advection in WEIO and vertical advection in WAS. The initiation process by ocean advection occurred about one-two month before MLT anomaly presents in WIO.

5. References

[1] Shukla J 1975 Effect of Arabian sea-surface temperature anomaly on Indian summer monsoon: A numerical experiment with the GFDL model *J. Atmos. Sci.* **32** pp 503–11

[2] Izumo T, Montegut C D, Luo J J, Behera S K, Masson S, and Yamagata T 2008 The role of the western Arabian Sea upwelling in the Indian monsoon rainfall variability, *J. Clim* **21** pp 5603–23

[3] Shaji C, Iizuka S, Matsuura T 2003 Seasonal Variability of Near-Surface Heat Budget of Selected Oceanic Areas in the North Tropical Indian Ocean *Journal of Oceanography* **59** pp 87

[4] Smith T M and Reynolds R W 2004 Improved extended reconstruction of SST (1854–1997) *J. Clim* **17** pp 2466–77

[5] Kalnay E The NCEP/NCAR 40-year reanalysis project 1996 *Bull. Amer. Meteor. Soc.* **77** pp 437-70

[6] Ng B, Cai W, Walsh K, Santoso A 2015 Nonlinear processes reinforce extreme Indian Ocean Dipole events *Sci. Rep.* **5** pp 11697

[7] Fathrio I , Manda A, IIzuka S, Kodama Y, Ishida S, 2017 Evaluation of CMIP5 models on sea surface salinity in the Indian Ocean *IOP Conf. Ser.: Earth Environ. Sci.* **54** 012039

[8] Fathrio I, Iizuka S, Manda A, Kodama Y, Ishida S, Moteki Q, Yamada H, Tachibana Y 2017 Assessment of western Indian Ocean SST bias of CMIP5 models *J. Geophys. Res. Oceans* **122** pp 3123–40

[9] Kawamura R 1998 A possible mechanism of the Asian summer monsoon-ENSO coupling *J. Meteor. Soc. Japan* **76** pp 1009-027

[10] Webster P J and Yang S 1992 Monsoon and ENSO: Selectively interactive systems *Quart. J. Roy. Meteor.Soc.* **118** pp 877-926

[11] Montégut C B, Vialard J, Shenoi S S, Shankar D, Durand F, Ethé C, and Madec G 2007 Simulated Seasonal and Interannual Variability of the Mixed Layer Heat Budget in the Northern Indian Ocean *J. Climate* **20** pp 3249–3268

[12] Shankar D, Vinayachndran P N, and Alallat U 2002 The monsoon currents in the northern Indian Ocean *Prog. Oceanogra*. **52 (1)** pp 63–120

[13] Yoko T, Tozuka T, and Yamagata T 2008 Seasonal Variation of the Seychelles Dome *J. Climate* **21** pp 3740–3754

[14] Shen X, and Kimoto M 1999 Influence of El Niño on the 1997 Indian summer monsoon *Journal of the Meteorological Society of Japan* Ser. II **77** (5) pp 1023-1037

[15] Yang W, Seager R, Cane M A, and Lyon B 2015 The rainfall annual cycle bias over East Africa in CMIP5 coupled climate models *J. Climate* **26** pp 9789–9802

IOP Conf. Series: Earth and Environmental Science **149** (2018) 012015 doi:10.1088/1755-1315/149/1/012015

Land use change impacts on discharge analysis using SWAT model at Ciherang Pondok DAM catchment area

M A Utamahadi, N H Pandjaitan, M I Rau
Department of Civil and Environmental Engineering, Faculty of Agricultural Technology, Bogor Agricultural University, IPB Darmaga Campus, PO BOX 220, Bogor 16002, Indonesia

Email : norahp@apps.ipb.ac.id

Abstract. The prompt increase of population influenced the requirement for new regions to fulfill people's primary needs. Its increased land use change and caused many impacts on the environment, including watersheds as well. Ciherang Pondok DAM catchment area is part of Cisadane watershed and was selected as the research area. This research aimed to analyse the water supply and water discharge change caused by the Urban Planning (RTRW) in 2020. The analysis was conducted using soil and water assessment tools (SWAT) model. Stages of this research were catchment area delineation, HRU identification, calibration and validation of models, and prediction of discharge and water demand. The result showed that RTRW of 2020 increased the maximum discharge of 1.6 m^3/s and decreased the minimum discharge of 0.01 m^3/s, hence the maximum and minimum discharge ratio increased 0.26% from 2016. Output discharge in 2020 at Ciherang Pondok Dam Catchment Area was classified as well, with discharge of 6.72 – 126.2 m^3/s, and could fulfil water demand. For the best result, it is better to use climate data from weather stations inside the study area and it is required an improvement in data archiving system.

1. Introduction

The prompt increase of population influenced the requirement for new regions to fulfill people's primary needs. This concern lead to a condition where lands were developed inappropriately and had themselves beyond their carrying capacity [1]. Consequently, this will affect the hydrologic system of a certain watershed [2-4]. In 1984, from 458 watersheds in Indonesia, there were 20 watersheds indicated as super priority and this condition keeps increasing to 60 watersheds in 1999, one of such is Cisadane Watershed [5-6]. From the review result by the Ministry of Forestry of Indonesia [7], in 2015, the handling of critical lands in Cisadane Watershed was not much, so that interferences and damages increased were caused by numerous human activities, especially within the upstream region of the basin.

Focusing in a narrower region, one of the areas influenced in the upstream region of Cisadane Watershed is Ciherang Pondok Dam Catchment Area, which is the main water resources for the local communities and the provision of raw water by water supply company (PDAM) Tirta Pakuan Bogor. This research is aimed to undertake analysis of water availability and the change of river discharge at Ciherang Pondok Dam Catchment Area using geographic information system (GIS)-based software, i.e. soil and water assessment tools (SWAT). The SWAT model has been utilized and proved as an effective yet efficient tool for undertaking research in analyzing the response between hydrological cycle and land use change [8-11].

2. Methodology

The research was undertaken in May 2017 in Ciherang Pondok Dam Catchment Area, Central Management of Regional River Flow (BPDAS) of Bogor Regency, and Development Planning Agency (BAPPEDA) of Bogor Regency. Tools used within this research include laptop with Microsoft Office, SWAT, ArcGIS 10.1 with ArcSWAT 2012 plug-in, Google Earth, SWAT-CUP 5.1.6, minitab, calculator, and global positioning system (GPS). Material used were digital elevation model (DEM) of Cisadane Watershed with 30x30 m resolution, river map of Cisadane Watershed, boundary map of Cisadane Watershed, soil type of Bogor Regency, administrational area of Bogor Regency, landuse map of Bogor Regency, 2020 RTRW of Bogor Regency, daily climate data and rainfall of 2007-2016, water level data of Cisadane River for 2012-2016, outlet point coordinate, and population number per Bogor District for 2010-2016. Steps undertaken for this research included data acquisition, catchment area delineation, establishment of hydrological response unit (HRU), calibration and validation model, and river discharge and water demand prediction. Detail research steps are given in Figure 1.

Within the simulation process, DEM SRTM map was initially projected from geographical coordinates to universal transverse mercator (UTM) WGS 48S projected system using ArcGIS 10.1. The data for the delineation of the catchment area used was the river and coordinate point of Ciherang Pondok Dam Catchment Area. Afterwards, HRU establishment was done by inserting 2007-2016 climatic and rainfall data to SWAT 2012, where the results were the daily simulations with warming up period of 2 years.

Prior undertaking calibration process, Nash-Sutcliffe (NS) and R^2 value are required to be attained from the SWAT daily discharge simulation. NS is defined as the objective function and can be calculated using Equation (1) [12]. The simulation result could be categorized as a good performance if NS > 0.75 and satisfactory performance if 0.36 < NS < 0.75 [13]. However if NS < 0.36, it is then categorized as unsatisfactory performance and needed to be re-calibrated. On the other hand, to portray the pattern accuracy of the model output with the observation values, coefficient or linear R^2 equation was used (Equation (2)). If R^2 reaches 1, this condition indicates a close relationship pattern between the resulted model predictions with the field observation result.

$$NS = 1 - \left[\frac{\Sigma(y-\hat{y})^2}{\Sigma(y-\bar{y})^2}\right] \quad (1)$$

$$R^2 = \left[\frac{\sum_{i=1}^{n}(Qobs,i-\bar{Q}obs,i)(Qcal,i-\bar{Q}cal,i)}{\sum_{i=1}^{n}(Qobs,i-\bar{Q}obs,i)^2\sum_{i=1}^{n}(Qcal,i-\bar{Q}cal,i)^2}\right]^2 \quad (2)$$

Calibration and validation were undertaken using SWAT-CUP. The SWAT model used more than 500 hydrological parameters, however, not all parameters were used. The selection of parameters was done by studying the literature from various publications and sorting the most frequent parameters used in the SWAT model. Afterwards, the sensitive parameters were selected using p-value analysis. The smaller the value of p-value, the parameters became more sensitive to changes in the value of NS and R^2. Calibration was done by inputting the discharge data of 2009-2016 SWAT simulation result and observation discharge data from the conversion result of 2015 water level data, whilst validation was done using the 2016 water level data. Then, trial and error was undertaken to obtain the best parameter values, indicated by NS and R^2. Finally, re-simulation was done using 2020 RTRW with additional input of the best parameter value from validation result on SWAT editor, in order to attain the change of discharge value and water availability as an outcome from the 2020 RTRW.

IOP Conf. Series: Earth and Environmental Science **149** (2018) 012015 doi:10.1088/1755-1315/149/1/012015

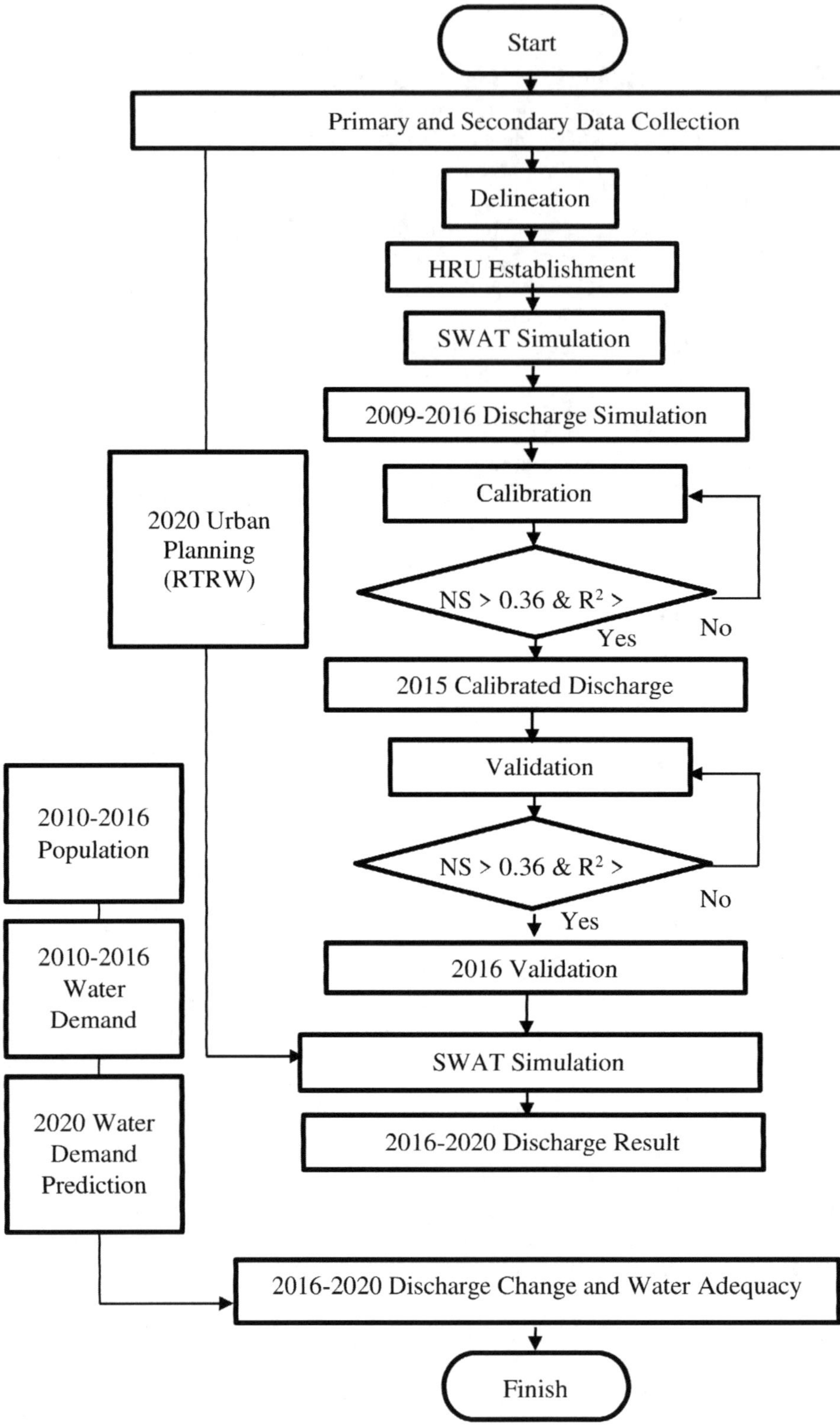

Figure 1. Research methodology steps.

3. Results and Discussions

3.1. General

Cisadane Watershed starts from Salak Mountain at the southern region of Bogor Regency towards Java Sea with length of 137.8 km. Cisadane River is one of the main rivers in Banten and West Java Province. Ciherang Pondok Dam Catchment Area is part of the upstream area of Cisadane Watershed. Ciherang Pondok Dam Catchment Area is within the humid tropical rainy climate type in the Koppen climate classification [14]. The selected outlet was the raw water tapping building or Ciherang Pondok intake, located at 06.68° South Latitude and 106.81° East Longitude with an elevation of 380 meters above sea level. Furthermore, some weather stations and rain gauges are likely to affect the hydrology system in the watershed. There are some rain gauges that enter the research area but due to constraints, the data was then not used. The general solution used was to utilize the nearest station data. This means that the rainfall data from the closest station to the research location was used for the analysis [15]. The weather stations used in this study were Dramaga Weather Station, Katulampa Rain Gauge, Citeko Weather Station, and Gunung Mas Rain Gauge.

3.2. Watershed Delineation and HRU Establishment

Watershed Delineation basically divided the Sub-Basin into several hydrological catchment areas. Based on the delineation result, the area of Ciherang Pondok Dam Catchment Area was 17,066.97 ha or about 20% of the area of Upper Cisadane Watershed (85,256.19 ha) and about 11.03% of Cisadane Watershed (154,654 ha). The catchment area was subdivided into 3 sections with an area of 804.08 ha, 6,024.71 ha, and 10,238.17 ha respectively. The research outlet went to Ciherang Pondok Dam Catchment Area section 1, so that the discharge value is considered only in that particular area. The delineation result map is presented in Figure 2.

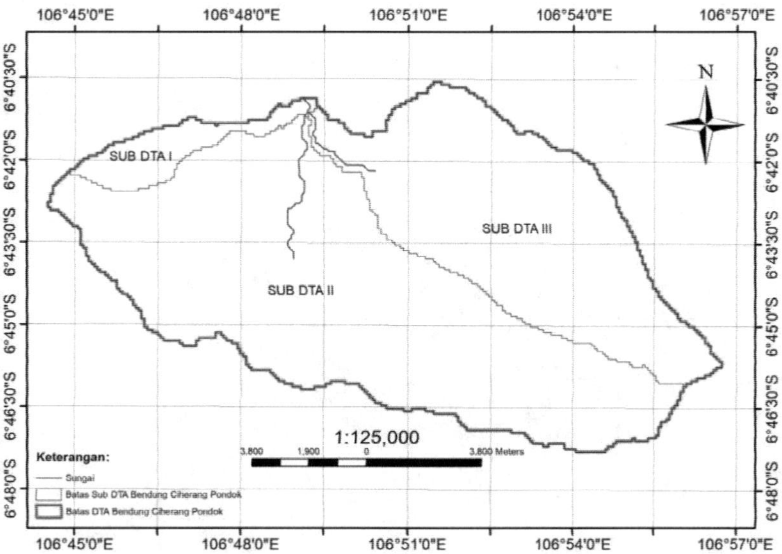

Figure 2. Delineation result of Ciherang Pondok Dam Catchment Area.

Within HRU establishment step, DEM, landuse map, soil type map, and slope classification data were all overlain using threshold by percentage method. For land use, 10% threshold was used, 5% threshold for soil type and 5% threshold for slope in accordance with Cisadane Watershed's land rehabilitation plan and soil conservation. HRU establishment can also be used to assess the effectiveness of a reforestation / rehabilitation effort as well as other efforts to improve the quality of water resources and the recharging area of a specific region.

Land use at Ciherang Pondok Dam Catchment Area was divided into 7 types, among which the largest was 9192.27 ha (53.86%) of agricultural land, secondary forest, forest, settlement, rice field, and the smallest among others was shrubs of 6.83 ha (0.04%). The landuse map is presented in Figure 3.

Based on the land distribution map in Figure 4, there were only 2 types of soil at Ciherang Pondok Dam Catchment Area. Most of the soil type in this area was brook andosol association of 9,605.57 ha (56.28%) and the rest was brook latosol of 7,461.41 ha (43.72%).

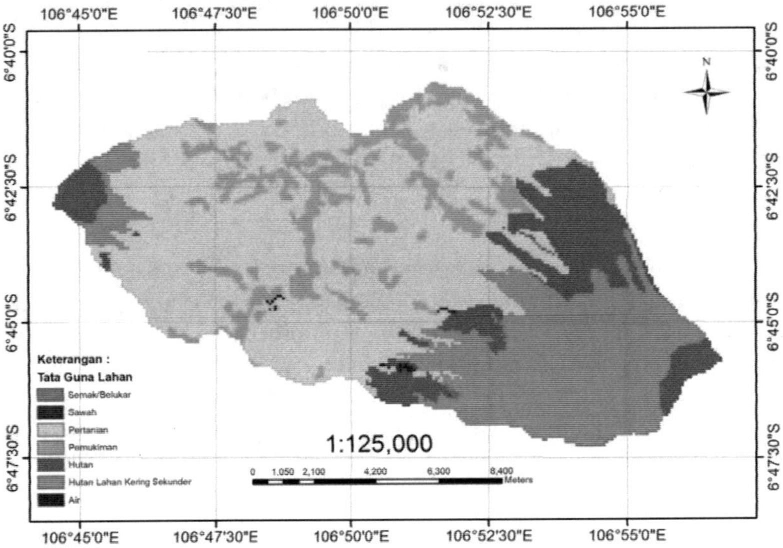

Figure 3. Landuse map of Ciherang Pondok Dam Catchment Area in 2016.

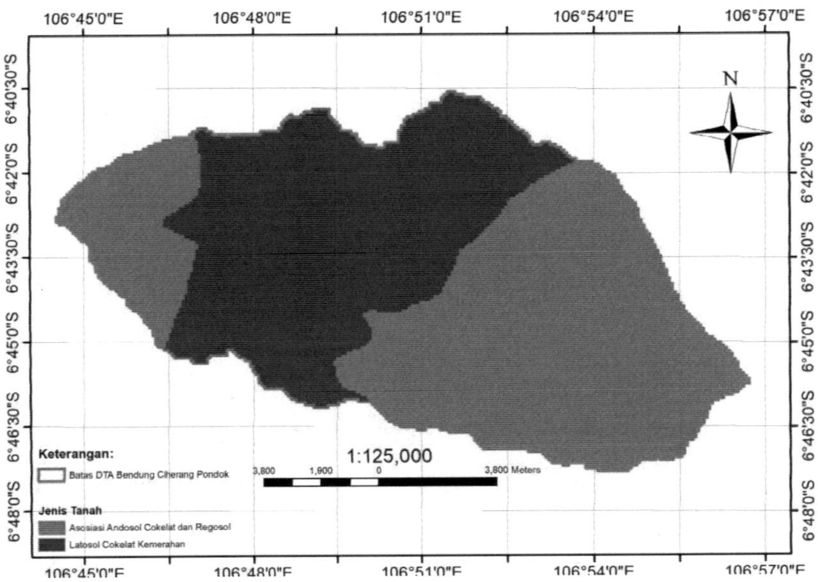

Figure 4. Soil type map of Ciherang Pondok Dam Catchment Area.

For this research, the slope was divided into 5 classes, i.e. 0-8% (flat), 8-15% (declivous), 15-25% (rather steep), 25-45% (steep), and >45% (very steep) [16]. The position of the area distribution is given in Figure 5. The result of the HRU establishment provided information regarding landuse, soil, slope, total area, and percentage of HRU within a sub-basin. In this research, 150 HRUs within 3 sub-basin were gathered.

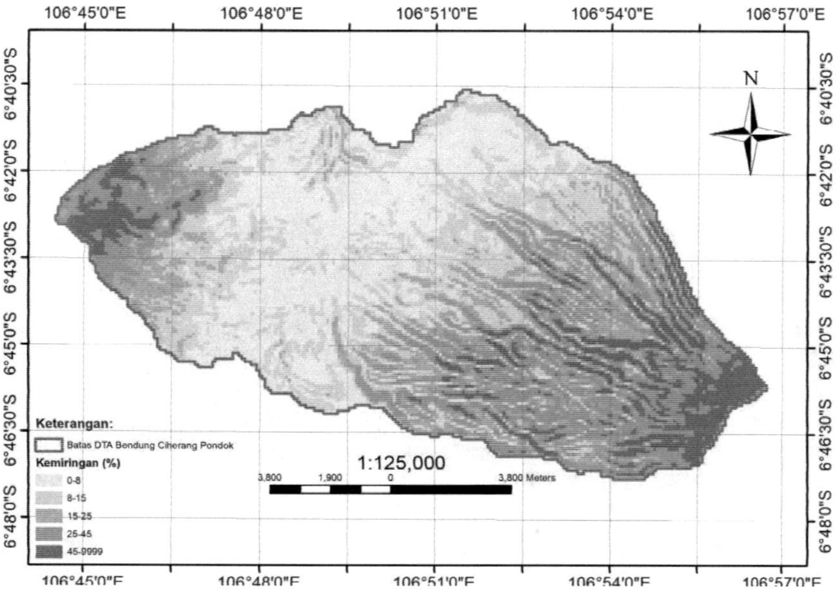

Figure 5. Slope map of Ciherang Pondok Dam Catchment Area.

3.3. SWAT Simulation and Validation Model

Based on simulation results with landuse data in 2016, the maximum daily discharge which occurred was 216.70 m^3/s, with minimum discharge of 0.61 m^3/s, and an average discharge of 18.59 m^3/s. Ciherang Pondok Dam Catchment Area water level data was initially converted into discharge value, in order to determine the value of NS and R^2 simulation results. Based on the data calculation results, NS value was -4.05, whereas R^2 was 0.20. The value was not satisfactory, hence, calibration was required.

The calibration and validation process were done using Sequential Uncertainty Fitting (SUFI 2) method with daily observation data in 2015 (for calibration) and 2016 (for validation). Calibration and validation were done as much as 32 times of iteration and simulation of 50-1000 times for each iteration. The number of final parameters used in this stage were 10 parameters, i.e. ALPHA_BF.gw, GW_DELAY.GW, GWQMN.gw, ESCO.hru, OV_N.hru, SLSOIL.hru, SLSUBBSN.hru, GW_REVAP.gw, REVAPMN.gw, and SURLAG.bsn.

Based on the calibration result, R^2 and NS value were 0.53 and 0.50, respectively, whist the value of R^2 and NS validation were 0.60 and 0.54, respectively. The values were eligible and it could be concluded that the data were valid and satisfactory. The calibration and validation results are shown in Figures 6 and 7.

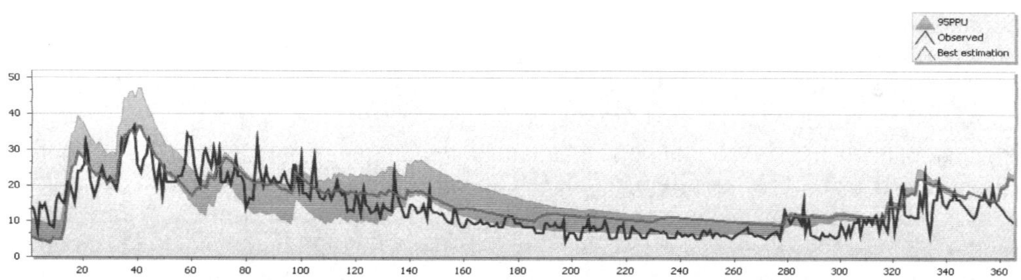

Figure 6. Calibration result of daily discharge in 2015.

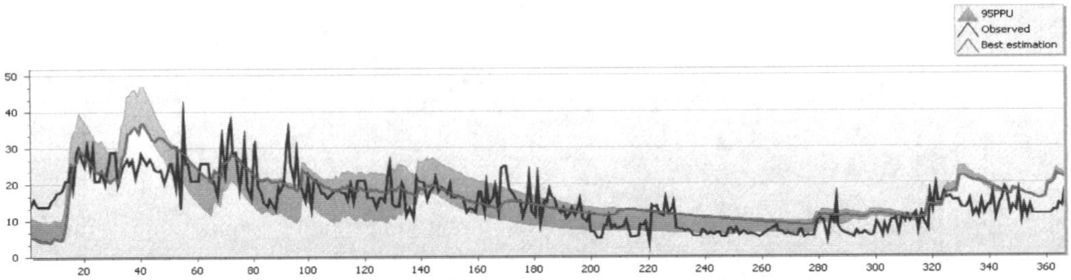

Figure 7. Validation result of daily discharge in 2016.

3.4. *Effect of Landuse Change on River Discharge*

The landuse change used was in accordance with the 2020 RTRW, which is presented in Figure 8. In the 2020 RTRW, forest land was reduced by 481.29 ha (2.82%) and agricultural land area of 2,392.79 ha (14.02%) from 2016. Most of the lands were transformed into residential and industrial land. Settlement area increased by 2,481.54 ha (14.54%) and there was land conversion into industrial land covering 421.55 ha (2.47%) from 2016. Paddy field was transformed into residential area according to 2020 RTRW.

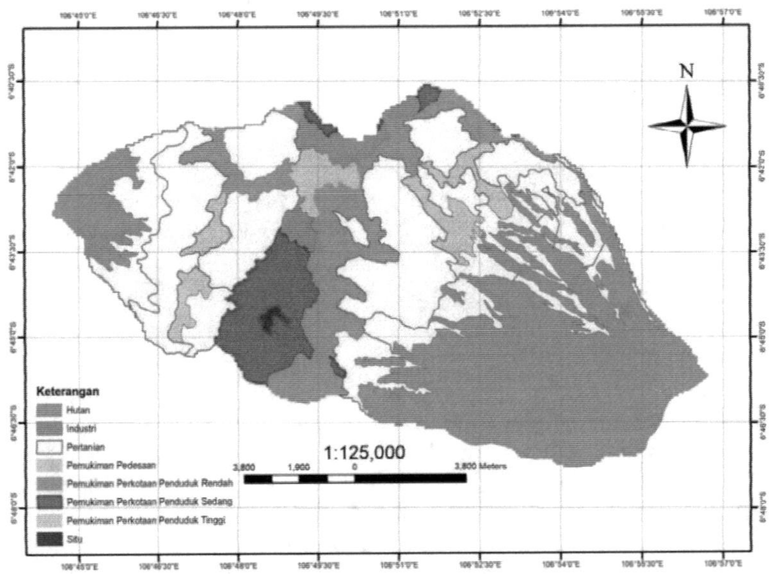

Figure 8. Landuse map in 2020.

Afterwards, SWAT simulation re-modelled with 2020 landuse and the sensitive parameters of validation result. Based on the results of data processing using the SWAT model, the value of discharge in 2020 was obtained. Based on the simulation results in Table 1, the maximum discharge of 2020 rose by 1.6 m^3/s, while the minimum discharge decreased by 0.01 m^3/s from 2016.

To assess the level of water availability based on the 2020 RTRW, predictive information of water demand was required. The number of population in 2010-2016 was used to obtain the water requirement . It was assumed that water requirement were 150 m^3/person/day with duration of water use of 9 hours/day. The prediction was done using minitab application. The utilization of raw water by Tirta Pakuan PDAM was 1.8 m^3/s and the value of water demand in 2020 was 5.74 m^3/s. Based on analysis results, the discharge of Ciherang Pondok Dam Catchment Area in 2020 could still meet 100% of PDAM water needs and was classified as uncritical (IK < 50%) [17].

3.5. Strategic Environmental Assessment (SEA)

In this research, the performance of catchment area was performed by parameter of river regime coefficient (KRS) and vegetation cover in accordance with the Forestry Minister Decree No P. 61 2014. Assessment results could be input into evaluation of government policies and planed as a function of strategic environmental studies, so that environmental impacts can be minimized [18].

The coefficient of river regime (KRS) showed fluctuations between the maximum and minimum discharge on the river. The sharp fluctuation of the discharge indicated that the watershed function was not performing well, which meaned that the quality of the watershed was low [19]. Based on the calculation data, the value of both KRS in the study sites were included in the criteria very well, according to the specific flow discharge criteria [20]. There was an increase in the value of KRS from 2016 to 2020 by 0.26%. However, it did not mean much since the value was very small but still consider as a very good category. The calculated data were presented in Table 1.

Table 1. KRS criteria at Ciherang Pondok Dam Catchment Area.

Parameter	2016	Criteria	2020	Criteria
Q max (m^3/det)	124.6	Very Good	126.2	Very Good
Q min (m^3/det)	6.73	Very Good	6.72	Very Good
Q ave (m^3/det)	20.21	Very Good	20.20	Very Good
Qmax/Qmin Ratio	18.51	Very Good	18.77	Very Good

Based on 2020 RTRW, the vegetation cover of Ciherang Pondok Dam Catchment Area decreased by 16.69% in 2016, but was still classified as good with 70%. Most of the vegetation land were transformed into residential and industrial land. The impact of this on a large scale will increase the fluctuation of river discharge in the rainy season but river discharge will be very low in the dry season [21].

4. Conclusion

Land-use changes based on the 2020 RTRW resulted a maximum discharge increase of 1.6 m^3/s and a minimum discharge decrease of 0.01 m^3/s, hence the ratio of the maximum-minimum discharge value of 2020 increased by 0.26% from 2016. The value of output discharge in 2020 at Ciherang Pondok Dam Catchment Area was still categorized as good, i.e. 6.72 - 126.2 m^3/s, and met 100% of PDAM water needs.

References

[1] Symeonakis E, Calvo-Cases A, Arnau-Rosalen E 2007 Land use and land degradation in southeastern Mediterranean Spain *J. Environt. Manag.* **40** 80–94
[2] Ferijal T 2012 Prediksi Hasil Limpasan Permukaan dan Laju Erosi Dari Sub Das Krueng Jreu Menggunakan Model SWAT *J. Agr.* **16** 29-38
[3] Pervez M S, Henebry G M 2015 Assessing the impacts of climate and land use and land cover change on the freshwater availability in the Brahmaputra River basin *J. Hydrol.: Reg. Stud.* **3** 285-311
[4] Welde K, Gebremariam B 2017 Effect of land use land cover dynamics on hydrological response of watershed: Case study of Tekeze Dam watershed, northern Ethiopia *Int. Soil Water Conser. Res.* **5** 1-16
[5] Arsyad S 2006 *Konservasi Tanah dan Air* Bogor: IPB Press
[6] Wibowo S 2004 *Masalah Degradasi Lahan dan Upaya Rehabilitasi Hutan dan Lahan. Prosiding Seminar Degradasi Lahan dan Hutan* Yogyakarta: Universitas Gadjah Mada and Department of Forestry
[7] Ministry of Forestry 2009 *Monitoring dan Evaluasi Pengelolaan DAS. Peraturan Menteri*

Kehutanan No P. 61 /MenhutII/2014 Jakarta: Ministry of Forestry
[8] Wu Y, Liu S, Yan W, *et al* 2016 Climate change and consequences on the water cycle in the humid Xiangjiang River Basin, China *Sto. Env. Res. Risk Ass.* **30** 225-235
[9] Kou J, Liu H, Zhang H, Zhou Q Hydrological response to climate change in taoer river basin based on SWAT model *Wat. Res. Pow.* **34** 12-16
[10] Zhang L P, Qin L L, Hu Z F, Zeng S D 2010 Simulated hydrologic responses to climate change of water source area in the Middle Route of South-to-North Water Transfer Project *J. Hydr. Eng.* **41** 1261-1271
[11] Liu J, Zhang C, Kou L, Zhou Q 2017 Effects of Climate and Land Use Changes on Water Resources in the Taoer River *Adv. in Met.* **1** 1-13
[12] Neitsch S L, Arnold J G, Kiniry J R, Srinivasan R, William J R 2004 *Soil and Water Assessment Tools Input/Output File Documentation Version 2005* Texas: Agricultural Research Service
[13] Van L M W, Garbrecht J 2003 Hydrologic simulation of the Little Washita river experimental watershed using SWAT *J Am. Water Resour. Assoc.* **39** 413-426
[14] Departement of Agriculture 1992 *Laporan Hasil Penelitian Daya Dukung Pertanian Lahan Kering di Daerah Aliran Sungai (DAS)* Jakarta: Pusat Penelitian Tanah dan Agroklimat
[15] Estiningtyas W 2012 *Penetapan Wilayah Cakupan Indeks Untuk Penerapan Asuransi Iklim* Bogor: IPB press
[16] Department of Settlement and Regional Infrastructure 2007 *Penataan Ruang: Undang-Undang RI N0.26 Tahun 2007* Jakarta: Departemen Permukiman dan Prasarana Wilayah
[17] Martopo S 1984 *Keseimbangan Ketersediaan Air di Pulau Bali* Yogyakarta: UGM
[18] Ministry of Environment 2009 *Pedoman Pelaksanaan Kajian Lingkungan Hidup Strategis. Peraturan Menteri Negara Lingkungan Hidup Nomor 27 Tahun 2009* Jakarta: Kementerian Lingkungan Hidup
[19] Suripin 2002 *Pelestarian Sumberdaya Tanah dan Air* Yogyakarta: Andi
[20] Handayani W, Indrajaya Y 2011 Analisis Hubungan Curah Hujan dan Debit Sub Sub Das Ngatabaru, Sulawesi Tengah *J. For. Nat. Cons. Res.* **8** 143-153
[21] Asdak 2007 *Hidrologi dan Pengelolaan Daerah Aliran Sungai* Yogyakarta: UGM Press

IOP Conf. Series: Earth and Environmental Science **149** (2018) 012016 doi:10.1088/1755-1315/149/1/012016

The use of remote sensing to estimate changes of seagrass extent and biomass in Cockburn Sound, Western Australia

S Vidyan
The University of Western Australia, Perth 6009, Australia

E-mail:Salira.vidyan@gmail.com

Abstract. The extent of seagrasses in Cockburn Sound was examined using Nearmap images of year 2010, 2012, 2014, and 2016 to be compared to the last assessment in 1999. It was identified that the seagrass coverage has increased by 231 Ha since 1999, with most of the growth occurred in the southern part. While the water quality in Cockburn Sound has improved, it is believed that there are other pressures affecting the slow growth rate of the seagrasses. Seagrass biomass was also evaluated using Landsat images of year 1994, 1999, 2010, 2012, 2014, and 2016 in addition to a field survey data of leaf biomass in 2016. Despite its increasing extent, seagrass in Cockburn Sound indicated a declining biomass since 1994, which is believed due to the changing nutrient content.

1. Introduction

Seagrass meadows play an essential role in climate change for their high contribution of carbon sequestration and storage. Carbon stocks of seagrass (1-meter depth) ranges between 10 – 829 Mg/Ha, with a mean of approximately 108 Mg/Ha [1]. Meanwhile, another study shows that the global carbon burial rate of seagrass ranges between 48 – 112 Tg C per year [2]. These numbers are comparable to the temperate forest (53 Tg C per year), tropical forest (78.5 Tg C per year), and boreal forest (49.3 Tg C per year) [3]. Besides, the longevity of carbon sequestration in seagrass ecosystems are higher compared to terrestrial soils [3].

Despite their importance, however, seagrass meadows have encountered extensive loss and degradation globally. It is suggested that approximately 30% of global seagrass area has been deteriorated, with an accelerating loss of less than 0.9% per year in the 1970s to more than 7% per year since 2000 [4]. This number is relatively higher compared to the global loss rate of other vegetated coastal areas in the same time frame, e.g. mangroves (1 – 2% per year) and salt marshes (1 – 3% per year).

Cockburn Sound in Western Australia is one of many sites where extensive seagrass loss and degradation has been recorded. Based on a mapping study by Kendrick, there has been a decline of more than 77% seagrass coverage between 1967 and 1999, which happened due to poor water quality [5]. Cockburn Sound has been used for various activities, including industrial, recreational, commercial and cultural heritage use, which resulted in changing water quality [6]. Since then, the industrial and agricultural activities in Cockburn have been restricted to improve the water quality [6]. However, recent field surveys have reported no significant expansion of seagrass meadows despite the improving water quality [7].

As the loss of seagrass meadows can be ecologically and socioeconomically significant, monitoring is essential. There have been several reports about seagrass health and water quality in Cockburn

Sound to date, such as [8] and [9]. However, there has been no recent studies which analyze the whole study area to look at the change in the spatial extent of seagrass habitat.

Remote sensing and Geographic Information Systems (GIS) is one of many methods that can be used to monitor ecological features, including seagrasses. Despite the absence of a consensus on a standard approach, remote sensing and GIS allow the study to be more holistic and cheaper due to the less field survey needed. It also enables the acquisition of long term data which generate highly accurate results, such as the mapping of seagrasses in Moreton Bay between 2004 and 2013 [10].

This study was conducted to evaluate the long-term change in seagrass spatial extent after the last assessment of seagrass coverage in 1999, using the combination of satellite and field data, and relate this to the potential seagrass carbon biomass production and overall carbon storage capacity of Cockburn Sound. Remote sensing and GIS methods were used in this study and the outcome is expected to illustrate the feasibility of concurrent methods in the identification of seagrass feature and the estimation of biomass.

2. Methodology

This study relies on satellite imagery as the main data. Seagrass feature was classified using *Nearmap* imagery. *Nearmap* is a high-resolution image of up to 0.075 m resolution with three bands of Red, Green, and Blue available from the year 2007 to present time. In addition to its high resolution, *Nearmap* imagery was used due to the easily accessible nature of the data. *Nearmap* images were collected in parts (scenes) to maintain the same resolution of 0.2 m. This study utilized the *Nearmap* images of year 2010, 2012, 2014, and 2016. The classification of previous years was taken from the results of a study by [5]. Meanwhile, the estimated carbon biomass were derived from *Landsat* images. In the case of *Landsat* image, additional image from the year 1994 and 1999 was used because the carbon biomass has not been discussed in the previous study [5]. Similarly, the older years were not assessed due to the unavailable data. The images' date acquisitions of both *Nearmap* and *Landsat* were varied from February to May in the same year depend on the quality and the availability of the images. These months were chosen as the occurrence of seagrass meadows are expected to be at their peak.

Field data of carbon biomass was used as the base of the statistical equation to derive the estimated above ground biomass. The field data was taken from a field survey, where seagrass samples from 12 sites were examined to acquire the shoot density, root biomass, rhizome biomass, leaf biomass, epiphytes, and productivity [7]. This study used the leaf and epiphytes biomass, as appear in table 1, to generate the above ground biomass of the study area. They were chosen as they represent the above ground biomass more than the root and rhizome biomass which are situated mostly below the ground.

Table 1. Above ground biomass of seagrass across sites in Cockburn Sound (Fraser & Kendrick, 2016).

Site	Leaf biomass ($g\ m^{-2}$)	Epiphytes biomass ($g\ m^{-2}$)	Mean ($g\ m^{-2}$)
GI 2 m	241.3	39.4	140.35
GI 3 m	88.9	16.6	52.75
GI 5 m	188	26.7	107.35
GI 7 m	104.7	22.6	63.65
Jervois Bay	94.2	37.5	65.85
Kwinana	133.9	46.4	90.15
Owen Anchorage	162.5	35.0	98.75
Woodman Point	154.5	37.2	95.85
WS 2 m	58	31.1	44.55
WS 3 m	24.7	16.7	20.7
WS 5 m	105.8	49.3	77.55
WS 7 m	89.9	28.3	59.1

IOP Conf. Series: Earth and Environmental Science **149** (2018) 012016 doi:10.1088/1755-1315/149/1/012016

2.1. *Seagrass classification*

Even though eight species (including *A. antarctica, A. griffithii, P. australis, P. coriacea, P. sinuosa, H. ovalis, H. tasmanica and S. isoetifolium*) could be identified in Cockburn Sound, it is difficult to differentiate each species in this study due to the necessity of examining the rhizome and structure of the seagrasses. Therefore, this study only classified the images into seagrass and non-seagrass class.

The identification of seagrass feature in this study was made using supervised classification method. The Fisher and MAXLIKE classifier in TerrSet software were utilized with the help of training sites consisting of seagrass and non-seagrass classes which were created manually. Fisher classifier was developed from a linear discriminant analysis similar to multivariate linear regression, where the equations maximize the variance between classes and minimize the variance within classes [11]. Meanwhile, the MAXLIKE classifier is based on the Bayesian probability theory, where it uses the mean and variance/covariance data of the signature to estimate the posterior probability that a pixel belongs to each class [11]. The maps of seagrass meadows each year would be presented as a result.

2.2. *Biomass calculation*

In this study, the carbon is represented by the above ground biomass. The approach to determine the seagrass' above ground biomass was adopted from a study which integrate the reflectance index with the in-situ biomass data [12].

The blue band, green band and red band of the *Landsat* images were used in this study. They would later be compared to find the best result with the closest resemblance to the in-situ data. The reflectance index of each band was extracted with reference to the sample sites and combined with the field data of above ground biomass. Software R was then used to calculate the linear regression, resulting in the equation of each band. Each equation was then applied accordingly to the images in ArcGIS to acquire the biomass in the study area. The maps of biomass would be presented as a result.

2.3. *Accuracy analysis*

In order to verify the results of seagrass classification, the images from Google Earth were used in place of the ground-truth survey due to time and budget constraints. Based on another study, Google Earth can derive land cover information with high accuracy [13], thus using the information to verify the results might be acceptable. 100 random points were created and converted into kmz format using ArcGIS software. The random points were imported to Google Earth to be used as the observation points. Seagrass feature was examined by its presence or absence in both the *Nearmap* classification and Google Earth each year. The results were exported again to ArcGIS. Percent accuracy was calculated by comparing the number of seagrass and non-seagrass classification in the results of *Nearmap* (predicted) and Google Earth (observed) with the help of two-way tables. Also, Kappa analysis was used to measure the difference between the observed and predicted maps, following the formula below. Where K is Kappa coefficient; Observed is the total number of correct observed sample; Expected is the total number of correct expected sample.

$$K = \frac{Observed - expected}{1 - expected}$$

The means difference between the results of biomass estimation and the in-situ biomass was also assessed using the pair-t-test in R software.

3. Results and Discussion

3.1. *Results*

3.1.1. *Seagrass extent*

Overall, the results show that the seagrass meadows in Cockburn Sound have been increasing from 688 Ha in 1994 to 918 Ha in 2016. The most significant increase occurred in the southern part, where more than 30% new seagrass area was formed (figure 1). Although not as significant as the southern part, the western and eastern part had dynamic changes, where it decreased until 1999, increased in 2014 and decreased again in 2016 (table 2, figure 1).

In 2010 and 2012, only the southern part data was available. The results show that there was only 1% growth appeared between the two years. However, compared to the year 1999, there was a 15% increased of seagrass coverage in the southern part.

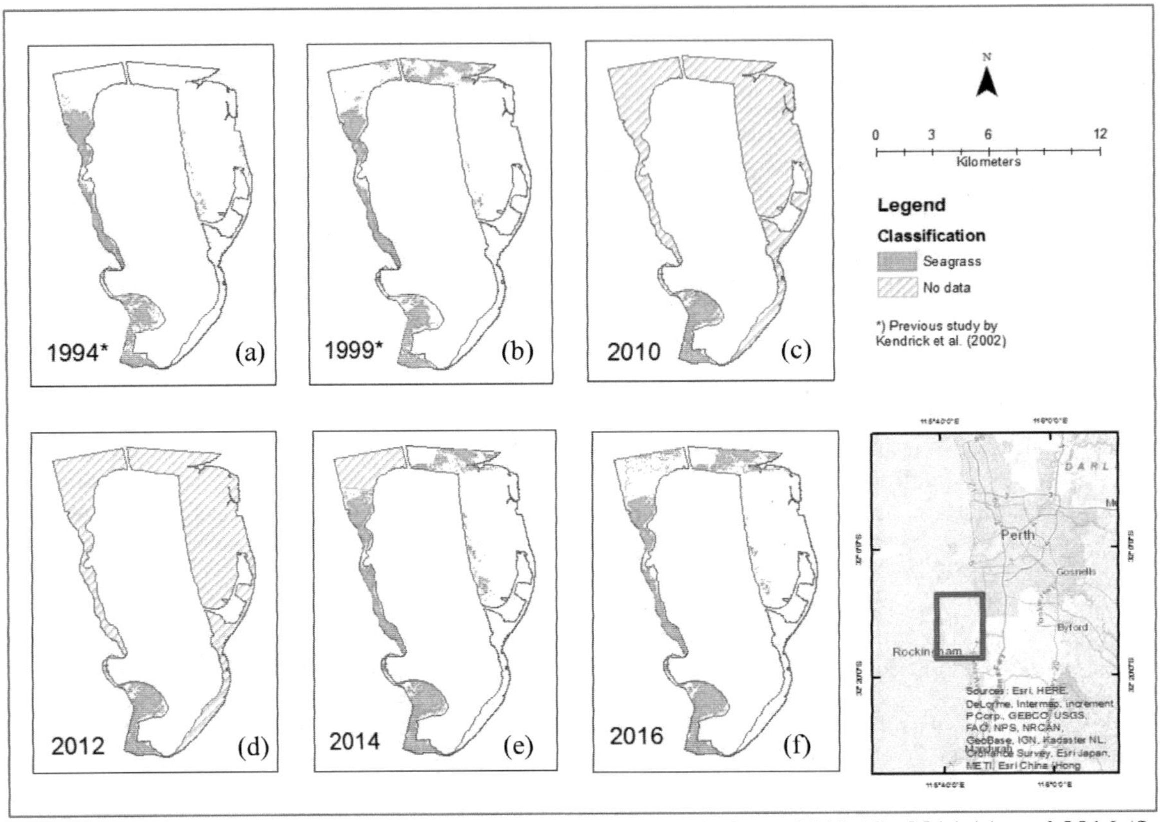

Figure 1. Map of seagrass extent in year 1994 (a), 1999 (b), 2010 (c), 2012 (d), 2014 (e), and 2016 (f) in Cockburn Sound.

Table 2. Change in seagrass coverage for the year 1994, 1999, 2010, 2012, 2014, and 2016.

Year	Area of Seagrass (*Ha*)				Percentage of seagrass area		
	West	East	South	Total	West	East	South
1994*	374.32	25.74	288.21	688.27	29.97	0.80	51.21
1999*	381.96	13.43	291.57	686.97	30.58	0.42	51.81
2010	No data	No data	376.79	376.79	No data	No data	66.96
2012	No data	No data	378.12	378.12	No data	No data	67.19
2014	421.41	42.05	395.35	858.81	33.74	1.31	70.26
2016	394.14	24.05	500.39	918.59	31.56	0.75	88.92

**Kendrick, et al (2002)*

A total of 858.8 Ha seagrasses was formed in 2014, with the most occurrence of 421 Ha in the western part, followed by the southern part with 395 Ha and the eastern part with 42 Ha. Compared to the previous year assessed (the year 2012), there was an increase of approximately 3% in the southern part. Meanwhile, compared to the year 1999, there was a 0.7% increase of seagrass meadows in the eastern part, as well as a 3% increase in the western part.

In the year 2016, seagrasses occupied a total of 918.6 Ha of sandbanks in Cockburn Sound, which is the greatest amount compare to the other years. In 2016, the southern part area had the most coverage of seagrass with 88.92% coverage. Meanwhile, the eastern part had the least coverage of only 0.7%. Between 2014 and 2016, seagrass cover increased by approximately 60 Ha. The most significant growth can be seen in the southern part, where it rose almost 19%, from previously 70% coverage in the year 2014 to 88.9% in the year 2016.

The accuracy assessment derived a 93% accuracy for correctly classifying seagrass cover. Meanwhile, Kappa coefficient of 0.4 obtained from the calculation of the results shows a fair agreement between the observed and expected classification.

3.1.2. Above ground biomass

The estimated biomass was obtained from the calculation using reflectance value of the image and the in-situ biomass. The reflectance value ranges from 0 to 1. The reflectance value of the sample sites was collected and used to calculate linear regression. The linear regression of blue, green, and red band resulted in three equations with different coefficients, as appears in table 3. The equation was then applied to the entire image with reflectance value. Based on the results, the blue band has the least gap of maximum and minimum estimated biomass, which is represented by the p-value. Also, the blue band has better water penetration ability, which could result in a better representation of seagrass biomass. Therefore, the equation derived from the blue band was chosen for the other years, despite the fact that it has the highest RMSE that may result in a significant difference between the estimated and in-situ biomass.

Table 3. The results of linear regression of each band.

Band	R	RMSE	P-value	Intercept	Coefficient	Equations
Blue	-0.099	61.97	0.94	114.44	232.41	114.44 + 232.41 BRI
Green	-0.047	60.48	0.49	93.17	1460.29	93.17 + 1460.29 BRI
Red	0.015	58.67	0.31	28.21	6328.47	28.21 + 6328.47 BRI

The results indicate a fluctuated carbon biomass between 1994 and 2016 (figure 2). Most of the highest biomass can be found in the Western and Eastern regions of Cockburn Sound. Meanwhile, the southern part in most years has the lowest biomass. The mean has decreased from 150 g/m^2 in the year 1994 to 145 g/m^2 in 1999 (table 4). However, the maximum biomass increased from 182 g/m^2 to 219 g/m^2, while the minimum value decreased from 145 g/m^2 to 140 g/m^2. The mean of biomass decreased in the year 2010, as well as the maximum and minimum value. It then increased again in 2012, where the mean was 154 g/m^2, the maximum value was 175 g/m^2, and the minimum value was 140 g/m^2. In 2014, the mean and minimum value decreased again, while the maximum value increased. Finally, all of them reduced and in the lowest state in 2016. The highest biomass identified was 219.3 g/m^2 in the year 1999, while the lowest was 118 g/m^2 in the year 2016. Meanwhile, the highest mean was identified in 2012, with 154.39 g/m^2. However, it has to be noted that the data available for seagrass classification in 2012 as well as 2010 was only covering the southern part. Therefore, it may not well-represent the whole study area.

The accuracy of the results was obtained from the comparison between the estimated and in-situ biomass presented in figure 3. Only year 2016 was assessed due to the data availability. The comparison shows a significant difference between the estimated and in-situ biomass. While the in-situ biomass ranges from below 100 g/m^2 to more than 200 g/m^2, the estimated biomass only has a range of 118 g/m^2 to 167 g/m^2.

The paired t-test shows that the mean of difference between the estimated and in-situ biomass are -32.81. With 95% confidence, the interval of the two is -61.12 to -4.49. Referring to the first step of the calculation, these numbers are related to the RMSE of the linear regression. While most of the estimated biomass have a difference of less than 32.81 g/m^2 with the in-situ biomass, some of them also have up to 62 g/m^2 difference.

The linear distribution of in-situ and estimated biomass shows a negative relationship. It is supported by the results of Pearson's correlation test between the two that show a correlation of -0.166 with a p-value of 0.72 and 95% confidence interval of -0.81 to 0.67.

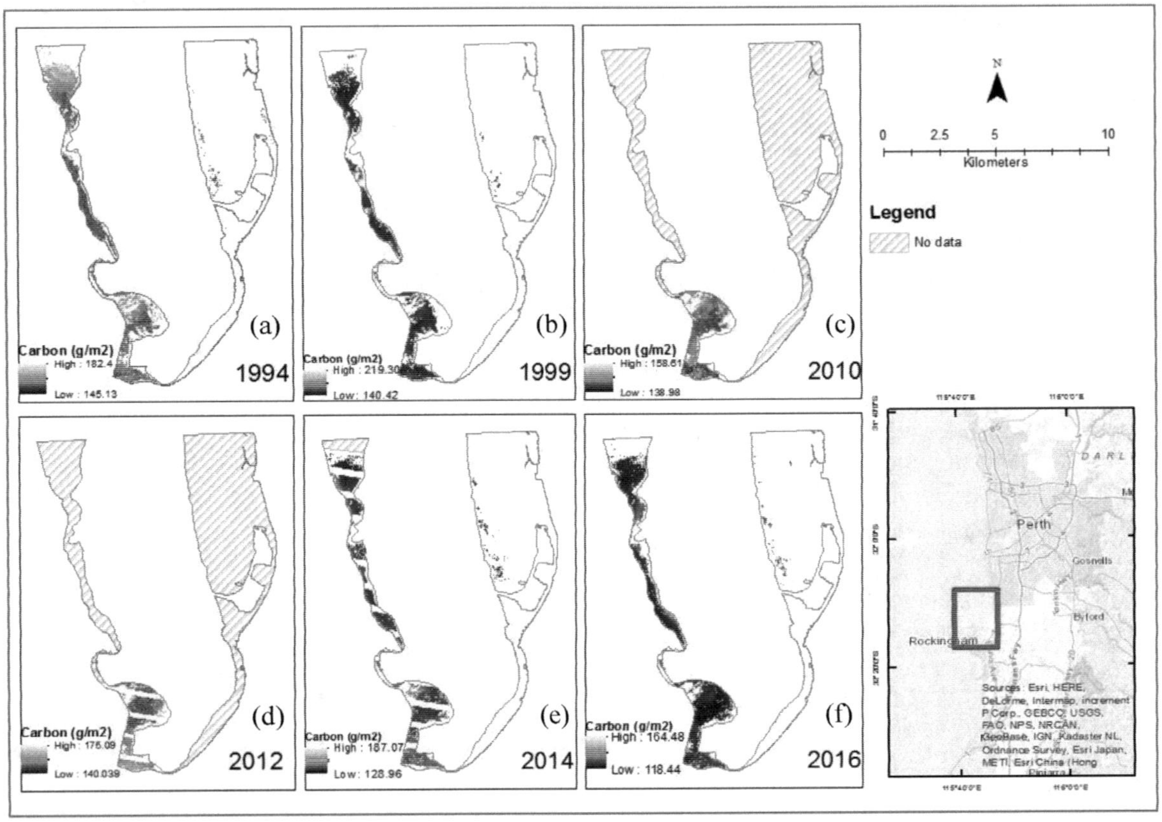

Figure 2. Map of biomass of year 1994 (a), 1999 (b), 2010 (c), 2012 (d), 2014 (e), and 2016 (f) in Cockburn Sound.

Table 4. Detail results of above ground biomass calculation

Year	Carbon biomass (g m^{-2})		
	Mean	Maximum	Minimum
1994	150.53	182.40	145.13
1999	145.12	219.30	140.42
2010	142.68	158.61	138.98
2012	154.39	175.09	140.04
2014	128.25	187.07	128.96

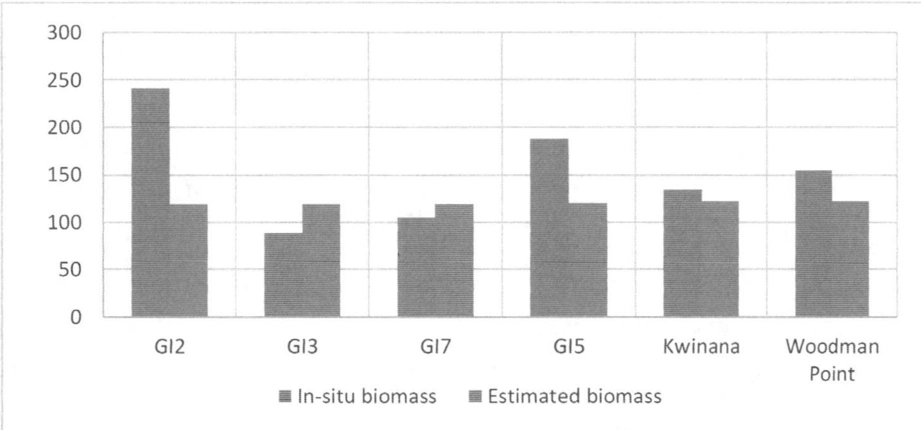

Figure 3. Graphic of comparisons between in-situ and estimated biomass in 2016.

3.2. Discussion

This study has identified a general trend of increasing seagrass coverage from 1994 to 2016 in Cockburn Sound, which was found mostly in the southern part. The fact that most of the industries are located in the eastern part indicates the growth rate of seagrasses can be influenced by the distance from the pollution sources. It is implied by the results that show the least significant increase of seagrass meadows in the eastern part, where the sources are concentrated, compared to the southern and western part. This result is relevant to a study which suggests that the close distance between aquaculture ponds and the shore in NE Hainan has reduced the light penetration and affected the seagrass' health [14]. It is also supported by another study that identified the contribution of *P. oceanica* in the evaluation of metal contaminations in Italy, where the rate of contamination was lower in the area further from the sources [15]. Based on previous assessments, the loss of seagrasses in the Southern Cockburn Sound was majorly due to the altered sediments impacted by the construction of causeways [16]. Thus the completion of the construction might be the main key in slowing down and even regenerate new seagrasses. On the other side, the seagrass meadows in the Western and Eastern Cockburn Sound have been depleted more because of the urchin overgrazing, shallower water, and nutrient enrichment [16], which would need more active management to improve the condition.

Aside from maintaining the water quality, several attempts of restorations, which have been concluded not practical, were expected to encourage seagrasses regeneration. A replantation was conducted in 2000, resulting in a survival of 40% plant units and confirmed that the Eastern and Western Cockburn Sound were the most suitable sites for rehabilitation using plug transplantation technique as the most efficient method [17]. Other transplant trials were conducted between 2004 and 2008 in the southern part of Cockburn Sound, which suggested a successful restoration, where a healthy meadow have grown and filled the gaps [18]. While it is possible for the successful replantation to keep expanding to date, which may explain the high growth rate in the southern part was beyond the influence of water quality, there has been no monitoring site in the southern part. Hence, no data recorded of the water quality which can provide a more detailed explanation of the reasons why the southern part has faster growth in comparison with the eastern and western part. However, similar to the previous restoration projects, the replantation was conducted on a small scale due to the expensive cost, intensive labor, and the possibility of severe impacts to the donor sites [18]. Furthermore, the re-colonization of *P. sinuosa* as the dominant species in Cockburn Sound is slow, due to the needs of modifications of C, N, and P content which may not fulfil the demand even one year after the transplantation [19]. While it has been more than 10 years since the replantation in the eastern and western part, the results of this study reflect an insignificant growth in the two areas. Thus, even though the replantation might have helped the Southern Cockburn Sound, it is not likely to be conducted on a large scale in the eastern and western part where the loss occupied a larger area due to

IOP Conf. Series: Earth and Environmental Science **149** (2018) 012016 doi:10.1088/1755-1315/149/1/012016

the inefficiency. Further assessment is necessary to indicate the reasons why the restoration might have worked more efficiently in the southern part and to identify the factors which interfere with the growth rate in the eastern and western part.

Meanwhile, the long-term assessment of seagrass' above ground biomass shows a decreasing pattern between 1994 and 2016, which may be associated with the declining density shoot. As mentioned previously, the density shoot in Cockburn Sound has decreased despite the increasing extent of seagrass meadows estimated in this study. A spatial expansion of seagrass meadows does not necessarily lead to a higher seagrass biomass, as biomass is a reflection of the investment of its resources [20]. Thus it is possible to get contradicting results of the seagrass extent and biomass. On the other hand, it is suggested that the carbon accumulation rates are higher in the older meadows [21]. The restored seagrasses had almost the same rate of carbon accumulation as the naturally growing seagrasses [21]. Accordingly, the formation of new seagrasses with lower biomass in Cockburn Sound might have affected the average biomass.

The changes of nutrient in Cockburn Sound might also affect the decreasing biomass. Contrary to the conventional paradigm regarding the excessive nutrient's detriment to the health of seagrass meadows, some studies, such as a reported positive correlation between seagrass biomass and nutrient availability [22]. The water movement was suggested to affect the seagrass biomass of tropical area in supplying the nutrients [20], which was in line with the findings for the temperate seagrasses. Similarly, seagrasses in Cockburn Sound also occupy a sheltered environment which preserves the meadows from the effect of storms [9]. The swell waves mainly come from the south - southwest in the summer and from the west – southwest in the winter [9]. While the water movement in Cockburn Sound is considerably low, occasional climatic events such as storm may influence the nutrient distribution, leading to a changing biomass despite the improving water quality.

This study also suggests the feasibility of using satellite data to properly assess the extent and the above ground biomass of seagrasses. While it has a high resolution with a relatively low cost compared to the other high-resolution images, *Nearmap* is rarely used to classify environmental features. It might be because it consists of only three bands (RGB), whereas many studies require more bands to identify particular features especially marine organisms like seagrasses. Therefore, this study could potentially be the first one using *Nearmap* image to determine the extent of seagrass meadows. However, the accuracy is still uncertain, given this study did not undertake a ground-truth survey. Although an accuracy percentage has been calculated with reference to expert annotation and Google Earth and resulted in an accuracy of more than 90%, field surveys are still necessary to obtain a more robust accuracy because the random points used in the assessment could appear in a stable area with not many changes. Thus they do not represent the accuracy of the more dynamics area. On the other side, *Landsat* images have been used in many studies to quantify the biomass of seagrasses which generate high accuracy results concurrent to the field verifications. While most of those studies assessed the seagrass biomass of recent years, it is difficult to conduct a long-term assessment like this study due to the limited in-situ data from the past. Thus, the results would be hard to be verified.

While remote sensing is an efficient method to assess an extensive area, the availability of the high-quality images also plays a major role in the results. The assessment of a particular event which may differ during a certain time of the year has to pay more attention to the date acquisition. The data, however, is not always proper for use on the preferred date. Hence, the results may not depict the condition very well. A thorough understanding of the mapping approaches is necessary to obtain a robust mapping result of inter-annual variation in seagrass distribution, as well as the impacts of episodic disturbance events [23].

4. Conclusion

This study has suggested the feasibility of using remote sensing in assessing seagrass extent. While the accuracy analysis showed the results have a high accuracy, field survey is still necessary. Nonetheless, the implications of this study are relevant for rehabilitation of seagrasses in Cockburn Sound. The seagrass extent has been expanding from the year 2010 to 2016, especially in the southern part, although still far from the initial state. Therefore, more focus should be given to the western and eastern part with the least significant growth, in addition to observing the potential factors that have supported the growth in the southern part. While water quality has been the focus of most restoration projects, its relation with the increasing seagrass extent is still uncertain. Additionally, given the low growth rate amidst the improved water quality, the pressures of seagrass' growth in Cockburn Sound remains unclear. Therefore, focusing on a more physical feature observation and restoration, such as transplantation, may be more effective despite it being inefficient. The absence of sample sites in the southern part has also made it difficult to compare the different growth rate spatially. Thus, adding a monitoring site in the southern part would enable the evaluation of the pressures in a diverse growing state.

5. References

[1] Intergovernmental Panel on Climate Change 2013 Coastal wetlands *2013 Supplement to the 2006 IPCC guidelines for National Greenhouse Gas Inventoried* eds D Alongi, A Karim, H Kennedy, G Chen, G Chmura, S Crooks et al. (Switzerland: IPCC) Chapter 1 pp 1-17

[2] Duarte C M, Middelburg J J and Caraco N 2005 Major role of marine vegetation on the oceanic carbon cycle *Biogeosciences* **2** pp 1–8

[3] McLeod E, Chmura G L, Bouillon S, Salm R, Bjork M, Duarte C M, Lovelock C E, Schlesinger W H, Silliman B R 2011 A blueprint for blue carbon: toward an improved understanding of the role of vegetated coastal habitats in sequestering CO2 *Frontiers in Ecology and the Environment* **9** pp 552-560

[4] Collier C J and Waycott M 2014 Temperature extremes reduce seagrass growth and induce mortality *Marine Pollution Bulletin* **83** 483-490

[5] Kendrick G, Aylward M J, Hegge B J, Cambridge M L, Hillman K, Wyllie A, Lord D A 2002 Changes in seagrass coverage in Cockburn Sound, Western Australia between 1967 and 1999 *Aquatic Botany* **73** pp 75-87

[6] Cockburn Sound Management Council 2005 *Environmental Management Plan for Cockburn Sound and its Catchment* (Perth, Australia: Department of Environment) pp 1-91

[7] Fraser M W, Kendrick G A, Perez A Z 2016 *Drivers of Seagrass Decline in Cockburn and Warnbro Sound* (Western Australia: Department of Environment Regulation) pp 1-35

[8] Keesing J K, Greenwood J, Donn M J, McFarlane D J 2016 *Spatial and Temporal Analysis of Water Quality Monitoring Data Collected from Cockburn Sound and Warnbro Sound between 1982/83 and 2013/14* (Report to the Cockburn Sound Management Council and the Western Australian Department of Water) (Australia: CSIRO)

[9] Mohring M and Rule M 2013 *Long-term Trends in the Condition of Seagrass Meadows in Cockburn and Warnbro Sounds* (Technical report to the Cockburn Sound Management Council)

[10] Roelfsema C M, Lyons M, Kovacs E M, Maxwell P, Saunders M I, Samper-Villarreal J, and Phinn S R 2014 Multi-temporal mapping of seagrass cover, species and biomass: A semi automated object based image analysis approach *Remote Sensing of Environment* **150** pp 172–187

[11] Eastman J R 2016 *TerrSet: Geospatial Monitoring and Modelling System* (The United States of America: Clark Labs)

[12] Misbari S and Hashim M 2016 Change detection of submerged seagrass biomass in shallow coastal water *Remote Sensing* **8** pp 1-29

[13] Tilahun A and Teferie B 2015 Accuracy assessment of land use land cover classification using Google Earth *American Journal of Environmental Protection* **4** pp 193-198

[14] Herbeck L S, Sollich M, Unger D, Holmer M and Jennerjahn T C 2014 Impact of pond aquaculture effluents on seagrass performance in NE Hainan, Tropical China *Marine Pollution Bulletin* **85** pp 190-203

[15] Lafabrie C, Pergent G and Pergent-Martini C 2009 Utilization of the seagrass Posidonia oceanica to evaluate the spatial dispersion of metal contamination *Science of the Total Environment* **407** pp 2440-2446
[16] Cambridge M L, Chiffings A W, Brittan C, Moore L and McComb A J 1986 The loss of seagrass in Cockburn Sound, Western Australia II: Possible causes of seagrass decline *Aquatic Botany* **24** pp 269-285
[17] Paling E I, van Keulen M and Tunbridge D J 2007 Seagrass transplanting in Cockburn Sound, Western Australia: A comparison of manual transplantation methodology using Posidonia sinuosa Cambridge et kuo *Restoration Ecology* **15** pp 240-249
[18] Sinclair E A, Verduin J, Krauss S L, Hardinge J, Anthony J and Kendrick G A 2013 A genetic assessment of a successful seagrass meadow (Posidonia australis) restoration trial *Ecological Management & Restoration* **14** pp 68-71
[19] Gobert S, Cambridge M L, Velimirov B, Pergent G, Lepoint G, Bouquegneau J-M, Dauby P, Pergent-Martini C, Walker D I 2006 *Biology of Posidonia* eds. Larkum T, Orth R J and Duarte C M *Seagrasses: Biology, Ecology and Conservation* (The Netherland: Springer) pp 387-408
[20] Rasheed M A and Unsworthy R K 2011 Long-term climate-associated dynamics of a tropical seagrass meadow: Implications for the future *Marine Ecology Progress Series* **422** 93-103
[21] Greiner J T, McGlathery K J, Gunnell J and McKee B A 2013 Seagrass restoration enhances "Blue Carbon" sequestration in coastal waters *PLoS ONE.* **8** pp 1-8 e72469
[22] Heck Jr K L, Pennock J R, Valentine J F, Coen L D and Sklenar S A 2000 Effects of nutrient enrichment and small predator density on seagrass ecosystems: An experimental assessment *Limnology Oceanography* **45** pp 1041-1057
[23] Roelfsema C M, Kovacs E M, Saunders M I, Phinn S and Lyons M 2013 Challenges of remote sensing for quantifying changes in large complex seagrass environment *Estuarine, Coastal and Shelf Science* **133** pp 161-171

GIS-based approach for quantifying landscape connectivity of Javan Hawk-Eagle habitat

C Nurfatimah[1], Syartinilia[1], Y A Mulyani[2]
[1]Landscape Architecture Department, Faculty of Agriculture,Bogor Agricultural University (IPB), Jl. Meranti, Kampus IPB, Darmaga - Bogor 16680, Indonesia.
[2]Department of Forest Resource Conservation and Ecotourism, Faculty of Forestry, Bogor Agricultural University (IPB), Jl. Ulin, Kampus IPB, Darmaga – Bogor 16680, Indonesia.

Email : cicinurfatimah@gmail.com

Abstract. Javan Hawk-Eagle (Nisaetus bartelsi; JHE) is a law-protected endemic raptor which currently faced the decreased in number and size of habitat patches that will lead to patch isolation and species extinction. This study assessed the degree of connectivity between remnant habitat patches in central part of Java by utilizing Conefor Sensinode software as an additional tool for ArcGIS. The connectivity index was determined by three fractions which are infra, flux and connector. Using connectivity indices successfully identified 4 patches as core habitat, 9 patches as stepping-stone habitat and 6 patches as isolated habitat were derived from those connectivity indices. Those patches then being validated with land cover map derived from Landsat 8 of August 2014. 36% of core habitat covered by natural forest, meanwhile stepping stone habitat has 55% natural forest and isolated habitat covered by 59% natural forest. Isolated patches were caused by zero connectivity (PCcon = 0) and the patch size which too small to support viable JHE population. Yet, the condition of natural forest and the surrounding matrix landscape in isolated patches actually support the habitat need. Thus, it is very important to conduct the right conservation management system based on the condition of each patches.

1. Introduction

Javan Hawk-Eagle is an endemic raptor species from Java Island [1]. The habitat preferences used by juvenile and immature differ from mature Javan Hawk-Eagle. The adult eagle selects the evergreen forest and secondary forest. Juvenile eagles tend to scatter out of their home territory and appear in less suitable areas as their habitat, whereas the immature eagles tend to choose habitats similar to adult eagles as a way to build new breeding territories [2]. These requirements are currently experiencing a major problem that threatens the sustainability of this species that is uncontrolled environmental damage. Currently only 675.000 hectares natural forest remains by 2013 in Java Island, a decrease by 32% from 2009. Moreover the remaining forest only cover 5% area compared to non-natural forest area [3]. This resulted in the predicted Javan Hawk-Eagle habitat being in a location far away from each other [4].

Connectivity plays an important role in the management of conservation of natural resources and wildlife [5]. Landscape connectivity has an understanding as the degree of ability of a landscape in facilitating the movement or dispersal of species between habitat patches [6,7]. While ecological connectivity can be interpreted as a connection or relationship that occurs between different landscape elements, ranging from energy or information, nutrient cycles, the spread of pollen to the movement of flora and fauna and metapopulation [8]. Connectivity is dependent on the structure of the landscape and the species ability to disperse [9]. The main components of connectivity comprise the spatially-oriented components or functional features and functional components that are species behavior and habitat preferences [10].

The movement of species is the main thing in population dynamics so connectivity becomes decisive in conservation [11]. Habitats that are not connected will become isolated patches and increase the potential for inbreeding. Interconnected habitats with high degree of connectivity are able to avoid the negative impacts of fragmentation such as habitat isolation, diminishing patch size, to decreased genetic quality due to inbreeding [12-15]. Connectivity also allows an individual to use some habitat fragments even though they are small in size so that they do not have to rely on certain large fragments and patches that benefit not only a group of species but also other species [16].

Previous study on Javan Hawk-Eagle have been focused on determining the breeding biology, habitat use, home range and predicted population density on specific study site [1, 2, 4, 5, 24]. There has been no attempt to see the connectedness between dispersed habitats. Therefore, in this study we evaluated the presence or absence of a link between fragmented habitats. The aims of this study were to: (1) analyze the available connectivity between patches, (2) evaluate the contribution of each patch to the habitat network, and (3) identify the conservation priority that need to be conducted based on the characteristic of each patch.

2. Method

2.1. Study Area

Central part of Java Island of about ±37300 km^2 consists of Central Java Province and Special Region of Yogyakarta Province. It has three national park which are Merbabu National Park and Merapi National Park on the mainland. And also Karimun Jawa National Park as small archipelago located at Java Sea. Central part of Java has the least suitable habitat for Javan Hawk-Eagle compared to western part and eastern part of Java [4]. This region also suffered by environment degradation due to the loss of 80% of natural dryland forest which result in severely fragmented habitat [17]. Habitat suitability model based on 2014 Landsat 8 OLI/TIRS in 250 m resolution concluded that only 19 suitable patches remain for Javan Hawk-Eagle [18]. Those patches located far from each other scattered all over central part of Java.

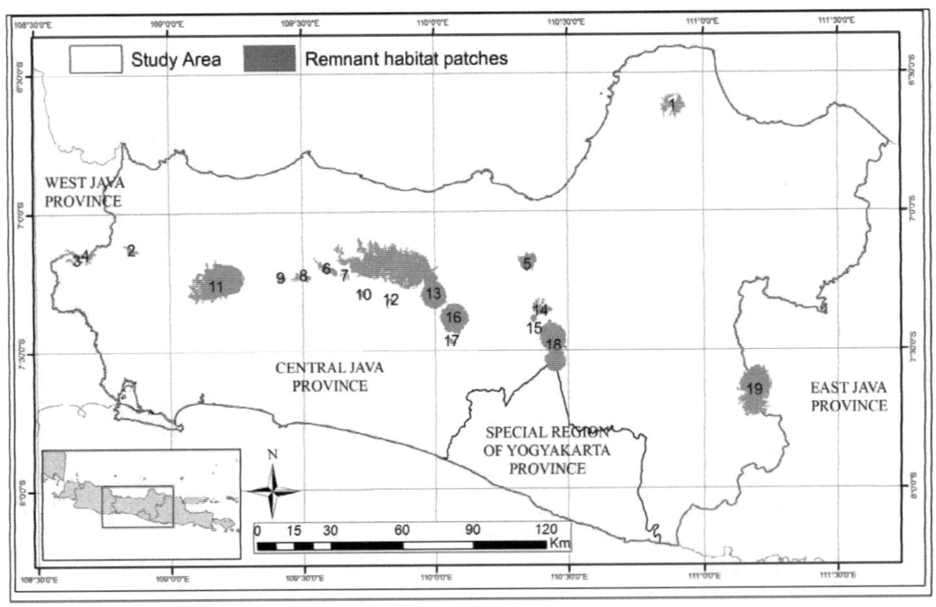

Figure 1. Study site and location of Javan Hawk-Eagle remnant habitat patches.

IOP Conf. Series: Earth and Environmental Science **149** (2018) 012017 doi:10.1088/1755-1315/149/1/012017

2.2. Analysis Tools

To perform connectivity analysis, this study used software called Conefor Sensinode 2.6 [5]. This software can be downloaded for free at http://www.conefor.org. Inside the software we use Probability of Connectivity (PC) index which is the probability or probability of two animals randomly placed on a landscape capable of reaching or reaching each habitat (interconnected) in a collection of habitat n patches that have a linkage level in it [1]. The index formula (1) is as follows:

$$PC = \frac{\sum_{i=1}^{n}\sum_{j=1}^{n} a_i a_j p_{ij}^{*}}{A_L^2} \quad (1)$$

Where a_i and a_j are the areas in the patch habitats i and j, the A_L is the total area of the study area landscape consisting of habitat and non-habitat patches. The result of the probability of a patch or path (where a path is made of a set of steps that does not pass the same patch twice) is the result of a pij at each step in the path. While $p*_{ij}$ is defined as the maximum probability outcome of all paths appearing between patches i and j (including single-step paths). If patch i and j are close enough then the maximum path probability result is just a step (direct movement between patch i and j). If the distance between patch i and j farther, then the maximum probability path will consist of several steps through stepping stone enclosing $p_{ij}^{*} > p_{ij}$.

The information required as input in this software is the node data contains a list of exisiting habitat nodes on the landscape along with its attributes. The habitat node data present in this method uses the first-stage patch analysis results. Attributes on the node are characteristics of the node itself considered to be related to the analysis to be used such as habitat area, habitat quality, and other attributes. Connection data contains the information needed to characterize connections between two nodes in the landscape. The data included in this study is Euclidean distance between each patch. The output of this stage is a patch function determined by three fractions ie *d*PCintra, *d*PCflux, and *d*PCconnector [20]. This fraction shows the contribution of each patch to the connectivity and overall landscape. *d*PCintra shows the contribution of the patch to the intrapatch connectivity or the patch weight itself if it connected to other patches. So although isolated the patch still has the weight. *d*PCflux shows the patch capability level in its role as the initial patch or final patch in the species dispersion process. *d*PCconnector shows the contribution of the patch as a connecting element or stepping stone between patch habitats.

Habitat availability and distance between patches were required to produce the degree of connectedness between patches. Only a patch of habitat remnant at a resolution of 250 m is used in the connectivity analysis. At this stage an edge-to-edge Euclidean distance method is used to obtain distance information between patches. Use of Euclidean distance assuming that birds can fly freely and move between habitat patches without taking into account the obstacles posed by different types of land cover. The dispersal distance used uses the median assumption of a home range of 2523 m from a maximum home range of 2000 ha [4]. The result of this stage is then validated with a land cover map to find any land cover characteristics on each patch contribution to the landscape habitat.

3. Result

3.1. Connectivity of Javan Hawk-Eagle remnant habitat patches in central part of Java

Nineteen patches generated from habitat suitability modeling can be ranked by the degree of importance of each patch in the overall habitat management (Figure 2). Patch 13 becomes the most important patch in the Javan Hawk-Eagle habitat in central Java (*d*PC = 72.2%, *d*PCflux = 21.5%, *d*PCintra = 49.9%), followed by patch 16 (*d*PC = 19.9%, *d*PCflux = 17.3%, *d*PCintra = 2.2 %), patch 11 (*d*PC = 11%, *d*PCflux = 0.24%, *d*PCintra = 10.8%), patch 19 (*d*PC = 7.1%, *d*PCflux = 0.001%, *d*PCintra = 7.1%) and patch 18 (dPC = 6%, *d*PCflux = 1%, *d*PCintra = 4.9%). These five patches are the largest (*d*A> 10%) patch of all Javan Hawk-Eagle remnant habitat patches in central Java and can be the core habitat. The next order of the degree importance are patch 6, patch 7, patch 14, patch 8, patch 17, patch 1, patch 9, patch 5, patch 10, patch 12, patch 15, patch 4, patch 3 and the lowest rank patch is patch 2.

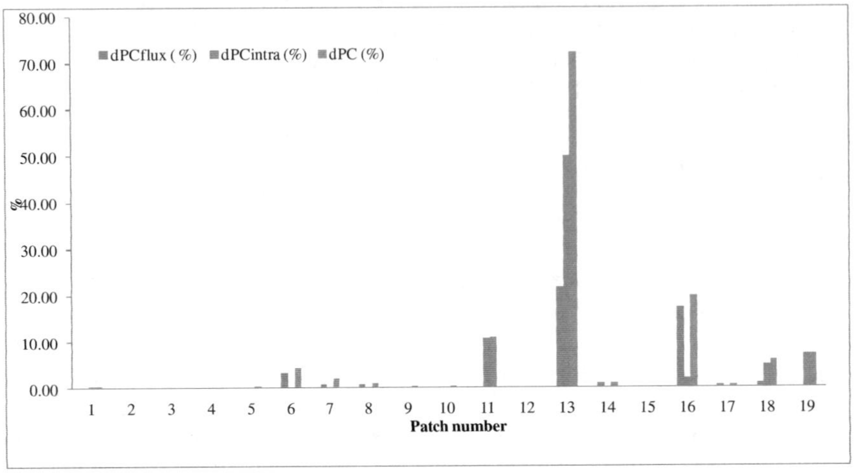

Figure 2. Graph of *dPC*flux, *dPC*intra and *d*PC index value.

Among the 19 patches there are 6 patches isolated from other patches because they have *d*PCconnector = 0 (Figure 3). These patches include patch 19, patch 17, patch 1, patch 10, patch 12, patch 3. Although the 19 patch is isolated but has an importance value on the 4th order of all patches. This indicates that there is a need for new habitat patches between patch 19 and the nearest patch which serves as a stepping stone or ecological corridor to help the Javan Hawk-Eagles move from patch 19 to other patches.

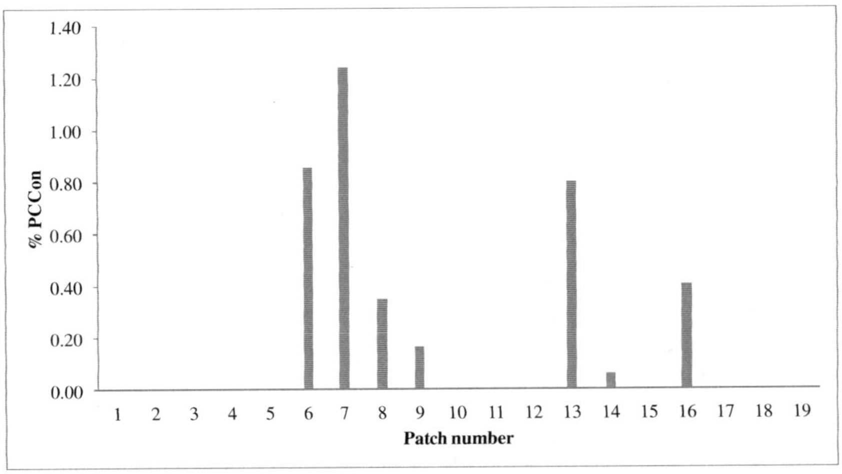

Figure 3. Graph of *d*PCconnector index value.

By combining the three connectivity fraction we able to determined the patches categories. Core patch defined by medium to high *d*PCflux, high *d*PCintra and low to high *d*PCconnector. Stepping stone patch defined by medium to high *d*PCflux, medium *d*PCintra and medium *d*PCcon. Meanwhile isolated patch defined by medium to high *d*PCflux, medium *d*PCintra and low *d*PCcon. This categorization was based on the fraction ranking of importance and overall connectivity. The distribution of patches category shown in Figure 4. The location of all patch defined by Table 1.

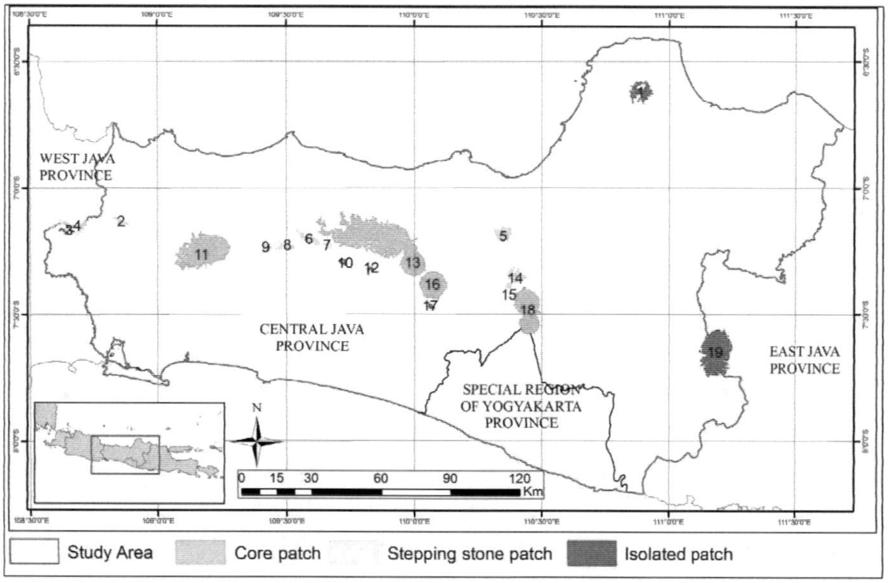

Figure 4. Patch categorization of Javan Hawk-Eagle remnant habitat patches.

Table 1. List of patch and its category.

Patch number	Location	Patch size (km^2)	Patch Category
Patch 1	Mt. Muria	46.36	Isolated
Patch 2	Mt. Sagara	9.00	Stepping stone h
Patch 3	Mt. Palasari	6.61	Isolated patch
Patch 4	Mt. Subang	23.8	Stepping stone
Patch 5	Mt. Ungaran	37.16	Stepping stone h
Patch 6	Mt. Imus, Mt. Beser, Mt. Langit, Mt. Igirtipis, Mt. Bongkok,	22.25	Stepping stone
Patch 7	Mt. Pesangon	5.19	Stepping stone
Patch 8	Mt. Cupu, Mt. Kraton, Mt. Jaha	13.25	Stepping stone
Patch 9	Mt. Beser, Mt. Butak, Mt. Malang, Mt. Tumbang, Mt. Bopong	6.13	Stepping stone
Patch 10	Mt. Pogong	5.09	Isolated
Patch 11	Mt. Slamet	238.2	Core
Patch 12	Mt. Pandan	4.50	Isolated
Patch 13	Mt. Lumping, Mt. Lebakbarang, Mt. Kemulan, Mt. Sindoro	511.2	Core
Patch 14	Mt. Telomoyo	26.3	Stepping stone
Patch 15	Mt. Andong	5.64	Stepping stone
Patch 16	Mt. Sumbing	107.6	Core
Patch 17	Southern hill of Mt. Sumbing	4.43	Isolated
Patch 18	Mt. Merapi dan Mt. Merbabu	160.7	Core
Patch 19	Mt. Lawu	193.8	Isolated

Patch source : [10]

3.2. Core patch

Each patch in the core category has different land cover types. The largest land cover on patch 11 is 68.93% (30.6 km^2) of forest cover. While for shrub, dry farmland and built areas are 13.45% (32.04 km^2), 8.14% (19.40 km2) and 7.94% (18.92 km^2) respectively. Patch 18 has 34% (52.67 km2) forest cover and 33% (51.26 km^2) of field cover. Patch 18 is a patch located in the area of Mount Merapi, Mount Merbabu and its surroundings. As many as 57.1% (91.71 km^2) of patch 18 (160.7 km^2) belonging to the area of Mount Merapi National Park and Mount Merbabu National Park. Patch 13 and patch 16 each have dry farmland as the most dominant cover. Patch 13 (Dieng Plateau and its vicinity) has a farmland cover of 56% and only 18% of forest land cover. In patch 16 (Mount Sumbing) dry farmland cover is 40% and forest cover is only 24%.

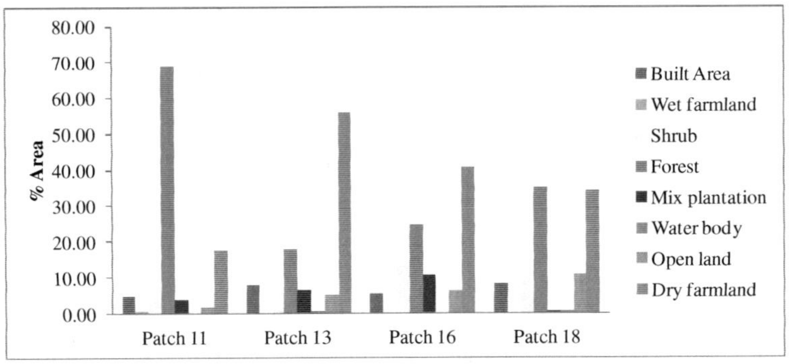

Figure 5. Landcover classification of core patch.

3.3. Stepping stone patch

The land cover on the stepping stone patch is dominated by forest cover (Figure 6). Patch 4 has a forest cover of 76.59% (18.22 km2), patch 5 has forest cover 56.05% (20.84 km2), patch 6 has forest cover 50.49% (11.23 km2), patch 7 has forest cover 55.79% (2.89 km2), patch 8 has a forest cover of 81.76% (10.85 km2), patch 9 has 79.08% forest cover (4.84 km2) and patch 15 has forest cover 32.45% (1.83 km2). Patch 2 which has a shrub cover as a dominant land cover of 40.22% (3.62 km2). Patch 14 has a dominant mixture of mixed plantation of 35.53% (9.34 km2).

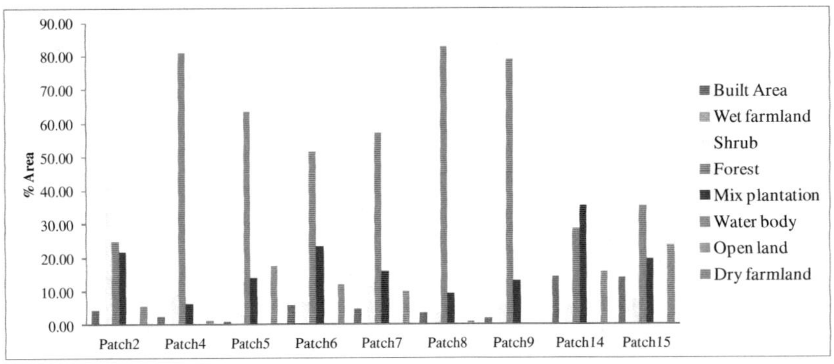

Figure 6. Landcover classification of stepping stone patch.

3.4. Isolated patch

Land cover on isolated patches is as shown in Figure 7. Patch 1 has a forest cover of 66.12% (30.6 km2), patch 3 has 67.12% forest cover (4.43 km2), patch 10 has forest cover 54.22% (2.75 km2) , patch 12 has forest cover 38.44% (1.73 km2), patch 17 has forest cover 73.81% (3.27 km2) and patch 19 has forest cover 37.42% (75.84 km2). Other landscape mosaics found in isolated patches include mixed plantation, shrubs, dry farmland, and open land.

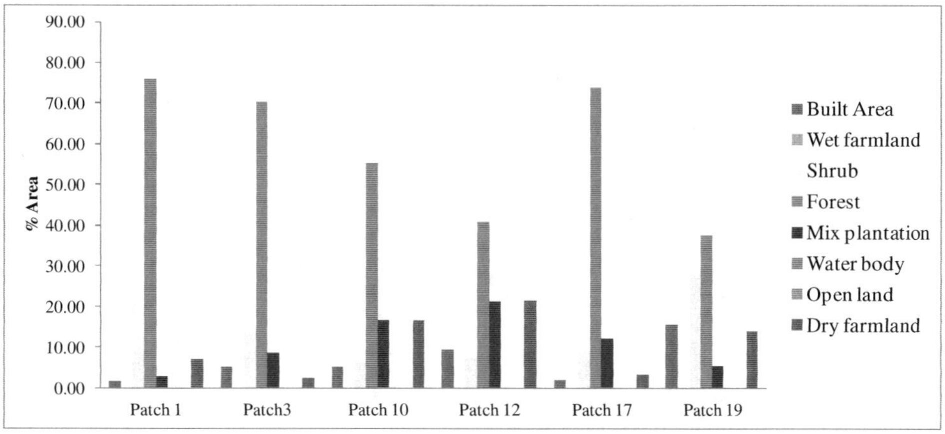

Figure 7. Landcover classification of isolated patch.

4. Discussion

Based on the results of the connectivity analysis, the Javan Hawk-Eagle remnant habitat patches in central part of Java is categorized into 4 core patches, 9 patch stepping stones and 6 isolated patches. One of the goals set by Indonesian Ministry of Environment and Forestry in the eagle conservation action plan is to increase in the population of these rare species. In relation to population increase, habitat function generated by degree of connectedness can be adapted to Metapopulation theory. Metapopulation can be interpreted as a genetic and recolonisation flow system carried out by an individual when the rate of extinction is high [21].

4.1. Core patch

As a patch which supporting the availability of habitats for breeding and increasing the population, core patches must be of high quality [22]. The core patch is expected to have a broad and patch quality that is capable of supporting significant numbers of Javan Hawk-Eagle populations to help increase the dispersal current toward surrounding patches. Using the median home range of Javan Hawk-Eagle (n = 12 km^2), it is estimated that the number of Javan Hawk-Eagles that can be supported by the core patch is 84 individuals or 42 pairs.

For raptors the success of nesting is limited to the availability of nesting sites [23]. The Javan Hawk-Eagle is a raptor that relies heavily on natural forests to build its nest [24]. With the wide size and different land use, not all patches have forest cover dominated by forests as habitat need for nesting.

Patch 11 and patch 18 have large availability of forests and adjacent landscape mosaics in the vicinity of agricultural areas that can serve as secondary habitats [24]. Yet, this condition is different from that occurred in patches 13 and 16. Based on interviews with sources from Raptor Indonesia, patch area 13 and 16 now has never been found again sightings of Javan Hawk-Eagle. This is consistent with the fact that the availability of natural forests is narrow and isolated from each other by expansion of horticultural and tobacco farming activities. The destruction of these forests has not only affected the habitat of the Javan Hawk eagle but also the environmental conditions. In patch 13, patch 16 and surrounding areas there is a decrease in water quality due to excessive use of pharmaceuticals and fertilizers for agriculture, severe droughts in the dry season and very high soil erosion [25]. Natural forest destruction on patch 13 and patch 16 began in the late 1980s and since 1997 the damage has been sharply driven by the expansion of tobacco farming into upland areas [26].

To avoid the isolation of the remaining natural forest patches, landscape restoration should be required on patch 13 landscape and 16 patches. Landscape restoration will produce sustainable and healthy mosaic landscapes, manage plant and animal diversity, protect ecosystems and community ecology, maintain genetic quality the best species, improved water quality, to reduce the level of soil erosion [27]. As a species indicator of the stability and sustainability of the ecosystem, the availability of a landscape structure that supports the habitat for Javan Hawks simultaneously supports the sustainability of other animal habitats.

Public awareness to maintain the sustainability of the Javan Hawk-Eagle in the core patch area is equally important. Javan Hawk-Eagle became one of the rare exotic animals whose level of illegal trade is very high. In the period of January to December 2015 there are 127 individuals of Javan Hawk-Eagles offered through online pages [28]. In addition to strong law enforcement, proper conservation education in the wider community can help to educate people not to hunt and maintain the Javan Hawk-Eagles.

4.2. Stepping stone patch

The main factor in the metapopulation system is the movement of species that is affected by quality, quantity and time [29]. The genetic flow rate is strongly influenced by the distance between patches so patches that are too far apart must be connected by corridor or stepping stone [30, 31]. Stepping stone can be defined as small vegetation patches scattered throughout the landscape [32].

In the remnant patches of Java Hawk-Eagle habitat there are 9 patches that serve as stepping stone with various width sizes. The smallest stepping stone patch is patch 7 cover 5.19 km^2 areas and the biggest patch is patch 14 measuring 26.30 km^2. The smallest stepping stone patch size meets the minimum width of the Javan Hawk-Eagle home range. So that the size of the stepping stone patch on Javan Hawk-Eagle remnant habitat in central Java meet the minimum size for conservation [33].

The Javan Hawk-Eagle is a species dependent on forests for nesting and protection. But the Javan Hawk-Eagle also requires the existence of landscapes outside the forest area to survive. Young Javan Hawk-Eagle is known to be able to scatter out of its original territory to other landscapes such as agricultural land, rice fields, tea plantations, fragmented small forests, up to stands of production trees [2]. Then by adulthood they are still looking for forests to be used as new territory breeding territory. The forest cover owned by each patch on the stepping stone patch can be a new shelter and nesting site. Farming areas and mixed gardens can be used as prey hunting sites. However, to be able to support the site as a new nesting place, the wide area becomes a factor that must be considered to avoid edge-effect and competition between species. One of the stepping stone patch that is estimated to become a nesting site is patch 5 (Mount Ungaran). Based on observations by Raptor Indonesia in 2016 there were sightings of some young eagle individuals in the patch 5. This indicates that as a stepping stone, patch 5 has the habitat quality and width which suitable for building breeding territory. Thus the diversity of landscape mosaics in the stepping stone patch is sufficiently appropriate to assist the dispersal of juvenile Javan Hawk-Eagle to find new mates and nesting places.

4.3. Isolated patch

There are 6 patches which not connected to other patches: patch 1, patch 3, patch 10, patch 12, patch 17 and patch 19. The metapopulation system will not work if the patch is isolated because there is no genetic flow of incoming movement. The cause of a patch being isolated is the distance between patches that goes beyond species dispersal capability and loss of patches around the isolated patch [17] [18]. Patch 1 (Mount Muria) and patch 19 (Lawu Mount) is a patch that has a great distance from the nearest patch. The closest patch of patch 1 is patch 5 ± 79 km distance. The closest patch of patch 19 is patch 18 ± 73 km. The distance exceeds the maximum range of Javan Hawk that is 20 km^2. In between patch 1-patch 5 and patch 18-Patch19 there is no stepping stone patch as dispersal facility. Patch 3, patch 10, patch 12 and patch 17 are detected as isolated patches in PC analysis because each patch has a small patch size (mean = 5.16 km^2). This means the patch is too small and unable to facilitate movement to many different patches or dispersal paths are disconnected.

The availability of large enough forest relative to the area of each patch means that the landscape on the patches is actually suitable for Javan Hawk-Eagle habitat. Especially on patch 19 as one of the patches with high importance in remnant habitat of Javan Hawk-Eagle in central Java. Patch 19 is administratively located on the border of Central Java Province and East Java Province. Ecologically patch 19 plays an important role as a hub with habitat in eastern Java. Patch 3 also has the same functionality as patch 19. Patch 3 becomes the central Java liaison with western Java. However patch 19 and patch 3 require further connectivity review because patch 19 may not necessarily be isolated if it has connectivity with one patch in eastern Java. Patch 1, patch 10, patch 12 and patch 17 can serve as a stepping stone toward the core patch. To connect the patches and minimize the impact of patch isolation it is necessary to construct an ecological corridor [29].

5. Conclusion

Through this study we show how appropriate software utilization can assist in wildlife habitat conservation activities. The use of connectivity index may indicate the connectedness of a fragmented habitat network. The index values of the fractions contained in the connectivity are able to help in determining the relationship of each patch with its ability as a facilitator of population dispersal. By applying this method into the remaining habitat patches of Javan Hawk-Eagle we able to specified the role or contribution of each patches in the habitat network which are core patch, stepping-stone patch and isolated patch.

This approach is very useful in determining appropriate conservation measures by adjusting to the conditions of each patch. Landscape restoration is necessary to recover and improve the quality of heavily degraded habitat at core patches. At stepping stone patches, maintaining the diversity of landscape mosaic is important to support its function in assisting the population dispersal of Javan Hawk-Eagle in a very disperse habitat. Meanwhile, to avoid isolation effect, ecological corridors are needed to build connectivity between patches. By utilizing this approach, managers and policy makers can decide which conservation priorities should be applied.

Acknowledgments

The authors would like to thank the Ministry of Research, Technology and Higher Education of Republic Indonesia under "Penelitian Unggulan Perguruan Tinggi" grant in fiscal year of 2016. Also thanks to Raptor Indonesia especially Asman A Purwanto for providing us the local information and guiding us during field survey.

References

[1] Nijman V and Sözer R 1998 Field identification of Javan Hawk-Eagle Spizaetus Bartelsi *Forktail* 13-16

[2] Nijman V and van Balen S 2003 Wandering star: age related habitat use and dispersal of Javan Hawk-Eagle (Spizaetus bartelsi) *Journal Ornithology* **144** 451-458

[3] Purba C P P, Nanggara S G, Ratriyono M, Apriani I, Rosalina L, Sari N A and Meridian A B 2014 The state of the forest Indonesia period of 2009-2013 *Forest Watch Indonesia*

[4] Syartinilia, Tsuyuki S and Lee J S 2009 GIS-based habitat model of Javan Hawk-Eagle (Spizaetus bartelsi) using inductive approach in Java Island, Indonesia Wildlife: Destruction, *Conservation and Biodiversity* ed J D Harris and P L Brown New York: Nova Science Publisher, Inc 301-312

[5] Saura S and Pascual-Hortal L 2007 A new habitat availability index to integrate connectivity in landscape conservation planning : Comparison with existing indices and application to a case study *Landscape and Urban Planning* **83** 91-103

[6] Taylor P D, Fahrig L, Henein K and Merriam G 1993 Connectivity is a vital element of landscape structure *OIKOS* **68** 571-573

[7] Clergau P and Burel F 1997 The role of spatio-temporal patch connectivity at the landscape level: an example in a bird distribution *Landscape and Urban Planning* **38** 37-43

[8] Marulli J and Mallarach J M 2005 A GIS methodology for assessing ecological connectivity: application to the Barcelona Metropolitan Area *Landscape and Urban Planning* **71** 243–262

[9] Xun B, Yu D and Liu Y 2014 Habitat connectivity analysis for conservation implications in an urban area *Acta Ecologica Sinica* **34** 44-52

[10] Boitani L, Falcucci A, Maiorano L and Rondinini C 2007 Ecological network as conceptual frameworks or operational tools in conservation *Conservation Biology* **21** 1414-1422

[11] Nikolakaki P 2004 A GIS site-selection process for habitat creation:estimating connectivity of habitat *patches Landscape and Urban Planning* **68** 77-94

[12] Keller L F and Waller M D 2002 Inbreeding effects in wild populations *TRENDS in Ecology dan Evolution* **17** 230-241

[13] Minor E S and Urban D L 2008 A graph-theory framework for evaluating landscape connectivity and conservation planning *Conservation Biology* **22** 297-307

[14] Coulon A, Cosson J F, Angibault, Cargnelluti, Galan M, Morellet N and Hewison J M 2004 Landscape connectivity influences gene flow in a roe deer population inhabiting a fragmented landscape: an individual–based approach *Molecular Ecology* 2841-2850

[15] Baranyi G, Saura S, Podani J and Jordan F 2011 Contribution of habitat *patch*es to network connectivity: Redundancy and uniqueness of topological indices *Ecological Indicators* **11** 1301-1310

[16] Martensen A C, Pimentel R G and Metzger J P. 2008. Relative effects of fragment size and connectivity on bird community in the Atlantic Rain Forest: Implications for conservation *Biological Conservation* **141** 2184– 2192

[17] Gunawan H, Prasetyo L B, Mardiastuti A and Kartono A P 2010 Fragmentasi hutan alam lahan kering di Provinsi Jawa Tengah *Jurnal Penelitian Hutan dan Konservasi Alam* **7** 75-91

[18] Nurfatimah C, Syartinilia and Mulyani Y A 2017 Potential habitat of Javan Hawk-Eagle based on multi-scale approach and its implication for conservation *IOP Conference Series: Earth and Environmental Science* **54** 012064

[19] Saura S and Torne J 2012 Conefor 2.6 (Manual Madrid: Universidad Politecnica de Madrid)

[20] Saura S and Rubio L 2010 A common currency for the different ways in which patches and links can contribute to habitat availability and connectivity in the landscape Ecography **33** 523-537

[21] Farina A 2006 *Principles and Methods in Landscape Ecology: Toward a Science of Landscape* Netherland: Springer 87-108

[22] Semlitsch R D and Jensen J B 2001 Core habitat, not buffer zone *National Wetlands Newsletter* **23** 5-7

[23] Newton I 1979 *Population Ecology of Raptor* London: T & AD Poyser Ltd

[24] van Balen S and Nijman V, Prins H H 2000 The Javan Hawk-Eagle: misconception about rareness and threat. *Biological Conservation* **96** 297-304

[25] Ngabekti S, Setyowati D L and Sugiyanto R 2007 Tingkat kerusakan lingkungan di dataran tinggi Dieng sebagai database guna upaya konservasi. *Jurnal Manusia dan Lingkungan* **14** 93-102

[26] Lavigne F and Gunnel Y 2006 Land cover change and abrupt environmental impacts on Javan volcanoes, Indonesia: a long-term persepctive on recent events *Regional Environmental Change* **6** 86-100

[27] Harker D, Libby G, Harker K, Evans S and Evans M 2000 *Landscape Restoration Handbook Second Edition* New York (US): Lewis Publisher

[28] Gunawan, Zulham, Pramono H, Djamaludin, Yuniar A, Hardina K, Mulyati S, Kuswandono and Kristiana I 2017 Release of confiscated raptors in Indonesia by Suaka Elang (Raptor Sanctuary): protocols and progress to date *BirdingASIA* **27** 88-93

[29] Hilty J A, Lidicker Jr W Z and Merelender A M 2006 *Corridor Ecology The Science and Practice of Linking Landscape for Biodiversity Conservation* (Washington: Island Press)

[30] Dramstad W E, Olson J D and Forman T T 1996 *Landscape Ecology Principles in Landscape Architecture and Land-use Planning* (Washington DC: Island Press)

[31] Uezu A, Metzger J P and Vielliard J M E 2005 Effects of structural and functional connectivity and *patch* size on the abundance of seven Atlantic Forest bird species *Biological Conservation* **123** 507-519

[32] Lindenmayer D B and Fischer J 2006 *Habitat Fragmentation and Landscape Change: An Ecological and Conservation Synthesis* (Washington: Island Press)

[33] Saura S, Bodin Ö and Fortin M 2014 Stepping stones are crucial for species' long-distance dispersal and range expansion through habitat networks *Journal of Applied Ecology* **51** 171-182

IOP Conf. Series: Earth and Environmental Science **149** (2018) 012018 doi:10.1088/1755-1315/149/1/012018

Carrying capacity of Peucang Island for ecotourism management in Ujung Kulon National Park

K H Wiyono[1,3], E K S H Muntasib[1,3], and F Yulianda[2,3]
[1] Forest Resources Conservation and Ecotourism Department, Faculty of Forestry
[2] Management of Marine Resource Department, Faculty of Fishery
[3] Bogor Agricultural University

E-mail:kukuhadii@gmail.com

Abstract. Peucang Island is one of island in Ujung Kulon National Park (UKNP), appointed as priority area and welcome area for tourism. This research aimed to calculate the carrying capacity of Peucang Island for ecotourism development (Study sites of this research are Karang Copong jungle trail and 8 sites of Peucangs beach). This research used observation method (wildlife exploration, measure the lenght of jungle track, and measure 10 parameters of beach), literature study and and interview method to collect data. The data of jungle track analyzed use Cifuentes's formula. The result showed that Karang Copong jungle trekking had 20,000 visitors/day for Physical Carrying Capacity (PCC), 4 838 visitors/day for Real Carrying Capacity (RCC), and 6 visitors/day for Efective Carrying Capacity (ECC). Observation of biological aspect showed that there were some damages of vegetation along the track, and the changes in animal behavior. The data of beach carrying capacity analyzed use Yulianda's formula that measured with the suitability map approach. Based on the suitability map, two beaches were classified in suitable category, while six beaches) were classified in highly suitable category for tourism activities. All of the beaches had different number of carrying capacity, specifically there are 70 visitors/day in highly suitable beach and 27 visitors/day in suitable beach. The number of visitor nowadays still not exceed from carrying capacity number of PCC, RCC of jungle trails and carrying capacity of the beach area, but the number has exceeded from the ECC numbers.

1. Introduction

Peucang Island is the one of island in Ujung Kulon National Park (UKNP) that has a managed accessibility, and facilities, and as welcome area for tourism visitor. Accessibility to go to Peucang Island reachable via sea travel (boat fast / slow boat) and takes about 2-6 hours of Carita or 3.5 hours from Labuan by slow boat, and 3 hours from Tamanjaya [1]. Physical and biological objects of natural resources in Peucang Island become a tourist attraction that can be utilized for ecotourism development in UKNP. Objects tourist attraction in Peucang include white sand and blue sea water for coastal tourism activities, and the attractions of wildlife such as wild boar, deer, and the long-tailed monkeys are often found around Karang Copong jungle trekking trails. The number of visitors increased since 2011 until 2015. Tourist activities in the protected area have a direct and indirect impact on the natural environment, as it relates to the growth in the number of visitors that can affect ecological integrity [2]. The increasing of visitor number should be considered to keep the natural sustainability in UKNP, then controlling and management of natural resources should be done effectively. Carrying capacity concept is the one of basic for sustainable concept to protect and develop area. Peucang Island as the one of priority area for ecotourism development in UKNP should be managed to keep the sustainability of natural resources which as tourism object are Karang Copong Jungle track and Peucang Beach area. This research aimed to calculate the carrying capacity of Peucang Island for ecotourism development. This study calculated the carrying capacity on Peucang

Island in study side are Karang Copong jungle trails and beach by using the physical aspect, biological, social and management.

2. Methods

2.1 Study Sites
Peucang Island is located in the northwest of the island of Java, in Sumur and Cimanggu District, Pandeglang with an area of ± 450 Ha. It takes 2-6 hours by slow or fast boat from Carita, 3.5 hours from Labuan, and 3 hours by slow boat from Tamanjaya to Peucang Island. This reseach was carried out at 9 study site, which are Karang Copong jungle trails and 8 beaches point around the Peucang Island.

2.2 Materials and Tools
This research used tourism map of Peucang Island, tally sheet, measuring tape, camera, GPS (Global Positioning System), sechhi disk, interview guidance and used ArcGIS 10.3 software to make widllife encounter map and suatibility map for beach areas.

2.3 Data Collection

2.3.1 Karang Copong Jungle Trails
We collected the data of physical, biological, social, management aspect. We used field observation to collect the data of biological aspect. We explored the diversity of flora and fauna around the track and 10 meters in the right and left side from the track. Flora observation, we listed the dominant species and the damage of flora around the track by tourism activities. Fauna observation, we listed the species, the quantity, activities, and time. It used the assumtion that we did observation when a track was empty and when tourism activities took a place in a track. We marked the point where we found all the animals with the GPS to make wildlife encounter map. Then physical aspect, the data that collected were the length and width of the track, land slope, dan type of soil by measured. The data used to count PCC and RCC. For social aspect, the data were amount of visitor number, group, visitors in a group. Management aspects, the data were amount of staff and management. The data of social and management aspect collected by interviewed to visitors and manager.

2.3.2 Peucang Beach.
We used biological and physical aspects as parameters to categorize the suitabilily of the beach areas for tourism development. Biological aspects, we used field observation to collect dangerous marine species which could be hazardous for beach tourism activities. Physical aspects, we also did field observation to determine the condition of some aspects such as the type of beach, wide beaches, water base material, the slope of the beach, the beach land cover, depth, current speed, wave height, water transparency and the availability of fresh water. All parameters has each score that according to matrix of land suitability for recreational beach tourism category [3]. The score used to count tourism suitability index (TSI), then we categorize the area by TSI score.

2.4 Data Analysis

2.4.1 Carrying capacity of Karang Copong Jungle track
It was analyzed by Cifuentes's formula [4]. There are three kind of carrying capacity number which are physical carrying capacity (PCC), Real Carrying Capacity (RCC) and Effective Carrying Capacity (ECC).

2.4.1.1 Physical Carrying Capacity (PCC)
PCC is defined as the maximum number of users that can physically fit into, or onto, a specific area.

$$PCC = A \ x \ \frac{V}{a} \ x \ Rf$$

A : Available area for public use (trail distance)
$\frac{U}{a}$: Area required per user to walk comfortably (1 visitor per m²)
Rf : Rotation factor (number of visits/day).

In order to measure PCC, the following assumptions were done:

- Horizontal space required to allow people to move freely on the trekking path that is 1 visitor / m^2
- Size available (A) is determined by the particular circumstances at the sites. Limitation of space for trekking path indicated by the size of tour groups and the required distance between the groups.
- Rotational Factor (Rf) is the daily number of visits allowed to a location

$$\text{Rf} = \frac{Open\ Period}{Avarage\ time\ of\ visit}$$

2.4.1.2 Real Carrying Capacity (RCC)

RCC is the maximum allowable number of users to the Karang Copong jungle trails, once the correction factors (*Cf*) derived from the particular characteristics of the site have been applied to the PCC. For the calculation of RCC, PCC was modified by a series of correction factors, such as precipitation (*Cf* 1), ground slope (*Cf* 2), the sensitivity of the soil (*Cf* 3) and social (*Cf* 4).

$$\text{RCC} \quad = \text{PCC}\ x \frac{100-Cf1}{100} x \frac{100-Cf2}{100} x \frac{100-Cfn}{100} x \frac{100-Cfn}{100}$$

To measure RCC, the following assumptions must be done:

- The correction factors are obtained by considering the environment, biophysical and social factors.
- These factors are closely linked to the specific conditions and characteristics of each site or activity.
- *Cf* is expressed as follows :

$$Cfn = \frac{Mn}{Mt}\ x\ 100\%$$

*M*n : limiting magnitude of variable
Mt : total magnitude of variable.

2.4.1.3 Effective Carrying Capacity (ECC)

ECC is the optimum number of tourists so that the area can hold based on consideration of the Manager. To find out the value of the effective support resource [4]. ECC calculations are as follows:

$$ECC = \frac{ICxMC}{RCC}\ x\ 100\%$$

ECC : Effective Carrying Capacity
MC : Management capacity
RCC : Real Carrying Capacity
IC : Infrastructure Capacity.

2.4.2 Suitability index and carrying capacity of beach

It was analyzed with tourism suitability index approach and after know where the beach that categorized as highly suitable and suitable, the carrying capacity would be counted.

2.4.2.1 Suitability of tourism analysis

The formula that is used for recreational beach tourism suitability Yulianda's formula [3] as follows:

$$\text{IKW} = \sum \left(\frac{Ni}{N\ max}\right)\ x\ 100\%$$

where TSI: Tourism Suitability Index, Ni : Parameter score, Nmax: Maximum score of tourism category. Nmax score for recreational beach tourism is 84. The suitability category which are highly suitable (75-100%), suitable (50-<75%), conditional suitable (25-<50%), and not suitable (<25%) [3].

We mapped beach areas that included as highly suitable and suitable category. Those area will be recomended to UKNP as recreational beach tourism area.

2.4.2.2 Carrying capacity of beach
The formula that is used to analysis is Yulianda's formula [3] as follows:

$$DDK = K\ x \frac{Lp}{Lt} x \frac{Wt}{Wp}$$

where:

CC : Carrying Capacity
K : Ecological potential of visitor per area unit (person)
Lp : Length of area for tourism activity (1 person/m^2)
Lt : Length of area for a person (meters)
Wt : Time period in one day (hours)
Wp : The time spent for each activities (hours).

Ecological potential visitors (K) is determined by the condition of ecological resources and the types of activities that will be developed. According to Yulianda [3], K score is 1 , and Lt is 50 meters. It means that 1 person need 50 meters of length of beach.

3. Results and Discussion

Peucang island has a very unique ecosystem with a wide diversity of plants and satwaliar so it was referred to as a miniature of tropical lowland rainforest in Ujung Kulon National Park. The object of attraction which are Peucang island that has a white sand beach, Karang Copong, kiara, wildlife diversity such as deer (*Cervus timorensis*), wild boar (*Sus scrofa*), long-tailed monkeys (*Macaca fascicularis*), and various types of bird, in addition there are coral reefs and marine life that became an object of attraction for snorkeling and diving activities in some spots [1]. Management based on carrying capacity is an effective management. According to Wearing and Neil [5], carrying capacity determine by depending on the season, time, some factors such as the visitors behaviours, facilities design and management, the dynamic characterisitics of the environment, and changes the behavior of the local community. There are three important aspects of carrying capacity i.e. biophysical, social, and cultural facilities related to visitor experience.

3.1 Karang copong jungle trails
There are two types of carrying capacity. At first we count number of carrying capacity based on physical, social and management aspects, and we also describe carrying capicity based on biological aspects.

3.1.1 Carrying capacity based on physical, social, and management aspects

3.1.1.1 Physical Carrying Capacity (PCC)
The calculation of PCC or physical carrying capacity aims to calculating the maximum number of visitors who can physically meet a predetermined space, at a certain time [4]. PCC value calculation results in Table 1 indicates that the line can accommodate as many as 20 000 visitors in one day.

Table 1. Physical carrying capacity.

The area of track (hectare)	Average time of visit (hours)	Open period (hours)	Area required per user (m^2)	Rf	PCC (Visitors/day)
3	1.5	10	1	6.67	2000

3.1.1.2 Real Carrying Capacity (RCC)

RCC used four correction factors in calculating the value consisting of the precipitation (CF1), land slope (CF2) and soil sensivity to erosion (CF3), and social (CF4). Correction also used by Ortega et al [6] in his research in calculating the value of the real carrying capacity of tourist Marietas Island in the Gulf of California that has been designated as National Parks using a correction factor, namely erodibility environment, accessibility, rainfall, floods, biological and vegetation. And Lucyanti [7] in her research used biophysical aspects are considered as a limiting factor of environmental support resources include: precipitation (Cf_1), kelerengan (Cf_2), erosivitas (Cf_3), vegetation (Cf4), and bird fauna (Cf5). The results of analysis of the four correction factor was found that each correction factor has a value which is then used in the calculation of the value of RCC. Precipitation correction factor (CF1) is values obtained by calculating the correction factor index values of precipitation in Peucang Island. Based on data from Meteorology and Geophysics Agency (BMKG), Pondok Betung Climatology Station found that the monthly precipitation in Pandeglang district is dominated by the wet months of precipitation value was over 100 mm, whereas in humid (60-100 mm rainfall value) most rare occurred during the year 2012 to 2016. The results of the precipitation correction factor (Cf1) in the calculation of the value of the RCC is 0.84.

Peucang topography is generally flat with a slope of 0-8% and some small ramps with a slope of 8-15% and undulating with a height of 0-17 m above sea level. Based on field observation results obtained reef slope trekking path Copong (CF2) tend to be flat. Results of the analysis conducted using ground slope index by decree. Minister of Agriculture No. 837/KPTS/UM/11/1980 obtained a flat slope is a first class number of land slope that has score 20. Subsequent analysis results is the score of land slope correction factor (Cf2), whereby a flat trekking has a score of 20, so that the obtained score of 0.80 (Cf2).

Soil sensitivity correction factor we used the data of soil types. Based on the type of soil maps Banten province that Peucang composed of alluvial soil types. Alluvial soil formation process depends on the origin of soil parent material factors and topographic factors [8]. By decree of Minister of Agriculture No. 837/KPTS/UM/11/1980, alluvial soil type is one type of land belonging to class 1 soil classification based on the level of sensitivity. Class 1 level is a type of soil which is not sensitive to erosion. The types of soil found along the path of trekking in the form of alluvial soil types that included the grade 1 soil type has a score is 20, and then the score of the Cf3 is 0.80.

Social correction factor (CF4), the results obtained from observations that the average number of people in a group that is 15-30 people. The number of visitors to each of the groups has been managed, 10 people for each the groups, and the distance of time for each of the groups is 30 minutes. Queiroz et. al. [8] explained that considerations of distance used each troupe is 250 meters with a total of 15 people in each group. Based on the data of visitors number (2016), the highest number of people visitors occured at August (999 visitors), and the lowest number of visitors occur in March (214 visitor) (figure 1). All the aspect of social correction factor counted, then correction factors social (Cf4) has a score is 0.55.

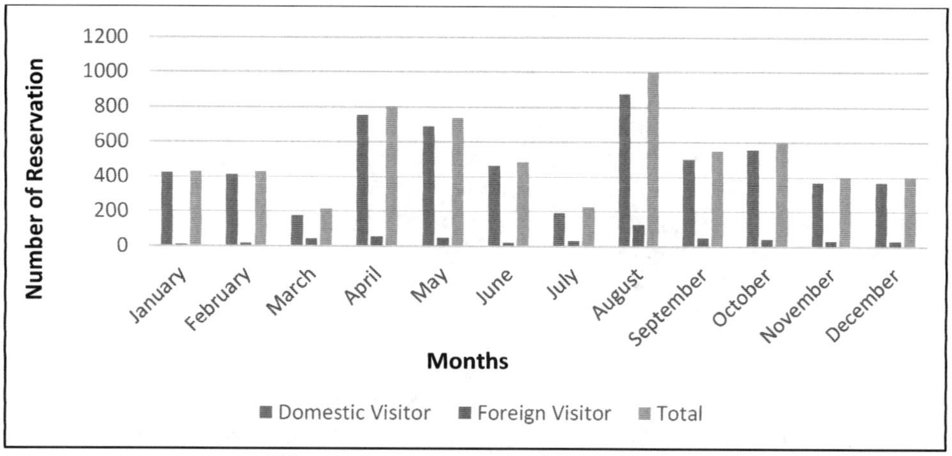

Figure 1. The number of reservation.

IOP Conf. Series: Earth and Environmental Science **149** (2018) 012018 doi:10.1088/1755-1315/149/1/012018

The results of RCC calculation is 4,838 people/day which is about 24.19% of the value of the PCC as shown in table 8. RCC will be lower than PCC because of a RCC value derived from the results of PCC has been reduced by the correction factors that influence as limiting factor [7].

Table 2. Real carrying capacity.

PCC (visitor/day)	Cf_1	Cf_2	Cf_3	Cf_4	RCC (visitor/day)
20 000	0.84	0.80	0.80	0.55	4 838

3.1.1.3 Effective Carrying Capacity(ECC)

ECC considered by availability of capacity management (Management Capacity/MC) and infrastructure capacity (IC/Infrastructure Capacity). Management capacity obtained by knowing the number of staff, and the number of staff needed for effective management. According to Lucyanti [7] increased capacity of the officer Manager in serving visitors need to be supported by capacity management based on variables such as the basic law, policies and regulations, equipment, personnel, financing, infrastructure and facilities. The number of human resources (HR) in Peucang Island resort are eight people assessed is still not adequate when the number of visits increased at certain times such as weekend time, and peak seasons. They need three more personnel assessed to improve the management of ecotourism, and it would be more effective. The results of the calculation of the MC (Management Capacity) that the value of 80%. The results of the calculation of the IC, namely by 20. The value of MC and IC who have obtained and then multiplied and divided by the value of the RCC 5,376, so it brings the power of effective support (ECC) for 6 as shown in table 3.

Table 3. Effective carrying capacity.

Open period (minutes)	Time to through the track (minutes)	Number of trail	The main manager	Bussines lisence holder	IC	ECC (visitors/day)
600	90	1	1	1	20	6

Sari [10] also researched about carrying capacity in mangrove forest in Sari Ringgung beach. Carrying capacity of mangrove trail in Saring Ringgung beach is 87 visitors/day, that is lower than PCC and RCC but more than ECC of Karang Copong jungle trail. Sari only use the biological aspect which are the species of bird and mangrove to count the carrying capacity because those were as the tourism attraction there. Total of Peucang Island visitor numbers at 2016 are 6 261 people. The avarage were 522 visitors per month or 17 visitors per day. That number visitor still lower than PCC and RCC, but has exceed ECC. The number of visitors in a protected area should be noted because of the growth in the number of visitors can affect to ecological integrity [2].

3.1.2 Carrying capacity based on biological aspects

3.1.2.1 Plant Conditions

The damage found in plants which are around line trails that were broken, plants saplings in addition plastic waste such as drinking bottles and glasses of mineral water as well as other items such as sandals were found on the trail. The damage caused by the behavior of visitors broke and trampling saplings that grow around the trail to avoid the slippery track and waterlogged. Wearing [5] describes the trampling of tourist activity on vegetation can cause bad impact for the vegetation such as damage and wounds on the network, the reduced power and vigor regeneration plants, loss of land as well as the cumulative impact of closure that can occur is a change in the composition of species where only species with high resistant power able to survive in areas that received pressure from tourist activities.

3.1.2.2 Wildlife condition

Wildlife observed on the time when the trails is assumed to be empty. There are ten species of wildlife found around the trail that mapped shown in figure 2. We marked the point where we found the animals with GPS, then mapped as wildlife encounter map used ArcGIS 10.3. The wildlife encounter map will help manager to develop ecotourism program which is animal watching program. The ecotourism should considere where is the area that the wildlife use for their activity, then the manager can set the facilities which is like canopy trail that will not be disturb the wildlife activity, and visitor also can watch animal behaviour in the wild.

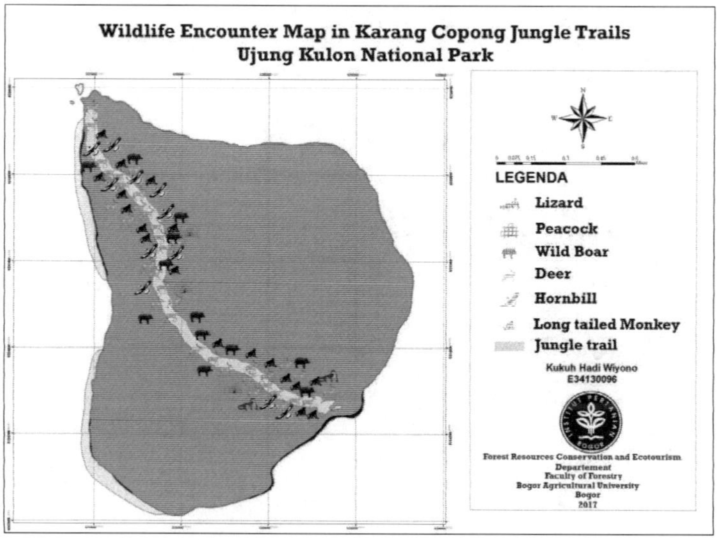

Figure 2. Wildlife encounter map.

Based on the results of observations made that reduction occurring type of wildlife that is found when an increase in the number of visitors on the track. A reduction in the number of types of wildlife can be caused by voice disturbance from the activities of visitors while on the trail. The observation result in table 4 can be explained that visitors will find species most numerous when the number of visitors as many as 1-10 people. So the number of visitors to this class is appropriate for visitors expected the diversity of species of wildlife around line trail as the main attraction.

Table 4. Species that are found on the class number of visitors.

The number of visitors	The number of wildlife species	Species
1—10	6	Deer, wild boar, peacock, lizard, hornbill, long-tailed monkey
11—20	3	Deer, wild boar, long-tailed monkey
21—30	2	Deer, long-tailed
31—40	1	Deer

There are some bad impact to wildlife. Wildlife behavior change has occurred as feeding behavior on wild boar, deer and long-tailed monkeys. Based on the results of observations made that look a few monkeys feeding around visitors hostel, deer are also seen eating junk, and the wild boar which ate the amenities. Rabbany [11] explained that the activity of visitors can cause changes in patterns of feeding behavior and eating habits directly and indirectly. Wildlife's diet can be changed directly due to the activity of feeding by visitors, and indirectly because of littering that effect wildlife to searching for food at dump. The long-tailed monkeys is now more aggressive. The long-tailed monkeys become more daring to take the food directly from visitors. According to Djuwantoko [12] that long-tail monkeys will be increasingly unattractive to become tourism attraction, if visitors are increasingly alarmed by the aggressive behavior of monkeys.

3.2 The suitability index and carrying capacity of beach

The beach carrying capacity on Peucang Island was analyzed by tourism suitability index (TSI). Calculation of the TSI can describes the category of the suitability of each location, and mapped into spatial map of suitability. It show the spatial locations of the beach on the island of Peucang that could developed for tourism activities. Then we count the carrying capacity at eight study site shown in figure 3.

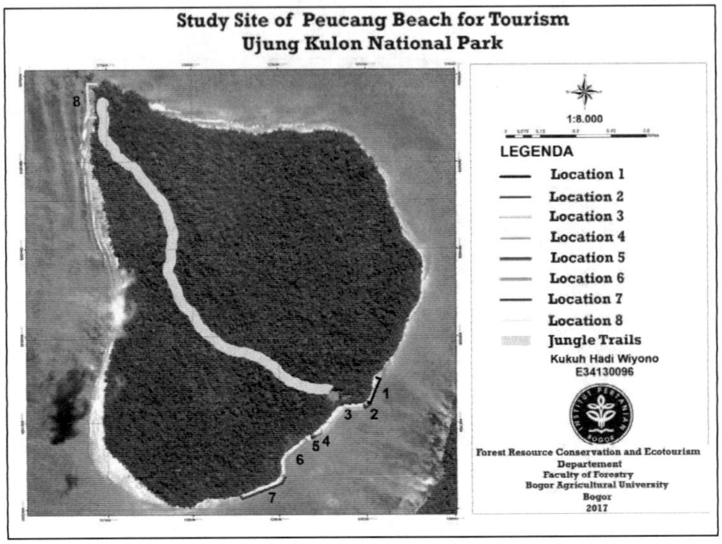

Figure 3. The study site for beach.

3.2.1 The suitability of resources for beach tourism

The results of tourism based on the suitability assessment index that used 10 parameters (biological and physical aspects) is 8 coastal location on the island of Peucang IKW has value for coastal recreational activities. It will be different if we count the TSI for marine tourism like Ketjulan [13] researched in Hari Island. He used 6 parameter, which are the brightness of the waters, cover of the reef community, types of life forms, types of reef fish, current speed, and depth of coral reefs [13]. Based on the classification of suitability with TSI that those are 2 locations that include suitable category and 6 categories include location higly suitable for beach recreation activities (table 5). Location 3 is the location that has the highest level of suitability than the seven other locations i.e. amounting to 94%. While the location 8 be the location that has the lowest value of the TSI. Parameters of became the dominant parameter of location 3 that lead location 3 as a location has a higshet TSI score are because it has the most extensive beach width, type of basic material in the form of white sand waters, as well as the location of the 3 is the beach closest to the source fresh water. While the dominant factor that affects location 8 has the lowest TSI, because it has the material basis of the waters in the form of sand and coral, the location of the most current speed 8 has great distance and a source of fresh water that is very far from the beach.

3.2.2 Carrying capacity of peucang beach.

The result of the assessment of Tourism Suitability Index that six location categorized as highly suitable and two beaches that are suitable shown at figure 4. Carrying Capacity of the eight location also should be assesed because according to Wunani [14] assessment of a carrying capacity is considered important because to know the maximum number of visitors that can fit in 1 day tour activities in order not to be cause interference both of visitor and the environment, so the utilization of sustainable coastal tourism. We mapped the beach area as highly suitable or suitable for beach recreational tourism depend on the TSI score that we count used 10 parameters.

Table 5. Tourism suitability of Peucang Beach.

Location	Total Parameter Score	TSI	Suitable Category
1	61	73%	Suitable
2	74	88%	Highly Suitable
3	79	94%	Highly Suitable
4	74	88%	Highly Suitable
5	69	82%	Highly Suitable
6	71	85%	Highly Suitable
7	68	81%	Highly Suitable
8	57	68%	Suitable

Figure 4. Suitability Map of Peucang Beach.

Beach Carrying Capacity calculated using several parameters such as the ecological potential visitors per unit area (K), long beach that utilized (Lp), the unit area for a specific category (Lt), the time provided for in a single day (Wt), and the time spent for each visitor activity (Wp). The other parameters in the form of the ecological potential visitors per unit area (K) it is assumed that one person using a length of 50 m beach for beach recreation activities. Yulianda [3] states that the average use of time for recreational beach is 3 hours. The time available to perform the activities of shore excursions for 10 hours. Carrying capacity of eigth location shown in table 6.

Location 5 has a lowest number of carrying capacity i.e. by 1 person, while location 7 has the highest number of carrying capacity i.e. 23 people. Based on observations on the visitors that every visitor who comes to Peucang Island will do such beach recreation enjoy the beautiful beaches of Peucang, enjoy views of the Ujung Kulon Peninsula, swim, sunbath on the beach and photo session. Total numbers of beach carrying capacity in Peucang Island i.e. 97 people per day. That number is far from the average number of visitors per day by 17 people per day. So Beach Peucang is still capable of accommodating the larger number of visitors with extent carrying capacity which has been obtained. The number of visitor on the beach should be considered by rotation factor as corection factor to count carrying capacity of the beach , because although amount of tourist can fit in at the beach, realilty shows a different situaion, becasuse as the number of beach tourist increases, visitor feel that the destination is losing quality [15]. Based on observations made against interview to visitors that there are some things that also need to be taken care of to ensure your comfort and safety related visitors with support for peak season. The existence of the ship which rests lined around the dock rated reduce the comfort of visitors do swim around the beach. In addition the visitor from plastic waste found at the base of the waters is also a threat Peucang Beach sustainability.

Table 6. Carrying capacity of Peucang Beach.

Location	K	Lp (m)	Lt (m)	Wt (Hours)	Wp (Hours)	DDK
1	1	192	50	10	3	13
2	1	23	50	10	3	2
3	1	306	50	10	3	20
4	1	61.85	50	10	3	4
5	1	11.3	50	10	3	1
6	1	310	50	10	3	21
7	1	340	50	10	3	23
8	1	212	50	10	3	14
Total						97

4. Conclusion

The result of carrying capacity based on the physical, social and management aspects consists of three values: PCC of 20 000 people/day, the value of RCC for 4838 people/day and ECC for 6 people/day. Carrying capacity of jungle trail obtained by the biological aspects has been damage to plants around the trail and change wildlife behavior because of visitor activities. Results of the determination of the suitability index of beach tourism that there are six beaches including highly suitable category and two beaches are categorized as suitable category. Peucang Beach carrying capacity of 97 people / day, with 70 people on the beach with a highly suitable category and 27 people on the beach with suitable category. The number of visitor nowadays still not exceed from carrying capacity number of PCC, RCC of jungle trails and carrying capacity of the beach area, but the number has exceeded from the ECC numbers. Management capacity should be developed which is the number of human resources for tourism management in Peucang Island should be more, because it was the one of limitation when visitor number increasing at peak season.

5. References

[1] [BTNUK] Balai Taman Nasional Ujung Kulon 2015 *Rencana Pengelolaan Jangka Panjang Taman Nasional Ujung Kulon* (in Bahasa) (Banten, Indonesia: BTNUK)
[2] Gurung H B 2010 Trends in Protected Areas (Queensland: CRC for Sustainable Tourism Pty Ltd)
[3] Yulianda F 2007 Ekowisata Bahari sebagai alternatif pemanfaatan sumberdaya pesisir berbasis konservasi Seminar Sains 21 Feb 2007 (Bogor: Fakultas Perikanan dan Ilmu Kelautan)
[4] Cifuentes M A 1992 Determination carrying capacity of tourism in protected area *(CTIE paper)* p 1-19
[5] Wearing S, and Neil J 2009 Ecotourism; Impact, Potentials and Possibilities 2nd Ed (Sydney: University of Technology, School of Leisure, Sport and Tourism) pp 78
[6] Ortega J L C, Dagostino R M C, and Magana A L C 2011 Estimating carrying capacity in a natural protected area as a conservation strategy *Conference Proc. Impact Assessment and Responsible Development for Infrastructure, Business and Industry* (Mexico, 28 May- 4 June 2011) pp 1-6
[7] Lucyanti S, Hartanto B, dan Izzati M 2013 Penilaian daya dukung wisata di obyek wisata bumi perkemahan palutungan taman nasional gunung ciremai propinsi jawa barat (in Bahasa) *Prosiding Seminar Nasional Pengelolaan Sumberdaya Alam dan Lingkungan 2013* (Indonesia, 27 August 2013) pp 234-240
[8] Tufaila M, dan Alam S 2014 Karakteristik tanah dan evaluasi lahan untuk pengembangan tanaman padi sawah di Kecamatan Oheo Kabupaten Konawe Utara *Agriplus* **24** (2) pp 184-194

[9] Queiroz R E, Ventura M A, Guerreiro J A, and Cunha R T 2014 Carrying capacity of hiking trails in Natura 2000 sites: a case study from North Atlantic Islands (Azores, Portugal) *Journal of Integrated Coastal Zone Management* **14** (2) pp 233-242
[11] Rabbany M G, Afrin S, Rahman A, Islam F, and Hoque F 2013 Environmental effects of tourism *American Journal of Environment, Energy, and Power Research* **1** (7) pp 117-130
[10] Sari Y, Yuwono S B, and Rusita 2015 Analysis of potential and carrying capacity along the track of mangrove forest in the coastal ecotourim Sari Ringgung beach, Pesawaran regency, Lampung *Jurnal Sylva Lestari* **3** (3) pp 31-40
[12] Djuwantoko, Utami R N, dan Wiyono 2008 Perilaku agresif monyet, *Macaca fascicularis* (Raffles, 1821) terhadap wisatawan di hutan wisata Alam Kaliurang, Yogyakarta (in Bahasa) *Jurnal Biodiversitas* **9** (4) p 301-305
[13] Ketjulan R 2010 Daya dukung perairan Pulau Hari sebagai objek wisata bahari (in Bahasa) *Journal Paradigma* **14** (2) pp 195-204
[14] Wunani D, Nursinar S, dan Kasim F 2014 Kesesuaian lahan dan daya dukung kawasan wisata Pantai Botutonuo Kecamatan Kabila Bone Kabupaten Bone Bolango (in Bahasa) *Jurnal Ilmiah Perikanan dan Kelautan* **1** pp 18–22
[15] Zacaria D A, William A T, and Newton A 2011 Recreation carrying capacity estimations to support beach management at Praia de Faro, Portugal Applied geography 31 pp 1075-108

IOP Conf. Series: Earth and Environmental Science **149** (2018) 012019 doi:10.1088/1755-1315/149/1/012019

Spatial distribution of ozone over Indonesia (Study case: Forest fire event 2015)

Sri Muslimah[1,2], Muhamad Buce Saleh[3], Rahmat Hidayat[4]
[1]Department of Information Technology for Natural Resources Management, Faculty of Mathematics and Natural Sciences, Bogor Agricultural University (IPB)
[2]Center for Environmental Research, Bogor Agricultural University (IPB)
[3]Department of Forestry Management, Faculty of Forestry, Bogor Agricultural University (IPB)
[4]Department of Geophysics and Meteorology, Faculty of Mathematics and Natural Sciences, Bogor Agricultural University (IPB)

E-mail: ms.srimuslimah@gmail.com

Abstract. Tropospheric ozone is known as surface ozone and caused several health impact. The objective of this study was to analysis spatial distribution of tropospheric ozone over Indonesia case study forest fire event in 2015. Monthly observation measured by Ozone Monitoring Instrument (OMI) have been analysed from January – December 2015 to study spatial distribution of tropospheric ozone related to forest fire event 2015. The study discovered high level of tropospheric column ozone (TCO) from October to November 2015. The result shows increasing average of TCO from September to October almost 6 DU. Meanwhile, monthly number of hotspot is higher in September 2015 with total number 257 hotspot which is acquired by Moderate Resolution Imaging Spectrometer (MODIS) Terra version 6.1 with confidence level same or more than 90%. The hotspot distribution compared with spatial TCO distribution and shows interesting time lag with respect to hotspot distribution, one month. Further study for daily comparison of TCO and forest fire event needed. This result suggested that the tropospheric ozone over the Indonesian region increases in 2015 were remarkable and corresponded to forest fire event.

1. Introduction

The abundance of ozone over world is 10% in troposphere, meanwhile good ozone in stratosphere is 90% of total abundance. Tropospheric ozone is considered to be important because of their impacts for human health and loss growth [1,2]. In Indonesia, national air quality standard of tropospheric ozone is 120 part per billion (ppb) [3]. Previous study mention about biomass burning does contribute 4-5% increases of global tropospheric ozone in tropic, which is Indonesia fires contribute during El Nino, as indicated of regional effect within and near burning area in troposphere. Actually, the biomass burning not directly produce tropospheric ozone, however produce volatile organic compounds (VOCs), CH_4, and CO in the presence of NOx and Sunlight which is ozone reaction in troposphere [4].

Tropospheric ozone concentrations measured in Indonesia have two location, first is in Kototabang, West Sumatera and secondly in Kemayoran, Jakarta [3]. Spatially tropospheric ozone does not represent from two location observation, need satellite observation for generate tropospheric ozone abundance. Many techniques deriving tropospheric column ozone (TCO) globally and have some improvement. The ozone monitoring instrument (OMI) and Microwave Limb Sounder (MLS) combination of Aura derived TCO globally using tropospheric residual method [5].

This paper presents the analysis of tropospheric column ozone from OMI measurements for the period January-December 2015 to identify their spatial variations related to hotspot, as indicator of forest fire, over Indonesia.

2. Data and Analysis

In the present study, monthly mean tropospheric column ozone (TCO) data for the period January-December 2015 were used. The TCO over Indonesia with spatial resolution 1 degree latitude by 1.25 degree longitude was derived from Ozone Monitoring Instrument (OMI) observations. The explanation method to generate tropospheric column ozone by subtracting Microwave Limb Sounders (MLS) stratospheric column ozone from OMI total column ozone data can be seen in previous study [5]. The daily hotspot data for the same period as TCO data were used and was obtained from Moderate Resolution Imaging Spectrometer (MODIS) on board the National Aeronautics and Space Administration's (NAS) terra satellite platforms version 6.1 with 1 km x 1 km spatial resolution (url: http://firms.modaps.eosdis.nasa.gov) [6].

Indonesia lying between the Indian Ocean and Pacific Ocean. The present study focuses on the tropospheric O_3 over Indonesia (12°N - 15°S and 90° - 150°E). Mapping the tropospheric column ozone data using Matlab 2017a and hotspot distribution is mapped by ArcGIS 10.5.

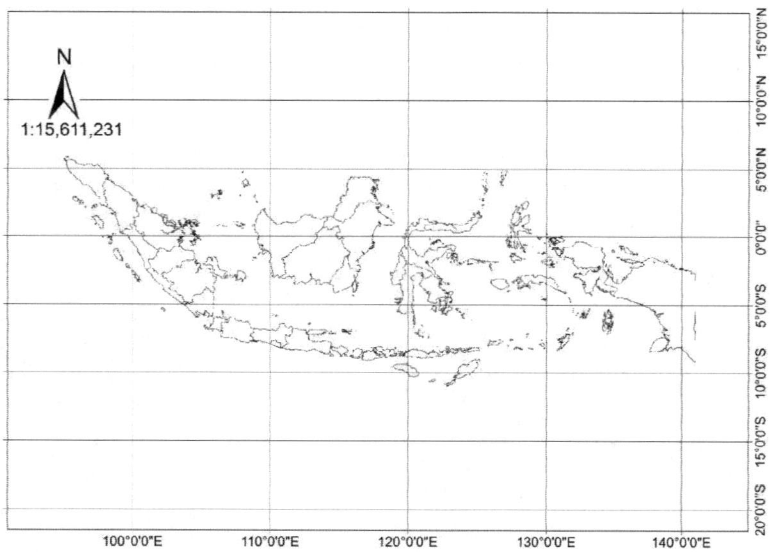

Figure 1. The locations used for the study.

3. Result and Discussion

The variation of TCO from January to December in 2015 (12°N-15°S and 90°-150°E) is illustrated in figure 2-4. There is gradual increase in spatial distribution of TCO from east toward the south west of Indonesia from January to December 2015. October illustrated a very high TCO distribution with November follows close behind. This progression coincides with the number of hotspot in September until November 2015. The TCO unit is Dobson Unit (DU). The interval of TCO value is 5 DU with minimum value 10 DU and maximum 50 DU (which have gradation color dark blue to dark red).

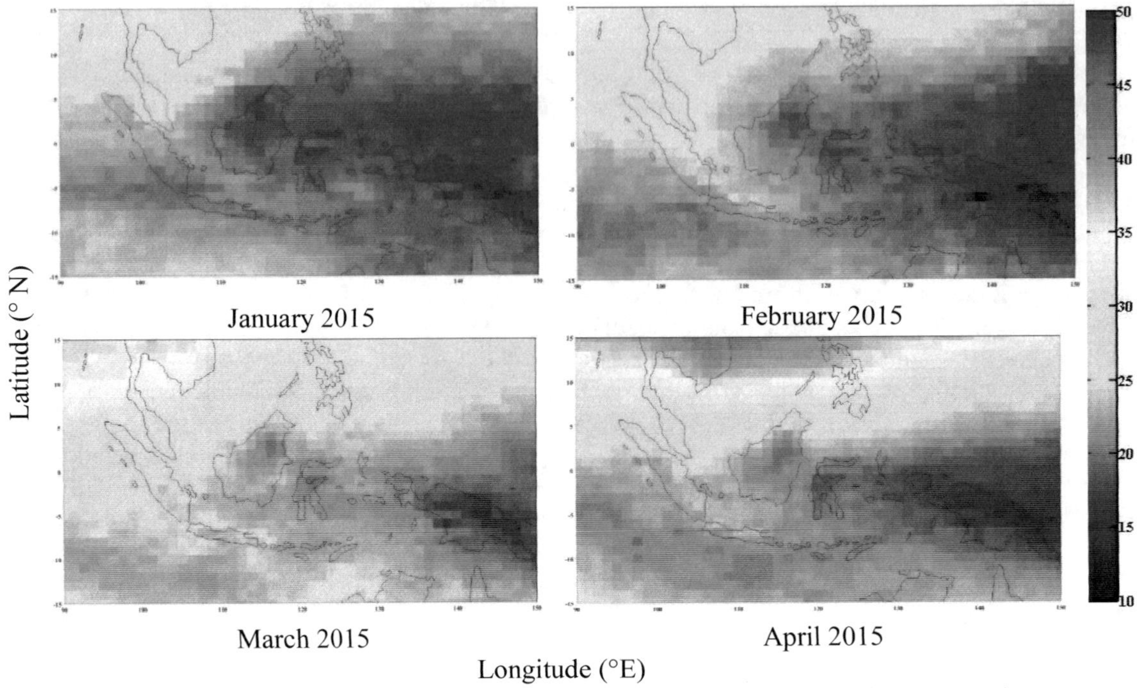

Figure 2. Spatial distribution of tropospheric ozone from January to April 2015 (DU) over Indonesia.

The spatial variations of tropospheric ozone from January to April 2015 is shown in figure 2. The highest value, which have red color, show in April 2015 but that is not Indonesia region. Highest value in north of Indonesia affected by global pattern of tropospheric ozone. Based on figure 3, the gradual variation of tropospheric ozone is higher in west of Indonesia and lowest in east part of Indonesia.

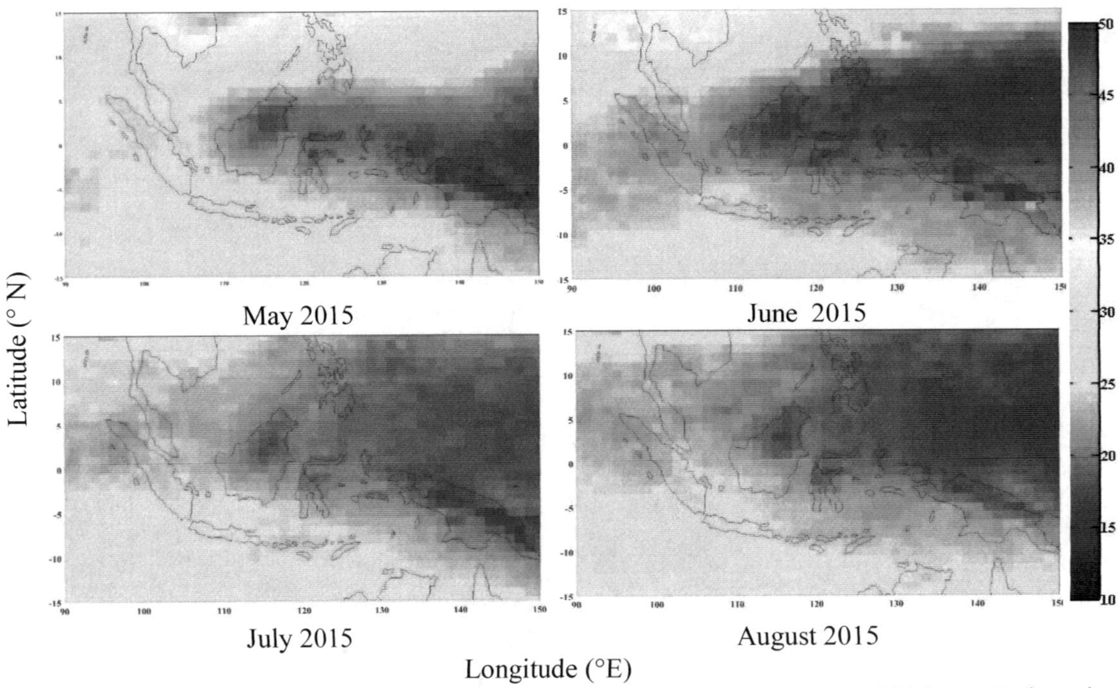

Figure 3. Spatial distribution of tropospheric ozone from May to August year 2015 over Indonesia.

In figure 3, spatial distribution of tropospheric ozone over Indonesia from May to August 2015 is shown. The average of tropospheric ozone in May, June, July, and August are 23.1 DU, 20.8 DU, 22.1 DU and 21.8 DU. The deviation standards from May to August 2015 are 2.9, 3, 2.7, and 3.1. It means average value can be representative of data distribution. Figure 4 show the spatial distribution of tropospheric ozone over Indonesia from September to December 2015. Based on figure 4, the highest concentration shown in October around Southern Sumatera Island, Indonesia and South of Indonesia (e.g. Java, Bali, Nusa Tenggara).

The maximum concentration of tropospheric ozone in September, October, November, and December are 32 DU, 39.7 DU, 36.7 DU and 30.6 DU and the average concentration each month are 24.5 DU, 30 DU, 26.8 DU and 24.5 DU, respectively.

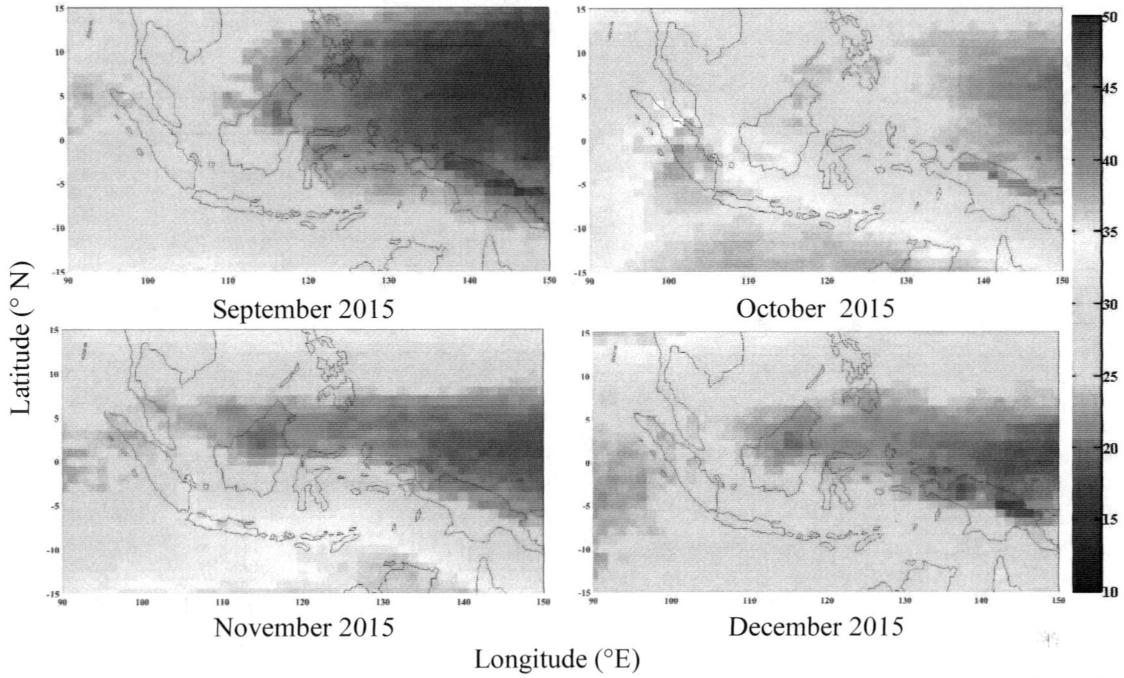

Figure 4. Spatial distribution of tropospheric ozone from May to August 2015 (DU) over Indonesia.

Compared with the TCO values of 20.5-39.7 DU in October for years 2015 and 16.3-33.7 DU in October for years 2014, the increases of TCO were about 4.2-6 DU. We suggested that the TCO increase could be attributed to the extensive forest fires in Sumatera with total forest fire area is 30,984.98 Ha in 2015 and largest forest fire area is in Center Kalimantan of 122,882.90 Ha [7].

Actually, the forest fire reported to data and information of Indonesian disaster reveals that highest number of forest fire in October with total event is 26 event [8]. The higher hotspot number in Indonesia also illustrated in figure 5. Here is hotspot distribution over Indonesia with confidence level 90% capture by Terra MODIS C6 (figure 5). We suggested there is correlation with highest values of TCO in October 2015. The highest number of hotspot in September and October is in Central Kalimantan, 87 and 73 hotspot. As mention in previous study, there are carbon and nitrogen which is found in biomass burning laboratory result of experiment and it leads to produce ozone in troposphere [9]. The duration after biomass burning and producing of tropospheric ozone is one month. We recognized that because when highest number of hotspot in September 2015, the tropospheric ozone increased one month after.

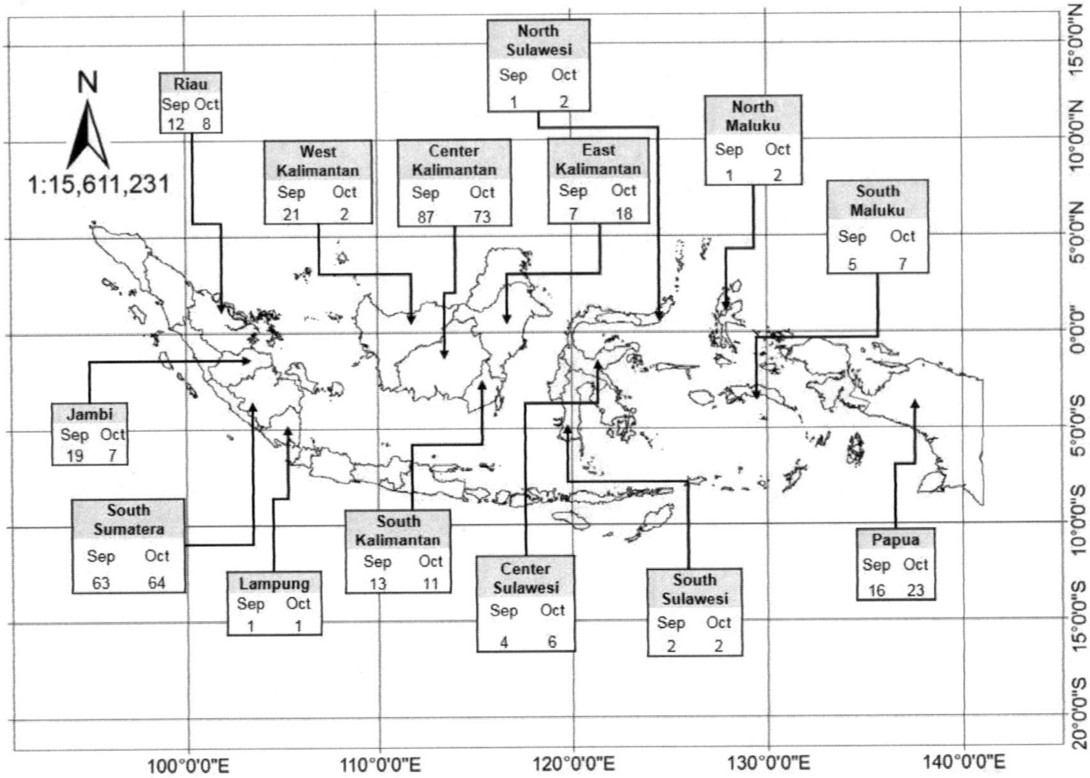

Figure 5. Hotspot distribution over Indonesia from September – October 2015 captured by Terra MODIS C6 with confidence level 90%.

4. Conclusion

We have studied the effects of biomass burning on tropospheric ozone in Indonesia derived from Aura OMI/MLS measurements for the time period January-December 2015. Biomass burning is interpreted by hotspot as indication of a fire event. Forest fire was increased substantially in Indonesia in year 2015 compared to previous years. The OMI measurements show sizeable increases from September to October and decreases from November to December 2015. Comparing the TCO distribution with hotspot reveals that there is time lag when hotspot is higher the TCO increases one month after that event. Having average of TCO over 12°N-15°S and 90°-150°E monthly for year 2015, it shown increases almost 5.9 DU from September to October 2015. Highest decreasing of average TCO about 3.7 DU occur in period October into November 2015.

References

[1] Chieppa J, Art C and Lori E 2015 Effects of tropospheric ozone on loblolly pine seedlings inoculated with root infecting ophiostomatoid fungi *Envir. Poll.* **207** 130-137

[2] Sun J, Feng Z and Ort D R 2014 Impacts of rising tropospheric ozone on photosynthesis and metoblite level on field grown soybean *Plant Scien.* **226** 147-161

[3] Meteorology, Climatology, and Geophysic Agency of Indonesia 2013 *Buku informasi perubahan iklim dan kualitas udara di Indonesia. Pusat Perubahan Iklim dan Kualitas Udara* (Jakarta: Climatology Division, Meteorology, Climatology, and Geophysic Agency of Indonesia)

[4] Ziemke J R, S Chandra, B N Duncan, M R Schoeberl, O Torres, M R Damon and P K Bhartia 2008 Recent biomass burning in the tropics and related changes in tropospheric ozone *geophy. Res. Let.* **36** 1-10

[5] Ziemke J R, Chandra S, Duncan B N, Froidevaux L, Bhartia P K, Levelt P F and Waters J W 2006 Tropospheric ozone determined from Aura OMI and MLS: evaluation of measurements and comparison with the global modeling initiative's chemical transport model *Journal of Geophysc. R.* **111** 1-18

[6] Louis G 2015 *Modis collection 6 active fire product user's guide revision A* (Department of Geographical Sciences University of Maryland)

[7] SIPONGI 2017 Rekapitulasi luas kebakaran hutan dan lahan (Ha) per provinsi di Indonesia tahun 2011-2016 accessed in http://sipongi.menlhk.go.id/hotspot/luas_kebakaran

[8] National Board for Disaster Management 2015 Data and information of Indonesian disaster accessed in http://dibi.bnpb.go.id/data-bencana

[9] Levine J S 2003 Biomass burning: the cycle of gases and particulates from the biosphere to the atmosphere *Treatise on Geochem.* **4** 143-158

Tourism hazard potentials in Mount Merapi: how to deal with the risk

J Muthiah, E K S H Muntasib, R Meilani
Forest Resources Conservation and Ecotourism Department, Faculty of Forestry
Bogor Agricultural University, Indonesia

E-mail:jadda_muthiah@yahoo.com

Abstract. Mount Merapi as one of the most popular natural tourism destination in Indonesia, indicated as disaster prone area. Hazard management is required to ensure visitors safety. Hazard identification and mapping are prerequisite in developing proper hazard management recommendation. This study aimed to map hazard potentials' geographical positions obtained with geographical positioning system and to identify the hazard management being implemented. Data collection was carried out in Mei – June 2017 through observation and interview. Hiking trail and Lava tour area was selected as the study site, since the sites are the main areas for tourism activities in Mount Merapi. The type of hazards found in the area included lava, tephra, eruption cloud, ash, earthquake, land slide, extreme weather, slope and loose rock. Early warning system had been developed in this area, however the mechanism to regulate tourism activities still had to be improved. Local tourism entrepreneurs should be involved in the network of early warning system stakeholders to ensure tourist safety, and their capacity should be improved in order to be able to perform the measures needed for handling accident and disaster occurrences. Interpretive media explaining hazard potentials may be used to improve visitors' awareness and ability to cope with the risk.

1. Introduction

There is a wide understanding that Indonesia is a mega biodiversity country and possesses exceptional natural beauty. This richness has become its main tourist attraction. But behind its charm, many tourism areas in Indonesia are actually disaster prone areas, being located between the Ring of Fire and the Alpide Belt [1].

Mount Merapi, as one of the most active volcanoes in Indonesia, erupted in 2010 with Volcanic Explosivity Index (VEI) 4. As a comparison, Krakatau eruption in 1883 was 100 times bigger with VEI 6 and Tambora Eruption in 1815 had VEI 7. However, a VEI 4 eruption is already a strong eruption with volume of erupted tephra amounted 10 Km^3 and the height of eruption column ranging from 10 -25 Km [2]. Mount Merapi is located in Special Region of Yogyakarta and Central Java Provinces. Special Region of Yogyakarta with its cultural uniqueness is one of the main tourist destinations in Indonesia. In 2015, there were approximately 308,485 international tourists and 3,814,450 domestic tourists occupying hotels in Yogyakarta [3]. On the other hand, Yogyakarta is a prone area for many natural disasters such as tsunami, earthquake, flood, extreme weather, land slide, drought, and volcano eruption. Disaster management has become increasingly crucial with the increase number of people visiting this area, since there will be more and more people involved in the disaster.

Like any other economic sectors, the tourism industry is highly vulnerable to disaster and its impact [4]. Natural hazards, such as the volcano eruption, earthquakes, landslides, ice storms in high mountain areas, high waves, torrential rains and storms in coastal areas, may affect tourist activities and tourist destinations, and can also cause significant change to the landscape of a region [5]. Hazards are commonly found in tourism attraction [6]. Considering that visitor safety is one of the most important aspects in ensuring the satisfaction of the visitors, managers are responsible for hazard management actions to reduce the likelihood of accident and/or the severity of the loss [6].The purpose of Disaster Risk Management is to reduce the underlying factors of risk and to prepare for and initiate an immediate response when disaster occur [7]. It also plays an important role in tourism marketing [8].

Identifying the current condition of disaster management may help improving the management to achieve the best model. Therefore this research aimed to identify the existing condition of disaster management in Mount Merapi tourism and provide recommendation for improvement.

2. Methodology

2.1. Study area

This research was conducted in Mount Merapi, Indonesia. Mount Merapi, located on two provinces in Indonesia, i.e. Central Java and Special Region of Yogyakarta, is managed under the Mount Merapi National Park. The mountain is one of the most active volcanoes in Indonesia, which had erupted regularly. The most recent eruption occurred in 2010, causing nearly 300 fatalities and destruction of thousands of houses in dozens of villages.

Like other tourists who only visited Yogyakarta, most of the tourists visiting Mount Merapi also used Yogyakarta international airport as their transportation gate and visited other attraction, such as cultural attraction, on its vicinity. While Special Region of Yogyakarta is a natural disaster prone area [9] it also is one of Indonesia main tourism destinations. Since the eruption of Mount Merapi affected a broader area, which include Special Region of Yogyakarta and its vicinity, tourism hazard management of mount Merapi should be integrated with Yogyakarta disaster management system.

The study was focused on two hiking trail and lava tour area in Mount Merapi. The first trail, Sapuangin hiking trail, is located within the coordinates S7° 34.067' E110°28.086' and S7°32.345' E110°27.494'. The second trail, Selo hiking trail, is located within S7°30.949' E110°27.162' and S7° 32.384' E110° 26.817'. Lava tour area is located on S7°41' E110° 25 and S7°34' E110°26'.

2.2. Data collection

Data on types and forms of hazards, tourism activities, occurrences of accidents, the likelihood and severity of the accidents, the existing management action, management plan, and stakeholders involved and their role in hazard management were collected in this study. Data were collected in May – June 2017 through field observation and interview. Field observation was carried out in Mount Merapi tourism area by participating in tourism activities in two hiking trails and lava tour. Interview with the manager of Mount Merapi National Park, local communities, visitors, local government, and local tourism agency were also conducted.The tool used in this research was digital camera, Global Positioning System (GPS), recording device, stationery, disaster prone area map, interview guide, and ArcGIS 10.3 software.

2.3. Data processing

Data of geographical position of each hazard was extracted from the GPS to prepare it for mapping of hazard potentials. Based on interview with area manager, local government, local community, visitors, and field observation, a list of identified hazards in Mount Merapi were developed. Information from the interview were used to decide on the likelihood and severity of each hazard potentials. There are five classes of likelihood (table 1) based on the probability of occurrences or how often each hazard

IOP Conf. Series: Earth and Environmental Science **149** (2018) 012020 doi:10.1088/1755-1315/149/1/012020

can occur, and five classes of severity (table 2) based on how severe the impact, lost or injury that can happened if one type of hazard occur.

2.4. Data analysis

The map of hazard potentials' in Mount Merapi was generated using Landsat 8 satellite imagery as basic map. Geographical position of each hazard potential was mapped using the ArcGIS 10.3 software. The potential hazards, which had been classified into its likelihood and severity classes, were then mapped into a qualitative matrix of risk analysis (table 3) to determine risk level for each hazard.

Table 1. Likelihood classification for risk assessment [10-11].

Likelihood	Description
Almost certain	Is expected to occurs in most circumstances
Likely	Will probably occurs in most circumstances
Possible/ Moderate	Frequently occur at some time
Unlikely	Could occur at some time
Rarely	May occur in exceptional/extreme circumstances

Table 2. Severity classification for risk assessment [10-11].

Severity	Description
Catastrophe	Fatalities/death
Major	Extensive injuries, with occasional death
Moderate	Seriously (Permanent) injury, medical treatment required
Minor	Minor consequences/injury: first aid treatment
Insignificant	Insignificant consequences, no injury

Table 3. Level of risk and management option needed [11].

Likelihood	Severity				
	Insignificant	**Minor**	**Moderate**	**Major**	**Catastrophe**
Almost certain	H [b]	H [b]	E [a]	E [a]	E [a]
Likely	M [c]	H [b]	H [b]	E [a]	E [a]
Possible/ Moderate	L [d]	M [c]	H [b]	E [a]	E [a]
Unlikely	L [d]	L [d]	M [c]	H [b]	E [a]
Rarely	L [d]	L [d]	M [c]	H [b]	H [b]

[a] Extreme risk, immediate action require. Management option: **avoid risk**
[b] High risk, senior management attention needed. Management option: **transfer risk**
[c] Moderate risk, management responsibility must be specified. Management option: **reduce risk**
[d] Low risk: Manage by routine procedures. Management option: **accept tolerable risk**

Risk level of hazards ranged from low risk to extreme risk, which requires different management option.Management option is mitigation in risk management strategy. The risk level will determined the priories of management option. Four options were available for tourism hazard management, i.e. accept tolerable risk, reduce risk, transfer/sharing risk, and avoid/terminate risk, which could be explained as follow [9, 10, 12]:

1. Tolerate: accept the risk, when it is not possible or practical to respond to the risk by the other strategies, or a response is not warranted by the importance of the risk, or the risk is not significant.
2. Reduce: reduces the probability and/or impact of an adverse risk event to an acceptable threshold. Even if this is costly but is often more effective than trying to repair the damage.
3. Transfer or sharing: let someone else take share of the risk (by insurance or passing responsibility for the risk to a contractor). The consequence is there are cost that need to be paid.
4. Avoid/Terminate: agree that the risk is too high and do not proceed with the activity.

3. Result and Discussion

3.1. Mount Merapi
Mount Merapi is divided into 3 area*s* based on their vulnerability for eruption, i.e. Disaster Prone Area level I, II, and III. Disaster Prone Area level III is a highly potentially devastated area of hot clouds, lava flows and lava, poisonous gases, bursts of rock (incandescent) and heavy ash rains (figure 1). There is no house allowed in this area, only temporary activities such as tourism, collecting grass for livestock, and collecting firewood were allowed. Community activities in disaster-prone areas of Mount Merapi are arranged according to the level of threat of eruption. There are four levels of volcano activities, as follow:

1. Level 1: Communities in disaster-prone areas III, II, I can perform daily activities. Especially for activities in the peak areas, the center of eruption, communities should remain vigilant and comply with local government regulations according to technical advice from the Center for Investigation and Technology Development of Geological Disaster.
2. Level 2: Communities in Disaster Prone Areas III, II, I can perform daily activities. Particularly for activities in disaster-prone areas III, communities must remain vigilant and comply with local government regulations according to technical advice from the Center for Investigation and Technology Development of Geological Disaster.
3. Level 3: Communities in Disaster Prone Areas III and II must prepare to evacuate by orders from local government according to technical advice from the Center for Investigation and Technology Development of Geological Disaster.
4. Level 4: Communities in Disaster Prone Area III must have been evacuated. In case of escalation of the threat of eruption, people in the Disaster Prone Area II should also evacuate, following the orders of the local government in accordance with technical advice from the Center for Investigation and Technology Development of Geological Disaster. Communities in Disaster Prone Areas I who live close to the river to further increase his vigilance against lava threats in case of rain.

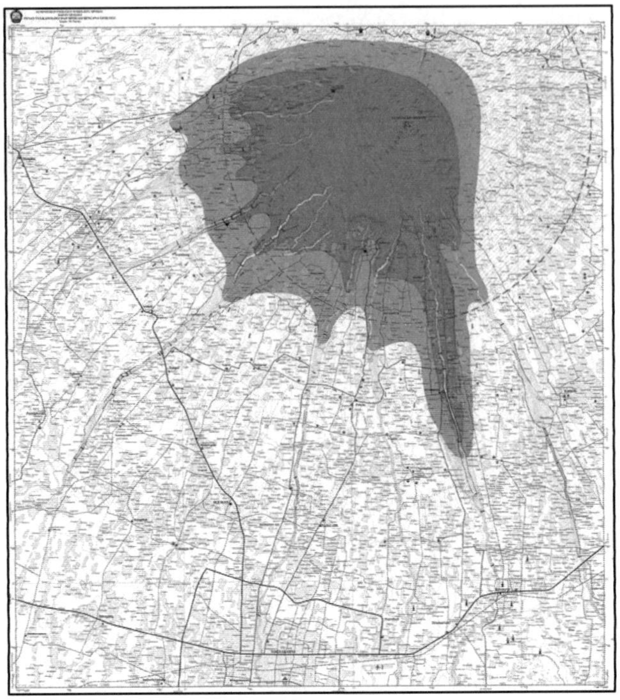

Figure 1. Mount Merapi disaster prone area [13].

3.2. Tourism in Mount Merapi and the hazard potentials

Generally, there were two tourism activities in Mount Merapi, which included hiking and jeep tour around the remaining of volcano eruption that known as'Volcano tour or Lava tour'. After 2010 eruption, only 2 hiking trails were opened, i.e. Selo and Sapuanging (Deles) hiking trail. Kinahrejo trail, an older trail severely damaged by the 2010 eruption, had been closed for hiking. However, because this trail was used for traditional ceremony called "Labuhan", local community had started cleaning the area. Therefore, even though it was not an official trail, some hikers were still found going through this trail. There was a path in Kinahrejo trail called the Sand Bridge, where the path that must be passed is a narrow sandy road with 8-10 meters-deep cliff on the right and left side of the trail (figure 2).

Figure 2. Example of hazard potential in hiking trail (from left to right: sand bridge, the summit).

Thirteen locations had been identified as having potential hazards along the trails. The type of hazards found in the area included lava, tephra, eruption cloud, ash, earthquake, land slide, extreme weather, slope and loose rock (figure 3).There were two cases of hypothermia, but they were not severe cases. Pasar Bubrah, and places higher than it, were forbidden area because of the loose rock, but some hiker were still going up, trying to get to the summit. The Garuda summit was destroyed in 2010 eruption, and considering the hazards along the trail from Pasar Bubrah to Garuda Summit, the reckless action of visitors forcing their way to the summit is very dangerous. This dangerous behavior is related to visitor perception toward the activity [14] when the motivation to do the activities is bigger that their caution for hazard.

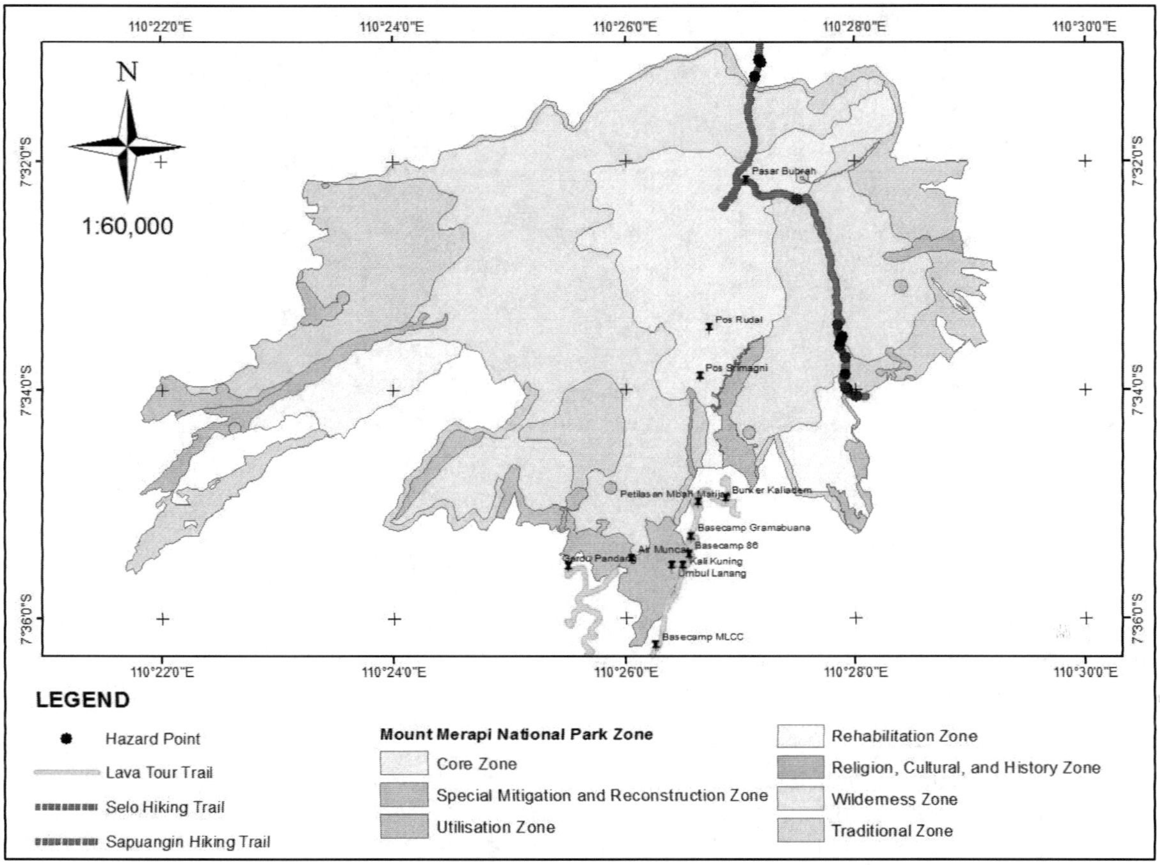

Figure 3. Mount Merapi tourism hazard point.

The remains of 2010 eruption were being used as tourist attractions in a tour program known as lava tour. This destination is an example that risk can be portrayed as part of the manifest attraction of certain activities [15]. There were 34 Jeep Communities found in Kaliadem and Kaliurang area. The jeeps were private property of the local community who also drives the jeep for the tour. The whole communities were incorporated in the Merapi Hill Tourism Jeep Association (*Asosiasi Jeep Wisata Lereng Merapi*). The jeep base camps were located in Kinahrejo and Kepuharjo village, which were included in the **level I** disaster prone area, while the tours took place in Kepuharjo village, Kaliadem, which is included in the **level III** disaster prone area, the most dangerous area. The attraction of this tour are the remains of 2010 Merapi eruption, such as house debris, lava path in river, bunker, Mbah Maridjan's (Mount Merapi spiritual caretaker) grave, etc (figure 4). Every driver know about early warning system, evacuation trail and emergency action needed, but they had not yet shared this information with the tourist in their interpretation along the tour.

Figure 4. Lava tour destination (From left to right: 'my remaining property' museum, Gendol river, bunker, Mbah Maridjan grave).

Accident was rarely happened in this Volcano tour. The record showed only one incident when a loose rock caused the passing jeep slightly slipped, but there was no one hurt. Until now, there had not been any exploration for new paths to develop the Volcano tours. When there is flood threat, there will be a warning from Regional Agency for Disaster Management in the form of siren and all Volcano tour activity will be put on hold in save area.

3.3. Existing Hazard Management in the Tourism Area of Mount Merapi

Management option for tourism hazard management in mount Merapi were classified from avoid risk, transfer risk, and reduce risk based on the hazard type. This management option is taken based on the classification of severity and likelihood of every hazard potential being identified (table 4.)

Table 4. Hazard management option.

No	Hazard	Severity	Likelihood	Risk Level	Management Option
1	Eruption	Major	Likely	Extreme	Avoid Risk
2	Land slide	Moderate	Possible/ Moderate	High	Transfer Risk
3	Earthquake	Minor	Possible/ Moderate	Moderate	Reduce Risk
4	Flood	Minor	Likely	High	Transfer Risk
5	Hiking accident	Minor	Likely	High	Transfer Risk

Management option for extreme risk level is to avoid the risk. Avoiding risk should be taken because the risk is so great that it is better to get people out of dangerous area and/or to prevent people to go in. Mount Merapi National Park handled extreme risk level hazard of eruption by applying entry prohibition for visitors in accordance with technical advice from the Geology Disaster Technology Development and Research Center. This action can be classified as **avoid the risk**. Hazard potential with high risk level should be handled with transferring the risk to the other party. Of course there are systems that make it work such as insurance etc. The hiking suggestion limitation to Pasar Bubrah emphasized that the summit is off limit and not covered by insurance. This action can be classified as **transfer the risk**. The patrol and installation of warning board in disaster prone area can be classified as **reduce the risk**.

Since some of the tourism areas were located outside the national park, more stakeholders were involved in hazard management of the area. The stakeholders that had been identified in this research included Mount Merapi National Park, Special Region of Yogyakarta Government, Central Java Government, National Agency for Disaster Management, Regional Agency for Disaster Management, The Geology Disaster Technology Development and Research Center, local community, and tourist. Each stakeholder had their own role in hazard and disaster management in Mount Merapi. Center for Investigation and Technology Development of Geological Disaster provided daily data of mount Merapi activity. They also collaborated with the Regional Agency for Disaster Management, who was responsible for community safety when natural disaster happened. This regional agency had a search and rescue team and a number of volunteers in every villages. If an emergency occurred, the volunteer that live near the site would perform the rescue operation in cooperation with the other local communities.

Yogyakarta special region have an early warning system involving government agencies, village officers, handy talkie community, and residents (figure 5). Early warning system is a set of functions required to provide warning information to enable people in a hazard-threatened area to take appropriate action to minimize risk [16].

Figure 5. Mount Merapi early warning system [17].

This system had already been running effectively, with each stakeholder fulfilling their role. This running system can be used not only when disaster occurred, but also when there were cases of accident with the opposite bottom-up flow. This mechanism is possible because Yogyakarta's early warning system was classified as community based early warning system. This kind of early warning system involved community driven collection and analysis of information that enable warning messages to help a community to respond to a hazard and reduce the resulting loss or harm [18].

Early warning system in Mount Merapi was divided into hard early warning system and soft early warning system. Hard early warning system took form of siren, sound of *kentongan* (wooden alarm devise), handy talkie, mosque speaker, radio and television. Soft early warning system took shape of disaster prone map, warning signs and evacuation tracks (figure 6), disaster education and training, volcano activities daily reports, etc. This soft early warning system helped community and tourists to be more aware of the hazard helped people to find direction during the fright when disaster occurred.

Figure 6. Evacuation direction and warning board.

Regional Agency for Disaster Management conducted training program to prepare the community, which is called Destana (*Desa Tangguh Bencana*/disaster resilient village).The people were trained to understand what they must do when disaster occur: to give first aid, to manage the crowd moving to a safer place, to know who they have to call, etc. Even studentswere prepared to know what they have to do in emergency situations. In order for the educational activities in school to be able to continue running during the time of disaster, the school have a "sister school" where students can join the educational activities until their school situation is safe.

3.4. Recommendation of Hazard Management Improvement

The existing hazard management in Mount Merapi has been well conducted. They had integrated early warning system and because of the last eruption, most of the stakeholders were aware of their role and had high willingness to help others. Yogyakarta local communities already had important role in collecting and analysing natural phenomenon to determine the hazard. However, improvement may be carried out, which include increasing their knowledge in designing, monitoring and managing early warning system, to make them the basis for the system. A community-based early warning system is a system in which the communities became the active participants in the design, monitoring and managing the system [19].

Visitor is also important stakeholder that should participate in the early warning system of Mount Merapi. However, they were not yet involved in the hazard management of the area. High number of visitors may either become a weakness or strength to Mount Merapi hazard management, depending on their understanding of emergency situation. If visitors had low understanding of emergency situation, they might hinder the action taken to minimize risk of the hazard, but if they had high understanding, they might be able to provide assistance in the action. Therefore, improving their capability in terms of hazard management may be necessary. There are several ways to achieve it, i.e. developing interpretation program that extend the basic knowledge of hazard for visitors, provide brief explanation for the hikers about the potential hazards and how to cope with them before they do their activities, and developing an integrated tourism information system. Using realistic visual media can improve tourist understanding of hazard and disaster that can happen [20], better tourist maps with information on potential hazard spots may reduce overall risks [21]. Jeep operators may take the role of educational agent for the tourists in lava tour. Financial contribution may influence participation

[22] and for this reason, engaging the jeep operators who generate their income from the tourist may be more helpful. As for hiking, the national park ranger can give the information and warning.With this improvement, the tourists are expected to be able to make the right action.

Summit restriction was still more of a request rather than a strict rule, because Mount Merapi National Park did not have the resources to make sure that the hikers obey the restriction. Therefore, providing the hikers with the necessary information, thus making them aware of the consequences of their action and how to avoid the risk, may help in reducing the risk for hikers.

4. Conclusion

Hazard management in Mount Merapi had applied both preventive and repressive actions, which included early warning system, evacuation scenario, evacuation, and rescue system. The hazard management had involved various stakeholders, including local residents, handy talkie community, and government institutions, leaving out the tourist from the system. The system was also focused more on the hazard of eruption, and less on other tourism hazard. To fill this gap, increasing tourist awareness of hazard management may become important. This may be done by giving tourist crucial information on hazard potentials (eruption and other hazards) using various instruments, such as warning sign, siren, and briefs on evacuation plan before they undertake tourism activities.

Acknowledgments

This publication is part of the research funded by Ministry of Research, Technology and Higher Education of the Republic of Indonesia. The author sincerely thank M. Asad, Veronica Anglisa Putri, and Grama Buana Adventure, who assist in data collection.

5. References

[1] UNDP [United Nation Development Programme] Indonesia 2014 *Institutionalizing Post-Disaster Recovery*. (Indonesia UNDP) p 6

[2] Wheeling Jesuit University / Center for Educational Technologies 2006-2018 Volcanic Explosivity Index (VEI) Washington Ave http://ete.cet.edu/gcc/?/volcanoes_explosivity/

[3] Dinas Pariwisata Daerah Istimewa Yogyakarta 2016 *Statistik Kepariwisataan tahun 2016* (Yogyakarta Indonesia: Dinas Pariwisata Daerah Istimewa Yogyakarta) p 19

[4] Me´heux K and Parker E 2006 Tourist sector perceptions of natural hazards in Vanuatu and the implications for a small island developing state *Tourism Management* **27** (2006) pp 69–85

[5] Ilieş DC, Dehoorne O and Ilieş A 2011 Some exe[a]mples of natural hazards affecting Geosites and tourist activities *Geo Journal of Tourism and Geosites* **7** (1) pp 33-38

[6] Jubenville A, Twight B Wand Becker RH 1987*Outdoor Recreation Management, Theory, and Aplication*. (Oxford Circle United Kingdom: Venture Publishing, Inc).

[7] Baas S, Ramsamy S, de Prick JD, Battista F 2008 *Disaster Risk Management System Analysis* (Rome: Food and Agriculture Organization of the United Nations)

[8] Rittichainuwat BN 2013 Tourists' and tourism suppliers' perceptions toward crisis management on tsunami *Tourism Management* **34** (2013) pp 112-121

[9] Badan Nasional Penanggulangan Bencana 2010 *Disaster Risk Index Map in DI Yogyakarta Province* (Jakarta Indonesia: BNPB)

[10] UNEP [United Nation Environment Programme] 2008 *Disaster Risk Management for Coastal Tourism Destinations Responding to Climate Change. A Practical Guide for Decisions Marker* (France: UNEP)

[11] Wilks J and Moore S 2004 *Tourism risk management for the Asia Pacific region: an authoritative guide for the managing crises and disasters* (Gold coast: CRC for Sustainable Tourism Pty Ltd.)

[12] Cater CI 2004 Playing with risk? participant perceptions of risk and management implications in adventure tourism *Tourism Management* **27** (2006) pp 317–325

[13] Balai Penyelidikan dan Pengembangan Teknologi Kebencanaan Geologi 2010 *Peta Kawasan Rawan Bencana Gunung Berapi* (Yogyakarta, Indonesia: BPPTKG)
[14] Gstaettner AM, Rodger K and Lee D 2017 Visitor perspectives of risk management in a natural tourism setting: an application of the Theory of Planned Behavior *Journal of Outdoor Recreation and Tourism* **19** pp 1–10
[15] Enterprise Risk Management Program 2012 *Guide to Risk Assessment and Response* (USA: The University of Vermont)
[16] UNDP 2010 Community-based Best Practice for Disaster Risk Reduction (Maputo Mozambique: UNDP and EUROPEAN Commission Humanitarian Aid)
[17] Mei ETW and Lavigne F 2012 *Influence of the institutional and socio-economic context for responding to disasters: case study of the 1994 and 2006 eruptions of the Merapi Volcano, Indonesia* (London: Geological Society) pp 171-186
[18] Macherera M and Chimbari M J 2016 A review of studies on community based early warning systems *Jàmbá: Journal of Disaster Risk Studies* **8** (1)
[19] Smith P J, Brown S, and Dugar S 2017 Community-based early warning systems for flood risk mitigation in Nepal *Natural Hazards and Earth System Sciences* **17** pp 423–437
[20] Yang B 2016 GIS based 3-D landscape visualization for promoting citizen's awareness of coastal hazard scenarios in flood prone tourism towns *Applied Geography* **76** pp 85–97
[21] Becken S, Hughey K F D 2013 Linking tourism into emergency management structures to enhance disaster riskreduction *Tourism Management* **36** (2013) pp 77-85
[22] Ofuoku AU 2011 Effect of community participation on sustainability of rural water projects in Delta Central agricultural zone of Delta State, Nigeria *Journal of Agricultural Extension and Rural Development* **3** (7) pp 130-136

Reducing the risk of potential hazard in tourist activities of Mount Bromo

R Meilani*, J Muthiah and E K S H Muntasib

Department of Forest Resources Conservation and Ecotourism, Faculty of Forestry, Bogor Agricultural University

E-mail: restimeilani@apps.ipb.ac.id

Abstract. Mount Bromo has been crowned as one of the most beautiful mountains in the world, having a particular landscape uniqueness. Not only volcano, Bromo also has savanna, sea of sands, and culture of Tengger tribe. Its panoramic landscape has attracted a large number of tourists, both domestic and foreign, despites the threat of eruption. To ensure tourists safety and satisfaction, the potentials hazard, both from eruption and other features should be managed carefully. The study objective was to identify and map hazard potentials and identify the existing hazard management. It was carried out in Mei – June 2017. Lava, tephra, eruption cloud, ash, earthquake, land sliding, extreme weather, slope, transportation modes (jeep, motorcycle, and horse), human, and land fire were found as potential hazards in Mount Bromo. Five locations had been identified as hazard area in the tourism areas, i.e. savanna, sea of sand, Bromo caldera and Pananjakan I trail and viewing point. Early warning system should be developed as part of hazard management in the area. Capacity building of local stakeholders and visitors would be needed to reduce risk of the hazard.

1. Introduction

Hazard is defined as any source that has the potential to cause harm, accident or adverse health impact on something or someone [1]. An older definition states hazard as physical events, phenomenon, or human activity with destructive potentials that may cause loss of life or injury, property damage, social and economic disturbance or environmental damage [2]. There are two causes of hazard: (1) nature, which include lightning, cliff, landslide, vegetation, wildlife, volcano activities, waves, current, rapids, troughs, and floods, and (2) human activities, which include unmaintained buildings, design of roads/lanes, and landscapes that have been changed [3]. Natural hazards such as volcano eruption, earthquakes, landslide, avalanche in high mountainous area, high waves, torrential rain and storm in coastal areas, can affect tourists' activities and tourism destinations, and can substantially changes the landscape of an area [4]. Hazard potentials in tourism destination may also cause risk for destination manager and visitors.

Risk is defined as possible adverse consequences or predicted loss (death, injury, financial loss, disturbance on livelihood and economic activities, and environmental damage) resulted from interaction between natural or human induced hazard and vulnerability [2] or the combination of hazard and vulnerability [5]. Risk can be calculated as the product of disaster likelihood multiplied by vulnerability, and divided by capacity to respond adequately [2].

Recreational opportunity in tourism destination is generally contain hazard, and therefore the manager should be responsible to carry out hazard management actions in reducing the likelihood of accidents or loss occurrence on visitors [3]. Such condition can also be found in Mount Bromo, an active volcano that recently erupted in 2016. Mount Bromo area is part of Bromo Tengger Semeru National Park located in East Java Province. The site is acknowledged as one of the most beautiful mountains in the world. It has various attractions, which include the scenic mountainous view, sunrise view, sea of sands, savanna, volcano and the culture of Tengger Tribe. Its panoramic and scenic beauty has attracted high number of visitors, both domestic and foreign, over the years. However, behind its magnificent beauty lies potential hazards threatening the safety of tourists visiting the site. Natural hazard in tourism area of Mount Bromo include volcanic activities of Mount Bromo, gorge/cliff, landslide, weather, fire, lightning, wind, earthquake, and various diseases associated with the respiratory tract [6].

During January 2016 – April 2017, eighteen cases of accidents in Mt. Bromo area has been recorded for insurance claims, and more might have happened that had not been reported. Visitors who had negative experience will have lower satisfaction and lower option of return visit [7]. There is significant relation between visitors' safety and destination's image, and that the image of the destination affects visitors' motivation to visit the destination [8]. While visitors to protected area should be limited based on its carrying capacity, a decrease of visitor number caused by such bad image would not be acceptable. Therefore, hazard/risk management is required. Risk management should be developed based on different types of client and travel [9]. Risk management stages includes planning risk management, risk assessment, risk control, risk financing, and risk monitoring and evaluation [10]. There are at least two approaches in developing hazard/risk management plan, i.e. job safety analysis/JSA [11] and tourism risk assessment developed by Wilks and Moore for APEC in 2004 [12], which later was also adopted by UNEP [2]. The APEC/UNEP approach is widely used in tourism hazard management researches. This research aimed at identifying and assessing potential tourism hazard and risk in Mount Bromo and providing its management recommendation, using the Wilks and Moore/APEC/UNEP approach.

2. Method

2.1. *Location and Time*

This research was carried out in Mount Bromo. The mountain is part of Bromo Tengger Semeru National Park, which is located on 7°54' - 8°55'13" South Latitude and 112°51' - 113°04'45'' East Longitude. Administratively, Mount Bromo is located on four districts, i.e. Probolinggo, Pasuruan. Lumajang. dan Malang Districts, East Java Province. Interview to residents was carried out to residents living in two adjacent villages, i.e. Ngadas Village, Malang District, and Ngadisari Village, Probolinggo District. The field survey was conducted on May 2017.

2.2. *Data Collection Method*

Survey method, which include field observation and interview with Mount Bromo manager, local community, and visitors, was employed in collecting data on hazard potentials, tourism activities, accidents occurrence, and existing management actions and hazard management plan. Geographical location of potential hazards or source of hazards were marked using the Geographical Positioning System (GPS) for later used in mapping the hazard. In addition, literature and document study was also carried out to obtain the data.

Interview with the manager of Mount Bromo, local community, visitors, and local tourism agency were carried out to obtain data on hazard potentials, its risk, severity and likelihood, accidents occurrence, existing management actions, and community and visitors perception toward hazard and visitors safety services. Management key informant, i.e. the officers in charge of visitor safety services and hazard management, were selected as respondent in the interview with the manager. Interview was also carried out with 30 conveniently sampled local people/residents from Ngadas and Ngadisari Villages located adjacent to the destination. A total of 30 visitor respondents, selected using convenient

IOP Conf. Series: Earth and Environmental Science **149** (2018) 012021 doi:10.1088/1755-1315/149/1/012021

sampling, were also interviewed. Included in the questionnaire were respondents' assessment on hazard likelihood and severity, using qualitative measures (table 1 and table 2).

2.3. Data Processing
Lists of identified hazards in Mount Bromo were developed from field observation, interview with area manager, interview with residents, and interview with visitors. The lists were then combined to develop a complete list of identified hazards in Mount Bromo, which included the likelihood and severity of each hazard potential. Since quantitative data on the frequency of occurrence and the risk incurred is lacking, qualitative data was used to explain the measures of likelihood (table 1) and severity (table 2), based on information provided by the area manager, residents, and visitors. The highest number of responses from the three groups of respondents were used in classifying hazard's likelihood and severity, in order to minimize bias (under or over estimation of likelihood and severity).

Table 1. Qualitative measures of likelihood [12][2]

Descriptor	Description
Almost certain	Is expected to occur in most circumstances
Likely	Will probably occur in most circumstances
Possible/Moderate	Frequently occur at some time
Unlikely	Could occur at some time
Rare	May occur in exceptional/ extreme circumstances

Table 2. Qualitative measures of severity [12][2]

Descriptor	Description
Insignificant	Insignificant consequences, no injury
Minor	Minor consequences/injury: first aid treatment
Moderate	Serious (permanent) injury, medical treatment required
Major	Extensive injuries, with occasional death
Catastrophe	Fatalities/death

2.4. Data Analysis
Potential hazards found in Mount Bromo tourism destination was mapped using Arcmap 10.3 software. The map was derived from map of zonation of Bromo Tengger Semeru National Park and hazard geographical position obtained using GPS in field observation through digitation, data extraction using DNR Garmin 5.4.1 software and map generation process (figure 1).

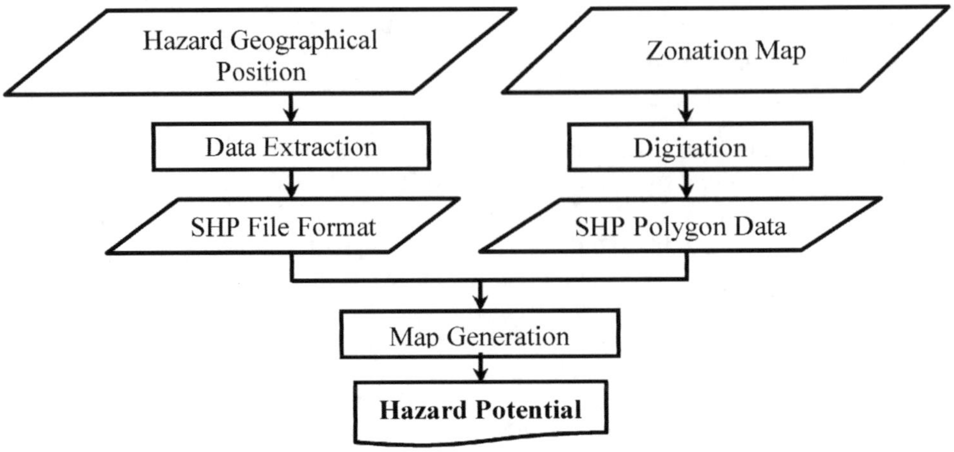

Figure 1. Hazard potential map generation process.

Risk assessment includes risk identification to determine possible risk, analysis/evaluation to determine the magnitude/level of risk, and value judgment/appraisal to determine risk's acceptability [10]. Risk assessment involves assessment to the identified risk based on the impact, separating the minor and acceptable risk to major risk by considering the likelihood (frequency of probability) of risk occurrence, and the impact (consequences) of the risk [12][2]. Hazard found in the destination was ranked based on its likelihood from rare to almost certain (table 1). Rank of impact severity was also assigned to each potential hazard, ranging from insignificant to catastrophe (table 2). The rank of likelihood and severity of each type of hazard was obtained from the highest number of responses given by respondents (managers, residents, and visitors).

After the rank of likelihood and severity was assigned to each type of hazard, the likelihood and severity of each hazard was then mapped into a qualitative matrix of risk analysis (table 3). Thus, a level of risk can be assigned to each potential hazard, ranging from low risk (acceptable risk) to extreme risk. For example, when the likelihood is moderate, and the severity is major, then the level of risk would be extreme, which would require immediate action.

Table 3. Qualitative Risk Analysis Matrix – Level of Risk [12]

LIKELIHOOD	**SEVERITY/CONSEQUENCES**				
	Insignificant	**Minor**	**Moderate**	**Major**	**Catastrophe**
Almost Certain	H	H	E	E	E
Likely	M	H	H	E	E
Moderate/Possible	L	M	H	E	E
Unlikely	L	L	M	H	E
Rare	L	L	M	H	H

Legend: E: Extreme risk, immediate action required; H: High risk, senior management attention needed; M: Moderate risk, management responsibility must be specified; L: Low risk, manage by routine procedures.

After the level of risk were identified, the management option could be determined. For tourism operators and destinations, management option for treating risk include accepting, avoiding, transferring or reducing risk (figure 2). After determining the management option, management actions could also be developed.

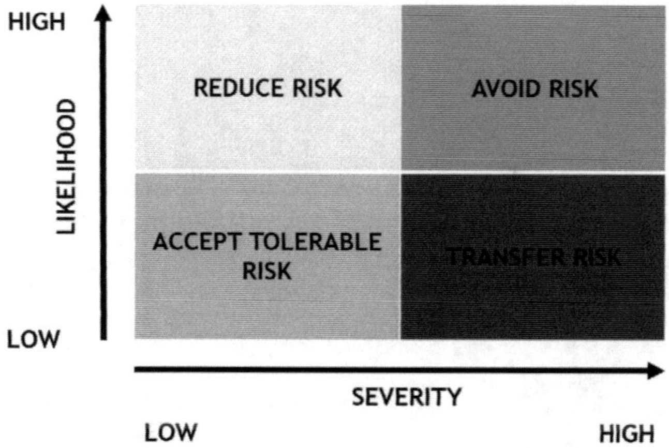

Figure 2. Risk evaluation matriks [12][2].

3. Result and Discussion

3.1. General Condition of Mount Bromo Tourism Area

Mount Bromo is located on the north part of Bromo Tengger Semeru National Park, in the use zone of Cemorolawang-Mentigen. Mount Bromo lies on the caldera of Tengger Mountainous Area, which most people known as Mount Bromo Area, along with several other mountains. Mount Bromo is surrounded by extensive sea of sand (around 5,290 ha) hemmed by high caldera wall forming high and steep cliff. The view of this caldera with sea of sand, Mount Bromo and other Mountain inside of it has attracted high number visitors to this area. Visitors watch the view from Mount Pananjakan, and recently new sites (Bukit Kingkong/kingkong hill and Bukit Cinta/love hill), had been developed to increase the capacity of the destination to accommodate visitors.

In addition to the view, the area also has savanna, various flora and fauna, and cultural attraction of Tengger tribe. Tengger people also plays important role in providing accommodation and in site transportation for the visitors, i.e. horses and jeep. The people firmly hold their tradition, and still continue to carry out traditional ceremonies such as *Yadnya Kasada* in which the people offers sacrifices into the crater of Mount Bromo. Other ceremonies include *Karo, Unan-unan, Entas-entas, Pujan Mubeng*, etc. The ceremonies often also attracts visitors to the area. Visitors were mostly involved in recreational activities such as enjoying the scenery, capturing sunrise from Mount Pananjakan, exploring the sea of sand, hiking the trail to Mount Bromo crater, and enjoying the scenery in Teletabies savanna and hill. They also took pictures in the objects. In order to reach the objects, visitors used rent jeep, rent horse, or motorcycles.

3.2. Hazard Potentials in Mount Bromo Tourism Area

Hazard potentials found in Mount Bromo Tourism Area included physical, biological, and socio-cultural hazards. Physical hazard potentials included cliff and steep topography, land/soil condition, fog, low temperature, wind, volcanic ash, fire, and eruption (figure 3; wind, fog and temperature were identified in the map as physical hazard). Biological hazard found were horses used as transportation mode for the tourist, while socio-cultural hazard included visitor behavior, residents'/local people's behavior, and local myth which causes accidents (figure 3).

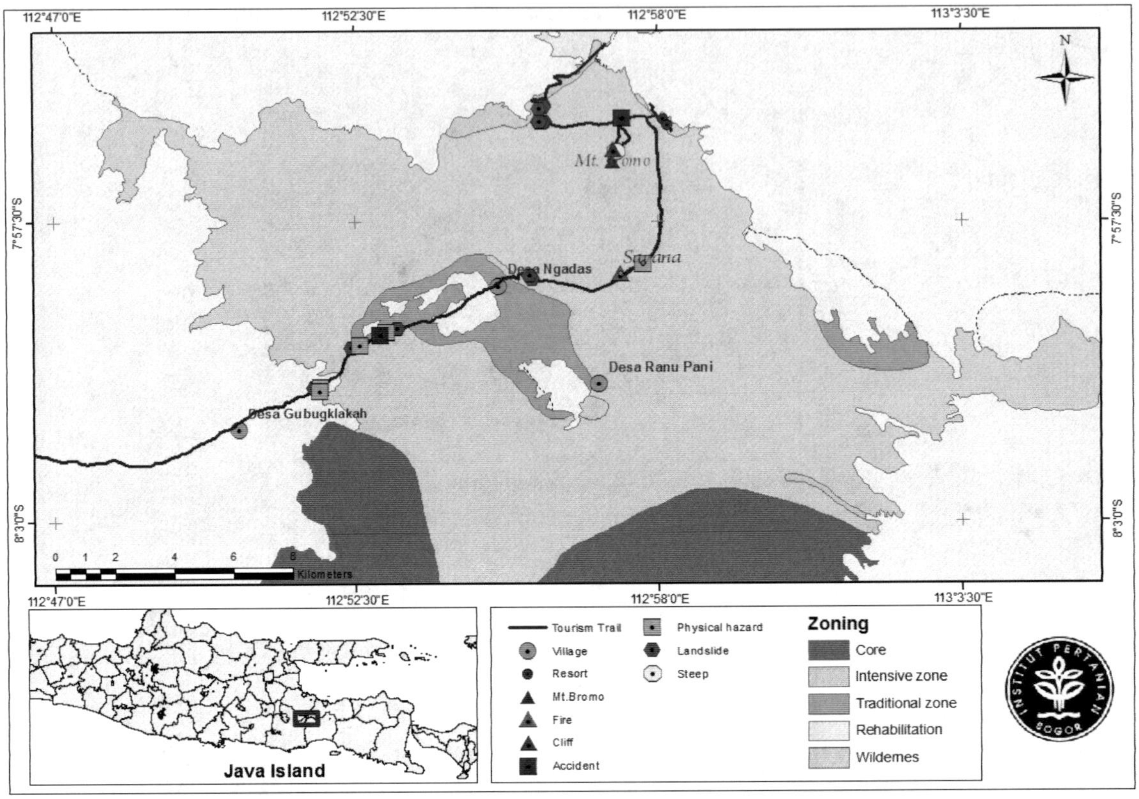

Figure 3. Potential hazard in Mount Bromo Tourism Area.

3.2.1. Cliff and Steep Topography. Cliff and steep topography could be found in several location in Mount Bromo Tourism Area, particularly in Mount Bromo itself (figure 3). Visitors might fall off the cliff during their activities. However, this hazard would pose higher risk when combined with human negligence/behavior. Steep and winding roads also posed threat to visitors who bring their own vehicles without proper knowledge of the area. Accidents often occurred on the road from the nearest village to the area.

3.2.2. Soil/land condition. Eroded area with landslide potentials could be found in several locations (figure 3). Landslide might harm and cause accidents for both visitors and the locals. The sea of sands was also identified as hazard potentials in this category. Frequent changes to soil condition in the sea of sands caused by water streams occurred during heavy rainfall might harm the visitors who were not familiar with the area. Visitors traveled across the sea of sand using jeep, horse, or motorcycle. Those who used their own vehicles and unaccustomed to the area might skid, slip and fall down that caused minor to major injuries to fatalities. The flood in sea of sands could also dragged the vehicles down, causing accidents.

3.2.3. Fog. Fog, which occurred almost every morning and afternoon in the area, limits the visibility of the area for the visitor. It would be difficult to see the road ahead, which might cause accidents, such as lost, fall, slip and stuck, and collision.

3.2.4. Low temperature. Air temperature in Mount Bromo Tourism Area was varying from a minimum range of 3°C- 5°C and maximum range of 20°C- 22°C. In mountain slopes, mountain peaks, and highland area, the temperature was relatively low and the air was thin. Visitors with rather poor physical condition or those who did not wear proper clothing might exposed to the risk of hypothermia.

3.2.5. Wind. Wind added to the risk of hypothermia in addition to structural damage toward physical and biological features of the area. The wind brought chill to the visitors. There were only two occurrence of physical and biological damaged caused by strong wind during 2012 – 2016 [13].

3.2.6. Volcanic ash. Volcanic ash is one of parent material of soil mass in sea of sand. The ash might cause disturbance to visitors' respiratory system, eye symptoms, skin irritation, and other indirect effect [14].

3.2.7. Fire. Fire occurred in the Teletabies savanna might cause harm to visitors. Fire usually occurred in dry season, mostly caused by human negligence.

3.2.8. Eruption. Mount Bromo is one of the most active volcanoes in Indonesia, considering its periodical eruption [15]. The most recent eruption occurred in 2016, when small eruption occurred. Apparently, despite its potential risk, the eruption also provided attraction for tourists, and benefit for the local people. Eruptions were usually accompanied by earthquake.

3.2.9. Horse. Horses were used as transportation mode for the visitors visiting Mount Pananjakan, Sea of Sand, and Mount Bromo crater. Horse might cause accidents for those who could not ride well, and risk of being kicked and collision. Poorly managed feces were still found in the area, which caused the feces fell in to the land, and though the feces might smell bad, it did not pose real threat to human health [16] unless the feces entered human body [17].

3.2.10. Human negligence: visitors' behavior, residents' behavior, and local myth. Visitors were often observed showing behavior that put risk to their own safety, such as climbing and trespassing safety fences put up by the area management to prevent visitors from falling off of the cliff. Such behavior is related to visitor perception toward the activity [18]. They perceived the activities as having more benefit than risks. By climbing and trespassing the fence, visitors might have a better look on the Mt. Bromo scenery, which they might think as the benefit of their action. Human behavior/negligence may also be associated with the lack of their knowledge of the environment. This was often observed in visitors' behavior in using transportation modes and other tourism activities. An increase number of accident cases occurred to visitors using automatic transmission motorcycle. There was a case of accident that cause death of visitor, in which a visitor with acute asthma neglected the hiking requirements. In addition, high number of visitors and local people that packed the crater's edge of Mount Bromo during Kasada ceremony posed risk of falling into the crater and gas suffocation, although the occurrence was low. Local people/the residents were involved in tourism activities, which include, but not limited to jeep driver, horse guide, food and souvenirs vendors. Their tradition to wear sarong as part of their attires might be risky during jeep driving. The sarong might get tangled, making the vehicles lose control and causing accidents. However, a more harmful threat would be from lack of maintenance of the vehicle and lack of safety equipment of the jeep. Lack of maintenance might cause engine breakdown or other damages, which would cause accidents when happened in steep road or vehicles-packed road, such as the road to Mount Pananjakan. The local people believed in the myth that anyone who visits Mount Bromo should be polite, both in their behavior and speech, and those who violated the rules might be suffering some consequences. There were two foreign tourists who got lost after carelessly urinated in the area. They could not find their way back, and just walked around in circles until they were found by the search team. The locals believed that their careless behavior caused the incident.

3.4. Residents' and Visitors' Perception of Hazard

Residents/local people and visitors had different perception of hazard potentials in Mt. Bromo. Hazard perceived by the residents include water stream/rain, fog, steep slope combined with human negligence, eruption, sea of sands, landslide, and human negligence. Quite a high percentage of residents' (28%) perceived steep slope combined with human negligence as potential hazards in this area, while the lowest percentage was sea of sands. The residents knew the area very well, that only a lower percentage of them perceived water stream/rain, fog, eruption, sea of sands and landslide as hazard. Apparently, the interaction between the local people and the volcanic environment is multifaceted and complex, in which people choose to live with the hazard, and exposed to its adverse impact but also enjoyed the benefits from the environment [15]. There were a higher number of hazard perceived by the visitors. Most visitors perceived human negligence as the cause of accidents, however there were still a large percentage of visitors (more than 40%) who confessed their lack of knowledge about accidents or hazard. Their lack of knowledge might be the cause of their negligence toward hazard. This led to a conclusion of the need of visitors' education about hazard.

3.5. Risk Level of Hazard in Mount Bromo

Risk assessment was carried out to identify the risk that might occurred from each hazard potentials based on its likelihood of occurrence and severity. Qualitative measures were used in this research due to lack of quantitative data (such as accidents data). In this research, respondents were asked for their assessment of hazards' likelihood and severity. The highest number of responses were used to classify hazard's likelihood and severity. Risk level for each type of hazard was then determined based on the likelihood and severity. For example, even though cliff and steep topography are unlikely to cause accidents (could occur at some times), when it does cause accidents, the consequences would be major (extensive injuries with occasional death), and therefore the level of risk is high according to the qualitative risk analysis matrix. Indeed, falling off of a cliff would cause extensive injuries, and might even cause death.

The assessment toward each potential hazards in Mount Bromo showed that the level of risk from hazard found in this area ranged from low to high risk, without extreme/very high risk. Cliff and steep topography, fog, low temperature, and human negligence had high risk, while soil condition, volcanic ash and fire had moderate risk, and the other hazards had low risk (table 4).

Table 4. Risk level of potential hazards found in Mount Bromo Tourism Area

No.	Hazard	Severity	Likelihood	Risk Level
1	Cliff and steep topography	Major	Unlikely	**High**
2	Soil condition/landslide potential	Minor	Moderate	**Moderate**
3	Fog	Minor	Likely	**High**
4	Low temperature	Minor	Likely	**High**
5	Wind	Minor	Rare	**Low**
6	Volcanic ash	Minor	Moderate	**Moderate**
7	Fire	Moderate	Rare	**Moderate**
8	Eruption (and earthquake)	Minor	Unlikely	**Low**
9	Horse	Minor	Rare	**Low**
10	Human negligence: visitors' behavior, residents' behavior, and myth	Major	Unlikely	**High**

3.6. Existing Hazard Management in Mount Bromo Area

The manager of Mount Bromo Tourism Area, the Office of Bromo Tengger Semeru National Park, had already implemented hazard management action. The management had already implement various

IOP Conf. Series: Earth and Environmental Science **149** (2018) 012021 doi:10.1088/1755-1315/149/1/012021

hazard management effort, which included the provision of insurance for visitors, routine patrol to control fire and monitor visitors' activities, cooperation with various stakeholders in handling of cases, increasing residents' and visitors' awareness of hazard, documenting accident cases. Insurance fee was included in the entrance ticket fee paid by the visitors. Investment in insurance is usually performed by business operators who understand the potentials of loss that may be incurred by hazards, but do not have the physical resources needed to overcome the hazard [19]. The national park management indeed had limitation of resources and lack of flexibility in managing financial source in implementing hazard management in particular, and tourism management in general.

Community and visitor education programs included extension program for the residents in relation to fire management, the use of volcanic status stamp to notify visitor of alert status regarding volcano, announcement made by ticketing officer to notify visitor about fog occurrence, and putting up written announcement about landslide occurrence and web announcement of automatic transmission motorcycle prohibition to the area. Despite all measures that had been taken by the management, there was still lack of hazard warning signs in the area. A study in the Alpen Mountainous area in Austria found that most of its visitors perceived that the lack of risk information means that there was not any risk in the area [20]. This indicated the need of hazard information services for the visitors.

3.7. *Hazard Management Recommendation*

There are four management options for tourism management, i.e. (1) retain/accept tolerable risk, when both severity and likelihood is low; (2) reduce risk, when the severity is low, but the likelihood is high; (3) transfer risk, when the severity is high, but the likelihood is low, and; (4) avoid risk, when both severity and likelihood is high, causing extreme risk/very high risk [2] [12]. Considering that the risk of hazards found in Mount Bromo were low to high level, without very high/extreme level risk, then there are three management options that can be applied in hazard management of Mount Bromo Tourism Area, i.e. accepting tolerable risk, reducing the risk, and transferring the risk. For low risk hazard, such as wind, horse and eruption which rarely to unlikely occured, accepting tolerable risk would be the option (table 5). For moderate risk hazard, such as fire, soil condition/landslide potentials, and volcanic ash which had rare to moderate likelihood, the option was reducing risk (table 5). The high risk of cliff and steep slope, fog, low temperature, and human negligence should be transferred (table 5).

Management actions for low risk hazard should include provision of information services (putting up written and oral announcement, warning signs, etc.) that may help the visitors avoiding the areas during hazard occurrence, and patrol to make sure the visitors took heed of the warnings. In cooperation with the local community involved in tourism activity, visitors should also be informed to be careful around horses and other hazardous objects/areas.

In order to reduce the risk, routine monitoring of hazard potentials should be carried out. In addition, visitor education, providing information services and establishing a standard operational procedures (SOP) for hazard management, and standard requirement for visitors entering the area, are also suggested to reduce the risk. Tourism manager should pay attention to 'best practice' initiatives such as having written policies and procedures, staff training, signage, visitor briefings, and monitoring of industry standards in reducing risk [12].

In cases where visitors being injured or killed, or loss of properties incurred, tourism managers traditionally relies on insurance to transfer their financial risk to the third party [12]. This measure had also been applied by Mount Bromo manager. Therefore, transferring risk using insurance should be maintained by the management in transferring financial loss to other party, the insurance company. However, considering visitors lack of knowledge and negligence toward hazards, visitor education should be included in the hazard management action. Interpretive program and media may be the options for visitor education. Visitors showed better cognitive, affective, and behavioral outcome with the increase exposure toward interpretive media [21]. Realistic visualization of hazard can help people to develop a better understanding of the hazard [22]. This measure should also be applicable in reducing the risk of moderate level risk. As an addition, educating the local community/the residents on how to help alerting visitors of hazard potential may also be helpful in reducing risk for the visitors.

Table 5. Hazard management option and action

No.	Risk Level	Hazard	Management Option and Action
1	High	Cliff and steep slope, fog, low temperature, and human negligence	Transfer risk: Insurance + visitor education (interpretive program and media), community education,
2	Moderate	Soil condition/landslide potential, volcanic ash, fire	Reduce risk: Routine monitoring of hazard potentials, visitor education (installing hazard signs/interpretive media), information services, set up SOP
3	Low	Wind, eruption (and earthquake), and horse	Accept tolerable risk: Information services, post-occurrence handling

4. Conclusion

Hazards found in Mt. Bromo tourism area had low to high risk. Three risk management options, i.e. accept tolerable risk, reduce risk, and transfer risk, should be applicable in this area. Areas such as the steps to Mount Bromo crater, and the steep trail to Pananjakan I viewing area, had high risk for the visitors. In occurrence of fog, the whole area of Mount Bromo, from the entrance gate to sea of sand and trail to Pananjakan I, posed high risk of accidents to visitors, and therefore the management should pay extra attention during fog. The officer should already provide warning starting from the entrance gate. In case of heavy fog, a temporary closing may be applicable to reduce the risk. Considering visitors lack of knowledge and negligence, hazard management should include raising visitors' awareness toward hazard and how to avoid and handle the risk. The management should consider to develop and implement interpretive program for visitor, install more signs and labels in the area to prevent accidents.

Acknowledgement

This paper is a part of a research entitled 'Tourism Hazard Management in Disaster-Prone Areas in Indonesia, funded by the Ministry of Research, Technology and Higher Education through Leading Strategic Research Scheme of University Leading Research, contract number: 011/SP2H/LT/DRPM/IV/2017 date April 20th, 2017, 2017. The authors are grateful for the help of students, Inggita Nuri and Sofia Ucu Utami, in collecting data and for all the parties that had supported this research.

References

[1] Canadian Centre for Occupational Health and Safety 2017 OSH answers fact sheet: hazard and risk. Available at: https://www.ccohs.ca/oshanswers/hsprograms/hazard_risk.html [2 Dec 2017].

[2] [UNEP] United Nation Environment Programme 2008 *Disaster Risk Management for Coastal Tourism Destinations Responding to Climate Change A Practical Guide for Decisions Maker* (Paris: UNEP)

[3] Jubenville A, Twight B W and Becker R H 1987 *Outdoor Recreation Management, Theory and Aplication* (Oxford Circle: Venture Publishing, Inc.)

[4] Ilieş DC, Dehoorne O and Ilieş A 2011 Some exe[a]mples of natural hazards affecting Geosites and tourist activities *GeoJournal of Tourism and Geosites* **7** (1) 33-38

[5] Kumpulainen S 2006 Vulnerability concepts in hazard and risk assessment Schmidt-Thomé P (Ed.) Natural and technological hazards and risks affecting the spatial development of European regions. *Geological Survey of Finland, Special Paper* **42** 65–74

[6] Rachmawati E, Faizah N K and Muntasib E K S H 2007 Potensi bahaya di kawasan wisata Gunung Bromo, Resort Tengger Laut Pasir, Taman Nasional Bromo Tengger Semeru, Jawa Timur *Media Konservasi* **12** (3) 140 – 144
[7] Howard R W 2009 Risky business? Asking tourist what hazards they actually encountered in Thailand *Tourism Management* **30** 359–365
[8] Suharto 2016 Studi tentang keamanan dan keselamtan pengunjung hubungannya dengan citra destinasi (Studi kasus Gembira Loka Zoo) *J. Media Indonesia* **14** (1) 287-304.
[9] Department of Labour Te Tari Mahi 2010 *Review of risk management and safety in the adventure and outdoor commercial sectors in New Zealand 2009/10* [Final Report] (Wellington: Department of Labour)
[10] Department of Environmental Affairs and Tourism 2006 Risk Management *Integrated Environmental Management Information Series* **23** (Pretoria: Department of Environmental Affairs and Tourism)
[11] Canadian Centre for Occupational Health and Safety 2017 OSH answers fact sheet: Job Safety Analysis. Available at https://www.ccohs.ca/oshanswers/hsprograms/job-haz.html [14 Jan 2018]
[12] Wilks J and Moore S 2004 *Tourism risk management for the Asia Pacific region: an authoritative guide for the managing crises and disasters* (Gold coast: CRC for Sustainable Tourism Pty Ltd.)
[13] Balai Besar Taman Nasional Bromo Tengger Semeru 2017 *Statistik Balai Besar Taman Nasional Bromo Tengger Semeru tahun 2016* (Malang: BBTNBTS)
[14] International Volcanic Health Hazard Network (IVHHN), Cities and Volcanoes Commission, GNS Science and the United States Geological Survey (USGS) 2007 *The Health Hazards of Volcanic Ash: A Guide for the Public* [pamphlet] https://www.civildefence.govt.nz/assets/Uploads/publications/volcanic-ash-health.pdf [2 *Dec 2017]*
[15] Bachri S, Stötter J, Monreal M and Sartohadi J 2015 The calamity of eruptions, or an eruption of benefits? Mt. Bromo human–volcano system a case study of an open-risk perception. *Nat. Hazards Earth Syst. Sci.* **15** 277–290, doi:10.5194/nhess-15-277-2015
[16] Quinn A 2001 Does Horse Manure Pose A Significant Risk to Human Health? *EnviroHorse paper* 2001. www.bayequest.com/static/pdf/manure.pdf
[17] Kitcehener N 2017 *6 Diseases You Can Catch From Your Horse* https://www.horse-canada.com/magazine_articles/6-diseases-you-can-catch-from-your-horse/ [2 Dec 2017]
[18] Gstaettner A M, Rodger K and Lee D 2017 Visitor perspectives of risk management in a natural tourism setting: an application of the Theory of Planned Behavior *Journal of Outdoor Recreation and Tourism* **19** 1–10
[19] Cioccio L and Michael E J 2007 Hazard or disaster: Tourism management for the inevitable in Northeast Victoria *Tourism Management* **28** 1–11
[20] Pröbstl-Haider U, Dabrowska K and Haider W 2016 Risk perception and preferences of mountain tourists in light of glacial retreat and permafrost degradation in the Austrian Alps *Journal of Outdoor Recreation and Tourism* **13** 66–78
[21] Weiler B and Smith L 2009 Does more interpretation lead to greater outcomes? An assessment of the impacts of multiple layers of interpretation in a zoo context *Journal of Sustainable Tourism* **17** (1) 91–105
[22] Yang B 2016 GIS based 3-D landscape visualization for promoting citizen's awareness of coastal hazard scenarios in flood prone tourism towns *Applied Geography* **76** 85–97

Relationship between land use and water quality in Pesanggrahan River

Hefni Effendi[1,2], Sri Muslimah[2,3] and Prita Ayu Permatasari[2]

[1]Department of Aquatic Resources Management, Faculty of Fisheries and Marine Science, Bogor Agricultural University (IPB)
[2]Environmental Research Center, Bogor Agricultural University (IPB)
[3]Department of Information Technology for Natural Resources Management, Faculty of Mathematics and Natural Sciences, Bogor Agricultural University (IPB)

E-mail: hefni_effendi@yahoo.com

Abstract. Pesanggrahan River watershed has several activities such as residential and commercial area in its catchment area. The purpose of this study was to analyse water quality related to spatial land use in Pesanggrahan River using GIS Analysis. River water quality in some locations, did not meet water quality standard of class III. From pollution load estimation it was revealed that segment 2 (Bogor City) has the highest BOD, COD, and TSS of 15,043 kg/day, 25,619 kg/day, and 18,104 kg/day respectively. On the other hand, the most developed area in Pesanggrahan Watershed is located in segment 7 (24.5%). Hence, it can be concluded that although an area has a fairly small developed area, high urban activity can cause high BOD, COD, and TSS.

1. Introduction

Pesanggrahan Watershed has an area of 55,141.84 ha with the river upstream in the city of Bogor which then flows into West Jakarta. Segmentation of Pesanggrahan Watershed area is divided based on administrative boundaries covering areas of Bogor City, Bogor District, Depok, South Tangerang, Tangerang City, South Jakarta and West Jakarta. The population within the Pesanggrahan Basin is different for each segmentation, in 2012 there are mostly in South Jakarta area of 710,406 population [1]. Based on the study of phytoplankton diversity in Pesanggrahan River both upstream and downstream showed moderate pollution, where the results of regression analysis showed that the nitrate content of waters significantly influenced the abundance of phytoplankton and zooplankton [2].

The contribution of pollutant source of Pesanggrahan River comes from household waste which is a waste from the tributaries obtained from spatial analysis. Previous studies have proven that the pollution load capacity in Pesanggrahan River needs to be reduced because it has exceeded the threshold [3]. It is important to set up the water class first in determining the pollution load capacity so that Pesanggrahan river can be utilized in accordance with its designation.

The importance of water for life such as supporting the economy, utilization as drinking water and washing make humans must be willing to preserve the environment [4]. The level of river water pollution of Pesanggrahan is crucial to be identified therefore it can be utilized in accordance with the designation of river class. The government regulates the classification of river classes into four classes in Government Regulation No. 82 of 2001, namely classes 1 (water use for drinking water), 2 (water use for tourism), 3 (water use for fisheries and livestock farming), and 4 (water use for crop irrigation).

Determination of water pollution level using the calculation method of storet index and pollution index, as done by [5,6,7,8,9].

Research on river water classes has been largely performed one of them is in Southern Iraq using water quality indexes [10]. Based on that study, the water quality index is able to classify water use. The purpose of this study was to analyze water quality related to spatial analysis of land use in Pesanggrahan River using GIS Analysis.

2. Methodology

2.1. Study area

The monitoring locations of Pesanggrahan watershed are located in seven regencies/cities (Bogor Regency, Bogor City, Depok, South Tangerang City, Tangerang City, South Jakarta, and West Jakarta) with total area of watershed is 55,141.84 ha. The upperstream area is located in Bogor Regency and flows to West Jakarta through Depok, South Tangerang, Tangerang City, and South Jakarta. Pesanggrahan Watershed have been reported has several activities such as residential and commercial area. Land use of Pesanggrahan River can be seen in figure 1. While table 1 showed area percentage of every land use.

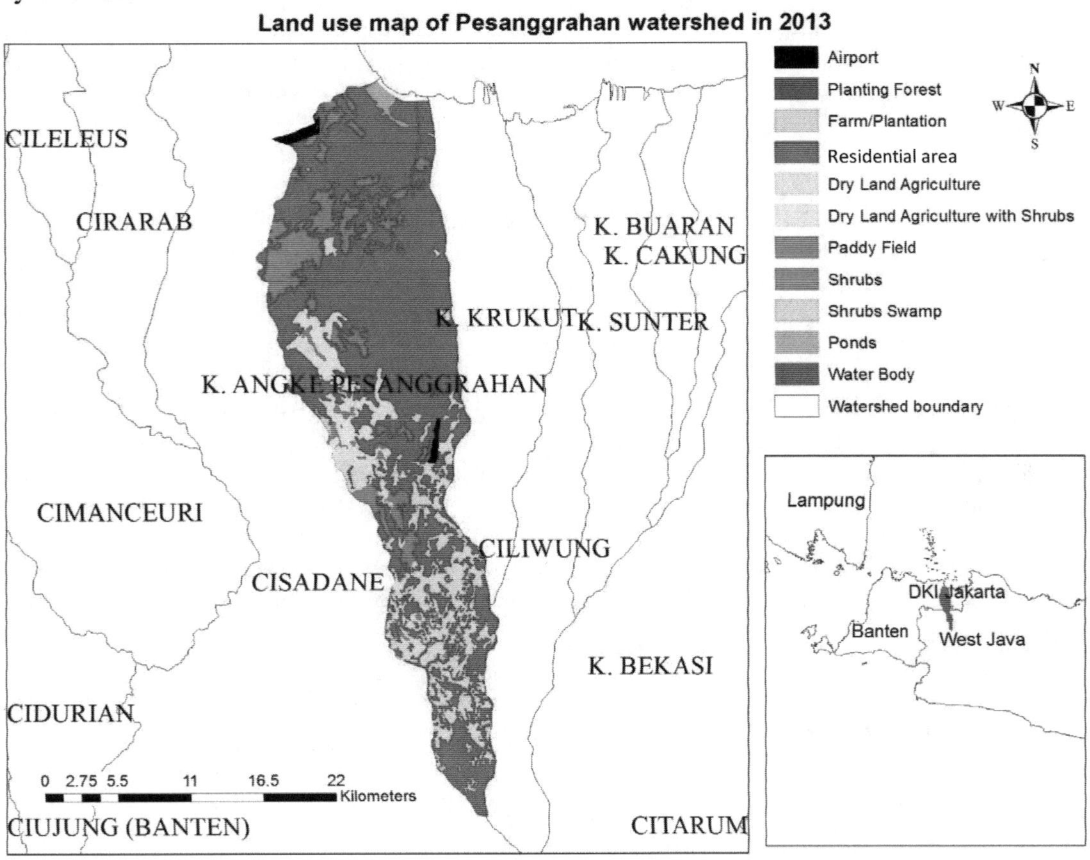

Figure 1. Land use map of Pesanggrahan Watershed in 2013.

IOP Conf. Series: Earth and Environmental Science **149** (2018) 012022 doi:10.1088/1755-1315/149/1/012022

Table 1. Land use and its percentage area in Pesanggrahan Watershed.

		Percentage Area (%)					
No	**Area**	**Built area**	**Forest**	**Plantation**	**Agriculture**	**Shrub**	**Water Bodies**
1	Bogor City	6.01	100.00	0.00	3.48	0.00	0.00
2	Bogor Regency	1.27	0.00	0.00	29.21	0.00	10.17
3	Depok	10.82	0.00	0.00	13.59	0.00	8.91
4	South Tangerang	11.59	0.00	100.00	30.55	93.34	6.41
5	Tangerang City	23.64	0.00	0.00	19.41	0.00	3.10
6	South Jakarta	22.17	0.00	0.00	0.10	0.00	0.00
7	West Jakarta	24.50	0.00	0.00	3.66	6.66	71.41
	Total	100.00	100.00	100.00	100.00	100.00	100.00

2.2. Tools and materials

The data used for analysis is water quality monitoring data from Balai Besar Wilayah Sungai Ciliwung-Cisadane (BBWS-CC) and Environmental Status Book (SLHD) from 2013 to 2014 (every year has different availability of monitoring data). Another data that used for analysis are number of population, administration map, Pesanggrahan watershed boundary map, water requirement every segment, planning official map and landuse landcover map. The tools used for research are ArcGIS 10.5 and microsoft excel.

2.3. Methods

Representative water quality data per year was concluded by averaging the data in that year. The parameters were total dissolved solids (TDS), total suspended solids (TSS), temperature, pH, biochemical oxygen demand (BOD), chemical oxygen demand (COD), and dissolved oxygen (DO) based on availability of data in every station. There were six monitoring stations for water quality observation, namely Jl. Ciputat Pasar Jum'at Lebak Bulus, Jl. H. Kelik, city park, Sawangan, Tanah Kusir, and Kebun Jeruk.

The potential polluted load in Pesanggrahan watershed was estimated by secondary data from SLHD report of Bogor City, Bogor Regency, Depok, South Tangerang, Tangerang City, South Jakarta, and West Jakarta. Because all of pollutant sources were non point source, it's difficult to conduct direct measurement. Emission factor pollution load from industry, mining, residential area, agriculture, farms, fishing, hotel, hospital, and waste was estimated. The potential polluted load parameters was counted for BOD, COD, and TSS (kg/day). The formula for estimating potential polluted load can be seen in equation 1-10. While emission factor for each parameter can be seen in table 2, table 3, table 4, table 5, table 6, and table 7.

2.3.1. Pollutant estimation for industry

$$Pollutant\ (\frac{Kg}{year}) = \frac{Ci\ \times V\ \times OpHrs}{1{,}000{,}000} \tag{1}$$

Ci = Concentration on wastewater (mg/L)
V = Velocity of wastewater (L/hour)
OpHrs = operational hours (hours/year)
1,000,000 = Conversion factor from mg to kg

2.3.2. Pollutant estimation for household

$$Pollutant\ (kg/day) = \frac{Population\ \times Emission\ factor\ \times equivalence\ ratio \times\ \alpha}{1000} \qquad (2)$$

Emission factor:
1. BOD = 40 gr/person/day
2. COD = 55 gr/person/day
3. TSS = 38 gr/person/day

Urban equivalence ratio:
1. Urban area = 1
2. Suburb area = 0.8125
3. Countryside = 0.625

Alpha (α): Load transfer coefficient (0.3 - 1)
α = 1 for location 0-100 meter from river
α = 0.85 for location 100-500 meter from river
α = 0.3 for location >500 meter from river

2.3.3. Pollutant estimation for livestock farming

$$Pollutant\ (kg/day) = \frac{number\ of\ livestock\ \times Emission\ factor\ \times 0.2}{1000} \qquad (3)$$

Table 2. Livestock emission factor.

Parameter	Unit	Buffalo	Cow	Horse	Pig	Sheep	Goat	Chicken	Swan	Duck
BOD		207	292	226	128	55.7	34.1	2.36	2.46	0.88
COD	gr/livestock	530	717	558	362	136	92.9	5.59	6.67	2.22
N Total	/day	2.6	0.93	38.08	4.62	0.278	1.62	0.002	0.061	0.001
P Total		0.39	0.15	0.31	0.31	0.063	0.11	0.002	0.006	0.005

2.3.4. Pollutant estimation for agriculture

$$Pollutant\ (kg/day) = \frac{crop\ area\ \times emission\ factor \times 0{,}1}{number\ of\ days\ of\ planting\ season} \qquad (4)$$

Table 3. Agriculture emission factor.

Pollutant (Kg/Ha/Planting Season)	Paddy	Secondary crop	Plantation
BOD	225	125	32.5
N	20	10	3
P	10	5	1.5
TSS	0.46	2.4	1.6
Pesticide	0.16	0.08	0.025

IOP Conf. Series: Earth and Environmental Science **149** (2018) 012022 doi:10.1088/1755-1315/149/1/012022

2.3.5. Pollutant estimation for hotel and hospital

$$Pollutant\ (kg/day) = \frac{number\ of\ room \times emission\ factor\ \times \%\ occupancy\ \times equivalence\ ratio\ \times Runoff\ Ratio}{1000} \quad (5)$$

$$Pollutant\ (kg/day) = \frac{number\ of\ bed\ \times emission\ factor\ \times \%\ occupancy\ \times equivalence\ ratio\ \times Runoff\ Ratio}{1000} \quad (6)$$

Table 4. Emission factor of hotel and hospital.

Type of Pollutant Source	Emission factor (gr/day)		
	BOD	**COD**	**TSS**
Hospital (per bed)	123	169.125	116.85
Hotel (per room)	55	75.625	52.25

2.3.6. Pollutant estimation for fisheries

$$Pollutant\ (kg/day) = \frac{fisheries\ production\ (kg/year) \times Emission\ factor}{365} \quad (7)$$

Table 5. Quantity of pollutant load per 1 kg of aquaculture production (kg/year).

BOD	COD	Nitrogen Total	Phosphor Total
0.0966	0.1449	0.0572	0.0131

2.3.7. Pollutant estimation for domestic waste

$$Pollutant\ (kg/day) = unhandled\ domestic\ waste\ (kg/hari)\ \times Emission\ factor \quad (8)$$

Table 6. Domestic waste emission factor.

Emission factor	Emission (Kg/day)
BOD	0.00282
COD	0.003878
TSS	0.002679

2.3.8. Pollutant estimation for small enterprise

$$Pollutant\ (kg/day) = \frac{Production\ (kg/day) \times Emission\ factor\ \times 0{,}3}{1000} \quad (9)$$

Table 7. Small enterprise emission factor.

No	Type of Small Enterprise	Parameter (Kg/day)		
		BOD	COD	TSS
1	Soybean processing	50	110	80
2	Tapioca processing	3.34	10.30	4.67

2.3.9. Total pollutant estimation

$$\begin{aligned} Total\ Water\ Pollution\ Load \\ &= Industry\ pollutant + Household\ pollutant + Livestock\ pollutant \\ &+ Agriculture\ pollutant + Hospital\ pollutant + Hotel\ pollutant \\ &+ Fisheries\ pollutant + Domestic\ waste\ pollutant + Small\ enterprise\ pollutant \end{aligned} \quad (10)$$

The landuse map was generated from planology agency of Indonesia. The classification of landuse was grouped into eleven classes such as airport, planting forest, farm/plantation, residential area, dry land agriculture, dry land agriculture with shrubs, paddy field, shrubs, shrubs swamp, ponds, and water body. The watershed boundary was generated from Indonesian watershed data. Quality standards of TDS, TSS, temperature, pH, BOD, COD, and DO based on Goverment Regulations No. 82 year 2001 can be seen in table 8.

Table 8. Quality standards based on government regulation No. 82 year 2001 for each water use class.

Parameters	Unit	Quality standards			
		Class I	Class II	Class III	Class IV
TDS	mg/l	1000	1000	1000	2000
TSS	mg/l	50	50	400	400
Temperature	°C	-	-	-	-
pH	-	6 – 9	6 – 9	6 – 9	5 – 9
BOD	mg/l	2	3	6	12
COD	mg/l	10	25	50	100
DO	mg/l	6	4	3	0

3. Results and discussion

All monitoring locations were located in downstream area which water use is for freshwater fish farming and husbandry (Class III). The averaged water quality data in 2014 is shown in table 9. The values in grey shading means that they didn't meet the quality standards of Class III. The high value of BOD usually indicates high residential and commercial lands [11]. According to Li's research [12] the sampling time also affects the water quality. BOD and COD concentrations tend to be higher during the dry season.

Table 9. Averaged water quality in 2014.

Parameter	Jl. Ciputat Pasar Jum'at Lebak Bulus	Jl. H. Kelik	Taman Kota	Sawangan	Tanah Kusir	Kebun Jeruk
TDS	107.7	133.2	143.3	99.0	140.0	194.0
TSS	28.3	57.0	42.5	112.0	71.0	43.0
Temperature	27.6	29.2	28.6	27.5	28.0	28.0
pH	7.4	7.4	7.4	7.5	7.7	7.7
BOD	3.8	9.6	8.6	2.7	2.5	3.5
COD	40.0	34.7	40.0	48.8	60.6	68.8

Parameter	Jl. Ciputat Pasar Jum'at Lebak Bulus	Jl. H. Kelik	Taman Kota	Sawangan	Tanah Kusir	Kebun Jeruk
DO	4.6	1.1	3.0	4.4	2.0	1.1

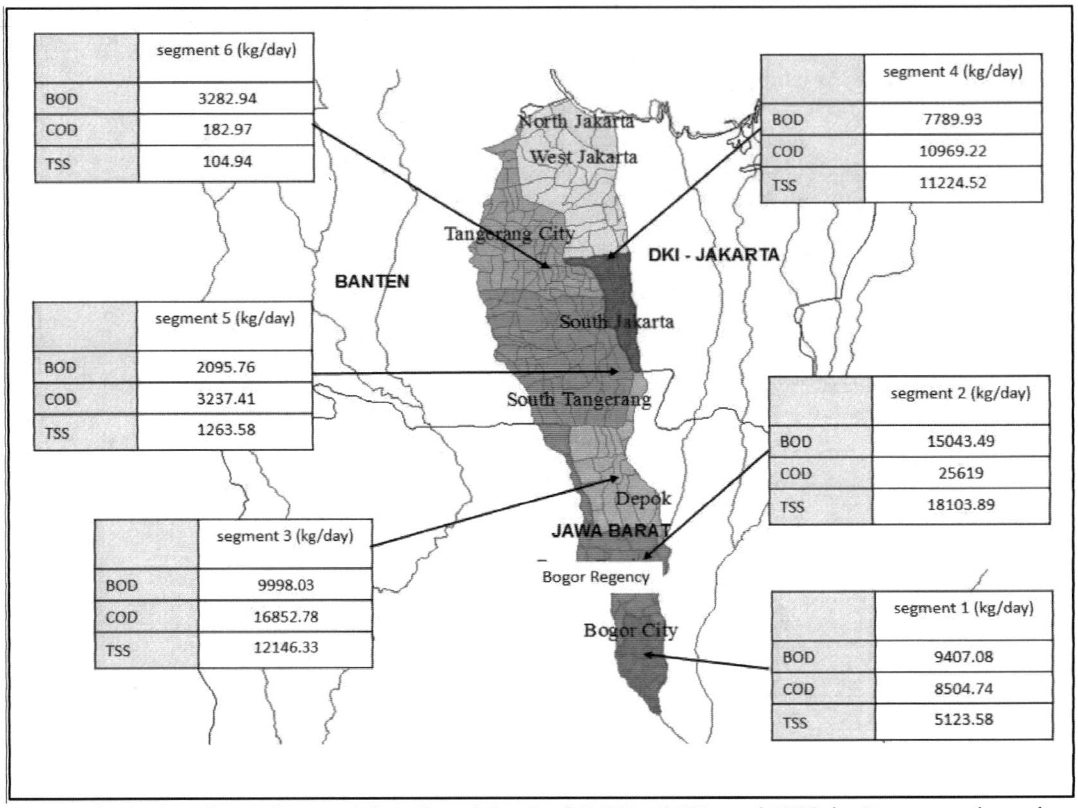

Figure 2. Estimation of potential polluted load of BOD, COD and TSS in Pesanggrahan river from 2012 – 2014 (kg/day).

Figure 2 illustrated the estimation of potential pollution load in Pesanggrahan river from 2012 to 2014. Segmentation division based on administrations boundary mentioned in table 1. The highest potential pollution load of BOD, COD, and TSS is in segment 2. Based on analysis, residential area contributed highest source of pollution load of BOD. Meanwhile, small and medium enterprise contributed highest source of COD and TSS in Pesanggrahan River.

Land use in the Pesanggrahan watershed is divided into eleven kinds of land utilization comprising airports, crops, plantations, residential area, dryland farming, dryland farming with shrubs, paddy fields, shrubs, swamps, mines, water bodies. Residential area has an area of almost 90% of the entire Pesanggrahan River Watershed. The contribution of this residential area to changes water quality is household waste. According to Downing *et al* [13], land convertion from forest to developed area or plantations is more common in tropical countries. The land cover in the built area is dominated by pavement, allowing sediment deposits to water bodies [14-16].

The highest developed area is located in segment 7 (24.5%). However, area with the highest BOD, COD, and TSS values is located in segment 2. The concept of BOD came after cities started dumping their waste water into the water bodies and this led to depletion of oxygen. Thus, although an area has a fairly small developed area but high urban activity, it may cause the BOD value to increase. Urbanization increases the high value of pollutants such as ammonia, chloride, and coliform [17]. On the other hand, in areas with urban land cover and accompanied by high agricultural activity tend to have high nitrate concentration. This is related to the use of fertilizers and pesticides [18]

4. Conclusion

The high concentration of BOD, COD, and DO that have been obtained from the sampling result shows high urban activity in some segments. It may cause water sample can not meet water quality standard of Class III. Based on the estimation of potential pollutant load, the highest BOD, COD and TSS values are located in segment 2. However, the percentage of developed area is located in segment 7. It shows that in segment 2 there is high urban activity and produce BOD, COD and TSS that exceed other segments. Although land cover has influence on water quality, land cover is not the only one factor that affected water quality. Domestic wastes are generally main cause of river water pollution in almost developing countries. Therefore, further research needed to ensure that land cover has important impact on water quality.

References

[1] Statistics Indonesia 2013 *South Jakarta in Figures* (Jakarta: BPS)
[2] Mohammad FF 2012 *The structure of the plankton community in Pesanggrahan River from the upstream (Bogor, West Java) to the downstream (Kembangan, DKI Jakarta)* (Depok: University of Indonesia)
[3] Komarudin M, Dayyadi S, Kurniawan B 2015 Analisis daya tampung beban pencemaran sungai Pesanggrahan (Segmen Kota Depok) dengan menggunakan model numerik dan spasial *Jurnal Pengelolaan Sourcedaya Alam dan Lingkungan* **5** 121-132
[4] Zhao YW, ZF Yang, F Xu 2007 Theoretical framework of the urban river restoration planning. *Environmental Informatics Archives* **5** 241-247
[5] Effendi H 2003 Telaah kualitas air. Kanisius, Jogjakarta, 258 p (In Indonesian).
[6] Effendi H, Romanto, Wardiatno Y 2015 Water quality status of Ciambulawung River, Banten Province, based on pollution index and NSF-WQI. *Procedia Environmental Sciences* **24** 228-237
[7] Effendi, H. 2016 Evaluation of water quality status of Ciliwung River based on pollution index. Proceedings IAIA 16 Nagoya, 11-14 May 2016. Japan.
[8] Effendi, H 2017 River water quality preliminary rapid assessment using pollution index *Procedia Environmental Sciences* **33** 562 -567
[9] Effendi H, Sabila M F, Setiawan Y 2018 Correlation between water quality and land use change in Ciliwung watershed *Nature Environment and Pollution Technology* **17**(1)
[10] Salam HW, Salwan AA 2017 Water quality index for Al-Gharraf river, Southern Iraq *Egyptian Journal of Aqua. Res.* **43** 117-122
[11] Tu Q 1990 *The Researches on the Eutrophication in Chaohu Lake* Heifei (China: University of Science and Technology of China Press) (Chinese)
[12] Li S, Gu S, Liu W, Han H, Zhang Q 2008 Water quality in relation to land use and land cover in the upper Han River Basin, China *Catena* **75** 216-222
[13] Downing J A, McClain M, Twilley R, Melack J M, Elser J, Rabalais N N, Lewis W M Jr, Turner R E, Corredor J, Soto D 1999 The impact of accelerating land-use change on the N-cycle of tropical aquatic ecosystems: current conditions and projected changes *Biogeochemistry* **46** 109 – 148
[14] Banner E, Stahl A, Dodds W 2009 Stream discharge and riparian land use influence in-stream concentrations and loads of phosphorus from central plains watersheds *Journal of*
[15] Tong S, Chen W 2002 Modeling the relationship between land use and surface water quality *Journal of Environmental Management* **66** 377-393
[16] Brabec E, Schulte S, Richards P 2002 Impervious surfaces and water quality: A review of current literature and its implications for watershed planning *Journal of Planning Literature* **16** 499514

IOP Conf. Series: Earth and Environmental Science **149** (2018) 012022 doi:10.1088/1755-1315/149/1/012022

[17] Sliva L, Williams D D 2001 Buffer zones versus whole catchment approaches to studying land use impact on river water quality *Water Research* **35** 3462 – 3472
[18] Hooda P S, Edwards A C, Anderson H A, Miller A 2000 A review of water quality concerns in livestock farming areas *Science of the Total Environment* **250** 143 – 167

Water quality of Cisadane River based on watershed segmentation

Hefni Effendi[1,2*], Prita Ayu Permatasari[1], Sri Muslimah[1], Mursalin[2]

[1] Environmental Research Center, Bogor Agricultural University, Bogor 16680, Indonesia
[2] Department of Water Resource Management, Faculty of Fisheries and Marine Science, Bogor Agricultural University, Bogor 16680, Indonesia

E-mail: hefni_effendi@yahoo.com

Abstract. The growth of population and industrialization combined with land development along river cause water pollution and environmental deterioration. Cisadane River is one of the river in Indonesia where urbanization, industrialization, and agricultural are extremely main sources of pollution. Cisadane River is an interesting case for investigating the effect of land use to water quality and comparing water quality in every river segment. The main objectives with this study were to examine if there is a correlation between land use and water quality in Cisadane River and there is a difference in water quality between the upstream section of Cisadane River compared with its downstream section. This study compared water quality with land use condition in each segment of river. Land use classification showed that river segment that has more undeveloped area has better water quality compared to river segment with developed area. in general, BOD and COD values have increased from upstream to downstream. However, BOD and COD values do not show a steady increase in each segment Water quality is closely related to the surrounding land use.Therefore, it can not be concluded that the water quality downstream is worse than in the upstream area.

1. Introduction

River is an important natural features of ecological life and really related with human activities such as industry, agriculture, and others [1]. Some of important function of river include the transport of deposits and nutrients that support biological community [2]. Rivers and other water bodies have own an important function for urban water resource and waste disposal. This role has changed over time during urban development [3]. River provides many ecological service functions for human life such as water source, ecological protection, etc. It also socially, economically, and environmentally provide urban area life [4].

Since 2008, most of of the world's population lives in urban area and it is estimated only one-third of the world's population lives in rural areas [5]. Population growth and an increase of human activities along watershed lead to an increased water pollution and environmental damage [6]. Nowadays, urbanization has caused some significant changes in watershed. Some of drainage system in the cities has been altered by human through channelization, containment, diversion, and burial [7]. Previous study also found the effect of urban development to rivers as well as to the effects of urban development on landscapes and river systems [8].

The amount of organic contamination which occurs due to high populations of microorganisms, has typically been monitored by Biochemical Oxygen Demand (BOD) and Chemical Oxygen Demand (COD) testing methods. A high level of BOD decreases river water quality due to decomposition of biodegradable organic matter that can be biologically oxidised, while COD represents organic compounds can be chemically oxidised than biologically oxidised [9]. Human economic activities such as animal husbandry and agriculture play higher position in supplying pollution of river water compared to others pollutants [10]. Wastewater of animal husbandry contains high concentrations of organic and inorganic nitrogen compound, pathogenic bacteria, and ammonia nitrogen [11]. In more rivers in urban areas, levels of phosphorus, sulfur, nitrogen, and other nonmetallic metals tend to be higher and affect the water quality [12].

Cisadane River is one of the river in Indonesia where urbanization, industrialization, and agricultural are extremely main sources of pollution. Cisadane River has a very important function as a source of raw water for the people of Bogor City. Cisadane River crosses several administrative areas ranging from Bogor Regency to Tangerang City with high urban activity that cause the river receives high enough pollutant load. Similar to Ciliwung River, Cisadane River is one of the main rivers in Jabodetabek area (Jakarta, Bogor, Depok, Tangerang and Bekasi) and has great influence for those areas. Cisadane River is an interesting location for examining the effect of land use to water quality and comparing water quality in every river segment.

2. Research method

2.1. Study area

Cisadane is a 138 km long river. The upstream of Cisadane River are coming from Gede-Pangrango and Salak Mountains and flow into Java Sea through Teluk Naga coastal waters. This river has some important role in agricultural, domestic, and industrial needs [13]. For convenience of analysis the river channels are divided into 7 segments. Segments were delineated beginning at upstream to the end of downstream of the stream reach of Cisadane River watershed through some administrative areas such as Bogor Regency, Bogor City, South Tangerang City, Tangerang Regency, and Tangerang City. In this study only 6 segments that will be discussed since there is no data availability in 7th segment. River segmentation from segment 1-6 based on administrative area can be seen in table 1 and map of study location can be seen in figure 1.

Table 1. River segments based on administrative area.

Segment Code	Description
1	Bogor Regency 1
2	Bogor City
3	Bogor Regency 2
4	South Tangerang City
5	Tangerang Regency
6	Tangerang City

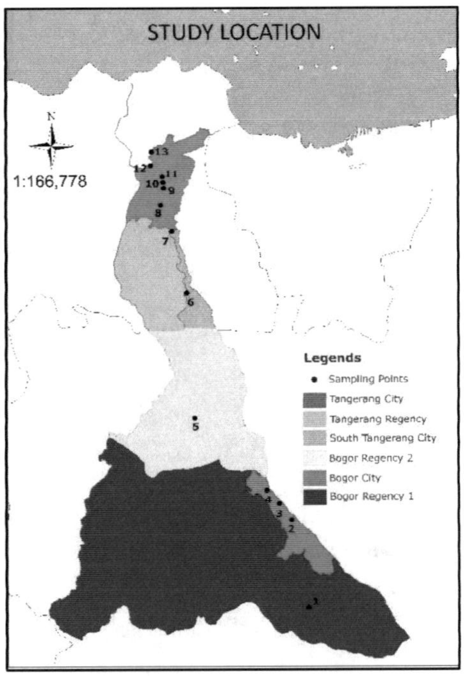

Figure 1. Map of study area.

2.2. Analytical methods

The thirteen location were sampled along the primary downstream segment of Cisadane River. Water sampling was conducted in rainy season (April) and dry season (September) in 2014. Water quality data consists of BOD and COD are measured and studied along these segments. These two variables were selected because they provide an index to measure the effect discharged wastewater on environment. The 13 sampling points will be grouped and averaged according to their segment. Sampling data will be used to relate the water quality and land cover in each segment of Cisadane River. Water quality data were attained from Ministry of Environment and Forestry Republic Indonesia. Sampling points location on Cisadane River can be seen in figure 1.

2.3. Spatial analysis
This research was using 30 m resolution of Landsat Thematic Mapper (TM) satellite imagery 2014. This data was attained from Indonesian Planology Agency. The land use classes were produced using a supervised classification technique with a maximum likelihood algorithm in the Erdas Imagine 9.2 (ERDAS Corporation, Norcross, GA, USA). The Landsat images has been corrected using aerial photographs and field survey data. Interpretation accuracy of the land use categories reached 88% according to the field survey (i.e., 85% more accurate than a random classification) [14]. The land use classes were classified into five groups: built area, bare land, agricultural land, forest, and water body. Built area consists of settlement and airport. Bare land consists of abandoned area and mining area. Agricultural land consists of plantation, rice field, shrubs, and secondary crops field. While water bodies including lake and pond.

3. Result

3.1. Land use classification on each river segment
Cisadane Watershed are grouped into five classes: built area, forest, agricultural land, water bodies, and bare land. These five classes are representatives of the actual landscape in the study region and could be determined using a moderate resolution Landsat image. Land use classification from segment 1-6 can be seen in figure 2 and table 2.

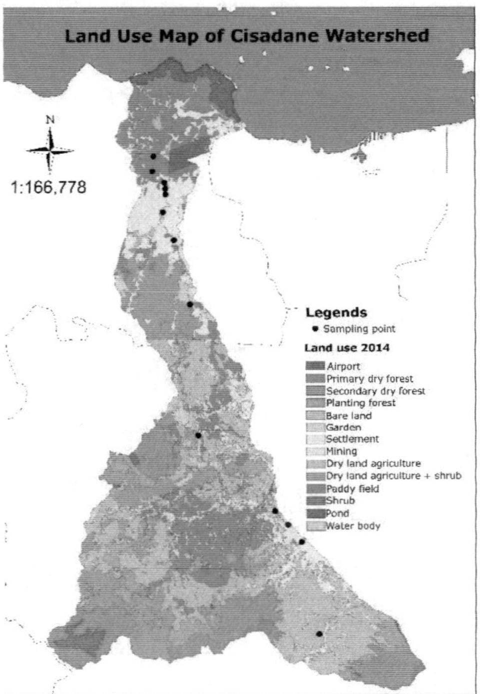

Figure 2. Land use map of Cisadane Watershed.

Table 2. Land use classification on segment 1-6 Cisadane River.

Segment	Land use	Total Area (Ha)	Percentage
1. Bogor Regency 1	Forest	6,662	32.8%
	Agricultural Land	12,431	61.2%
	Built area	1,188	5.9%
	Water bodies	26	0.1%
	Total	**20,307**	**100.0%**
2. Bogor City	Forest	109	2.2%
	Agricultural Land	2,176	44.1%
	Built area	2,649	53.6%
	Water bodies	5	0.1%
	Total	**4,939**	**100.0%**
3. Bogor Regency 2	Forest	19,453	22.4%
	Agricultural Land	58,754	67.8%
	Built area	7,922	9.1%
	Water bodies	559	0.6%
	Bare land	30	0.0%
	Total	**86,718**	**100.0%**
4. South Tangerang	Forest	-	0.0%
	Agricultural Land	1,563	55.8%
	Built area	1,156	41.3%
	Water bodies	83	3.0%
	Total	**2,802**	**100.0%**
5. Tangerang Regency 1	Forest	-	0.0%
	Agricultural Land	9,526	78.5%
	Built area	2,521	20.8%
	Water bodies	91	0.7%
	Total	**12,138**	**100.0%**
6. Tangerang City	Forest	-	0.0%
	Agricultural Land	1,849	21.9%
	Built area	6,488	76.7%
	Water bodies	118	1.4%
	Total	**8,455**	**100.0%**

Table 2 show that in the segment 1 the built area has a smaller area than the forest, agriculture, and water bodies. The amount of built area is quite high in segment 2 considering the urban activity in Bogor City is higher than the upstream river area in Puncak. In segment 3, agricultural land dominate the area and while built area is less than 10%. In segment 4, built area has a fairly high amount even though the agricultural area still dominates the segment. In segment 5, the number of built areas decreases and agricultural land still dominates the segment. In segment 6, built area dominates land use. In the last three segments of the study area there was no forest area in land use classification. This is due to the higher economic activity in the downstream area, which requires more land to be built than open areas.

3.2. *Water quality*

According to Government Regulation No. 82/2001, the water quality classification in Indonesia includes four quality classes based on its designation:

Table 3. Water quality of classification in Indonesia.

Class	Designation	Standard (mg/L)	
		BOD	COD
I	Drinking water	2	10
II	Water recreation facilities	3	25
III	Freshwater fish farming and husbandry	6	50
IV	Crops irrigation	12	100

Based on its allocation for drinking water source, the water quality of Cisadane River should be in class II. Based on BOD measurements in each segment, water quality in Cisadane River for BOD and COD parameters is quite bad. BOD value in six segments range from 12-18 mg/L. BOD values in some segments are still suitable for class IV while others do not even meet the criteria of class IV. For COD parameters, segments 1-3 still meet class III criteria while the downstream area only meets the criteria of class IV (figure 3).

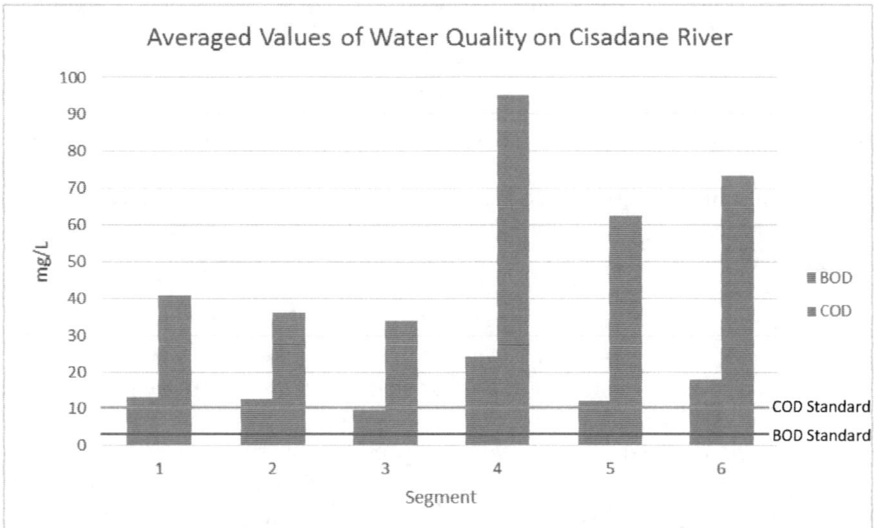

Figure 3. Averaged values of water quality on Cisadane River.

Water quality from upstream to downstream shows very fluctuate results. BOD and COD values from segment 1 to segment 3 show a decrease. In segment 4 both variables increase, then show a decrease in segment 5 and rise again in segment 6.

4. Discussion

4.1. *Land use effect on water quality*

Undeveloped areas such as natural forest area has positive correlation with good water quality while developed area such as settlement and commercial area usually give negative impact on water quality [15]. Land use classification showed that river segment that has more undeveloped area has better water quality compared to river segment with developed area. The high value of BOD in each segment shows the high development along the river. Tu *et al.* [16] explained that BOD had a significant positive correlation with developed areas such as residential and commercial lands. Huang *et al.* [17] also said that the areas of the forest land and grassland has negative correlation to COD. Thus, development areas in Cisadane River really give negative impact on water quality.

4.2. Water quality differences in upstream and downstream area
Based on figure 3, in general, BOD and COD values have increased from upstream to downstream. However, BOD and COD values do not show a steady increase in each segment. A study of water quality differences in upstream and downstream has been conducted before. The study revealed that upstream and downstream water qualities of River Tano in Ghana, do not differ much in terms of temperature, pH, sulphide, sulphate, ammonia, nitrite, manganese, and total alkalinity [18]. Water quality is closely related to the surrounding land use. Therefore, it cannot be concluded that the water quality downstream is worse than in the upstream area.

5. Conclusion
This study showed that developed area was positively correlated to the deterioration of water quality, While, forest, agricultural area, and water area were negatively correlated to the deterioration of water quality. While, segment location doesn't affect water quality. Based on these results, it is necessary to increase the area of green open space (forest and agricultural land) and water body in the local land use planning of Cisadane Watershed. It is therefore recommended that the relevant stakeholders should come together to enforce the environmental laws regarding protection of water bodies.

References
[1] Lansing JS, Lansing PS, Erazo JS 1998 The Value of a River *Journal of Political Ecology* **5**
[2] Das J, Acharya BC 2003 Hydrology and assessment of lotic water quality in Cuttack City, India *Water, Air, and Soil Pollution* **150** 163–175
[3] Gierlinger S, Haidvogl G, Gingrich S, Krausmann F 2013 Feeding and cleaning the city: the role of the urban waterscape in provision and disposal in Vienna during the industrial transformation *Water Hist* **5** 219–239
[4] Zhao YW, ZF Yang, F Xu 2007 Theoretical framework of the urban river restoration planning. *Environmental Informatics Archives* **5** 241-247
[5] DESA 2014 *World Urbanization Prospects The 2014 Revision* (New York: United Nations)
[6] Sumok P 2001 River water quality monitoring : sharing sarawak experience *Proceeding 6th SITE Research Seminar*, 13-14 September 2001
[7] Ceola S, Laio F, Montanari A 2015 Human pressure on rivers is increasing worldwide and threatens water security. Pages 109–110 *in* Christophe Cudennec, Siegfried Demuth, Anil Mishra, and Gordon Young, editors. Proceedings of the 11th Kovacs Colloquium, June 16–17, 2014, Paris. Proceedings of the International Association of Hydrological Sciences, 366. Copernicus GmbH, Gottingen, Germany
[8] Zaharia L, Toroimac GI, Cocoş O, Ghiţă FA, Mailat E 2016 Urbanization effects on the river systems in the Bucharest City region (Romania) *Ecosystem Health and sustainability* **2**
[9] Hur J, Cho J 2012 Prediction of BOD, COD, and Total Nitrogen Concentrations in a Typical Urban River Using a Fluorescence Excitation-Emission Matrix with PARAFAC and UV Absorption Indices *Sensors* **12**
[10] DOE 2001 *Malaysia Environmental Quality Report* Department of Environment, Ministry of Science, Technology and Environment (Kuala Lumpur: Maskha Sdn Bhd)
[11] Gasim MB, Jamil MM, Rahim SA, Toriman ME 2009 Water-quality assessment of the Langat River at kilometre 7, Jalan Kajang-Bangi, Selangor, Malaysia. *Arab World Geographer* **12** 188–198
[12] Sunyer J, Spix C, Quénel P, Ponce-de-León A, Pönka A, Barumandzadeh T, Touloumi G, Bacharova L, Wojtyniak B, Vonk J, Bisanti L, Schwartz J, Katsouyanni K 1997 Urban air pollution and emergency admissions for asthma in four European cities: the APHEA Project

Thorax **52** 760-765

[13] Congalton RG 1991 A review of assessing the accuracy of classifications of remotely sensed data *Remote Sens. Environ.* **37** 35–46

[14] Siahaan R, Indrawan A, Soedharma D, Prasetyo L B 2012 Keanekaragaman makrozoobenthos sebagai indikator kualitas air Sungai Cisadane, Jawa Barat-Banten *Jurnal Bioslogos* **2** 1-9

[15] Bu H, Meng W, Zhang Y, Wan J 2014 Relationships between land use patterns and water quality in the Taizi River basin, China *Ecol. Indic.* **41** 187–197

[16] Tu Q 1990 *The Researches on the Eutrophication in Chaohu Lake* (China: University of Science and Technology of China Press) (Chinese)

[17] Huang J, Zhan J, Yan H, Wu F, Deng X 2013 Evaluation of the Impacts of Land Use on Water Quality: A Case Study in The Chaohu Lake Basin *The Scientific World Journal* **2013**

[18] Adiyiah J, Aboagye H, Acheampong MA 2013 Comparative Assessment of the Upstream and Downstream Water Qualities of River Tano in Ghana *Journal of Environmental Science and Engineering* A **2** 283-292

Characteristics of the oceanic MCC, continental MCC, and coastal MCC over the Indonesian maritime continent

Trismidianto
Center of Atmospheric Science and Technology, National Institute of Aeronautics and Space, Jakarta, Indonesia

E-mail: trismidianto@lapan.go.id

Abstract. This study explains the comparison of mesoscale convective complexes (MCC) characteristics in the oceans, land and in the coast over Indonesian maritime continent (IMC). MCCs were identified and tracked during 15-years (2001-2015) over IMC by infrared satellite imagery using an algorithm that combines criteria of cloud coverage, eccentricity, and cloud lifetime. Infrared satellite imagery was obtained from Himawari generation satellite data. This study showed most of the continental MCC found near the mountains and the high elevation areas. The frequency of MCC occurrences was larger over the land than over the ocean. The oceanic MCCs, which lasted for more than 12 hours, were longer-lived than the continental MCCs. The MCCs with small size most frequently occurred in the continent, in contrast, the MCC with the medium and large size were most concentrated over the ocean. Generally, the continental and coastal MCC initiation occurs in the late afternoon and reach maximum size around midnight before decaying the next morning. In contrast, the oceanic MCC dominantly develops in midnight, and reach maximum size in the morning and then MCC decayed and dissipated from noon until afternoon. The evolution of MCC development in the ocean, land, and in the coast has almost the same stages and ways.

1. Introduction

Maddox [1] defined Mesoscale Convective Complexes (MCC) is a very large of Mesocale Convective System (MCS) observed around the globe. MCS is generally defined as an organized cluster of convective clouds that is larger than an individual cloud, but smaller than cyclone. MCC is a particular type of MCS; an MCC is a large, circular, long-lived cluster of the cloud systems identified by satellite. It often emerges out of other storm types during the late-night and early morning hours. MCCs can cover an entire state. Maddox [1] divided MCS into two types as linear and circular type, and MCC include as the circular type that found in midlatitude and tropics region included IMC. This system grows upscale from convective towers to a convective-stratiform coupled vertical circulation, ultimately meeting its demise as only a stratiform rain region [2].

MCC phenomenon has been studied by researchers for more than a decade in the world including IMC after Maddox [1] introducing the concept of the MCC. Laing and Fritsch [3] have studied the characteristics of MCCs globally. They found that more than 400 MCCs occurred at various locations around the globe each year, but mostly in tropical and mid-latitude locations, with approximately 66 percent of the occurrences in the Northern Hemisphere. They also found that 91% of the events occurred over continents and the rest occurred over the ocean. In general, they came to a conclusion that, the life-cycle of MCCs in globe about 10 hours and most of them occurred in the night. In principally, MCCs usually occurred over land or close to the coast. However, they are not described in detail the MCC characteristics in the ocean, inland, and in the coast. MCC does not only occur in the middle latitude region but also occurs in tropical regions including IMC [1].

Ismanto [4] and Trismidianto [5] have reported that there are many MCCs occurs in the Indonesian Maritime Continent, especially in the Indian Ocean and over Kalimantan Island. One example of the case of MCC event that occurred in ocean has been reported by Trismidianto [7]. In another paper, Trismidianto [6] have analyzed the MCCs that occurred on the Kalimantan Island. However, that papers did not describe in detail the MCC characteristics in the ocean, inland and in the coast over IMC. So that, the MCC characteristics that occurred on the ocean and inland over IMC have not been clearly reported. Therefore, the purposes of this paper were to explain in detail the MCC characteristics that occur in the oceans, inland, and along the coast.

2. Data and Method

This research utilized brightness temperature (T_{BB}) from the Himawari geostationary satellites, provided by the Japan Meteorological Agency (JMA). In detail, the data consists of Himawari-5/GMS-5 for data from January 2001-April 2003, Pacific GOES/GOES-9 for data from May 2003-June 2005, Himawari-6/MTSAT-1R for data from July 2005 - June 2010, Himawari-7/MTSAT-2 for data from July 2010 - June 2015, and Himawari-8 for data from July 2015 - December 2015. This satellite data has spatial and time resolutions $0.05^{o}\times0.05^{o}$ and one hour, respectively that could be downloaded from http://weather.is.kochiu.ac.jp.

MCC identified by input the temperature, latitude, and longitude values for each cloud shield pixel that obtained from infrared (IR) to a modified version of a computerized MCC program using MATLAB based on characteristics of MCCs [1] as shown in Table 1. This method was adapted the method from Ismanto [4] and Trismidianto [5] that based on the maximum spatial correlation tracking technique (MASCOTTE) method by Carvalho and Jones [8].

This program results in the MCC information consisted of MCC location, time of MCC, area for cloud shield and interior cloud, the eccentricity of cloud shield and interior cloud, and then the life-cycle of MCC found by the manual checking and tracking. The MCC information used to explain the MCC characteristics over the ocean, inland, and coastal regions. In this study, we defined the oceanic (continental) MCCs as a system that reaches maximum extent while positioned over the ocean (land) as the following method by Morel and Senesi [9] and Blamey and Reason [10]. This research follows the method by Ogino [11] and Mori [12] to define the coastal region. The coastal MCC is defined the MCCs that occurred near the coastal at between ocean and continent during MCC reaching maximum extent or mature stage.

Table 1. Physical characteristics of MCCs [1].

Size:	A-Cloud shield with continuously low $T_{BB} \leq$ - 32°C (241 K) must have an area $\geq$ 100,000 km^2 B-Interior cold cloud region with $T_{BB} \leq$ - 52°C (221 K) must have an area $\geq$ 50,000 km^2
Initiate:	Size definitions A and B are first satisfied
Duration:	Size definition A and B must be met for period of $\geq$ 6 hours
Maximum extent:	Contiguous cold cloud shield ($T_{BB} \leq$ - 32°C (241 K)) reaches a maximum size
Shape:	Eccentricity (minor axis/major axis) $\geq$ 0.7 at time of maximum extent
Terminate:	Size definitions A and B no longer satisfied

3. Result and Discussion

3.1. The frequency of MCC occurrences

This study has been found around 1028 MCCs during fifteen years from 2001 to 2015 as shown in figure 1a. The MCCs found separate in several areas, some of them were over coastal and ocean, and others were in high elevation areas over land. Figure 1a shows that the MCCs that occur in the mainland are frequently found in the high elevation areas or near the mountainous area, such as Central Kalimantan, East Kalimantan, Western Sumatra and Central Sulawesi. This result is similar with Ashley [13] and Laing and Fritsch [14] who reported that the mean location for the development of the MCCs in America was found on the lee side of the mountain. Morel [9] also reported that a large number of MCS over several areas of Europe found at mainly continental, strongly related with orography and the effect of topography. A mountain breeze and a valley breeze are two related, localized winds that occur one after the other on a daily cycle which also give a contribution to convective activity over the IMC due to IMC is a unique geographical region that composed of a complex system of mountainous islands. The interaction between the topography with the westerly wave propagation is assisted in the development of convective storm or MCC in Africa [15].

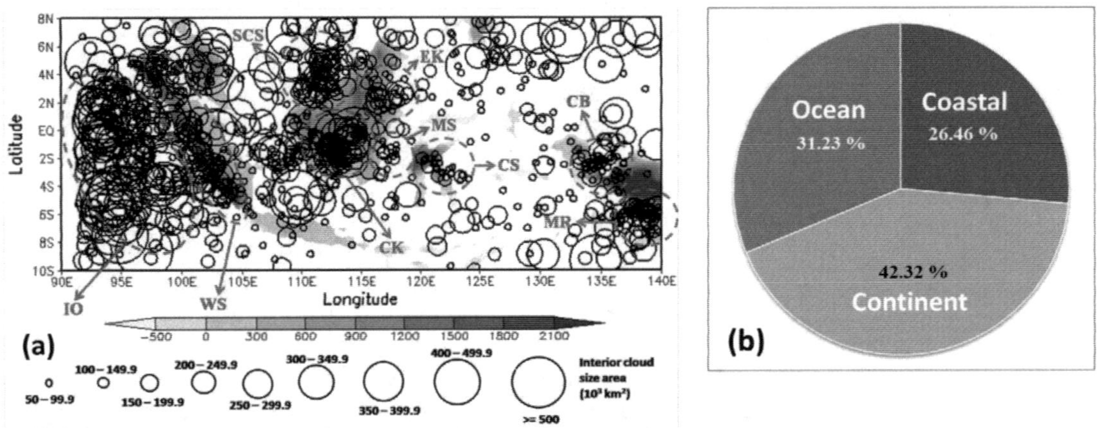

Figure 1. (a) Geographical distribution of the MCCs in the IMC in 15-years (2001-2015). The locations of MCC determined when the interior cloud size area of MCC reach maximum extent. The circles representing the interior cloud size area (10^3 km^2). The red broken line indicated the MCCs concentration region in IO (Indian Ocean), WS (Western Sumatra), SCS (South China Sea), CK (Central Kalimantan), EK (East Kalimantan), MS (Makassar Strait), CS (Central Sulawesi), CB (Cendrawasih Bay), and MR (Merauke). The color shaded refers to the elevation of the unit in meters (m). (b) Percentage of the MCC occurrences over the IMC in 15-years (i.e., 2001-2015) over the ocean, the coastal, and the continent.

The oceanic MCC mostly concentrated over the Indian Ocean, whereas the coastal MCCs concentrated over the Makassar Strait, Cendrawasih Bay, and the South China Sea near Kalimantan Island. Most of the oceanic MCCs concentrated over the Indian Oceanhad large interior cloud size over 300,000 km^2 up to 500,000 km^2 since this area contained one of the biggest heat sources as the driving force for the global circulation in the tropics [16]. This area was also one of the regions where deep cumulus convection and heavy rainfall occurred most frequently in the tropics [17]. The oceanic MCCs with large size also occurred in the Pacific Ocean and the Java Sea. Moreover, an oceanic area with the lower MCC occurrences was found in the Molucca Sea and the Banda Sea.

IOP Conf. Series: Earth and Environmental Science **149** (2018) 012024 doi:10.1088/1755-1315/149/1/012024

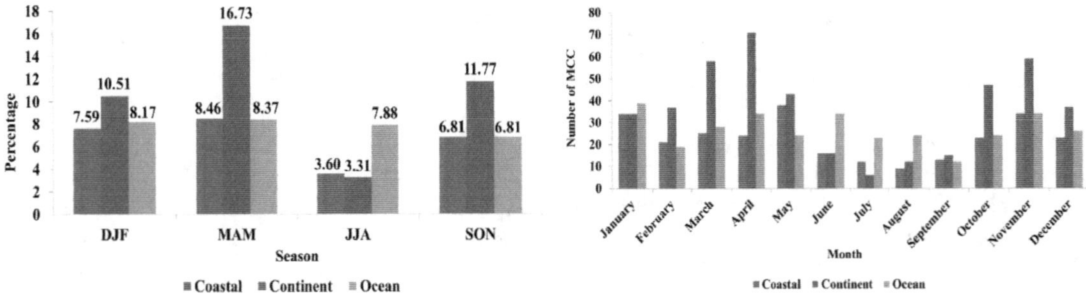

Figure 2. (a) Percentage of the MCC occurrences in coastal, continent and ocean over the IMC in 15-years (2001-2015) for each season in DJF (December, January and February), MAM (March, April and May), JJA (June, July and August), and SON (September, October and November). (b) A total of MCC occurrences over the IMC during 15-years (2001-2015) for each season in coastal, continent and ocean.

This result also supported by previous research, among other i.e., Trismidianto [5] that stated there were many MCC occurrences over the Indian Ocean near Sumatra and Ismanto [4] stated that one of concentration area for oceanic MCC was the Indian Ocean. Moreover, Mori [12] stated that IMC consists of many islands with very long coastlines, many narrow straits control the global (Pacific to Indian) ocean circulation and one of the major contributors to changes in tropical rainfall was the deep convection [18]. Land-sea breeze and mountain breeze play an important role in the tropical precipitation, include IMC so that most of the precipitation in IMC concentrated over the continent than over the ocean [19]. The rainfall outside the coastal region accounts for 52% and 14% on the ocean and the land, respectively. However, on the inside of the coastal region accounts for approximately 34% of the total over the entire tropics [11].The percentage of MCC occurrences in the IMC was found greater over the continent (42.32%) than over the ocean (31.23%) and over the coastal region (26.46%) (figure 1b). Laing and Fritsch [3] found that most of the MCCs was located in continents (almost of 91%), however, it can occur over oceans.

Due to IMC is one of monsoon region, the seasons of IMC divided into two seasons. i.e., wet seasons on December, January and February (hereafter DJF) and dry seasons on June, July and August (hereafter JJA), however, there are two transition seasons. i.e., March, April, and May (hereafter MAM) as transition seasons from wet to dry seasons and September, October and November (hereafter SON) as transition seasons from dry to wet seasons. Figure 2a shows the comparison of the coastal, oceanic and continental MCCs, where 7.59%, 10.51% and 8.17% of MCCs occurred in the coastal, continent and ocean during DJF, respectively. The percentage of the MCC occurrence increased in MAM around 8.46%, 16.73% and 8.73% of MCCs found in coastal, continent and ocean, respectively. In contrast, during JJA, most of MCCs occurrence were the oceanic MCC with 7.88% events, while the percentage of the MCC which occurred in the continent and coastal are 3.31% and 3.60%, respectively. The percentage of MCC which occurred in the continent around 11.77% in SON, while the MCC which occurred in the oceanic and coastal have the same percentage around 6.81%.

The frequency of MCC occurrence in the ocean, land and in the coastal per month was shown in Figure 2b. As we can see that the peak of MCC occurrence in the land found in March, April, and November. It was possible due to likely linked to the synoptic-scale environment, which was more baroclinic in nature compared to the late summer months. It was similar to Tyson and Preston-Whyte [20] that stated the peak of MCC activity in southern Africa mostly occurred during the early before summer months. However, the MCC that occurred in the ocean was frequently found in June, July, and August. This result similar to Velasco and Fritsch [21] that reported the longer MCC season may be due to the influence of the oceans on the relatively smaller landmass. However, some of the previous paper stated that there is the effect of ENSO on MCC activity in Africa [22] and in America [23]. The frequency of MCC events in the coastal was frequently found in January, May, and November, but less than the number of the continental MCC in the similar month.

3.2. Morphological characteristics of MCC

Figure 3a shows the greatest interior cloud size of MCC more than 1000 x 10^3 km^2 occurred in April. The MCC with large size also occurred in March, July, September, and November with interior cloud size area more than 600 x 10^3 km^2. Moreover, most of the MCC with medium size around 200 x 10^3 km^2 occurred in July. The MCC with small size most frequently occurred in the land (continent), in contrast, the MCC with the medium and large size are most concentrated over the ocean as shown in figure 3b. The averaged of cloud shield size of MCC over ocean, land and in the coast were 387,865 km^2, 275,880 km^2, and 293,054 km^2, respectively. While the averaged of interior cloud size of MCC over the ocean, continent and in the coast were 188,590 km^2, 122,550 km^2, and 135,720 km^2, respectively. MCC with area size more than 500,000 km^2 frequently occurred in the ocean.

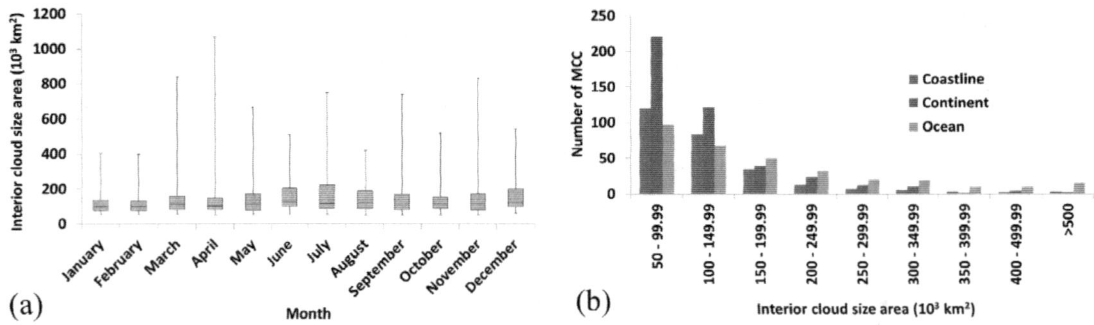

Figure 3. (a) Box and Whisker plot of interior cloud size area of MCC frequency by month during 15-years. (b) Frequency distribution of the MCCs interior cloud maximum area each month during 15-years.

Most climatologies of MCSs concern MCCs, and hence make use of a measure of cloud-shield shape with calculating eccentricity. The eccentricity is very important to distinguish MCC with the other type of MCS. Figure 4 displays the distribution of eccentricity at the time of maximum extent during mature stage. The eccentricity defined as the ratio of the minor axis to the major axis of the MCC best-fitting ellipse at the time of maximum extent.

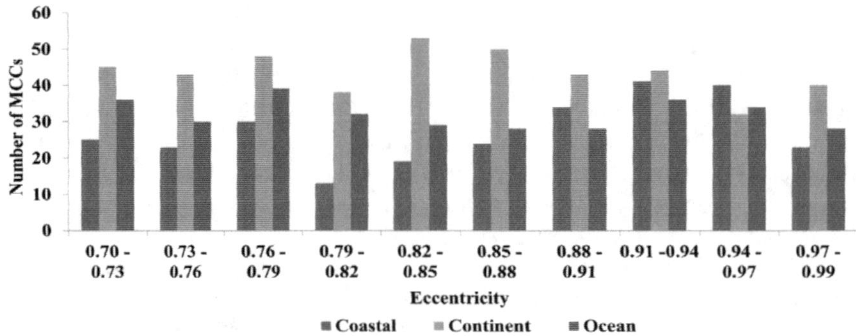

Figure 4. Number of MCC eccentricity at the time of maximum extent during 15-years (2001-2015).

This research found that averaged of eccentricity for all of the MCCs over the IMC was 0.85, with eccentricity averaged for the MCC which occurred in the continental, ocean and in the coast were 0.84, 0.85 and 0.86, respectively. It indicated that majority of MCCs over the IMC was more circular which similar to MCC over the USA [24]. In general, most of the MCCs has eccentricity between 0.91 - 0.94 and between 0.76 - 0.79, most of them occurred over the ocean as shown in figure 4. Most of the MCCs with large eccentricity more than 0.82 found over the ocean, while the majority of the small eccentricity occurred in the coastal region.

3.3. Duration and life-cycle of MCC

Figure 5 shows the comparison of the duration time of MCC which occurred in the continent, coastal, and the oceans. Most of the MCC with duration around 8 - 10 hours occurred on the land. The long-lived MCCs were most frequently concentrated in the ocean, while the short-lived MCCs were most frequently found in the land. Most of the coastal MCCs have duration around 8 -13 hours. This result similar to the global MCCs by Laing and Fritsch [3], where they stated that the oceanic MCCs have been found to be slightly larger and last longer than the continental MCCs. The average duration of the coastal, continental and oceanic MCCs was approximate ~11.5 hours, ~10 hours and ~12.5 hours, respectively. However, this averaged duration less than average for oceanic MCCs in southern Africa (~14 h; [10]).

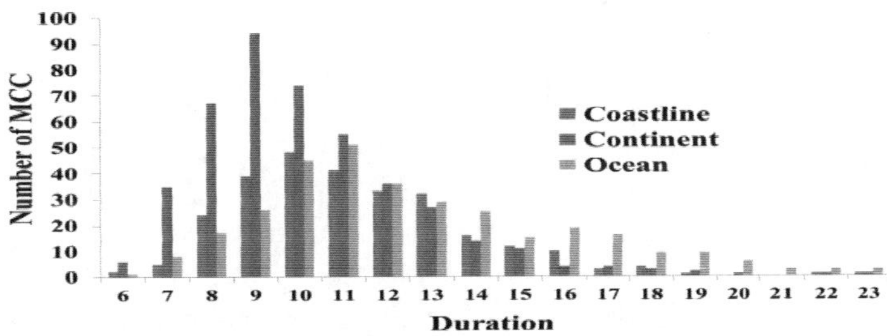

Figure 5. Frequency distribution of duration time of MCCs over the coastal, ocean, and continent during 15-years.

The comparison of MCC life-cycle in the ocean, land and in the coast was shown in figure 6. The MCC which occurred on the coast was usually started in the night to midnight around 1900 - 0000 LT and reached maximum size in the early morning around 0300 - 0700 LT. This system decayed in the morning, several hours after maximum extent, and then dissipated in the afternoon. While the MCC which occurred on the continent was also started in the night. The maximum size of the continental MCC is also predominantly in the early morning, but later than coastal around 0000 - 0400 LT. This system decayed and dissipated in the morning and daytime, respectively. In contrast, the frequency of MCC occurrences in the ocean for initiation, mature, decay and dissipation/post-MCC between 0000 - 0400 LT, 0400 - 0700 LT, 1000 - 1200 LT and 1400 -1800 LT, respectively. It indicated that most of the oceanic MCCs end much later in the morning than the continental MCCs. It is similar with Blamey and Reason [10].

3.4. The evolution of MCC development based on the case studies

The first case study in this research was the continental MCC as shown in figure 7. The MCC occurred on 14 – 15 April 2012 in transition seasons (MAM). Figure 7 shows that the MCC was generated in the Central Kalimantan is triggered by merging of the orographic cloud that migrated southward toward to the MCC area with the other group of the cloud that grew in near the MCC area. At 2100 LT, during the growth stage, this merger cloud expanded southward became grew with the larger size until reaches maximum extent in the midnight (2300 LT). This MCC started decaying in the morning around 0300 - 0700 LT. During decay stage, the new convective system was generated in the surrounding of the MCC area. The MCC was dissipated in the morning at 0700 - 0900 LT, and then the new convective systems migrated to Sumatra, northern Kalimantan, and Java Island. A few hour after the MCC was dissipated, several cloud system appears over Java, eastern Sumatra, Malaysia and Brunei in northern Kalimantan. It indicated that the new convective systems and squall line induce the other cloud in its surrounding area during migration.

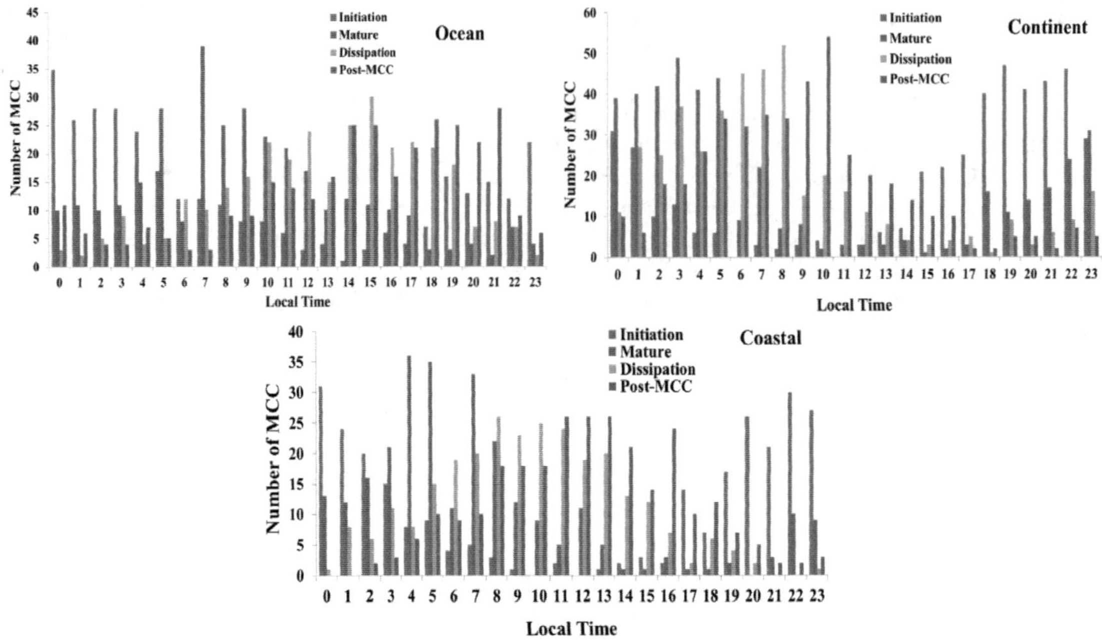

Figure 6. Frequency distribution of the oceanic, continent and coastal MCCs during critical stage (initial, mature, decay and dissipation or post-MCC stage).

The second case study in this research was the oceanic MCC in the Indian Ocean near Sumatra on 27–28 October 2007 and the third case study was the coastal MCC that occurred in the coastal of Sulawesi on 22 - 23 October 2011 as shown in figure 8. The second case study has been analyzed in detail by Trismidianto [7]. MCC in the second case began develops triggered by the cloud from the west and the coastal of the Sumatra in the midnight 0100 LT. The MCC reached maximum size in around the morning. The MCC was decayed and then dissipated in the afternoon. During the MCC dissipated, seem the cloud migrated to the west and predominantly to the east toward over the Sumatra. The MCC in the third case began from the small cloud in the coastline of Sulawesi in the midnight that continues grew rapidly to reaches maximum size in the morning. MCC decayed after several hours and then dissipated in the noon or afternoon. The MCC that occurred over the coastal of the Sulawesi tends to propagate to the Kalimantan Island after the MCC dissipated. It influenced the convective activity and rainfall over the Kalimantan.

This study has found around 1028 MCCs during fifteen years from 2001 to 2015. The greatest frequency of MCC occurrences was found in the continent that reached 42.32% from all events. Most of them were found in the Central Kalimantan. The MCC which occurred in the ocean and coastal was around 31.23% and 26.46%, respectively. The Indian Ocean is the favored region for the oceanic MCCs due to this area is one of the heat sources that interpreted as the driving force for the global circulation in the tropics. There are several favored region for the coastal MCCs. i.e., the coast of western of Sumatra, the South China Sea near the northern of Kalimantan, and the coastal of Merauke. The MCC with small size most frequently occurred in the continent. In contrast, the MCC with the medium and large size was most concentrated over the ocean. The MCC systems are longer lasting in the western Indonesian than in the central Indonesian and eastern Indonesian. Most of the oceanic MCC concentrated over the Indian Ocean. Most of them have large interior cloud size over 300,000 km^2 even much going MCC which has a scale of 500,000 km^2. The majority of MCCs over the IMC was more circular due to the average of eccentricity for all of the MCCs over the IMC are 0.85. The oceanic MCCs are slightly larger and last longer than the continental and coastal MCCs. Almost half of MCC occurrences during MAM occurred on the continent. In contrast, during JJA, most of MCCs occurrences were the oceanic MCC.

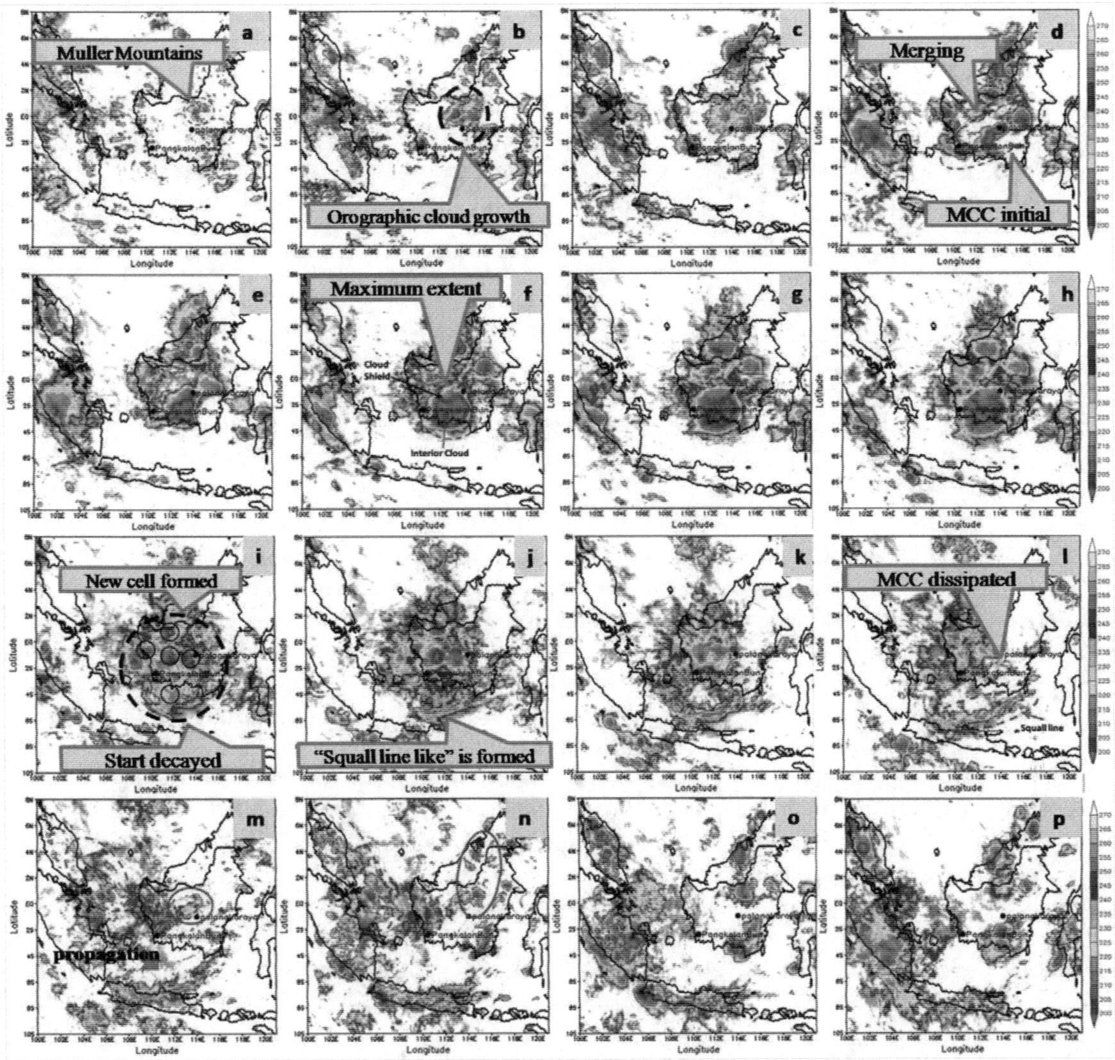

Figure 7. Horizontal distribution of T_{BB} for MCC criteria from infrared data obtained by MTSAT-1R over Kalimantan Island on 14-15 April 2012, showing the eight stages of MCC evolution: (a) MCC-12h or first storm stage (1300 LT), 14 April 2012; ((b),(c)) pre-MCC stage (1500 LT to 1700 LT), 14 April 2012; (d) initial stage (1900 LT), 14 April 2012; (e) growth stage (2100 LT), 14 April 2012;((f),(g),(h)) mature stage (2300 LT, 0000 LT, 0100 LT)), 14 April 2012; ((i),(j) and (k)) decay stage (0300 LT, 0500 LT and 0700 LT), 15 April 2012; (l) dissipation stage (0900 LT), 15 April 2012; (m) post-MCC stage (1300 LT) 15 April 2012; ((n),(o) and (p)), A few hours after MCC occur (1500 LT, 1700 LT and 1900 LT), 15 April 2012. Unit for T_{BB} is Kelvin.

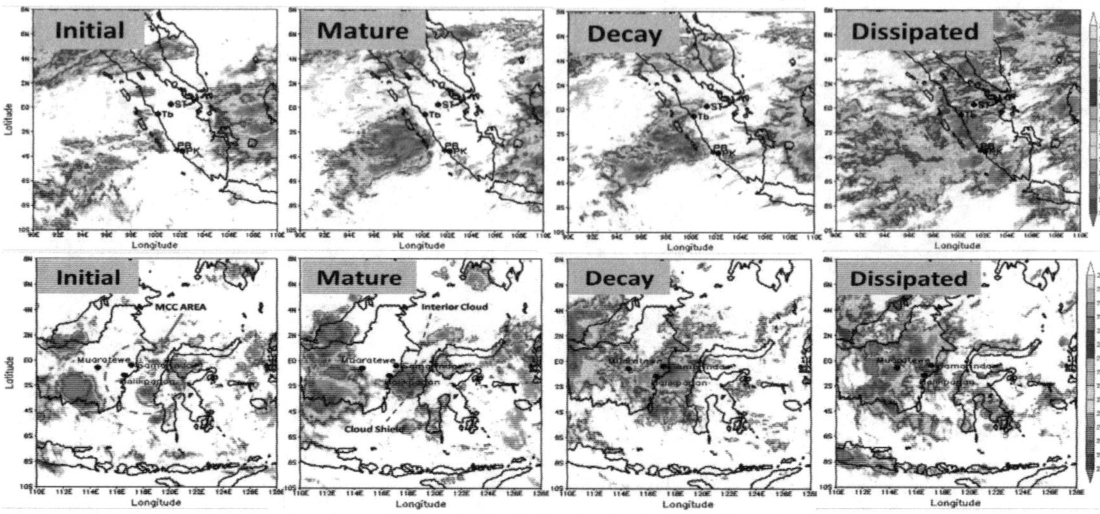

Figure 8. Horizontal distribution of T_{BB} for MCC criteria from infrared data obtained by MTSAT-1R; (Top) over the Indian Ocean near Sumatra on 27–28 October 2007 during initial stage (0100 LT), mature stage (0700 LT), decay stage (1300 LT) and dissipation stage (1600 LT), 28 October 2007. (Bottom) in the coastal of Sulawesi on 22 - 23 October 2011 during initial stage (0100 LT), mature stage (0600 LT), decay stage (1100 LT) and dissipation stage (1300 LT), 23 October 2011.The color indicates T_{BB} with unit Kelvin.

In general, the MCC started to begin occurring in the late afternoon and reached maximum size around midnight before decaying the next morning. The MCC was dissipated from the noon until the afternoon. The characteristics of the evolution of MCC in the coastal and continental was relative similar than the general MCC. In contrast, the oceanic MCC dominantly develops in midnight and reached maximum size in the morning and then MCC decayed and dissipated from noon until afternoon. The evolution of MCC development in the ocean, land and in the coast has almost the same stages and ways. MCC started to develop from the small cloud that continues grew rapidly to reach maximum size and then decayed after several hours. During decay stage, the new convective system was generated in the surrounding of the MCC area, after that, the MCC dissipated. The new convective systems propagated to its surrounding area and influenced the convective activity and rainfall over its surrounding area

4. Conclusion

A total of 1028 MCCs were identified and tracked during fifteen years for the 2001 – 2015 period. The largest frequency of occurrences of the MCC over the IMC was found over the continental (42.32%). While the MMCs frequency over the oceanic and the coastal regions were of 31.23% and 26.46%, respectively. The favored region for development of the continental MCC was found in the Central Kalimantan. The Indian Ocean was the favored region for the oceanic MCCs since this this area was one of largest heat sources functioned as the driving force for the global circulation in the tropics. There were several favored regions for the coastal MCCs. i.e., the coastal of western of Sumatra, the South China Sea near the northern of Kalimantan, and the coastal of Merauke. The MCC with small size most frequently occurred over the continent. Meanwhile, the MCC with the medium and large size were most concentrated over the ocean. The MCC systems were longer lasting in the western Indonesian than in the central Indonesian and eastern Indonesian. Most of the oceanic MCC was concentrated over the Indian Ocean with large interior cloud size over 300,000 km^2 up to 500,000 km^2. The majority of MCCs over the IMC were more circular due to the average of eccentricity for all of the MCCs over the IMC of 0.85. The oceanic MCCs were slightly larger and last longer than the continental and coastal MCCs. Almost half of MCC occurrences during MAM occurred on the continent. In contrast, during JJA, most of MCCs occurrences were the oceanic MCC

In general, MCC initiation occurred in the late afternoon or early evening and reach maximum size around midnight before decaying the next morning. The systems dissipated from the noon until the afternoon. The coastal and continental MCC was relative similar than the general MCC. In contrast, the oceanic MCC dominantly developed in midnight, and reach maximum size in the morning and then MCC decayed and dissipated from noon until afternoon. The evolution of MCC development in the ocean, land and in the coast had almost the same stages and ways. MCC development began from the small cloud that continues grew rapidly to reache maximum size, and then decayed after several hours. During decay stage, the new convective system was generated in the surrounding of the MCC area, after that, the MCC dissipated. The new convective systems propagated to its surrounding area and influenced the convective activity and rainfall over its surrounding area.

References

[1] Maddox R A 1980 Mesoscale Convective Complexes *Bull. Amer. Meteor. Soc.* **61** 1374-1387

[2] Houze R A Jr 2004 Mesoscale Convective System *Review of Geophysics* **43**

[3] Laing A G and Fritsch J M 1997 The Global Population of Mesoscale Convective Complexes *Q. J. R. Meteorol. Soc* **123** 389-405

[4] Ismanto H 2011 *Characteristics of Mesoscale Convective Complexes over Maritime Continent* (Magister Thesis) (Bandung: FITB)

[5] Trismidianto 2012 *Study of Mesoscale Convective Complexes in the Indian Ocean to Weather Pattern above Sumatra* (Magister Thesis) (Bandung: FITB)

[6] Trismidianto, Yulihastin E, Satyawardhana H, Ishida S 2017 A composite analysis of the Mesoscale Convective Complexes (MCCs) development over the Central Kalimantan and its relation with the propagation of the rainfall systems IOP Conference Series: Earth and Environmental Science **54**

[7] Trismidianto, Hadi T W, Ishida S, Moteki Q, Manda A and Iizuka S 2016 Development Processes of Oceanic Convective Systems Inducing the Heavy Rainfall over the Western Coast of Sumatra on 28 October 2007 *SOLA12* 6–11

[8] Carvalho L M V and Jones C 2001 A Satellite Method to Identify Structural Properties of Mesoscale Convective Systems Based on the Maximum Spatial Correlation Tracking Technique (MASCOTTE) *J. of Applied Meteorology American Meteorology Society* **40** 1683-1701

[9] Morel C and Senesi S 2002 A climatology of mesoscale convective systems over Europe using satellite infrared imagery II: Characteristics of European mesoscale convective systems *Q. J. R. Meteorol. Soc.* **128** 1973-1995

[10] Blamey R C and Reason C J C 2012 The role of Mesoscale convective complexes in Southern Africa Summer Rainfall *J. Clim* **26**

[11] Ogino S Y, Yamanaka M D, Mori S and Matsumoto J 2016 How much is the precipitation amount over the tropical coastal region? *J. Clim.* **29** 1231–1236

[12] Mori S, Hamada J I, Tauhid Y I, Yamanaka M D, Okamoto N, Murata F, Sakurai N, Hashiguchi H and Sribimawati T 2004 Diurnal land-sea rainfall peak migration over Sumatra Island, Indonesian Maritime Continent, observed by TRMM satellite and intensive rawinsonde soundings *Mon. Wea. Rev* **132** 2021-2039

[13] Ashley W S 2003 A Distribution of Mesoscale Convective Complexe Rainfall in the *United State American Meteorology Society* **131** 3003 – 301

[14] Laing A G and Fritsch J M 2000 The Large Scale environment of the Global Population of Mesoscale Convective Complexes *Monthly Weather Review* **128** 2756 – 2776

[15] Garstang M, Kelbe B E, Emmitt G D, and London W B 1987 Generation of convective storms over the escarpment of northeastern south africa *Monthly weather review* **115** 429–443

[16] Ramage C S 1968 Role of a tropical "maritime continent" in the atmospheric circulation *Mon. Weather Rev.* **96** 365–369

[17] Yamanaka M D, H Hashiguchi, S Mori, P Wu, F Syamsudin, T Manik, Hamada J-I, M K Yamamoto, M Kawashima, Y Fujiyoshi, N Sakurai, M Ohi, R Shirooka, M Katsumata, Y Shibagaki, T Shimomai, Erlansyah, W Setiawan, B Tejasukmana, Y S Djajadihardja, and J T Anggadiredja 2008 HARIMAU radar-profiler network over Indonesian maritime continent: A GEOSS early achievement for hydrological cycle and disaster prevention *J. Disaster. Res.* **3** 78-88

[18] Tan J, Jakob C, Rossow W B and Tselioudis G 2015 Increases in tropical rainfall driven by changes in frequency of organized deep convection Nature **519**

[19] Qian J H 2008 Why precipitation is mostly concentrated over islands in the Maritime Continent *J. Atmos. Sci.* **65** 1428–1441

[20] Tyson P D and Preston-Whyte R A 2000 The Atmosphere and Weather of Southern Africa *Oxford University Press.* **396**

[21] Velasco I and Fritsch J M 1987 Mesoscale Convective Complexes in the America *J. Geophysic. Res.* **92** 9591-9613

[22] Mason S J and Jury M R 1997 Climate variability and change over southern Africa: A reflection on underlying processes *Prog. Phys. Geogr.* **21** 23–50

[23] Durkee J D, Mote T L and Shepherd M J 2009 The Contribution of Mesoscale Convective Complexes to Rainfall across, Subtropical South America *Int J. Clim* **22**

[24] Augustine J A, and K W Howard 1989 Mesoscale convective complexes over the United States during 1985 *Mon. Wea. Rev.* **116** 685–701

Mesoscale Convective Complexes (MCCs) over the Indonesian Maritime Continent during the ENSO events

Trismidianto and H Satyawardhana
Center of Atmospheric Science and Technology, National Institute of Aeronautics and Space, Jakarta, Indonesia

E-mail: trismidianto@lapan.go.id

Abstract. This study analyzed the mesoscale convective complexes (MCCs) over the Indonesian Maritime Continent (IMC) during the El Niño/Southern Oscillation (ENSO) events for the the15-year period from 2001 to 2015. The MCCs identified by infrared satellite imagery that obtained from the Himawari generation satellite data. This study has reported that the frequencies of the MCC occurrences at the El Niño and La Niña were higher than that of neutral conditions during DJF. Peak of MCC occurrences during DJF at La Niña and neutral condition is in February, while El Niño is in January. ENSO strongly affects the occurrence of MCC during the DJF season. The existences of the MCC were also accompanied by increased rainfall intensity at the locations of the MCC occurrences for all ENSO events. During JJA seasons, the MCC occurrences are always found during neutral conditions, El Niño and La Niña in Indian Ocean. MCC occurring during the JJA season on El Niño and neutral conditions averaged much longer than during the DJF season. In contrast, MCCs occurring in La Niña conditions during the JJA season are more rapidly extinct than during the DJF. It indicates that the influence of MCC during La Niña during the DJF season is stronger than during the JJA season.

1. Introduction

IMC lies at the heart of tropical region's warm pool. It is one of the world's most convectively active areas and thereby plays an important role in the variability of the tropical and global climate system. It is located between two continents (Asia and Australia) and between two oceans (Indian and the Pacific Ocean) [1]. Convection over the IMC consists of various spatial and temporal scales ranging from local circulation to large-scale disturbance in the tropical oceans near of IMC [2]. One of the types of convection system over IMC is the Mesoscale Convective System (MCS) as an organized cluster of convective clouds that is larger than an individual cloud, but smaller than cyclone. A very largest of MCS that are widely observed around the globe is defined as Mesoscale Convective Complex (MCC), which represents a class of MCS with a huge, near-circular, long-lasting cold cloud shield [3].

Houze [4] stated that MCCs have the important role in the rainfall changes in the tropics include IMC. MCCs are primary contributors to tropical rainfall [5]. Trismidianto [6] found that the contribution of MCC to total rainfall over the IMC during fifteen years reached 20%. Most studies have also reported that MCCs producing the heavy rainfall as a cause of the flooding, localized flash floods, mudslides, and property and crop damage (e.g., [3,7]). Trismidianto [8] have also reported that MCC in the Indian Ocean near of Sumatra Island influencing weather pattern in several regions over Sumatra Island.

The rainfall of IMC has also been considered to be closely linked to El Niño-Southern Oscillation (ENSO) in the Pacific Ocean. For example, the transition season anomalies in IMC region were found to be related with a drier condition over most part of the country during El Niño or the warm phase of ENSO [9,10]. ENSO is a naturally occurring phenomenon involving fluctuating ocean temperatures in the central and eastern equatorial Pacific, coupled with changes in the atmosphere. Indonesia is strongly influenced by the ENSO on rainfall in dry season and weakly influenced in wet season [11].

The linked of ENSO and MCC has been reported by some previous researchers but has not been described in detail, for example; Velasco and Fritsch [12] suggested that El Niño might play a role in MCC activity. Laing and Fritsch [13] also suggested the possibility of a relationship between El Niño and MCCs. Although there have been assertions in the literature of a connection between MCCs and El Niño, there has not been a general study directly addressing the changes in MCC related to the ENSO. Given that the climate of IMC is influenced by ENSO, while MCC are primary contributors to tropical rainfall. For that reason, it is important to know how they change between El Niño and La Niña. So that, this study focuses on the role of ENSO in modulating the spatial and temporal distribution of MCC events.

2. Data and method

This research utilized brightness temperature (T_{BB}) from the Himawari geostationary satellites, provided by the Japan Meteorological Agency (JMA). In detail, the data consists of Himawari-5/GMS-5 for data from January 2001-April 2003, Pacific GOES/GOES-9 for data from May 2003- June 2005, Himawari-6/MTSAT-1R for data from July 2005 - June 2010, Himawari-7/MTSAT-2 for data from July 2010 - June 2015, and Himawari-8 for data from July 2015 - December 2015. This satellite data has spatial and time resolutions $0.05^o \times 0.05^o$ and one hour, respectively that could be downloaded from http://weather.is.kochiu.ac.jp. This study used Nino 3.4 index to the analysis of the ENSO. This index can be downloaded free from https://www.esrl.noaa.gov/psd/gcos_wgsp/Timeseries/nino34/. The estimated rainfall data, corresponding to the MCCs and ENSO in this research, were obtained from the Real-Time TRMM TMPA (TMPA-RT) 3B41RT v7 dataset, which has hourly temporal resolution and $0.25° \times 0.25°$ spatial resolution. These data could be downloaded free from ftp://disc2.nascom.nasa.gov/data/TRMM/Gridded/3B41RT/.

MCC identified by input the temperature, latitude, and longitude values for each cloud shield pixel that obtained from infrared (IR) to a modified version of a computerized MCC program using MATLAB based on characteristics of MCCs [3]. He stated that MCC have cloud shield with continuously brightness temperature or black body temperature (hereafter T_{BB}) $\leq$ -32°C (in this study converted to 241 K) and must have an area $\geq$ 100,000 km^2. The interior cold cloud has $T_{BB} \leq$ -52°C (in this study converted to 221 K) and must have an area $\geq$ 50,000 km^{2}for over six hours, although the cloud shield does not have to maintain an eccentricity of $\geq$ 0.7 for its entire life cycle. The eccentricity is also important to distinguish MCC from the other MCS type, for example; squall line, bow echo, CC and the others. The areal extent and eccentricity values used to determine the duration and life cycle of the MCCs, while the centroid position indicated the path of propagation of the systems.

This method adapts the method from Ismanto [14] and Trismidianto [8] that based on the maximum spatial correlation tracking technique (MASCOTTE) method by Carvalho and Jones [15]. This program results in the MCC information consists of MCC location, time of MCC, area size for cloud shield and interior cloud and the eccentricity of cloud shield and interior cloud, and then by the manual checking and tracking, we get the life cycle of MCC.

MCC events are grouped according to the ENSO events that occurs. The ENSO event is divided into three categories based on the SST anomalous value for the Nino 3.4 area, ie neutral conditions, El Niño and La Niña. The division of the ENSO category based on the anomalous SST Nino 3.4 is described in table 1. The anomalous value of SST Nino 3.4 has a monthly temporal resolution, whereas the analysis performed in this study is three monthly or seasonal, hence the average of three months before the value of anomalous SST Nino 3.4. Due to IMC is one of monsoon region, the seasons of IMC divided into two seasons. i.e., wet seasons on December, January and February (DJF) and dry seasons on June, July and August (JJA), however, there are two transition seasons. i.e., March, April, and May (MAM) as transition seasons from wet to dry seasons and September, October and November (SON) as transition seasons from dry to wet seasons.

Table 1. The 3-monthly mean of the sea surface temperature anomaly in Nino 3.4 region, which are: green represents La Niña, yellow represents neutral and red representing the El Niño condition.

Year/Season	DJF	MAM	JJA	SON
2000	-1,7	-1,0	-0,6	-0,6
2001	-0,8	-0,4	-0,1	-0,3
2002	-0,2	0,2	0,7	1,1
2003	1,0	0,0	0,0	0,3
2004	0,3	0,0	0,4	0,7
2005	0,5	0,3	0,1	-0,2
2006	-0,8	-0,4	0,2	0,8
2007	0,6	-0,2	-0,4	-1,3
2008	-1,7	-0,9	-0,2	-0,3
2009	-0,9	-0,3	0,6	1,0
2010	1,5	0,5	-0,9	-1,6
2011	-1,5	-0,8	-0,4	-0,9
2012	-0,9	-0,3	0,4	0,3
2013	-0,3	-0,2	-0,3	-0,1
2014	-0,3	0,2	0,3	0,6
2015	0,6	0,8	1,6	2,3

3. Result and Discussion

3.1. The Influence of the MCCs to rainfall while ENSO occured in December-January-February (DJF) Period

A climatological study of MCCs during 2001 - 2015 over IMC has been identified using IR satellite imagery with creating an algorithm that combines criteria of cloud coverage, eccentricity, and cloud lifetime. This study used the MCC data which has been found by Trismidianto [6] around 1028 MCCs during fifteen years from 2001 to 2015. This research was a continuation of the research conducted by Trismidianto [6] with grouping and analyzing the MCC occurrences when the ENSO phenomena during 2001 - 2015 periods in DJF and JJA season.

Table 2. The occurances of MCC according to the three conditions of ENSO.

Month/ENSO condition	Neutral	El Niño	La Niña
December	27	24	26
January	26	35	30
February	35	32	35
Total MCC	**88**	**91**	**91**

Variations in actual MCC totals per ENSO phenomenon during DJF season has been observed with a peak of MCC activity occurred during La Niña and El Niño condition which has the same

IOP Conf. Series: Earth and Environmental Science **149** (2018) 012025 doi:10.1088/1755-1315/149/1/012025

number of events that is about 91 events as shown in table 2. While the MCC occurrences at the time of neutral condition are 88 events. The peak of MCC occurrences during DJF season at the time of La Niña and the neutral condition is in February, while El Niño is in January. It is due to the peak of diurnal convective activity and rainy season over IMC at the time of DJF is in February, where according to Trismidianto [16] that diurnal convective is also associated with MCC formation above IMC

Table 3. The physical parameters of MCC in three conditions of the ENSO during December-January-February for 2000 – 2015 periods.

	Eccentricity	Cloud Shield Area Size (km^2)	Interior Cloud Area Size (km^2)	Duration (Hours)
Neutral	0.84	371009.5	157434.7	11.5
El Niño	0.85	257748.6	119463.3	11.2
La Niña	0.86	321722.6	145679.6	11.3

During the DJF season, the number of MCCs occurring at the time of the neutral conditions is less than the number of MCC that occurred during El Niño and La Niña condition, but MCCs that occur during neutral conditions have larger average sizes than El Nino and La Nina as shown in table 3. The average of cloud shield and interior cloud area size of MCCs is around 371009.5 km^2 and 157434.7 km^2, respectively. It is more than MCC that occurred during El Niño and La Niña condition with the average of cloud shield area size of MCCs is 257748.6 km^2 and 321722.6 km^2, respectively. This indicates that the El Niño phenomenon at DJF has an effect on the MCC physical characteristics and does not significantly affect the number of MCC events in IMC.

The MCC that occurred during El Niño condition is slightly less than with global MCC by Laing and Fritsch [13] that reported the MCC for the global population is 354,000 km^2. However, the MCCs in IMC is more than from MCC in the western Pacific region that has the greatest frequency of MCC between 2 x 105 km^2 until 3 x 105 km^2 as reported by Miller and Fritsch [17]. This MCC is also more than with MCC over subtropical South America that has cloud size area 256,500 km^2 [18]. Although some MCCs produce cold cloud shields that extend over a million km^2, the size, and duration of MCSs and MCCs are positively correlated, i.e., larger systems tend to last longer [19,20,13].

Most climatologies of MCSs concern MCCs, and hence make use of a measure of cloud-shield shape with calculating eccentricity. The eccentricity is very important to distinguish MCC with the other type of MCS. The eccentricity defined as the ratio of the minor axis to the major axis of the MCC best-fitting ellipse at the time of maximum extent. This research found that average of eccentricity for all of the MCCs over the IMC during neutral, El Niño, and La Niña condition is 0.84, 0.85 and 0.86, respectively as shown in table 3. It indicated that majority of MCCs over the IMC are more circular which similar to MCCs over the USA [21]. In general, most of the MCCs has eccentricity between 0.91 - 0.94 and between 0.76 - 0.79.

In this study, we define the MCC duration as the duration between the time of initial stage and the time of dissipation stage. The frequency of distribution of the duration of all MCC systems that occurred at the ENSO event is shown in table 3. The average duration of the MCC that occurred during neutral, El Nino and La Nina condition was approximate 11.5 hours, 11.2 hours and 11.3 hours, respectively. It is more than average for MCCs in southern Africa which average duration around 9.5 hours [22]. Also, it is slightly longer than the global average around 10 hours [13] and much less than those found in subtropical South America which duration around14 hours [18].

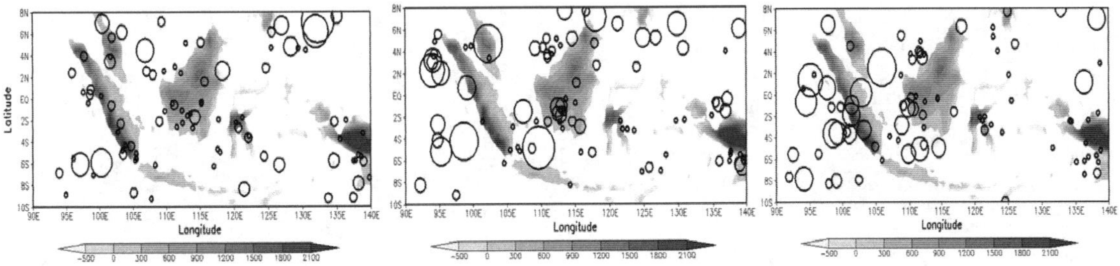

Figure 1. Frequency distribution of the MCC occurrences over IMC during fifteen years from 2001 to 2015 in wet seasons or DJF, for; (a) El Niño, (b) La Niña, and (c) Neutral condition. The circles show the size and location of MCC when the mature stage at the time of the interior cloud size area of MCC reached the maximum extent (10^3 km^2). The color shaded refers to the elevation of the unit in meters (m).

Figure 1 shows that the ENSO influenced on MCC events in different places, such as in Kalimantan, Sumatra, Java, and Papua. At the time of El Niño, the MCC occurrences spread more in Sumatera and Kalimantan, while during La Niña, MCCs tended to spread in the Indian Ocean and the Java Sea, but some of MCC concentrated in Central Kalimantan. In the neutral condition, MCC spread in the ocean and continent. The MCC occurrences in Java only occurred at the time of neutral and La Niña conditions. At the time of El Niño, MCCs that occur in IMC is not as large as MCCs that occur during La Niña and neutral conditions. It is due to the reduced supply of water vapor to become rain-producing convective clouds in IMC. It appears that the largest MCCs at the time of DJF El Niño occurred in the southwestern waters of Sumatra (the eastern Indian Ocean), which is the most frequently formed area of large-scale MCC and the Pacific Pole (warm pole). What is interesting is the area with the small and small number of MCCs at the time of La Niña and Neutral, such as in the Banda Sea, Timor Sea and the Arafura Sea, precisely during El Niño many MCCs occur and have a larger scale. Walker's circulation shift at the time of El Niño occurs, possibly causing shifts and changes in the scale of MCCs further east. In addition, during the neutral DJF and La Niña periods, MCCs spreads mostly in land and sea, whereas in the El Niño DJF period, MCCs occur on land and very few MCCs are found in the oceans.

Figure 2 shows that the MCCs affects rainfall in some areas for example in the Indian Ocean that has high rainfall during neutral and La Niña conditions. MCC events during normal and La Niña also caused high rainfall in the Eastern Indian Ocean, while MCC events at the time of El Niño affected more rainfall in the mainland such as southern Sumatra, Central Kalimantan, Southeast Sulawesi and some area in Papua. In addition, El Niño conditions also reduce the number of MCC occurrences in the ocean which resulted in low rainfall in the area, such as in the Java Sea and the western coast of Sumatra. It results reinforces the results of research that has been reported by Trismidianto [6,16] that MCC is very important to contribute to the rainfall that occurs in IMC. MCC also affects not only the existing rainfall at the MCC location, but also affects rainfall around the MCC location through the formation of new convective cells when MCC begins to become dissipated. Numerous studies have demonstrated that ENSO can modulate the synoptic and mesoscale environments South America, which in turn can influence convective activity and precipitation patterns across the tropical and subtropical regions (e.g., [23,24,12]). Specifically, Velasco, and Fritsch [12] found that the number of MCCs doubled during the 1982/83 El Nino event. However, this research showed no apparent relationship between MCC frequency and ENSO that similar with Durkee and Mote [18] that analysis of ENSO and MCC cloud-top characteristics, longevity, and the distribution of MCC rainfall show no apparent relationship.

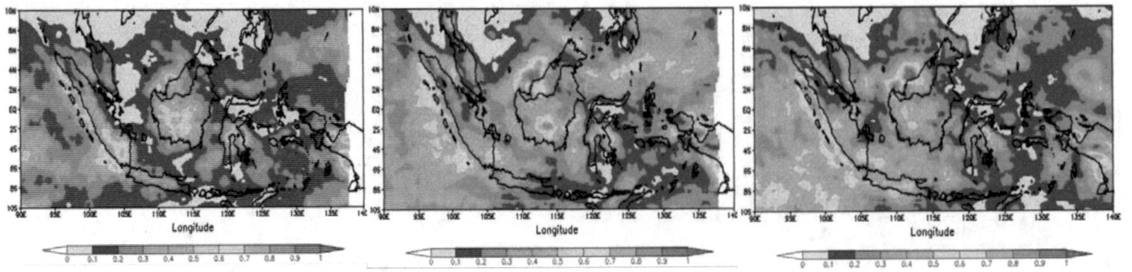

Figure 2. Rainfall accumulation per MCCs occurrences number in the IMC during fifteen years from 2001 to 2015 on DJF, for; El Niño (left), La Niña (middle), and Neutral condition (right). The color shaded refers to rain rate of the unit in millimeters (mm).

3.2. The influence of the MCCs to rainfall while ENSO occured in June-July-August (JJA) period

Variations in actual MCC totals per ENSO phenomenon during JJA season has been observed as shown in table 4 with the peak of MCC activity occurred at the time of neutral condition with the number of MCC is 95 events, while the number of MCC that occurred during JJA at the time of La Nina is 30 events. MCC occurs at least few at the time of El Nino during the JJA season with about 19 events. The peak of MCC occurrences during JJA season during neutral, El Nino and La Niña is in June compared to July and August. This result almost similar with Tyson and Preston-Whyte [25] that stated the peak of MCC activity in southern Africa is mostly occur during the early before summer months due to likely linked to the synoptic-scale environment, which is more baroclinic in nature compared to the late summer months. Velasco and Fritsch [12] that reported the longer MCC season may be due to the influence of the oceans on the relatively smaller landmass. However, some of the previous paper stated that there is the effect of ENSO on MCC activity in Africa [26] and in America [27]. Comparison to DJF, the number of MCC events at JJA during neutral condition occurs more frequently. However, at the time of JJA, the phenomena of El Niño and La Niña had the effect of reducing the number of MCC occurrences in IMC, which at the time of the DJF was not so influential.

Table 4. Same as table 2, but for JJA.

Month/ENSO condition	Neutral	El Niño	La Niña
June	48	8	21
July	23	4	5
August	24	7	4
Total MCC	**95**	**19**	**30**

During the JJA season, the number of MCC events occurring at El Niño was smaller than the number of MCCs in neutral and La Niña conditions, however, the average of MCC that occurred at the time of El Niño larger than during neutral and La Niña conditions. The average of cloud shield and interior cloud area size of MCCs at the time of El Niño is around 444089.4 km^2 and 213422.4 km^2, respectively as shown in table 5. It is more than MCC that occurred at the time of the neutral and La Niña condition with the average of cloud shield area size of MCCs is 348253.1 km^2 and 253462.5 km^2, respectively. This indicates that the El Niño phenomenon at DJF has an effect on the MCC physical parameters and does not significantly affect the number of MCC events in IMC. MCC that occurs during El Nino is the large MCC compared to MCC in some regions of the world and more than with global MCC by Laing and Fritsch [13] that stated the mean cloud-top area for the global population is 354,000 km^2.

MCC occurring during the JJA season on El Niño and neutral conditions averaged much longer than during the DJF season, where the duration of MCC life-cycle when El Niño and neutral conditions were 12.9 and 11.6 hours, respectively. MCCs that occur during the JJA season at the time of El Niño conditions is the long-lived MCC types. It indicates that El Niño also influences the development of MCC in IMC. In contrast, MCCs occurring in La Niña conditions during the JJA season are more rapidly extinct than during the DJF season with an average life-cycle duration of MCC ranging from 10.2 hours. It indicates that the influence of MCC during La Niña during the DJF season is stronger than during the JJA season. Almost similar with DJF season, this research also found that average of eccentricity for all of the MCCs over the IMC during neutral, El Niño, and La Niña condition during JJA seasons is circular types.

Table 5. Same as table 3, but for JJA.

	Eccentricity	Cloud shield area size (km^2)	Interior cloud area size (km^2)	Duration (Hours)
Neutral	0.88	348253.1	162834.7	11.6
El Niño	0.83	444089.4	213422.4	12.9
La Niña	0.84	253462.5	115142.8	10.2

Figure 3 shows that at the time of the neutral condition during JJA season, the MCC most frequently occurs if compared with El Niño and La Niña condition. However, the MCC, which is centered in the Indian Ocean at the time of JJA, is not affected by ENSO since MCC events are always found in neutral, El Niño and La Niña, although during El Niño and La Niño the number of MCC occurrences in the Indian Ocean is less. But in Indonesia, the influence of La Niña and El Niño is very strong at the time of JJA. This is indicated by the decreasing number of MCCs compared to the neutral condition. At the time of JJA in the IMC, MCC occurs mostly around the equator and north of equator while in the southern part of Indonesia there are few MCC events. In general, MCC at El Niño and La Niña are not as large as in neutral conditions.

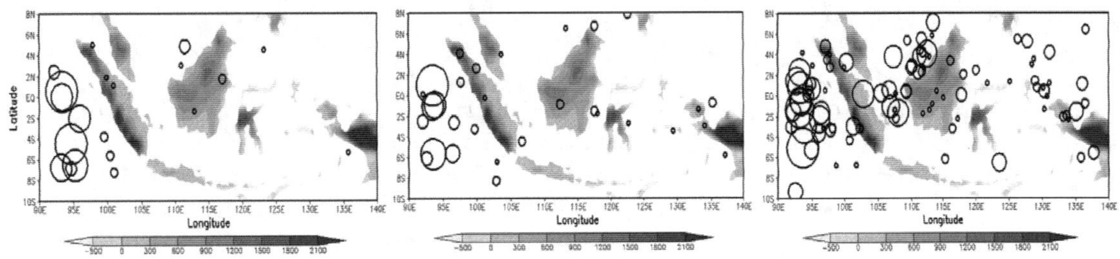

Figure 3. Same with figure 1, but for JJA.

During JJA period, rainfall under ENSO neutral and La Niña conditions are higher and more prevalent in IMC compared to El Niño. The highest rainfall is on the western coast of Sumatra, the western coast of Kalimantan and Papua. While low rainfall is seen in the southern region of Indonesia. However, at the time of El Niño, high rainfall occurred in the East Indian Ocean and the western coast of Sumatra while MCCs formed, comparing to La Niña and neutral conditions. While the reduction of rainfall at the time of El Niño occurred in South Kalimantan, Sulawesi, Maluku Islands and Papua. This is because the MCC occurrences at the time of El Niño is not so much, but it produces very high rainfall in the area. This is reasonable considering that MCCs that occur on JJA El Niño has a large scale (size) in the eastern Indian Ocean. But unlike the northern region of Kalimantan, because of high rainfall occurs despite the number of incidents MCCs are small and small scale. This indicates at the time of JJA El Niño, the presence of MCCs causes high rainfall in the area. While in the area of Papua, at the time of La Niña happened, rainfall in Papua area is very high although MCCs that happened little and with the small scale. This indicates the magnitude of the influence of MCCs at the time of JJA La Niña in the area. Even the effect is greater when compared to the JJA neutral which has a large number of occurrences and scale of MCCs. In general, the MCC occurrences at JJA neutral has an effect to more widespread rainfall when compared to La Niña and El Niño which have a high rainfall effect focused on a particular area.

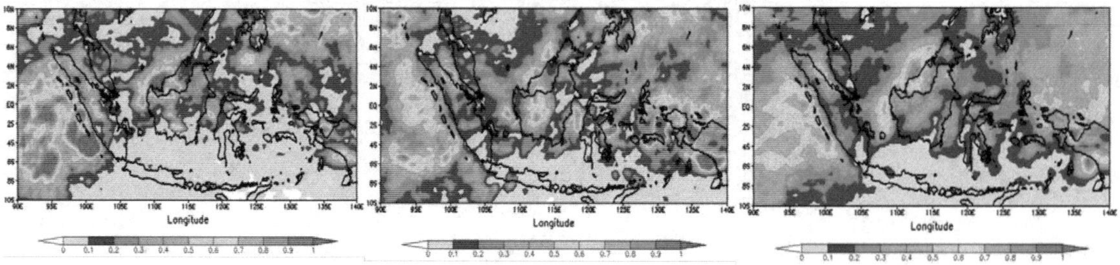

Figure 4. Same with figure 2, but for JJA.

4. Conclusion

During DJF seasons, the frequencies of the MCC occurrences during El Niño and La Niña conditions was higher than that of neutral conditions, but the size of MCCs at Neutral condition was larger than the other conditions. At the time of El Niño condition, the MCC occurrences frequently found spread in the southern of Sumatra and the Central Kalimantan. At the La Niña, the MCCs tended to spread in the Indian Ocean and the Java Sea, but also found concentrated in the Central Kalimantan. While at the neutral conditions, the MCC spread evenly in the oceans and the mainland of IMC. The MCC occurrences in the Java Sea and Java Island are just found in the neutral and La Niña conditions. It is generally concluded that ENSO strongly affects the occurrence of MCC during the DJF season. The existences of the MCC were also accompanied by increased rainfall intensity at the locations of the MCC occurrences for all ENSO events. The MCC occurrences at neutral and La Niña caused high rainfall in the Eastern Indian Ocean, whereas the MCC occurrences at El Niño affected more rainfall in the mainland of southern Sumatra. During JJA seasons, the MCCs which concentrated in the Indian Ocean are not affected by ENSO, whereas the MCC occurrences are always found during neutral conditions, La Niña and El Niño. The influence of MCC during La Niña during the DJF season is stronger than during the JJA season. Comparison to DJF, the number of MCC events at JJA at the time of neutral condition occurs more frequently. However, at the time of JJA, the phenomena of La Niña and El Niño had the effect of reducing the number of MCC occurrences in IMC, which at the time of the DJF was not so influential. This research found that average of eccentricity for all of the MCCs over the IMC at the time of neutral, La Niña, and El Niño condition during JJA and DJF seasons is tend to circular types.

References

[1] Ramage C S 1968 Role of a tropical “maritime continent” in the atmospheric circulation. *Mon Weather Rev* **96** 365–369

[2] Nakazawa T 1988 Tropical super clusters within intraseasonal variations over the western Pacific *J. Meteor. Soc* Japan **66** 823-839

[3] Maddox R A 1980 Mesoscale Convective Complexes *Bull. Amer. Meteor. Soc* **61** 1374-1387

[4] Houze R A Jr 1977 Structure and Dynamics of a Tropical Squall Line System *Mon. Wea. Rev* **105** 1540 – 1567

[5] Nesbitt S W, Cifelli R, and Rutledge S A 2006 Storm Morphology and Rainfall Characteristics of TRMM Precipitation Features *Montly Weather Review* **134** 2902- 2721

[6] Trismidianto, E Yulihastin, H Satyawardhana, J T Nugroho, S Ishida 2017 The Contribution of the Mesoscale Convective Complexes (MCCs) to total rainfall over Indonesian Maritime Continent *IOP Conference Series: Earth and Environmental Science* **Vol. 54**

[7] McAnelly R L and Cotton W R 1989 The Precipitation life cycle of Mesoscale Convective Complexes over the Central United States *Mon. Wea. Rev* **117** 784 – 808

[8] Trismidianto 2012 Study of Mesoscale Convective Complexes in the Indian Ocean to Weather Pattern above Sumatra *Magister Thesis* FITB ITB
[9] Hendon H H 2003 Indonesian rainfall variability: Impacts of ENSO and local air–sea interaction. *J. Clim* **16** (11), 1775-1790
[10] Aldrian E and Susanto D 2003 Seasonal variability of Indonesian rainfall in ECHAM4 simulations and in the reanalyses: The role of ENSO. *Int. J. Clim.* **87** 41-59
[11] Haylock M, McBride J. 2001 Spatial coherence and predictability of Insonesian wet season rainfall *Journal of Climate* **14** 3882-3887
[12] Velasco I and Fritsch J M 1987 Mesoscale Convective Complexes in the America *J. Geophysic. Res.* **92** 9591-9613.
[13] Laing A G and Fritsch J M 1997 The Global Population of Mesoscale Convective Complexes *Q. J. R. Meteorol. Soc* **123** 389-405
[14] Ismanto H 2011 Characteristics of Mesoscale Convective Complexes over Maritime Continent *Magister Thesis* FITB ITB.
[15] Carvalho L M V and Jones C 2001 A Satellite Method to Identify Structural Properties of Mesoscale Convective Systems Based on the Maximum Spatial Correlation Tracking Technique (MASCOTTE) *J. of Applied Meteorology. American Meteorology Society* **40** 1683-1701
[16] Trismidianto, Hadi T W, Ishida S, Moteki Q, Manda A and Iizuka S 2016 Development Processes of Oceanic Convective Systems Inducing the Heavy Rainfall over the Western Coast of Sumatra on 28 October 2007 *SOLA* **12** 6−11
[17] Miller D and Fritsch J M 1991 Mesoscale Convective Complexe in Western Pasific Region. *Amer. Meteor. Soc* **119** 2978 – 2992
[18] Durkee J D and Mote T L 2009 A Climatology of Warm Season Mesoscale Convective Complexes in Subtropical South America. *Int J. Clim.* **12 pp**
[19] Tollerud E I, Augustine J A and Jamison B D 1992 Cloud top characteristics of mesoscale convective systems in 1986 Pp. J3–J7 in preprint volume of the 6th conference on satellite meteorology and oceanography, *Am. Meteorol. Soc*. Atlanta, USA
[20] Laurent H 1996 Tracking of convective cloud clusters from Meteosat data *Tenth Meteosat Scientific Users' Conference*
[21] Augustine J A, and K W Howard, 1989 Mesoscale convective complexes over the United States during 1985 *Mon. Wea. Rev.* **116** 685–701
[22] Blamey R C and Reason C J C 2012 The role of Mesoscale convective complexes in Southern Africa Summer Rainfall *J. Clim* **26**
[23] Curtis S 2008 The El Nino–Southern Oscillation and global precipitation. *Geogr. Compass* **2** 600–619
[24] Grimm V R Barros and M E Doyle 2000 Climate variability in southern South America associated with El Niño and La Niña events. *J. Climate,* **13** 35–58
[25] Tyson P D and Preston-Whyte R A 2000 The Atmosphere and Weather of Southern Africa *Oxford University Press.* **396**
[26] Mason S J and Jury M R 1997 Climate variability and change over southern Africa: A reflection on underlying processes *Prog. Phys. Geogr.* **21** 23–50
[27] Durkee J D, Mote T L and Shepherd M J 2009The Contribution of Mesoscale Convective Complexes to Rainfall across, Subtropical South America *Int J. Clim.* **22**

Predicting drought propagation within peat layers using a three dimensionally explicit voxel based model

A A Condro*, H Pawitan and I Risdiyanto
Department of Geophysics and Meteorology, Faculty of Mathematics and Natural Sciences, Bogor Agricultural University

E-mail : aacondro@gmail.com

Abstract. Peatlands are very vulnerable to widespread fires during dry seasons, due to availability of aboveground fuel biomass on the surface and belowground fuel biomass on the sub-surface. Hence, understanding drought propagation occurring within peat layers is crucial with regards to disaster mitigation activities on peatlands. Using a three dimensionally explicit voxel-based model of peatland hydrology, this study predicted drought propagation time lags into sub-surface peat layers after drought events occurrence on the surface of about 1 month during La-Nina and 2.5 months during El-Nino. The study was carried out on a high-conservation-value area of oil palm plantation in West Kalimantan. Validity of the model was evaluated and its applicability for disaster mitigation was discussed. The animations of simulated voxels are available at: goo.gl/HDRMYN (El-Nino 2015 episode) and goo.gl/g1sXPl (La-Nina 2016 episode). The model is available at: goo.gl/RiuMQz.

1. Introduction

Contribution of tropical peatland is of importance to the global climate, mainly due to its regulating services by storing a huge amount of carbon. Tropical peatland has been estimated as storing carbon of about 88.6 Gt C (range 81.7-91.9 Gt C), equal to 15-19% of the of the global peat carbon pool, where 68.5 Gt C of this is distributed in Southeast Asian region (77%), equal to 11–14% of the global peat carbon pool, with the largest share from Indonesia of about 57.4 Gt C or 65% of the tropical peat carbon pool, followed by Malaysia of about 9.1 Gt C or 10% of the tropical peat carbon pool [1].

Nevertheless, tropical peatland ecosystem is fragile, particularly to profound perturbations [2]; [3]. Stability of tropical peatland ecosystem is strongly determined by its hydrological system [2]; [1]; [4]; [5]. Drainage system applied to peatland can perturb its ecosystem functions, particularly on its hydrological function [4]. At sub-surface water levels of less than 40 cm below the peat surface and surface water levels of less than 100 cm above the peat surface, peatland is vulnerable to subsidence and fire [5].

Moreover, according to [6], CO_2 emitted from decomposition of drained peatlands was between 355 Mt year^{-1} and 855 Mt year^{-1} in 2006 of which 82% came from Indonesia, largely Sumatra and Kalimantan. Meanwhile, widespread peat fires in insular Southeast Asia resulted C emissions, with a conservative estimate of around 0.1 Gt year^{-1} [3]. In addition, [7] estimated losses of tree species in peatland area of Central Kalimantan, Indonesia, of about 23% to 92% due to peat fires.

Meanwhile, monsoon plays an important role in determining seasonality of rainfall in humid tropical region, such as Indonesia, which inter-seasonal variability is associated with El-Nino Southern Oscillation (ENSO). During El-Nino years, the region can experience a longer dry season, while during

La-Nina years, the onset of the monsoon is usually earlier than normal [8]. Furthermore, increasing temperature during El-Nino years can potentially trigger wildfires.

The objectives of the study are: (1) to predict drought regime propagation within peat layers using three dimensionally model of peatland hydrology (the PEAT VOXEL model); and (2) to analyze vertical peat water behaviors under ENSO conditions.

2. Methods

2.1. Study area

The model was evaluated in a peatland restoration zone of oil palm plantation, located in Nataikuini Village, Kendawangan Sub-district, Ketapang District, West Kalimantan Province, Indonesia (110.896° -110.943° E, 2.831°-2.794° S) with total area of about 900 ha (Figure 1). Four land-cover types were identified: unburnt forest, shrub, burnt forest, and bareland (Table 1). We applied object-based image classification method using multi-resolution segmentation algorithm [9] to Landsat 8 imagery. Accuracy of the land-cover map was assessed using 38 ground truthing points with kappa accuracy of 0.89.

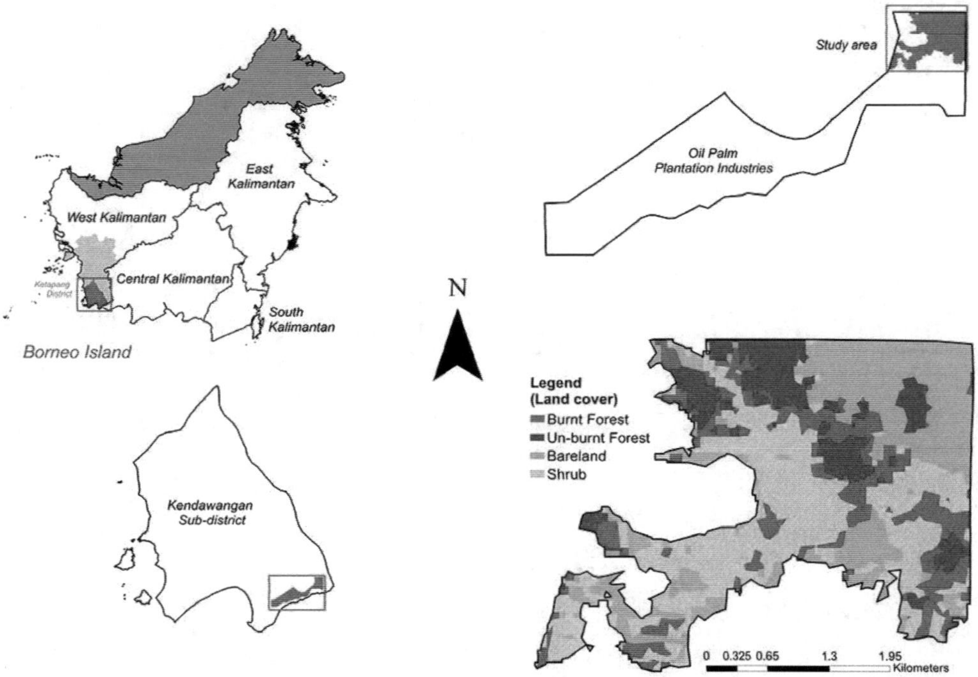

Figure 1. Map of the study area in a peatland restoration zone of oil palm plantation in Ketapang, West Kalimantan, Indonesia and the land cover map used in this study (the lower right corner).

Table 1. Description of land cover categories used in the study.

Land cover
Unburnt forests dominated by vascular plants: *Shorea balangeran* (endangered according to IUCN Red List), Alstonia trees, and *Combretocarpus rotundatus*.
Shrub dominated by fern and other vascular plants (at seedling stage).
Bareland refers to non-vegetated fire-climax areas, mostly resulted by catastrophic wildfire in 2015 (based on reports by the oil palm plantation company).
Burnt forests refer to forest regrowth after experiencing wildfires in 2013.

2.2. Data

There are two data types used for parameterizing the model: spatial and tabular (Table 2). Tabular data consist of: (i) meteorological parameters, (ii) peat parameters, and (iii) land cover parameters. Spatial data consist of: (i) land cover map, classified from Landsat 8 imagery of path/row 120/62 with acquisition date of February 20th 2017, (ii) digital elevation model (DEM) based on SRTM, and (ii) peat thickness map, based on surface interpolation of ground measurement points. In this case, the spatial data were resampled into 60-m.

Table 2. Input parameters of the model.

Type	Input parameters		Source
Tabular	Meteorological parameters	Daily precipitation	Measurement by IOI Group from 1 rain gauge station using ombrometer[a]
		Potential evapotranspiration[b]	NASA POWER data
	Peat properties	Overland flow velocity	[10]
		Hydraulic conductivity by land cover type	[11]
		Field capacity by land cover type	[11]
		Saturated capacity by land cover type	Authors' defined
	Land-cover properties	Potential evapotranspiration correction factor	[12]
		Drought resistance	[12]
		Interception effect on transpiration	[12]
Spatial	DEM		SRTM
	Land-cover map	4 categories	Landsat 8 OLI 120/62
	Peat thickness map		IOI Group

[a]Since the study area is relatively flat, time series of daily precipitation based on measurement from 1 rain gauge station should represent the whole area [13].

[b]Potential evapotranspiration was estimated using Penman-Monteith method [14].

To validate the PEAT VOXEL model, five instruments for water level measuring were constructed using Arduino micro-controller (Nano ATmega328) and ultrasonic HC-SR04 sensor, installed and distributed purposively to represent each land cover type in the study area, and recorded daily surface and sub-surface water levels for 2 weeks (30-March-2017 to 12-April-2017). Principally, the sensors transmit ultrasound pulse beams and calculate the return time of the signals from the peat surface level, thus depression of water level can be measured. The model goodness of fit [15] between daily simulated data and measured data was about 87%.

2.3. Model description

The PEAT VOXEL model was developed using NetLogo 3D agent-based modelling tool [16]. It captures spatio-temporal dynamics of surface water due to land-atmospheric interactions based on GenRiver conceptual framework [12] and spatio-temporal dynamics of sub-surface water due to water diffusion within peat layers based on Darcy's Law, which is three-dimensionally explicit. Temporal resolution of the model is daily, and three-dimensional spatial resolution of the model (xyz) is 60 m x 60 m x 0.1 m.

Assumptions of the model are: (i) surface water velocity is constant at 30 m/s [10]; (ii) total amount of sub-surface peat water within the entire peat layers is at equilibrium condition, i.e. water sinks from and leakages to outside (surrounding peat layers not under consideration) are assumed to be at equal rates [17]; and (iii) hydraulic conductivity and water retention capacity of peat layers are the main determinants of water diffusion within peat layer.

A diplotelmic peat characterization system, i.e. peat is categorized into acrotelm-catotelm layers ([18]; [19]) was used in this study. The acrotelm layer is the upper part of peat layers, with thickness of about 25 cm and relatively high hydraulic conductivity due to biological activities within the layers, such as vegetation rooting and microbial activities; while the catotelm layer is the deeper and more dense part of peat layers, with relatively low hydraulic conductivity due to low biological activities at its anaerobic condition [20].

In the PEAT VOXEL model, the movement of water on the surface is based on elevation-driven pixels neighbourhood rule, incorporating 8 pixel-neighbours, while the diffusion of water within peat layer is based on Darcy's Law-driven voxels neighbourhood rule, incorporating 6 voxel-neighbours.

2.4. Mathematical model of PEAT VOXEL

Land-atmospheric interactions within the PEAT VOXEL model was conceptualized based on generic water balance model [12], incorporating precipitation, evapotranspiration, interception and infiltration, using the following equations.

$$I_a = I_p\,(1 - exp(\frac{P_g}{I_p})) \tag{1}$$

Actual interception (I_a) in mm is defined as water loss due to water evaporation from each land cover type to the atmosphere, determined by gross precipitation (P_g) in mm and potential interception of each land cover type (I_p) in mm.

$$Inf = P_g - I_a\,, Inf < \min(SC - H) \tag{2}$$

Infiltration (*Inf*) is defined as water penetration from the surface into the sub-surface (peat layers), determined by net precipitation (P_g - I_a) in mm, saturated capacity of peat layers (*SC*) in mm, and actual water content of peat layers (*H*) in mm. In this study, *SCs* of acrotelm and catotelm were parameterized to 200 and 150 mm per voxel, respectively.

$$ETA = \beta\,[(ETP \times k_o) - (c \times I_a)] \tag{3}$$

$$\beta = \frac{\overline{H}}{R_d \times \theta_{fc}} \tag{4}$$

Actual evapotranspiration (*ETA*) in mm contributes to water losses from peat surface and acrotelm layers. This parameter is determined by potential evapotranspiration (*ETP*) in mm, evapotranspiration coefficient of each land cover (k_o) in mm, actual interception (I_a) in mm, interception effectiveness to transpiration (*c*), and relative water content (β) as averaged water content within acrotelm (*H*) in mm, drought-resistance of each land cover type (R_d), and field capacity (θ_{fc}) in mm. In this case, c, β and R_d are fractional parameters, which values ranging from 0 to 1.

Sub-surface water diffusion within peat layers (acrotelm-acrotelm) is conceptualized based on Darcy's Law, modified with regards to target voxels, incorporating 6 ordinal voxel neighbours of source voxels, using the following equation:

IOP Conf. Series: Earth and Environmental Science **149** (2018) 012026 doi:10.1088/1755-1315/149/1/012026

$$\left(\frac{dH}{dt}\right)_i = -K\frac{(H - h_i)}{L_i} \tag{5}$$

Water diffusion within peat layers increases at higher hydraulic conductivity (K), higher gradient of water between source voxels and their target voxels (H - h_i), and shorter inter-voxel distance (L_i). In this case, i denotes the 3D ordinal axes (xyz) of each voxel.

$$w_i = \frac{(\frac{dH}{dt})_i}{\sum_{\mathrm{i}}^{6}(\frac{dH}{dt})_i} \tag{6}$$

Outflows from source voxels are weighted (w_i) with regards to actual water contents of the target voxels.

$$\theta = \sum_{i}^{6} - \left(\frac{dH}{dt}\right)_i, \theta < H \tag{7}$$

Total outflows from source voxels to target voxels (θ) is therefore limited by actual water content of the source voxels.

$$\gamma_i = \theta\, w_i \tag{8}$$

$$H_{t+\Delta t} = H_t - \sum_{i}^{6} \gamma_i + \sum_{\mathrm{i}}^{6} \alpha_i\,, \alpha_i < H_{max} \tag{9}$$

Equation (8) describes the weighted outflows from target voxels (γ_i). Furthermore, equation (9) describes peat water level of source voxels at time $t + \Delta t$, where the total inflows from neighbouring voxels are limited by maximum capacity of the source voxels (H_{max}).

2.5. Drought propagations and vertical peat water distributions
Vertical peat water distributions within peat layers was evaluated using ratio of peat water content (mm) and saturated capacity (mm) of each layer. We visualized the ratio of peat water using filled contour plot that represented the temporal dynamics of peat water content within peat layers. Furthermore, drought propagations under different ENSO scenario was evaluated using standardized data [21] of precipitation, surface, and sub-surface peat water in acrotelm and catotelm zone so that we can predict the time travel of the deficit shifting throughout peat layers.

3. Results and Discussions

3.1. Peatland drought assessments
Based on the simulation results, peat water content ratio across peat layers was calculated as the voxel-averaged ratio between actual peat water content relative to its saturated water capacity, ranging from 0 to 1, indicating the wetness of the peat voxels (Figure 2). In addition, drought propagation from incoming precipitation to surface and peat layers [21] was evaluated based on standardized hydrological parameters (Figure 3).

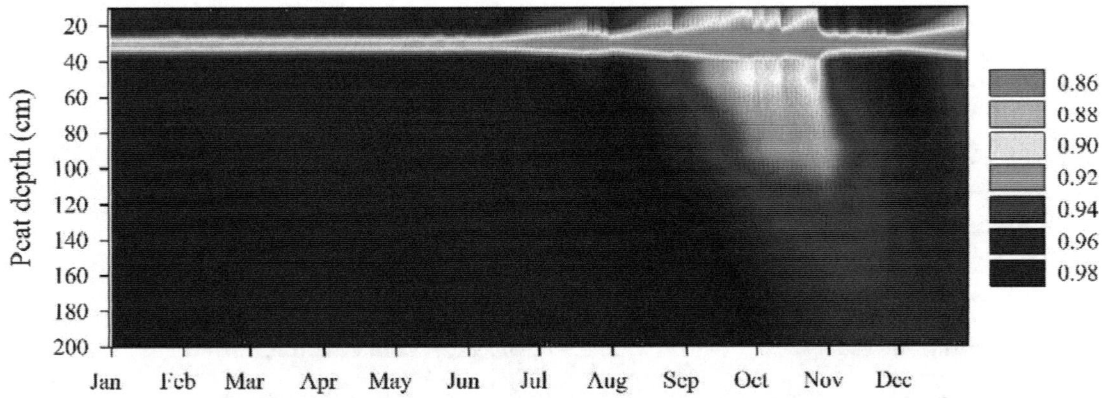

(a)

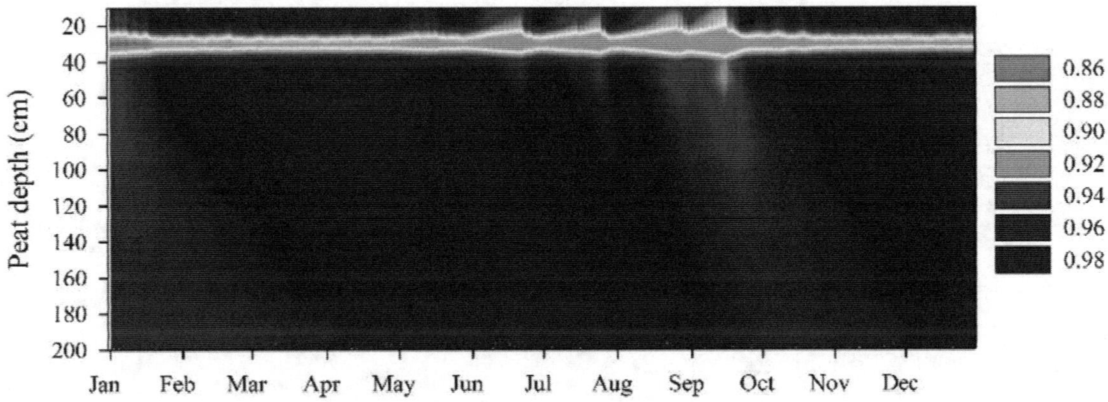

(b)

Figure 2. Water content ratio within peat layers during: (a) 2015 El-Nino episode; and (b) 2016 La-Nina episode.

Based on water content ratio (Figure 2), relatively dry peat voxels during 2015 El-Nino episode were found at maximum peat thickness of ~80 cm from July to October; while during 2016 La-Nina episode, relatively dry peat voxels were found at maximum peat thickness of around 20-40 cm from July to September (on dry season). This peat water depression ignited wildfires in the study area during September to October 2015[1].

[1] Based on Environmental, Health, and Safety Department wildfires report in 2015

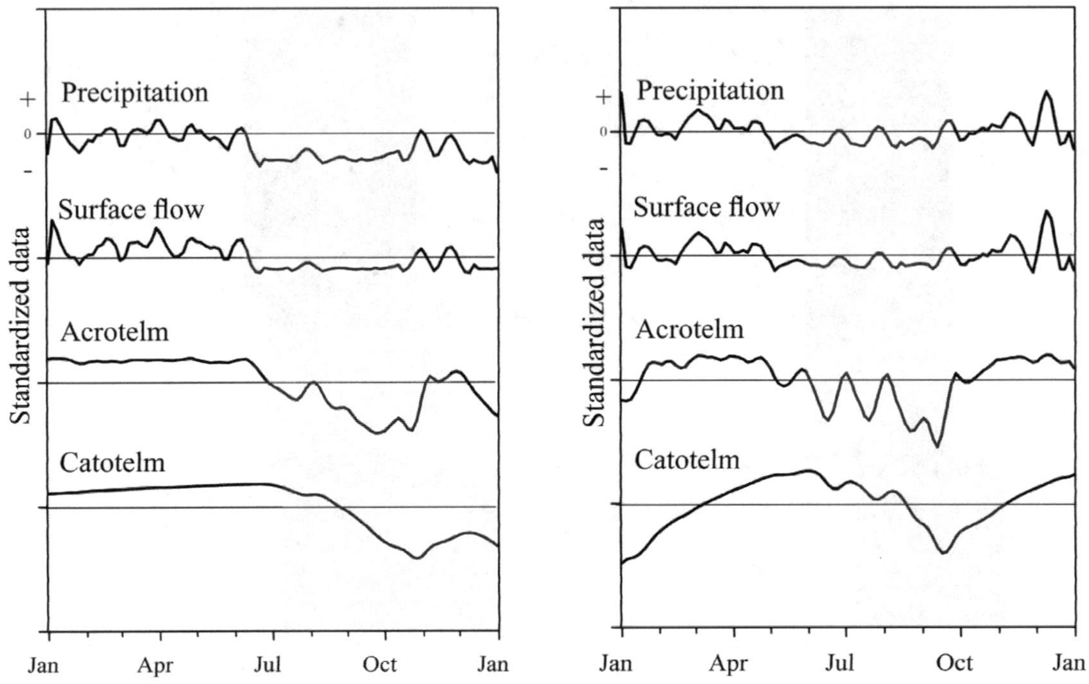

(a) (b)

Figure 3. Drought propagation from incoming precipitation to surface and peat layers during: (a) 2015 El-Nino episode; and (b) 2016 La-Nina episode. Yellow highlights indicate the deficiency period of hydrological parameters.

The study also found that drought propagation time lags into sub-surface peat layers after drought events occurrence on the surface were about 2.5 months during 2015 El-Nino episode and 1 month during 2016 La-Nina episode (Figure 3). Drought propagation explains drought transformation processes from meteorological drought (i.e. precipitation deficiency) to hydrological drought (i.e. catotelm water deficiency). Assessment of drought propagation in this study was carried out based on methods developed by [21].

4. Conclusion

The PEAT VOXEL model captures the spatio-temporal dynamics of surface and sub-surface peatland hydrology with relatively high performance with average of about 87%. Results from scenario-based simulations suggest that drought propagation from surface into sub-surface of peat layers had a time lag of about 2.5 months during El-Nino period and 1 month during La-Nina.

Acknowledgements

We are grateful for the financial assistance from IOI Group (http://www.ioigroup.com/) and Meteorology and Atmospheric Pollution Laboratory, Department of Geophysics and Meteorology, Faculty of Mathematics and Natural Sciences, Bogor Agricultural University (http://geomet.ipb.ac.id/). We would like to express our gratitude towards: (1) Uriel Joseph Wilensky and the NetLogo development team (http://ccl.northwestern.edu/netlogo/) for providing open source modelling platform; and (2) the founders of Arduino (https://www.arduino.cc/) for providing open source microcontroller platform.

References

[1] Page S E, Rieley J O, and Banks C J 2011 Global and regional importance of the tropical peatland carbon pool *Glob. Chang. Biol* **17** (2) 798–818

[2] Page S E and Baird A J 2016 Peatlands and Global Change: Response and Resilience *Annu. Rev. Environ. Resour.* **41** (1) 35–57

[3] Page S E and Hooijer A 2016 In the line of fire: the peatlands of Southeast Asia *Philos. Trans. R. Soc. B Biol. Sci.* **371** (1696) 20150176

[4] Ritzema H 2008 The Role of Drainage In The Wise Use of Tropical Peatlands *Restor. Trop. Peatlands* 78–87

[5] Wösten J H M , Clymans E, Page S E, Rieley J O, and Limin S H Peat-water interrelationships in a tropical peatland ecosystem in Southeast Asia *Catena* **73** (2) 212–224

[6] Hooijer A, Page S, Canadell J G , Silvius M, Kwadijk J, Wösten H, and Jauhiainen J 2010 Current and future CO2 emissions from drained peatlands in Southeast Asia *Biogeosciences* **7** (5) 1505–1514

[7] Shiodera S, Atikah T D, Apandi I, Seino T, Haraguchi A, Rahajoe J S, and Kohyama T S 2015 Peat-fire impact on forest structure in peatland of central Kalimantan *Tropical Peatland Ecosystems* 197–212

[8] Hamada J I, Yamanaka M D, Matsumoto J, Fukao S, Winarso P A, and Sribimawati T 2002 Spatial and Temporal Variations of the Rainy Season over Indonesia and their Link to ENSO *J. Meteorol. Soc Japan* **80** (2)285–310

[9] Blaschke T 2010 Object based image analysis for remote sensing *ISPRS J. Photogramm. Remote Sens* **65** (1) 2–16

[10] Holden J,Kirkby M J, Lane S N, Milledge D G, Brookes C J, Holden V, and McDonald A T 2008 Overland flow velocity and roughness properties in peatlands *Water Resour. Res.* **44** (6) 1–11

[11] Mustamo P, Hyvärinen M, Ronkanen A K, and Kløve B 2016 Physical properties of peat soils under different land use options *Soil Use Manag* **32** (3) 400–410

[12] van Noordwijk M, Farida A, Suyamto D, Lusiana B, and Khasanah N 2015 Spatial variability of rainfall governs river flow and reduces effects of land use change at landscape scale: GenRiver and SpatRain simulations *Modsim 2003 Int. Congr. Model. Simulation (***1-4)** 572–577

[13] Dingman S L 2015 *Physical hydrology* (Ilinois US: Waveland Press, Inc.)

[14] Allen R G, Pereira L S, Raes D, Smith M, and a B. W 1998 Crop evapotranspiration - Guidelines for computing crop water requirements - FAO Irrigation and drainage paper 56 *Irrig. Drain* 1–15

[15] Costanza R 1989 Model goodness of fit: A multiple resolution procedure *Ecol. Modell* **47 (**3–4) 199–215

[16] Wilensky U and Rand W 2015 *An Introduction to Agent-Based Modeling*

[17] Xie Z H, Di Z H, Luo Z D, and Ma Q 2012 A Quasi-Three-Dimensional Variably Saturated Groundwater Flow Model for Climate Modeling *J. Hydrometeorol.* **13 (**1) 27–46

[18] Holden J 2005 Peatland hydrology and carbon release: why small-scale process matters *Philos. Trans. R. Soc. A Math. Phys. Eng. Sci.* **363** (1837) 2891–2913

[19] Holden J and Burt T P 2003 Hydrological studies on blanket peat: The significance of the acrotelm-catotelm model *J. Ecol.* **91 (**1) 86–102

[20] Ballard C E, McIntyre N, Wheater H S, Holden J, and Wallage Z E 2011 Hydrological modelling of drained blanket peatland," *J. Hydrol.* **407** (1–4) 81–93

[21] Van Loon A F 2015 Hydrological drought explained *Wiley Interdiscip. Rev. Water.* **2 (**4) 359–392

Determining critical groundwater level to prevent degraded peatland from severe peat fire

E I Putra[1,2*], M A Cochrane[2], Y Vetrita[3,4], L Graham[5] and B H Saharjo[1]

[1]Faculty of Forestry, Bogor Agricultural University, Bogor, Indonesia
[2]University of Maryland Center for Environmental Sciences, Frostburg, MD USA
[3]Geospatial Sciences Center of Excellence (GSCE), South Dakota State University, Brookings, SD USA
[4]Indonesian National Institute of Aeronautics and Space (LAPAN), Jakarta, Indonesia
[5]Borneo Orangutan Survival Foundation (BOSF), Palangkaraya, Indonesia

E-mail: eindraputra@yahoo.com

Abstract. Peat fires have been a severe recurrent problem for Indonesia, but droughts due to prolonged dry season aggravate burning conditions. To get a better understanding of this issue, we studied fire conditions in a portion of the ex-Mega Rice Project (MRP) area, Central Kalimantan. To examine fire season and hydrology factors affecting peat fires we analyzed daily TRMM data, Nino 3.4 SST Anomalies, and changing groundwater levels (GWL) from 300 dipwells. Our results quantify time-lags between the period of lowest precipitation and the lowest GWL; providing some ability to predict fire risk in advance of the lowest GWL. The rise of Nino 3.4 SST anomalies is significant risk factors for peat fire as they signify dry months which may yield large fire occurrences. GWL in 2011 was lower than in 2012, but fires were more frequent in 2012, indicating that low precipitation amounts in the wet season of 2011/2012 left the peat in a dry condition early in 2012. Most of the fires occurred in areas with GWL less than -30 cm, powerfully illustrating the importance of maintaining GWL at more than -10 cm, to prevent degraded peatlands from experiencing surface and deep peat fires.

1. Introduction

Peat fire have been a severe recurrent problem in Indonesia, but our prolonged observations indicate that recent Indonesian peat fires tend to occur every year in the dry season, even in non-El Nino years. The country then suffered from severe peat fires episode when it experienced droughts due to long dry season, where burning conditions provoke. To get a better understanding of this critical issue, we studied the recent fire situation and hydrology factors affecting peat fires in a portion of the Mega Rice Project (MRP) area, Central Kalimantan. A giant tropical peat swamp-forest conversion [1] has converted 9,191 km^2 of peatland in the area to the fields for rice cultivation and promote transmigration [2], but it just left huge drained peatland area mainly due to the built of over 4,600 km of channels to open-up the area and make it suitable for cultivation. Without sufficient consideration of peatland hydrology and the area topography [3], this project just exacerbated the problem when the over drained peatland became more vulnerable to fire. Fires commonly occur in this area annually during the dry season (July-September), but burning conditions aggravated during pronounced El Niño years when the peatlands there are subjected to extremely severe dry conditions.

Groundwater level (GWL) has used widely as a critical indicator for predicting fire occurrences [4,5,6]. Our previous study [4] revealed that the fires in the MRP area were mostly, over 99%, occurred

IOP Conf. Series: Earth and Environmental Science **149** (2018) 012027 doi:10.1088/1755-1315/149/1/012027

when the GWL was below the soil surface with the peak fire period started in the middle of August when daily precipitation reached the minimum and GWL tent to decreased steeply. Recurrent fires in the Ex-MRP area may result to the decreasing of peatland capacity for retaining and absorbing water from precipitation, lowering groundwater levels further and making the area more susceptible to fire, particularly in the dry season. Here, we examine the situation by linking the GWL, precipitation and fire tendency in the area.

2. Methods

In this study, we are using monthly GWL data from 300 dipwells in the northern part of Block A and southern part of Block E of the MRP area (figure 1). These dipwells established in 2010 within the Kalimantan Forest and Climate REDD+ demonstration project (KFCP). The GWL were monitored monthly during January 2010-January 2013 using the blow-straw method [7]. We conducted static DGPS Survey to provide exact peat surface elevation for these 300 dipwells [8].

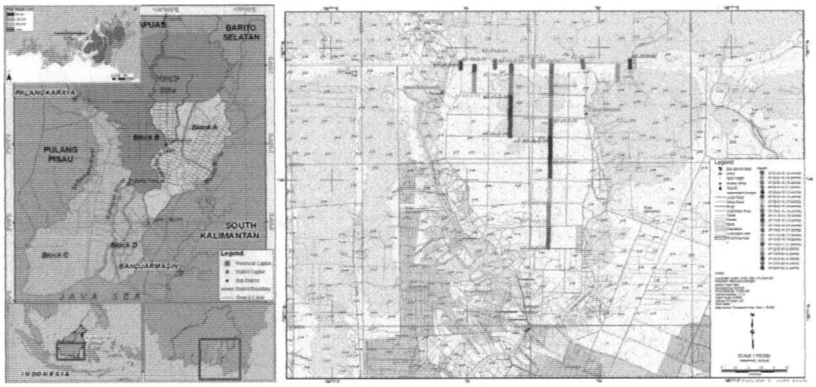

Figure 1. Map of study area [7] (left) and location of dipwells [8] (right)

We examine the relationship between groundwater level, a rate of precipitation and fire occurrences during 2010-2012. IDW (Inverse Distance Weighting) multivariate interpolation technique is used to figure the groundwater level surrounding each deep well point. Monthly TRMM (Tropical Rainfall Measuring Mission) data derived from GES-DISC NOAA (http://mirador.gsfc.nasa.gov/) is used to illustrate precipitation patterns in the study area. As the study area covers four TRMM pixels, we have four different precipitation regions for the study area (figure 2). We used Terra/Aqua MODIS active fire data from NASA-EOSDIS (https://earthdata.nasa.gov/) to explain fire occurrences and tendency in the area.

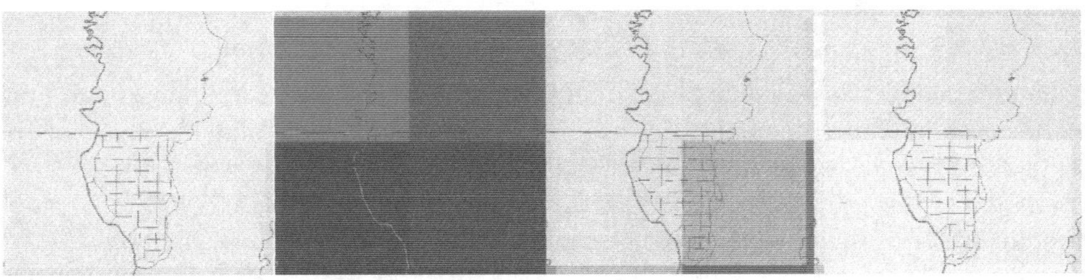

Figure 2. Monthly precipitation surrounding area, July – October 2012. The graduated color indicates precipitation rates; dark blue for lowest precipitation (0-10 mm), yellow to moderate rain (30-40 mm); light blue and green indicates precipitation rate between the lowest and moderate.

3. Results and Discussion

Monthly precipitation is commonly used to define the dry and wet season in Central Kalimantan [3,4,8]. A dry month is determined by Mackinnon et al. [9] as a month with less than 100 mm mean monthly rainfall, and a wet month when the mean monthly rainfall exceeds 200 mm. All regions experienced similar U-shaped monthly precipitation patterns with peak dry season between July and September and a trough reaching a minimum in August (figure 3). As the same findings revealed in the previous study [10], then we can point June-October as the dry season and November-May as the wet season in the MRP area; provides some ability to predict fire risk in advance. Low precipitation in the dry season dries the degraded peatland in the MRP area, increasing its susceptibility to fire, and creating suitable conditions for peat fires to ignite.

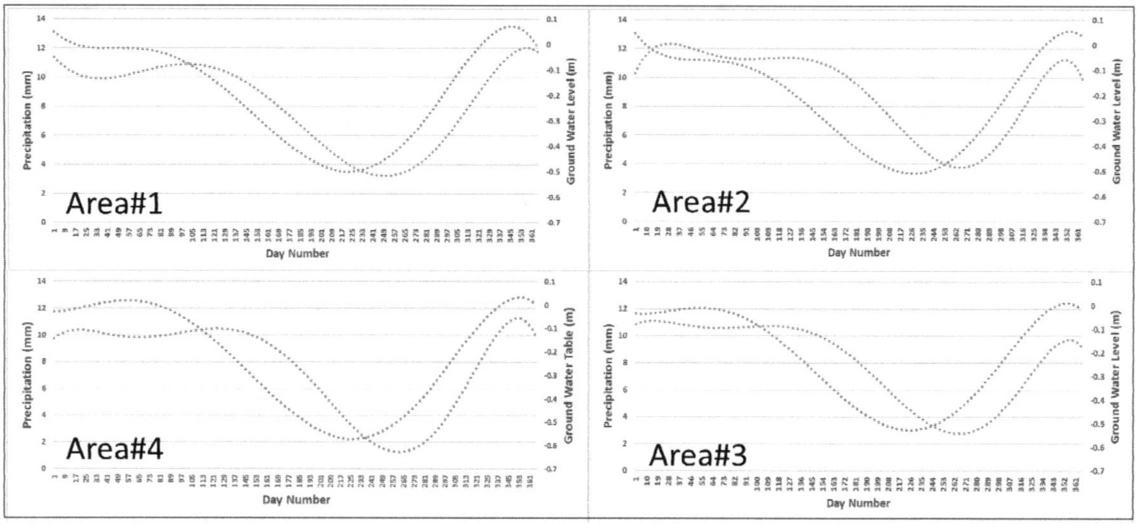

Figure 3. Daily groundwater level and precipitation tendency at four TRMM pixels in the study area

In its natural condition, peat swamp forest always inundated with water [11]. The peat acts as a natural sponge and normally waterlogged already (UNDP, 2006). GWL in MRP area peatlands previously remained positive above the peat surface in wet season [10]. However, GWL in the study area now remains in deficit for the whole of the year (figure 3), which can reflect the severe dry condition of peat in the area. This situation may also be considered unnatural for the system and can explain the degraded status of the peatland in the study area where the peat has lost its ability for absorbing and storing water, brings the high risk of future fires in the area. Peatland in the study area experienced severe dry conditions during the dry periods of July – October, as shown by the red and orange colors in figure 4, resulting in the frequent peat fire occurrences here. The study area has suffered from almost yearly peat fire occurrences since 2001, except 2008 (figure 5).

We observed a time-lag between the lowest precipitation and the lowest GWL for a month in 2011 and two-month in 2012 (figure 6). Lowest precipitation is found in August for both 2011 and 2012, while the lowest GWL occurred in September (2011) and October (2012). GWL starts to increase after having continuous high precipitation. These time-lags strongly indicates that peatlands here have lost the capacity to effectively absorb and retain water from rainwater droplets, suggesting that groundwater levels in the study area will recover only after having plenty water supplies from continuous precipitation. Less precipitation rate in 2012 compared to 2011 brings longer time-lag in 2012 compared to 2011.

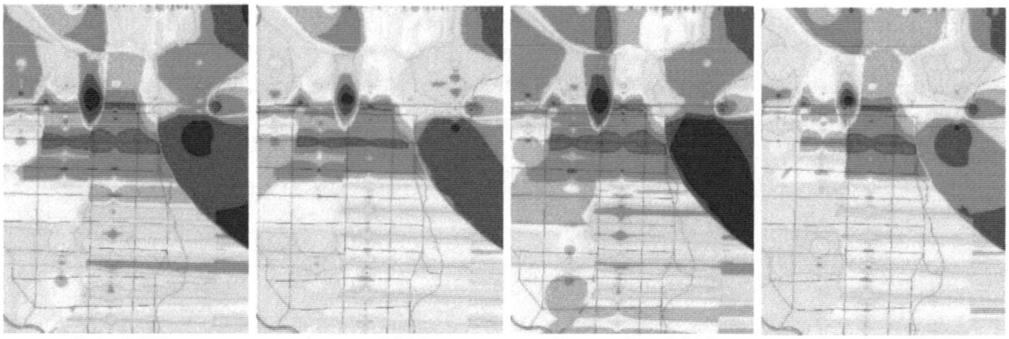

July 2012 August 2012 September 2012 October 2012

Figure 4. Change of groundwater level from July – October 2012. Red colour indicates deep GWL (more than 30 cm below peat surface); dark blue colour for positive GWL; light blue, yellow and orange represent GWL between 0 (peat surface) and -30 cm (30 cm below peat surface).

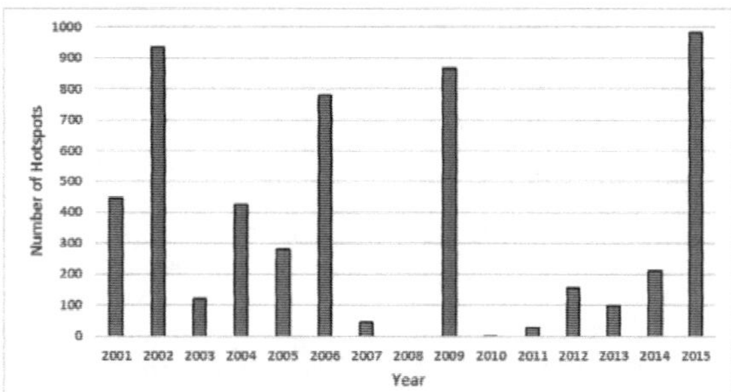

Figure 5. Number of hotspots detected by Terra/Aqua MODIS Satellite in study area, 2001 - 2015

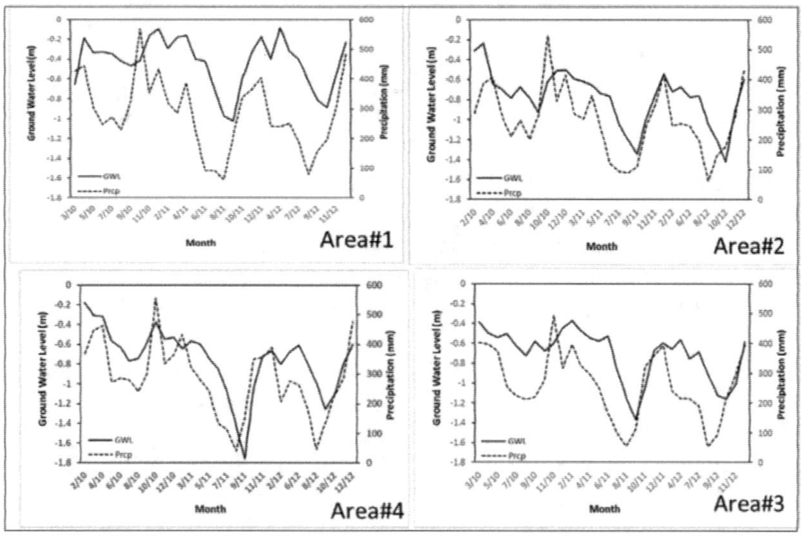

Figure 6. Groundwater level and precipitation patterns at four TRMM pixels in the study area, 2011 – 2012. Solid lines represent the GWL and dot lines represent precipitation.

Figure 7 (left) shows the coincidence of peat fire occurrences with the precipitation curve. Fires usually occur following precipitation patterns with little or no rainfall during the dry period. Large number of peat fires tends to start three months after the start of the decreased of precipitation. However, GWL in 2011 is lower than 2012 (figures 7 and 8), but fires were more frequent in 2012.

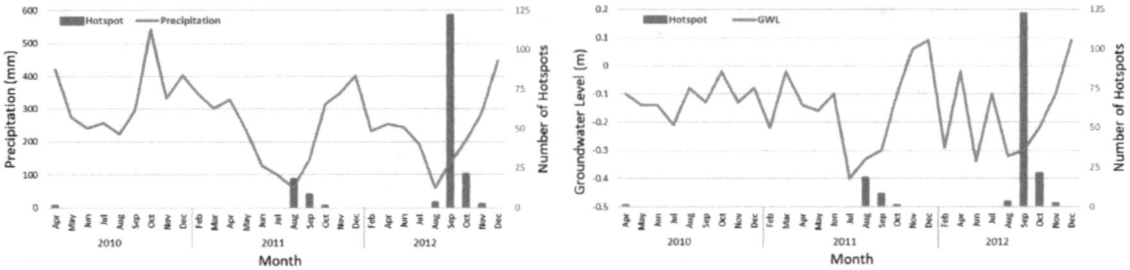

Figure 7. Relationship between hotspot occurrences in study area with monthly precipitation (left) and groundwater level (right), 2010-2012

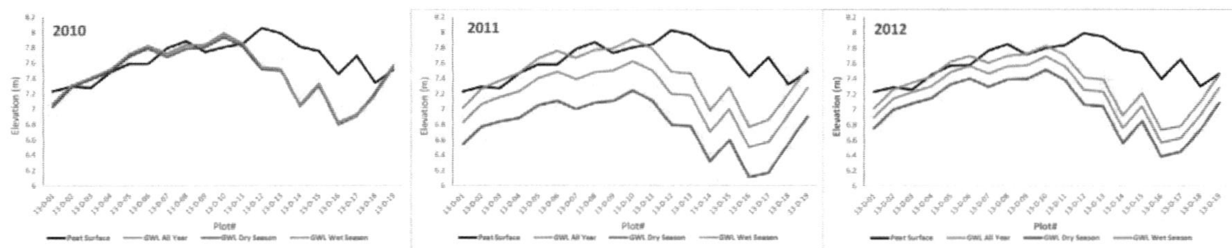

Figure 8. Comparison between groundwater level in dry season and wet season at some dipwells, 2010-2012. The red lines represent GWL in the dry season, blue lines represent GWL in the wet season, black lines represent peat surface, and the orange lines represent the average of GWL.

In 2011, precipitation reached the lowest point in August after experiencing a gradually decreased from May. This condition creating the necessary dry conditions for fires to ignite. Equally, large number of fires occurred in August and September. In 2012, large numbers of fires started during August, three months after the commencement of decreasing precipitation patterns during June 2012. This phenomenon may indicate that wet season precipitation (November 2011-May 2012) was not sufficient to raise the water table high enough to rewet the degraded peat in the study area, leaving the peat in a dry condition early in 2012 and effectively extending the fire season when peat burning was possible. Wet season usually started from November, but the fires occurrences in November 2012 explain the dry condition of the peat was still existing in the beginning of the wet period after suffering from severe dry conditions for prolonged dry periods, maintaining the appropriate conditions where ignition may readily occur [14].

Sea surface temperature (SST) anomalies is found greatly responsible for severe drought occurrences in years coinciding with intense El Nino events [13,15], but our previous work [16] suggested SST Anomalies as a useful index to determine drought in Palangka Raya instead of the El-Nino event. This study notices a significant decrease of precipitation in dry season is following the rise of SST anomalies (figure 9), bring the dry condition in the area, and thus allowing the fire to ignite in ease.

Fires in 2011 and 2012 started when the area suffer from lack accumulated precipitation for less than 5 mm in a week. However, the rising of SST Anomalies in 2012' dry season to high positive value may bring less precipitation and more prolonged drought to the area. Thus, fires in 2012 were much higher compared to 2011. Fires in 2012 began in early August when the SST Anomalies was +0.6^{0}C and there was no rainfall during a week. Numbers of fires in 2012 significantly expanded to more than 100 in September and October when the area subjected the low accumulated weekly precipitation of less than

25 mm and the increasing of SST Anomalies to +0.8^{0}C. Therefore, we strongly suggest considering the rising of Nino 3.4 STT Anomalies to any positive value for a significant risk factor for fire prevention efforts as it may initiate the driest months, likely to yield large fire occurrences.

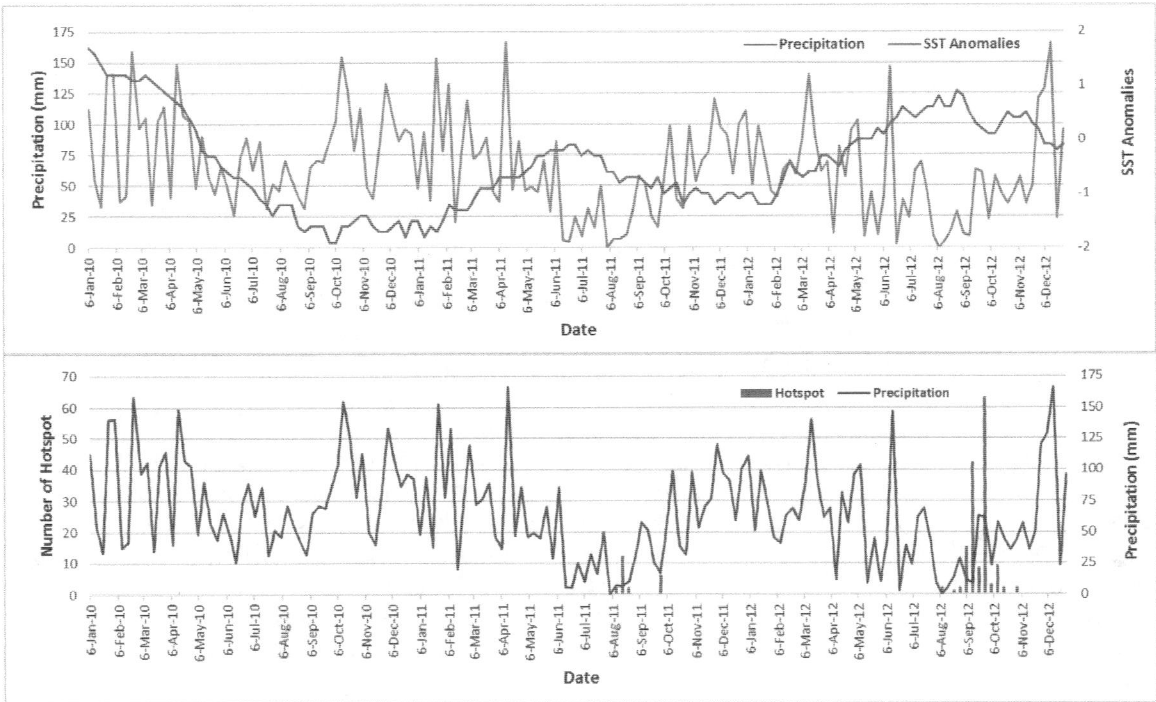

Figure 9. SST Anomalies, weekly precipitation and hotspot occurrences in study area, 2010-2012

Relationships between fire occurrences and GWL (figures 7 and 10) may suggest that the low GWL accelerated conditions where fires ignite with ease. In 2011, large number of fires started occurring in August following the drop of the mean GWL to – 33 cm, while in August 2012 fires began with a mean GWL of -34 cm (Fig. 7). Usup et al. [5], Wosten et al. [6], Putra and Hayasaka [12], Susilo *et al.* [15] and DeVries [17] suggested critical groundwater level of 40 cm below peat surface to prevent fire. However, our findings suggest that shallower GWL below peat surface should be maintained to prevent peat fire occurrences in dry-degraded peatlands.

Most of the fires in the study area occurred with GWL conditions of 30 – 39 cm below the peat surface (figure 10), but fire occurrences with GWL of less than 10 cm below peat surface may strongly suggest that degraded peatlands are very vulnerable to fires even under relatively moist conditions. Therefore, degraded peatlands should be maintained in wet conditions with critical GWL of less than 10 cm below peat surface, to prevent the area from experiencing surface peat fires. Dry conditions of degraded peatland create a suitable condition for the fire to burn downward into deeper peat layers and ignite deep peat fires, resulting in devastating peat fires in the area.

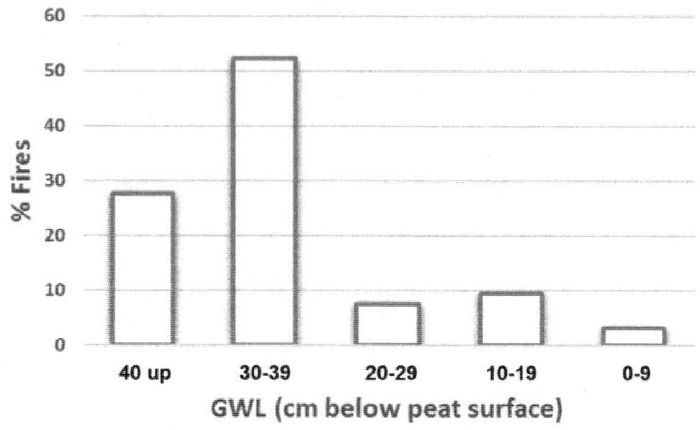

Figure 10. Percentage of peat fires occurrences on each GWL depth, September 2012

4. Conclusions

This study reveals that the degraded peatlands in the northern part of Block A and southern part of Block E of Ex-MRP have lost its capacity in absorbing and retaining water from rainwater droplets, keeping them in drier conditions than natural for most of the year, and therefore are very vulnerable to fire. The rising of Nino 3.4 STT Anomalies should be considered to predict significant fire risk as it may yield large fire occurrences. Time-lags between lowest precipitation and lowest GWL may provide some abilities to predict fire risk in advance of the lowest GWL. We propose the critical GWL of less than 10 cm below peat surface to prevent the degraded peatland from experiencing surface peat fires that may escalate to devastating deep peat fires.

Acknowledgements

The research was funded by NASA (NNX13AP46G) under UMCES/IPB-NASA Tropical Peat Fire Research Project. The authors wish to thank KFCP for sharing the hydrology data from the project, and the KFCP Hydrology field team for its collection.

References

[1] Notohadiprawiro T 199. Mega-project of Central Kalimantan wetland development for food crop production, belief and truth *Proc. of the International Peat Symposium the Spirit of Peatlands* Jyvaskyla, Finland, 7 – 9 Sept. 1998.

[2] Sabiham S, Winarna, Pulunggono HB, Novarina D and Sahari B 2016 What is the way forward on Indonesian peatlands? *15th International Peat Congress* 2016 Malaysia

[3] Siegert FH, Boehm DV, Rieley JO, Page SE, Jauhiainen J, Vasander H and Jaya A 2001 Peat fires in Central Kalimantan, Indonesia: fire impacts and carbon release *Proc. of the International Symposium on Tropical Peatlands* 142-154

[4] Putra EI, Hayasaka H, Takahashi H, and Usup A 2008 Recent Peat Fire Activity in the Mega Rice Project Area,Central Kalimantan, Indonesia *Journal of Disaster Research* **3** (5): 1-8

[5] Usup A, Hashimoto Y, Takahashi H, and Hayasaka H 2004 Combustion and thermal characteristics of peat fire in tropical peatland in Central Kalimantan, Indonesia *Tropics* **14** (1):1-19

[6] Wosten JHM, Clymans E, Page SE, Rieley JO and Limin SH 2008 Peat-water interrelationships in a tropical peatland ecosystem in Southeast Asia *Catena* **73**: 212-224

[7] Ichsan N, Hooijer A, Vernimmen R, Applegate GB 2014 *KFCP Hydrology and Peat Monitoring Methodology* Scientific Report, Kalimantan Forests and Climate Partnership, Indonesia.

[8] *Final report for GPS Positioning Services for peatland surface elevation stdiy in Block A. PT*

Geoindo Giri Jaya December 2016
[9] Aldrian E and Susanto RD 2003 Identification of three dominant rainfall regions within Indonesia and their relationship to sea surface temperature *International Journal of Climatology* **23**: 1435-1452
[10] MacKinnon K, Hatta G and Mangalik A 1997 *The Ecology of Kalimantan* Oxford University Press
[11] Ritzema H and Wosten H 2002 *Hydrology of Borneo's peat swamps* STRAPEAT – Status Report Hydrology The Netherlands
[12] Putra EI and Hayasaka H 2011 The effect of the precipitation pattern of the dry season on peat fire occurrence in the Mega Rice Project area, Central Kalimantan, Indonesia. *Tropics* **19** (4): 145 – 156
[13] Murdiyarso D and Adiningsih E 2006 Climate anomalies, Indonesian vegetation fires and terrestrial carbon emissions *Mitig Adapt Strat Glob Change* 101 – 112
[14] Putra EI, Cochrane MA, Vetrita Y, Graham LLB and Saharjo BS 2016 Degraded peatlands, Ground Water Level and severe peat fire occurrences *15th International Peat Congress* 2016. Malaysia
[15] Susilo GE, Yamamotoa K and Imaia T 2013 Modeling groundwater level fluctuation in the tropical peatland areas under the effect of El Nino The 3rd International Conference on Sustainable Future for Human Security SUSTAIN 2012 *Procedia Environmental Sciences* **17**: 119 – 128
[16] Putra EI 2010 *Recent peat fire trend in the Mega Rice Project (MRP) area, Central Kalimantan, Indonesia*. PhD Thesis. Hokkaido University, Japan
[17] DeVries B 2010 *Monitoring the Effects of Hydrological Restoration Efforts in Degraded Tropical Peatlands* MSc Thesis Wageningen University Germany

IOP Conf. Series: Earth and Environmental Science **149** (2018) 012028 doi:10.1088/1755-1315/149/1/012028

Contribution of land use changes to meteorological parameters in Greater Jakarta: Case 17 January 2014

D E Nuryanto[1,2], H Pawitan[1], R Hidayat[1], and E Aldrian[3]
[1]Department of Geophysics and Meteorology, FMIPA Bogor Agricultural University, Kampus IPB Darmaga, Bogor 16680, West Java, Indonesia
[2]Research and Development Center, The Indonesian Agency for Meteorology Climatology and Geophysics, Jl Angkasa I No 2 Kemayoran Jakarta Pusat 10620, Indonesia
[3]Agency for The Assessment and Application of Technology, Jl MH Thamrin No 8 Jakarta 10340, Indonesia

Email : danang.eko@bmkg.go.id

Abstract. The impact of land use changes on meteorological parameters during a heavy rainfall event on 17 January 2014 in Greater Jakarta (GJ) was examined using the Weather Research and Forecasting (WRF) model. This study performed two experimental simulation methods. The first WRF simulation uses default land use (CTL). The second simulation applies the experiment by changing the size of urban and built-up land use (SCE). The Global Forecast System (GFS) data is applied to provide more realistic initial and boundary conditions for the nested model domains (3 km, 1 km). The simulations were initiated at 00:00 UTC January 13, 2014 and the period of modeling was equal to six days. The air temperature and the precipitation pattern in GJ shows a good agreement between the observed and simulated data. The results show a consistent significant contribution of urban development and accompany land use changes in air temperature and precipitation. According to the model simulation, urban and built-up land contributed about 6% of heavy rainfall and about 0.2 degrees of air temperatures in the morning. Simulations indicate that new urban developments led to an intensification and expansion of the rain area. The results can support the decision-making of flooding and watershed management.

1. Introduction

An exceptionally heavy rainfall is a serious threat to human life and property. Even with the recent rapid improvements in numerical weather forecasting, the forecast of heavy rainfall events remains a problem for atmospheric modeling communities. Greater Jakarta (GJ) is a potentially active region for heavy rainfall events that can trigger flash floods [1-5]. In addition, the heavy rainfalls are particularly devastating, especially if they last a long time. This is also due to a densely populated and low-lying coast. It has a significant impact on the socio-economic conditions of the GJ bordering areas. The events that imply heavy flooding in GJ could be caused by the activity of mesoscale convection systems (MCSs).

Previous studies on MCS activity in Indonesia Maritime Continent (IMC) were performed using satellite observation data [6-11]. But it is still not fully understood about the predictability of the MCS. A recent study by Trismidianto [8] discusses MCC and is associated with its contribution to precipitation in IMC, but there is still no clarity of heavy or light rainfall. Meanwhile, Putri *et al.* [11]

have only discussed seasonal variations of MCS in IMC where MCS on land tend to show a lower TBB minimum, larger size, and higher rainfall compared to MCS at sea.

In this study, the mesoscale model Weather Research and Forecasting - Advanced Research WRF (WRF-ARW) is used to simulate the meteorological features of a heavy rainfall event over GJ on January 17, 2014, and its sensitivity to study the effects of land use on predictive patterns. The sensitivity of the model simulations to the initial conditions is also examined on the basis of land use changes. A brief description of the model as well as the parameterization schemes and data used in the study are presented in Section 2. Model-simulated results together with the assessment of the performance of the model for different land uses are presented in Section 3 with the conclusions in Section 4.

2. Data and Methods

Version 3.5 of the non-hydrostatic regional numerical weather forecast system WRF-ARW was used for simulations. The WRF model is initialized and the boundary conditions (with the interval of 6 hours) are specified using the Global Forecast System (GFS) data. The GFS data have high spatial (0.5°) resolution for periods from 13 to 19 January 2014.

Three nested grids with one-way nesting were used, extending over 1,179 km x 1,179 km; 363 km x 363 km and 151 km x 151 km at 9; 3 and 1 km resolution as shown in Figure 1. The simulations were done for 27 vertical levels, the modeling period was 6 days, and the temporal resolution of the output data was 1 hour. The WRF model used the following schemes of parameterization of sub-grid processes: Lin *et al.* [12] for cloud microphysics; the RRTM scheme for longwave radiation [13]; the Dudhia scheme for shortwave radiation [14]; the BMJ scheme for the cumulus option [15-17] and the scheme of Yonsei University [18] for the boundary layer (Table 1).

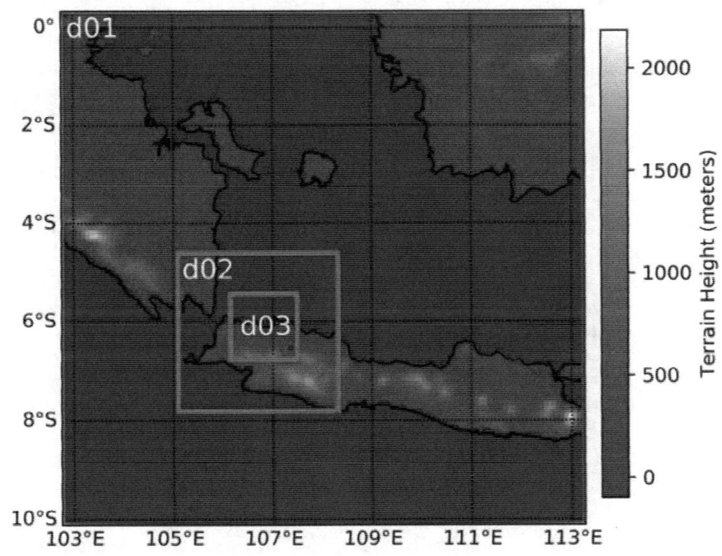

Figure 1. Map with the topography of the grid domain (d01 denotes a resolution of 9 km, d02 denotes a resolution of 3 km and d03 a resolution of 1 km). Note that the map is the computational grids used from the WRF simulations.

The Lin *et al.* [12] scheme is a single moment scheme in which only the mass mixing ratio of each hydrometeor species is transmitted with the number density estimated using an exponential size distribution function. Most cumulus parameterizations should not be used for grid resolutions less than 10 km. But in this case, we used the BMJ scheme because BMJ has significantly increased the coverage of lighter rainfall [19]. In addition, the reason for choosing the YSU boundary layer scheme is that the YSU tends to give the deepest PBL.

The impact of land use changes on patterns of meteorological parameters on 17 January 2014 in GJ are examined using the Weather Research and Forecasting (WRF) model. This study conducted two different land uses for experimental methods of simulation. The Global Forecast System (GFS) data is applied to the first model domain (9 km) with the purpose of provide a more realistic and desirable initial and boundary conditions for the nested model domains (3 km, 1 km). The reference experiment (CTL) was chosen as the standard model configuration described in the previous. Experiment SCE is the same as CTL configuration, except for the land use index modification (Figure 2). The SCE experiment is carried out by changing the land use index in most GJ urban and built-up land. This is done to see how urban heat islands affect local meteorological processes.

Table 1. Details of the WRF-ARW model specifications.

Configurations	**Domain 1**	**Domain 2**	**Domain 3**
Center point longitude	106.749		
Center point latitude	-6.168		
Horizontal grid dimension X	131	121	151
Horizontal grid dimension Y	131	121	151
Horizontal grid resolution	9 km	3 km	1 km
Number of vertical level	27	27	27
Microphysics	Lin *et al.*	Lin *et al.*	Lin *et al.*
Planetary boundary layer	YSU	YSU	YSU
Cumulus parameterization	BMJ	BMJ	BMJ
Longwave radiation scheme	RRTM	RRTM	RRTM
Shortwave radiation scheme	Dudhia	Dudhia	Dudhia

To evaluate the WRF performance, observations from 5 surface weather stations were used in the urban and surrounding GJ (indicated by a blue triangle in Figure 2). The data are available in the Database Center of the Indonesian Agency for Meteorological Climatology and Geophysics (BMKG). The available data are hourly observed surface wind speed and wind direction. In this work, a statistical approach is used to obtain a quantitative comparison. The following equation is the root mean square error (RMSE) and standard deviation of the error (SDE) as the intended method:

$$RMSE = \left(\frac{1}{n} \sum_{i=1}^{n} (x_i - y_i)^2 \right)^{1/2} \quad (1)$$

$$SDE = \left(\frac{1}{n-1} \sum_{i=1}^{n} (e_i - \bar{e})^2 \right)^{1/2} \quad (2)$$

$$e_i = x_i - y_i \quad (3)$$

with:

n = number of samples
x_i, y_i = individual samples with index i
$\bar{e}$ = error mean

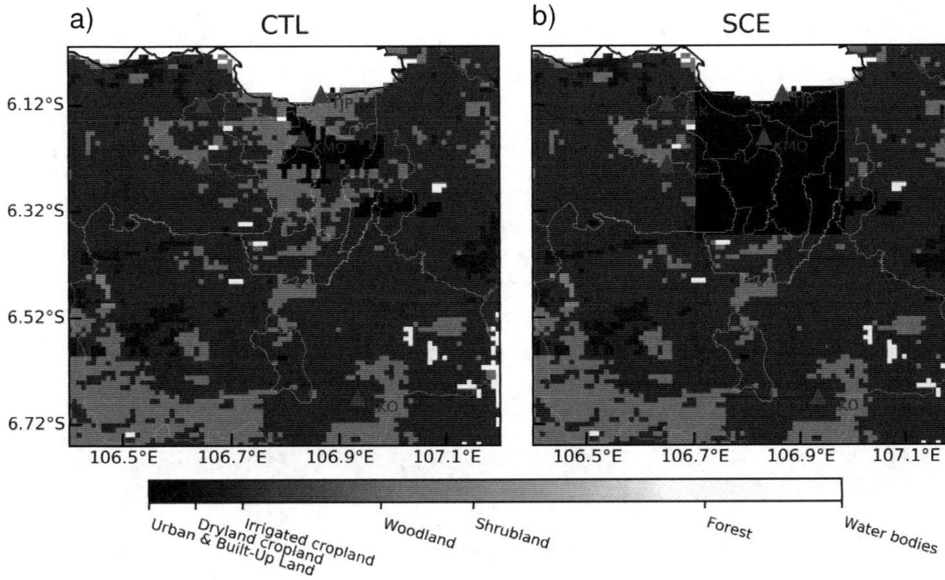

Figure 2. Shading plots showing the reference experiment (a) and the SCE experiment (b), with different land use index in most of GJ. Blue triangles denote stations: Tanjung Priok (TJP), Soekarno-Hatta Cengkareng Airport (CGK), Kemayoran (KMO), Curug (CRG) and Citeko (CKO) for the inner modeling domain. Note that the maps in CTL and SCE are the computational grids used from the WRF simulations.

To diagnose the impact of land use on meteorological features, we plot the spatial map of surface temperature at 0700 LT on January 17, 2014. This technique is conducted by recording the spatial value of the surface temperature of CTL, SCE, and difference (CTL-SCE). It is commonly used to know the difference of a particular phenomenon in a particular area. In this case, this is done in a specified latitude range, i.e. 6.01°S - 6.76°S.

3. Results and Discussion

Figure 3 shows the simulated and observed surface air temperature for the five stations (blue triangles in Figure 2). The simulations capture the diurnal variation including the maximum and minimum temperatures during much of the period of the heavy rainfall event within approximately 0.2 °C of the different temperature. On day 2 of the simulation, a maximum temperature reaches about three hours later in the simulation than in the observed time. There is not a tendency of WRF simulation results to detect the peak of air temperature. In the startup period of the simulation on the fifth day, the simulated air temperature indicates an overestimation, except at Tanjung Priok Station. The temperature difference between the simulation result and the observation is less than 2 °C for the six simulation days (Table 2).

Figure 4 shows the simulated and observed sea level pressure (SLP) for the five stations. The simulations also capture the diurnal variation of SLP during the heavy rainfall event. On day 4 of the simulation, a maximum SLP reaches the urban area about 3 hours earlier in the simulation before the observed. There is a tendency of WRF simulation results to detect the peak of SLP lower than observed. In the starting period of the simulation on the fifth day, the simulated SLP indicates an underestimation. The SLP difference between the simulation result and the observation is less than 2 for the six simulation days (Table 2).

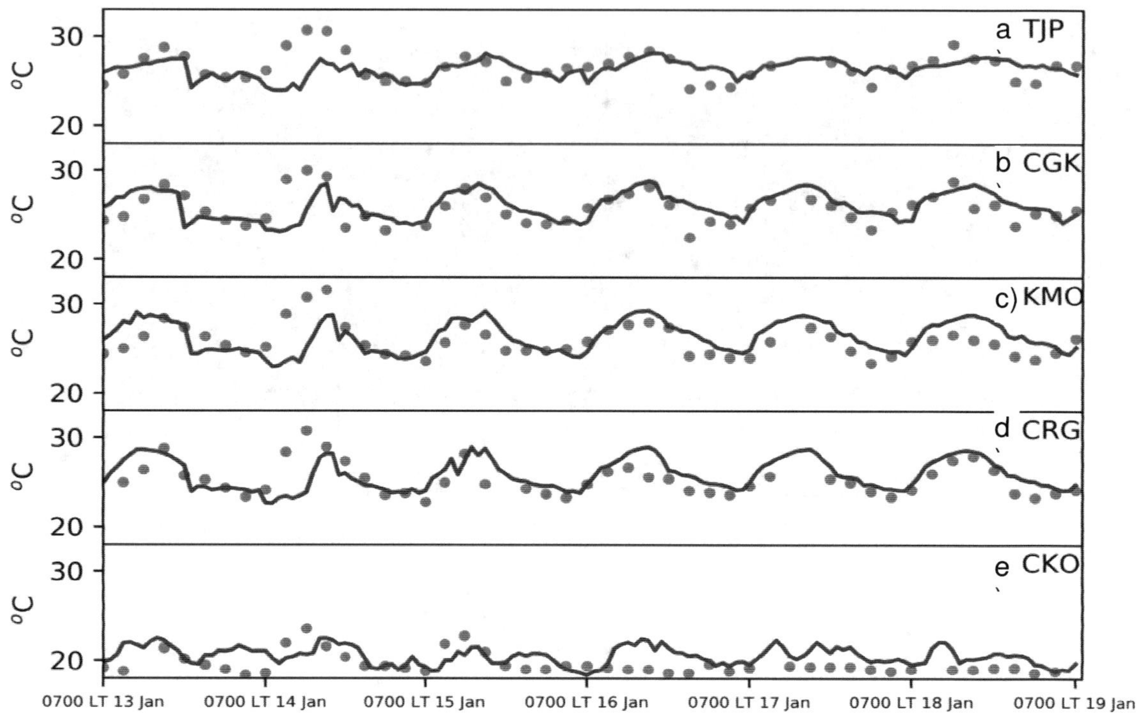

Figure 3. Simulated and observed of surface air temperatures at stations a) Tanjung Priok (TJP), Soekarno-Hatta, b) Cengkareng (CGK), c) Kemayoran (KMO), d) Curug (CRG) and e) Citeko (CKO) (observed: red dots, simulated: blue line). The data refer to the 13th - 19th of January 2014.

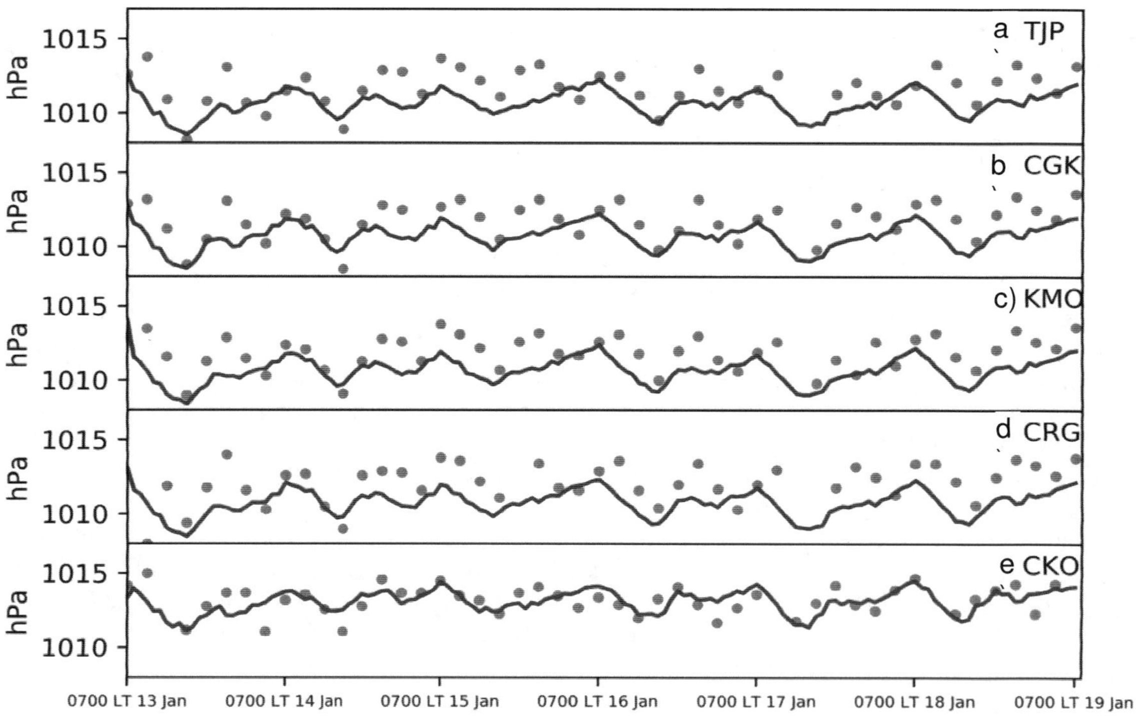

Figure 4. As in Figure 3 but for sea level pressure.

Table 2. Statistics showing root-mean-square error (RMSE) and standard deviation of the error (SDE) for each station between the simulation and observation results.

	Air temperature		Sea level pressure	
	RMSE	SDE	RMSE	SDE
Tanjung Priok	1.6	1.6	1.5	1.1
Cengkareng	1.7	1.7	1.4	1.0
Kemayoran	1.8	1.8	1.5	1.0
Curug	1.8	1.8	1.7	1.2
Citeko	1.7	1.4	0.8	0.8

Furthermore, the average diurnal temperature in CTL, SCE and CTL - SCE has been showed. There is indicated a contribution of land use change in air temperature increase in GJ at 01:00 WIB (Fig. 5). This is shown by the dominance of negative values the difference of CTL-SCE. The maximum difference is about 0.6 degrees and in GJ about 0.2 degrees. This means that land use change results in a temperature increase of about 0.2 degrees with simulated at 01:00 LT

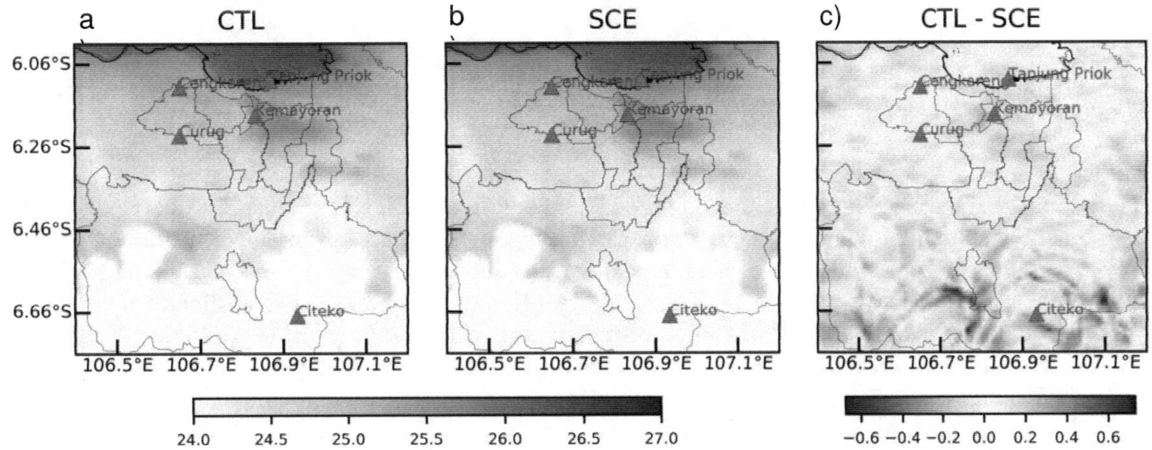

Figure 5. Diurnal means of air temperature for a) CTL, b) SCE and c) differences of CTL – SCE simulations at 0100 LT.

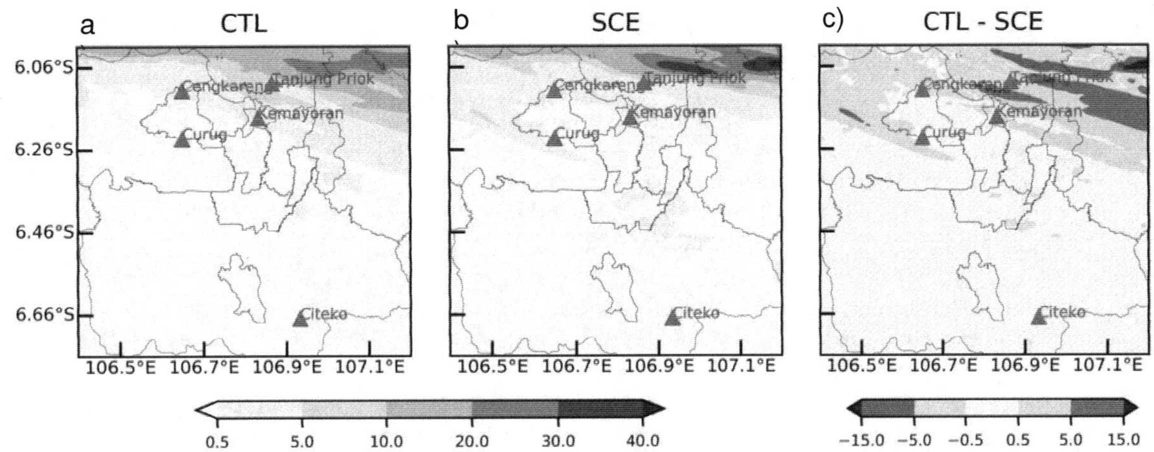

Figure 6. As in Figure 3 but for accumulated rainfall at periods 2200 – 0700 LT on 17 January 2014.

Figure 6. shows an accumulated rainfall for CTL and differences of CTL – SCE simulations periods 2200 – 0700 LT on 17 January 2014. There is a difference of up to 15 mm in the changed land use. The northern GJ has higher precipitation in the land-use change scenario compared to the south. The results suggest new urban developments caused an intensification and expansion of the area experiencing rainfall but mainly influenced early morning rainfall with increase of up to 15 mm.

4. Conclusions

The WRF simulations directed for 6 days from 13 January 2014 00 UTC until 19 January 2014 00 UTC. Commonly, measured air temperature and rainfall pattern in GJ region show a good promise between observed and simulated data. The results also indicate consistent significant contribution of urban development and associated land use changes to air temperature and rainfall. According to the model simulation, urban and built up land contributed about 6% in heavy rainfall in the morning and about 0.2 degrees of air temperature different. Simulations suggest new urban developments caused a strengthening and development of the area experiencing rainfall but mainly influenced morning rainfall with increase of up to 15 mm.

Acknowledgments

This research was supported by Department of Geophysics and Meteorology, Bogor Agricultural University and the Indonesian Agency for Meteorology Climatology and Geophysics through the Graduate Study Program in Applied Climatology.

5. References

[1] Wu P, Hara M, Fudeyasu H, Yamanaka M D, Matsumoto J, Syamsudin F, Sulistyowati R and Djajadihardja Y S 2007 The impact of trans-equatorial monsoon flow on the formation of repeated torrential rains over Java Island *SOLA* **3** 93–96

[2] Wu P, Arbain A A, Mori S, Hamada J, Hattori M, Syamsudin F and Yamanaka MD 2013 The Effects of an Active Phase of the Madden-Julian Oscillation on the Extreme Precipitation Event over Western Java Island in January 2013 *SOLA* **9** 79 - 83

[3] Trilaksono N J, Otsuka S and Yoden S 2011 Dependence of model-simulated heavy rainfall on the horizontal resolution during the Jakarta flood event in January–February 2007 *SOLA* **7** 193–196

[4] Trilaksono N J, Otsuka S and Yoden S 2012 A time-lagged ensemble simulation on the modulation of precipitation over West Java in January–February 2007 *Mon. Wea. Rev.* **140** 601– 616

[5] Siswanto, van Oldenborgh G J, van der Schrier G, Lenderink G and van den Hurk B 2015 Trend in High-Daily Precipitation Events in Jakarta and the Flooding of January 2014 *Bull. Amer. Meteor. Soc.* **96** 131-135

[6] Ismanto H. 2013. Distribusi Spasial dan Temporal Mesoscale Convective Complexes di Benua Maritim. *Bull Megasains* **4** 74 – 81

[7] Trismidianto, Hadi TW, Ishida S, Manda A, Iizuka S and Moteki Q 2016 Development processes of oceanic convective systems inducing the heavy rainfall over the western coast of Sumatra on 28 October 2007 *SOLA* **12** 6–11

[8] Trismidianto, Yulihastin E, Satyawardhana H, Nugroho JT, Ishida S 2017a The Contribution of the Mesoscale Convective Complexes (MCCs) to total rainfall over Indonesian Maritime Continent. *IOP Conf Ser: Earth Environ Sci.* **54** 012027. doi:10.1088/1755-1315/54/1/012027.

[9] Trismidianto, Yulihastin E, Satyawardhana H, Ishida S. 2017b A composite analysis of the Mesoscale Convective Complexes (MCCs) development over the Central Kalimantan and its relation with the propagation of the rainfall systems. *IOP Conf Ser: Earth Environ Sci.* **54** 012036. doi:10.1088/1755-1315/54/1/012036

[10] Nuryanto D E, Pawitan H, Hidayat R, Aldrian E 2017 Propagation of convective complex systems triggering potential flooding rainfall of Greater Jakarta using satellite data. *IOP Conf Ser: Earth Environ Sci.* 54 012028. doi:10.1088/1755-1315/54/1/012028

[11] Putri NS, Hayasaka T, Whitehall K D 2017 The Properties of Mesoscale Convective Systems in Indonesia Detected Using the Grab 'Em Tag 'Em Graph 'Em (GTG) Algorithm. J Meteor Soc Japan. DOI:10.2151/jmsj. 2017-026

[12] Lin Y-L, Farley R D, Orville H D 1983 Bulk parameterization of the snow field in a cloud model, *J Clim Appl Meteor.* **22** 1065–1092

[13] Mlawer E J, Taubman S J, Brown P D, Iacono M J, Clough S A 1997 Radiative transfer for inhomogeneous atmosphere: RRTM, a validated correlated-k model for the longwave. *J Geophys Res.* **102** 16663–16682

[14] Dudhia J 1989 Numerical study of convection observed during the winter monsoon experiment using a mesoscale two-dimensional model. *J Atmos Sci.* **46** 3077–3107

[15] Betts A K 1986 A new convective adjustment scheme. Part I: observational and theoretical basis. Quart J Roy *Meteor Soc.* **121** 255–270

[16] Betts A K and Miller M J 1986 A new convective adjustment scheme. Part II: single column tests using GATE wave, BOMEX, and arctic air-mass data sets. Quart J Roy *Meteor Soc.* **121** 693–709

[17] Janjic Z I 1994 The step-mountain eta coordinate model: further developments of the convection, viscous sublayer and turbulence closure schemes. *Mon Weather Rev.* **122** 927–945

[18] Hong S Y, Noh Y, and Dudhia J 2006 A new vertical diffusion package with an explicit treatment of entrainment processes. *Mon Weather Rev.* **134** 2318–2341

[19] Jankov I, Gallus W Jr, Segal M, Shaw B, Koch S 2005 The impact of different WRF model physical parameterizations and their interactions on warm season MCS rainfall. *Wea Forecasting.* **20** 1048–1060

IOP Conf. Series: Earth and Environmental Science **149** (2018) 012029 doi:10.1088/1755-1315/149/1/012029

Analysis of climate change impact on rainfall pattern of Sambas district, West Kalimantan

Sinta Berliana Sipayung, Amalia Nurlatifah, Bambang Siswanto, and Lilik Slamet S
Centre of Atmospheric Science and Technology, National Institute of Aeronautics and Space

E-mail: s_berlianasipayung@yahoo.com

Abstract. Climate change is one of the most important issues being discussed globally. It caused by global warming and indirectly affecting the world climate cycle. This research discussed the effect of climate change on rainfall pattern of Sambas District and predicted the future rainfall pattern due to climate change. CRU and TRMM were used and has been validated using in situ data. This research was used Climate Modelling and Prediction using CCAM (Conformal Cubic Atmospheric Model) which also validated by in situ data (correlation= 0.81). The results show that temperature trends in Sambas regency increased to 0.082°C/yr from 1991-2014 according to CRU data. High temperature trigger changes in rainfall patterns. Rainfall pattern in Sambas District has an equatorial type where the peak occurs when the sun is right on the equator. Rainfall in Sambas reaches the maximum in March and September when the equinox occurs. The CCAM model is used to project rainfall in Sambas District in the future. The model results show that rainfall in Sambas District is projected to increase to 0.018 mm/month until 2055 so the flow rate increase 0.006 m^3/month and the water balance increase 0.009 mm/month.

1. Introduction

Climate Change is one of the events on earth that is caused by rising air temperature (global warming) and can cause changes in the earth in many aspects. One effect that can be caused by climate change is the change in rainfall patterns. Previous research has shown that in recent years climate change has resulted in changes in rainfall patterns of different regions [1-3].

Sambas is one of the areas in West Kalimantan that has an important role in the availability of food, especially in the province of West Kalimantan. Rainfall is certainly an important aspect in meteorological agriculture that can support water supply in West Kalimantan and can affect food availability. Therefore, it is necessary to ascertain how the effect of climate change on the availability of water in Sambas so that in the future if any event of extreme dry or flood in Sambas will soon be mitigated.

Other studies about the effects of climate change on water availability and agricultural conditions have been conducted by researchers in various parts of the world. This study has been conducted in China and stated that climate change may affect the frequency of extreme weather events such as drought and flood [4]. Other research cites the need for adaptation and mitigation related to climate change impacts on agricultural conditions and water availability in China [5]. Climate change in China itself is the impact of human activity, and can change hydrological cycle and spatio temporal distribution in China [6].

The aim of this study is to determine the impact of climate change on water availability in Sambas and forecast the future projection of water availability in Sambas due to climate change.

2. Material and Method

The research stage used in this research is the first to analyze the rainfall pattern based on in situ data. The in situ data used is the in situ data based on BMKG station measurement. Further analysis of climate change conditions based on CRU data and TRMM satellite data is done. Simulation of rainfall by CCAM model was done for the next step. Both satellite data and CCAM model data are validated by observational data. If the validation result is good, it means that the model is quite representative in describing the climatic conditions, then the monthly rainfall projection by the CCAM model can be done. If rainfall data based on the CCAM model has been simulated and projected, then the projection of Sambas River discharge condition by ArcSWAT model based on CCAM rainfall input data can be done. The last stage is the projection of the Sambas River discharge condition by the ArcSWAT model based on CCAM rainfall input data.

2.1. Data

2.1.1. Insitu Data

The insitu data used in this research is monthly BMKG rainfall data from 2010-2015.

2.1.2. GFS Data

GFS (Global Forecast System) data is a global datasets used for CCAM (Conformal Cubic Atmospheric Model) input. This data were downscaled to 14 km resolutions.

2.2. Methods

2.2.1. Rainfall Estimation with TRMM Satellite Data and CRU

The Tropical Rainfall Measuring Mission (TRMM) is a satellite that used in precipitation measurement, TRMM 3B43 have 0.25^{o} spatial resolution and monthly temporal resolution [7].

2.2.2. Climate Modeling Methods with CCAM

The Conformal-Cubic Atmospheric Model (CCAM) has been developed at CSIRO [8, 9] and includes a fairly comprehensive set of physical parameterizations, mostly similar to those of the Mk 3 GCM [8]. Conformal-Cubic Atmospheric Model (CCAM) is an atmospheric numerical model which developed by CSIRO, Australia. This model previously developed for the limited-area in Division of Atmospheric Research Model (DALRAM). CCAM is a global model which have variable-based conformal cubic grid resolution. CCAM use Schmidt transformation for regional and local forecasts with multiple nesting techniques for downscaling and having topographic and land use data that has been integrated with the system [10].

2.2.3. Hydrological Modeling Methods with ArcSWAT

The SWAT (Soil and Water Assessment Tool) is a driver basin model that can predict the impact of land use and water management to watershed condition. This model use HRUs data such as topography data, soil type data, land-use data, and weather data to predict and determine the watershed condition (discharge, nutrition, etc.) [11].

3. Results and Discussion

Based on figure 1 it can be seen that generally Sambas District has flat topography. The lower altitude reaches 37 meters above mean sea level. While the plateau with 730-1.407 meters above mean sea level is located only in the north eastern district of Sambas. Topography certainly has big influence for drainage of river water [12].

IOP Conf. Series: Earth and Environmental Science **149** (2018) 012029 doi:10.1088/1755-1315/149/1/012029

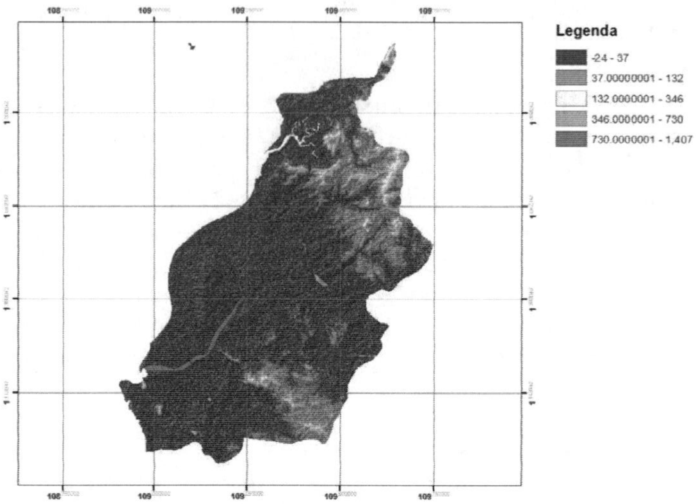

Figure 1. Topography Maps of Sambas District.

If viewed from the location of Sambas that Sambas is close to the equator, then there is a hypothesis that the rainfall pattern of Sambas District is equatorial. However, this is not simply assured. Based on rainfall observation data, rainfall composite in 2010-2015 shows that rainfall in Sambas Regency tends to be flat. However, the maximum value of rainfall are in December and January (figure 2).

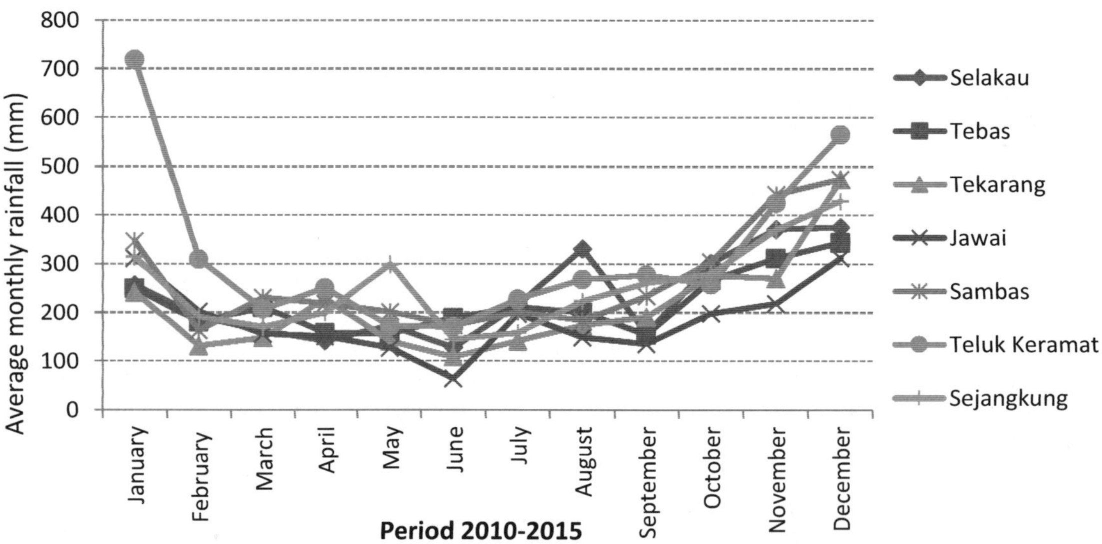

Figure 2. Rainfall pattern from insitu Data.

The TRMM 1998-2009 data shows that the rainfall fluctuation in Sambas is not so fluctuate (figure 3). The standard deviation value is only reached 73 with a minimum value is about 154 mm. TRMM rainfall data shows the maximum rainfall pattern in Sambas is on December while in February to September is tend to ble flat.

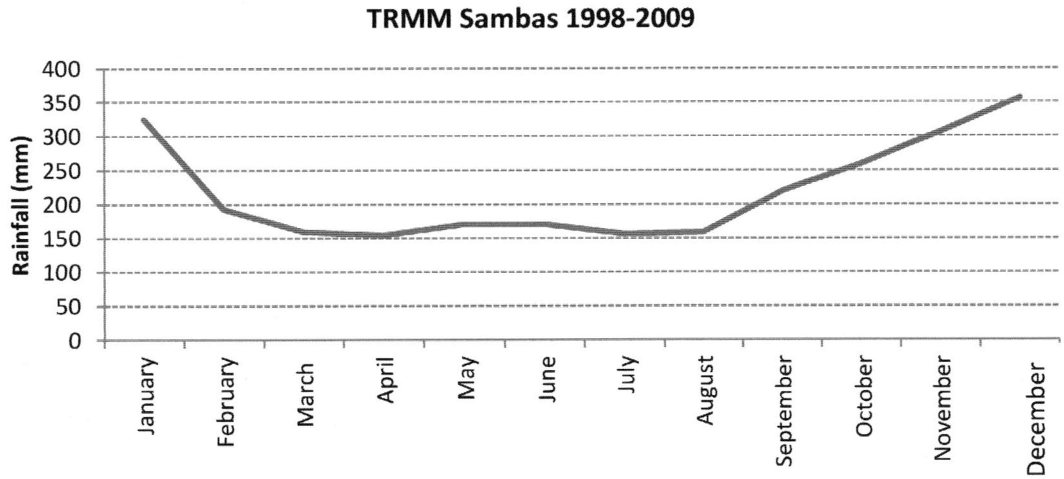

Figure 3. Rainfall pattern from TRMM Data.

Similar to observation data, CCAM simulation results show that rainfall in Sambas is fluctuate even though the maximum value is seen in March, August, and November (figure 4). However, although there are significant differences in patterns, but the correlation between observation data and CCAM has high value around 0.81. This can be interpreted that CCAM simulation can adequately describe the trend or rainfall pattern in Sambas District.

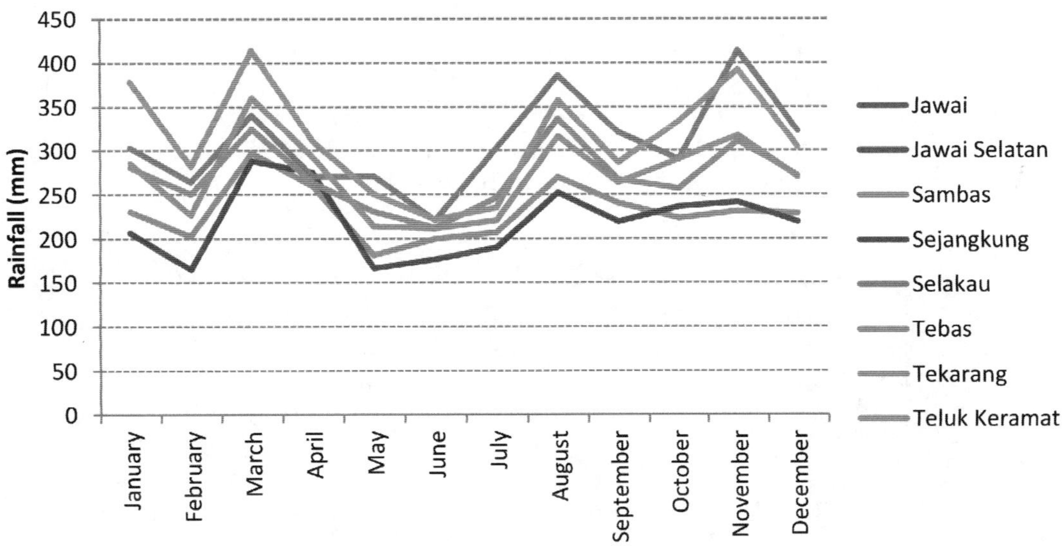

Figure 4. Rainfall pattern from CCAM Data.

The pattern of rainfall itself has an important role in the analysis of water availability in a region. This is due to rainfall patterns can regulate the condition of ground water and river water discharge in a river. Rainfall, topography, and vegetation can also affect run-off conditions in an area [13]. Based on the latest condition analysis, monthly rainfall in Sambas in a year tends to be equally distributed and reach maximum in December and January. However, it is not possible that there are other factors such as climate change which can affect the rainfall patterns so the water availability also can change.

Based on the graph 4, the CRU data shows that there is temperature increase about 0.25°C per 30 years. 1991-2014 became the hottest year with 27-28°C (figure 5). This can be indicated as the occurrence of global warming in Sambas District with an temperature increase around 0.3-1°C in the last 100 years.

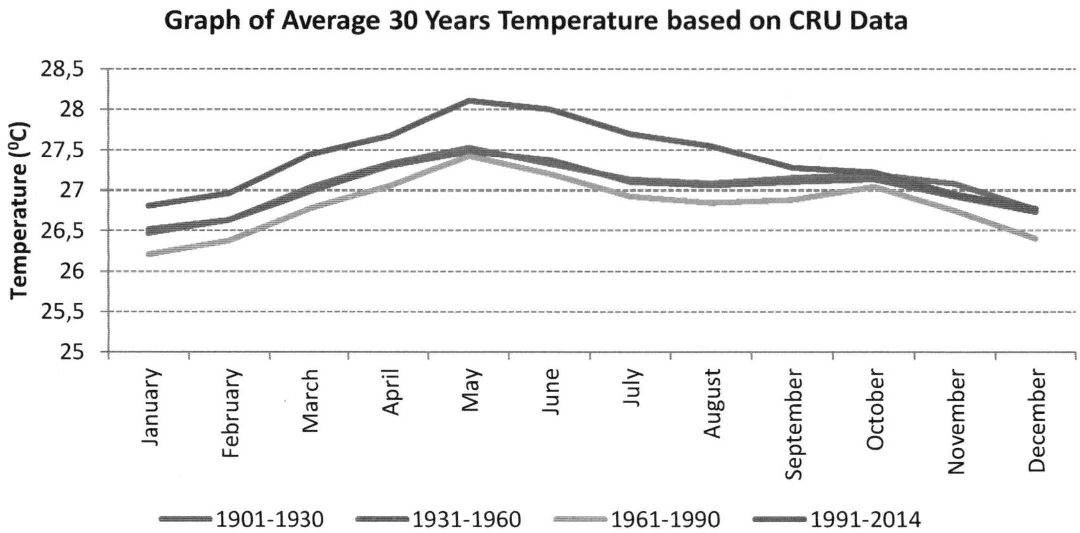

Figure 5. Graph of average 30 months Temperature based on CRU Data.

The global warming can impact the agricultural welfare and water availability [14]. This is due to climate change can also affect the rainfall pattern in an area [15]. This also applies in Sambas district.

The CCAM temperature simulation plot states that in 1971-2000 temperatures in Sambas reached 300-302 ⁰K (figure 6a). In 2040-2069 there is an increase up to 302.5-304.5 ⁰K (figure 6b). This increase in temperature ranges from 1.9-2.4 ⁰K across Sambas (figure 6c).

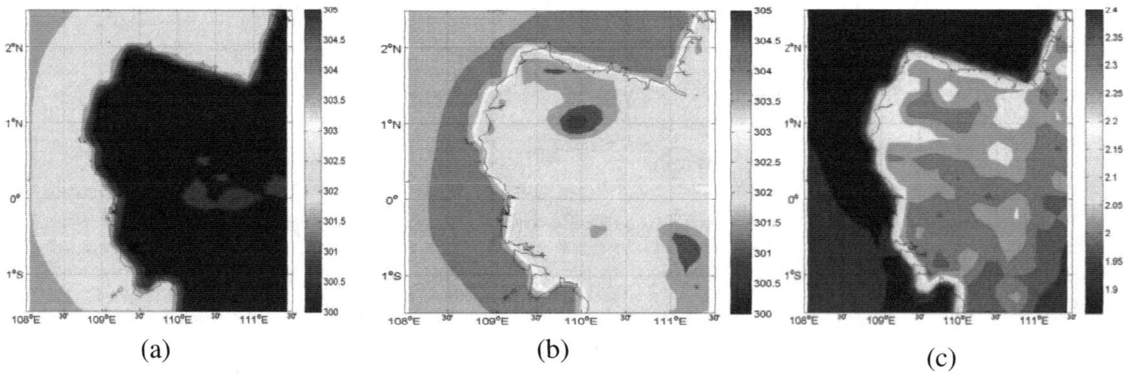

Figure 6. (a) Plot of average temperature in 1971-2000, (b) Plot of average temperature in 2040-2069 (c) Plot of temperature anomaly.

The CCAM simulation states that in the future, in 2040-2069 there will be an increase of rainfall in Sambas regency around 250-400 mm (figure 7b), while 1971-2000 reaches only 200-350 mm (figure 7a). This increase averages around 10-40 mm in each region (figure 7c).

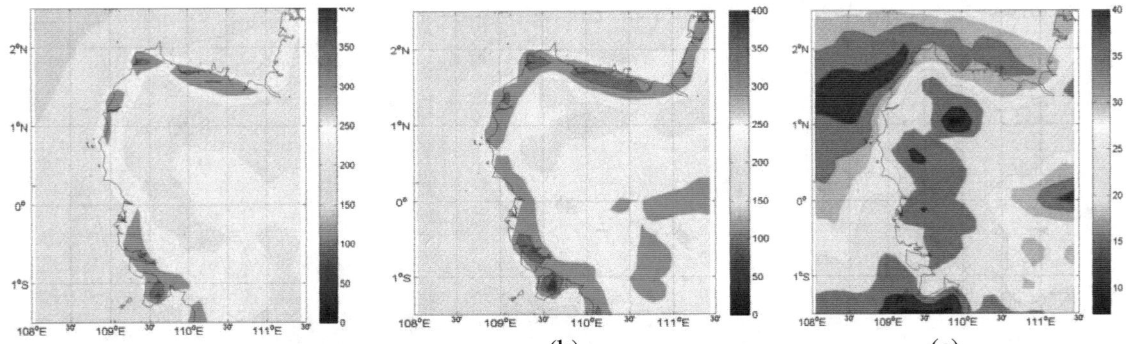

Figure 7. (a) Plot of average rainfall in 1971-2000, (b) Plot of average rainfall in 2040-2069 (c) Plot of rainfall anomaly.

Rainfall increase also occurs in rainfall projections as seen in monthly rainfall plot of temperatures from 2006-2055 that expected to be increase (figure 8). This increase is estimated to occur around 0.018 mm/month based on the linear regression that has been done.

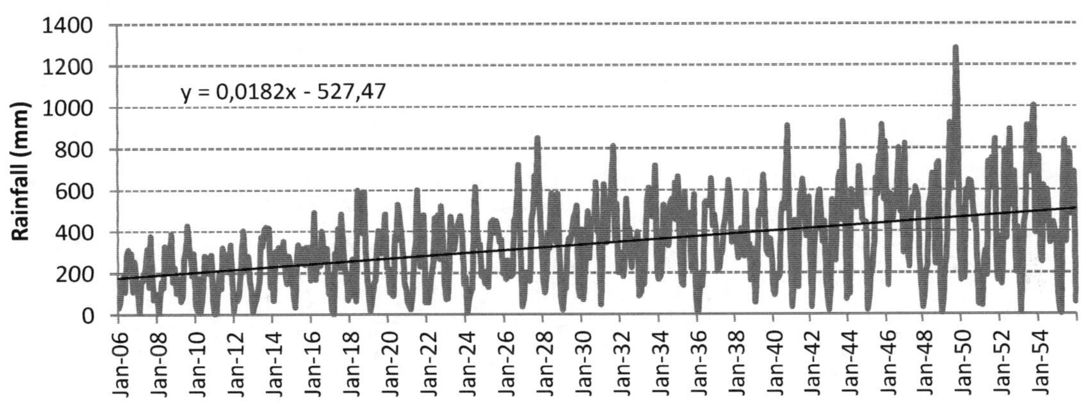

Figure 8. Graph of rainfall projection on Sambas River based on CCAM Model.

Changes in rainfall can lead to change in water availability. In this case, the water availability is measured as river flow or water balance. The simulation of discharge conditions is done in Sambas River using ArcSWAT model based on climate data input which obtained from CCAM output. ArcSWAT simulation results show that the rainfall increase in Sambas Regency is also expected to have an impact to the increase of flow rate in Sambas River (figure 9). The increase reached 0.006 m^3/month based on the linear regression plot in 2006-2055.

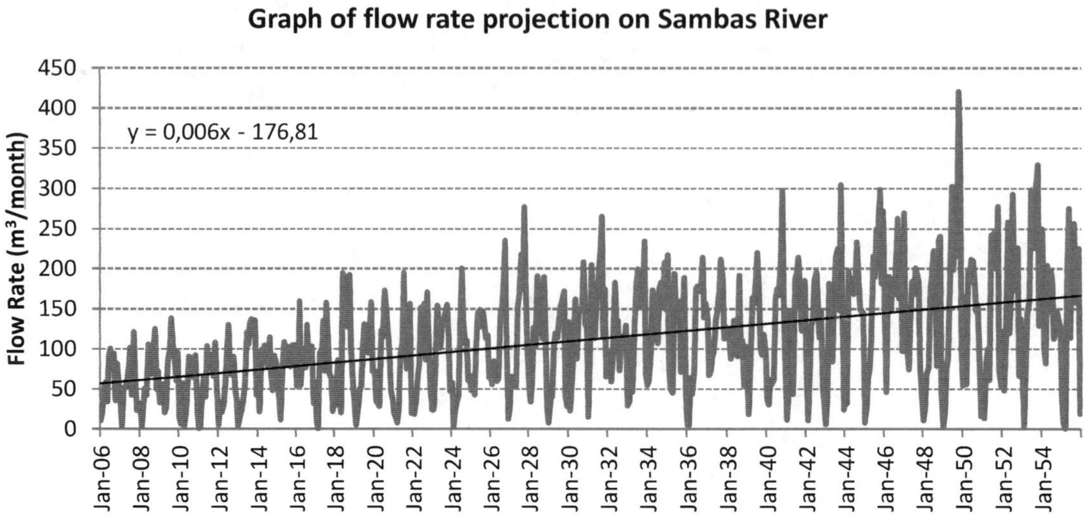

Figure 9. Graph of flow rate projection on Sambas River.

Water availability is also measured from the existing water balance conditions in Sambas District. Based on the simulation of water balance conditions, it is estimated that in the future the water balance conditions in the Sambas River will also increase (figure 10). This is reinforced by a linear regression analysis which states that the water balance in Sambas River will increase by 0.009 mm/month until 2055.

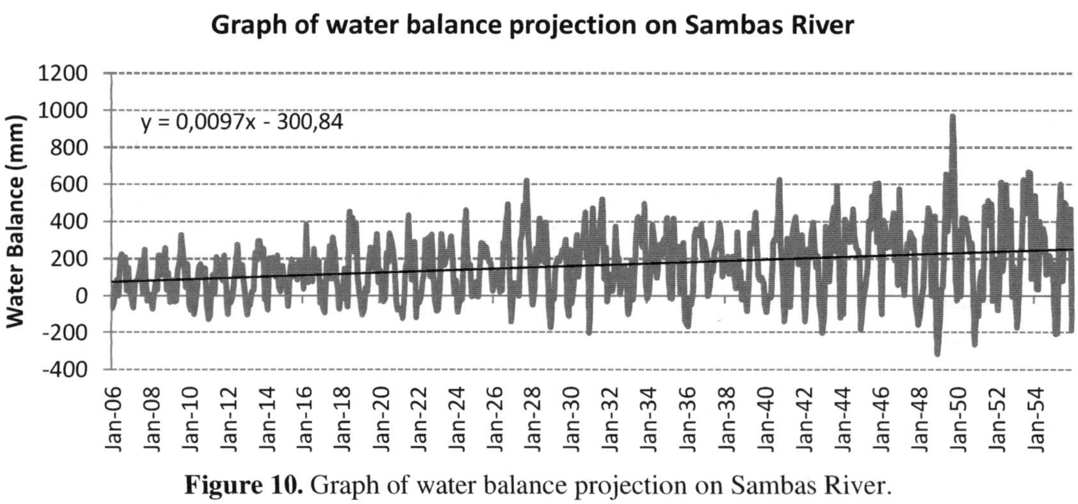

Figure 10. Graph of water balance projection on Sambas River.

Based on analysis of flow rate and water balance in Sambas River, it is estimated that in the future, water availability in Sambas Regency (based on CCAM climate model simulation for SWAT input data) will increase. This is due to climate change can make rainfall increase so that flow rate and water balance will increase. High rainfall and increasing water supply conditions are expected to cause flood. Therefore, prevention of flood potency should always be held.

4. Conclusions

Observation data state that climate change has occurred in the Sambas region and causing its temperature to rise in the last 30 years. Climate change is expected to lead to rainfall increase in Sambas regency around 0.018 mm/month until 2055. Increased rainfall causes increased Sambas River flow around 0.006 m^3 / month and increased water balance around 0.009 mm / month. Knowing these conditions, it is hoped that this will become the foundation in the regional planning system, starting from infrastructure planning and community social system as a form of climate change mitigation and adaptation.

Acknowledgments
The authors would like to thank the Atmospheric Science and Technology Centre which have provided moral and material support so that this research can be realized.

References

[1] Storch H V, Zorita E, Cusbasch U 1993 Downscaling of global climate change estimates to regional scales: an application to Iberian rainfall in wintertime *Journal of Climate* **61** 161–1171
[2] Trenberth K E 1998 *Atmospheric Moisture Residence Times and Cycling: Implications for Rainfall Rates and Climate Change* (Berlin: Springer)
[3] Williams S E, Bolitho E E, Fox S 2003 *Climate Change in Australian Tropical Rainforests: An Impending Environmental Catastrophes* (Royal Society Publishing)
[4] Jun X, Qing-Yun D, Yong L, Zheng-Hui X, Zhi-Yu L, Xing-Guo M 2017 Climate Change and Water Resources: Case Study of Eastern Monsoon Region of China *Advances in Climate Change Research* **8** 63-67
[5] Xing-Guo M, Shi H, Zhong-Hui L, Su-Xia L, Jun X 2017 Impacts of Climate Change on Agricultural Water Resources and Adaptation on the North China Plain *Advances in Climate Change Research* **8** 93-98
[6] Yu-Jie W, Da-He Q 2017 Influence of Climate Change and Human Activity on Water Resources in Arid Region of Northwest China: An Overview *Advances in Climate Change Research* **8** 268-278
[7] https://mirador.gsfc.nasa.gov/collections/TRMM_3B43__007.shtml accessed 28 September 2017
[8] McGregor J L 2005 CCAM: Geometric Aspects and Dynamical Formulation (CSIRO)
[9] McGregor J L, Dix M R 2008 *An Updated Description of The Conformal Cubic Atmospheric Model* (New York: Springer)
[10] Thatcher M, McGregor J L 2009 Using a scale-selective filter for dynamical downscaling with the conformal cubic atmospheric model *Monthly Weather Review* **137** 1742-1752
[11] Arnold J G, Moriasi D N, Gassman P W, Abbaspour K C, White M J, Srinivasan R, Santhi C, Harmel RD, van Griensven A, Van Liew MW, Kannan N, Jha MK 2012 SWAT: model use, calibration, and validation *American Society of Agricultural And Biological Engineers* **55** 1491-1508
[12] Price K 2011 Effects of Watershed Topography, soils, land use, and climate on baseflow hydrology in humid regions: A review *Progress in Physical Geography* **35** 465-492
[13] Dunne T, Zhang W, Aubry B F 1991 Effects of rainfall, vegetation, and microtopography on infiltration and runoff *Water Resources Research* **27** 2271-2285
[14] Vorosmarty C J, Green P, Salisbury J 2000 Global Water Resources: Vulnerability from Climate Change and Population Growth *Science* **289** 284-288
[15] Udayanshankara T H 2016 Impact of climate change on rainfall pattern and reservoir level *Journal of Hydrology: Regional Studies* **3** 473-493

IOP Conf. Series: Earth and Environmental Science **149** (2018) 012030 doi:10.1088/1755-1315/149/1/012030

Seasonal variability of convectively coupled equatorial waves (CCEWs) in recent high-top CMIP5 models

Dzaki Zakaria[1], Sandro W. Lubis[2] and Sonni Setiawan[1]
[1] Department of Geophysics and Meteorology, Bogor Agricultural University (IPB), Indonesia
[2] Department of Geophysical Sciences, The University of Chicago, USA

E-mail: dzakizakaria28@gmail.com

Abstract. Tropical weather system is controlled by periodic atmospheric disturbances ranging from daily to subseasonal time scales. One of the most prominent atmospheric disturbances in the tropics is convectively coupled equatorial waves (CCEWs). CCEWs are excited by latent heating due to a large-scale convective system and have a significant influence on weather system. They include atmospheric equatorial Kelvin wave, Mixed Rossby Gravity (MRG) wave, Equatorial Rossby (ER) wave and Tropical Depression (TD-type) wave. In this study, we will evaluate the seasonal variability of CCEWs activity in nine high-top CMIP5 models, including their spatial distribution in the troposphere. Our results indicate that seasonal variability of Kelvin waves is well represented in MPI-ESM-LR and MPI-ESM-MR, with maximum activity occurring during boreal spring. The seasonal variability of MRG waves is well represented in CanESM2, HadGEM2-CC, IPSL-CM5A-LR and IPSL-CM5A-MR, with maximum activity observed during boreal summer. On the other hand, ER waves are well captured by IPSL-CM5A-LR and IPSL-CM5A-MR and maximize during boreal fall; while TD-type waves, with maximum activity observed during boreal summer, are well observed in CanESM2, HadGEM2-CC, IPSL-CM5A-LR and IPSL-CM5A-MR. Our results indicate that the skill of CMIP5 models in representing seasonal variability of CCEWs highly depends on the convective parameterization and the spatial or vertical resolution used by each model.

1. Introduction

Convectively coupled equatorial waves (CCEWs) are one of the most prominent atmospheric disturbances in the tropics. These disturbances substantially control the tropical weather system periodically from daily to subseasonal time scales [1-5]. CCEWs are generated by large-scale convection in the equatorial region around 20° N – 20° S [5–8]. They are including atmospheric Kelvin wave, MRG wave, n=0 ER wave and TD-type waves [9, 7]. Kelvin waves propagate eastward with a period of approximately 12 – 20 days and are influenced by El-Nino phenomenon [10]. MRG waves propagate westward with a period of approximately 4–5 days and are significantly influenced by La-Nina phenomenon [10]. TD-type and ER waves have similar characteristics as MRG. These waves propagate westward and have period of approximately 3–5 days for TD-type waves and approximately 10 – 48 days for ER waves. According to [2], MRG and TD-type waves have an important role to generate tropical cyclone in west northern Pacific (WNP). The CCEWs activity has a strong influence on the variability of tropical weather. Moreover, vertical propagation of CCEWs activity has a significant impact on the dynamical coupling between troposphere and stratosphere [3,11,9].

Previous studies have shown that CCEWs are not well simulated in general circulation models (GCMs) in the twentieth century experiment from the Coupled Model Intercomparison Project phase 3 of the World Climate Research Program (CMIP3) [15]. The biases in simulating CCEWs in these models are associated with the use of convection parameterization in each model. However, the representation of CCEW's seasonal variability in high-top CMIP5 model has remained elusive. Therefore, it is worth investigating how the recent high-top CMIP5 models represent the CCEWs' variability and also to understand what causes the discrepancies among the model simulations. The purpose of this study is to examine the seasonal variability of CCEW activity in recent high-top CMIP5 models, with the main focus on their spatial distribution in the troposphere.

2. Data and Methods

2.1 Data
CCEWs activity was identified using daily-mean outgoing longwave radiation (OLR) from nine CMIP5 and NOAA reanalysis (observation) from January 1975 to December 2005 (~30 years). This study focused on the equatorial latitudinal bands, between 20° N – 20° S with resolution of 2.5° x 2.5°. The nine models from CMIP5 were chosen, including MPI-ESM-LR, MPI-ESM-MR, IPSL-CM5A-MR, IPSL-CM5A-LR, CanESM2, HadGEM2-CC, MIROC-ESM, MRI-CGCM3 and CMCC-CESM. These models have the same component and complexity (atmosphere, land surface, ocean and sea-ice), but with different spatial resolution [2]. They also used different convective parameterization.

2.2 Space Time Spectral Analysis (STSA)
The zonal propagation of CCEWs activity was analysed by using STSA technique similar to [5]. This method is used to elucidate the wave propagation as a function zonal wavenumber (k) and frequency (ω). A brief formulation can be expressed as follow:

$$x(\lambda, t) = \Sigma_k xk\,(\lambda, t) \tag{1}$$

$$x(\lambda, t) = Re\ \Sigma_{k,\omega}[W_{k,\omega}e^{i(kx+\omega)} + W_{k,-\omega}e^{i(kx+\omega)}] \tag{2}$$

$$x_k(\lambda, t) = \sum_k \quad \sum_{\pm\omega} W_{k,\pm\omega} \cos(k\lambda \pm \omega t + \varphi_{k,\pm\omega}) \tag{3}$$

In this study, we used a modified version of STSA analysis, in which the raw datasets were not partitioned into equatorial symmetric and anti-symmetric components (no symmetry constraints).

2.3 Spatial Distribution
This method is used to study the seasonal variation of CCEWs activity. We use Hovmoller technique in order to understand the seasonal variation of the filtered fields. First, the modelled OLR dataset were interpolated to the fixed horizontal grid (2.5° x 2.5°) as in the observed OLR dataset, by using a bilinear interpolation. Then we calculate the monthly variance as well as the seasonal variance in order to encapsulate the amplitude of the waves.

3. Results

Figure 1 shows the space-time spectra diagram of CCEWs activity based on OLR daily data from the observation. The different colours in the polygon of STSA diagram indicate the observed frequency-zonal wavenumber bands for the Kelvin wave (red), MRG wave (magenta), ER wave (cyan), TD-type wave (green) and MJO (blue). We can see from this diagram that Kelvin wave propagate eastward with a period of 2.5 – 17 days. On the other hand, both MRG wave and ER wave propagate westward with period of 2.2 – 3.6 and 10 – 48 days, respectively. TD-type wave propagate westward with a period of 2.5 – 5 days. These filter bands are used as a wave domain to isolate the spatial and seasonal variability of CCEWs in some high tops CMIP5 models. The results from the model simulations will be compared to the observation, therefore, the analysis from the NOAA dataset is considered as a reference observed values.

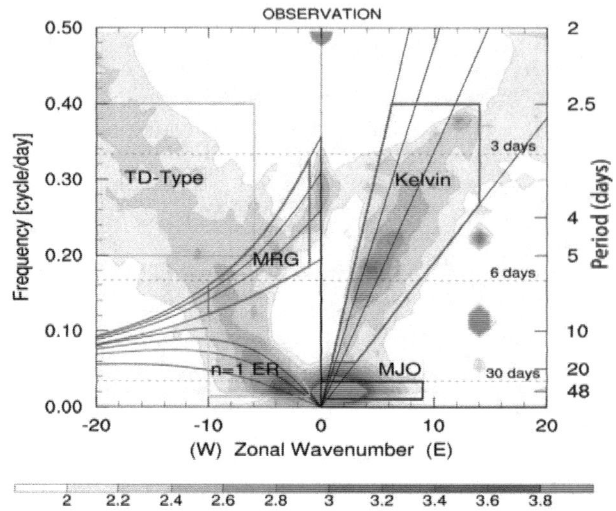

Figure 1. Zonal wavenumber frequency spectra of the anomalies OLR from observation (NOAA data reanalysis).

Figure 2 shows the seasonal evolution of Kelvin wave activity and its corresponding spatial distribution during JJA (June, July, August). We focus on JJA, since it is the season where the Kelvin wave activity is strongly observed in the observation and the model simulations. From the Hovmöller diagram, we can see that the observed Kelvin wave is strongly observed in the Northern Hemisphere (NH) equator (0-10°N) from April to July, with amplitude around 120-160 $(W/m^2)^2$. The peak is observed in May (during boreal spring and summer). On ther other hand, in the Southern hemisphere (SH), the strongest activity is observed from June to December, with amplitude around 30-40 $(W/m^2)^2$ (Fig. 2, top). A seconadry active periode of Kelvin wave appears in October to November from boreal fall or early winter. Almost all CMIP5 models in Fig. 2, except MPI-ESM-LR and MPI-ESM-MR, unreasonably describe these main characteristics of the seasonal cycle. Nevertheless, both models underestimate the amplitude of the Kelvin waves in the equator. Other best candidates are CanESM2, HadGEM2-CC and MIROC-ESM, however the magnitude is still very low compared to that in the observation.

Figure 3 shows the seasonal evolution of MRG wave activity and its corresponding spatial distribution during JJA (June, July, August). Similar to Kelvin waves, we focus on JJA since it is the season where the MRG wave activity is strongly observed in both observation and model simulations. The results show that the MRG wave is strongly observed in the NH equator (0-10°N) from December to April, with amplitude around 50-60 $(W/m^2)^2$. The peak is observed from June to August (JJA). In the SH, the MRG activity maximizes from April to January, however the amplitude is somewhat weaker compared to that in the NH, around 30-40 $(W/m^2)^2$. The best models that can simulate the seasonal cycle of MRG waves close to the observation are IPSL, CanESM2 and HandGEM2-CC.The observed MRG waves are stronger during JJA over the Northern Pacific to the Northern Atlantic, and they persists from May to December around with amplitude about 50–60 $(W/m^2)^2$ (Fig. 3). The IPSL, CanESM2 and HandGEM2-CC models are relatively good in representing the spatial distribution of MRG waves, however the magnitudes are underrepresented (negative bias) over the northern part of the Pacific Ocean around 10 N and are overrepresented in the off-equatorial regions, around 20N.

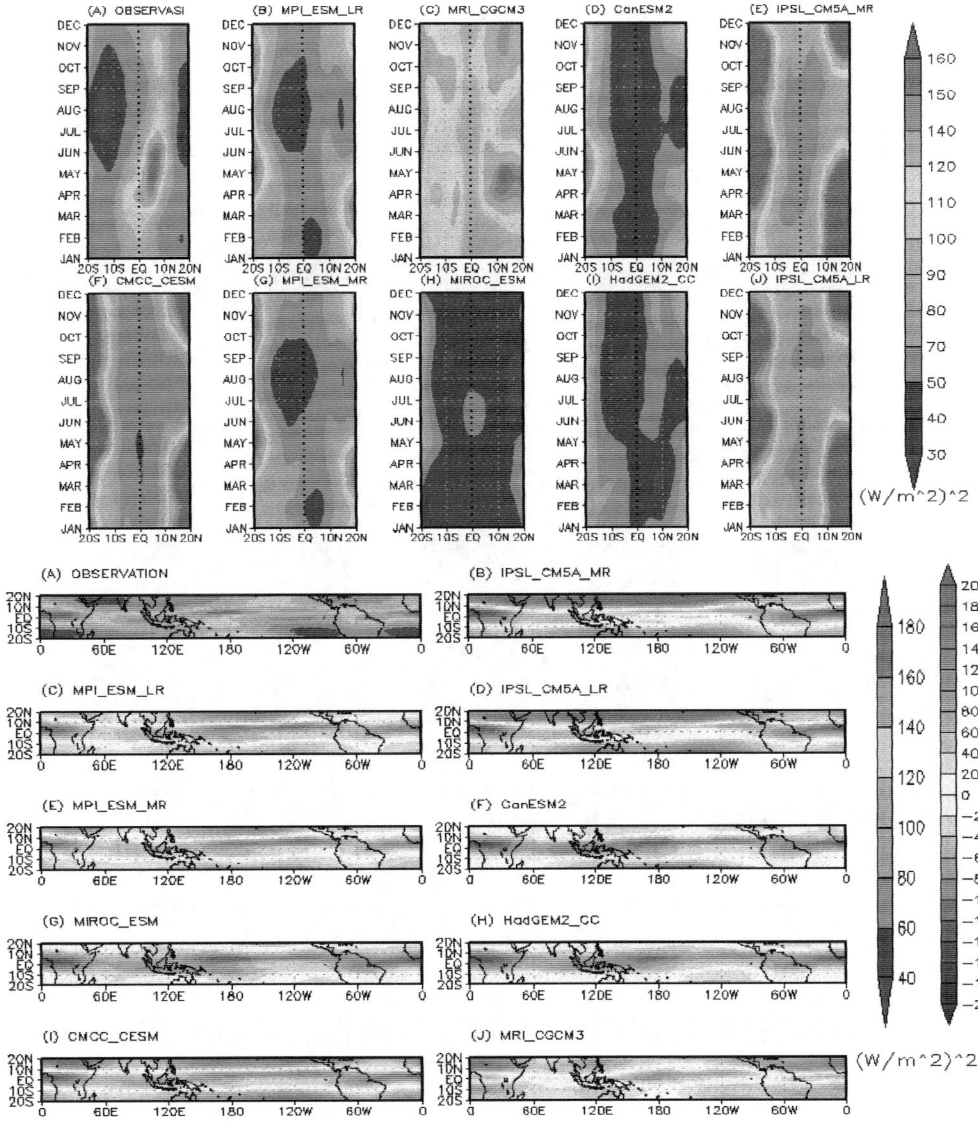

Figure 2. Seasonal cycle of Kelvin wave activity and its spatial distribution in JJA (June, July, August) from the observation (a) and CMIP5 models (b-j). The values in Figs. 2b-j (bottom) indicate residual values (model minus observation).

Figure 4 shows that seasonal cycle of ER waves as well as their spatial distributions in the observation and CMIP5 model simulations. The ER wave activity is observed from boreal spring to boreal winter (from December to April) with amplitude 130–150 $(W/m^2)^2$. The peak is observed in August. A secondary active period of ER wave appears in the SH, from January to March. The amplitude is much weaker compared to those in the observation, about 50 – 70 $(W/m^2)^2$. The CMIP5 models that exhibit a similar seasonal cycle of ER waves as in the observation are CanESM2, HadGEM2-CC, IPSL-CM5A-LR and IPSL-CM5A-MR. These models are able to simulate double peaks (in the NH and SH) of ER wave seasonal cycle as in the observation, but with weaker amplitudes. Furthermore, focusing on the spatial distribution of ER waves, it is well established that the ER wave activity maximized over the northern and southern parts of the maritime continent [5].

In particular, the ER waves are strongly observed over the western north Pacific basin and along the Atlantic basin. Both IPSL-CM5A-MR and IPSL-CM5A-MR can mimic the spatial pattern of ER waves in JJA, but tend to be overestimate. In particular, the amplitude of ER waves is much stronger in the model around the ITCZ regions. In IPSL-CM5A-MR, the model also underestimates the amplitude of ER waves over the northwestern part of the tropical Pacific basin.

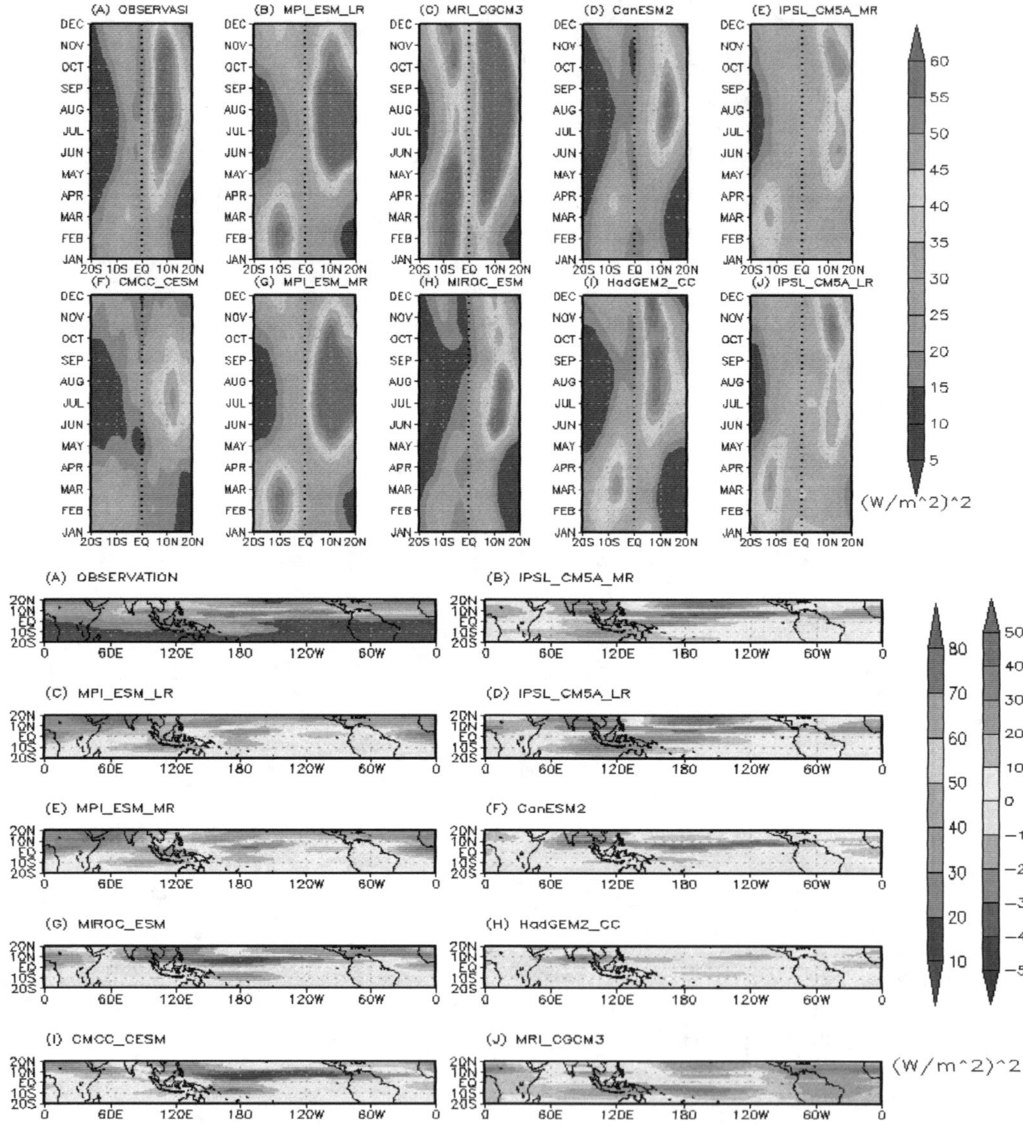

Figure 3. Seasonal cycle of MRG wave activity and its spatial distribution in JJA (June, July, August) from the observation (a) and CMIP5 models (b-j). (b-j). The values in Figs. 3b-j (bottom) indicate residual values (model minus observation).

IOP Conf. Series: Earth and Environmental Science **149** (2018) 012030 doi:10.1088/1755-1315/149/1/012030

Finally, the seasonal cycle of TD-type waves and their spatial distribution in the model and the high-top CMIP5 models are presented in Fig. 5. The TD-type waves, in general, have similar seasonal cycle as the ER waves. The MRG wave activity is strongly observed from boreal spring to boreal winter (from December to April) with amplitude 110–130 $(W/m^2)^2$, with the peak occurring in August. A secondary active period of TD-types waves are observed from January to March in the SH. These characteristics are consistent with the results found by Huang *et. al* 2013 and they argued that TD-type waves are intimately associated with the tropical cyclone (TC) geneses/activities. Furthermore, the high-top CMIP5 models that successfully represent the seasonal cycle of TD-type wave are CanESM2, IPSL-CM5A-MR, IPSL-CM5A-LR and HadGEM2-CC. These models are similar with the models that are able capturing the MRG wave activity, suggesting that the MRG and TD-type waves are interacting and/or coupled. These models, however, have lower amplitude compared to the observation than others.

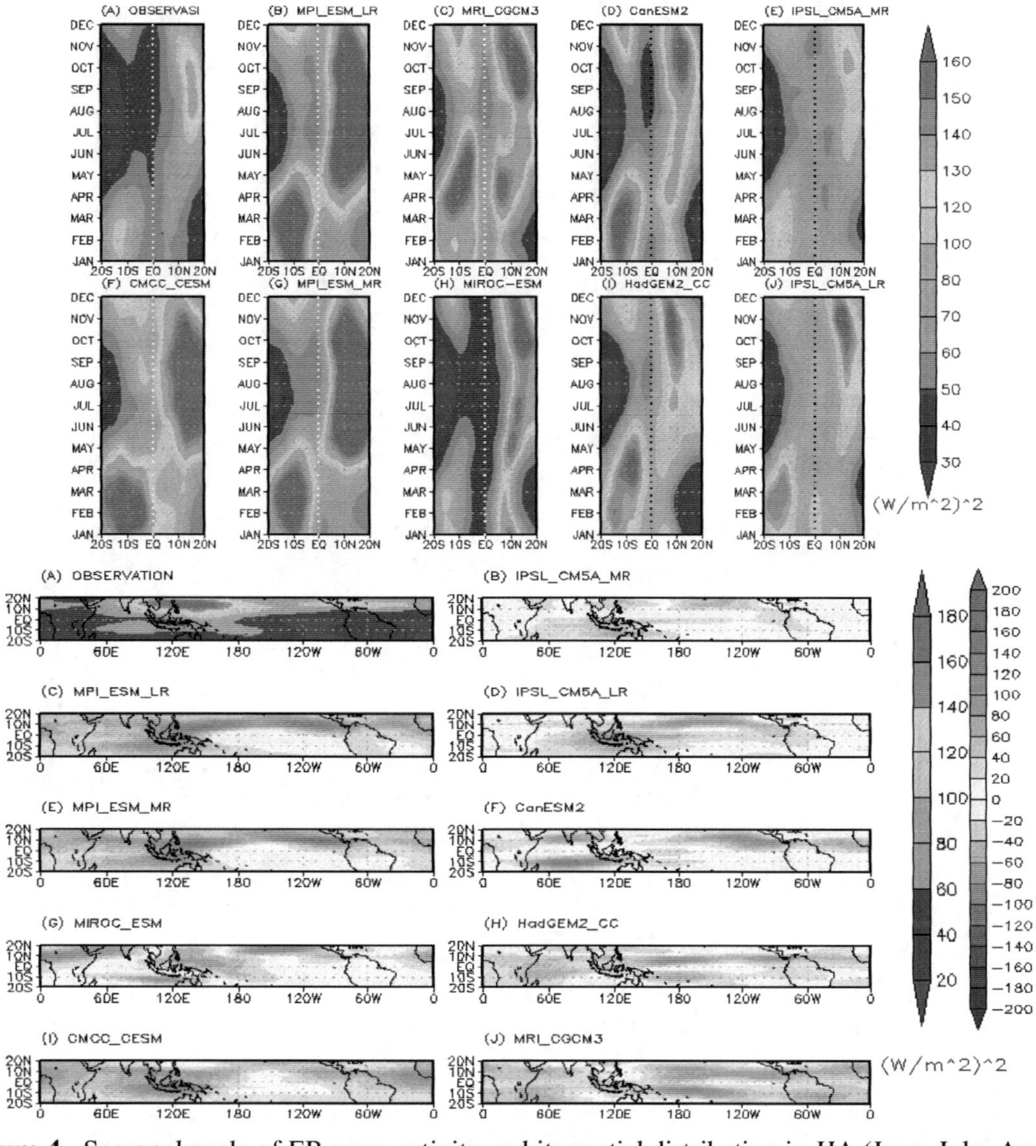

Figure 4. Seasonal cycle of ER wave activity and its spatial distribution in JJA (June, July, August) from the observation (a) and CMIP5 models (b-j). (b-j). The values in Figs. 2b-j (bottom) indicate residual values (model minus observation).

IOP Conf. Series: Earth and Environmental Science **149** (2018) 012030 doi:10.1088/1755-1315/149/1/012030

In terms of spatial distribution, there are two regions with strongest biases; firstly, the positive biases are observed over the Indian Ocean, central Pacific Ocean and Atlantic Ocean. Secondly, the negative bias (underrepresented amplitude of TD-type waves) are observed over the northwestern Pacific ocean, indicating that most high-top CMIP5 models still underestimate the frequency of TC activities compared to the observation.

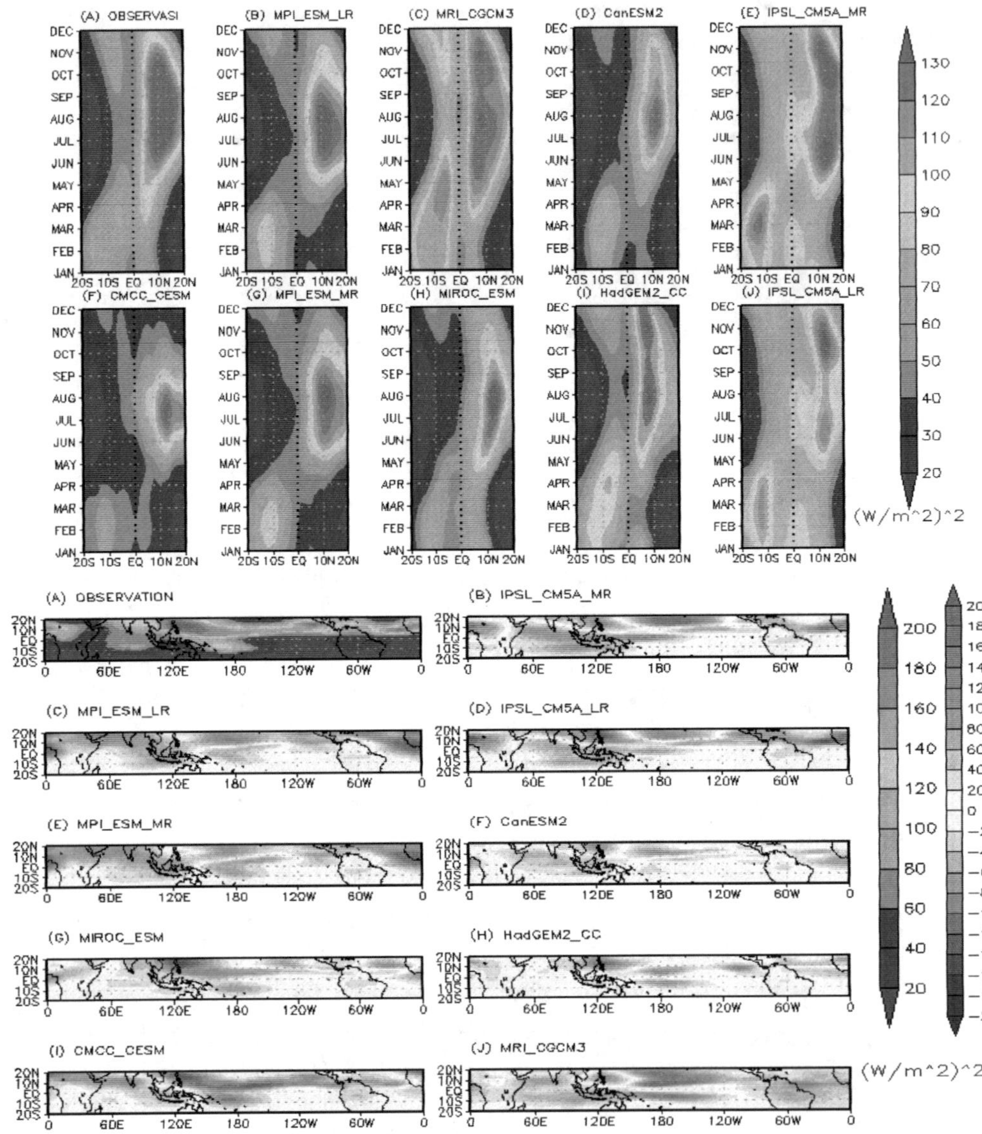

Figure 5. Seasonal cycle of TD-type wave activity and its spatial distribution in JJA (June, July, August) from the observation (a) and CMIP5 models (b-j).

According these result, model that can simulate CCEWs tend to have an intermediate complexity and resolution, such as IPSL, CanESM2 and HandGEM2-CC. We also aware that the types of the convective parameterization e.g., Convective Available Potential Energy (CAPE) used by the model will results in different CCEWs activity [4,13,14,1]. We found that model that has complex convective parameterization (rather than the simplified version) tend to better simulate the seasonal variability of

CCEWs. Further studies are required to understand the extent to which the convective parameterization affects the variability and amplitude of CCEWs in coupled GCM simulations.

4. Summary

This study evaluates some characteristics of CCEWs, including Kelvin waves, mixed Rossby-gravity waves, equatorial Rossby waves, and tropical depression (TD)-type waves, in some high-top CMIP5 models. The seasonal variability of each CCEW type was analyzed by using space-time spectral analysis (STSA). The key results of the current study are summarized as follow:

1) Most high-top CMIP5 models are capable in simulating CCEWs, as can be seen from the STSA analysis. The seasonal variability of Kelvin waves is well represented in MPI-ESM-LR and MPI-ESM-MR.

2) The seasonal variability of MRG waves is well simulated in CanESM2, HadGEM2-CC, IPSL-CM5A-LR and IPSL-CM5A-MR.

3) The seasonal variability of ER waves is well represented by CanESM2, HadGEM2-CC, IPSL-CM5A-LR and IPSL-CM5A-MR and maximized during boreal spring.

4) The seasonal variability of TD-type waves is well represented in CanESM2, HadGEM2-CC, PSL-CM5A-LR and IPSL-CM5A-MR, with maximum activity observed during boreal summer.

5) Our results indicate that the capability of CMIP5 models in representing seasonal variability of CCEWs highly depends on the convective (cumulus) parameterization and the resolution used by the models.

Further studies are required to understand what causes the discrepancies among the model simulations. We should note that without a realistic Kelvin wave and TD-type wave, the MJO and tropical cyclone (TC) genesis would not be able to be simulated in GCMs since the structure and propagation of the MJO and TC is similar to the convectively coupled Kelvin wave and TD-type waves.

Acknowledgment

We thank the reviewers for their constructive comments and WCRP for providing CMIP5 datasets.

References

[1] Alsepan G, Lubis S W, Setiawan S 2016 Analysis of the equatorial lower stratosphere Quasi-Biennial Oscillation (QBO) using ECMWF-Interim Reanalysis Data Set. *Earth Environ. Sci* 31 : 1-14.

[2] Fatullah N Z, Lubis S W and Setiawan S 2017 Characteristics of Kelvin waves and Mixed Rossby-Gravity waves in opposite QBO phases. *Journal IOPscince* doi.org/10.1088/1742-6596/755/1/011001

[3] Faturochman I, Lubis S W and Setiawan S 2017 Impact of Madden-Julian Oscillation (MJO) on global distribution of total water vapour and column ozone. *Journal IOP Science* doi.org/10.1088/1755-1315/54/1/012034.

[4] Lubis S W, Setiawan S 2010 Identification of Equatorial Atmospheric Kelvin Waves based on NCAP/NCAR Reanalysis I Data. Journal of Physics 12:71-82. ISSN 0854-3046.

[5] Lubis S W, Jacobi C 2015 The modulating influence of convectively coupled equatorial waves (CCEWs) on the variability of tropical precipitation. *J.Climtol.* **7**(35):1465-1483.

[6] Rakhman S, Lubis S W, Setiawan S 2017 Impact of ENSO seasonal variation of Kelvin wave and mixed Rossby-Gravitu waves. *Journal IOPscince*. doi:10.1088/1755-1315/54/1/012035.
[7] Flato G, J Marotzke, B Abiodun, P Braconnot, S C Chou, W Collins, P Cox, F Driouech, S Emori, V Eyring, C Forest, P Gleckler, E Guilyardi, C Jakob, V Kattsov, C Reason and M Rummukainen 2013. Overview: *Evaluation of Climate Model* In: Climate Change: The Physical Science Basis. Contribution of Working Group I to the Fifth Assessment Report of the Intergovernmental Panel on Climate Change 741-866.
[8] Takayabu Y N 1994 Large-scale cloud disturbances associated with equatorial waves. I:Spectral features of the cloud disturbances. *Journal of the Meteorological Society of Japan*. **72**(3):433–449.
[9] Kawatani Y, Takahashi M, Sato K, Alexander S P, Tsuda T 2009 Global distribution of atmospheric waves in the equatorial upper troposphere and lower stratosphere: AGCM simulation of sources and propagation. *J. Geophys. Res.* **114** doi: 10.1029/2008JD010374.
[10] Lot F, Denvil S, Butchart N, Cagnazzo C, Giorgetta M A, Hardiman S C, Manzini E, Krismer T, Duvel J P, Maury P, Scinocca J F, Watanabe S, Yukimoto S 2014 Kelvin and Rossby-gravity wave packets in the lower stratosphere of some high-top CMIP5 model. *Journal of Geophysical Research: Atmospheres*. doi:10.1002/2013JD020797.
[11] Tiedtke M 1989. A comprehensive mass flux scheme for cumulus parameterization in large-scale models. *Mon. Wea. Rev*. (117): 1779–1800.
[12] Emanuel K A 1991: A scheme for representing cumulus convection in large-scale models. *J. Atmos. Sci*. **48:**2313–2329.
[13] Kiladis G N, Matthew C, Wheeler, Patrick T, Katherine H, Straub, Paul E Roundy 2009 Convectively coupled equatorial waves. *Reviews of Geophysics*. **47**(2):1–42.
[14] Taylor K E, Stouffer R J, Meehl G. 2007 A Summary of the CMIP5 Experiment Design. World 1–33.
[15] Huang P, Chou C, Huang R 2013 The activity of convectively coupled equatorial waves in CMIP3 global climate models. *Theoretical and Applied Climatology*. 112(3–4): 697–71.
[16] Nordeng TE 1994 Extended versions of the convective parametrization scheme at ECMWF and their impact on the mean an transient activity of the model in the tropics. ECMWF Research Department Tech. European Centre for Medium-Range Weather Forecasts.
[17] Wheeler M, Kiladis G N 1999 Convectively coupled equatorial waves: analysis of cloud and temperature in the wavenumber-frequency dominan. *J Atmos Sci*. **56**(3):374-339.
[18] Wheeler M, Kiladis G N, Webster P J 2000 Large-scale dynamical fields associated with convectively coupled equatorial waves. *J Atmos Sci*. **57**(5): 613-640.
[19] Yang G Y, Hoskins B J, Slingo J M. 2011. Equatorial waves in opposite QBO phases. *J. Atmos. Sci*. **68**: 839–862, DOI: 10.1175/2010JAS3514.1.
[20] Zhang G J, McFarlane N A 1995 Sensitivity of climate simulations to the parameterization of cumulus convection in the Canadian Climate Centre general circulation model. Atmos-Ocean **33**(3):407–446.

Spatial Conflict of Mining Land in Tolitoli District -Province of Central Sulawesi

Y Suwarno and R Windiastuti

Geospatial Information Agency
Jl Raya Jakarta Bogor KM 46 Cibinong, Jawa Barat, 16911, Indonesia

Email : yatinsuwarno@yahoo.com

Abstract. Spatial planning is supposed to be applied in the use of space, so there will be no overlapping space utilization. In fact, there are still overlapping uses of land, between the area of mining and plantation, as well as with forest areas. The purpose of this study was to find out the conflicts that occured due to overlapping permits given to mining and plantation companies, and also to forest status. The method used was by overlaying the maps of Mining Business Permit with that of Plantation Business Permit, and also with Forest Area Map. In Tolitoli District there were 23 mining business permit holders with 7 types of mining commodities, covering total areaof 81,503.54 Hectare. In addition, there were 5 companies holding plantation business permits, mostly on palm oil, and only 2 companies with rubber and sengon wood business commodities, with a total area of 80,005.35 Hectare. From the result of spatial analysis, it was found that there was an overlapping area of 22,869.70 Hectare, while the area of 118,072.93 Hectare did not overlap. The Mining Business Permit overlapped with the Plantation Business Permit covering an area of 18,853.32 Hectare, and 4,301.77 Hectare were located in Forest Protected Area and Nature Reserve.

1. Introduction

Spatial planning of a district is made by considering the physical condition that is vulnerable to disaster; the potential of natural resources, human resources, and artificial resources; economic, social, cultural, political, legal, defense, security, environmental and scientific and technological conditions as a whole; as well as geostrategic, geopolitics, and geo-economics (Chapter 3, article 6) [1] In reality, there is still a lot of overlap in the use of lands due to conflict ofinterests byeach sector.

The Spatial Planning Map of Tolitoli District has been made for 20 years period (2012-2032), and has been ratified in Tolitoli District Regulation Number 16/2012 [2]. The Sspatial Planning Map consisting of spatial pattern and spatial structure maps is composed of various thematic maps, which sometimes still have problems. These problems include overlapping between mining and plantation permits, and overlapping between each permit with forest areas. The land overlap is thought to be caused by the use ofdifferent sources and scales of base mapswhen the license was assigned. The occurrence of overlapping land is predicted due to several reasons such as different map sources, different base map scales, and different map production years.

Spatial problems such as overlapping multiple land uses can be solved by spatial analysis in Geographic Information System. Spatial analysis is an approach in geography and discipline associated with it by using statistical methods to simplify spatial patterns. Spatial analysis is a technique or process that involves a count and evaluation of logic being done in order to seek or discover potential relationships or patterns that may exist between geographic elements (contained in digital data with boundaries of certain study area).In spatial analysis, spatial data is needed, i.e. data that has a geo-reference in which various attribute data lies in various spatial units.

1.2 Objective

The objective of this research was to find out overlapping of land use permits in Tolitoli District, the factors that caused it, and the solution and its recommendation to minimize the potential of spatial conflicts.

2. Materials and method

2.1. Material and Tool

The materials used were mining permit map, plantation permit map, and forest status map. All the aps were in shapefile format. The hardware used was a personal computer or notebook, while spatial data processing software used was ArcGIS.

2.2. Analysis Method

The analytical method was the overlay (union) in the geographic information system among the land use permit maps to determine the spatial relationships among several maps [3]. Union is a spatial analysis function used to aggregate some of the spatial elements presented in a selected theme to become an element [4,5].

2.3 Analysis Framework

The analytical framework for the evaluation of land use permit conflicts in this study is presented in Figure 1.

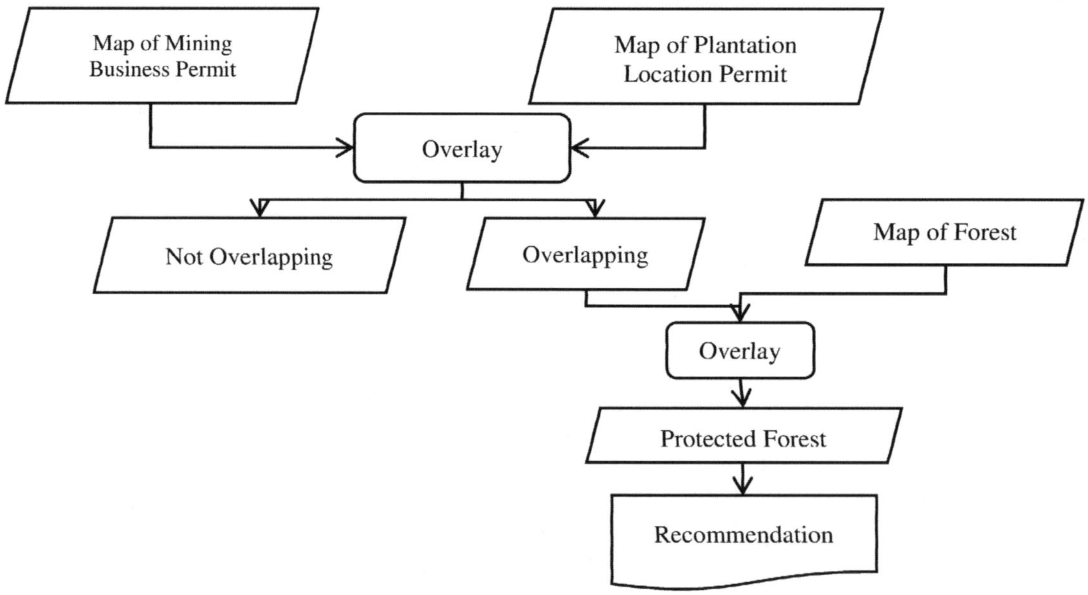

Figure 1. Spatial Analysis for Evaluation of Space Use Permit.

3. Results and discussion

3.1. Mining Business Permit

In Tolitoli District there were 25 metal mineral mining business license holders, consisting of 3 gold companies, 12 molybdem companies, 6 copper companies, 2 iron ore companies and one manganese and galena [6,7]. Location map of the distribution of mining business permits in Tolitoli District is shown in Figure 2, while the name of the company and its type of business is presented in Table 1.

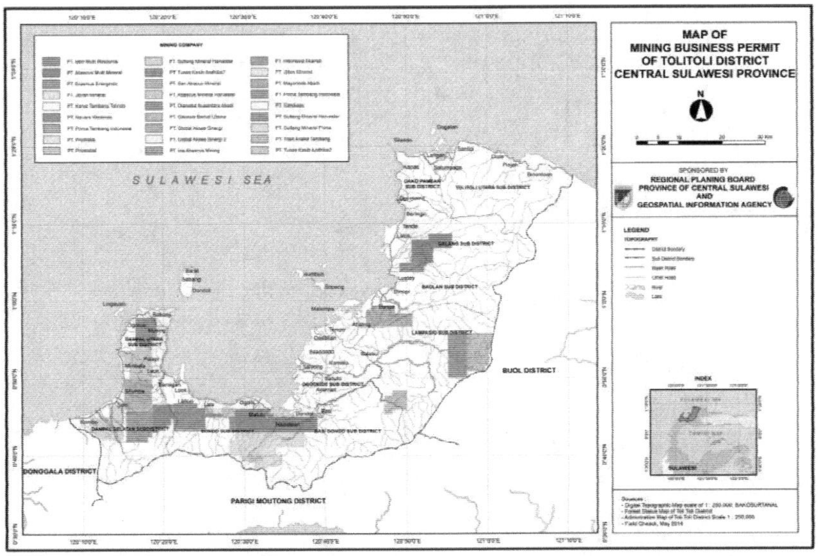

Figure 2. Map of Mining Business Permit in Tolitoli District.

According to data from the Mining Office of Tolitoli District (2014), it has legalized 7 mining business licenses, comprising of gold, copper, molybdenum, iron ore, galena, manganese and granite. All types of minerals under 23 companies hold mining business licenses with an area of each commodity for each company as in Table 1.

3.2 Plantation Location Permit

In Tolitoli District, there were 5 plantation permit companies, mostly palm oil commodities, and two rubber and sengon wood commodity companies (Table 2) [8]. Bukit Berlian Persada Company with oil palm commodity had 7 locations of business license (Table 2). Location map of plantation permits in Tolitoli District is presented in Figure 3.

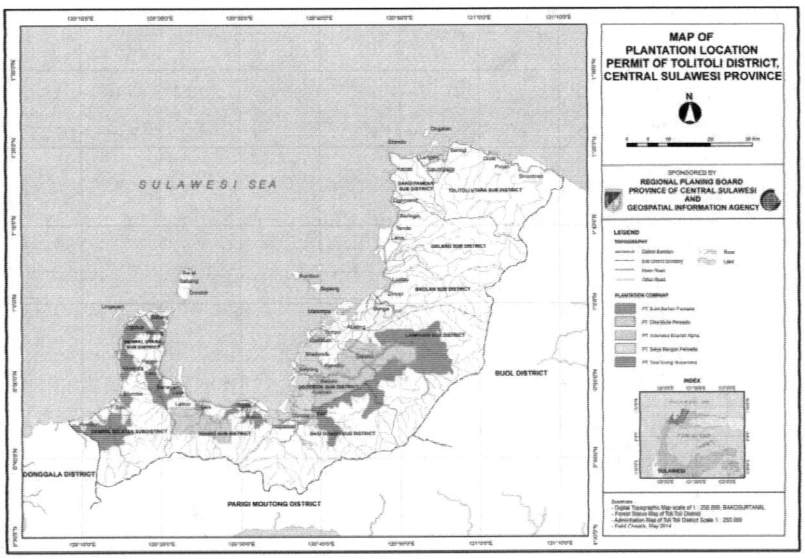

Figure 3. Map of Plantation Location Permit in Tolitoli District.

IOP Conf. Series: Earth and Environmental Science **149** (2018) 012031 doi:10.1088/1755-1315/149/1/012031

Table 1. Mining Business Permit in Tolitoli District.

Num.	Name of Company	Mineral Commodity	Area (Ha)
1	Abascus Mineral Harvester	Molybdem	4,219.97
2	Abascus Mineral Harvester	Cooper	4,169.48
3	Dianelsa Nusantara Abadi	Molybdem	5,371.95
4	Erasmus Energindo	Galena	1,330.25
5	Genesis Berkat Utama	Mangaan	2,279.74
6	Global Akses Sinergi 2	Cooper	6,795.71
7	Ina Abascus Mining	Molybdem	4,486.20
8	Indonesia Ekaristi	Molybdem	69.28
9	Jibra Mineral	Molybdem	9,846.77
10	Karya Tambang Tolindo	Iron ore	5,162.03
11	Mayorindo Abadi	Granite	179.42
12	Mbh Multi Resource	Gold	132.30
13	Mbh Multi Resource	Molybdem	0.27
14	Mbh Multi Resource	Cooper	2,412.18
15	Navara Westindo	Molybdem	8,312.93
16	Prima Tambang Indonesia	Gold	4,658.01
17	Promistis	Molybdem	1,035.24
18	San Diego	Cooper	2,738.39
19	Sulteng Mineral Harvester	Molybdem	3,497.15
20	Sulteng Mineral Prima	Iron ore	4,864.89
21	Titan Aneka Tambang	Gold	4,714.13
22	Tunas Kasih Andhika2	Molybdem	593.47
23	San Abacus Mineral	Molybdem	4,633.81
		Total	**81,503.54**

Source: Mining and Industry Service Division of Tolitoli District. 2014

In addition, in TolitoliDistrict there were individual plantations (such as coconut, cloves, cocoa, nutmeg, etc.) [2]. Data of exact areas for all of these individual estates were not available, so they did not appear on the map.

Table 2. Plantation Permit in Tolitoli District.

Num.	Name of Company	Plantation Commodity	Area (Ha)
1	Bukit Berlian Persada	Palm Oil	18,470.23
2	Citra Mulia Perkasa	Rubber and Sengon wood	19,952.34
3	Indonesia Ekaristi Alpha	Palm Oil	8,486.74
4	Setya Bangun Persada	Palm Oil	12,547.14
5	Total Energi Nusantara	Rubber and Sengon wood	20,548.89
		Total	**80,005.35**

Source: Plantation Service Division of Tolitoli District. 2014

IOP Conf. Series: Earth and Environmental Science **149** (2018) 012031 doi:10.1088/1755-1315/149/1/012031

3.3 Spatial Conflict of Mining Land

The local government of Tolitoli District has issued permits to utilize the space (land) of its territory. In this analysis only 2 (two) permits of space utilization were used, that were Mining Business Permit and Plantation Location Permit. Space utilization permit analysis was conducted to determine whether there was any overlaps or not, between permits of space utilization for mining and plantation. The Map of Space Utilization Permit was an overlay between the Map of Mining Business Permit and Map of Plantation Location Permit (Figure 4). The results of the widespread calculation with overused GIS results are presented in Table 3and Table 4.

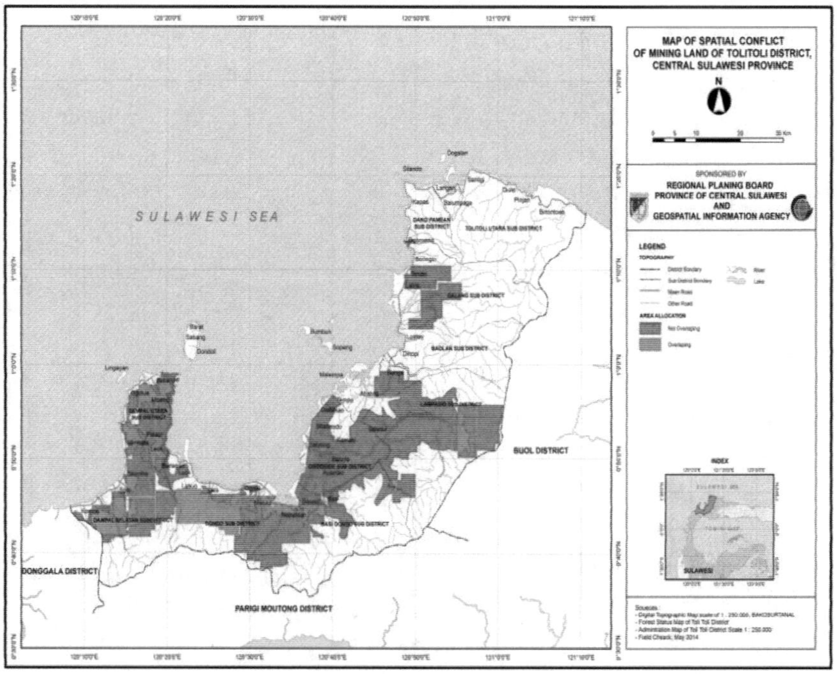

Figure 4. Map of Space Utilization in Tolitoli District.

Table 3. Spatial Utilization Conflict in Tolitoli District.

Num.	Utilization	Area (Ha)	Area %)
1	Not Overlapping	118,072.93	83.77
2	Overlapping	22,869.70	16.23
	Total	**140,942.63**	**100.00**

Table 4 shows the condition of space utilization in Tolitoli District in general. Overlay results indicated that there wereoverlapping permits for the utilization of space for mining and plantations, covering area of 22,869.70 Ha (16.23%).Table 5 presents a more detailed overlap, between each company. Permits for Bukit Berlian Persada (palm oil) overlapped with 10 (ten) mining companies with a total area of 12,828.80 Ha. Permits for Indonesia Eucharistic Alpha (oil palm) overlapped with 4 (four) mining companies with a total area of 5,241.09 Ha. Permits for Total Nusantara Energy (rubber and sengon wood) overlapped with 3 (three) mining companies, covering 1,532.53 Ha. Permits for Setya Bangun Persada (oil palm) overlapped with 2 (two) mining companies, covering 1,540.26 Ha. In addition to granite, most of the mining businesses in Tolitoli District were on a type of metal mineral (gold, copper, molybdem, etc.), which were still in the exploration stage. However, when they come into exploitation stage later, some problems may arise, because all of these mineral minings would be open pits [9,10].

Table 4. Overlap area between Mining Business Permit and Plantation Location Permit.

Mining Company	Plantation Company	Sub District	Area (Ha)
Abascus Mineral Harvester	Bukit Berlian Persada	Dampal Selatan	1,088.43
		Dampal Utara	196.09
Dianelsa Nusantara Abadi	Bukit Berlian Persada	Dondo	93.28
Dianelsa Nusantara Abadi	Indonesia Ekaristi Alpha	Dampal Selatan	230.02
		Dondo	4,914.79
Genesis BerkatUtama	Total Energi Nusantara	Basi Dondo	24.35
Global Akses Sinergi 2	Bukit Berlian Persada	Dampal Utara	850.04
Global Akses Sinergi	Total Energi Nusantara	Lampasio	1,088.43
Ina Abascus Mining	Bukit Berlian Persada	Dampal Selatan	517.00
Indonesia Ekaristi	Indonesia Ekaristi Alpha	Dondo	6,61
Jibra Mineral	Setya Bangun Persada	Basi Dondo	21.63
		Dondo	24.81
Mbh Multi Resource	Indonesia Ekaristi Alpha	Dampal Selatan	54,468
		Dondo	46.05
Navara Westindo	Setya Bangun Persada	Basi Dondo	1,315.69
		Dondo	1,484.73
Prima Tambang Indonesia	Total Energi Nusantara	Lampasio	1,352.29
Promistis	Bukit Berlian Persada	Dondo	273.59
San Diego	Bukit Berlian Persada	Dampal Utara	656.44
		Dondo	1.45
Sulteng Mineral Harvester	Bukit Berlian Persada	Dampal Utara	1,455.88
Sulteng Mineral Prima	Bukit Berlian Persada	Dampal Selatan	3,918.87
San Abacus Mineral	Bukit Berlian Persada	Dampal Selatan	93.80
		Dampal Utara	77.67
		Total	**18,853.32**

3.4 Evaluation of Space Utilization Permit

Table 5 presents the area and location of mining land that overlapsed with plantation land in Protected Forest area. There were 6 mining companies in Tolitoli District that overlapped with 4 plantation companies. Figure 5 presents an evaluation of the use of space in protected forest area.

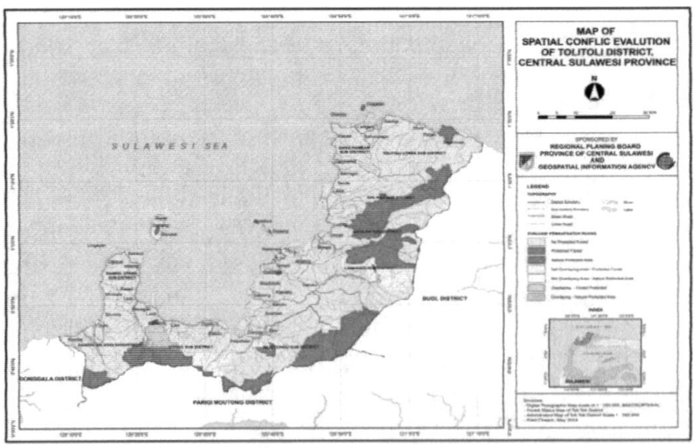

Figure 5. Map of Space Utilization Evaluation Permit of Tolitoli District.

Table 5. Overlap area of mining and plantation lands in Forest Protected Status (Num. 635).

Num.	Plantation Company	Mining Company	Sub District	Area (Ha)
Protected Forest	Bukit Berlian Persada	Dianelsa Nusantara Abadi	Dondo	0.91
Protected Forest	Bukit Berlian Persada	Dianelsa Nusantara Abadi	Dondo	0.65
Protected Forest	Bukit Berlian Persada	Dianelsa Nusantara Abadi	Dondo	0.11
Protected Forest	Indonesia Ekaristi Alpha	Mbh Multi Resource	Dondo	0.27
Protected Forest	Indonesia Ekaristi Alpha	Mbh Multi Resource	Dondo	12.52
Protected Forest	Indonesia Ekaristi Alpha	Dianelsa Nusantara Abadi	Dampal Selatan	76.69
Protected Forest	Indonesia Ekaristi Alpha	Dianelsa Nusantara Abadi	Dondo	3,611.53
Protected Forest	Indonesia Ekaristi Alpha	Jibra Mineral	Dondo	1.00
Protected Forest	Total Energi Nusantara	Prima Tambang Indonesia	Lampasio	0.17
Protected Forest	Total Energi Nusantara	Prima Tambang Indonesia	Lampasio	4.99
Protected Forest	Total Energi Nusantara	Global Akses Sinergi	Lampasio	9.76
Protected Forest	Total Energi Nusantara	Prima Tambang Indonesia	Lampasio	0.56
Protected Forest	Total Energi Nusantara	Prima Tambang Indonesia	Lampasio	0.63
Protected Forest	Total Energi Nusantara	Prima Tambang Indonesia	Lampasio	228.42
Protected Forest	Setya Bangun Persada	Navara Westindo	BasiDondo	3.17
Protected Forest	Setya Bangun Persada	Navara Westindo	BasiDondo	1.35
Protected Forest	Setya Bangun Persada	Jibra Mineral	BasiDondo	13.13
			Sub Total	3,965.85
Nature Reserve / Protection Area	Indonesia Ekaristi Alpha	Dianelsa Nusantara Abadi	Dampal Selatan	82.11
Nature Reserve / Protection Area	Indonesia Ekaristi Alpha	Dianelsa Nusantara Abadi	Dondo	253.81
			Sub Total	335.92
			Total	**4,301.77**

4. Conclusion

There were 22,869.70 Ha (16.23%) overlaps of space utilization for mining business permits and plantation site permits in Tolitoli District. The result of evaluation of space utilization permits in Protected Area showed that, on the area of 17,021.72 ha (39.17%) there is permit of space utilization for mining and/or plantation purposes. The result of spatial use permit evaluation in Production forest shows that, on an area of 3,335.28 Ha (9.72%) there is permit for space utilization for plantation that is 650.55 Ha (1.90%) for mining 2,682.99 Ha (7.82%), while overlapping plantations and mining area of 1.74 Ha (0.01%). The overlapping of space utilization in Tolitoli district was caused by 2 (two) possibilities, which were (1) different kinds and level of accuracy of the maps used or (2) lack of coordination between the licensing institutions.

Acknowledgements

Acknowledgments were conveyed to the Regional Planning and Development Agency of Central Sulawesi Province that have contributed to the financing of this research, the Regional Planning and Development Agency of Tolitoli District that has assisted in providing data, the entire leaders of the Geospatial Information Agency who have given trust and direction in this research. Special thanks to Mr. Sularsana for his assistance in the Geographic Information System works.

References

[1] Lembaran Negara Republik Indonesia Tahun 2007 No 68 *Undang-Undang Republik Indonesia No 26* tentang Penataan Ruang Jakarta
[2] Lembaran Daerah Kabupaten Tolitoli Tahun 2012 No 01 Tahun 2012 *Peraturan Daerah Kabupaten Tolitoli No 16* on spatial planning Tolitoli 2012-2032
[3] Burrough M D 1998 *Principals of Geographical Information Systems* Oxford: Oxford University Press 216
[4] ESRI 1997 *ArcView Spatial Analyst. Environmental Systems* Research Institute. Inc. Redlands: USA
[5] Prahasta, Eddy 2002 Sistem Informasi Geografis Tutorial ArcView Bandung: CV Informatika
[6] Lembaran Negara Republik Indonesia Tahun 2009 No 4 *Undang-Undang Republik Indonesia No 4* Concerning Mineral and coal mining
[7] Lembaran Negara Republik Indonesia Tahun 2010 No 28 *Peraturan Pemerintah Republik Indonesia No 22* on Mining Areas
[8] Central Bureau of Statistics Tolitoli District 2014 *Tolitoli District in Figures 2013* ISSN: 0794.0123 catalog of BPS **218**
[9] Mottana A Crespi R and Liborio G 1978 *Guide to Rocks and Minerals* New York: Simon and Schuster Inc.
[10] Peraturan Presiden Republik Indonesia Nomor 28 Tahun 2011 C*oncerning the use of protected forest for subterranean mining* 32

IOP Conf. Series: Earth and Environmental Science **149** (2018) 012032 doi:10.1088/1755-1315/149/1/012032

Detection of mesoscale convective complexes using multispectral RGB technique of Himawari-8 (Case Study: Jakarta, 20 February 2017)

Fatkhuroyan, T Wati
Indonesia Agency for Meteorology Climatology and Geophysics
BMKG, Kemayoran, Central Jakarta, Indonesia

E-mail:fatkhuroyan@bmkg.go.id

Abstract. Mesoscale Convective Complexes (MCC) is a well-organized convective cloud that has big size and long lifetime. The aim of the study is to detect and to monitor the development of MCC around Jakarta on 20th February 2017 using satellite Himawari-8. This study uses the analyzing method of the infrared channel and multispectral imagery RGB Technique to monitor the development of radiative, morphology and cloud position which describe the cloud top microphysics, structure and movement of the MCC. On 20th February 2017, the result from Himawari-8 shows that there are many dense-clouds with small ice particle and cloud top temperature could be < -50^{0}C which can be seen as red and yellow dot colour by RGB Technique. The MCC caused a severe storm at Jakarta and its surrounding area.

1.Introduction

Mesoscale convective complexes is a well-organized convective cloud that has a big size and long lifetime. The concept of MCC have been introduced in 1980 that has cloud shield with continuously low IR temperature $\leq$ -32^{0}C with area $\geq$100,000 km and has interior cold cloud region with temperature $\leq$-52^{0}C with area 50,000 km^2 [1]. The MCC also has the shape of the cloud with eccentricity $\geq$ 0.7 at time maximum extent and it can affect the severe weather [1,2]. Recently, the operational numerical weather-prediction model is not sufficient to represent the MCC and to forecast of such system is still difficult [3], mainly based on real-time observation and knowledge concerning the environmental conditions which are favourable for MCC triggering [4]. A possible way to improve MCC prediction is by multi-parametric tools and conceptual models based on average behaviour of several parameters coming from different observation such as satellite, images and model output fields [5].

Himawari-8 is a new Japanese geostationary meteorological satellite operated since July 7th 2015 with optical sensors significantly higher in temporal and spatial resolution than previous geostationary orbit such as MTSAT-2 [6]. There are 16 observation channels with 0.5 or 1 km spatial resolution for visible and near-infrared bands and 2 km for infrared bands. The shortened revisit times around 10 minutes for full disk provide new levels of capacity for identification and tracking of rapidly changing weather phenomena and for the derivation of quantitative products [7]. There are still a few previous studies about MCC at Maritime Continent. The spatial and temporal MCC over Indonesia was studied for 5 years (2001-2005) and concluded that MCC was frequently happened at Java Sea during March-April-May (MAM) and December-January-February (DJF) [8]. The MCC over the Maritime Continent have large lightning densities, while MCC in the oceanic convergence zones are more connected and have proportionately larger strati form rain fractions [9]. The aim of this study is to

IOP Conf. Series: Earth and Environmental Science **149** (2018) 012032 doi:10.1088/1755-1315/149/1/012032

detect and to monitor the development of MCC around Jakarta and other provinces near Jakarta on 20th February 2017 using satellite Himawari-8.

2.Data and methods

Himawari-8 satellite imagery was used for studying the structure and development of the MCC on 20th - 21st February 2017 and observation rainfall data from 13 stations in Sumatra and Java on 19th – 21st February 2017 for rainfall data verification. The locations of rain gauges sites is presented in Figure 1.

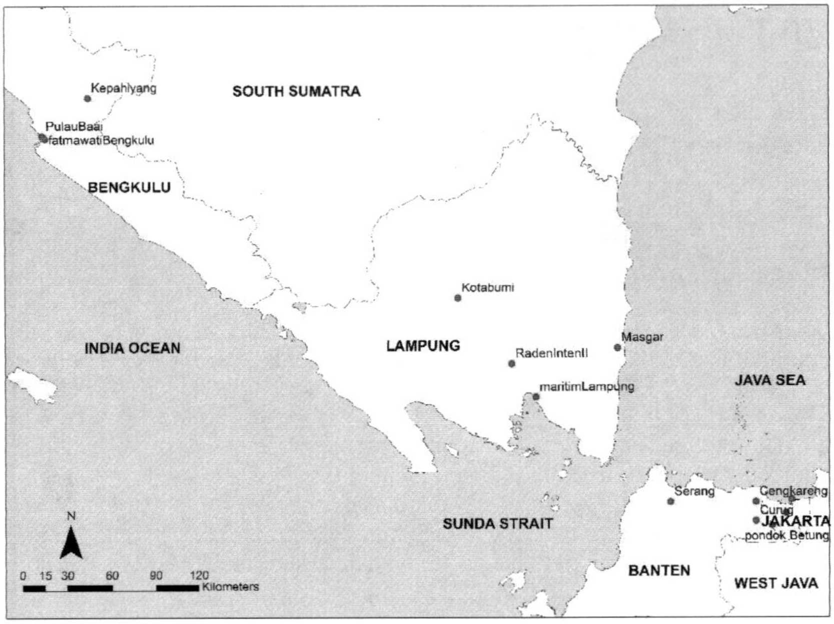

Figure 1. The Locations of rain gauge sites (red dots).

IR black body temperature (T_{BB}), derived from ten minutes data of Himawari-8 since Feb 20th 2017 10.00 UTC until Feb 21st 2017 03.00 UTC were used to identify and monitor the development the MCC and their physical characteristics based on Maddox (1980) as shown in table 1.

Table 1. MCC based upon analyses of enhanced IR satellite imagery.

	Physical Characteristics
Size:	A – Cloud shield with continuously low IR temperature $\leq$ - 32°C must have an area $\geq$ 100 000 km^2 B – Interior cloud could region with temperature $\leq$ - 52°C must have an area $\geq$ 50 000 km^2
Initiate:	Size definitions A and B are first satisfied
Duration:	Size definition A and B must be met for a period $\geq$ 6 h
Maximum Extent :	Continuous cold could shield (IR temperature $\leq$ - 32°C) reaches maximum size
Shape:	Eccentricity (minor axis/ major axis) $\geq$0.7 at time of maximum extent
Terminate:	Size definition A and B no longer satisfied

Studies of MCC structures and propagation using Himawari-8 Satellite consist of detection and monitoring of convective cloud cell use Infrared channel in order to monitor time evolution of radiative, morphology and location of clouds. It also describe cloud top microphysics, structure, growth and the movement of MCC.

Multispectral Himawari-8 imagery analysis of RGB (Red Green Blue) technical analysis [10] using SATAID software [11] for visualization of Himawari-8 data have conducted for investigating the organization and cloud top microphysics of every MCC cycle phase.

Red colour is displayed by compositing the differentiation of channel 12.3μm minus 10.4μm, green colour is from differentiation of channel 10.4μm minus 3.9μm and blue colour is from channel 10.4μm [12,13]. This RGB is effective to distinguish clouds with high cloud top (Cumulonimbus) and fog or low-level clouds because the difference of 3.9μm is included at this scheme.

3.Results and Discussion

Life-cycle of MCC will be shown by sequential infrared image (10.8μm) with blue cloud-shield and red interior cold cloud-shield and by multispectral imagery from a combination of some channels [10]. There are eight steps of MCC life-cycle, which consists of MCC-12h, pre-MCC, Initiation, Growth, Mature, Decay, Dissipation and post-MCC [14]. The development of MCC was started at 20th Feb 2017 10 UTC with the development of many embedded cumulonimbus clouds around west of Lampung and Bengkulu along the Indian Ocean (Fig. 2a). At 13 UTC, it reached initiate condition (Fig. 2b) which had interior cold-cloud region $\geq$ 50,000 km^2 (red colour) and cloud shield $\geq$ 100,000 km^2 (blue colour) in Lampung until the Indian Ocean. The MCC had growth bigger and cover all Lampung area (Fig. 2c) then it reached mature condition while it cover all Lampung area, some part of South Sumatera, Bengkulu, the Indian Ocean until the Sunda Strait (Fig.2d) at 18 UTC. The MCC decayed in the following day at 21st Feb 2017 01 UTC (Fig. 2e) when it extended across the Sunda Strait, the Indian Ocean, Banten, and Jakarta. It caused heavy rainfall and floods. In the end, The MCC dissipated at 03 UTC in the morning (Fig. 2f).

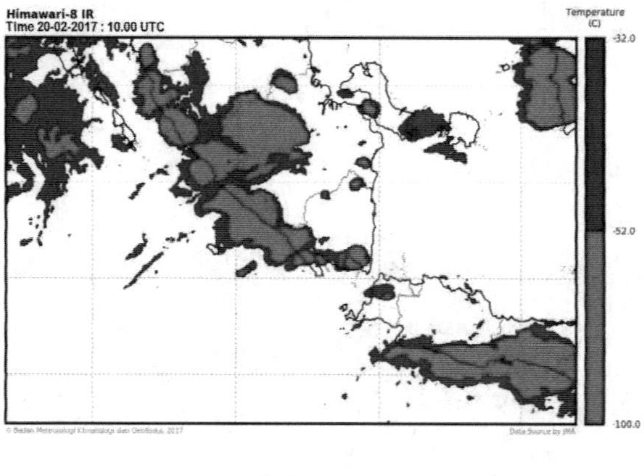

(a)

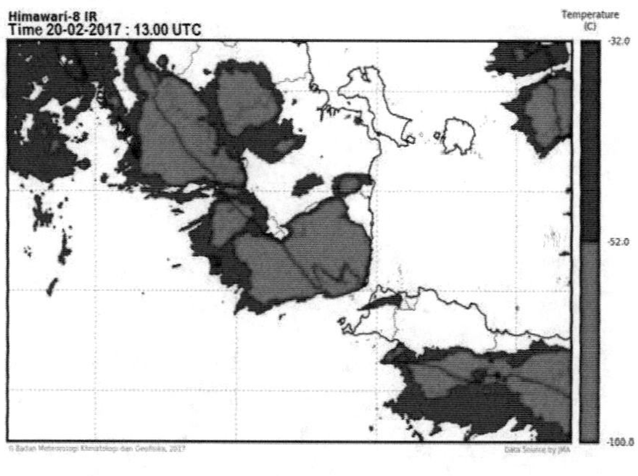

(b)

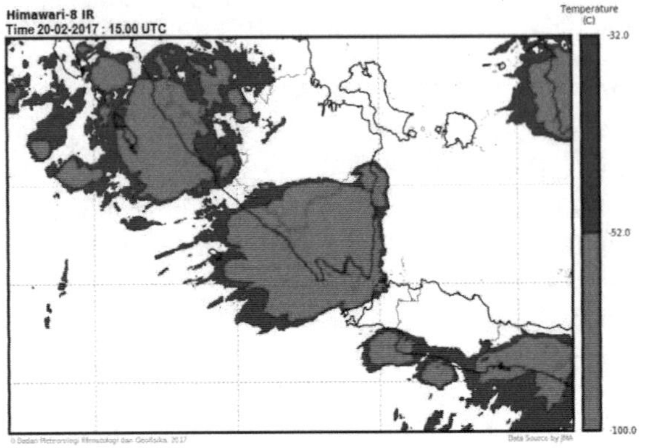

(c)

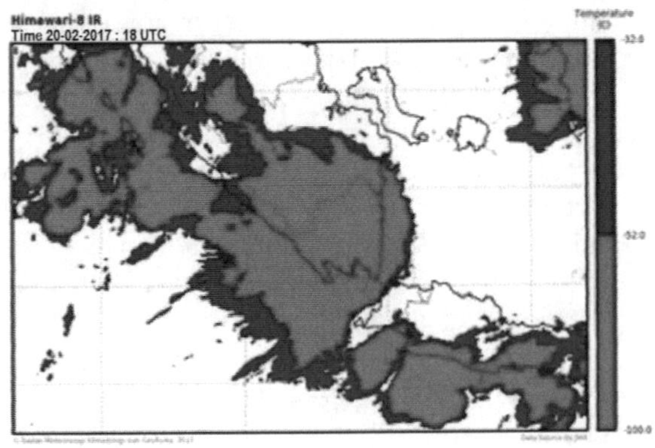

(d)

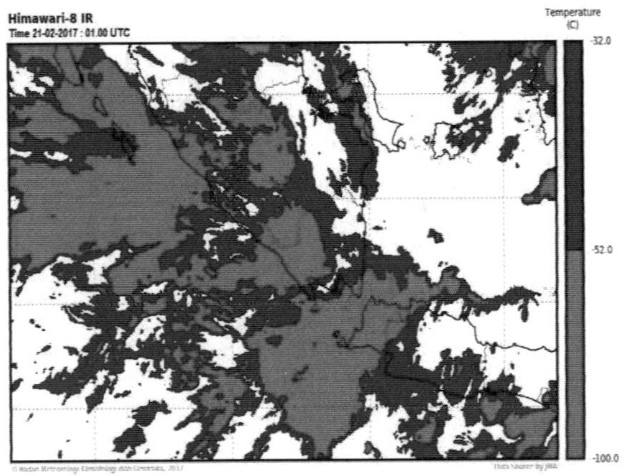

(e)

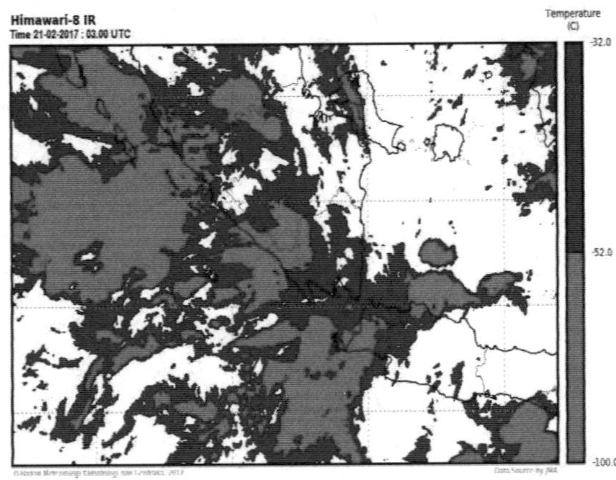

(f)

Figure 2. (a) Pre-MCC (b) Initial (c) Growth (d) Mature (e) Decay (f) Dissipation using IR channel.

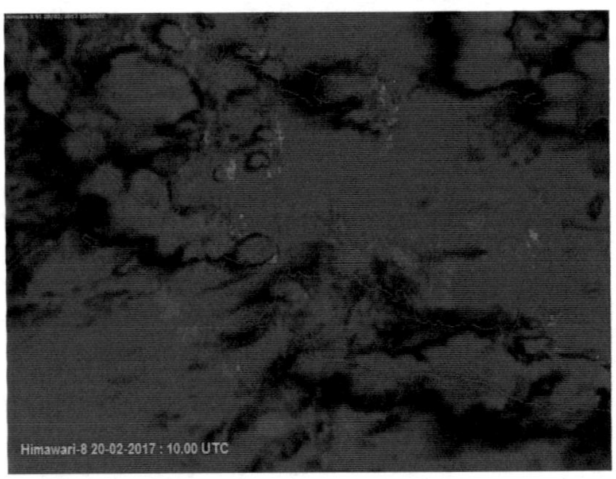

(a)

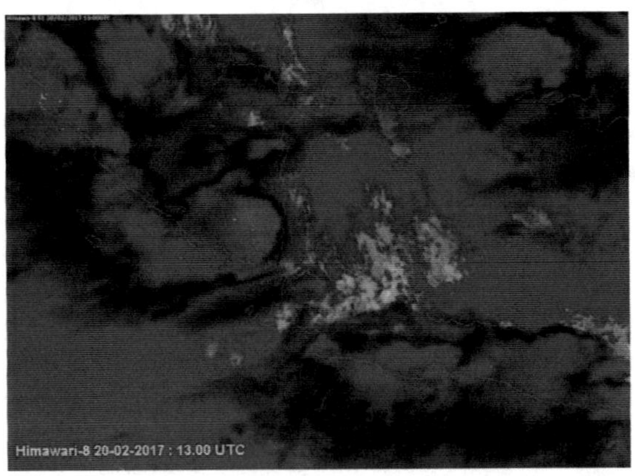

(b)

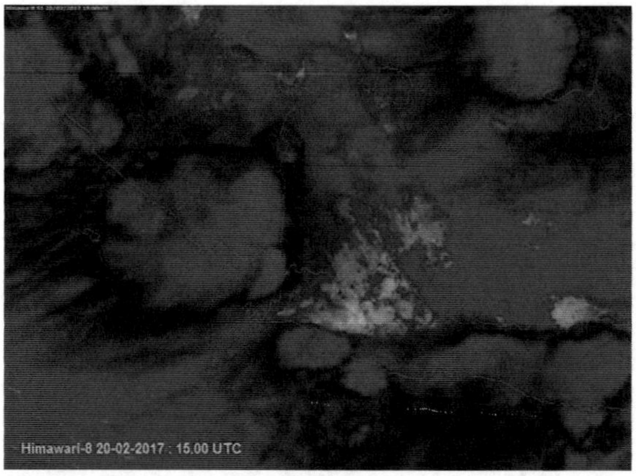

(c)

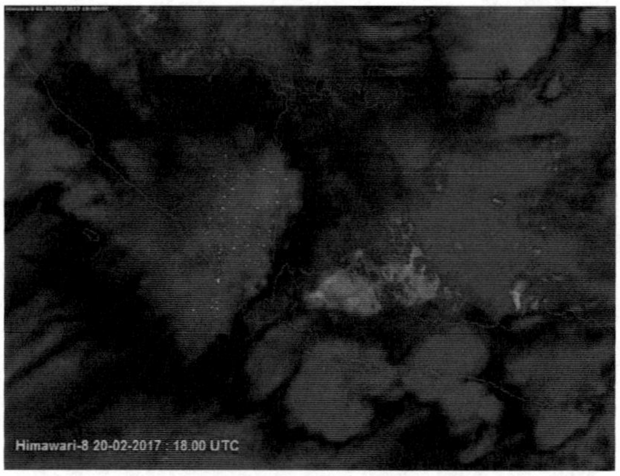

(d)

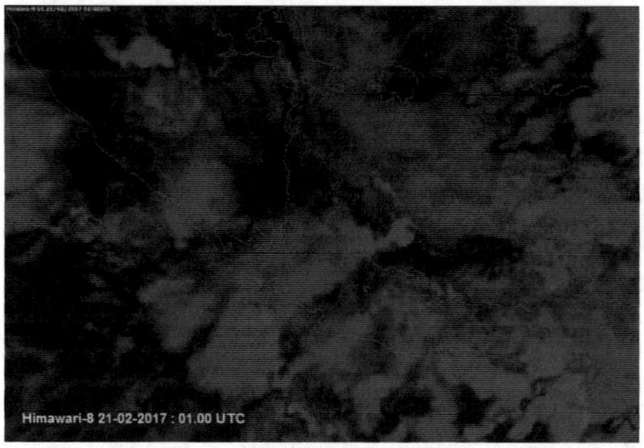

(e)

(f)

Figure 3. (a) Pre-MCC (b) Initial (c) Growth (d) Mature (e) Decay (f) Dissipation using Night Microphysics RGB.

The development of the MCC since Pre-MCC phase until dissipation phase also can be monitoring very well using multi-spectral night microphysics RGB as seen as in fig. 3(a-f). From multispectral imagery, it was clearly seen that there were cold-thick-high level clouds which displayed in red colour over Lampung, the Sunda Strait and the Indian Ocean during growth phase, Moreover, grey colour over the Java Sea, northen part of Jakarta indicated low-level cloud (Fig. 3c). Cold-thick clouds with ice particles in the top of high cloud with temperature at the top could be reach < -50^{0}C with small ice particles and cause storm severity were displayed by yellow dot red colour during mature phase of the MCC(Fig. 3d). Cold and thick high-level clouds still stayed during decay and dissipation phase wich caused heavy rainfall at western part of Java and Jakarta.

Table 2 and Table 3 represent the amount of observation rainfall from rain gauges in Sumatra and Java. On 19 February 2017, the rainfall which more than 50 mm is only in Serang, Banten, while on 20 February 2017, the heavy rain almost happened in Lampung (Masgar, Maritim Lampung and Raden Inten II), Banten (Pondok Betung, Serang and Cengkareng) and Jakarta (Tanjung Priok and Kemayoran). According to Indonesian National Board for Disaster Management (BNPB), on 20 February 2017 the flood because of heavy rain happened at 54 point in Jakarta and surrounding, the inundation recorded from 10 cm until 150 cm. The flood also happened in some part of Tangerang Banten (Pondok Ranji) on February 20, 2017. While in Bandar Lampung (Capital city of Lampung Province), the flood happened on 21 February 2017 with the inundation recorded 50 cm until 80 cm [15].

Table 2. The rainfall in Sumatra (Bengkulu and Lampung) in mm/day.

Date	Fatmawati	Pulau Baai	Kepahiyang	Masgar	Maritim Lampung	Raden inten II	Kotabumi
19-02-2017	17	7	1	6	1	10	7
20-02-2017	30	43	15	107	81	88	4
21-02-2017	5	6		8	5	6	27

Table 3. The rainfall in Java (Banten and Jakarta) in mm/day.

Date	Pondok betung	Serang	Curug	Cengkareng	Tanjung Priok	Kemayoran
19-02-2017	11	55	3	16	21	9
20-02-2017	93	50	33	73	116	180
21-02-2017	11	3	13	10	12	11

4.Conclusions

Himawari-8 imagery is able to detect and monitor the development, structure, growth and movement of the MCC on 20th February 2017. Multispectral imagery also revealed the organization, cloud top microphysics every phase of the MCC's life cycle appropriately and attractively in order to give better early warning about extreme rainfall and flood disaster. The MCC on 20 February 2017 has caused storm and heavy rains in Jakarta and Lampung causing flood and inundations in some locations on 20-21 February 2017, even though the inundation is also due to the poor drainage of the cities.

Acknowledgements
This study was funded by INSINAS program of the Ministry of Research, Technology and Higher Education of the Republic of Indonesia (No: 063/P/RPL-LIPI/INSINAS-2/VI/2017. PI: Dr. Apip, M.Eng (Research Center for Limnology-LIPI). The authors are also grateful to BMKG as Indonesian agency for Meteorology Climatology and Geophysics for Himawari-8 satellite and rainfall observation data.

References

[1] Madox, R.A 1980. Mesoscale Convective Complexes, *Bul. Amer. Meteor. Soc.* **61** 1374 -1387
[2] McAnelly R. L and Cotton W R 1989 The Precipitation life cycle of Mesoscale Convective Complexes over the Central United States. *Mon. Wea. Rev* **117** 784 - 808
[3] Kain J S,Weiss S J, Baldwin M E, Carbin G W, Bright D, J J Levit, and J A. Hart 2005 Evaluating high-resolution configurations of the WRF model that are used to forecast severe convective weather: The 2005 SPC/NSSL Spring Experiment. *Preprints, 21st Conf. on Weather Analysis and Forecasting and 17th Conf. on Numerical Weather Prediction, Washington, DC Amer. Meteor. Soc., CD-ROM, 2A.5*
[4] Ducrocq, V, Tzanos, D, and Senesi S 1998 Diagnostic tools using a mesoscale NWP model for the early warning of convection. *Meteorlol. Appl, 5, 329-349.*
[5] Roca, R., T Fiolleau, and D Bouniol, 2017 A Simple Model of the Life Cycle of Mesoscale Convective Systems Cloud Shield in the Tropics. *J. Climate. doi:10.1175/JCLI-D-16-0556.1*
[6] Nakazono, A 2005 Status of the MTSAT programme. *APSDEU-6,* Seoul-Korea
[7] Bessho K *et al* 2016 An Introduction to Himawari-8/9-Japan's New-Generation Geostationary Meteorological Satellites *Japan J. Meteor Soc* 94 151
[8] Ismanto, H. 2013 Distribusi spasial dan temporal Mesoscale Convective Complex di Benua Maritim. *Buletin Megasains.*
[9] Virts K S and Houze R A Jr 2015 Variation of lightning and convective rain fraction in

mesoscale convective systems of the MJO. *J. Atmos. Sci* **72** 1932–1944 doi: 10.1175/JAS-D-14-0201.1

[10] Shimizu A 2015 Outline of RGB Composite Imagery. Meteorological Satellite Center, Japan Meteorology Agency (JMA)

[11] Tanaka Y 2009 SATAID-Powerful Tool for Satellite Analysis RSMC Tokyo-Typhoon Center, Japan Meteorology Agency (JMA)

[12] BoM-Melbourne Vlab CoE 2016 Himawari-8 RGB Training Available from:www.virtuallab.bom.gov.au/training/hw-8-training/introduction-resources-and-case-studies/ [cited 2017 6th June]

[13] Kerkmann J 2012 Effective Training and Use of RGB Satellite Products for Forecasters. *Eumetsat RGB Satellite Products Workshop 2012*

[14] Cotton W R, Lin M S, McAnell R. L and Tremback C J 1989 A composite model of mesoscale convective complexes. *Mon. Wea. Rev* **117** 765-783

[15] Indonesian National Board for Disaster Management. 2017. News. Available on https://www.bnpb.go.id/ [cited 2017 18th November]

IOP Conf. Series: Earth and Environmental Science **149** (2018) 012033 doi:10.1088/1755-1315/149/1/012033

Assessment of the impact of climate change and land cover change on landslide in Tana Toraja district

M Ardiansyah[1,2], **A Rafiuddin**[1], **S D Jadmiko**[1] **and R Boer**[1,3]
[1]The Center for Climate Risk and Opportunity Management in Southeast Asia and Pacific, Bogor Agricultural University, Kampus IPB Baranangsiang, Bogor 16143, Indonesia
[2]Department of Soil Science and Land Resource, Faculty of Agriculture, Bogor Agricultural University, Bogor, Indonesia
[3]Department of Geophysics and Meteorology, Faculty of Mathematics and Natural Sciences, Bogor Agricultural University, Bogor, Indonesia

Email: ardysaja@gmail.com

Abstract. Landslides are the frequent and widespread climate hazard in Indonesia that cause loss of human life and damage to property. Tana Toraja District is one of the regencies in Indonesia with the highest number of landslide events. Throughout the year 2016 there were at least 3 (three) landslide incidents that caused casualties and disconnection of transportation access. The study aimed to assess the impact of climate change and land cover change on landslide hazard in the Tana Toraja District. The study showed that in 2014 the number of landslide area with high risk was 53.3% and very high risk 14.4%, while in 2031 the high risk 56.4% and very high risk 12.7%. Thus, in high climate risk area, landslide adaptation and risk reduction strategies in the framework of climate change are necessary.

1. Introduction

Climate variability and extreme events have become a serious conversation over the last few decades. The IPCC's Fifth Assessment Report stated that warming of the global climate is clear, and since the 1950s, many of the observed changes are unprecedented over decades to millennia [1]. The period from 1983 to 2012 was very likely the warmest 30-year period of the last 1400 years in the Northern Hemisphere. The globally averaged land and ocean surface temperature data show a warming of 0.85 (0.65 to 1.06)°C over the period 1880 to 2012. Climate change are heavily influenced by human activity and potentially causing a serious impact [1]. The impact of which is the increased frequency and intensity of extreme weather. In the hilly areas and steep slopes it caused landslide [2-3]. During 2016 there were 599 incidents of landslides in Indonesia causing 180 deaths and missing people, 107 injured and 37,748 displaced people [4].

The occurrence of landslide disaster is most prevalent in Java Island, while outside Java Island it is recorded to be mostly occurred in Sulawesi Island, especially South Sulawesi. Tana Toraja District is one of the regencies in South Sulawesi with the highest number of landslide events. In the hilly and mountainous terrain where the Tana Toraja located, it created a large number of landslides, while the plains were affected by overflowing rivers banks caused by floods. Throughout 2016, there were 3 landslide incidents causing 1 death, and 316 displaced people. This is the most frequent occurrence of landslides in South Sulawesi Province [4]. In most cases the main trigger for landslides are heavy or prolonged rainfall. In the future the frequency and intensity of increased rainfall due to climate change will trigger more landslides. The landslide is also triggered by forest cover change. With better forest cover, proportion of rainfall intercepted by the canopy will be high and this will be evaporated back to the atmosphere, so less portion of the rainfall reaches soil surface. Thus if the forest cover of an upstream area decreases, more part of the rainfall will mostly be a direct surface flow and can cause landslides in the upstream area. Therefore, we predict an increase in the number of area in Tana Toraja District exposed to landslide risk in the future. The objectives of this study were to assess the impact of climate change and land cover change on landslide hazard.

2. Material and Method

2.1. Study area

The study was conducted in Tana Toraja District, South Sulawesi Province, Indonesia, one of the provinces during 2016 often experienced landslide [4]. The district is centred around latitude 2° - 3° N and longitude 119° - 120° E, and located at an altitude of 400-3,075 masl. Tana Toraja District comprises 47 sub-districts and 112 villages [5], where is adjacent to Toraja Utara District and West Sulawesi Province in the north, to Enrekang and Pinrang District in the south, to Luwu District in the east, and to West Sulawesi Province in the west as shown in figure 1.

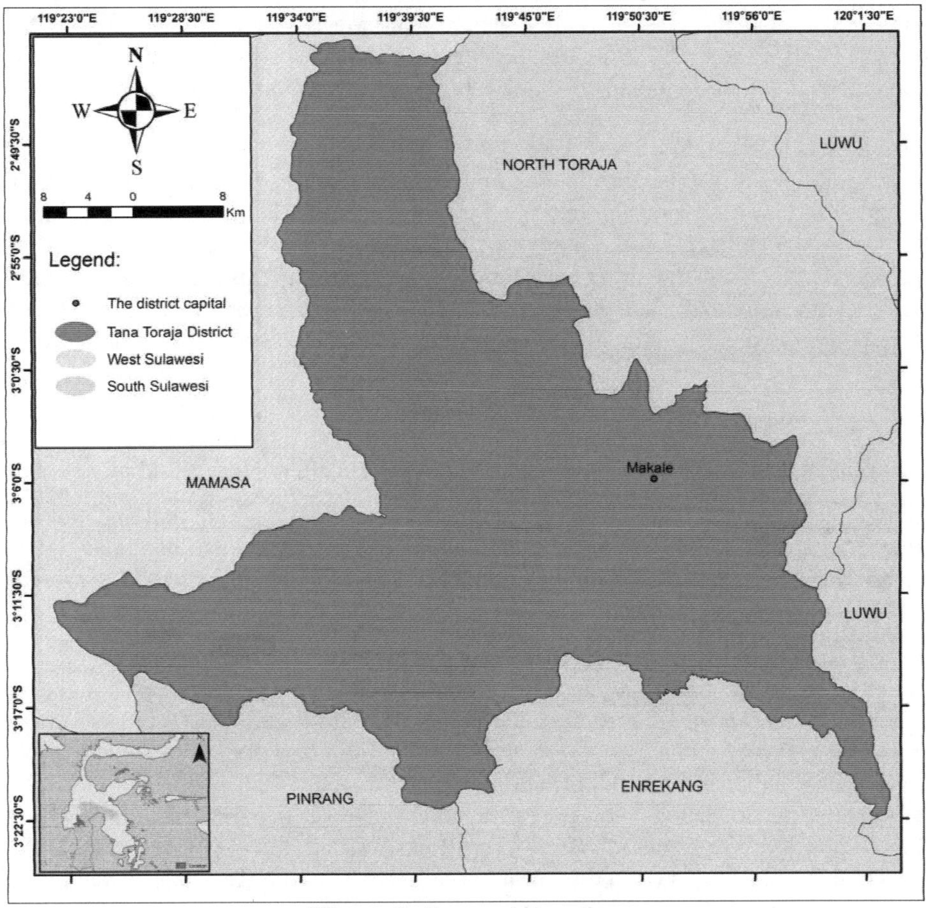

Figure 1. Research location.

2.2. Method

The method consists of 4 main stages: analysis of rainfall data, analysis of land cover and land cover change, analysis of current landslide, and analysis of landslide projection in 2031.

2.2.1. Analysis of rainfall data
The analysis of rainfall data comprises two stages: current and future rainfall analysis. Current rainfall analysis used historical rainfall data from Climate Hazards Group InfraRed Precipitation with Station (CHIRPS) output. Meanwhile, future rainfall analysis utilized rainfall projection 2021-2050 based on RegCM4.3 output with scenario RCP4.5 [6-8].

2.2.2. Analysis of land cover change
Land cover 2014 and projected land cover in 2031 was used to assess its impact to landslide. Landsat image of 2000 and 2014 was interpreted visually to generate land cover map for year 2000 and 2014. The results were validated using Google Earth. Land cover classification based on land classification prepared by Ministry of Environment and Forestry that are grouped into 8 classes for the study area: shrubs, forest, plantation, settlement, dryland agriculture, paddy, bare land, and water body.

Land cover projection until 2031 actually followed the spatial plans of Tana Toraja District. It is a result of a spatial land cover model. The modeling was carried out in 2 stages: the 2000 model and the 2014 model. The 2000 model was used to construct a land cover in 2014 and the 2014 model was a model used to projection land cover in 2031. The raster cell size analyzed was 100 m x100 m. The spatial model was built using CLUE-S software (Conversion of Land Use and its Effects at Small Regional Extent) [9,10]. The CLUE-S model incorporates 4 variables as a condition for analyzing land-use modeling, which includes demand for land, elasticity, location characteristic, and spatial policy [11].

2.2.3. Landslides analysis in 2014
The landslide model used in this study was a model developed by the Ministry of Agriculture [12], commonly used in Indonesia, especially in the agricultural sector. The key parameters used to generate landslide hazard were a slope, rainfall, land cover, lithology, soil texture, and elevation. The slope and altitude data were derived from the 30-meter shuttle radar topography mission (SRTM) data. Soil texture was taken from field sampling, and lithology data from the land system map. The weight of each parameter was determined using rational method [13] and calculated following equation [14]:

$$wij = \frac{n-rj+1}{\sum(n-rj+1)} \tag{1}$$

where: wij: normalized value, n: number of criteria (k = 1, 2,... n), rj: position of criteria sequence.

The rational method was done based on ordering the importance of parameters, in which all parameters used are given weight based on its influence, the greater the influence also the greater the weight. The parameter sequence and score of criteria based on the rational method is presented in table 1 [13].

Slope in the Tana Toraja District is steep until very steep and is the main factor triggering landslides, so this parameter ranked first with the highest weight. If the rainfall is high or prolonged, this became the next trigger factor for landslide prone area, the higher the rainfall it increase the potential of landslide hazard in the wavy and hilly areas, especially with slope of more than 15%. The bigger slope the greater the landslide potential caused by the slope stability was getting smaller.

The land cover, especially vegetation cover, such as forests reduce the impact of the rainfall, so an area with less vegetation cover caused rainwater will easily seep into soil layer making the soil layer becomes saturated and the landslide potential increased. Thus, land cover change that declining forest cover may increase the landslide potential.

IOP Conf. Series: Earth and Environmental Science **149** (2018) 012033 doi:10.1088/1755-1315/149/1/012033

Table 1. Sequence and score of parameter for landslide assessment.

No	Parameter	Score	No	Parameter	Score
1	Slope (%)	•	4	Soil texture	•
•	0 - 8	1	•	Sand	1
•	8 - 15	2	•	Silt	2
•	15 – 25	3		Silt loam	3
•	25 – 45	4	•	Clay loam	4
•	>45	5	•	Clay	5
2	Maximum monthly rainfall (mm/month)	•	5	Lithology	•
•	0 – 100	1	•	Alluvium recent riverine rocks	1
•	101 – 300	2	•	Sedimentary rocks	2
•	301 – 400	3	•	Breakthrough rocks	3
•	>400	4	•	Volcanic rocks	4
3	Land cover	•	6	Elevation	•
•	Water body	1	•	0 - 740	1
•	Forest	2	•	741 - 1280	2
•	Settlement	3	•	1281 – 1820	3
•	Shrub, Plantation	4	•	1821 - 2360	4
•	Bare land, Paddy, Dryland Agriculture	5	•	>2360	5

Lithology was the fourth parameter that influenced the occurrence of landslide, because the soil formed from weathering of rocks, and its material generally was eroded in the form of soil and rock. Furthermore, soil texture was the fifth parameter that triggered the occurrence of landslide. Soil texture is affected by the content of clay, silt, and sand and represent the type of soil that affect the landslide. The higher the clay content of a soil the more potential to form soil fractions into chunk, if it is saturated and located on a steep slope will cause landslides.

The last sequence was elevation parameters, because the higher of the altitude of a place the greater kinetic energy produced, so the potential for landslides will also be higher.

Based on the description above, the six hazard parameters were then sorted in order of importance and weight, to obtain the normalized weight, as presented in table 2.

Table 2. Weight of landslide hazard parameters for Tana Toraja District.

Parameter	Sequence of criteria (r_j)	Sequence of Weight (n - rj + 1)	Normalized Weight
Slope	1	6	28
Rainfall	2	5	24
Land cover	3	4	19
Lithology	4	3	14
Soil texture	5	2	10
Elevation	6	1	5

The landslide hazard is then estimated using following equation:

$$H = 28\,(S) + 24\,(Rf) + 19\,(Lc) + 14\,(Lt) + 10\,(Ts) + 5\,(e) \tag{2}$$

Where H: landslide hazard; (s) slope, (rf) rainfall; (lc) land cover; (lt) lithology; (ts) texture; (e) elevation.

2.2.4. Landslide analysis in 2031

The factors used to predict landslide hazard in Tana Toraja District were basically the same as the method used in landslide hazard analysis of 2014. Parameters for rainfall used projected rainfall for the period of 2021-2050, and land cover used land cover projection 2031, while slope, lithology, soil texture, and elevation remained the same as the current condition.

The results were further classified into 5 hazard classes, i.e. very low, low, medium, high, and very high hazard using following formula [14]:

$$\textit{Interval class} = (\textit{the highest score} - \textit{the lowest score})/\ \textit{number of classes}) \quad (3)$$

3. Result and Discussion

The rainfall conditions in Tana Toraja District between the historical rainfall and the projected rainfall of 2021-2050 increase in all areas including in the capital district of Makale and surrounding areas. Comparison of historical rainfall map and projection year period of 2021-2050 is shown in figure 2. The southern area of Tana Toraja has more rainfall in the future. Based on the trend analysis, the anomaly rainfall will increase both of RCP4.5 and RCP8.5 scenarios. The rainfall will increase between 7 to 9 percent in the 2021-2050 relative to baseline (figure 3). Probability of wet month in a year is shown in figure 4. The analysis indicated the number of wet month in the future has two conditions. The RCP4.5 shows the number of wet month will increase more than the historical condition, while the RCP8.5 indicated the frequency of wet month will increase about 5 percent than historical. The increase of rainfall in the future can increase the chances of landslide incident in Tana Toraja District.

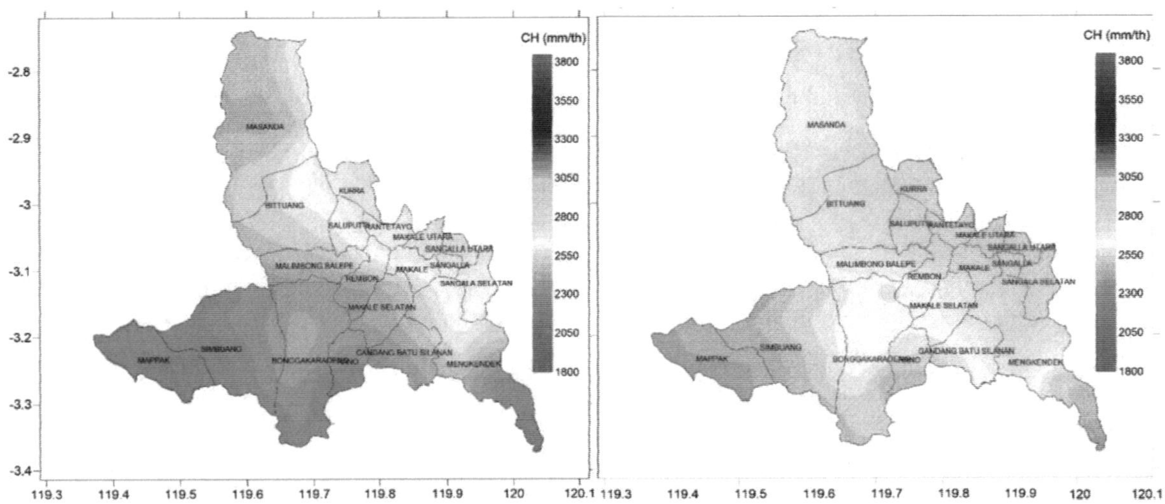

Figure 2. Maps of historical (left) and projected rainfall on 2021-2050 (right).

IOP Conf. Series: Earth and Environmental Science **149** (2018) 012033 doi:10.1088/1755-1315/149/1/012033

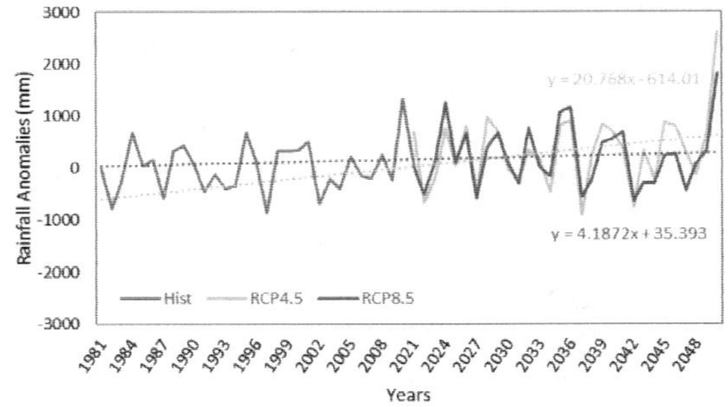

Figure 3. Trend of rainfall anomalies.

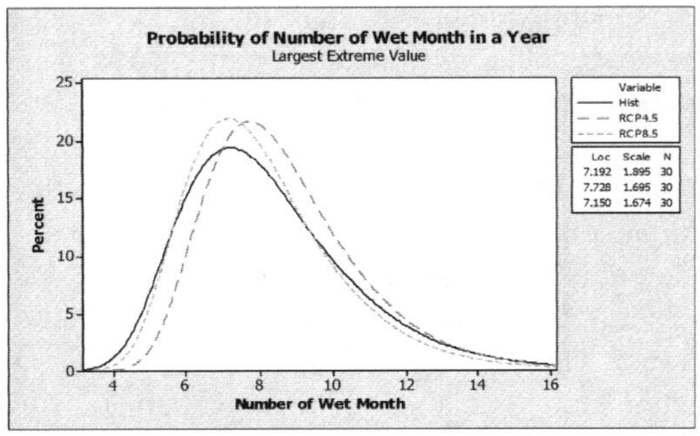

Figure 4. Probability of number of wet month in a year.

In 2014, land cover is dominated by shrub, forests, dryland agriculture, and bare land. Shrub was 61,640 hectares (ha) or 30.44% of the total area, forest was 43,859 ha, dryland agriculture was 32,350 ha (15.97%), and bare land was 28,911 ha (14.28%). Meanwhile, settlement and paddy was dominant land cover found in the plains, each covering only 988 ha (0.49%) and 17519 ha (8.65%). The land cover map of 2014 and projection of 2031 as shown in figure 5.

In 2031, land cover shows decreases in forest by 3.82%, shrubs by 13.49%, and plantations by 0.03%, while other land covers increase. Dryland agriculture is the most dominant land cover area of 48,421 ha (23.91%), then forests, bare land and shrubs are 17.84%, 17.70% and 16.95%, respectively. As depicted in figure 5, forest, shrubs and bare land located in hilly and steep area.

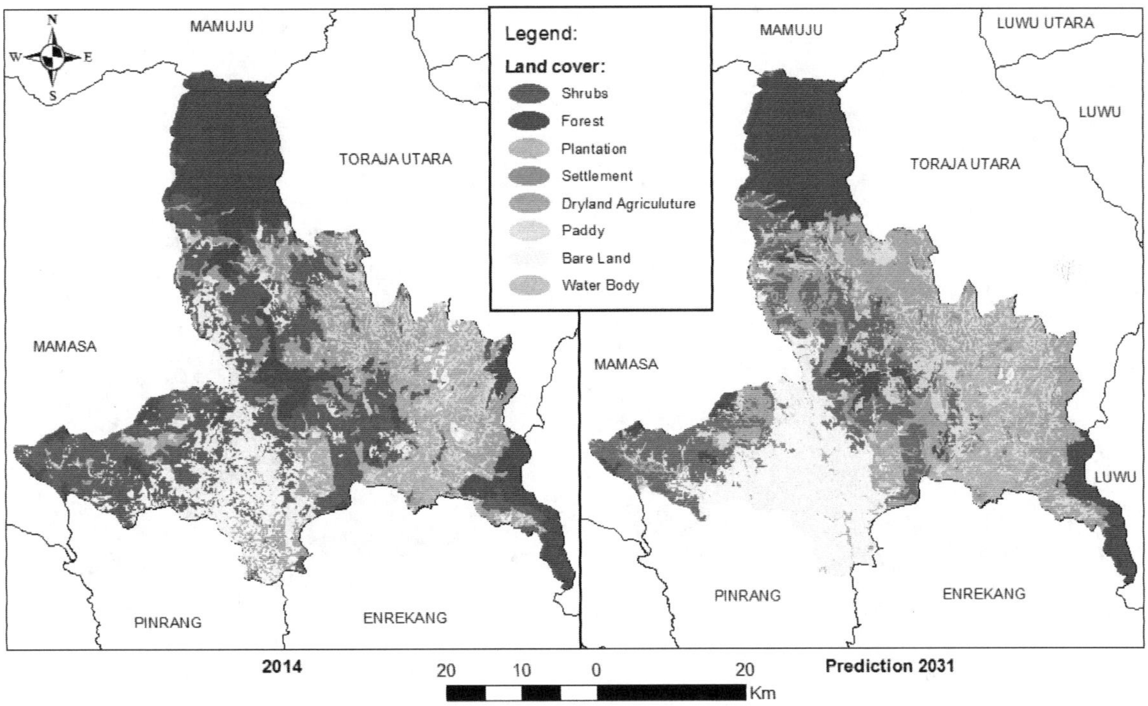

Figure 5. Maps of land cover 2014 (left) and projection in 2031 (right).

In the period from 2000 to 2014, the forest decreased by about 1.6% and it continued to decline in the period of 2014-2031 with a total loss of about 3.82%. In the area with better forest cover, the proportion of rainfall intercepted by the canopy will be high and this will be evaporated back to the atmosphere, so less portion of the rainfall reaches the soil surface. Various studies found that the canopy interception loss may account for 10-50% of season-long or annual rainfall. Thus, if the forest cover of an upstream area decreases, more part of the rainfall will mostly be a direct surface flow. As the result, it can cause landslides in the upstream area [15,16].

Increasing rainfall and changes in land cover in the steep and hilly mountainous area dominated by low vegetation cover such as bare land and dryland agriculture making Tana Toraja District more vulnerable to landslide hazard in the future. In 2014, the landslide hazard is dominated by high class of 107,938 ha (53.3%). Furthermore, the medium class is 53,995 ha (26.7%), very high 29,188 ha (14.4%), low and very low class of 10,910 ha (5.39%) and 481 ha (0.24%). In 2031, the high class of landslide hazard increases to 114,163 ha (56.4%), while the very high class slightly decreases to 25,719 ha (12.7%) as summarized in table 3 and depicted in figure 6.

Table 3. Area of landslide hazard in Tana Toraja District in 2014 and 2031

Level of Hazard	2014		2031	
	Area			
	(ha)	(%)	(ha)	(%)
Very low	481	0.24	139	0.07
Low	10,910	5.39	5,021	2.48
Medium	53,995	26.66	57,470	28.38
High	107,938	53.30	114,163	56.37
Very High	29,188	14.41	25,719	12.70
Total	**202,512**	**100.00**	**202,512**	**100.00**

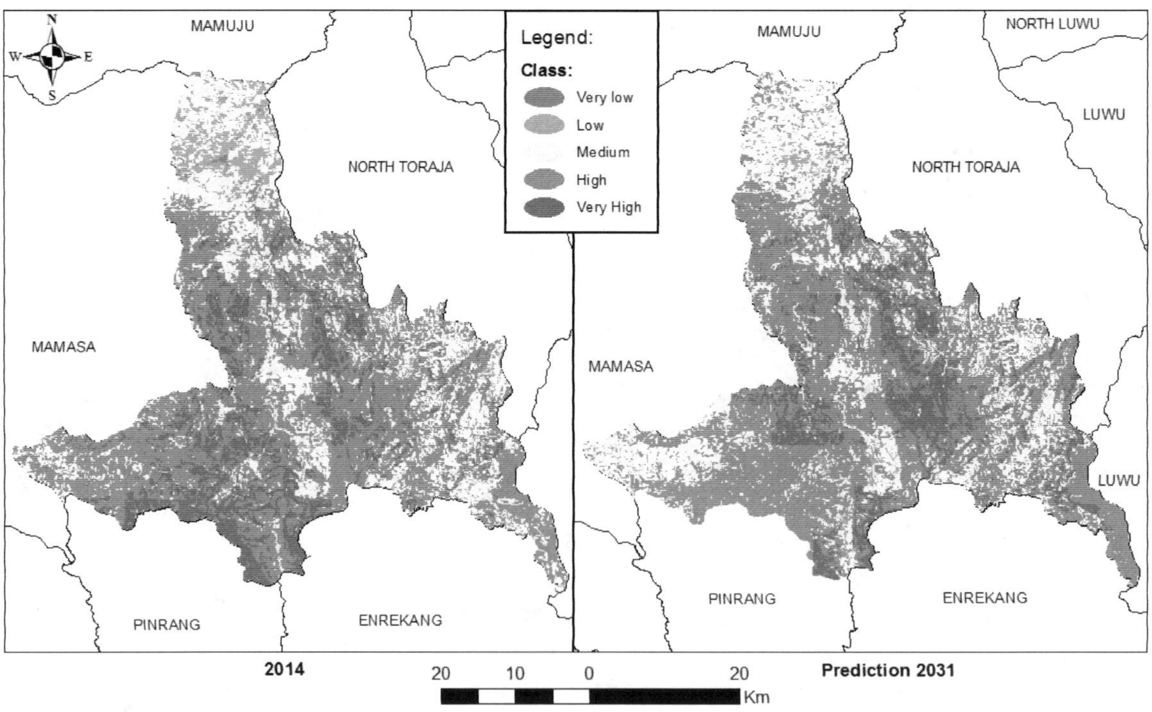

Figure 6. Maps of landslide hazard 2014 (left) and prediction 2031 (right).

The increase of landslide hazard in 2031 is influenced by increased rainfall and land use change. The projected high rainfall in the period of 2021-2050 compared to historical precipitation led to an increase in the amount of rainfall in the north and east. A high intensity of rainfall can trigger a landslide [17]. In addition, the condition of land use change in the Tana Toraja District on the steep slopes to very steep also trigger the occurrence of a landslide. High rainfall intensity and human activities on the use of dryland agriculture can trigger landslides. In recent decades population growth and climate change have triggered landslides [18].

Spatially, the landslide hazard under current condition and projection is dominantly in high and very high class that spread in hilly and mountainous area, where covered by low vegetation cover such as shrubs, bare land, and dryland agriculture. This happens because, in the high rainfall, those land use types are a type of land that is not strong enough to bind the soil, making the soil become soft and saturated water so these is easy to occur landslide [19]. Meanwhile the plains are dominated by very low to medium landslide hazard, where land cover is rice fields, settlements, and water body.

4. Conclusions

An increase in landslide hazard is generally expected to occur in Tana Toraja District for the future, especially in the rainy season. The study shows that in the 2014 the number of areas of high-level landslide hazard is 53.3% and very high class 14.4%, while in the 2031 high level 56.4% and very high level 12.7%. Thus, in high hazard areas, landslide adaptation and risk reduction strategies in the context of climate change are required.

References

[1] IPCC 2014 *Climate Change 2014, Synthesis Report Contribution of Working Groups I, II and III to the Fifth Assessment Report of the Intergovernmental Panel on Climate* eds Core Writing Team, R.K. Pachauri and L.A. Meyer (Geneva: Switzerland) p 151

[2] Gariano S L and Guzzetti F 2016 *J. Earth-Sci. Rev.* **162** 227–252

[3] Reder A, Rianna G, Mercogliano P and Pagano L 2016 *Proc. Earth and Planetary Sci.* **16** 166–76

[4] National Disaster Management Authority 2017 *Disaster information data of Indonesia* Accesed at http://bnpb.cloud/bnpb/xdibi/xdibi_list

[5] BPS 2016 *Center Statistic of Indonesia, Tana Toraja in Figure* 2016 (Tana Toraja: BPS)

[6] Huang Y F, Ang J T, Tiong Y J, Mirzaei M and Amin M Z M 2016 *Proc. Engineering* **154** 710–17

[7] Kwon T S and Lee C M 2015 *J. Asia-Pacific Biodiversity* **8** 49–65

[8] Lee C M, Kwon T S, Kim S S, Park G E and Lim J H 2016 *J. Asia-Pacific Biodiversity* **9** 116–37

[9] Zhang P, Liu Y, Pan Y and Yu Z 2011 *J. Math. and Comp. Model.* **58** 588–595

[10] Luo G, Yin C, Chen X, Xu W and Lu L 2010 *J. Ecological Complexity* **7** 198–207

[11] Verburg P 2010 *The CLUE Modelling Framework Course Material Institute for Environmental Studies, Course Material* (Amsterdam: Amsterdam University) p 53 Accesed at http://www.ivm.vu.nl/en/Images/Exercises_tcm234-284019.pdf

[12] PUSLITANAK 2004 *Activity report of assessment of drought, flood and landslide disaster potential in multi watershed area, West Java with Geographic Information System, Indonesian Agency for Agricultural Research and Development Ministry of Agriculture* (Bogor: Center of Soil and Agroclimate Research)

[13] Silviani R V 2013 *Analysis of landslide hazard and risk in the upstream of Ciliwung Watershed and its relationship with spatial planning, Theses, Graduate School* (Bogor: Bogor Agricultural University)

[14] Ikqra 2012 *Geomorphological study of ternate island and landslide risk assessments Theses, Graduate School* (Bogor: Bogor Agricultural University)

[15] Ollauri A G and Mickovski S B 2017 *J. of Hydrol.* **549** 374–387

[16] Chen C W, Chen H and Oguchi T 2016 *J. of Geomorph.* **273** 1–13

IOP Conf. Series: Earth and Environmental Science **149** (2018) 012033 doi:10.1088/1755-1315/149/1/012033

[17] Hasnawir 2012 *J of Forest. Research Wallacea* **1** 62–73
[18] Do X K, Kim M, Nguyen H P T and Jung K 2016 *Proc. Engineering* **154** 990–994
[19] Hidayah A, Paharuddin and Massinai M A 2017 *J. Geocelebes* **1** 1–4

Spatial model of land use change related to sediment yield (case study: Cipeles and Cilutung watershed, West Java)

D W Wulandari, E Kusratmoko and T L Indra
Magister Program Departement of Geography, Faculty of Mathematics and Science, Universitas Indonesia, Depok 16424

Email: dianwahyu21@gmail.com

Abstract. Land use changes (LUC) as a result of increasing human need for space are likely to destroy the hydrological function of the watershed, increase land degradation, stimulate erosion and drive the process of sedimentation. This study aimed to predict LUC during the period 1990 to 2030 in relation to sediment yield in Cilutung and Cipeles Watershed, West Java. LUC were simulated following the model of Cellular Automata-Marcov Chain, whereas land use composition in 2030 was predicted using Land Change Modeler on Idrisi Selva Software. Elevation, slope, distance from road, distance from river, and distance from settlement were selected as driving factors for LUC in this study. Erosion and sediment yield were predicted using WATEM/SEDEM model based on land use, rainfall, soil texture and topography. The results showed that the areas of forest and shrub have slightly declined up to 5% during the period 1990 to 2016, generally being converted into rice fields, settlements, non-irrigated fields and plantations. In addition, rice fields, settlements, and plantations were expected to substantially increase up to 50% in 2030. Furthermore, the study also revealed that erosion and sediment yield tend to increase every year. This is likely associated with LUC occurring in Cipeles and Cilutung Watershed.

1. Introduction

Land use (LU) is defined as a series of people's activities on a land, the purpose of these activities are to get product and/or benefit through the utility of natural resources of the land. Land use can be reflected human interactions with environment on earth [1] and the social-economy characteristic of people in a certain geographical space [2]. The increase of quantity of inhabitant and the economy demand on an area of earth make the change of rapid land use as one of the most environmental issue globally [3]. The land use change (LUC) is a complex interaction between structural factor and human behavior with technology, demand, and social interaction [4]. LUC in a watershed, especially showed by the decrease of forest area and the increase of rice fields, at one side, and variability of rain intensity on the other, these cause changes in hydrology balance in the watershed system that triggered land level increasing and soil erosion [5]. The results are the decreasing of land fertility, the extent of critical area, and degraded land in the watershed [6,7]. Spatially, the patterns of LUC on earth are variable and the dynamic in geographic scale is very complex. These also influence the natural process of atmosphere, lithosphere, hydrosphere, and biosphere difference in intensity between geographical spaces [3].

Spatial modeling the LUC on a watershed will assist the understanding of complicated system and produce valuable information about configuration of LU in the future [1]. These information definitely very important in watershed management, especially to land resource management. For examples cases in Cipeles and Cilutung Watershed, West Java, we analyze the LUC and simulate the change in the future with integration of Markov-Cellular Automata method. Then, we integrate with WATEM-SEDEM model (Water and Tillage Erosion Model and Sediment Delivery Model) to predict sediment yield and its spatial distribution. Markov-Celluar Automata method has been used widely as model and to predict LUC in numerous room scale [8-10].

Cipeles and Cilutung Watershed are part of subsystem of Cimanuk Watershed. Since Jatigede Dam operated on August 2015, Cipeles and Cilutung River have important role to maintain balance of sediment supply to the Java sea. Base on identification of BPDAS Cimanuk- Citanduy, in 2003 critical land in Cimanuk Watershed was dominated at outside of forest area about 132,665 hectares. This high level of critical area will trigger quality degradation of watershed environment. This will lead to increasing of soil erosion and sedimentation in the river, more over in the lake, in the dam and irrigation channels. In this study, we predict the LUC in 2030, which is the same time with the end of Rencana Tata Ruang Wilayah (RTRW) of Regencies in the study area.

2. Method

2.1. Study Area

The research location is situated in Cipeles and Cilutung watershed. Both are sub system of Cimanuk watershed. Geographically Cipeles and Cilutung lies between 6°38'09.5" – 7°04'04.6" South and 107°45'44.7" – 108°24'17.4" East. Administratively, Cipeles watershed is in Sumedang Regency, and Cilutung watershed is in Majalengka Regency.

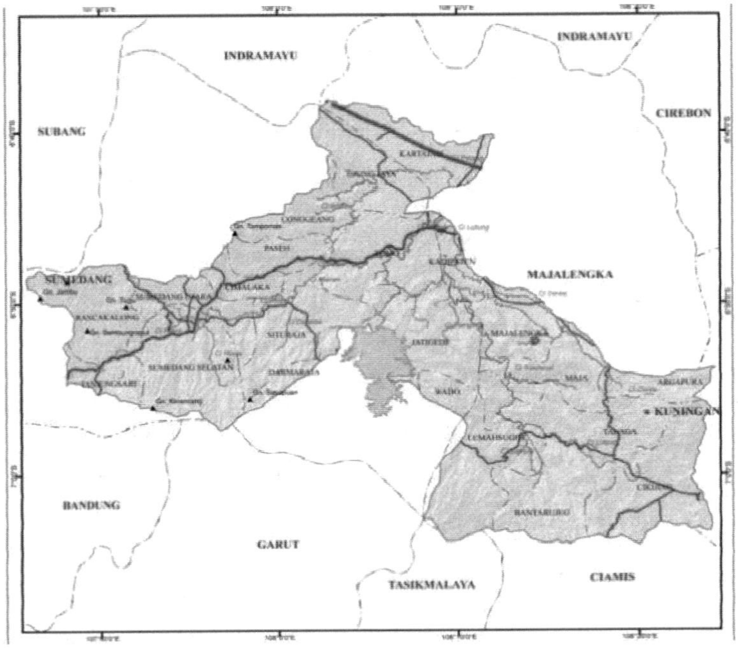

Figure 1. Map of the study areas (Cipeles and Cilutung Watershed).

2.2. Data and Materials

- Landsat images of path/row 121/065; Landsat 5 TM for the year 1990 and 2010, Landsat 7 ETM for the year 2000, and Landsat 8 OLI for the year 2016. The satellite acquisition date for the images was July to October.
- Digital Elevation Models Shuttle Radar Topography Mission (DEM SRTM).
- Land use of Sumedang and Majalengka Regency scale 1: 50,000 (year 1990, 2000, 2010, and 2016), obtained from Ministry of Agricultural dan Spatial.
- Annual Rainfall Data obtained from Meteorology, Geophysics, and Climatology Agency (BMKG)
- Soil Texture of West Java Province obtained from Puslitanak.
- Software: Arc GIS. 10.1, ENVI 5.1, Idrisi Selva, and WATEM/SEDEM version 2004.

2.3. Data Processing and Analysis

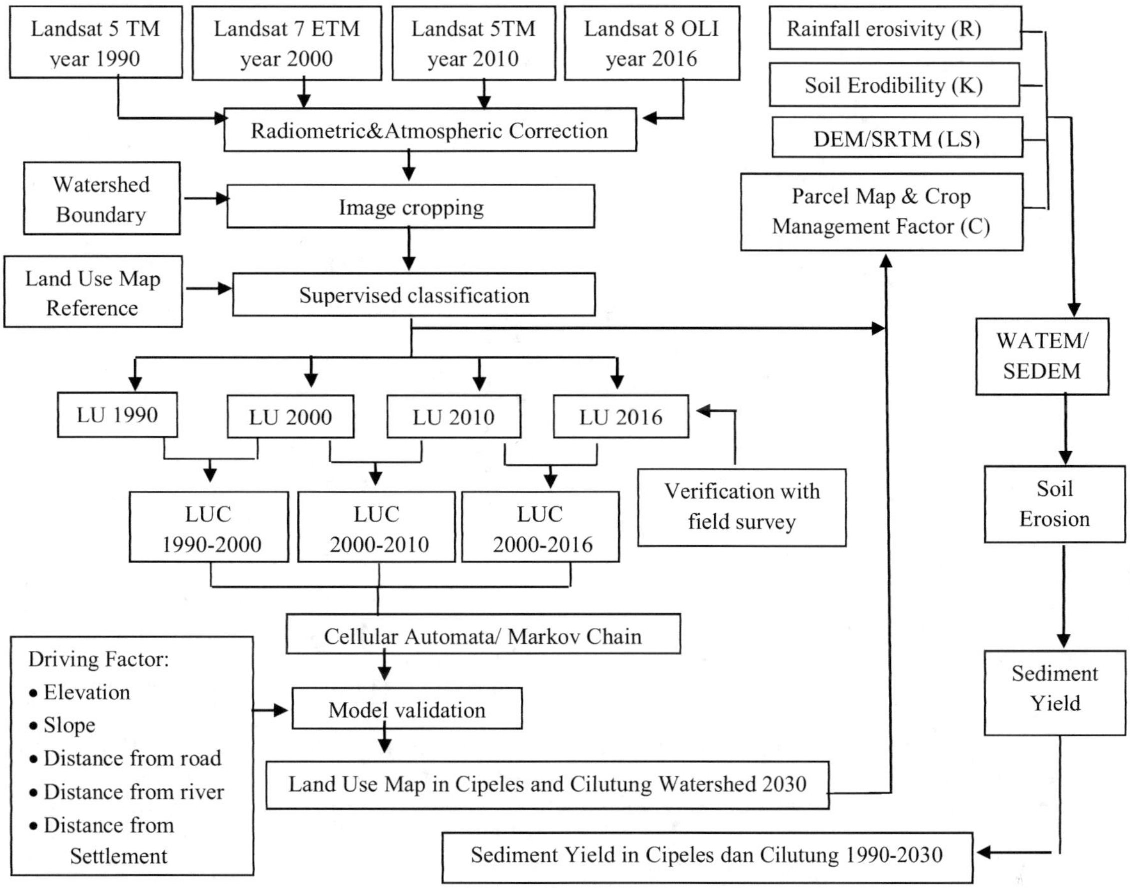

Figure 2. Landuse classification and sedimet yield analysis process.

LUC Analysis was done with menu land change modeler (LCM) on software Idrisi 17 Selva. The prediction of LU in 2030 was acquired by observing the LUC on year 2000 and 2016 as a reference years. Potential Transition Accuracy test was done to count how much influence of each driving factor to LUC. In this study driving factors were consider as the factors that influence the LUC which are consist of: the slope, elevation, distance from road, distance from river, and distance from settlement [8,9]. The prediction of LUC was done using Business As Usual scenario, which means this model calculate without obstructive and supportive factors that influence the LUC.

Analysis the effect of LUC on sediment yield was using WATEM-SEDEM model version 2004. The result of this modeling was a spatial imaging of soil erosion that can be used to detect part of watershed which the most endangered by soil erosion. This model consists of three stages which are: calculation of soil erosion with RUSLE equation, calculation the capacity of sediment transportation, and sediment flow to the river [11]. Software WATEM/SEDEM version 2004 was built based on RUSLE equation, which is $E = R \times LS_2D \times K \times C \times P$, where E represent annual soil erosion; R for rainfall erosivity; K for soil erodibility; C for crop management factors; P for erosion control practice; and LS_2D for two dimensional topography factor. The results are value and the spread of erosion, and sediment deposition. All input layers were generated in Idrisi Selva with a resolution of 20x20m. The resolution of 20x20m

IOP Conf. Series: Earth and Environmental Science **149** (2018) 012034 doi:10.1088/1755-1315/149/1/012034

was applied because the WATEM/SEDEM model was calibrated with resolution 20x20m. The input layer consisted of rainfall erosivity (R), soil erodibility (K), crop cover (C), and parcel map. In WATEM/SEDEM model, P factor was considered to be 1, because the conservation practices are mostly in small scales that cannot be detected by Landsat satellite images [12].

3. Result and Discussion

3.1. The Land Use Change in Cipeles and Cilutung watershed between 1990 to 2016

The land use in Cipeles and Cilutung watershed classified as water body, forest, plantation, rice fields, settlements, shrub, and non-irrigated fields. Each class of land use will change dynamically in extent, more or less in size. This land use dynamics were effects of human activities in the watershed. In detail, the extent of land use in 1990, 2000, 2010, and 2016 are reveals in Table 1.

Tabel 1. Land Use Change between 1990 to 2016.

No	Land Use	1990		2000		2010		2016	
		Hectares	%	Hectares	%	Hectares	%	Hectares	%
1	Water body	1,226	0.88	1,226	0.88	1,226	0.88	2,487	1.79
2	Forest	49,019	35.31	42,303	30.48	35,834	25.82	28,111	20.25
3	Plantation	14,422	10.39	14,578	10.50	18,825	13.56	18,235	13.14
4	Settelments	4,243	3.06	7,447	5.36	9,970	7.18	12,625	9.10
5	Rice Fields	28,284	20.38	42,682	30.75	43,820	31.57	49,585	35.72
6	Shurb	32,625	23.50	19,909	14.34	17,505	12.61	13,547	9.76
7	Non-irrigated fields	8,990	6.48	10,665	7.68	11,628	8.38	14,219	10.24
	Total area	138,809	100.00	138,809	100.00	138,809	100.00	138,809	100.00

In 1990, forest had dominated in Cipeles and Cilutung watershed with area of 49,019 Ha or 35.31% of the watershed. But in 2000, 2010 and 2016, the forest areas were reduced into 5% per 10 years. Shurb area was significantly reduced. In 26 years proportion of Shurb in Cipeles and Cilutung watershed had decreased from 23.50 % in 1990 became 9.76 % in 2016. The increasing of settlement area in 26 years (1990 to 2016) was 8,382 Ha. The water bodies (river, lake) had the area extent relatively the same from 1990 to 2010, which was 1.226 Ha or 0.88 % of the extent of Cipeles dan Cilutung watershed. But, in 2016 the extent was increase into 2,487 Ha or 1.79 % of watershed. This increasing was because of Jatigede Dam that had been operated since August 2015.

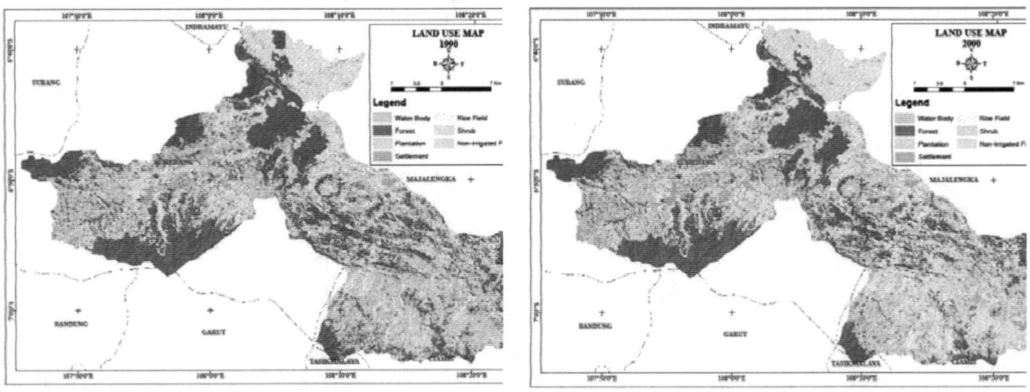

(a) (b)

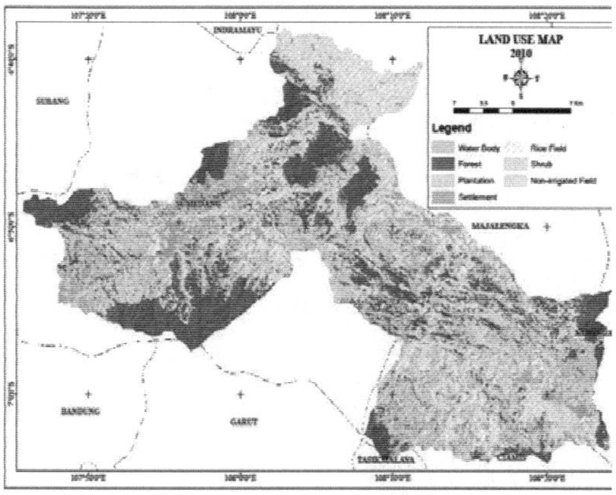

(c)

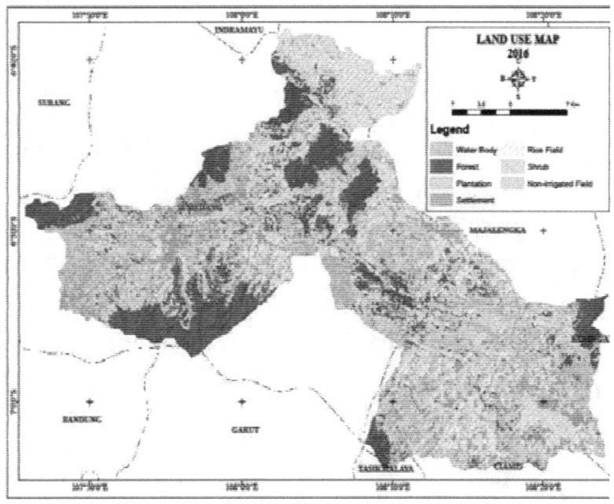

(d)

Figure 3. Land Use Change in Cipeles and Cilutung Watershed Map (a) 1990 (b) 2000 (c) 2010 (d) 2016

3.2. Factors that Affect Land Use Change

The driving factors of LUC that used in this study were the slope, elevation, distance from the road, distance from the river, and distance from the settlement. The level of strength of each driving factor to each utility was measured by Cramer's value. The higher of the Cramer's value, means the higher the effect of driving factors on LUC. Elevation had the highest Cramer's value of 0.246 (Figure 3). This showed that elevation factor had affect more to the chance of land use on Cipeles and Cilutung watershed. Commonly, the elevation of less than 500m had more likely to LUC. The second higher value (0.233) is the distance from the settlement. This showed that the closer to the settlement, the higher the possibilities of LUC. The change commonly were became settlement, rice field, and non-irrigated field. And the last one is the slope of mountain with Cramer's value of 0.216. The slope of mountain had become the border of LUC. The areas which were too steep had become the limit of development area.

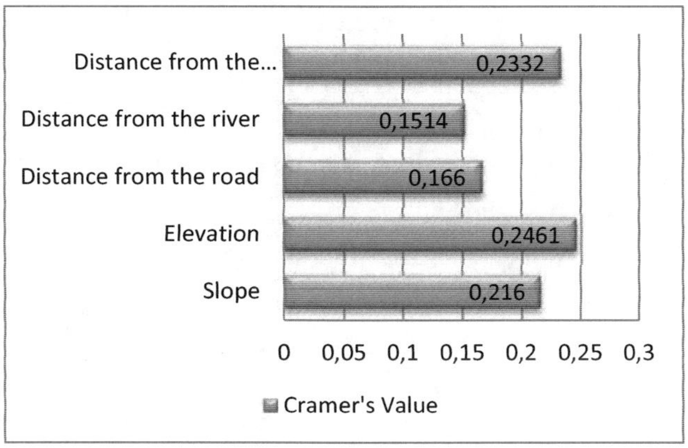

Figure 4. Cramer's Value for each Driving Factors

3.3. Prediction of Land Use in 2030

Kappa Value as results of cross tabulation between actual LU of 2016 with simulated LU of 2016 was 0.7256. This value reveals that the model that had been used had good synchronization. In other words, model of LUC that already arranged can be used to predict LUC in 2030. The result of LU in 2030 is shown in Figure 4. The use of water body, forest, shrub, and non-irrigated field were predicted will change into other form, this will decrease the extent in 2030. Compared with 2016, the extent of forest and shrub decrease each 108 Ha (29% of watershed) and 5,081 Ha (38%). Meanwhile, rice fields, settlements, and plantation were predicted will increasing in extent. The plantation increase into 4,446 Ha or increased proportionally into 24 % of the extent of the watershed in 2016. The settlement was predicted will be the most extensive change of 6,292 Ha or increase the proportion 50% of the extent of settlement in 2016.

Tabel 2. Land Use change between 2016 and 2030.

No.	Land Use	2016	2030	Change	
		Hectares	Hectares	Hectares	%
1	Water body	2,487	2,440	-47	-2%
2	Forest	28,111	20,003	-8,108	-29%
3	Plantation	18,235	22,681	4,446	24%
4	Settelments	12,625	18,917	6,292	50%
5	Rice Fields	49,585	53,419	3,834	8%
6	Shurb	13,547	8,466	-5,081	-38%
7	Non-irrigated fields	14,219	12,883	-1,336	-9%
	Total area	138,809	138,809		

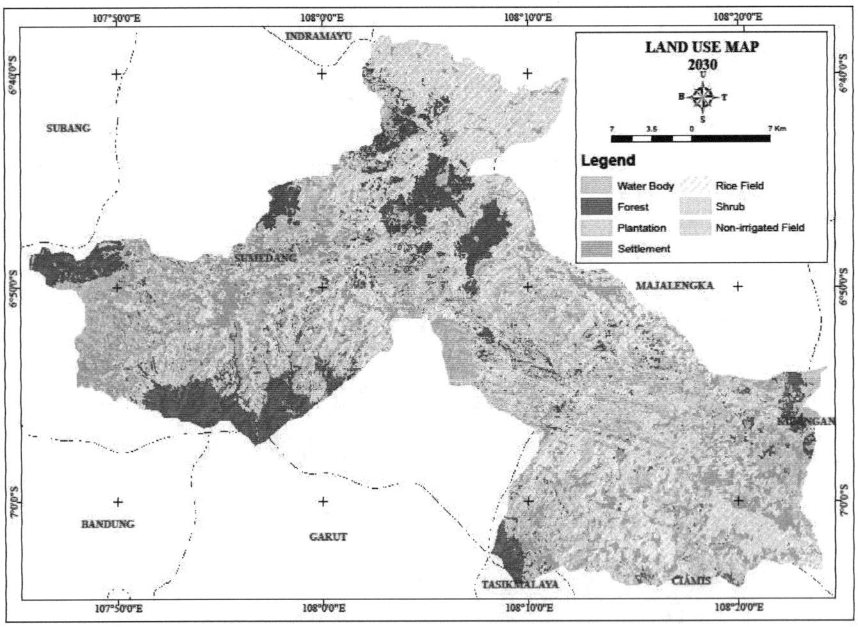

Figure 5. Land Use Change in Cipeles and Cilutung Watershed Map year 2030.

3.4. Model of Soil Erosion and the Sediment yield

Table 3 shown the summary of WATEM-SEDEM model results. Along with prediction of LUC, total of sediment yield in study area in 2016 was 10.51×10^6 tons, increased almost 24.38% from the LUC in 2010. The modeling results show that in 2030 total production of sediment yield predicted will be 13.1×10^6 ton, increase about 2.6×10^6 ton from year 2016. Most of sediment yield (75%) was deposit in the watershed and only 12% drifted by Cipeles and Cilutung River. The low value of sediment delivery ratio (SDR) affected by geomorphology of watershed [14]. Along with the increase of sediment production, the sediment yield that drifted by Cipeles and Cilutung river were also increasing. In 1990 total of drifted sediment was 454,269 ton and increased into 1,225,726 tons in 2016 and predicted will be around 1,542,140 tons in 2030.

Tabel 3. Summary of the WATEM/SEDEM model results.

WATEM/SEDEM output	1990 (ton/year)	2000 (ton/year)	2010 (ton/year)	2016 (ton/year)	2030 (ton/year)
Total sediment production	4,832,039	6,925,720	8,450,038	10,510,259	13,095,628
Total sediment deposition	3,705,019	5,092,698	6,317,254	7,854,479	9,818,185
Total sediment export	1,127,020	1,833,022	2,132,784	2,655,780	3,277,443
Total River export	454,269	820,642	1,007,341	1,225,726	1,542,140

3.5. Spatial Distribution of Soil Erosion and Sediment Deposition

The percentage of deposition in Cipeles and Cilutung watershed tends to increased (Table 4). Area with level of deposition 10 tons/Ha per year (classified as high – severe) has increased significantly. In 1990 the proportion was only 11% of the watershed, increased into 18% in 2016 and predicted will be 20% of the watershed in 2030. Spatially, the area with deposition category high and severe found in non-irrigated fields, plantation, and rice fields. This is because the management of non-irrigated fields, plantation, and rice field causes the capacity of soil particle transportation had become higher.

Tabel 4. The deposition classification.

Deposition rate (ton/ha/year)	Deposition classes	Coverage of the deposition affected area (%)				
		1990	2000	2010	2016	2030
0 s/d 2	Very Slight	1.84	2.36	2.53	2.70	2.84
2 s/d 5	Slight	0.06	0.09	0.11	0.14	0.19
5 s/d 10	Moderate	3.79	5.01	5.38	5.76	6.06
10 s/d 50	High	8.24	11.12	12.17	13.54	14.57
50 s/d 100	Severe	2.63	3.31	3.75	4.47	5.10
100 s/d 500	Very severe	0.06	0.08	0.11	0.13	0.19

Based on the erosion rate, study area dominated with the slight erosion class (less than 2 tons/Ha per year). However, along with LUC, the slight erosion class will getting smaller. In 1990 the proportion was 66% of the watershed, but in 2016 became 46% and predicted will be 41% in 2030. The decreasing of slight erosion class was followed by the increasing of area with level of erosion more than 10 tons/Ha per year (classified as high to severe). In 1990, it was only 9.6% of the watershed and will increasing into 17.3% in 2030. The more extensive erosion area with high to severe classification was due to the decreasing of forest and shrub area, converted into non-irrigated field.

Tabel 5. Soil erosion classification.

Erosion rate (ton/Ha/year)	Erosion classes	Coverage of the erosion affected area (%)				
		1990	2000	2010	2016	2030
-100 s/d -500	Very severe	0.48	0.60	0.70	0.89	1.11
-50 s/d -100	Severe	1.81	2.29	2.58	3.10	3.47
-10 s/d -50	High	9.63	12.90	14.29	15.82	17.27
-5 s/d -10	Moderate	5.03	6.55	7.00	7.42	7.77
-2 s/d -5	Slight	0.05	0.08	0.10	0.13	0.18
0 s/d -2	Very Slight	66.36	55.60	51.28	45.91	41.27

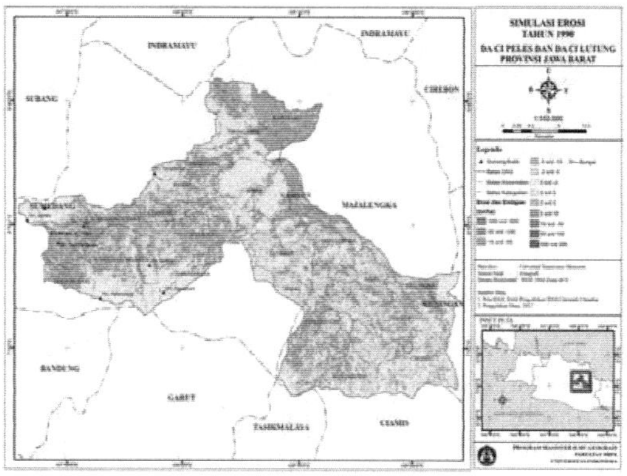

(a)

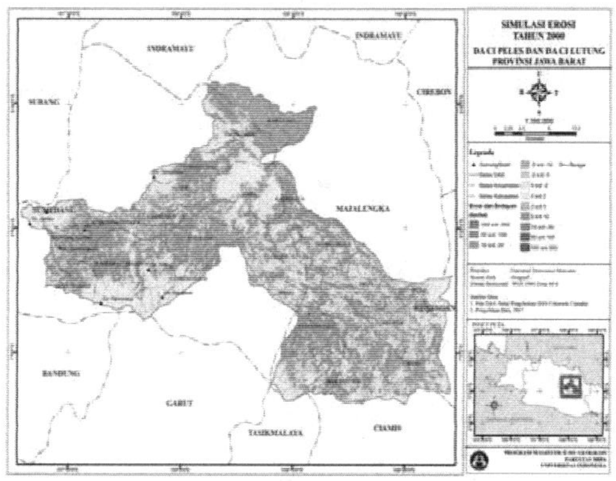

(b)

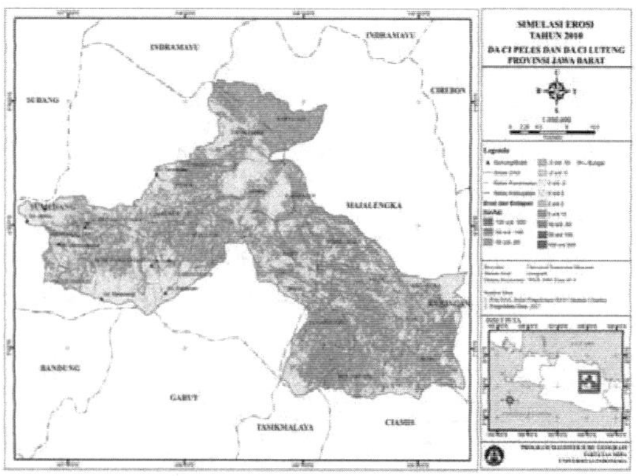

(c)

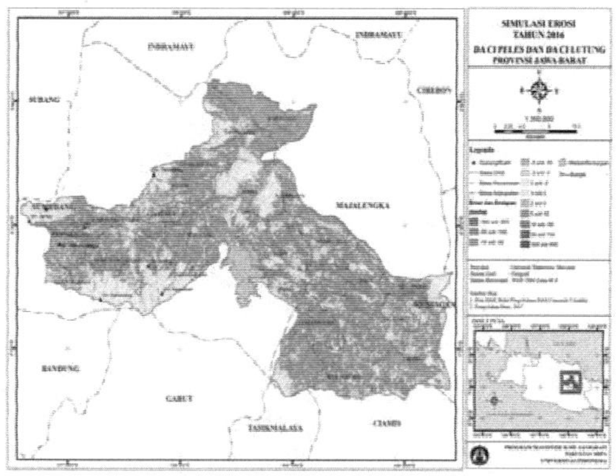

(d)

IOP Conf. Series: Earth and Environmental Science **149** (2018) 012034 doi:10.1088/1755-1315/149/1/012034

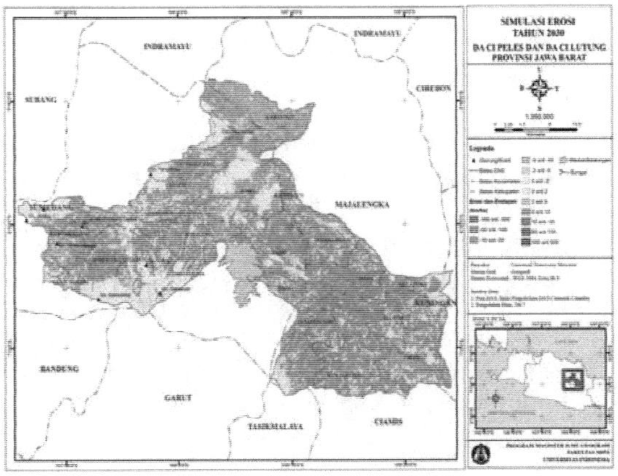

(e)

Figure 6. Erosion/deposition in Cipeles and Cilutung Watershed Map (a) 1990 (b) 2000 (c) 2010 (d) 2016 (e) 2030.

4. Conclusions

The land use change model using integration of Markov-Cellular Automata method with business as usual scenario, had been done successfully in Cipeles and Cilutung watershed. The LUC period from 1990 to 2016 showed that there were decreasing in forest and shrub area up to 5% every 10 years. Its commonly change into rice field, settlement, non-irrigated field, and plantation. These were caused by the increasing of human demand of field and settlement. The prediction of soil erosion and sediment yield in Cipeles and Cilutung watershed tend to increase every year due to the LUC. Spatial distribution showed that area with high erosion and sediment yield were in forest and shrub area which changes into non-irrigated field because the non-irrigated field is the source of soil erosion.

Reference

[1] Koomen E, Rietveld P, de Nijs T 2008 Modelling land-use change for spatial planning support *Ann Reg Sci* **42** 1–10

[2] Sandy I Made 1977 *Penggunaan lahan di Indonesia. Direktorat Tata Guna Tanah* Jakarta: Departemen dalam Negeri 115

[3] Lambin E F, Geist H J, Lepers E 2003 Dynamics of landuse and land-cover change in tropical regions *Annual Review of Environment and Resources* **28** 205–241

[4] Verburg P H, van Eck J R., de Hijs T C, Dijst M J, Schot P 2004 Determination of land use change patterns in the Netherlands *J. Environ.* **31** 125–150

[5] Alwi L O, Sinukaban N, Solahuddin S, Pawitan H 2011 Kajian dampak dinamika penggunaan lahan terhadap erosi dan kondisi hidrologi DAS Wanggu *J Hidrolitan.* **2**(2) 74-86.

[6] Kusratmoko E, Dayanti S T, Supriatna 2017 The critical land in Komering watershed as a result of land use changes from 2000-2016 period *IOP Conf. Series* **54** 012020

[7] Salsabilla A, Kusratmoko E 2017 Assessment of Soil Erosion Risk in Komering Watershed, South Sumatera, Using SWAT Model. *AIP Conference Proceedings* 1862 030192

[8] Kusratmoko E, Albertus S D Y, Supriatna 2017 Modelling land use/cover changes with markov-cellular automata in Komering Watershed, South Sumatera. *IOP Conf. Series: Earth and Environmental Science* **54**

[9] Marko K, Zulkarnain F, Kusratmoko E 2016 Coupling of Markov chains and cellular automata spatial models to predict land cover changes (case study: upper Ci Leungsi catchment area)

IOP Conf. Series: Earth and Environmental Science **47**
[10] Marwa W A H, Paul E G, Jeffrey A H, Boshra B S 2015 Land use/land cover change detection and prediction in the north-western coastal desert of Egypt using Markov-CA *Applied Geography* **63** 101–112
[11] Bezak, Rusjan N S, Petan S, Sodnik J, Mikos M 2015 Estimation of soil loss by the WATEM/SEDEM model using an automatic parameter estimation procedure *Environ Earth Sci.* **74** 5245–5261.
[12] Syahli F 2015 *The Impact of Land Use Change on Soil Erosion in Serayu Watershed, Case Study Merawu Watershed, Banjarnegara, Central Java* Yogyakarta: Thesis Graduate School Gajah Mada University Faculty of Geo-Information Science and Earth Observation University of Twente
[13] Alatorre L C, Beguería S, García-Ruiz J M 2010 Regional scale modeling of hillslope sediment delivery: A case study in the Barasona Reservoir watershed (Spain) using WATEM/SEDEM. *Journal of Hydrology* **391** 109–123

IOP Conf. Series: Earth and Environmental Science **149** (2018) 012035 doi:10.1088/1755-1315/149/1/012035

Identification and delineation of areas flood hazard using high accuracy of DEM data

B Riadi, B Barus, Widiatmaka, M J P Yanuar, B Pramudya
Bogor Agricultural University, Jalan Raya Baranangsiang Bogor

E-mail : bambang.riadi@big.go.id

Abstract. Flood incidents that often occur in Karawang regency need to be mitigated. These expectations exist on technologies that can predict, anticipate and reduce disaster risks. Flood modeling techniques using Digital Elevation Model (DEM) data can be applied in mitigation activities. High accuracy DEM data used in modeling, will result in better flooding flood models. The result of high accuracy DEM data processing will yield information about surface morphology which can be used to identify indication of flood hazard area. The purpose of this study was to identify and describe flood hazard areas by identifying wetland areas using DEM data and Landsat-8 images. TerraSAR-X high-resolution data is used to detect wetlands from landscapes, while land cover is identified by Landsat image data. The Topography Wetness Index (TWI) method is used to detect and identify wetland areas with basic DEM data, while for land cover analysis using Tasseled Cap Transformation (TCT) method. The result of TWI modeling yields information about potential land of flood. Overlay TWI map with land cover map that produces information that in Karawang regency the most vulnerable areas occur flooding in rice fields. The spatial accuracy of the flood hazard area in this study was 87%.

1. Introduction

Landforms has an important role for the study of flooding, because the land form is one of the vessels where the process of running water comes from the input of rain to the sea. Areas that are highly affected by the flood are areas with flat and sloping reliefs. Indicators of flood-prone landforms are floodplains, marine terraces, swamps and rear swamps [1], geomorphologically dangerous flooded areas characterized by a concave or flat morphology of landforms and their morphoaransemen, which are associated with rivers, with meander flow patterns and or braided. The location and location of the landform can be used as one of the potentially flood-potent and mapped areas [2]. Landforms in the coastal areas of Karawang regency are alluvial plains, areas with landforms of alluvial plains, alluvial valleys, river bends, and swamps that are susceptible to flooding because they are low areas or basins. The research location in Karawang regency was chosen because it met the criteria as flood hazard areas, located in coastal areas with slope characteristics $<2\%$ and had an average rainfall of ≥ 200 mm including in flood hazard class [3]. The characteristics of the flood hazard areas can be identified by the form of land that is the result of geomorphological phenomena, and the form of land may indicate a basin as a wetland area. The development of information technology raises the challenge in the distribution of spatial information of flood hazard areas that can be determined more quickly, precisely and accurately. Utilization of DEM and Landsat-8 Image data is expected to provide efficiency in the identification and delineation activities of flood hazard areas.

DEM data used in this research is TerraSAR-X Image 2011 with high accuracy of 1 m. TerraSAR-X image has a high spatial resolution so it can be an alternative in generating DEM data accurately, the utilization of DEM is expected to result in more accurate landform analysis. The high spatial resolution of the TerraSAR-X sensor provides access to surface heterogeneity on a better scale [4]. The approach in determining the drainage network for wetland identification is simpler, and directly obtains an interconnected runoff flow network. This approach is considered a better approach because it relies on the runoff analogy to establish the flow path [5]. Topographic Wetness Index (TWI) analysis method can give an illustration of an area having potential flood hazard or have a high chance of flooding.

The Landsat-8 image of October 2015 is a multispectral image that can provide the earth surface physical information, for the purpose of identifying wetland area using wetness parameter. The Tasseled Cap Transformation (TCT) method is a mathematical model formula for calculating brightness, greenness, and wetness [6], which will generate new digital value information from digital numbers in each band (band 1 to band 5 and band 7) on the Landsat image [7]. The use of spectral value transformation as an object sharpening effort in digital data satellite imagery that will generate new digital value information. Transformation of spectral values can be divided into two, namely (1) sharpen information, (2) summarize information from the number of channels that exist [8].

The purpose of this study was to identify flood hazard areas using high-resolution DEM data with Topographic Wetness Index (TWI) method. This method generates derived data from DEM data, using flow accumulation and slope functions to display potentially flooded basin areas [9]. Flood hazard map are generated from the TWI map overlay with a history of flood events and The Tasseled Cap Transformation (TCT) map. Image data processing in this study using software ILWIS 3.85 which is downloaded for free.

2. Method

2.1. Location and Time of Study

The research location in Karawang regency, located in the northern part of West Java Province, with geography position between 107°02 '- 107°40' East and 5°56 '- 6°34' South. The research and data processing activities were conducted from August 2016 to March 2017. Data analysis was conducted at the Geospatial Information Agency. Hypsography of study location is located at an altitude of 0 meters up to 80 meters above sea level, climatologically monthly rainfall data obtained from [10] and The Tropical Rainfall Measuring Mission [11] Karawang regency shows that bulk rain at low intensity levels.

2.2. Materials and tools

The materials used in this study consisted of the image and history of flood events presented in Table 1.

Table 1. Materials used in the study.

No.	Materials	Specification
1.	DEM/DTM TerraSAR-X years 2011, Otrhorectification	Elevation accuracy, 1 m
2.	Citra Landsat-8, recording of OctoberOctober 2015	Resolution 30 m
3.	Map of flood events Year 2015- BIG	Scale 1:60,000
4.	History of Floods – Agriculture Department of Karawang Regency	2016
5.	Topographic map – BIG	Scale 1:25,000

The tool used for spatial data processing is a set of computers with Microsoft Office 2010 software, Global Mapper V.10, ER Mapper 7, Ilwis 3.85 and ArcGIS 10.2.

2.3. Data analysis procedure

This research is done through several stages. Research stages generally consist of: (a) the preparation and data collection phase, (b) the data processing interpretation stage, (c) the field check stage, (d) the data analysis stage, and I the presentation stage of the results. The expected output of this study is the identification of flood hazard areas, with the classification of high levels of luminance, medium luminosity and low luminosity. The software used to process spatial data is Global Mapper 10, ER Mapper 7, Ilwis 3.85 and ArcGIS 10.1.

2. *Perparation*

At this stage, a review of the literature on study areas and literature related to research topics, TerraSAR-X data collection, Topographic map and Landsat-8 imagery was performed.

b. Processing and interpretation

b.1. Flood hazard mapping with Topographic wetness index (TWI)

Topographic Wetness Index (TWI) is a wetness index that can be used as an indicator of an area that has the potential for flooding. Topographic Wetness Index (TWI) is derived data generated from DEM data in a steady state –steady state – using flow accumulation and slope functions [12]. Wetness index is a key variable that controls hydrological and biogeochemical processes [13].

The accuracy of the TWI results relies heavily on the topographic surface, using high-resolution DEM data allowing the implementation of more detailed modeling. Therefore, the use of TerraSAR-X DEM data is necessary to develop the TWI method to be able to delineate more precise flood hazard areas [14]. TerraSAR-X signals have sensitivity to wet areas, thus there is an increase in wet area information. The results of the analysis will show the degree of wetness in an area and give an indication as the flood hazard area.

To DEM data is done Shaded Relief / Hillshade process with the intention to present a relief image of a region on raster data which is still in 2-D format (two dimension) by giving 3-D (three dimension) impression. Appropriate luminance and shading techniques will result in a 3-D display of DEM data. The availability of high-resolution DEM data will improve the detail of topographic surface reliefs that are the primary data for TWI modeling. The formula of the TWI method (Formula 1), developed by [15], in a run-off model and a topographic wetness index that has no units. The Topographic Wetness Index is a wetness index determined from the previously calculated surface variables with the following equation:

$$\text{Formula TWI} = ln \frac{\alpha}{\tan \beta} \qquad (1)$$

Where α = *Flow Accumulation* and tan β are slopes.

The TWI model describes water trends accumulated at certain points with slopes that indicate the effect of gravity on water movement [8]. A location that has a sloping slope will have a high TWI index value, thus potentially high water puddle. Conversely, a location that has a steep slope will have a low TWI index value, resulting in a low potency of water puddles. The data processing begins with DEM data recovery with the Fill Sink function, a process of eliminating depression or sink which is a condition where there is a striking elevation difference with very small coverage. For hydrological study the condition of this sink can interfere with the calculation, so it needs to be removed first sink. One of the keys in obtaining derivatives of surface hydrological characteristics is by obtaining flow direction on each raster data cell [16]. Because in the basic concept, flow direction is determined from a high value to a lower value [8].

The flow accumulation function is applied in modeling the amount of accumulated flow of water present in a particular region. As a result there will be an accumulated water value which is also identical to the actual river network in the field. Flow Accumulation serves to calculate the number of upstream areas or the countless stream of cells flowing into each cell. Theoretically, the process of river extraction is done by collecting pixels that have the same flow direction and accumulation trends and their adjacent locations [17]. Stream Networks is a function of making a network of rivers in a particular region. The information used is the flow accumulation described by determining the value of flow accumulation that is considered as a river based on the extraction and collection of pixels that have the same flow and accumulation tendency toward the adjacent location.

TWI in this research resulted from the processing of raster data using software Ilwis 3.85. The wetness index is grouped at each interval of 5 index scales, the high wet index value being indicative of the wetland region. The wetness index value on the index 15 scale is assumed to be the boundary of the flood hazard area. Implementation of boundary delineation of flood areas, done by vectorization method (raster to vector conversion) and or visual interpretation. In general delineation results need to be generalized or simplified by maintaining the general characteristics of the map. The methods include selection, simplification, merging, and magnification. Selection is done for objects that need to be removed and / or combined because they do not meet geometry specifications or do not fit into the classification of elements that can be displayed on the map scale. Selection is done by removing and / or aggregating segments of polygon at least 0.5 mm x 0.5 mm [18].

b.2. Mapping of flood hazard areas in paddy fields

Utilization of Landsat-8 Imagery, analyzed by The Tasseled Cap Transformation (TCT) method, as one method of identification of wetland area by transforming spectral value. The TCT method as an attempt to perform digital data object sharpening of Landsat satellite imagery will produce new spectral values or new information [20]. Land surface objects have varied spectral responses on wet / waterlogged and dry / unfertilized water.

The math calculation model of The Tasseled Cap Transformation (TCT) according to [15] is as follows (Formula 2),

$$
\begin{aligned}
& i1*0.1511+i2*0.1973+i3*0.3283+i4*0.3407+i5*{-0.7117}+i6*{-0.4559} \quad (\textit{wetness}) \\
& i1*{-0.2941}+i2*{-0.2430}+i3*{-0.5424}+i4*0.7276+i5*0.0713+i6*{-0.1608} \quad (\textit{greenness}) \qquad (2) \\
& i1*0.3029+i2*0.2786+i3*0.4733+i4*0.5599+i5*0.5080+i6*0.1872 \quad (\textit{brightness})
\end{aligned}
$$

Of the three above formulas of concern in this study is wetness, this formula indicates the area associated with the presence of water. The highest digital value indicates as the water surface area and the lowest as a dry area. Water gives the appearance of white or bright, and the appearance of the black indicates gray scale, Table 2. Is the new digital value of the classification wetness level. The software used in the processing of Tasseled Cap Transformation (TCT) Landsat-8 image is ERDAS-7.

Table 2. New digital figures showing the wetness of TCT analysis results.

Class	Digital value of TCT	Land cover
1	192 - 253	Wet areas, rice fields, ponds
2	160 - 191	Plantations with high and rare plants
3	76 - 159	Open area, gardens with rare plants
4	0 - 75	Fields, rice fields run out of harvest, open areas for construction

b.3. Delineate the distribution of flood areas

Mapping of flood hazard areas resulted from incorporation of hazard area map maps with historical data of flood events. Flood hazard maps of TWI results in overlay with historical data of flooding on paddy fields consisting of frequency of flood events, height of inundation, and duration of inundation (Figure 1). How to combine flood incident report data into the map is to add flood data as an administrative district map attribute. This map is then combined with a flood hazard map for flood risk analysis.

The criteria for the height of the flood waters in the wetlands will be very concerned with the height of the plant, the height of the rice plant 70 to 120 cm. Therefore, the criteria of flooded puddles with a height of > 70 cm indicate that paddy fields have high potential flood hazard. The application of the above criteria consider the condition of rice crops that when submerged in floods for a week will die [16], and if the rice fields are often submerged for several days under cropping conditions will cause crop damage and also decrease yield [21].

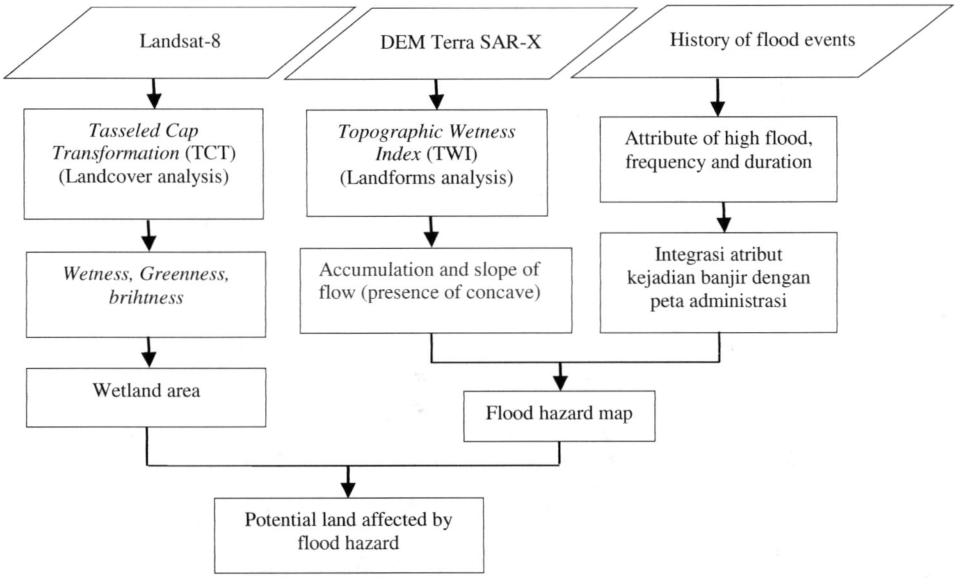

Figure 1. Flow chart for the preparation of flood hazard maps.

c. Field survey

The survey was conducted for primary data collection and secondary data as well as validation of TWI analysis result map, so it can be known the errors or deficiencies. The determination of the field was done randomly (purposive random sampling) and the flood event history was obtained from the [20] of Karawang Regency (Table 2).

d. Data analysisi, validation and presentation

The flood hazard areas were analyzed by overlaying between the TWI maps and the historical map of flood events in the district unit units. Potential of flood distribution is analyzed by overlaying between TWI maps and TCT maps. Subsequently validated using maps of flood events published by Geospatial Information Agency (2015). Test the accuracy of flood hazard analysis results with overlay method between TWI map with Flood Map of Year 2015 and TCT Map. The accuracy test using a wide parameter with a sample of 3 (three) locations, accuracy is presented in percentage with broad baseline reference on TWI Map.

3. Result and discussion

Flood hazard mapping with Topographic wetness index (TWI)

Index calculation with input fill sink and flow accumulation DEM produce wetness index (Figure 2). The TWI analysis is arranged sequentially with the Stream Networks analysis, arranging a network of streams that exist in a given region using flow accumulation information. Flow accumulation is explained by determining the value of flow accumulation considered as a river.

The red color on (Figure 2) has a high Index value more than 25 (>25). This region has a sloping slope so that the potential for high water puddle, green color has index value 15 in this study as an indication of boundary of the inundation area. The blue color indicates a location that has a steep slope and has a low TWI index value, potentially low water inundation occur.

IOP Conf. Series: Earth and Environmental Science **149** (2018) 012035 doi:10.1088/1755-1315/149/1/012035

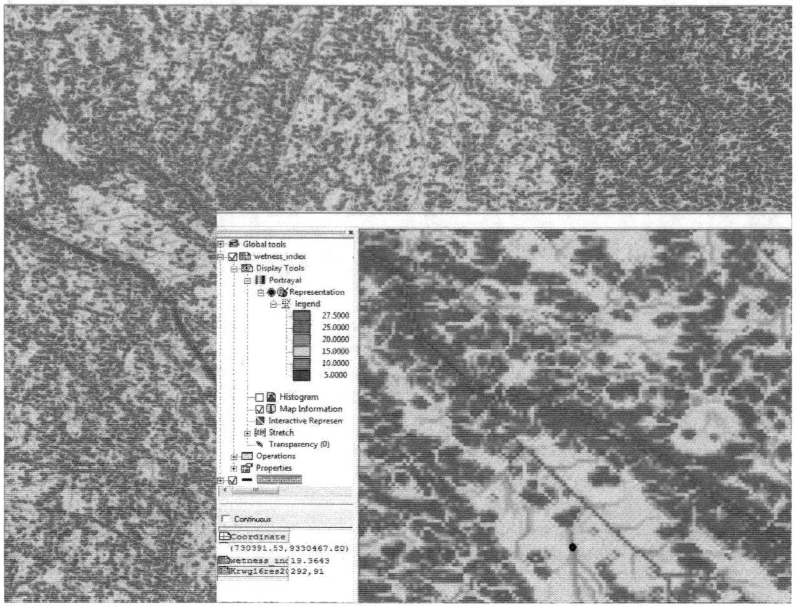

Figure 2. Wetness Index (WI>15 indicating wetlands).

Delineation of boundary areas indicating the danger of flooding is done through visual interpretation that shows the distribution of flood hazard areas (Figure 3). This condition is in accordance with the opinion of [21] that the main factors affecting the flood hazard in the field are flat topography, low elevation and soil type with fine texture.

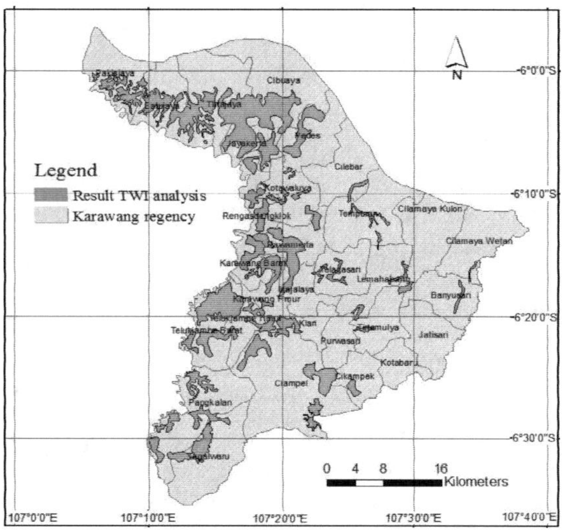

Figure 3. Topographic wetnees index map.

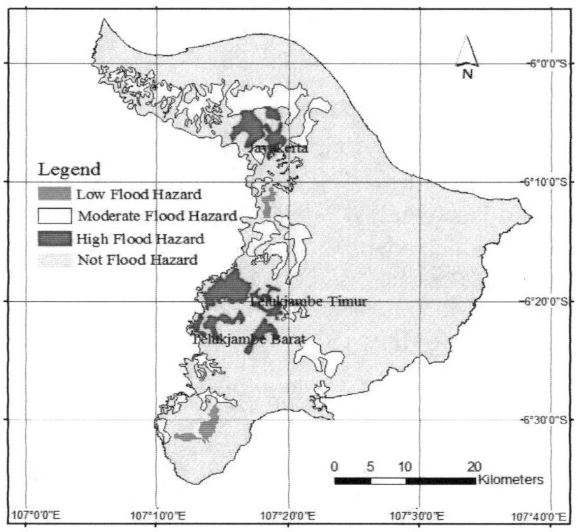

Figur 4. Flood hazard map.

Mapping of flood hazard areas in paddy fields

Data of flood occurrence frequency, height of puddle and duration of flood of rice field obtained from Agriculture Department of Forestry Plantation and Animal Husbandry of Karawang regency (Table 2). Height of puddle, central and southern regions have high grade with high puddle more than 70 cm (>70 cm), while for the north dominated by water level of puddle less than 70 cm (<70 cm) in medium category. High criteria for flooding more than 70 cm (>70 cm) indicate that the research area has high flood potential, especially in paddy field areas. Floods that occur annually, potentially become areas with high flood hazard if the average height of flood water more than 70 cm (>70 cm) and long puddle more than 7 days (>7days). The research area based on the criteria of flood hazard has high flood hazard potential due to fluvial land form in the northern region, in the middle of high flood hazard potential due to the flood plain located at the longitudinal bar of Citarum river and Cibeet river (Figure 4).

Table 3. Reports of flood events in rice fields during the planting season 2008/2009 -2013/2015.

No	Districts	Frequency	High flood average (cm)	Flood duration	Flood criteria
1.	Cikampek	1	60	6	low
2.	Rengasdengklok	1	70	7	low
3.	Tirtajaya	1	65	16	low
4.	West Karawang	2	100	15	medium
5.	East Karawang	2	100	15	medium
6.	Klari	2	100	5	medium
7.	Pangkalan	2	150	5	Medium
8.	Cilebar	2	70	16	Medium
9.	West Cilamaya	2	60	16	Medium
10.	East Cilamaya	2	60	16	Medium
11.	Ciampel	3	150	10	Medium
12.	Cibuaya	3	50	16	Medium
13.	Tempuran	3	70	15	Medium
14.	Jayakerta	4	80	16	High
15.	Rawamerta	4	70	15	medium
16.	East Telukjambe	5	100	15	High
17.	West Telukjambe	5	150	15	High
18.	Pakisjaya	4	70	16	medium
19.	Batujaya	5	60	16	medium
20.	Pedes	4	60	16	medium

Processed from data : [22]

Rainfall

The monthly rainfall data from The Tropical Rainfall Measuring Mission (TRMM) is presented in Table 3. Rainfall in Karawang regency at the low classification level because the monthly average is less than 100 mm, referring to the rainfall classification level criteria from the Bogor Climatology Station 2015. Flood incident in Karawang regency is not only caused by local rainfall which is extreme but also from flood of Citarum River, Cibeet River and other river that pass in Karawang regency because of high rainfall in upstream area.

Table 4. Average monthly rainfall in Karawang regency.

Month	Average monthly rainfall in the year (mm)						
	2010	2011	2012	2013	2014	2015	2016
January	16	6	8	18	24	13	7
February	8	5	7	8	20	15	17
March	8	5	5	8	12	8	10
April	5	6	8	8	9	6	9
May	5	5	4	8	5	4	5
June	7	4	2	3	4	1	7
July	5	1	0	9	6	0	3
August	5	0	0	1	0	0	5
September	9	0	0	0	0	0	7
October	10	3	2	3	0	0	9
November	8	8	8	6	8	10	8
Desember	6	6	10	15	8	8	3

Source: [23]

Delineation of flood area distribution with The Tasseled Cap Transformation (TCT)

Land surface objects have varying spectral responses to wet areas or inundated and dry or not waterlogged. The TCT method as an attempt to refine the digital data object of satellite imagery produces new information. The wetland area is indicated by white color, while the dry land area is dark (Figure 5).

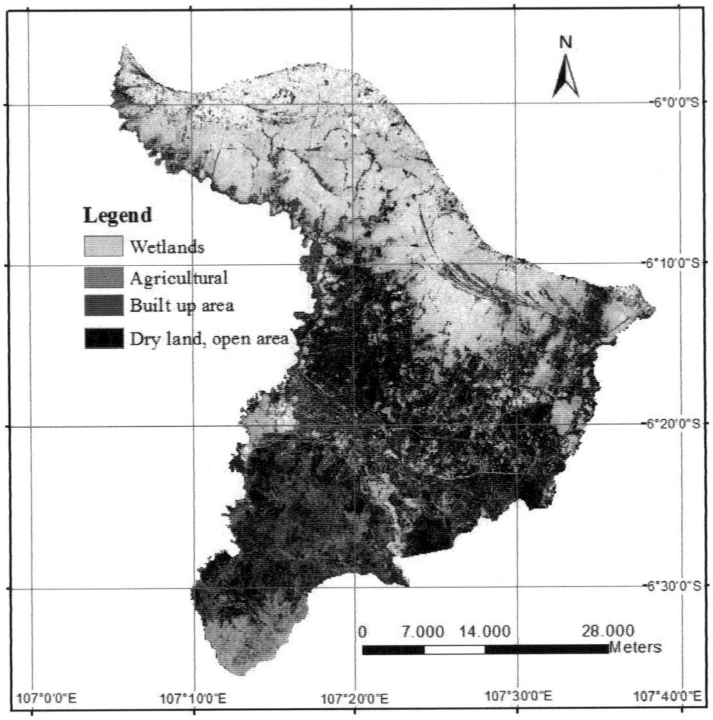

Figure 5. The Tasseled Cap Transformation.

Flood hazard areas with an area of 28,366 ha in Karawang regency mostly flooded rice fields. Results Overlay TCT map with TWI map (Figure 6) shows the distribution of flood hazard areas. Areas with high flood hazard covering 7,489 ha (7.61%), medium flood hazard covering 19,188 ha (19.49%) and low flood hazard covering 1,689 ha (1.72%). Potential flood hazard with long inundation more than 7 days and high water inundation over 70 cm can lead to crop failure.

Validation

Flood hazard analysis results in validation by overlaying between TWI maps and map of flood events of 2015 published by Geospatial Information Agency (Figure 7).

The overlay results show that the flood area is relatively the same as the map of flood event 2015 published by Geospatial Information Agency. The flood hazard area is calculated using TWI map with an area of 28,366 ha, while the calculation result of the map of flood incident that occurred in 2015 with an area of 25,168 ha. The accuracy test uses the broad parameter presented in Table 4. with an average of 87% accuracy.

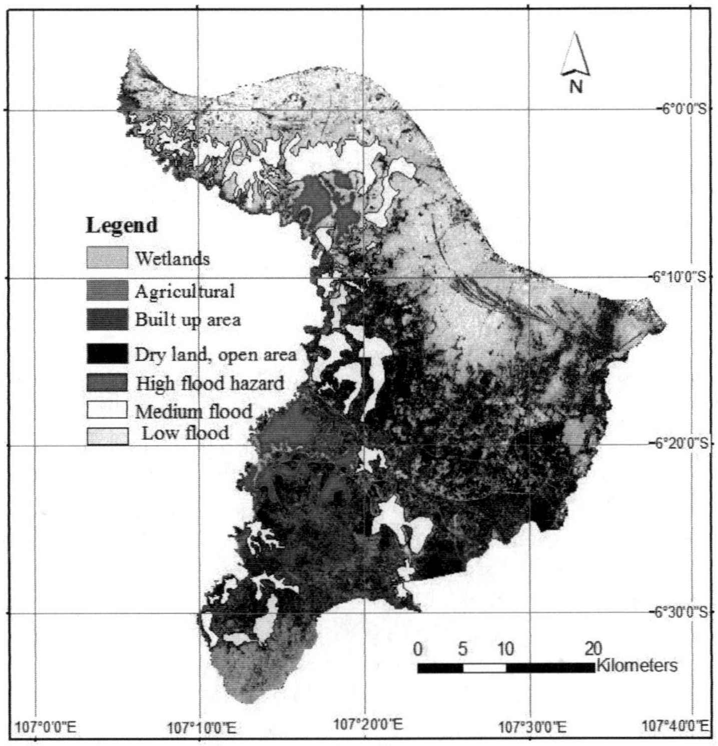

Figure 6. Overlay TWI dengan TCT.

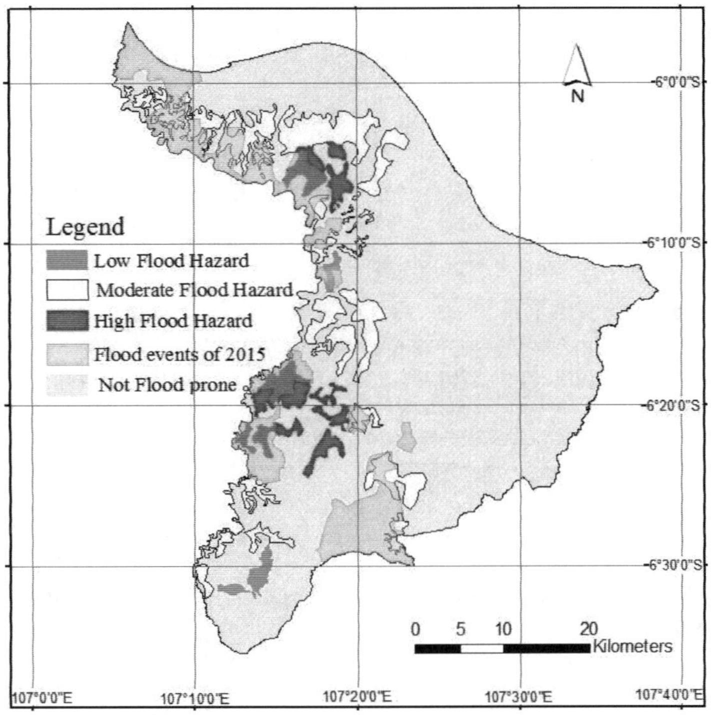

Figure 7. Map validation.

Table 5. Map test accuracy.

TWI (ha)	TCT (ha)	BIG (ha)	Accuracy %	Sample area
28,366	-	25,168	87%	Entire region
2,234	2,128	2,166	95%	Karangligar
2,857	2,695	6,813	94%	Ciampel/Cikampek
3,591	2,216	2,893	71%	Pangkalan
	Average		87%	

4. Conclusions

Flood hazard area mapping using high resolution DEM data and Topography Wetness Index (TWI) method is successful and gives good results. The wetness index is grouped at each of 5 index intervals, with the humidity index 15 being the boundary of the flood hazard area. The flood hazard areas resulting from TWI map analysis with a history of flood events indicate that there are areas with high flood hazard levels. High flood areas are indicated that the frequency of floods occurs more than once every year with depths of more than 70 cm and inundation more than 7 days. Some northern areas that are fluvial / lowland plains have high flood potential in paddy fields. In the middle of areas with high flood hazard are floodplains, encounter of Citarum river with Cibeet river, this area is paddy field. To recognize the Tasseled Cap Transformation (TCT) wetlands the Landsat-8 data was able to identify wetland and dryland cover. Delineation of flood spreading for TCT land cover analysis with TWI shows the distribution of flood areas that occurred in Karawang regency in paddy fields. Monthly rainfall in Karawang regency including low less than 100 mm / month, more flood events caused by the overflow of rivers due to high rainfall in the upstream area. Spatial studies indicate that high flood hazard is indicated in paddy fields with 7,489 ha of rice field, wetland with flood hazard is indicated with area of 19,188 ha and low flood hazard with 1,689 ha. The accuracy of spatial analysis in this study was 87%.

5. Acknowledgments

Thank you to the Geospatial Information Agency for assistace TerraSAR-X DTM data and USGS Website for the ease of data access Landsat and The Faculty of Geo-Information Science and Earth Observation (ITC) for open source ILWIS 3.8 software, so this article can be resolved

References

[1] Riadi B, Suriadi A B 2017 Identification of paddy field using Landsat image in Karawang Regency, West Java. *The 3rd International Symposium on LAPAN-IPB Satellite: For Food Security and Environmental Monitoring 2016 IOP Conf. Series: Earth and Environmental Science* **54** doi:10.1088/1755-1315/54/1/012014

[2] Pourali, S H, Arrowsmith C, Chrisman N, Matkan A.A and Mitchell D 2014 Topography Wetness Index Application in Flood-Risk-Based Land Use Planning. *Appl. Spatial Analysis* doi: 10.1007/s12061-014-9130-2.

[3] SNI 8197 [Standart Nasional Indonesia] 2015 Metode pemetaan rawan banjir skala 1:50.000 dan 1:25.000

[4] Somantri L 2008 Pemanfaatan Teknik Penginderaan Jauh Untuk Mengidentifikasi Kerentanan dan Risiko Banjir

[5] Hery T P 2013 *Ekstraksi Morfometri Daerah Aliran Sungai dari Data Digital Surface Model* Yogyakarta: UGM Press.

[6] Baghdadi N, Cerdan O, Zribi M, Auzet V, Darboux F, El Hajj M *et.al.* 2008 Operational performance of current synthetic aperture radar sensors in mapping soil surface characteristics: Application to hydrological and erosion modelling. *Hydrological Processes* **22**(1) 9–20.

[7] Ito O, Ella E, Kawano N 1999 Physiological basis of submergence tolerance in rainfed lowland rice ecosystem. *Field Crops*

[8] Kauth, R J, Thomas, G S 1976. The tasseled Cap – A Graphic Description of the Spectral-Temporal Development of Agricultural Crops as Seen by LANDSAT. *Proceedings of the Symposium on Machine Processing of Remotely Sensed Data, Purdue University of West Lafayette, Indiana* 4B-41 4B-51.

[9] Chapman G A, Young M A, Gray J M 2005 Using Compound Topographic Index to Deliniate Soil Landscape Facets from Digital Elevation Models for Comprehensif Coastal Assessment. The Modelling and Simulation Society of Australia and New Zealand Inc

[10] Buchanan B P, Fleming M, Schneider R L, Richards B K, Archibald J, Qiu Z, Walter MT 2014 Evaluating topographic wetness indices across central New York agricultural landscapes. *Hydrology and Earth System Science* 18:3279–3299

[11] The Tropical Rainfall Measuring Mission (TRMM) 2017Available at: http://trmm.gsfc.nasa.gov. [1 August 2017]

[12] Badan Meteorologi Klimatologi dan Geofisika - Stasiun Klimatologi Bogor 2015 Peta Prakiraan Curah Hujan Bulan Juni 2015 Kabupaten Karawang.

[13] Beven K and Kirkby M 1979 A physically based variable contributing area model of basin hydrology/ Un modèle à base physique de zone d'appel variable de l'hydrologie du bassin versant *Hydrological Sciences Journal* **24** (1)

[14] Murdiyanti, S R. 2010 Integrasi Transformasi Spektral Citra Landsat ETM Dan SIG Untuk Pemetaan Pola Rotasi Tanam Lahan Sawah Kabupaten dan Kota Semarang Serta Daerah Sekitarnya di Jawa Tengah. *Majalah Geografi Indonesia* (MGI) **24** (2):121-141

[15] Baig MHA, Zhang L, Shuai T, Tong Q 2015 Derivation of a tasseelled cap transformation based on Landsat 8 at- satellite reflectance. https:// www. researchgate. net/publication/ 262005316_Derivation_of_a_tasselled_cap_transformation_based_on_Landsat_8_at_satellit e_reflectance/figures

[16] Dinas Perkebunan Kehutanan Perkebunan dan Peternakan (Distanhut) 2016 Data Produksi dan Produktivitas Padi 2006 - 2016.

[17] Hartini S, Hadi M P, Sudibyakto S, Poniman A 2015 Risiko Banjir pada Lahan Sawah di Semarang dan Sekitarnya. Majalah Ilmiah Globe **17**(1):51-58.

[18] Badan Informasi geospasial – Pusat Pemetaan Rupabumi dan Toponim 2014 *Panduan Generalisasi Peta Rupabumi Indonesia untuk Peta Skala Sedang*. Bogor : Cibinong

[19] Ikhwani 2013. Ketahanan Varietas Padi Toleran Rendaman dan Responnya terhadap Pemupukan. *Jurnal Lahan Suboptimal* **2**(1) ISSN: 2252-6188

[20] Badan Pusat Statistik 2015 Kabupaten Karawang dalam Angka

[21] Hojati M 2016 Determination of a Topographic Wetness Index Using High Resolution Digital Elevation Models. *European Journal of Geography.* **7**(4):41 – 52 ©Association of European Geographers

[22] Dinas Perkebunan Kehutanan Perkebunan dan Peternakan (Distanhut) 2016 Laporan Kejadian Banjir di Lahan Sawah 2006 - 2015.

[23] The Tropical Rainfall Measuring Mission (TRMM) 2017 Available at http://trmm.gsfc.nasa.gov [cited 1 August 2017]

IOP Conf. Series: Earth and Environmental Science **149** (2018) 012036 doi:10.1088/1755-1315/149/1/012036

Landuse change identification using SPOT 6 for food security analysis of Denpasar City, Bali Province, Indonesia

N Supardan, M Panularsih and M Darmawan
Geospatial Information Agency, Jalan Raya Jakarta - Bogor KM 46 Cibinong, Bogor West Java, Indonesia

Email : niendya_salam@yahoo.co.id

Abstract. Land use change was common in Indonesia, including Bali, especially land conversion from paddy field to other uses. Based on statistics data of Bali 2015, the area of rice fields continues to decline every year, both irrigated and rain-fed rice fields. This study was conducted to analyze land use change and its impact on food security in Denpasar. Land use change dynamics can be observed by multi-temporal land use changes. GIS analysis was performed to obtain spatial land use changes in the study area by using topographic map 1: 25,000 with data sources 2002 and land cover classification of SPOT 6 image acquisition 2015. The results of GIS analysis indicate that settlement and built-up area increased by 1,736 hectares and the area of rice field decreased by 1,695 hectares during 13 years. The increasing of settlement rate was 133,5 ha / year and the decreasing of rice field was 130 ha / year. Meanwhile, the result of food availability analysis shows that the Denpasar City still deficit of rice amount 48,997 tons in 2014, and 69,175 tons in 2015 or only fulfill about 45% of food requirement.

1. Introduction

Landuse change or land conversion can be interpreted as a change in the function of part or the entire region of the original function into other functions that have a negative impact on the environment and the potential of land itself [1]. Conversion of land also means the conversion or transfer of land generally involves a transformation in the allocation of land resources from one use to another use [2]. Land conversion from agriculture to non-agriculture, especially paddy field to non-paddy field such as plantations, settlements or other awoke region is very common in Indonesia, including in Bali Province. Referring to the 2015-2019 RPJMN that food sovereignty is one priority sector development dimension. Geospatial Information Agency of Indonesia as the only organizer for basic geospatial information in Indonesia can play a role in providing basic geospatial information, remote sensing imagery and supporting spatial analysis of strategic issues that can support the development priorities. Study of the use of satellite imagery for land changes analysis [3] especially in agricultural land is highly relevant mainly carried out in which the results are used to support food security analysis of a region in terms of the spatial aspect. Denpasar Municipality is a selected sample, since this area is one of the tourist destinations, with the highest density in the province of Bali, and farmland tend to change the function.

Percentage of land conversion in Bali over the past 15 years, is the highest occurred in Denpasar, on average 1.30% per year to switch the function of agricultural land become non-agricultural land[4]. Therefore, in this study, Denpasar became the object of a sample to be examined regarding food security.

According to the World Health Organization (WHO), the definition of the main components of food security namely food availability, food access and food utilization [5]. Food availability is the ability of a person, group of people (households) has a number of food sufficient for basic needs [6]. Food access is the ability to have the resources economically and physically to get nutritious food. While the utilization of food is the ability to utilize food properly and proportionate. But in this paper is focused on the components of food availability in Denpasar based on spatial analysis and the use of the statistical data of the Regional Statistical Bureau. Research questions on this study is: how extensive the paddy field changes become non paddy field in Denpasar by using SPOT imagery 6 acquisitions 2015, and whether the existing paddy field is still able to meet food needs the population entire Denpasar.

This research was important to do despite the availability of food was not the only factors that determine the food security of an area, but there were other factors that will determine the achievement of food security [5]. However, due to conversion of agricultural land was a serious threat for security and food safety, it was necessary to study in this region. Conversion of fertile farmland have been less offset by a concerted effort to develop agricultural land through the use of marginal land. On the other hand, food agriculture land conversion leads to reduced tenure thus decrease the income of farmers. Therefore, the results of this study can be used as a warning for this local government to look back the land conditions, especially land changes from agricultural into non-agricultural which can cause reducing availability of food (rice) in this region. Besides this, with the results of this study can be used as the input of policy what will be done, such as controlling wetland conversion, creating new wetland or other policies that can be strengthening food security.

2. Data and Methodology

Land use change dynamics can be observed by multi-temporal land use changes. The data used in this study consisted of spatial and non-spatial data. Spatial data used topographic map scale of 1: 25,000 created 2002 and SPOT 6 acquisition 2015. While non-spatial data used statistical data from BPS [4] such as number of population, population projection, wetland area, the number of crops, etc. Land use information is the result of human activities in land, land use or land function, so it can't always be estimated directly from remote sensing images, but can be recognized from the land cover association. The procedure of processing and analysis of remote sensing imagery is done by beginning stages of pre-processing data. At this stage, the geometric process correction is needed in order to improve the appearance of the object distortion and geographic position within the limits of the study area. The radiometric and atmospheric correction to eliminate or reduce atmospheric interference is done by systematically process.

The land use classification is carried out by visual image interpretation. Visual interpretation made based on basic element interpretation approach such tone / color, texture, pattern, size, shape, shadow and association as a guidance for land use classes [7]. In this study, the land use are grouped into 12 classes: lake, ponds, mangroves, swamp forest/peat, grasslands, plantations/gardens, settlements and activities, swamps, paddy fields, scrub, moor/ fields, and non-cultivated vegetation.in The land use change were observed by overlaying the maps of land use in 2002 obtained from topographical maps and land use maps in 2015 from the results of SPOT 6 image interpretation.

Food self-sufficiency is the ability of a region to produce food that can ensure sufficient food needs until at an individual level. Food self-sufficiency can be calculated based on the carrying capacity of food in a region. Or consumption needs of the population to food, especially rice by 114 kg / capita / year [8]. If rice production in a district divided by the population exceeds the consumption of rice per capita per year, then the district can meet the needs for itself, even if production is large, then the district may send

IOP Conf. Series: Earth and Environmental Science **149** (2018) 012036 doi:10.1088/1755-1315/149/1/012036

its food production to other districts. Otherwise, the district can't be self-sufficient, so it must import food or rice from other districts. The ratio of consumption to the net production of rice normative per capita is one of the determinants of food insecurity [9]. Carrying capacity of the food in each municipality can be formulated as follows:

$$DDP = (L x Pp)/Pe > 114 \quad (1)$$

where:

DDP	= Food Carrying Capacity (kg/capita/year)
L	= Harvesting Area (ha)
Pp	= Rice Productivity (kw/ha)
Pe	= Number of Population (people)

The DDP index 114 explains that food consumption per capita was 114 kg /capita/year is a critical limit food consumption per person per year. If DDP $\leq$ 114, it means the food carrying capacity is deficit, that food production can't meet the needs of the population residing in this region. Otherwise, if the DDP> 114, then food production in the region could meet the food needs of the population residing in the territory, it means food surplus.

3. Result and Discussion

3.1. Image Classification

Remote sensing technology continues to evolve so as to produce image data of high resolution remote sensing satellite capable of recording an object with a size of less than one meter. This is certainly easier for users to recognize an object recorded by satellite imagery and provide more detailed information. This study used land use data obtained from topographic map scale of 1: 25,000 and the results of SPOT 6 image interpretation acquired 2015 which has a spatial resolution of 1.5 meters. Based on the topographic map scale of 1: 25,000 were created in 2002, land use / coverage area of Denpasar consists of lake, ponds, mangroves, swamp forests, grasslands, plantations, swamps, paddy fields, scrub, moor / fields, and non-cultivated vegetation. The largest land use in Denpasar City are the settlements and activities, fields, and plantations / gardens. Based on the results of rectified SPOT 6 image classification 2015, it can be seen that the largest land use is still the same, namely settlements and its activities, paddy fields, and plantations / gardens, but its wide has undergone changing. Figure 1 shows the SPOT imagery 2015, the research sites that used as a data source for / land use classification. Figure 2 shows the topographic map created in 2002 and Figure 3 shows the results of SPOT image classification map, 2015.

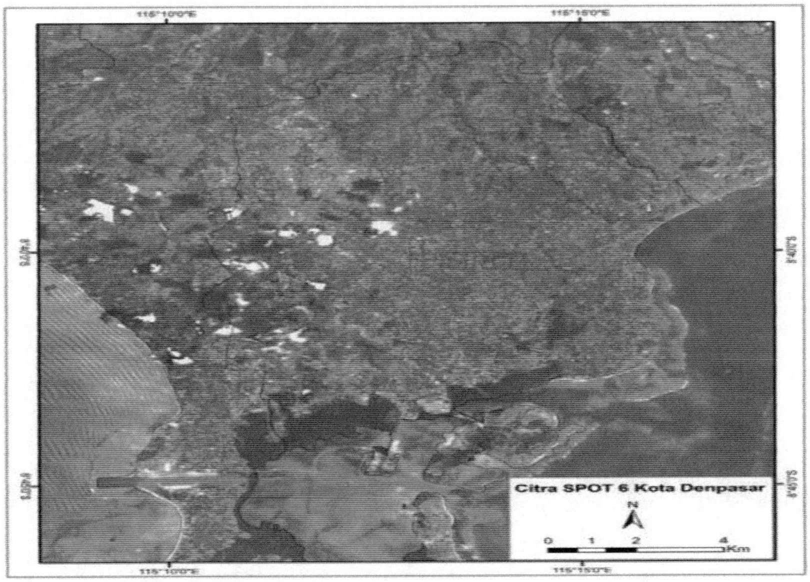

Figure 1. Orto rectified SPOT Image 6 acquired 2015 of Denpasar City.

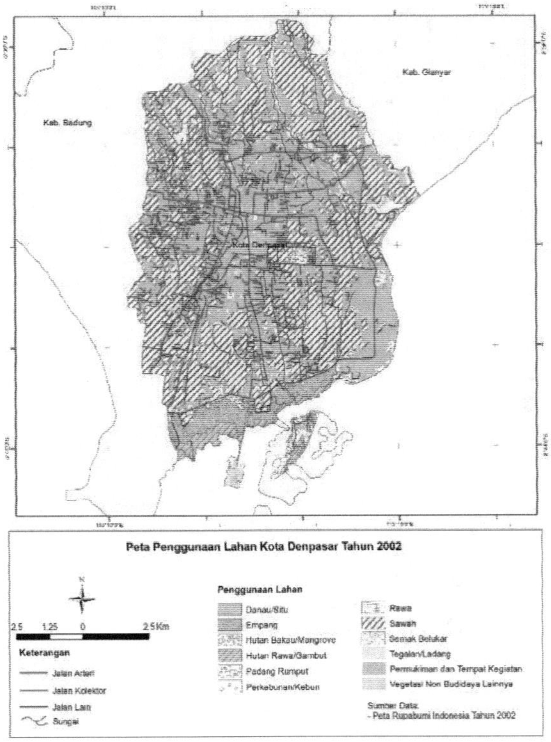

Figure 2. Landuse / land cover 2002
Source: Base map 1:25.000.

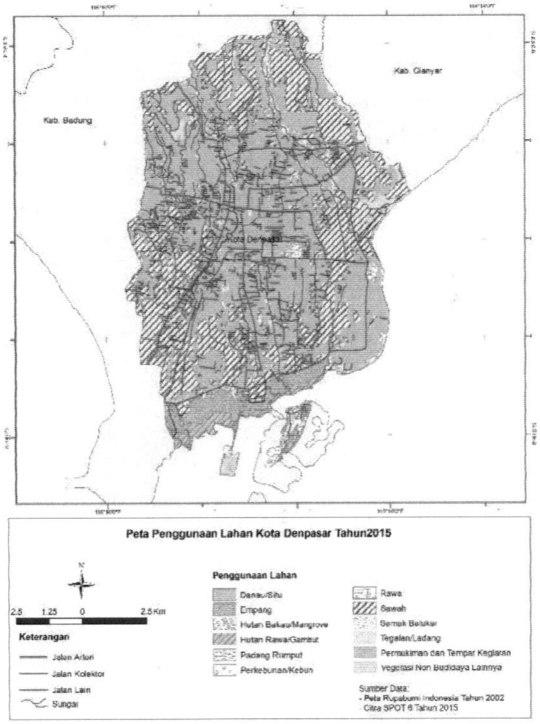

Figure 3. Landuse / land cover 2015
Source: SPOT 6 acquired 2015.

3.2. Land Use Change

Based on the land use changes analysis 2002 – 2015 was obtained the coverage of land use change, as shown in Figure 5. The map showed some land use changes, namely some grassland change into settlements, some plantations / gardens change into settlements, some plantations / gardens changes into paddy fields, some scrub changes into settlements and some paddy fields change into settlements. Settlement and its activities is the largest increased amount 1,735.58 ha for 13 years or increasing rate 133.5 ha/year, followed by paddy fields change decreased amount 1,694.81 Ha or decreasing rate 130 ha/year, and plantation change decrease amount 26.90 Ha or 2 ha/year. The increased in built up area was accompanied by a drop in agriculture areas. The land conversion is mainly for housings and settlement purposes [10]. For more details can be seen in Table 1 that inform land use change in Denpasar based on data from 2002 and 2015. Figure 4 shown illustrated with a graph.

Tabel 1. Land use change of Denpasar City 2002 – 2015.

Landuse/land cover	Topographic Map 2002 (Ha)	SPOT Imagery 2015 (Ha)	Land Use change (Ha)
Lake	10.33	10.33	0
Ponds	384.14	384.14	0
Mangroves Forest	2.27	2.27	0
Swamp forests/Peat	245.47	245.47	0
Grasslands	195.53	182.62	-12.91
Plantations/garden	687.68	660.78	-26.90
Settlement and its activities	8621.76	10357.34	1.735.58
Swamps	33.79	33.79	0
Paddy fields	5238.15	3543.34	-1.694.81
Scrub	20.5	19.98	-0.52
Moor / fields	111.23	110.79	-0.44
Non-cultivated vegetation area	32.53	32.53	0

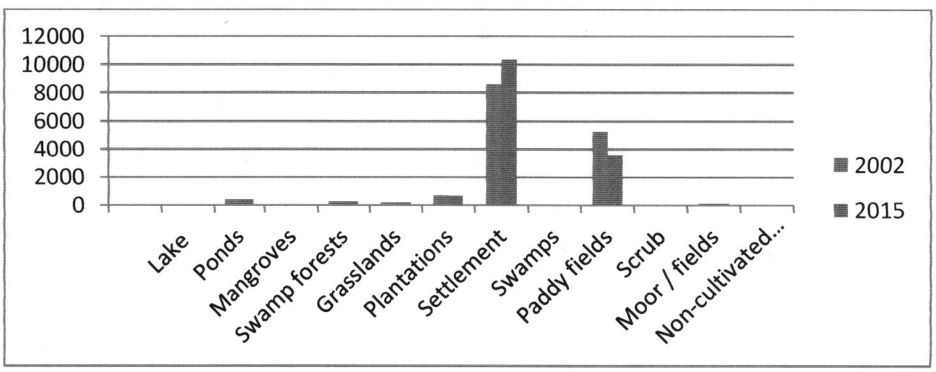

Figure 4. Land use change graph of Denpasar City 2002 – 2015.

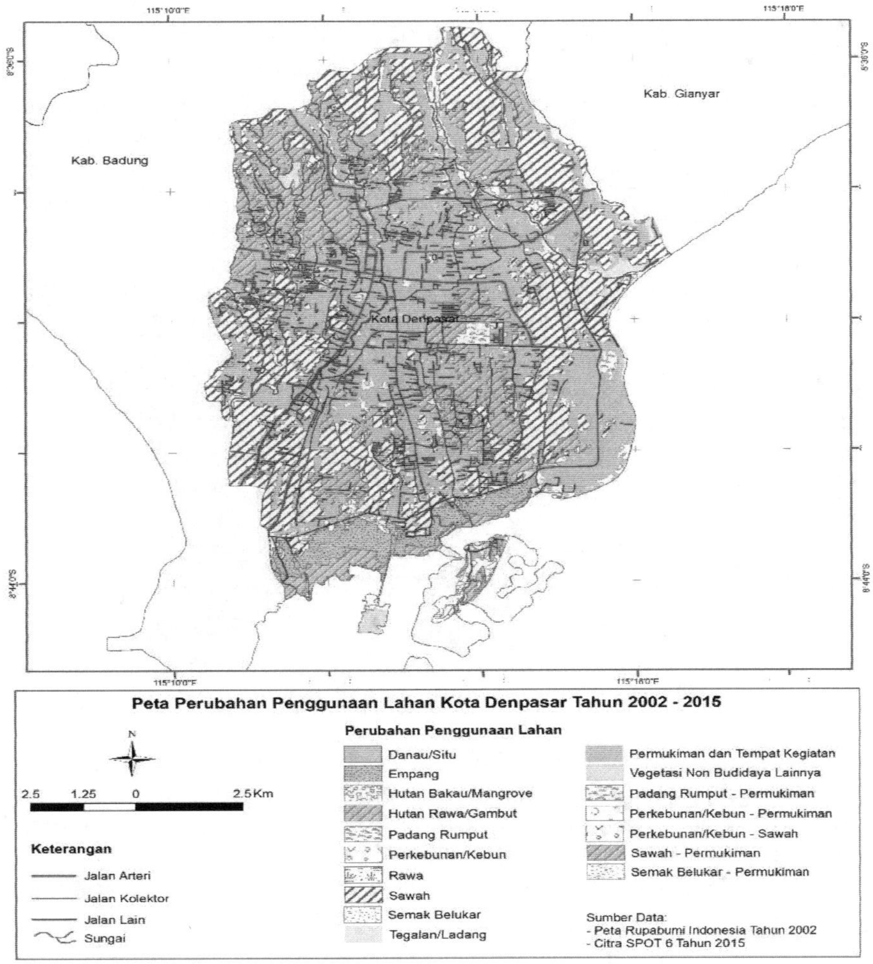

Figure 5. Land use change map of Denpasar City 2002 – 2015.

3.3. Food Security Analysis

For Indonesia, food is often associated with rice because this type of food is the main staple food. The impact of land conversion, especially paddy fields would eliminate the function fields as land intended for the production of rice [11]. With the conversion to other functions will reduce rice production. This could threaten food production in a region that led to the inability to meet the food needs of population in the region. For example, due to conversion of paddy fields in Java for a period of 18 years (1981-1998) accounted for accumulation has been loss of 50.9 million tons of grain, or approximately 2.82 million tons of grain per year. When calculated rice equivalent, then the loss of food production is approximately 1.7 million tons per year [12]. The calculation result of the food carrying capacity of Denpasar City in 2014 was 39.72 with harvested area was amount 4,232 ha. An average yield of 6.19 tons / ha, rice production reached 26,200 tons. Of the value of production will produce 16,558 tons of rice. The result of the calculation of food requirements in Denpasar with a population of 659,623 and an average consumption of 114 kg per capita per year reached 75,197 tons. The calculations show that the food security of Denpasar currently not able to feed themselves independently. There were deficit amount 48,997 tons of rice that can't be met by the Denpasar from existing rice crops.

Using the same formula for data 2015, based on data submitted by the Head of Department of Agriculture, Food and Horticulture of Denpasar as mentioned on Denpostnews.com stated that Denpasar on average produce 10-13 tons of rice per hectare during 2015, so the total rice that produced amount 31,213 tons. The projected population of the Denpasar city in 2015 is around 880,600. The food carrying capacity was 35.45, its means <114. The need total rice in 2015 was amount 100,388 tons, there were deficit amount 69,175 tons. It means the rice available from Denpasar City only about 45% of the need, it can't met the food need of entire population throughout the year, unless there were import policy of the local government.
Based on the wetland dynamics analysis, wetland in Denpasar continues to decline, while the population growth rate reached 1.96% per year. Population growth continues increasing so demand food availability is increased as well. With the wetland conversion rate is high enough, the paddy fields in the Denpasar city no longer be sufficient for current residents and future, can be seen from the above calculation. Denpasar dependence on other areas related to rice supply as a staple food even higher. Therefore we need appropriate government policies related with land conversion of agricultural land into non-agricultural [13]. The local government should be stipulate of the primary rice fields area which should not be conversion unless the reason for the public interest, with feasibility studies [14]. Need to apply rewards and punishments related to farmers who are able to maintain his farm and working creatively to increase the productivity of land, for example by applying appropriate agricultural and does not switch professions.

4. Conclusion

Conversion from agricultural into non-agricultural land in Denpasar City was quite high based on change identification by using SPOT imagery. Settlement and built-up area was the most increasing landuse, otherwise paddy field was the most decreasing landuse for period 2002 - 2015. Based on food security analysis, Denpasar city still deficit of rice amount 48,997 tons in 2014, and 69,175 tons in 2015 or only fulfill about 45% of food requirement. Moratorium Policy is needed for preventing land use change especially from paddy field to other usage.

Acknowledgments

Thanks to Geospatial Information Agency for supporting and facilitating of this research since preparation until publication.

References

[1] Muslikin M K 2015 Study Wetland Conversion to non Wetland and Rice Production Impact in Blora Regency 2000 - 2010 Departemen of Geography, Faculty of Sosial Science, Universitas Negeri Semarang

[2] Soemarno 2011 Konversi Lahan: Landuse Planning and Management Summary of Teaching Material Departement of Soil, Agricultural Faculty, University of Brawijaya Malang

[3] Butt A, Rabia Shabbir, Sheikh Saeed Ahmad, Neelam Aziz 2015 Land use Change Mapping and Analysis using Remote Sensing and GIS: A Case Study of Simly Watershed, Islamabad, Pakistan. *The Egyptian Journal of Remote Sensing and Space Sience* **18**(2) 251 - 259

[4] BPS Provinsi Bali 2015 Land Area According to Land Uses in Bali Province 2015 Publication Number: 51531.1603 ISBN : 2356-4334

[5] Mun'im, A 2012 An Analysis of factors influencing food security, access and absorption factor on food security in food-surplus regencies: a partial least Square path modeling approach *Journal of Agro Ekonomi* **30**(1) 41-58

[6] Burchi F and Pasquale de Muro 2015 *Food Availability to Nutritional Capabilities: Advancing Food Security Analysis. Food Policy* 60 10 – 19 Published by Elsevier Ltd

[7] Purwadhi S H, Rokhmatuloh F and Haryani N S 2013 *Application of Remote Sensing Technology for Regional Development* Jakarta: Departement of Geography, Faculty of Math and Science – University of Indonesia

[8] Mahbubi A 2013 Model dinamis supply chain beras berkelanjutan dalam upaya ketahanan pangan nasional *Jourrnal of Manajement & Agribisnis* **10**(2) 81-89. Publisher: Graduate Program of Management and Business Bogor Agricultural University

[9] Sulistyo W and Winarko E 2015 Spatial Autocorrelation Modeling for Food Security and Insecurity Condition in Klaten Regency. Proceeding of National Seminar of Information Technology and Communication. ISSN 2089-9815: 35-42

[10] Harini R 2007 Land Conversion and Driving Forces of Urban Fringe Area of Yogyakarta. *Proc. of Southeast Asean-German Summer School on Urbanization. Manila*

[11] Iqbal M, Sumaryanto 2007 Strategi Pengendalian Alih Fungsi Lahan Pertanian Bertumpu Pada Partisipasi Masyarakat Bogor: Center for Sosio Economic Analysis And Agricultural Policy, *Journal of Agricultural Policy Analysis* **5**(2) 167-182

[12] Irawan B, Friyatno S 2012 *Dampak Konversi Lahan Sawah di Jawa Terhadap Produksi Beras dan Kebijakan Pengendaliannya* Bogor: Center for Research and Development of Socio Economic Agricultural

[13] Nasoetion L I 2003 Konversi Lahan Pertanian: Aspek Hukum dan Implementasinya *Proceeding of Nasional Seminar "Multifungsi Lahan Sawah dan Konversi Lahan Pertanian"* Bogor: Center for R & D of Soil and Agroclimate, Agricultural Research and Development Agency

[14] Parwata IMO 2014 Lahan Pertanian Berkelanjutan Availabe at: http://distantp.baliprov.go.id/lahan-pertanian-pangan-berkelanjutan [Cited 20 January 2017]

IOP Conf. Series: Earth and Environmental Science **149** (2018) 012037 doi:10.1088/1755-1315/149/1/012037

Analysis of vegetation changes in Cidanau watershed, Indonesia

R N Khairiah[1], Y Kunihiko[2], L B Prasetyo[1], Y Setiawan[1,3]
[1]Department of Tropical Biodiversity Conservation, Faculty of Forestry, Bogor Agricultural University.
[2]Department of Biological and Environmental Engineering, the University of Tokyo.
[3]Center for Environmental Research, Bogor Agricultural University.

Email: rahmi_khairiah10@apps.ipb.ac.id

Abstract. Vegetation change detection is needed for conserve of quality and water cycle in Cidanau watershed. The NDVI was applied to quantify the vegetation changes of Cidanau watershed for three different years 1989, 2001, and 2015. Using NDVI we mapped the reflectance from chlorophyll and distinguished varying amounts of vegetation at the pixel level by index. In the present study, as a preliminary study, we proposed a vegetation change detection analysis based on the NDVI from 1989 through 2015. Multi-temporal satellite data i.e. Landsat imagery with 30 m spatial resolution are used in the present study. It is reported that agroforestry land exhibited the greatest reductions in highly dense vegetation class in 1989-2001 and also moderate vegetation class in 2001-2015. It's mean that amount of vegetation present in agroforestry land is getting lower year by year.

1. Introduction

Cidanau watershed is one of the priority watersheds in Indonesia. A rare ecosystems and the only one natural wetland that existing in Indonesia which appear in the mountain region, included nature reserve that contains the last lowland swamp forest in Java Island. The Rawa Danau reserve also plays an important role as a water catchment for the Cidanau River [1]. Vegetation change has played an important role in quality and water cycle condition which Cidanau watershed is the only one sources of freshwater supplies for domestic and industrial water in Banten Province, Indonesia.

Vegetation change detection can be done by remote sensing data through a process called change detection. It can simply be defined as the process of identifying differences in vegetation or land cover over time [2]. Variation in spectral response involves situations where the spectral characteristics of the vegetation in a given location change over time [3]. [4] Mentioned that the change detection approach by the temporal dynamics of vegetation over a long period is able to detect changes in the earth's surface, either the gradual change caused by the extreme climatic variations or changes dramatically due to human or natural. Some researchers also noted that human activities, especially some ecological construction projects, greatly influence vegetation change [5-9]. Quantitatively, NDVI will generally show the proper sense of change in greenness at the pixel level by index. NDVI is one of the most successful indices for simple and quick identification of vegetated areas and their condition [10]. Therefore, the aim of this study was to detect the vegetation changes based on the NDVI for three different years 1989, 2001, and 2015.

2. Methods

2.1. Study Area

Cidanau watershed is located in Serang District and Pandeglang District, Banten Province, Indonesia (060 07' 30"-060 18' 00" S and 1050 49' 00"-1060 04' 00" E). Cidanau watershed covers an area of 22620 Ha. [11]. As mostly in Indonesian islands, the Cidanau watershed has tropical climate with two seasons; dry and rainy seasons. In the dry season, rainfall is only 50-87 mm, which happens in June to October. In this season, relative humidity is 77.60-81.40% and shining intensity is 60.20-78.60%, and temperature is 26.14 °C-27.00 °C and wind speed is around 4.80-5.40 knot per hour. In the rainy season, rainfall is 133-346 mm, which happens in November to May. In the same season, the relative humidity is 79.60-84 % and shining intensity is 31.60-68.20 %, and temperature is 26.04-27.10 ^{0}C with wind speed 5.00-5.90 knot per hour [12]. Figure 1 shows the location of Cidanau watershed.

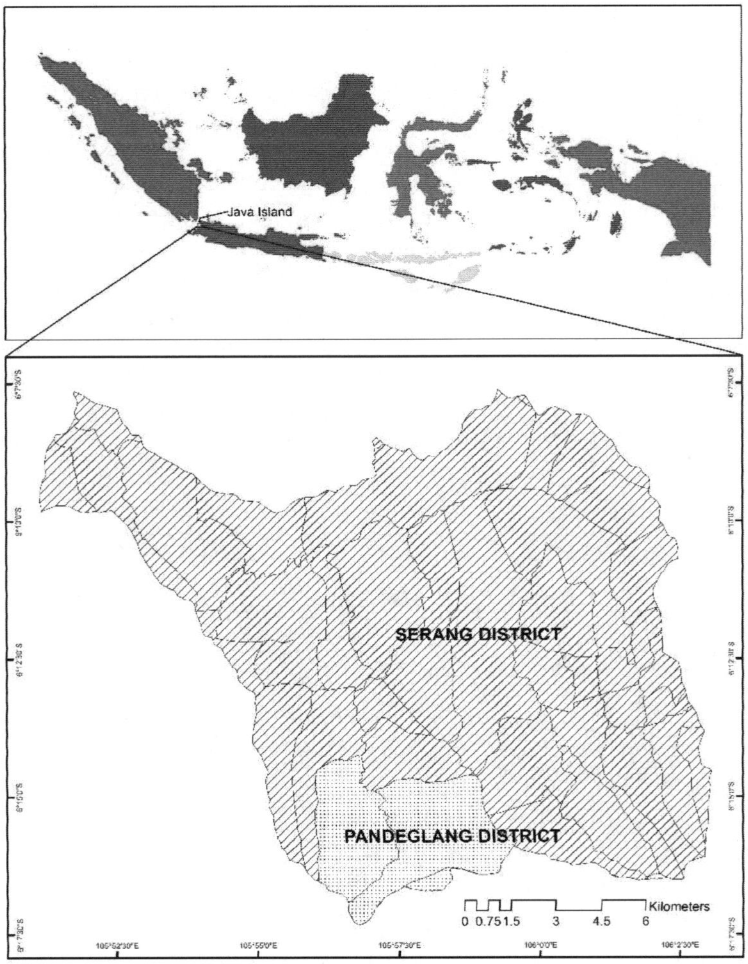

Figure 1. The map of study area.

2.2. Satellite image datasets

In order to quantitatively our predictions of vegetation change, we used Multi-temporal satellite data i.e. Landsat imagery with 30 m spatial resolution. The characteristics of Landsat imagery data that used in the present study are shown in Table 1. Cloud cover is a major problem in working with optical remotely sensed data sets in humid tropical forest environments such as Indonesia and Brazil [13-15]. Unlike Brazil, for example, Indonesia doesn't have a seasonally cloud-free window, requiring more data intensive methods to overcome persistent cloud cover [16].

IOP Conf. Series: Earth and Environmental Science **149** (2018) 012037 doi:10.1088/1755-1315/149/1/012037

Table 1. Acquired satellite remote sensing data.

Satellite sensor	Date of Acquisition	Spatial Resolution	Clouds Cover	Spectral Resolution
Landsat 5 TM	July 29, 1989	30 m	0.00%	Band 5,4,3
Landsat 7 ETM+	July 6, 2001	30 m	10.00%	Band 5,4,3
Landsat 8 OLI TIRS	July 5, 2015	30 m	2.66%	Band 6,5,4

*https://earthexplorer.usgs.gov/ (downloaded 2017)

2.3. Image processing

Before using data for detecting vegetation changes, pre-processing operations were performed on three images including geometric correction, atmospheric correction, cloud and shadow masking. The Dark object subtraction (DOS) method for atmospheric correction, was applied using QGIS software for digital image processing, remote sensing and GIS based software (ERDAS Imagine 9.1 and ArcGIS 10) were used for image processing, classification, analysis to achieve the objectives of the study. ERDAS Imagine is used to calculate NDVI for all of the three images (1989, 2001 and 2015). Then three individual NDVI was applied to detect areas of vegetation changes and derived quantitative data were generated and summarized using ERDAS Imagine, GIS software and spreadsheet. The NDVI derived map of 1989, 2001 and 2015 were crossed to generate the change map of vegetation in the study area. Very low values of NDVI (minus one (-1.00) to 0.10) correspond to non-vegetated surface likes barren areas of rock or sand, shrub and grassland (0.20 to 0.30), while high values indicate temperate and tropical rainforests (0.60 to 0.80) [17]. Based on this information, the three date NDVI images were classified into six ranges by fixing the thresholds for NDVI classification and the method used is equal interval (Table 2).

Table 2. Classification for vegetation coverage.

Classes	Value Intensity
No Vegetation	value < 0.19
Less Vegetation	value >0.20 and value <0.34
Less Moderate Vegetation	value >0.35 and value <0.49
Moderate Vegetation	value >0.50 and value <0.64
Dense Vegetation	value >0.65 and value <0.79
Highly Dense Vegetation	value >0.80 and value <0.94

Supervised classification has been widely used to detect land use types. In supervised classification, spectral signatures are collected from specified locations in the image by digitizing various polygons overlaying different land use types. Classes of the resulting image were recoded into 10 land-use classes. The accuracy of the classified land use was assessed using high resolution images Google Earth. The overall accuracy was computed by dividing the total correct by the total number of pixels in the error matrix. Another discrete multivariate technique of use in accuracy assessment is called Kappa [18].

3. Results

The NDVI is used in the present study to detect vegetation changes. The NDVI images representing the amount of vegetation present at each time in Figure 2, areas with healthy vegetation is white to grey, and the area where no vegetation is black. The white area which represents vegetated areas has stronger near-infrared reflectance. Conversely, negative NDVI values were recorded in a dark area. This is as a result of the fact that features reflect more in the visible band than they do in the near-infrared band, indicating regions of low vegetation, typical water, cloud, bare soil and rock [19]. The mean of NDVI for overall watershed were 0.75 (1989), 0.45 (2001) and 0.35 (2015). It shows that amount of vegetation present in 1989, getting lower in 2001 and also 2015. The mean NDVI between 1989, 2001 and 2015 is characterized by a general decrease.

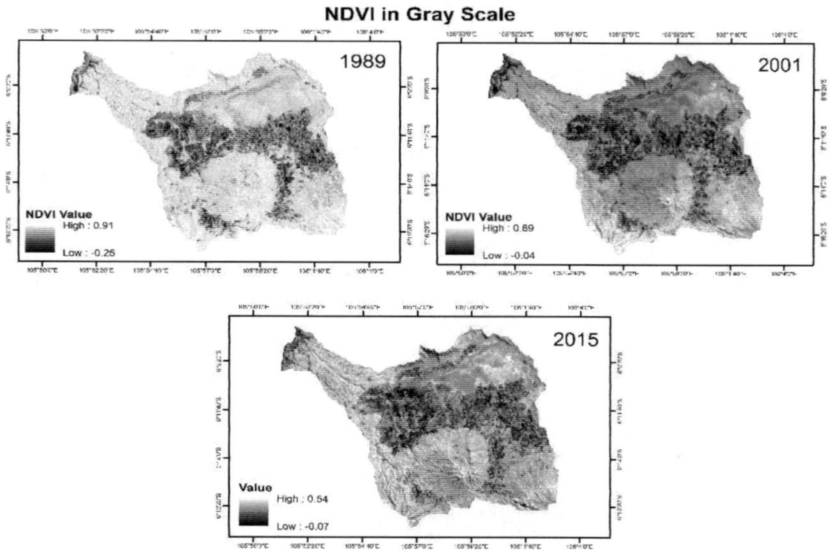

Figure 2. NDVI map for three different years (1989, 2001, and 2015).

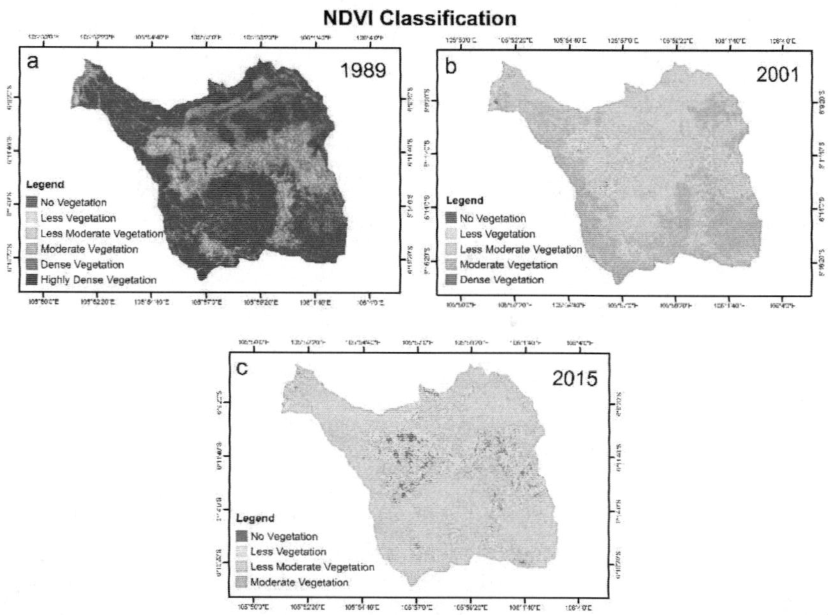

Figure 3. (a,b,c) NDVI derived classified map of study areas different years (1989, 2001, and 2015).

IOP Conf. Series: Earth and Environmental Science **149** (2018) 012037 doi:10.1088/1755-1315/149/1/012037

The change detection technique, which was employed in this study, was the post-classification comparison. In 1989, the site was dominated by 59.01% highly dense vegetation class. In 2001, it was dominated by 53.68% less moderate vegetation class. Meanwhile, in 2015 it was dominated by less moderate vegetation also about 67.00%. In 2001 and 2015, there are no site was categorized into classes highly dense vegetation. Are presented in Table 3&4 and Figure 3. The obvious difference was the area of highly dense vegetation class was high in 1989 but then disappeared in 2001 and 2015.

Table 3. Vegetation changes from 1989 to 2001.

Area Covered in 1989 (Hectare)	Area Covered in 2001 (Hectare)					
	No Vegetation	Less Vegetation	Less Moderate Vegetation	Moderate Vegetation	Dense Vegetation	Grand Total
No Vegetation	6.35	4.57	2.95	0.19	0.00	14.05
Less Vegetation	8.84	71.27	55.87	4.53	0.06	140.58
Less Moderate Vegetation	18.64	413.66	417.00	22.32	0.05	871.67
Moderate Vegetation	16.72	964.79	1381.35	82.16	0.24	2445.27
Dense Vegetation	8.76	831.12	4071.39	889.29	0.39	5800.95
Highly Dense Vegetation	1.84	131.99	6213.14	6993.31	7.19	13347.47
Grand Total	61.14	2417.41	12141.70	7991.80	7.95	22620.00

Table 4. Vegetation changes from 2001 to 2015.

Area Covered in 2001 (Hectare)	Area Covered in 2015 (Hectare)				
	No Vegetation	Less Vegetation	Less Moderate Vegetation	Moderate Vegetation	Grand Total
No Vegetation	26.52	29.84	4.78	0.00	61.14
Less Vegetation	350.06	1678.00	389.23	0.12	2417.41
Less Moderate Vegetation	325.52	4602.41	7210.38	3.39	12141.70
Moderate Vegetation	16.01	383.73	7552.68	39.38	7991.80
Dense Vegetation	0.00	0.00	7.35	0.60	7.95
Grand Total	718.11	6693.98	15164.42	43.49	22620.00

To explain this, we tried to observe the distribution with the land use of Cidanau watershed to find out the location of changes. The pattern of the land use changes for three different years, 1989, 2001 and 2015 are presented in Table 5 and Figure 4. It is reported that agroforestry land exhibited the greatest reductions in highly dense vegetation class in 1989-2001 and also moderate vegetation class in 2001-2015. It's mean that amount of vegetation present in agroforestry land is getting lower year by year. The economic crisis in Indonesia may induced on vegetation changes rapidly in 1989-2001. Many people who depend on natural resources for subsistence, by cutting down trees in the private landowners or illegal area. The lack of government monitoring, caused to the amount of vegetation present continues to decreased dramatically. [20] Explains that the economic crisis, has induced an unstable population flow between urban area and rural area. There are not enough lands to develop for new cultivation. These implies illegal activities by poor farmers and strangers (refugees) from other place.

Table 5. Land use derived changes statistics of the study area for three different years (1989, 2001, and 2015).

Land Use	Area (Hectare) in 1989	Area (Hectare) in 2001	Area (Hectare) in 2015
Secondary forest	925.27	899.98	927.20
Swamp forest	1110.22	955.57	946.47
Agroforestry	11839.19	12025.28	12300.08
Forest plantations	354.15	366.77	368.43
Shrubs	68.27	33.96	47.32
Grass marshes	1418.31	804.73	829.55
Paddy fields	6192.88	6814.60	6482.76
Open land	1.11	2.10	1.05
Housing	627.96	628.01	627.52
Water bodies	82.63	89.00	89.61
Total	22620.00	22620.00	22620.00

Land use changes in Table 5 above statistics shows that in each year agroforestry land and paddy fields dominated the area Cidanau watershed. According to [21], this is as a result of socio-economic conditions in the region Cidanau watershed communities who depend on agriculture and plantation sector. According to data from DAS Communications Forum Cidanau in 2013, the livelihoods of residents in the watershed Cidanau 33% are in the agricultural sector.

Table 6. Overall classification and kappa accuracy.

Accuracy Assesment	in 1989	in 2001	in 2015
Overall classification accuracy	89.00%	89.00%	92.73%
Overall kappa index	0.88	0.87	0.91

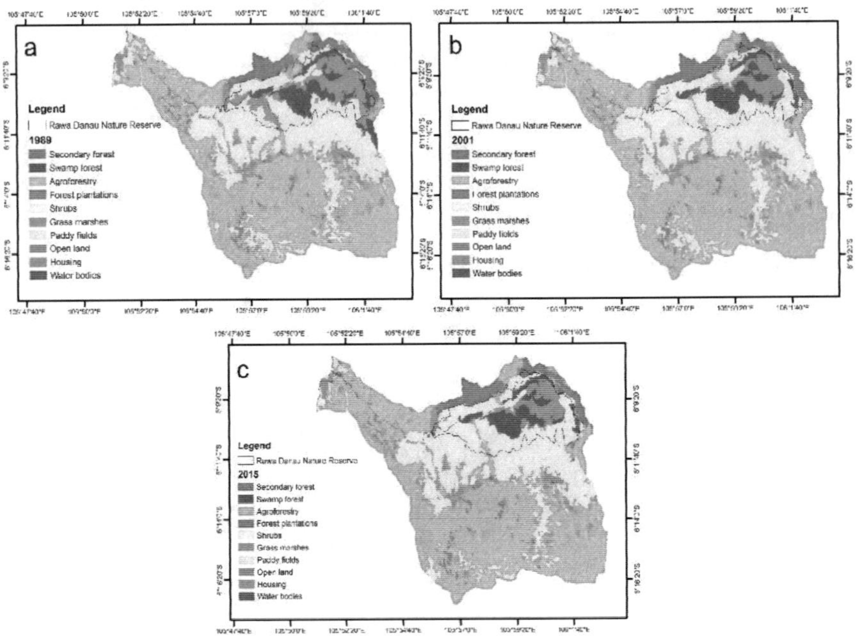

Figure 4. (a, b, c) Land use cover change map for three different years (1989, 2001, and 2015).

4. Conclusion

The study highlights was to try to understand the changes in vegetation by looking the land use distribution. It is reported that agroforestry land exhibited the greatest reductions in highly dense vegetation class in 1989-2001 and also moderate vegetation class in 2001-2015. It's mean that amount of vegetation present in agroforestry land is getting lower year by year.

References

[1] World Agroforestry Centre 2010 Rewards for, Use of and shared investment in Pro-poor Environmental Services schemes (RUPES-II) Where we work: brief profiles of RUPES action research sites

[2] Richards J A and Jia X 2006 Remote sensing digital image analysis: An introduction, Springer Verlag

[3] Hoffer R M, Noyer S C and Mroczynski R P 1978 A comparison of LANDSAT and forest survey estimates of forest cover Proceedings of the Fall Technical Meeting of the American Society of Photoqrammetry 221-231, Falls Church: Virginia

[4] Setiawan Y, Yoshino K 2011 Land use change detection by characterizing the vegetation dynamics: case study of Java Island, Indonesia. *Journal of the Japan Society of Photogrammetry and Remote Sensing* **50** (2) 96-103

[5] Cao X, Gu Z H, Chen J, Liu J, Shi P J 2006 Analysis of human-induced steppe degradation based on remote sensing in Xilin Gole Inner Mongolia China. *J. Plant Ecol.* **30** (2) 268-277

[6] Zhuo L, Cao X, Chen J, Chen Z X, Shi P J 2007 Assessment of grassland ecological restoration project in Xilin Gol grassland *Acta Geogr. Sin.* **62** (5) 471-480

[7] Wang J, Guo N, Cai D H, Deng Z Y 2009 The effect evaluation of the program of restoring grazing to grasslands in Maqu country. *Acta Ecol. Sin.* **29** (3) 1276-1284

[8] Li H, Cai Y L, Chen R S, Chen Q, Yan X 2011 Effect assessment of the project of grain for green in the karst region in Southwestern China: a case study of Bijie Prefecture. *Acta Ecol. Sin.* **31**(12) 3255-3264

[9] Huang SW, Li X S, Wu B F, Pei L 2012 The distribution and drivers of land degradation in the Three-North shelter forest region of China during 1982-2006. *Acta Geogr. Sin.* **67** (5) 589-598
[10] Vikram A, Bhamare S M 2012 Change detection of vegetation cover using Remote Sensing and GIS
[11] Khairiah R N, Prasetyo L B, Setiawan Y, Kosmaryandi N 2016 Monitoring model of payment for environmental service (PES) implementation in Cidanau watershed with stands density approach. *Procedia Environmental Sciences* **33** 269-278
[12] [Bappeda] Regional Development Planning Agency of Banten. 2002 Master Plan of Arrangement for Conservation and Development of the Cidanau watershed. Provincial Development Planning Board of Banten and PT Formation Four Patterns Consistent Consultants Serang
[13] Hoekman D H 1997 Radar monitoring system for sustainable forest management in Indonesia. Geoscience and remote sensing 1731-1733
[14] Asner G P 2001 Cloud cover in Landsat observations of the Brazilian Amazon. International *Journal of Remote Sensing* **22** 3855-3862
[15] Hansen M C, Stehman S V, Potapov P V, Loveland T R, Townshend J R G, DeFries R S, Pittman K W, Arunarwati B, Stolle F, Steininger M K, Carroll M and DiMiceli C 2008b Humid tropical forest clearing from 2000 to 2005 quantified by using multitemporal and multiresolution remotely sensed data. *Proceedings of the National Academy of Sciences* **105** 9439–9444
[16] Broich M, Hansen M C, Potapov P, Adusei B, Lindquist E, and Stehman S V 2011b Time-series analysis of multi-resolution optical imagery for quantifying forest cover loss in Sumatra and Kalimantan, Indonesia. *International Journal of Applied Earth Observation and Geoinformation* **13** 277–291
[17] Weier J and Herring D 2000 Measuring Vegetation (NDVI & EVI)
[18] Cohen J 1960 A coefficient of agreement for nominal scales. *Educational and Psychological Measurement* **20** (1) 37–46
[19] Lillesand T M, Kiefer R W, Chipman J W 2004 Remote Sensing and Image Interpretation, John Wiley and Sons: New York
[20] Yoshino K, Ishikawa M, Budi I S 200 Socio-economic causes of recent environmental changes in Cidanau watershed, West Java Indonesia. *Rural and Environmental Engineering* **44** 27-41
[21] Badadung G 2017 Changes on the surface of carbon stored in the Cidanau watershed Banten Province [Thesis] Bogor Agricultural University

Measuring deforestation using remote sensing and its implication for conservation in Gunung Palung National Park, West Kalimantan, Indonesia

N I Fawzi[1], V N Husna[2] and J A Helms[3]
[1]Conservation Program, Yayasan Alam Sehat Lestari Sukadana, West Kalimantan
[2]Environmental Research Center, Bogor Agricultural University
[3]Health in Harmony, Portland, Oregon, United States

E-mail: vinahusna@gmail.com

Abstract. Gunung Palung National Park (1,080 km^2, 1°3' – 1°22' S, 109°54' – 110°28' E) was first protected in 1937 and is now one of the largest remaining primary lowland mixed dipterocarp forests on Borneo. To help inform conservation efforts, we measured forest cover change in the protected area using 11 multi-temporal Landsat series images with path/row 121/61. Annual deforestation rates have declined since measurement began in 1989, to around 68 hectares per year in 2011 and 112 hectares per year in 2017. Halting deforestation in this protected area requires to tackle its underlying economic and social causes, and find ways for communities to meet their needs without resorting to forest clearing. Community empowerment, forest rehabilitation, and health care incentives as payment for ecosystem services can help reduce deforestation in Gunung Palung National Park. This becomes a positive trend which we must continue to always work in forest conservation. Future forest monitoring will be dependency with remote sensing analysis and open source remote sensing data such as Landsat and Sentinel data remain an important data source for historical forest change monitoring.

1. Introduction

Tropical deforestation is a major cause of global biodiversity loss and carbon emissions. The island of Borneo, including its Indonesian controlled territory known as Kalimantan, contributes to this problem. Between 1973 and 2010, Kalimantan lost 30.7% (123,941 km^2) of its forest, and Borneo as a whole lost 73.55% [1]. Industrial plantations were the biggest cause of deforestation since the 1970s [2,3], followed by agriculture, wildfires, and climate change [1]. Currently, less than 10% of forests in Southeast Asia are protected, and even protected areas are threatened with deforestation [4].

Gunung Palung National Park in West Kalimantan protects one of the largest remaining primary lowland mixed dipterocarp forests in Borneo. It is home to approximately 2,500 wild orangutans, or about 10% of the world's surviving population [5]. But the park suffers from high rates of deforestation, with 38% of its lowland forests lost between 1999 and 2002 [6]. This forest loss is partly due to a large number of people (up to 47-72% of households in some areas) who rely on illegal logging for their primary source of income [7]. To achieve zero deforestation, we must therefore not only protect the park, but also address the problem at its roots by empowering people to make a living without logging.

Here we describe deforestation using remote sensing and its implication to conservation efforts in Gunung Palung National Park. We measure forest cover using satellite remote sensing technology, because it is less costly and time intensive than field-based methods [8] and allows us to assess long term changes in landscape composition over the entire area of the park at multiple scales [9,10]. Maps produced from remote sensing techniques also provide information for conservation planning and forest management [11].

2. Methods

2.1. Study area

Gunung Palung National Park was first protected in 1937 with an area of 300 km^2, and later expanded to a total of 1,080 km^2. It is located in the regencies of North Kayong and Ketapang, in the province of West Kalimantan, Indonesia (1°3' – 1°22' S, 109°54' – 110°28' E, elevation 0 to 1,031 meters above sea level, figure 1). The park conserves many habitat types, including mangrove forest, peat swamp forest, lowland rainforest, and montane forest, and is home to several threatened species, such as Bornean orangutans (Pongo pygmaeus), proboscis monkeys (Nasalis larvatus), sun bears (Helarctos malayanus), and helmeted hornbills (Rhinoplax vigil).

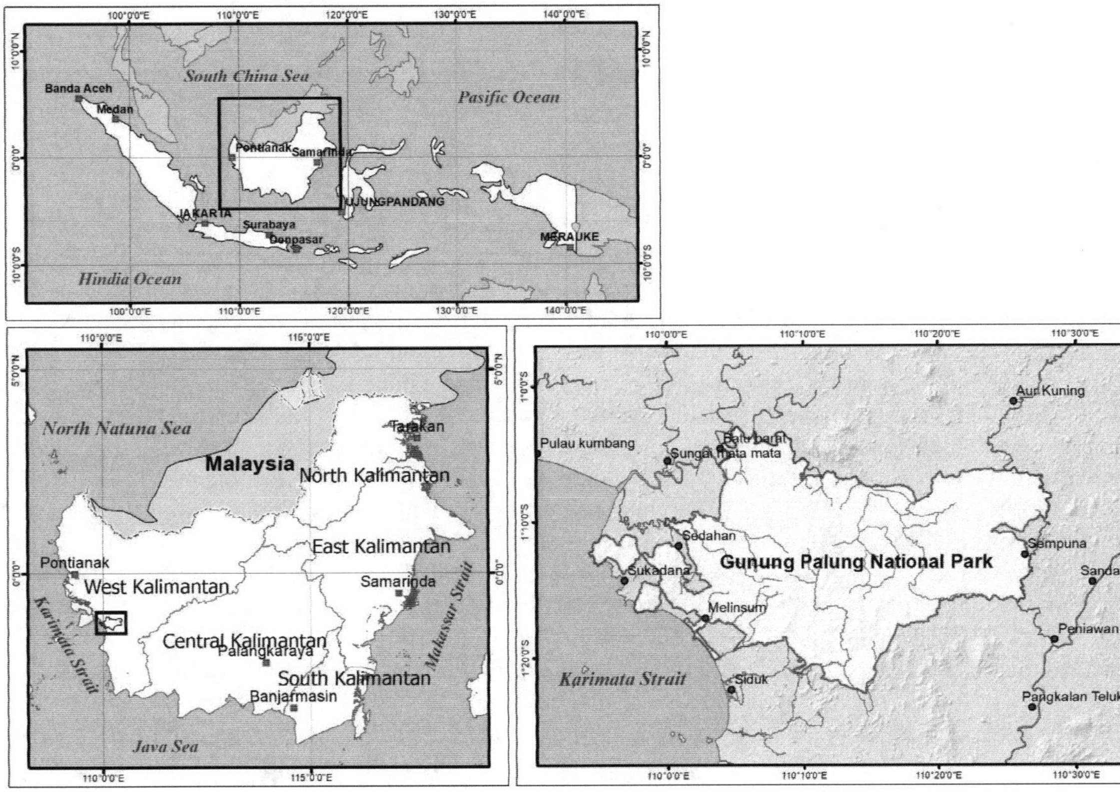

Figure 1. Study area in Gunung Palung National Park.

2.2. Data

We used the following 11 multi-temporal Landsat series images (path/row 121/61) to map forest cover over time and determine rates of forest loss in the park:

1. Landsat TM acquired on 12 September 1989 and 3 October 3 1989.
2. Landsat TM acquired on 21 July 21 1997, 22 August 1997 and 28 December 1997.
3. Landsat ETM+ acquired on 25 June 2002, 11 July 11 2002 and 28 August 28 2002
4. Landsat TM acquired on 7 July 2007.
5. Landsat TM acquired on 30 September 2011.
6. Landsat 8 (OLI) acquired on 22 March 2017.

IOP Conf. Series: Earth and Environmental Science **149** (2018) 012038 doi:10.1088/1755-1315/149/1/012038

The spatial resolution for Landsat images is 30 meters. The data were geometrically corrected and projected to the Universal Transverse Mercator zone 50S (WGS 1984 datum). We used the newest version of the national park's borders, issued by the Ministry of Environment and Forestry of Indonesia decree number SK.733/Menhut-II/2014.

2.3. Analysis

We classified land cover in the park using visual interpretation aided by maximum likelihood methods [12]. For classification classes, we used a modified version of the method of [13] which classifies land cover into: (1) montane forest, (2) peat forest, (3) lowland forest, (4) mangrove, (5) secondary forest, (6) young regenerating forest, (7) scrub or open land, (8) logged forest, (9) mixed agriculture, (10) mixed plantation, and (11) settlement. After conducting targeted ground checks for validation and to assess map accuracy, we found that our methods correctly classified land cover in over 95% of cases. Measuring annual deforestation rate we using Compound Interest Law were mathematically calculated as (equation 1):

$$r= 1/((t\_2-t\_1))\ \ln A\_2/A\_1 \qquad (1)$$

Where A_1 and A_2 are forest cover at time t_1 and t_2, respectively (unit % per year) [14].

3. Results and Discussions

3.1. Forest loss measurement

At the time of the first measurement in 1989, the park had already lost around 11,607 ha or 10.74% of its forest area (table 1, figure 2). From 1989 to August 1997, the annual deforestation rate was 0.87% or around 800 ha/year. In one four-month period in 1997, the park lost nearly 22.000 ha of forest or 20.3% of the park's area. By 2011 the deforestation rate had declined to around 68 ha/year, and was around 112 ha/year in 2017. That corresponds to an annual deforestation rate of 0.12% in 2011 and 0.21% in 2017. These rates are nearly 200 and 120 times less than the peak deforestation rate in December 1997, or 19 and 11 times less than rates in 2002. Corresponding with this decline in deforestation, since 2002 many logged areas in the park lowlands have been regenerating into secondary forest (figure 3). These patterns mirror global deforestation rates, which have decreased by half from the 1990s to 2010 [15]. But deforestation rates in Gunung Palung are still around three times as high as the global average of 0.08% per year [17].

This forest loss is related to a growing population and Indonesia's transmigration policy that encourages development [3]. The main causes of forest change in this park are illegal logging and conversion to community's plantations and agriculture. In low elevation areas near settlements, forests are converted to mixed agriculture, such as paddy farming or gardening [17]. In hilly areas far from settlements, forests are converted into mixed plantations of rubber, durians, chilies, or other products. In southern areas of the national park, forests were cleared and logged through a timber concession operated by PT. Marsela Wana Sekawan [18]. The timber concession exacerbated the logging problem and allowed deforestation to penetrate into the interior of the park. After logging, cleared areas around the park perimeter were colonized by alang-alang grass (Imperata cylindrica), which encourages and tolerates wildfires, creating large areas of anthropogenic grassland.

Table 1. Land cover change in Gunung Palung National Park from 1989 to 2017.

No	Landuse	1989	Aug-97	Dec-97	2002	2007	2011	2017
1	Lowland forest	48,741.55	46,921.43	28,360.32	24,060.45	19,579.92	19,436.93	19,125.77
2	Peat forest	24,561.19	19,892.65	17,244.91	14,085.23	13,366.52	13,265.14	12,951.78
3	Montane forest	22,790.20	22,790.04	22,200.86	22,149.86	22,126.82	22,126.82	22,126.82
4	Mangrove forest	311.91	311.91	213.78	213.78	196.96	169.62	118.04
5	Mixed agriculture	83.30	136.96	235.08	1,800.03	2,330.43	2,479.78	2,764.48
6	Mixed plantation	3,653.29	4,714.56	4,714.56	5,192.37	5,332.56	6,346.62	6,842.24
7	Scrub or open land	2,807.31	6,243.31	14,017.88	7,607.28	8,800.11	7,068.74	7,097.09
8	Degraded forest	5,063.86	6,889.85	20,913.32	26,523.77	29,821.87	28,756.91	11,107.40
9	Young regenerating forest	0.00	111.90	111.90	6,267.93	6,345.52	7,621.28	2,517.68
10	Secondary forest	0.00	0.00	0.00	111.90	111.90	740.76	23,330.37
11	Settlement	0.00	0.00	0.00	0.00	0.00	0.00	30.96
	Total	**108,012.61**	**108,012.61**	**108,012.61**	**108,012.61**	**108,012.61**	**108,012.61**	**108,012.61**

3.2. Implications for Conservation

To achieve zero deforestation we must address the causes of forest loss in Gunung Palung National Park. Conservation can encompass a range of activities, from total preservation of natural resources to sustainable extraction, and is not limited to public policy [19]. Forest monitoring and reporting are important to measure forest condition or natural resources for conservation and management. An up-to-date forest information need fast assessment and reliable method. The result from measuring forest cover change using remote sensing is we can know the annual rate of deforestation and its distribution. We can also differentiate between deforestation and forest degradation or partial canopy cover loss.

Satellite remote sensing area contribute to multi-temporal monitoring to better understanding impact of anthropogenic activities [20]. These including current information of habitat extend and its change, habitat condition, species diversity and the threats [21]. In recent years, the trend of logging activities are decreased and reforestation effort are increased [17]. Knowing how and why the changes over times is important especially in long-term forest loss or gained forest (e.g. afforestation). Knowing forest losses and gain help scientist to monitor carbon stock and gauge forest ability to reduce greenhouse gasses.

In Gunung Palung National Park, degraded forest are becoming young regenerating forest or secondary forest with rate up to 6.5% in 2011 and 18.8% in 2017. The gained forest due to increased law enforcement and global conservation effort to save remaining rain forest. This part also influenced by Reducing Emissions from Deforestation and Forest Degradation (REDD+) mechanism where work to slow rates of forest degradation. We also identified a fragmented forest where separate from main forest in Gunung Palung. In hilly areas inside the park around the settlement, from images we found reforested area. However, ground-checking reveal reforested area is durian or rubber tree plantation. Hence, ground-checking always important part in forest assessment using remote sensing.

IOP Conf. Series: Earth and Environmental Science **149** (2018) 012038 doi:10.1088/1755-1315/149/1/012038

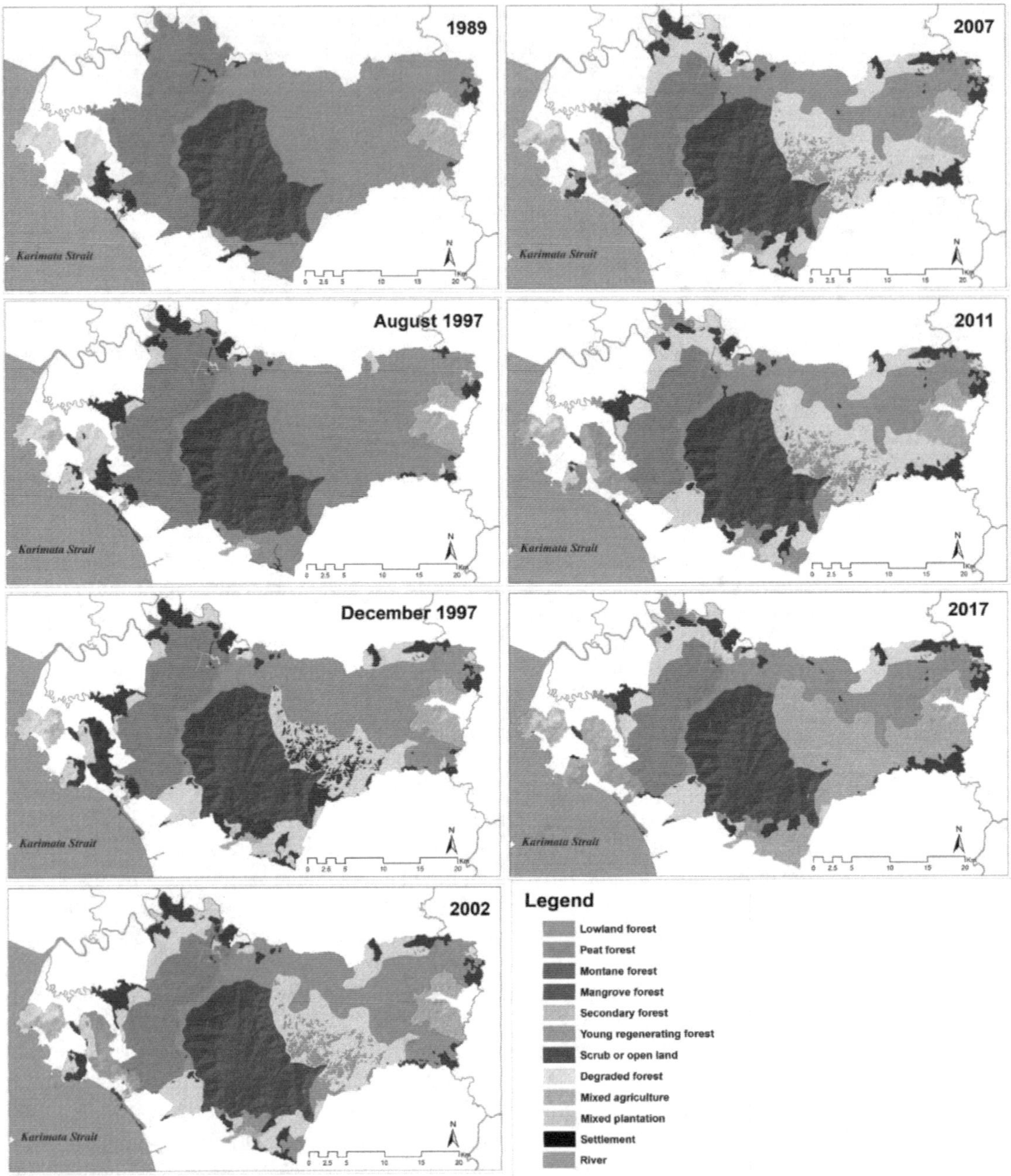

Figure 2. Land cover map of Gunung Palung National Park. Deforestation reached a peak in 1997, when a fifth of the park's area was cleared in just four months. Deforestation rates have declined since then, and many degraded areas are growing into secondary forest.

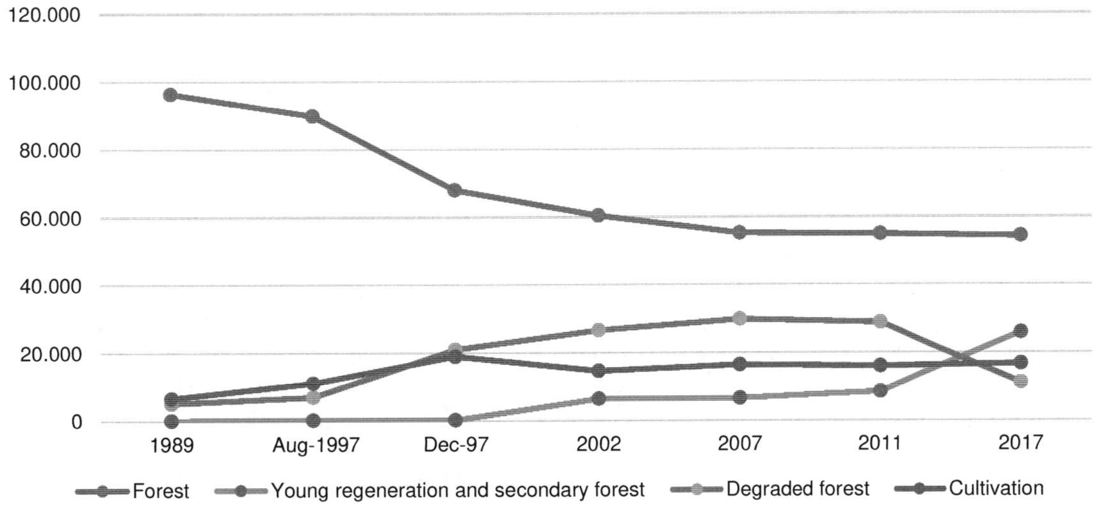

Figure 3. Pattern of forest loss in Gunung Palung national Park. Primary forest extent has declined every year, but some logged forests are beginning to regrow into secondary forest.

3.3. Discussion

National Park is closely with environmental, water, land use, educational, and social term that must be understood. Forest conservation is linked to many aspect such as environmental, economic and social, and government regulation. The base part of forest conservation in Gunung Palung National Park is save remaining forest with tackling its underlying economic and social causes. What is no conservation effort to save forest?

Biodiversity will remain decrease, fragmentation, and habitat for endangered species will loss. We also facing lower stock carbon if we still loss many forest. In the future, enhance forest management which good management base on social factor should be part of conservation. And remote sensing is an effective way to monitor forest cover with multi-temporal analysis [10]. The increase availability of remote sensing data including hyperspectral images, radar, LiDAR, and utilization of unmanned aerial vehicles (UAV), allow us to study broader forest monitoring [22]. Future forest monitoring will be dependency with remote sensing data and analysis [23]. We highlighted that free and open source remote sensing data such as Landsat and Sentinel data remain an important data source for historical forest change monitoring. The majority of conservation research have used Landsat images to assess change in protected area, with consideration of continues data [21]. But, we need better spatial resolution to get more detailed information, e.g. canopy density, tree height, vegetation health, anthropogenic threat assessment, etc. these information can give impact for conservation management in future.

In practical conservation program, people participation for conservation is important part. Halting deforestation in this protected area requires us to tackle its underlying economic and social causes, and find ways for communities to meet their needs without resorting to forest clearing. Increase willingness to participate in forest management and conservation with increasing people's knowledge about the benefits due to conservation and land use right security become positive thing in conservation [24]. Conservation efforts that have been done is community empowerment, forest rehabilitation, and health care incentives as payment for ecosystem services can help reduce deforestation in Gunung Palung National Park.

Failing to account for social factors around protected areas can lead to repeated deforestation in the future [25]. Poverty, for example, is a major cause of deforestation [26], as poor people living near protected areas often rely on livelihoods that cause forest loss, such as shifting agriculture, timber harvesting, or plantations. We should emphasis to solve land tenure problem to people who cultivated land inside national park. Maybe we should promote forest farming or agroforestry, a sustainable way to use forest resources as a non-timber forest products (NTFPs) as a more comprehensive concept of forestry land use [27]. The NTFPs product such as wild boars, deer, wild fruits, firewood, wild vegetables, and wild mushrooms; and valued range from 34 – 169 USD/ha [17]. Even tough, the NTFPs contribution from forest to poverty alleviation is minor at present [28], but still income from forest can give an average 5.8% contribution to total household income [29].

4. Conclusion

Our forest monitoring in Gunung Palung National Park using remote sensing show deforestation rate now reduce to 0.21% in 2017, lower up to 120 times in peak of deforestation in December 1997. We found remote sensing is the effective way to map forest change in Gunung Palung Park. To stop this deforestation in protected area, we should finding way for communities to meet their needs without resorting to forest clearing or we fight deforestation at its source by tackling its underlying economic and social causes. Community empowerment, forest rehabilitation, and provide health care as an environmental payment service is the conservation can do in Gunung Palung National Park to reduce deforestation. The increased prosperity, community loggers started to quit his job. This becomes a positive trend which we must continue to always work in forest conservation. The increase availability of remote sensing data including hyperspectral images, radar, LiDAR, and utilization of unmanned aerial vehicles (UAV), allow us to study broader forest monitoring and its conservation. We highlighted that free and open source remote sensing data such as Landsat and Sentinel data remain an important data source for historical forest change monitoring.

Acknowledgements

We would say thank you to Yayasan Alam Sehat Lestari board who gave us opportunity and funding to this research.

References

[1] Gaveau D L, Kshatriya M, Sheil D, Sloan S, Molidena E, Wijaya A, Wich S, Acrenaz M, Hansen M, Broich M, Guariguata M R, Pacheco P, Potapov P, Turubanova S, Meijaard E 2013 Reconciling Forest Conservation and Logging in Indonesian Borneo *PLoS ONE* **8**

[2] Kissinger G, Herold M and De Sy V 2012 *Drivers of Deforestation and Forest Degradation* A Synthesis Report for REDD+ Policymakers p. 48

[3] Tsujino R, Yumoto T, Kitamura S, Djamaluddin I and Darnaedi D 2016 *History of forest loss and degradation in Indonesia* Land Use Policy **57** 335–347

[4] Spracklen B D, Kalamandeen M, Galbraith D, Gloor E and Spracklen D V 2015 A global analysis of deforestation in moist tropical forest protected areas *PLoS ONE* **10**

[5] Johnson A E, Knott C D, Pamungkas B, Pasaribu M and Marshall A J 2005 *A survey of the orangutan (Pongo pygmaeus wurmbii) population in and around Gunung Palung National Park, West Kalimantan, Indonesia based on nest counts* Biological Conservation

[6] Curran L M, Trigg S N, Mcdonald A K and Astiani D 2004 Lowland Forest Loss in Protected Areas *Terra* **303** 1000–1003

[7] Hiller M A, Jarvis B C, Lisa H, Paulson L J, Pollard E H B and Stanley S A 2004 Recent trends in illegal logging and a brief discussion of their causes: a case study from Gunung Palung National Park, Indonesia *Journal of Sustainable Forestry* **19** 81–212

[8] Wang K, Franklin S E, Guo X and Cattet M 2010 Remote sensing of ecology, biodiversity and conservation: a review from the perspective of remote sensing specialists *Sensors* **10** 9647–9667

[9] Grecchi R C, Beuchle R, Edemir Y, Aragão L E O C, Arai E, Simonetti D and Achard F 2017 An integrated remote sensing and GIS approach for monitoring areas affected by selective logging : A case study in northern Mato Grosso, Brazilian Amazon *International Journal of Applied Earth Observation and Geoinformation* **61** 70–80

[10] Masek J G, Hayes D J, Hughes M J, Healey S P and Turner D P 2015 The role of remote sensing in process-scaling studies of managed forest ecosystems *Forest Ecology and Management* **355** 109–123
[11] Kraxner F, Schepaschenko D, Fuss S, Lunnan A, Kindermann G, Aoki K, Dürauer M, Shvidenko A, See L 2017 Mapping certified forests for sustainable management - A global tool for information improvement through participatory and collaborative mapping *Forest Policy and Economics* **83** 10–18
[12] Hagner O and Reese H 2007 A method for calibrated maximum likelihood classification of forest types *Remote Sensing of Environment* **110** 438–444
[13] Anderson J R, Hardy E E, Roach J T and Witmer R E 1976 A Land Use and Land Cover Classification System for Use with Remote Sensor Data *United States Geological Survey Professional Paper* **964**
[14] Puyravaud J P 2003 Standardizing the calculation of the annual rate of deforestation *Forest Ecology and Management* **177** 593–596
[15] Sloan S and Sayer J A 2015 Forest Resources Assessment of 2015 shows positive global trends but forest loss and degradation persist in poor tropical countries *Forest Ecology and Management*
[16] FAO 2015 *Global Forest Resources Assessment 2015* (FAO Forestry)
[17] Zamzani F, Onda N, Yoshino K and Masuda M 2009 Deforestation and agricultural expansion processes in Gunung Palung National Park, West Kalimantan, Indonesia *JMHT* **15** 24–31
[18] Ode M D L 2012 *Etnis Cina Indonesia Dalam Politik: Politik Etnis Cina Dan Singkawang Di Era Reformasi 1998-2008* Yayasan Pustaka Obor Indonesia
[19] Meinard Y 2017 What is a legitimate conservation policy? *Biological Conservation* **213** 115–123
[20] Buchanan G M, Brink A B, Leidner A K, Rose R and Wegmann M 2015 Advancing terrestrial conservation through remote sensing *Ecological Informatics* **30** 318–321
[21] Nagendra H, Lucas R, Honrado J P, Jongman R H G, Tarantino C, Adamo M and Mairota P 2013 Remote sensing for conservation monitoring: Assessing protected areas, habitat extent, habitat condition, species diversity, and threats *Ecological Indicators* **33** 45–59
[22] Sankey T, Donager J, McVay J and Sankey J B 2017 UAV lidar and hyperspectral fusion for forest monitoring in the southwestern USA *Remote Sensing of Environment* **195** 30–43
[23] Houborg R, Fisher J B and Skidmore A K 2015 Advances in remote sensing of vegetation function and traits *International Journal of Applied Earth Observation and Geoinformation* **43** 1–6
[24] Brännlund R, Sidibe A and Gong P 2009 Participation to forest conservation in National Kabore Tambi Park in Southern Burkina Faso *Forest Policy and Economics* **11** (7) 468–474
[25] Baird T D and Leslie P W 2013 Conservation as disturbance: Upheaval and livelihood diversification near Tarangire National Park, northern Tanzania *Global Environmental Change* **23** 1131–1141
[26] Zwane A P 2007 Does poverty constrain deforestation? Econometric evidence from Peru *Journal of Development Economics* **84** 330–349
[27] Sisak L, Riedl M and Dudik R 2016 *Non-market non-timber forest products in the Czech Republic-Their socio-economic effects and trends in forest land use* (Land Use Policy) pp. 390–398
[28] Sakai S, Choy Y K, Kishimoto-Yamada K, Takano K T, Ichikawa M, Samejima H, Kato Y 2016 Social and ecological factors associated with the use of non-timber forest products by people in rural Borneo *Biological Conservation* **204** 340–349
[29] Oli B N, Treue T and Smith-Hall C 2016 The relative importance of community forests, government forests, and private forests for household-level incomes in the Middle Hills of Nepal *Forest Policy and Economics* **70** 155–163

IOP Conf. Series: Earth and Environmental Science **149** (2018) 012039 doi:10.1088/1755-1315/149/1/012039

Hydrological similarity approach and rainfall satellite utilization for mini hydro power dam basic design (case study on the ungauged catchment at West Borneo, Indonesia)

W G Prakoso[1*], K Murtilaksono[2], S D Tarigan[2], Y J Purwanto[3]

[1]Graduate School Bogor Agricultural University, Jl. Raya Darmaga, Gedung Sekolah Pascasarjana IPB, Kampus IPB Darmaga Bogor 16680 - Jawa Barat, Indonesia
[2]Watershed Management, Departement Of Soil Science and Land Resources, Jl. Meranti Kampus IPB Darmaga, Bogor 16680
[3]Departement Of Civil and Environmental Engineering, Gedung FATETA Jl. Lingkar Akademik, Jawa Barat 16680, Indonesia

E-mail : wahyu_prakoso@apps.ipb.ac.id

Abstract. An approach on flow duration and flood design estimation on the ungauged catchment with no rainfall and discharge data availability was been being develop with hydrological modelling including rainfall run off model implemented with watershed characteristic dataset. Near real time Rainfall data from multi satellite platform e.g. TRMM can be utilized for regionalization approach on the ungauged catchment. Watershed hydrologically similarity analysis were conducted including all of the major watershed in Borneo which was predicted to be similar with the Nanga Raun Watershed. It was found that a satisfactory hydrological model calibration could be achieved using catchment weighted time series of TRMM daily rainfall data, performed on nearby catchment deemed to be sufficiently similar to Nanga Raun catchment in hydrological terms. Based on this calibration, rainfall runoff parameters were then transferred to a model. Relatively reliable flow duration curve and extreme discharge value estimation were produced with reasonable several limitation. Further approach may be performed in order to deal with the primary limitations inherent in the hydrological and statistical analysis, especially to give prolongation to the availability of the rainfall and climate data with some novel approach like downscaling of global climate model.

1. Introduction

Hydroelectricity potential in Indonesia reaches approximately 22 GW, however, the power utilization from hydroelectricity still as low as 6 % from it's potential in 2012. Hydroelectricity as renewable energy power source in Indonesia facing challenging condition from the design stage, e.g. causing by unavailable of critical data for design parameterization including rainfall, discharge and sedimentation at many region of Indonesia. Even when it's available usually not in well distributed condition, not long enough period, and or with lower reliability and quality.

Approach on Flow duration and flood design estimation on the ungauged catchment with no rainfall and discharge data availability are been being develop by researchers. In this case, approach which are used in the research with following some of the publication [1];[2];[3];[4]. Generally, discharge characteristic of ungauged watershed can be estimated by extrapolating and transferring discharge characteristic from hydrologically similar gauged watershed with regionalization process[5];[6];[7].

Further approach can by develop with hydrological model including rainfall run off model implemented with watershed characteristic dataset. Near real time Rainfall data from multi satellite platform e.g. TRMM, APHRODITE, and CMORPH can be utilized for regionalization approach on the ungauged catchment [8];[9];[10];[11];[12];[13];[14] .

The primary objective of this study are :
a. Determine / making Flow duration curve estimation for mini hydro power plant design on ungauged catchment
b. Determine/ making Design Flood estimation for mini hydro power plant design on ungauged catchment

Scope of works during this study are limited to :
a. Hydrologically similarity analysis for the ungagged- gauged watershed
b. Setup Model, Calibration, and validation for gauged watershed model (donor watershed)
c. Setup Model, FDC analysis and Flood Design estimation for ungauged watershed (target watershed) using transferred parameters from gauged donor watershed model.
d. TRMM Rainfall Satellite Data

2. Methods

2.1. Location
Mini hydro power site located on the Nanga Raun Watershed, upper Kapuas, West Borneo Indonesia

2.2. Data and Software
Data and Software which are used in the study

Table 1. Data and software

Data	Source	Note	Software	Version
DEM	SRTM		ArcGIS	10.1
River Polyline	SRTM, RBI		SAGA	
Rainfall Satellite Data	TRMM, GSMap	TRMM 3B42 Ver.7	DHI MIKE	11
Evaporation	BMKG		SMADA	
Raingauge	BWS, BMKG		MS Excell	2010
Discharge	BWS			
Landsystem	BIG	1: 250.000		
Land Unit	BBSDLP			
Groundwater	BGL ESDM	1: 250.000		

2.3. Flow Chart

Working flow of the study are as follows

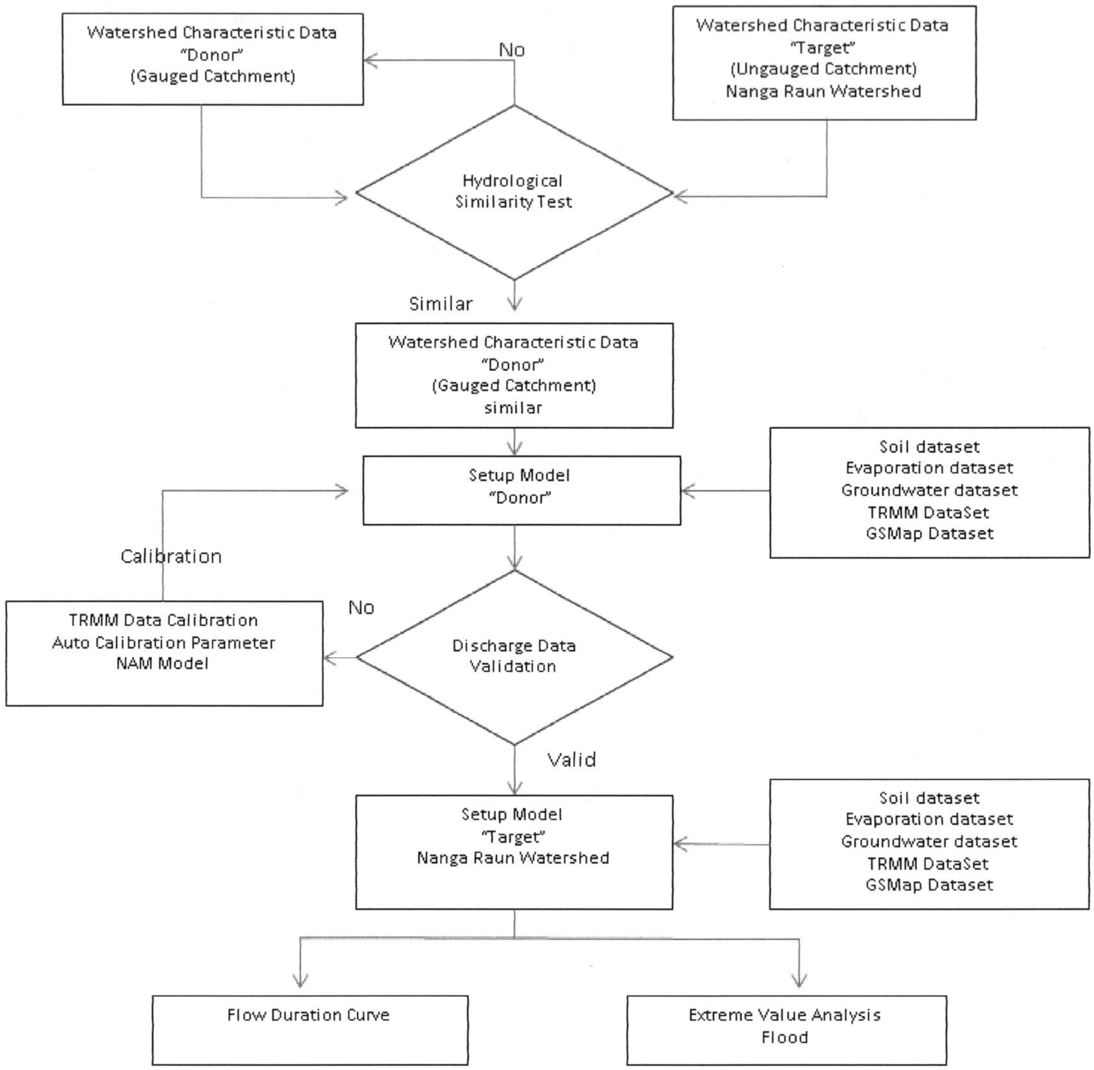

Figure 1. Research Flow Chart

3. Results/Discussion

3.1. Watershed Hydrologically Similarity

Watershed hydrologically analysis are including all of the major watershed in Indonesian Borneo Region which predicted to be similar to the Nanga Raun Watershed. Hydrologically similarity criteria are used in this study following the catchment similarity analysis protocol proposed by previous researcher [15];[16];[17];[18];[19];[20];[21]. Including :

a. Watershed area
b. GSMap Rainfall
c. Median TRMM Daily Rainfall
d. 95th percentile

IOP Conf. Series: Earth and Environmental Science **149** (2018) 012039 doi:10.1088/1755-1315/149/1/012039

e. Average TRMM Daily Rainfall
f. Percentage of TRMM Rainy days (percentage of rainy day TRMM, TRMM Rainfall > 0)
g. Available Water Content
h. Years of Data

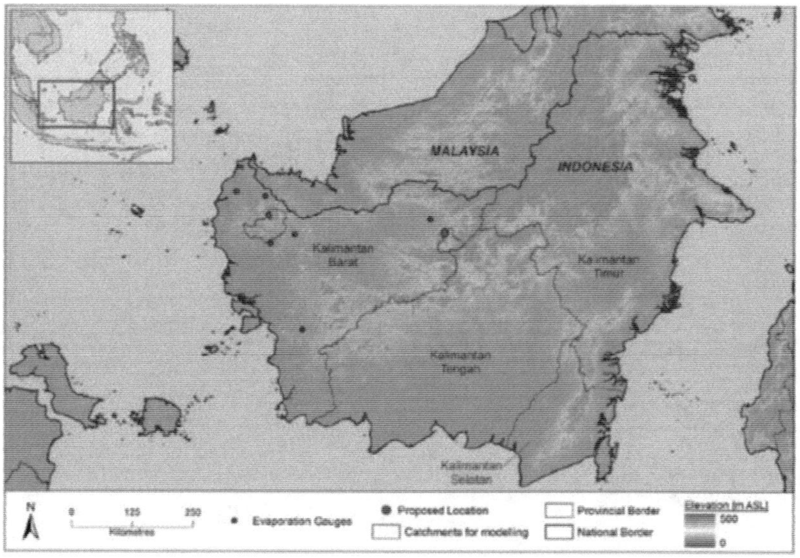

Figure 2. Similar Donor Watershed Gauge No. 03-044-01-04

3.2. Setup Model, calibration , and Validation Donor Watershed Model

Based on the hydrologically similar donor watershed, setup parameter in the NAM Mike 11 model are transferred to the target watershed (Nanga Raun Watershed) with replacing the watershed characteristic dataset, including :
a. DEM and Watershed boundary
b. Evaporation data set
c. Soil dataset
d. Groundwater dataset
e. TRMM Rainfall dataset

Running NAM – MIKE 11 Model are used for simulated discharge. Discharge data then will be calibrated using the observed discharge data from the automatic water level recorder available on donor watershed.

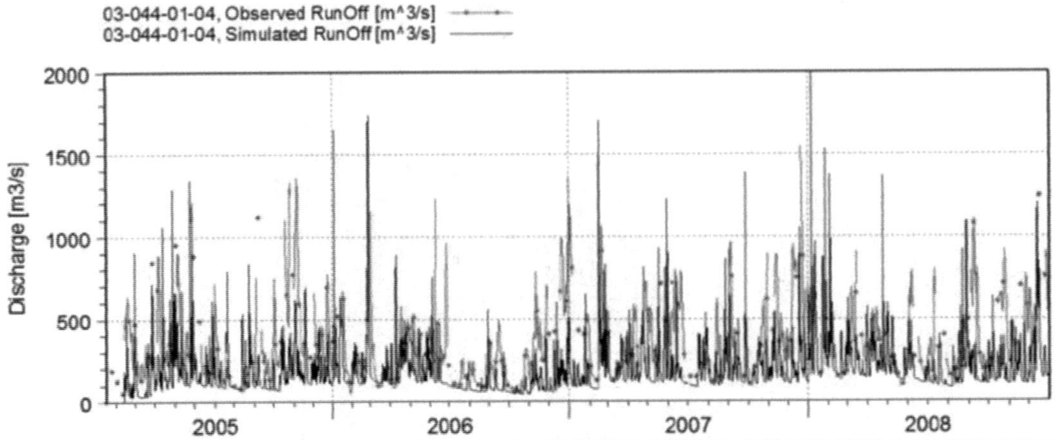

Figure 3. Comparison of observed (red) and simulated (black) discharge at 03-044-01-04, based on TRMM Rainfall Data

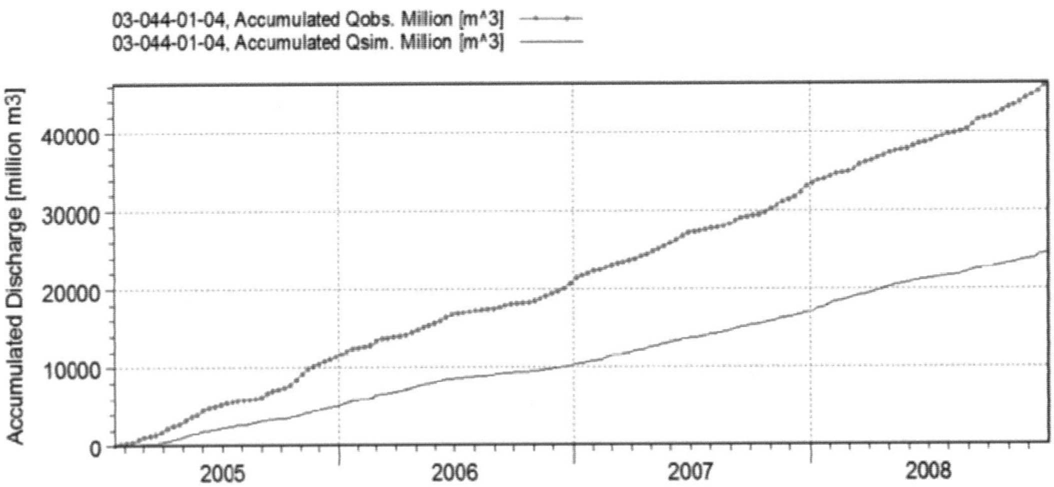

Figure 4. Comparison of observed (red) and Simulated (black) accumulated discharge, based on TRMM Rainfall Data

The simulated runoff does appear to be too low, as is confirmed by a comparison of the accumulated discharges, in Figure 4. This may indicate that the TRMM data underestimate the actual rainfall over the catchment. The TRMM rainfall data were increased by a factor of 1.4 (with trial-error process in order to match the observed water balance based on discharge data), and a series of auto-calibration routines was run in order to optimize the model parameters. Following this, the comparison of simulated discharge with the observed data is as shown below Figure 5. And Figure 6 Show the simulated versus observed discharge time-series, which indicates that a much closer fit has been achieved than previously. The difference in water volumes at the end of the simulation is approximately 2 %. Auto calibration NAM MIKE 11 Rainfall Runoff Model Parameters are including :

a. Maximum water content in surface storage (Umax)
b. Maximum water content in root zone storage (Lmax)
c. Overland Flow Runoff Coefficient (CQOF)
d. Time constants for Interflow (CKIF)
e. Time constants for Routing Overland flow (CK1,2)

f. Root zone Threshold value for overland flow (TOF)
g. Root zone Threshold Value for interflow (TIF)
h. Time constants for routing base flow (CKBF)
i. Root Zone Threshold value for Groundwater Recharge (Tg)

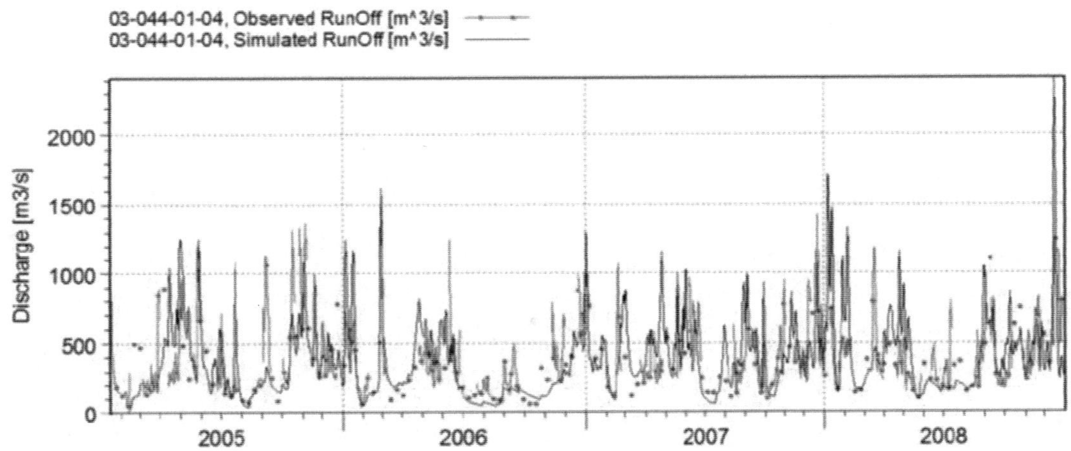

Figure 5. Comparison of Observed (red) and Simulated (black) discharge at 03-044-01-04, after a number of auto-calibrations have been run (using scaled TRMM Data)

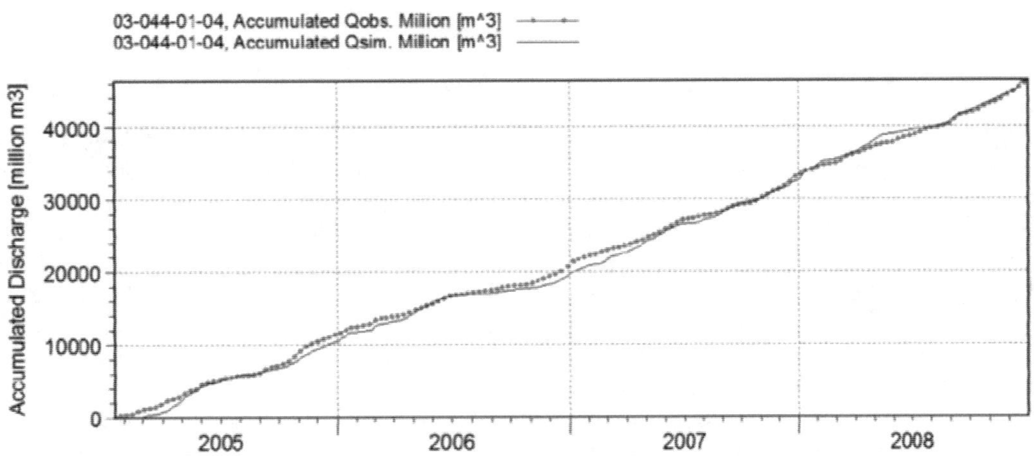

Figure 6. Comparison of Observed (red) and simulated (black) accumulated discharge, after a number of auto calibrations have been run (using scaled TRMM data)

While the time series shown above does not indicate a very close match between observed and simulated discharge for specific events. Instead, it is most important that the model be shown to capable of representing the basic hydrological regime of the river at the site location. For this reason, it is most interesting to calibrate the model based on comparison of Flow duration Curve (observed FDC Vs simulated FDC), this approach has previously been used by another researchers [14].

As shown in Figure 7, the final calibrated simulation appears to be capable of representing the observed Flow Duration Curve at 03-044-01-04 relatively close (with deviation below 5 %). This provide some confidence that the calibrated parameter set may be reasonably transferred to a model of the proposed site on the ungauged catchment like in Nanga Raun Catchment, West Borneo in order to simulate the hydrological regime at that location too.

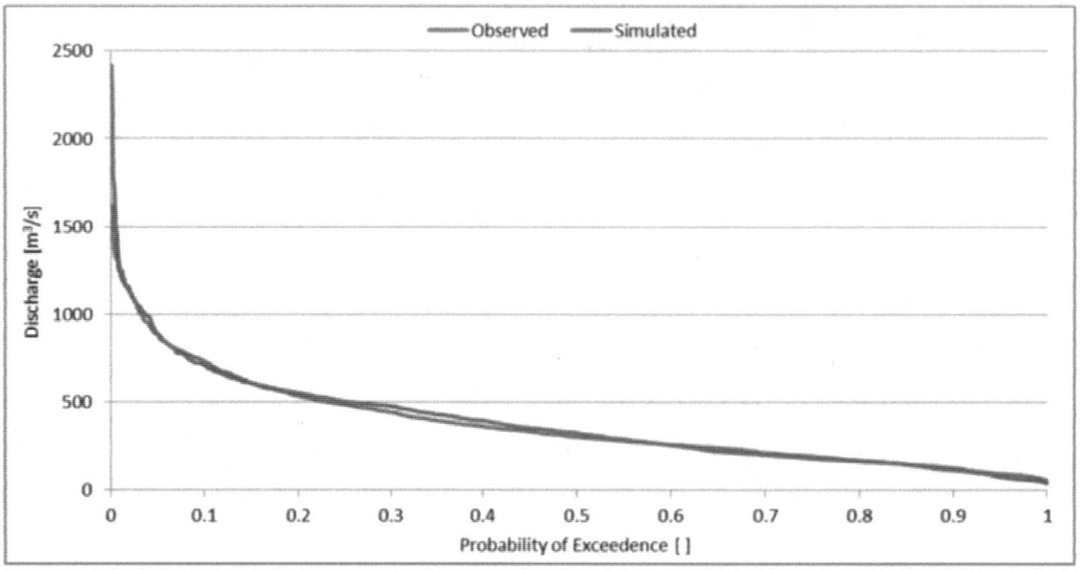

Figure 7. Comparison of observed (blue) and Simulated (red) Flow Duration Curve at Gauging Station 03-044-01-04, Based on TRMM Rainfall Data

3.3. Simulation of Nanga Raun Catchment (target/ site ungauged watershed) based on Calibrated parameters

Following on from the hydrological model calibrations presented in section b, the calibrated rainfall-runoff model parameters were then transferred to the models of the proposed site's catchment in Nanga Raun. It is assumed that these parameters will provide a reasonable assumption for the project catchment too, based on the hydrological similarity. The catchment averaged TRMM rainfall data were scaled up by a factor of 1,4 and used to run simulation from 2/3/2000 to 1/1/2013 (based on the availability of rainfall data). Note that evaporation data were only available from 1/1/2002 to 31/12/2008, with the period before and after that filled in based on the monthly averages of the available data. The flow duration curve and Extreme Value estimation (Design Flood) produce from these simulation result are presented in Figure 8, Figure 9 and Figure 10.

The extreme value analysis (Design Flood Estimation) processing was performed using simulated discharge data, from March 2000 – December 2012, based on scaled TRMM rainfall data from the nearest climatological station. Given that almost 13 years of data are available, this suggests that extreme events up to 30 year may be reliably predicted, with larger events becoming less and less reliable.

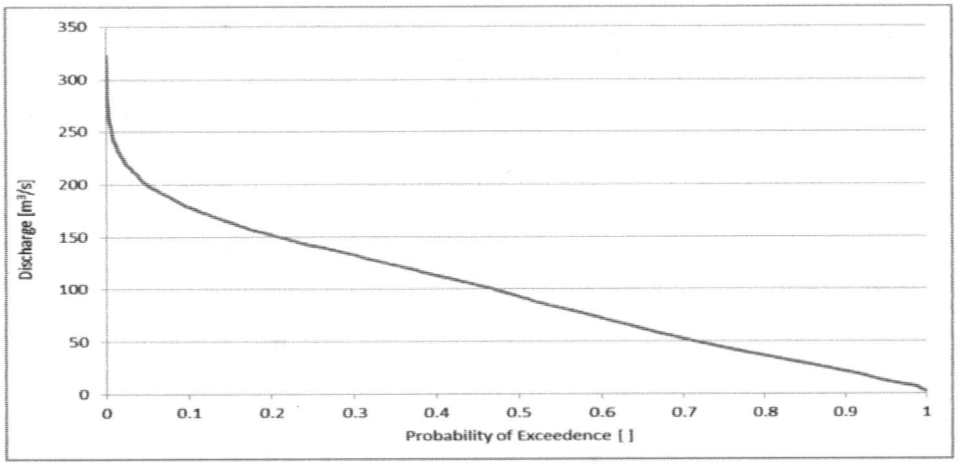

Figure 8. Flow Duration Curve for Proposed Mini hydro Power Site At Nanga Raun, Based on TRMM Rainfall Data

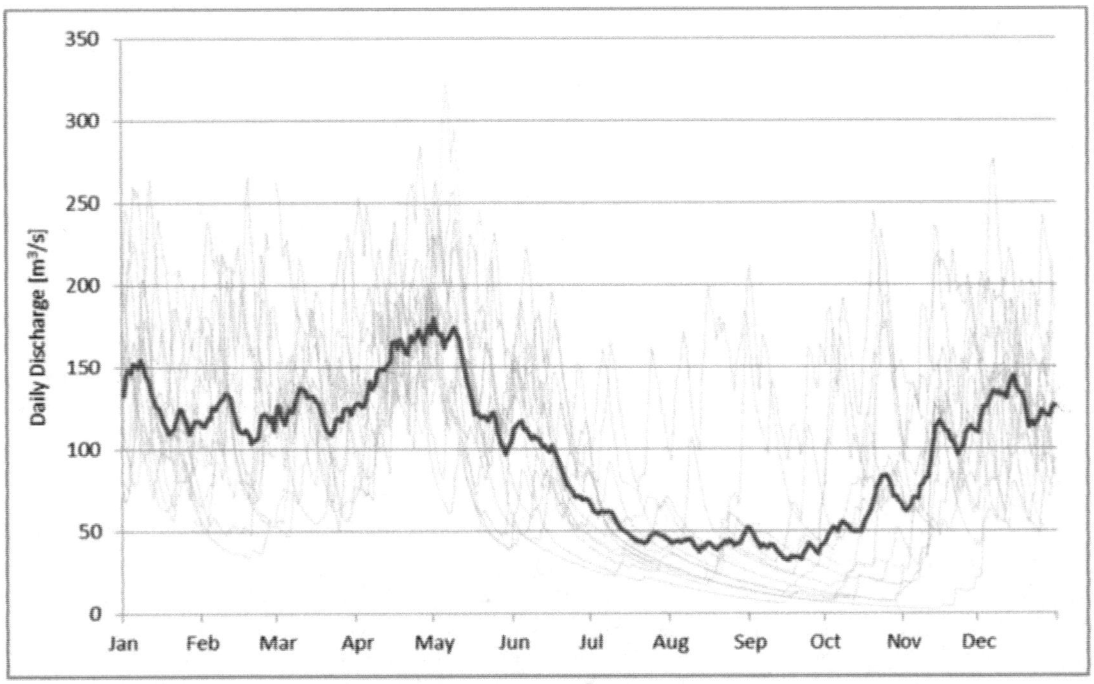

Figure 9. Average Daily Discharge Over the Calendar Year (red line), with individual years simulated shown by light grey lines.

Given the relatively short period of historical discharge data available (less than 13 years), the partial duration series (PDS) method of extracting extreme values was used, instead of annual maxima. It was found that PDS threshold value of 180 m3/s, slightly higher than 90th percentile of simulated discharge values, provided good results. In other words, measured discharge values greater than this value were considered to be extreme events and included in subsequent analysis. In addition, an inter event time of 15 days was used as independence criterion to ensure that the same extreme flood event was not counted more than once. A variety of extreme value distributions types were trialed in order to select that which appeared to fit the data best. It was found that a Log-normal distributions, using the method of moments for estimation Lower bound quantile method II provided the best fit.

The extreme value probability distribution fitted to the simulated discharge data sets is presented in Figure 10, showing a relatively close fit. While the fit is not as close for the larger flood events, this result is considered acceptably conservative. As noted previously, the limited data available only allow prediction of extreme events with return periods of up to approximately 30 years. However, it is possible to estimate the extreme events of return period 1,5,10,25,50,100, and 200 years with less and less reliable. Thus, the extreme value distributions fitted to the available data have been extended to allow predictions of these events. It should be stated very clearly that these predictions are based on insufficient data.

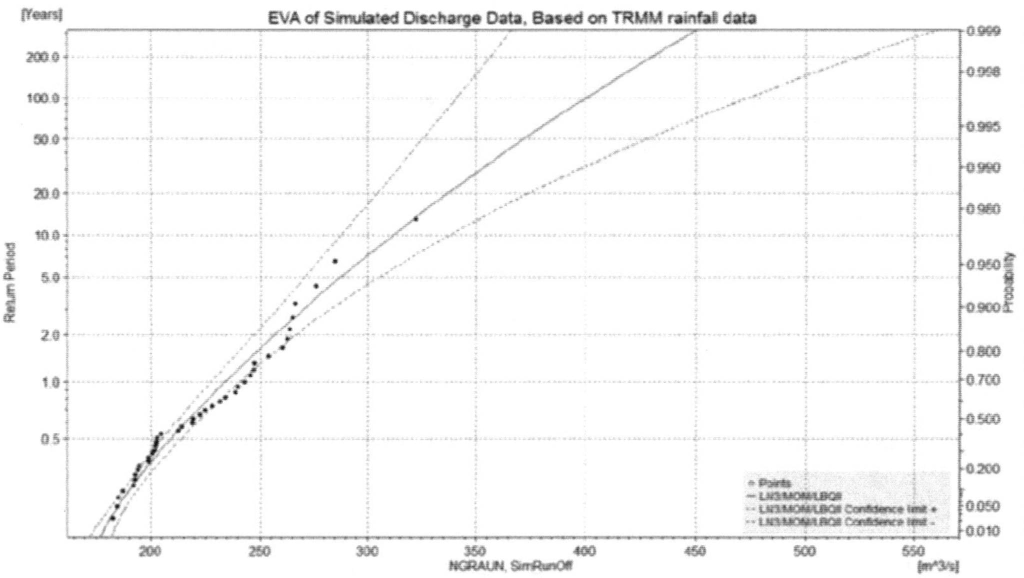

Figure 10. Probability distribution (Full Line) fitted to simulated discharge data (confidence intervals as dashed lines), assuming Log-normal distribution and method of moments for estimations

4. Conclusion

In conclusion, it was found that a satisfactory hydrological model calibration could be achieved using catchment weighted time series of TRMM daily rainfall data (after scaling), performed on nearby catchment deemed to be sufficiently similar to the proposed ungauged catchment in hydrological terms. Based on this calibration, rainfall runoff parameters were then transferred to a model of Nanga Raun catchment (as target ungauged catchment proposed as mini hydro site) and run to obtain a time series of simulated discharge. Relatively reliable flow duration curve and extreme value estimation (design flood) were produced with reasonable several limitation.

Further approach may be performed in order to deal with the primary limitations inherent in the hydrological and statistical analyses, especially to give prolongation to the availability of the rainfall and climate data with some novel approach like downscaling of global climate model to obtain either historical and future rainfall and climate data on the ungauged catchment.

References

[1] Ganora D, Claps P, Laio F and Viglione A 2009 An Approach To Estimate Nonparametric Flow Duration Curves In Ungauged Basins *Water Resource Research* **45** W10418
[2] Niadas IA 2005 Regional Flow Duration Curve Estimation In Small Ungauged Catchments Using

Instantaneous Flow Measurements And A Censored Data Approach *Journal of Hydrology* **314** 48-66.

[3] Post DA 2004 A New Method For Estimating Flow Duration Curves : An Application To The Burdekin River Catchment, North Queensland, Australia *CSIRO Land and Water, Davies Laboratory*

[4] Mazvimavi D 2003 Estimation Of Flow Characteristics Of Ungauged Catchment: Case Study In Zimbabwe [PhD Thesis] Wageningen University, ITC

[5] Nathan RJ and Mc Mahon TA 1990 Identification Of Homogenous Regions For Purposes Of Regionalization *Journal of Hydrology* **121** 217-238

[6] Bullock A and Andrews A 1997 Introduction To Southern Africa FRIEND Technical Document in Hydrology No.**15**. UNESCO: Paris

[7] Hall MJ and Minns AW 1999 The Classification Of Hydrologically Homogeneous Regions *Hydrological Sciences Journal* **44**(8) 693-704

[8] Adler RF, Huffman GJ, Bolvin DT, Curtis S and Nelkin EJ 2000 Tropical Rainfall Distributions Determined Using TRMM Combined With Other Satellite And Rain Gauge Information *Journal of Applied Meteorology* **39** 2007-2023

[9] Hughes DA 2006 Comparison Of Satellite Rainfall Data With Observations From Gauging Station Networks *Journal of Hydrology* **327** 399-410

[10] Mantas VM, Liu Z, Caro C and Pereira AJSC 2015 Validation Of TRMM Multi-Satellite Precipitation Analysis Products In Peruvian Andes *Atmospheric Research* **163** 132-145

[11] Pombo S and de Olieviera RP 2015 Evaluation Of Extreme Precipitation Estimates From TRMM In Angola *Journal of Hydrology* **523** 663-679

[12] Salio P, Hobouchian MP, Skabar YN and Vila D 2015 Evaluation Of High Resolution Satellite Precipitation Estimates Over Southern South America Using A Dense Rain Gauge Network *Atmospheric Research* **163** 146-161

[13] Semire FA, Mohd-Mokhtar R, Ismail W, Mohamad N and Mandeep JS 2012 Ground Validation Of Space Borne Satellite Rainfall Products In Malaysia *Advances in Space Research* **50** 1241-1249

[14] Westerberg IK, Guerrero JL, Younger PM, Beven KJ, Seibert J, Halldin S, Freer JE, and Xu CY 2011 Calibration of Hydrological Models Using Flow Duration Curves *Hydrol. Earth Syst. Sci* **15** 2205-2227

[15] Castellarin A, Galeati G, Brandimarte L, Montanari A and Brath A 2004 Regional Flow Duration Curve : Reliability For Ungauged Basins *Advances in Water Resources* **27** 953-965

[16] Merz R, Bloschl G and Parajka J 2006 Regionalization Method In Rainfall-Runoff Modelling Using Large Catchment Samples *IAHS Publ.* **307** 117-125

[17] Makungo R, Odiyo JO, Ndritu JG and Mwaka B 2010 Rainfall – Runoff Modelling Approach For Ungauged Catchments : A Case Study Of Nzhelele River Sub Quatemary Catchment. *Physic and Chemistry of The Earth* **35** 596-607

[18] Sawiez K, Wagener T, Sivapalan M, Troch PA and Carillo G 2011 Catchment Classification : Empirical Analysis Of Hydrologic Similarity Based On Catchment Function In The Eastern USA *Hydrol. Earth. Syst. Sci.* **15** 2895-2911

[19] Patil S and Stieglitz M 2012 Control On Hydrologic Similarity : Role Of Nearby Gauged Catchment For Prediction At An Ungauged Catchment. *Hydrol. Earth. Syst. Sci.* **16** 551-562

[20] Zelelew MB and Alfredsen K. 2014. Transferability Of Hydrological Model Parameter Spaces In Estimation Of Runoff In Ungauged Catchments *Hydrological Sciences Journal* **59**(8) 1470-1490

[21] Munoz E, Parra V and Arumi JL 2016 Caution In Using The Hydrologic Similarity Concept In Volcanic Areas. Practical Recommendations For Management Based On Andean Watershed. *Geophysical Research Abstract* **18** 2016-9177

IOP Conf. Series: Earth and Environmental Science **149** (2018) 012040 doi:10.1088/1755-1315/149/1/012040

Estimation of carbon emission from peatland fires using Landsat-8 OLI imagery in Siak District, Riau Province

Dinda Aisyah Fadhillah Hafni[1*], Lailan Syaufina[2], Nining Puspaningsih[3], Indah Prasasti[4]

[1] Graduate School of Bogor Agricultural University, Bogor 16680, Indonesia
[2] Department of Silviculture, Faculty of Forestry, Bogor Agricultural University, Bogor 16680, Indonesia
[3] Department of Forest Management, Faculty of Forestry, Bogor Agricultural University, Bogor 16680, Indonesia
[4] Remote Sensing Application Center, LAPAN, Jakarta 13710, Indonesia

E-mail: dfadhillahhafni@gmail.com

Abstract. The study was conducted in three land cover conditions (secondary peat forest, shrub land, and palm plantation) that were burned in the Siak District, Riau Province, Indonesia year 2015. Measurement and calculation carbon emission from soil and vegetation of peatland should be done accurately to be implemented on climate change mitigation or greenhouse gases mitigation. The objective of the study was to estimate the carbon emission caused peatland fires in the Siak District, Riau Province, Indonesia year 2015. Estimated carbon emissions were performed using visual method and digital method. The visual method was a method that uses on-screen digitization assisted by hotspot data, the presence of smoke, and fire suppression data. The digital method was a method that uses the Normalized Burn Ratio (NBR) index. The estimated carbon emissions were calculated using the equation that was developed from IPCC 2006 in Verified Carbon Standard 2015. The results showed that the estimation of carbon emissions from fires from above the peat soil surface were higher than the carbon emissions from the peat soil. Carbon emissions above the peat soil surface of 1376.51 ton C/ha were obtained by visual method while 3984.33 ton C/ha were obtained by digital method. Peatland carbon emissions of 6.6 x 10^{-4} ton C/ha were obtained by visual method, whereas 2.84 x 10^{-3} ton C/ha was obtained by digital method. Visual method and digital method using remote sensing must be combined and developed in order to carbon emission values will be more accurate.

1. Introduction

Peatland ecosystem plays an important role in maintaining environmental balance, including preventing floods during the rainy season and releasing moisture back into the air during the dry season, however, this ecosystem is very fragile. Peatland ecosystems disturbed by particular factors such as drying with the canalization system will gradually dry up and have vulnerability to fire. Organic materials in peatland ecosystems have a great potential in triggering the fires when it is supported by the suitable weather/climate conditions.

Peatland fire is categorized in the most dangerous types of fires [1]. Peatland fires are a major contributor to Indonesia's greenhouse gas (GHG) emissions. About 500 billion tons of carbon are stored in the vegetation that exists worldwide. The carbon element is the dominant compound in peatland fires, almost 45% of the plant's dry matter is carbon [2]. The increasing carbon compounds as greenhouse gases, the greenhouse effect is also increasing.

The emission calculations for tropical peatlands in relation to fires been based upon limited field data. A study of the West and East Kalimantan region conducted by the Forestry Research and Development Team and the Directorate General of Plants (2009) states that there are obstacles in implementing the IPCC guideline 2006, particularly the data limitations that result in low levels of detail. Data needed to calculate emissions using IPCC guideline 2006 are activity data and emission factor data or absorption. Activity data such as peatland fires are required in hectares, so that accuracy of peatland fires are urgently needed to produce accurate carbon emissions data.

Remote sensing and Geographic Information Systems (GIS) play an important role in the study of factors affecting fire occurrence and understanding fire behavior [6]. Previous studies have used satellite-based sensors AWiFS, LISS-III, ETM +, SPOT, AATSR, AVHRR, and MODIS [7]. Images from MODIS and Landsat TM /ETM + satellites are common sources of data and have the potential to monitor forest fires [7]. Several studies have studied the topic related to the use of remote sensing satellite imagery to identify forest and land fires, including the identification of burned areas, however, the research has not been widely developed in Indonesia. Researchers in Indonesia have been detection using Landsat TM multi-time satellite imagery, MOS-MESSR, and Landsat TM multisensor images, and using SPOT images [3], [4], [5]. Therefore, it is necessary to conduct the study on burned areas with the the aim was to test the reliability of quantitative methods (digital) and qualitative (visual) methods in order to estimate the burned area.

2. Research method

2.1. Study area

The research was conducted on May 2016 to April 2017. The study was conduced in Siak Regency. Data processing and analysis of imagery data were performed in Remote Sensing Laboratory and GIS, Forestry Faculty of IPB, Bogor, and Utilization Center of Remote Sensing LAPAN.

2.2. Tools and materials

The tools used in this study were a set of computers with some software such as Microsoft Excel 2016 for tabulation and graphics processing, digital camera, GPS (Global Positioning System), compass, oven, trash bag, plastic straps, phi band, tally sheet. The materials used in this study was primary and secondary data, namely: (1) top surface plat conditions (trees, poles, stakes, seedlings, necromass, dead woods) and peat soils, (2) the preliminary research data (burned area and land cover) [8]

2.3. Method and data analysis

2.3.1. Determination and preparation of research plots.

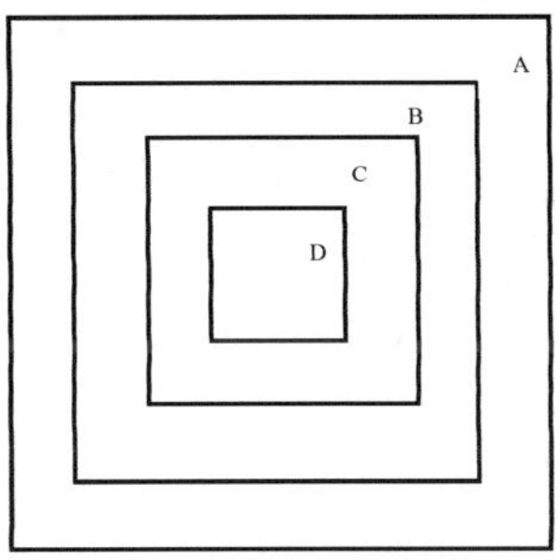

Where:
A : tree plot (20m × 20m);
B : pole plot (10m × 10m);
C : stake plot (5m × 5m);
D : seedling plot (2m × 2m).
The plot distance is 50 m.

Figure 1. Research plot design.

The sampling technique used is a randomly sampled from SNI 7724 year 2011 on measurement and calculation of carbon stock. The size of the largest plot is 20 m × 20 m (figure 1), whereas the plot

IOP Conf. Series: Earth and Environmental Science **149** (2018) 012040 doi:10.1088/1755-1315/149/1/012040

size for vegetations are: (a) seedling with a minimum area of 4 m^2, (b) stake with a minimum area of 25 m^2, (c) pole with a minimum area of 100 m^2, and (d) tree with a minimum area of 400 m^2.

2.3.2. Biomass data collection. The biomass data were collected in each plot. All sections were collected in each collection plot, then weighed the total wet weight and sampled each of ± 300 grams wet weight. The biomass samples were dried by oven in the laboratory with a temperature range of 70°C to 85°C until it reached a constant weight for 2 × 24 hours. Thus will be obtained dry weight (biomass).

2.3.3. Soil sampling. The soil samples were taken in a distrubed and intact soil. The disturbed soil was weighed to obtain the net weight while the intact soil was taken using a ring sampler to obtain the bulk density. Soil samples were taken as many as 3 samples in each plot 20m × 20m.

2.3.4. Calculation of biomass carbon stock. The tree biomass was estimated using allometric equations depending on DBH (Diameter at Breast Height) in cm and WD (Wood Density) in g cm^{-3} [9]:

$$AGB_{est}=0.242\times DBH^{2.473}\times WD^{0.736} \quad (1)$$

Other biomass data were calculated based on SNI 7724 year 2011. Then, these biomassa data were converted to carbon stock data.

2.3.5. Estimation of carbon emisions. Estimating carbon emissions from peatland fires used the IPCC 2006 method in the Verified Carbon Standards 2015 [10]. Estimating can be used the following equations:

$$E_{biomassburn,\,i,\,t}=\sum_{g=1}^{G}\left(\left(\left(A_{burn,\,i,\,t}\times B_{i,\,t}\times COMF_i\times G_{g,\,i}\right)\times 10^{-3}\right)\times GWP_g\right) \quad (2)$$

Where:

$E_{biomassburn,i,t}$	Greenhouse gas emissions due to biomass burning as part of deforestation activities in stratum i in year t of each GHG (CO_2, CH_4, CO) (t CO_2e)
$A_{burn,i,t}$	Area burnt for stratum i in year t (ha)
$B_{i,t}$	Average aboveground biomass stock before burning stratum i, year (t d.m. ha^{-1})
$COMF_i$	Combustion factor for stratum i (unitless)
$G_{g,i}$	Emission factor for stratum i for gas g (kg t^{-1} d.m. burnt)
GWP_g	Global warming potential for gas g (t CO_2/t gas g)
g	1, 2, 3 ... G greenhouse gases (unitless)
i	1, 2, 3 ...M strata (unitless)
t	1, 2, 3, ... t* time elapsed since the start of the project activity (years)

$$E_{peatburn,\,i,\,t}=\sum_{g=1}^{G}\left(\left(\left(A_{peatburn,\,i,\,t}\times P_{i,\,t}\times G_{peat,\,g,\,i}\right)\times 10^{-3}\right)\times GWP_g\right) \quad (3)$$

Where:

$E_{peatburn,i,t}$	Greenhouse emissions due to peat burning in stratum i in year t of each GHG (CO_2, CH_4, CO) (t CO_2e)
$A_{peatburn,i,t}$	Area peat burnt in stratum i in year t (ha)
$P_{i,t}$	Average mass of peat burnt in stratum i, year t (t d.m. ha^{-1})
$G_{peat,g,i}$	Emission factor in stratum i for gas g (kg t^{-1} d.m. burnt)
GWP_g	Global warming potential for gas g (t CO_2/t g)
g	1, 2, 3 ... G greenhouse gases unitless)

i 1, 2, 3 …M strata (unitless)
t 1, 2, 3, … t* time elapsed since the start of the project activity (year)

3 Result and discussion

3.1. Condition of measurement location for carbon stock

Map of research object in Dayun Village Dayun Subdistrict and Kotoringin Village Mempura Subdistrict of Siak District can be seen in figure 2. These locations are quite far from Siak City, which is about 15–30 km and could be reached by two-wheeled or four wheels vehicles. Mileage to the location of research about 1 hour. The location's coordinate of this measurement is the UTM 48 N (table 1).

Table 1. The Coordinate of sampling location.

Land Cover	Plot	Latitude (°LU)	Longitude (°BT)
Shrub	1	0.66724	102.10985
Shrub	2	0.66681	102.11005
Shrub	3	0.66624	102.11015
Oil palm plantation	1	0.87683	102.07080
Oil palm plantation	2	0.87666	102.07136
Oil palm plantation	3	0.87655	102.07176
Secondary forest	1	0.66904	102.09885
Secondary forest	2	0.66887	102.09888
Secondary forest	3	0.66880	102.09891

The location of carbon stock measurement is land under the control of Natural Resources Conservation Center (BBKSDA) of Riau Province. The condition of burning area consisted of 3 types of land cover namely shrub, oil palm plantation, and secondary forest. The topography of the three types of land cover is relatively flat. The thickness of peat in 3 land types was categorized as medium (1–2 m) for oil palm plantation and very deep category (>3 m) for shrub and secondary forest. Types of land cover are frequent fires due to dry or non-flooded peat conditions and the presence of many artificial canals nearby.

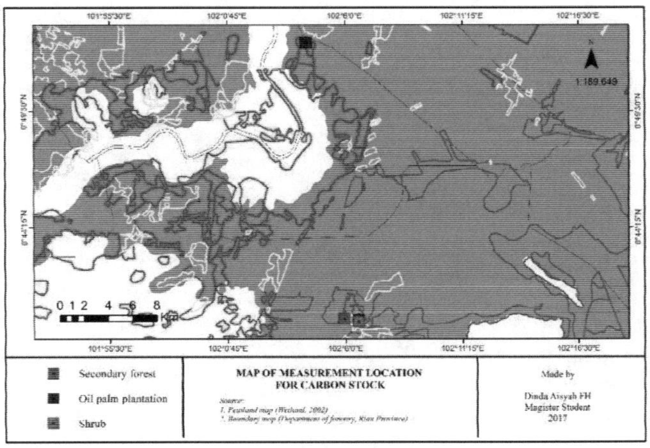

Figure 2. Map of measurement location for carbon stock.

3.2. Carbon stock at above ground of the burned area

This research used the allometric equation of Manuri *et al.* [9], Lubis [11], Hairiah *et al.* [12]. These equations were used to determine the carbon content at above ground. The results of carbon (ton/ha) contents calculations of the 9 sample plots consisting of different levels of regeneration, litter, and necromass are presented in table 2 and table 3.

Tables 2 and 3 show that the carbon content in secondary forest (104.72 ton/ha) was higher than that of oil palm plantation (42.47 ton/ha) and shrub (54.69 ton/ha). Carbon content in the swamp secondary forest area suggests the highest carbon content compared to other land cover types due to the highest tree density in the area in each diameter class. According to Syam'ami *et al.* [13], the amount of carbon stock in a region is closely related to the diameter growth and the quantity of tree biomass since most of the biomass is contained in tree trunks. The tree component has the largest percentage because the stem is the woody part and the largest storage of photosynthesis for growth. In secondary forest areas, it was obtained stands with a diameter almost 60 cm, this condition was not found in shrubs. The results of Indrajaya's research [14] showed that trees with stem diameters >60 cm, although the percentage was only 4.93%, has contributed 67% in storing carbon in the above-ground biomass.

Table 2. Total carbon content at above and below ground for the observation site of secondary forest and shrub.

Plot	Carbon of litter (ton/ha)	Carbon of understory Plants (ton/ha)	Carbon of dead tree (ton/ha)	Carbon of trees, poles, stakes (ton/ha)	Carbon of root (ton/ha)	Total carbon content (ton/ha)
Secondary forest						
I	3.02	0.41	0.45	82.93	11.35	98.16
II	1.13	0.28	0.62	137.14	18.77	157.95
III	2.69	0.06	0.66	48.08	6.58	58.06
Average	2.28	0.25	0.58	89.38	12.24	104.72
Shrub						
I	2.05	0.11	0.45	27.35	3.74	33.70
II	2.81	0.04	1.10	96.86	13.26	114.07
III	2.71	0.03	0.98	11.06	1.51	16.29
Average	2.52	0.06	0.84	45.09	6.17	54.69

Table 3. Total carbon content at the observation site of oil palm plantation.

Plot	Carbon of litter (ton/ha)	Carbon of understory plants (ton/ha)	Carbon of fronds (ton/ha)	Carbon of dead tree (ton/ha)	Carbon of tree (ton/ha)	Total carbon content (ton/ha)
I	0.00	2.09	57.26	0.00	4.16	63.51
II	7.34	1.99	0.00	0.00	9.48	18.82
III	0.59	1.63	37.84	0.73	4.30	45.09
Average	3.97	1.90	18.92	0.73	5.98	42.47

The root of tree was the second largest carbon stock after the tree trunk. Root carbon stocks are directly proportional to the above-ground carbon stocks [15]. The percentages of tree root biomass in secondary forest and shrubs were respectively by 14%. This result is in line with Indrajaya [14] research results in the protected forest in Ketrok Malinau District. Areas with healthy stands condition will be able to optimally store carbon at above and below ground.

Litter carbon stocks on land cover types of secondary forests and shrubs were smaller than that finding by Wahyuni *et al.* [16] but larger than Mahfud *et al.*[17]. This difference is caused by differences in the physical condition of the area and also the measurement time which conducted during the dry season. According to Liao *et al.* [18], litter production in the rainy and autumn seasons is high while that is moderate to low out of the seasons. However, the high litter production during the rainy season cannot survive for a long time on the forest floor due to the rain flows.

The carbon stock of understory plants was similar to Ariani *et al.* [19] result that is 0.13–0.53 tons C/ha but much lower than Azham [20] research results. The carbon stock of understory plants in this study is considered low due to human activity and canopy factor. The shrubs are frequently disturbed by surrounding communities in the form of land clearing for planting preparation or as an accessibility area, moreover, the secondary forest has a more dense canopy so the carbon stock is relatively low.

The carbon stock of small diameter dead tree was relatively low, this is also lower than that found by Wahyuni *et al.* [16], amounted to 1.64 tons C/ha. Carbon stock of dead tree is an indicator of low illegal logging and the relative good of stands health conditions.

The largest total carbon stocks from table 2 and 3 if sorted on each of the existing land covers were obtained in the secondary forest of 104.72 ton/ha, shrub of 54.69 ton/ha, then palm plantation of 42.47 ton/ha. The results of the previous research indicated that secondary forest carbon stocks are 83.49 ton/ha [21], shrub carbon stocks of 55.19 ton/ha [22], carbon stocks of oil palm plantation with age between 6–10 year of 34.16 ton/ha [23]. Several results of the previous research were not much different with the results of carbon stocks in this study. Furthermore, according to Hairiah *et al.* [24], the differences of carbon stocks amounts between types of land cover depend on the diversity and density of existing plants, type of soil, and management method of land. The more diverse plants provide larger carbon stocks.

3.3. *Physical and chemical properties of peat soil*

The results of physical and chemical properties analysis of peat soil prior to burning are presented in table 4. The bulk density of secondary forest (0.22 g/cm^3) was lower than that of shrubs (0.26 g/cm^3) and oil palm plantation (0.30 g/cm^3). The bulk density is affected by the depth of the groundwater level and affects the decomposition process. The difference of groundwater depth is caused by the land processing system, which is drainage or canalization channel. Differences in the value of bulk density are also influenced by the level of peat maturity. In addition, the bulk density also has a reciprocal relationship with porosity [25], in which soils with a relatively low bulk density generally have high porosity. Furthermore, oil palm plantations have high bulk density value compared to other types of land. This condition might be caused by the land preparation before planting in the form of drainage, soil grinding, and organic or inorganic fertilization which can change the nature of peat [26].

Above ground Organic-C of peat soils in secondary forests was higher than in other land types. The lower organic-C values on oil palm plantation cover due to decomposition activities by soil microorganisms, erosion, and subsidence. Drained peatland conditions will change the condition of peat from anaerobes to aerobes. Moreover, the results of Nugroho *et al.* [26] suggested that the c-organic value of peat soils will be degraded since the conversion of secondary peat forests into oil palm plantations.

Table 4. Physical and chemical properties of peat soil before burning in various types of land cover.

Variable	Land cover		
	Secondary forest	Shrub	Oil palm plantation
Bulk density (g/cm^3)	0.22 ± 0.02	0.26 ± 0.02	0.30 ± 0.02
Organic-C (%)	49.26 ± 0.40	49.22 ± 0.44	45.79 ± 0.11

Note: The numbers written after the ± sign are the standard deviations

3.4. *Carbon stock in top layer of peat soil*

Carbon stock in the top layer of Peat Soil (10 cm) is interpreted from bulk density and c-organic content of peatland (table 5). The highest carbon stock in the top of peat soil (10 cm) on oil palm plantation (138.542 ton/ha) was higher than that of secondary forest (108.63 ton/ha) and shrub (126.712 ton/ha). The carbon stock in the top of peat soil (10 cm) in land covered oil palm plantation is influenced by the value of bulk density which is relatively higher than other land covers.

In this study, sampling was done at the same depth because the data related the top layer of peat soil is required for the calculation of carbon emissions from peatland fires. The calculation of this carbon stock cannot be equated with the overall peat soil carbon stocks using peat thickness variables. Aswandi *et al.* [27] stated that the peat soil carbon stock in secondary forest is higher than that of oil palm plantations. This is due to variations of peat thickness variable and secondary forest land cover that have a deeper peat thickness than that in oil palm plantations. The results of Safitri's research [28] stated that peat soil carbon in oil palm plantation with a depth of 28 cm, 34 cm, and 146 cm were 266 tons / ha, 341 tons/ha and 2050 tons/ha respectively. The results of Safitri's research [28] can also be compared with the results of this study which suggested that the carbon stock of peat soil with a depth of 10 cm automatically lower than 266 tons/ha, this is in line with the results of the calculations presented in table 5.

Table 5. Carbon stock in the top layer of peat soil in research location of secondary forest, shrub, and oil palm plantation.

Land cover [8]	The depth of soil sample (cm)	*Bulk density* (g cm^{-3})	% C-Organic soil content	Carbon stock (ton/ha)
Secondary forest	10	0.219	0.493	108.363
Shrub	10	0.258	0.492	126.712
Oil palm plantation	10	0.302	0.458	138.542

3.5. Carbon emission from peatland fires

The forest and land fires cause biodiversity loss and will also affect the amount of carbon emissions generated from the burning areas. Lu *et al.* [29] stated that forest and land fires result in the release of carbonaceous compounds into the air. Subsequently, the increasing carbon compounds as greenhouse gases raise the greenhouse effect. The carbon is the dominant compound in forest fires since nearly 45% of the plant's dry matter is carbon [2]. Most of the carbon emitted into the air is in the form of CO_2, the excess is CO_2 and hydrocarbons (mainly CH_4 and smoke), while sulfur will be left as smoke and slightly formed into SO_2 and chlorine elements form the CH_3Cl compounds [30]. Carbon emissions from fires cause carbon stocks decline. Thus, an approach based on carbon stocks in the land that is assumed for unburnt is expected to provide information on carbon emissions occurred in burning areas.

The carbon stock of each land cover will be different, therefore in case of forest and land fires with high biomass to low biomass contents will impact on less carbon content and furtherly contribute to the value of carbon emissions. The data of Carbon stock is partly obtained from the result of previous research through a direct measurement in the field.

Table 6. The depth and mass of the peat burn.

Land cover [8]	The depth of peat burn (cm)	The mass of the peat burn (ton/ha)
Secondary forest	33.17 ± 3.65	73 x 10^{-7}
Oil palm plantation	25.00 ± 2.89	65 x 10^{-7}
Shrub	52.57 ± 2.25	158 x 10^{-7}

The calculation of carbon emissions from peat fires used the data of bulk density and the surface depth of the burning peat. Table 6 shows that the highest average of burning peat depth is found on the land cover of shrub. The highest level of peat depth in shrubs is caused by higher fire intensity than others. The depth of the peat burned in the secondary forest was about 15–74 cm, in the oil palm plantation of 15–30 cm, and in the shrub of about 30–72 cm.

According to Utami *et al.* [31], Indonesia has different fire patterns from other regions in Southeast Asia. Indonesia contributes to open burning biomass emissions resulting from burning of rice stem or straw, forest, and peatland fires. Indonesia contributes the fourth largest emissions of open burning biomass in Southeast Asia, however, considering the amount of emissions in peatlands, Indonesia contributes for about 84% of carbon dioxide emissions.

Table 7. Carbon emissions from forest and peatland fires.

Land cover	Carbon emission (ton C/ha)			
	Finding result		Reference [33]	
	Visual Method [8]	Digital Method (NBR model) [8]	Visual Method	Digital Method (NBR model)
Carbon emission at above ground of peat soil (A)				
Shrub	524.84	3024.09		
Secondary forest	230.69	553.67		
Forest Plantation	381.81	-		
Oil palm plantation	239.16	406.58		
Total	1376.51	3984.33		
Carbon emission of peat (B)				
Shrub	4.5×10^{-4}	2.57×10^{-3}	4.3×10^{-4}	2.45×10^{-3}
Secondary forest	5.0×10^{-5}	1.20×10^{-4}	4.7×10^{-5}	1.12×10^{-4}
Forest Plantation	8.0×10^{-5}	-	7.5×10^{-5}	-
Oil palm plantation	9.0×10^{-5}	1.50×10^{-4}	8.4×10^{-5}	1.42×10^{-4}
Total	6.6×10^{-4}	2.84×10^{-3}	6.3×10^{-4}	2.70×10^{-3}
Total carbon emission (A + B)	1376.51	3984.33		

Carbon emissions from peat fires can be seen in table 7. This carbon emission calculation is performed on the different land cover and is generated from CO_2, CO, and CH_4 emissions. However, Carbon emissions from peat fires are not known in detail. The results of Anita *et al.* [32], used a gas analyzer and gas chromatography, stated that CO2 emissions from shrubs, oil palm plantation, and secondary forest were 497.4 ppm; 523.2 ppm; 457.2 ppm respectively. In contrast to this study that used the Verified Carbon Standard [10], the results of this study showed that carbon emissions in the shrub were higher than others since the severity of the fires is also high. Peat soil carbon emission value was considered very low, however, a relatively high value of carbon emissions at above ground of peat soil is also obtained. The total carbon emissions resulting from the visual method are different from those of the digital method due to some constraints already mentioned in the previous chapter.

Stockwell *et al.* [33] stated that changes in the value of emission factors used in the calculation of greenhouse gas emissions from peat soil fires in tropical countries need to be obtained. This is because the default emission factor value published in IPCC 2003 is the result of a laboratory study which used only single samples. Furthermore, the CO_2 emission factor value changed from 464 g / kg to 1564 ± 77 g / kg, the CO changed from 210 g/kg to 291 ± 49 g/kg, and CH_4 changed from 21 g/kg to 9.51 ± 4.74 g/kg. Table 7 presents the differences between peatland carbon emissions calculations using IPCC [34] and emissions factor of Stockwell *et al.* [33]. The result of the final calculation for peat soil carbon emission from fire was 0.05% lower than using the value of emission factor of Stockwell *et al.* [33].

4 Conclusion

The estimation of carbon emissions from fires from above the peat soil surface were higher than the carbon emissions from the peat soil. Carbon emissions above the peat soil surface of 1376.51 ton C/ha were obtained by visual method while 3984.33 ton C/ha were obtained by digital method. Peatland carbon emissions of 6.6×10^{-4} ton C/ha were obtained by visual method, whereas 2.84×10^{-3} ton C/ha was obtained by digital method. Visual method and digital method using remote sensing must be combined and developed in order to carbon emission values will be more accurate.

IOP Conf. Series: Earth and Environmental Science **149** (2018) 012040 doi:10.1088/1755-1315/149/1/012040

References

[1] Wosten JHM, Van Den Berg J, Van Eijk P, Gevers GJM, Giesen WBJT, Hooijer A, Idris A, Leenman PH, Rais DS, Siderius C, Silvius MJ, Suryadiputra N, Wibisono IT 2006 Interrelationships between hydrology and ecology in fire degraded tropical peat swamp forests. *Water Resources Development* **22** 157–174

[2] Hao MW, Liu MH, Crutzen PJ. 1990. Estimate of annual and regional release of CO_2 and other trace gases to the atmosphere from fire in tropic based on the FAO statistics for the period 1975 and 1980. In Goldammer JG, editor. *Fire in Tropical Biota (Ecological Studies 84)* (New York : *Springer)* pp 440–460

[3] Jaya INS 2000a Detecting burnt forest damage using digital spot imagery *Tropical Forest Management* Journal **6** 7–23

[4] Jaya INS, M Ikhwan, Nurhendra 2000b Technique for detecting forest fire using mutitemporal images: a case study in South Sumatera and Riau Provinces *Tropical Forest Management Journal* **6** 25–37

[5] Jaya INS, Endang P, MB Saleh 2000c Detection of Post-Fire Forest Condition through MOS-MESSR and Landsat TM Multisensor Imagery: A Case Study in PT. MHP South Sumatra. *Journal of Tropical Forest Management* **6** 55–70

[6] Lentile LB, Holden ZA, Smith AM, Falkowski MJ, Hudak AT, Morgan P, Benson NC 2006 Remote sensing techniques to assess active fire characteristics and post-fire effects *International Journal of Wild Land Fire* **15** 319–345

[7] Verma S, Kuimi TV, Sathya M, Shanmuganathan J 2015 Monitoring changes in forest fire pattern in Mudumalai Tiger Reserve, Western Ghats India, using remote sensing and GIS *Global Journals Inc.* **15** 13–19

[8] Hafni DAF 2017 Estimation Burned Area and Carbon Emission from Peatland Fires in Siak District, Riau Province [thesis]. Bogor (ID): Bogor Agricultural University

[9] Manuri S, Cris B, Nunung PN, Kristell H, Nisa N, Helmut D, Louis V, Chandra ASP, Eka W 2014 Tree biomass equations for tropical peat swamp forest ecosystems in Indonesia *Forest Ecology and Management* **334** 241–253

[10] [VCS] Verified Carbon Standard 2015 VCS Module: Estimation of greenhouse gas emissions from biomass and peat burning (E-BPB) (Online) Available: http.//database.v-c-s.org/methodologies/redd-methodology-framework-redd-mf-v15

[11] Lubis AR 2011 Estimation of palm oil carbon stocks based on allometric equation in peatland garden meranti understand, PT. Perkebunan Nusantara IV, Labuhan Batu Regency, North Sumatera [undergraduete thesis] Bogor (ID): Bogor Agricultural University

[12] Hairiah K, Ekadinata A, Sari RR, Rahayu 2011 *Measurement of Carbon Reserves. From Land to Land Level.* Ed ke-2. (Indonesia : World Agroforestry Centre, ICRAF SEA Regional Office)

[13] Syam'ami, Arfa AR, Susilawati, Yusanto N 2013 Above-ground carbon stocks in various land cover systems in amandite sub-sub *Journal of Tropical Forests* **13** 148–158

[14] Indrajaya Y 2013 Protected forest carbon reserves Ketrok in Malinau District, East Kalimantan to support REDD + mechanism *Journal of Social Research and Economic Forestry* **10** 99–109

[15] Cairns MA. Brown S, Helmer EH, Baumgardner 1997 Root biomass allocation in the world's upland forest *Oecologia* **111** 1–11

[16] Wahyuni S, Chairul, Arbain A 2013 Estimated carbon stocks above ground level and plant species diversity in Bukit Tangan Island forest production area of PT. Kencana Sawit Indonesia, South Solok *Journal of Biology* **2** 18–26

[17] Mahfud A 2016 Potential carbon stocks in high conservation value areas in oil palm plantations in support of RSPO implementation [thesis] Bogor (ID): Bogor Agricultural University

[18] Liao JH, Wang HH, Tsai CC, Hseu ZY 2006 Litter production, decomposition and nutrien return of uplift coral reef tropical forest *Forest Ecology and Management* **235** 174–185

[19] Ariani, Sudhartono A, Wahid A 2014 The biomass and carbon plant under Tambling lake in Lore

Lindu National Park area *Warta Rimba.* **2** 16 –170

[20] Azham Z 2015 Estimated carbon stocks on secondary forest land cover, shrubs and bushes in Samarinda City *Agrifor* **14** 325–338

[21] Rochmayanto Y. Dudung D, Teddy R, Elias 2010 Change of carbon stock and its economic value on conversion of peat swamp forest to pulp plantation forest *Journal of Tropical Forest Management* **16** 18–26

[22] Manuri S, Lingenfelder, KH Steinmann 2011 Tier 3 Biomass Assessment for Baseline Emission in Merang Peat Swamp Forest. Paper Presented at Workshop on Tropical Wetland Ecosystem of Indonesia CIFOR 11-14 April 2011 Merang REDD Pilot Project. Palembang

[23] Yuliyanto 2015 Estimation of carbon stock stored in oil palm (Elaeis guineensis Jacq.) and analysis of soil fertility in PT Daria Dharma Pratama Ipuh Bengkulu plantation [thesis] Bogor (ID): Bogor Agricultural University

[24] Hairiah K, Rahayu 2007 *Carbon Measures Stored at Various Land Use* Indonesia: World Agroforestry Centre, ICRAF SEA Regional Office)

[25] Islami T 1995 *Land, Water and Plant Relations*. Semarang (Indonesia: IKIP Semarang Press)

[26] Nugroho KG, Widjaya A 1997 Soil hidroulic properties of Indoneisa Peatland. Proc. Tropical Petaland. Serawak. Malaysia

[27] Aswandi, Ronggo S, Haryono S, Hartono 2016 Carbon losses from drainage and tropical peatland degradation in Trumon and Singkil Aceh *J. Humans and the Environment* **23** 334–341

[28] Safitri I 2010 Determination of carbon stock of sapric, hemic, and fibric peat (case study at Lubuk Gaung palm oil plantation, Sungai Sembilan sub-district, Dumai) [undergraduete thesis] Bogor (ID): Bogor Agricultural University

[29] Lu A, Hanqin T, Mingliang L, Jiyuan L, Jerry MM 2006 Spatial and temporal patterns of carbon emissions from forest fires in China from 1950 to 2000 *J. Geophys. Res.* **111** (D05313). doi:10.1029/2005JD006198

[30] Syaufina L, Saharjo BH, Nurhayati AD. 2009. *Smoke Disaster Mitigation Through Dispersion Modeling and Content.* Bogor (ID): Fakultas Kehutanan IPB.

[31] Utami A, ChihHua C 2016 Air pollution emissions from open burning biomass on peatlands in Indonesia using GIS data and MODIS image *Journal of Environmental Science and Technology* **8** 1–9

[32] Anita S, Amri TA, Hanifah TA, Furnando E, Lukas A 2017 Carbon dioxide emissions due to forest fires in Bukit Batu Area, Bengkalis Regency, Indonesia. *IOP Conf. Series: Earth and Environmental Science* DOI: 10.1088/1755-1315/68/1/012002

[33] Stockwell CE, Thilina J, Mark AC, Kevin CR, Erianto IP, Bambang HS, Ati DN, Israr A, Donald RB, Isobel JS, Elizabeth AS, Robert JY 2016 Field measurement of trace gases and aerosols emitted by peat fires in Central Kalimantan, Indonesia, during the 2015 El Niño. *Atmos. Chem. Phys.* **16** 11711–11732 DOI: 10.5194/acp-16-11711-2016

[34] [IPCC] *International Panel on Climate Change* 2006 IPCC Guidelines for National Greenhouse Gas Inventory, National Greenhouse Gas Inventories Program. Japan

IOP Conf. Series: Earth and Environmental Science **149** (2018) 012041 doi:10.1088/1755-1315/149/1/012041

Identification of ex-sand mining area using optical and SAR imagery

Novie Indriasari[1,2], Eko Kusratmoko [1], Tito Latif Indra[1] and Atriyon Julzarika[2]

[1] Geography Department, Faculty of Mathematics and Natural Sciences, Universitas Indonesia

[2] Indonesian National Institute of Aeronautics and Space (LAPAN)

E-mail: indriasari.novie8@gmail.com

Abstract. Open mining activities in Sumedang Regency has been operated since 1984 impacted to degradation of environment due to large area of ex-mining. Therefore, identification of ex-mining area which generally been used for sand mining is crucial and important to detect and monitor recent environmental degradation impacted from the ex-mining activities. In this research, identification ex-sand mining area using optical and SAR data in Sumedang Regency will be discussed. We use Landsat 5 TM acquisition date August 01, 2009 and Landsat 8 OLI acquired on June 24, 2016 to identify location of sand mining area, processed using Tasselled Cap Trasformation (TCT), while the landform deformation approached using ALOS PALSAR in 2009 and ALOS PALSAR 2 in 2016 processed using SAR interferometry (InSAR) method. The results show that TCT and InSAR method can can be used to identify the areas of ex-sand mining clearly. In 2016 the total area of ex-mining were 352.92 Ha. The land deformation show that during 7 years period since 2009 has impacted to the deformation at 7 meters.

1. Introduction

Traditional activities of open mining in Sumedang regency has been operated since 1984, while the estate mining companies with heavy machine equipment to excavate the sand came in 1993 [1]. According the local news, both the traditional and the estate mining companies were mostly operated in illegally.

Implementation of remote sensing data to detect the ex-mining area as a prominent method. This due to the accurate, effective and cost efficiency [2]. Optical remote sensing imagery as the main source in detecting the ex-mining area provides two-dimension analysis resulted from the completeness of wave length channels. But the optical imagery difficult to detect the depth of landform deformation [3]. Moreover, inaccurate interpretation was interferes and mixed due to the fast growing of grasses and shrubs [4]. Therefore, SAR imagery were deployed to detect the landform deformation that caused the degradation of environment and wide area of abandoned land.

This study aims to identify an appropriate method to detect the ex-mining area using combination of optical satellite imagery and SAR imagery. Selection of appropriate methods for data processing will affect the results obtained. Several research conducted on detecting ex-mining using NDVI [5,6] and VIDN [7] to identify abandoned area of mining using optical satellite imagery. This research used the implementation of Tasseled Cap Transformation (TCT) method for optical imagery analysis and Interferometry Synthetic Aperture Radar (InSAR) in detecting landform deformation based on SAR imagery.

IOP Conf. Series: Earth and Environmental Science **149** (2018) 012041 doi:10.1088/1755-1315/149/1/012041

The successfully of landcover classification of TCT method due to the ability in detecting and differentiate the phenomenon of vegetation growth/phase [8,9]. The InSAR method were processed through combination of two/more of SAR imagery which known as a source to provide information of height or Digital Elevation Model (DEM) [10]. InSAR often to be utilised for research on landform deformation [2,11].

2. Materials and methods

2.1. Data

Optical remote sensing applied in this study were; Landsat 5 TM in August 01, 2009 and Landsat 8 OLI in June 24, 2016, while DEM derived from ALOS PALSAR in 2009 (level product 1.0) and ALOS PALSAR 2 in 2016 (level product 1.1). The ex-mining in Sumedang regency located at 6°46'21.27'S – 6°48'28.01"S, 107°55'59.17"E - 107°59'46.41"E (shown in figure 1).

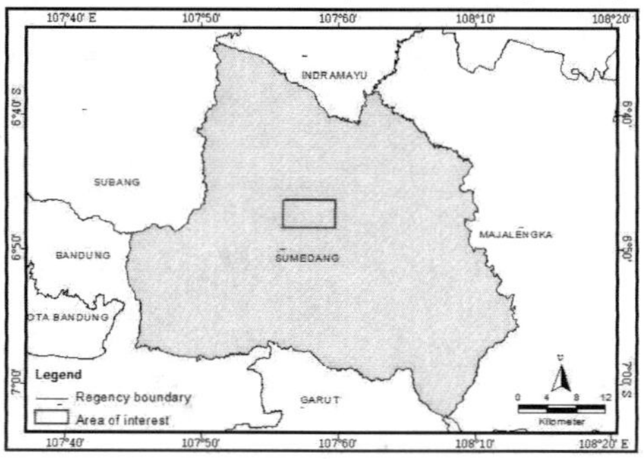

Figure 1. Study area.

2.2. Method

Location of ex-mining area were detected from Landsat imagery. The steps to detect the ex-mining area were firstly, the image pre-processing the TOA correction of Landsat imagery which normalized the digital number into reflectance and to avoid the illumination geometry [12], the next steps is the TCT process which resulted in a specific landcover classes physically. The scatterplot of TCT as a based to determine the suitable training sample were using the composite of Brightness and Greenness bands of TCT. Based on the result from the field surveying for verification the mapping result, it can be seen that only few of ex-mining area were processed reclamation and others were mostly abandoned and covered by grass and bush. Therefore in this study, the ex-mining area identified in 2009 will be added to the ex-mining area in 2016.

DEM creation were firstly processed by focusing method that synchronize between raw and metadata as to fulfil the requirement in interferometry processing. The result of focusing step was the Single Look Complex (SLC) that provide information of intensity value and result phase from acquisition. The second-step was the interferogram to obtained the fringe (contrast between dark and bright) from the different phase of master and slave imagery. The third-step was the flat earth correction method for the interferogram imagery that based on the earth surface measurement in Indonesia, and followed by filtering process order to enhanced the coherence of interferogram imagery. The final process in InSAR production was the phase unwrapping which corrected the cyclic phase of 2π to absolute phase (unwrap) before converted to DEM.

Landform deformation resulted from ex-mining area was measured using the subtraction of DEM in 2016 with the DEM in 2009. Research flow diagram can be seen in figure 2.

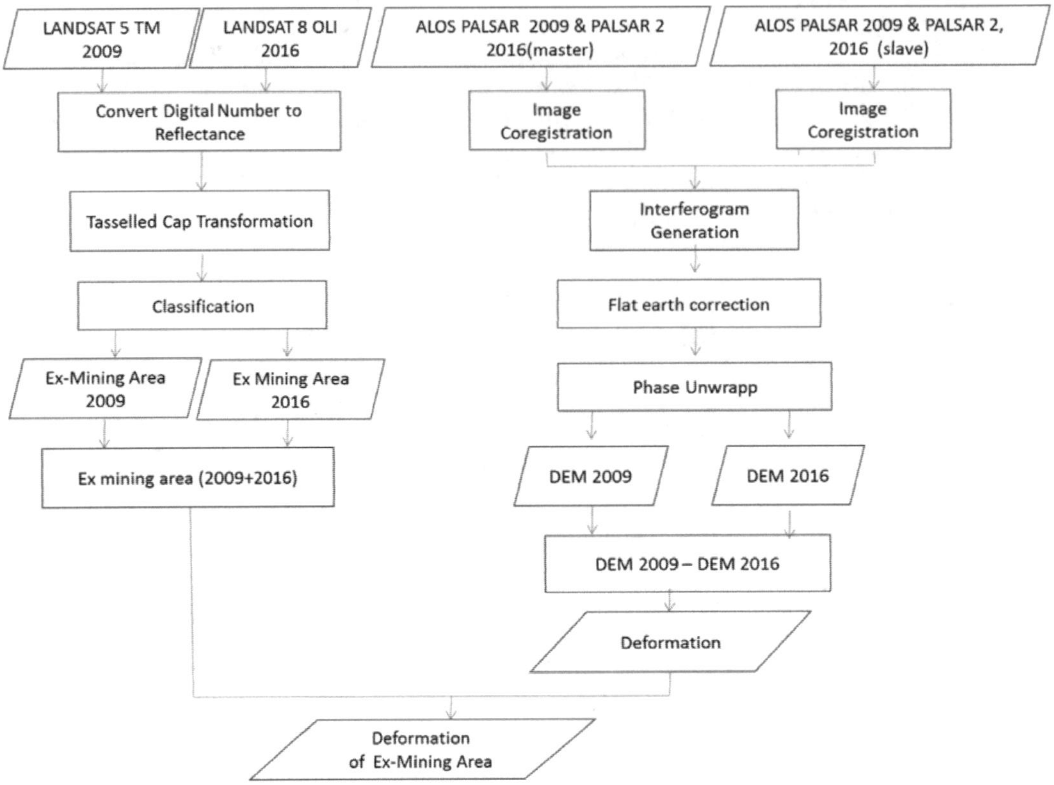

Figure 2. Flow diagram of research.

3. Results and discussion

The TCT method implemented for Landsat 5 TM and Landsat 8 OLI in Sumedang Regency resulted in a clear and detail differentiation for several land cover types where detected from Brightness channel for Red and Greenness channel for Green and Wetness channel for Blue (figure 3). For TCT Landsat 5 TM (figure 3a), it can be seen from the red to purple objects identified as soil surface that has vegetation patches and for bare land, cyan to dark blue use to detect vegetation, and building area depicted as an orange. Slight difference when viewing TCT Landsat 8 OLI (figure 3b) that the bare land look in dark purple.

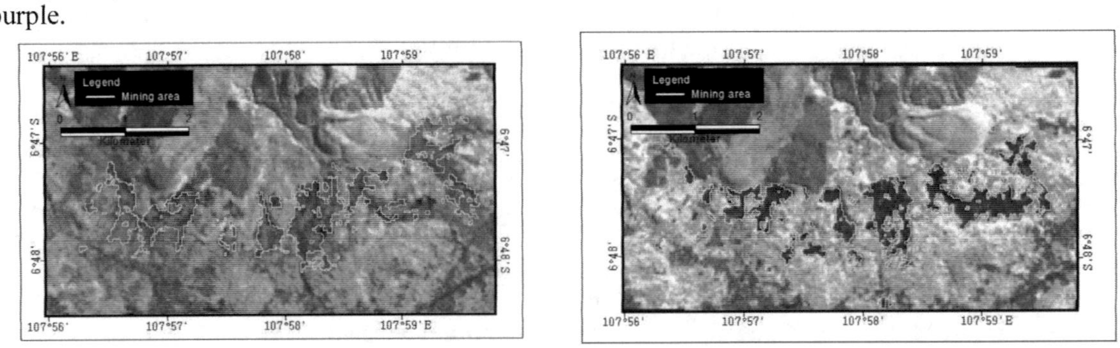

(a) (b)

Figure 3. TCT of (a) Landsat 5 TM and (b) Landsat 8 OLI.

The TCT Brightness obtained from the sum of six bands. The TCT Brightness beneficial for detecting soil character. This due to the transformation of the total soil reflectance. The transformation in Brighteness channel occurred when the dense canopy vegetation increase causes the increasing reflectance in NIR band and lower in the visible bands.

The Greenness shows the high absorbtion of visible band due to the vegetation pigment, such as chlorophyll and leaves structures (high reflectance in NIR band). The combination of Brightness and Greenness based on the Tasseled Cap developed by Kauth and Thomas applied in the Landsat MSS, shows that the method can provide and identified all of the landcover for Landsat 5 and Landsat 8. The accurate method to identify the bare land of ex-mining from TCT can be conducted by delineating the scatterplot as a part stage in the cap analysis (table 1).

Table 1. Land cover scatterplot.

Land Cover	**Image**	**Scatterplot**	**Result of Scatterplot Delineation**
Bare land	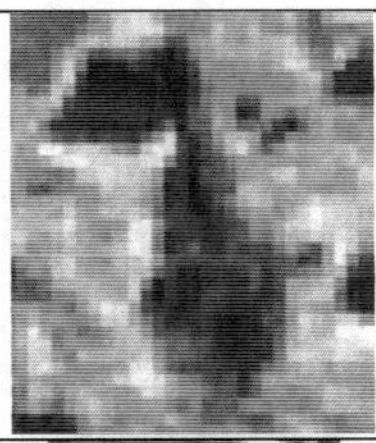	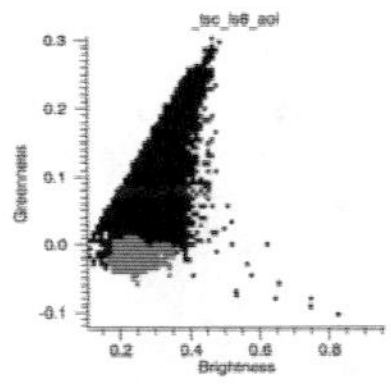	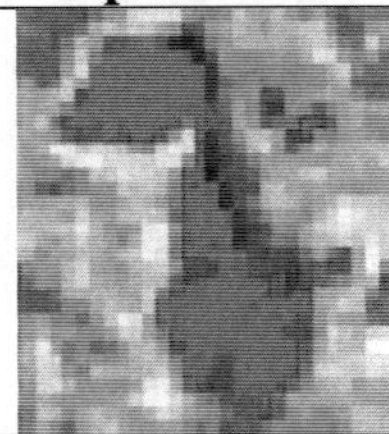
Forest		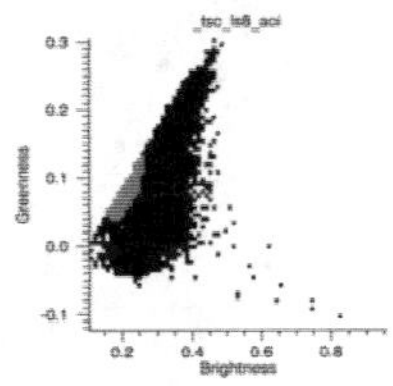	
Bush	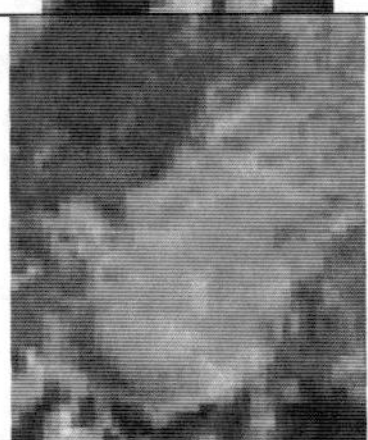	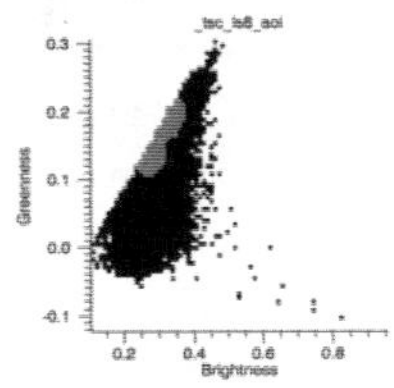	
Grass	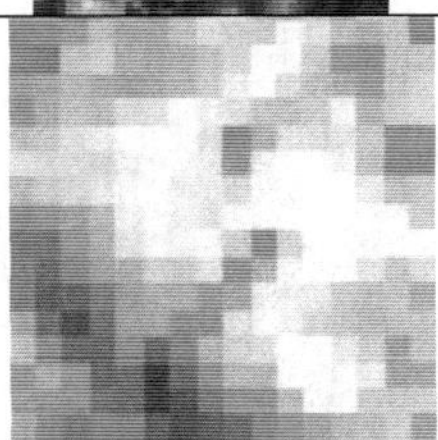	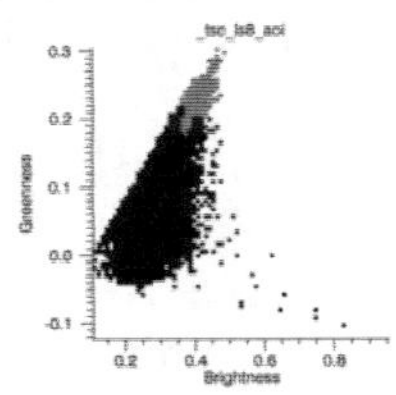	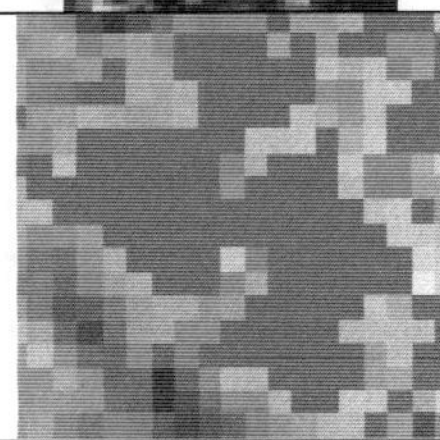
Building area	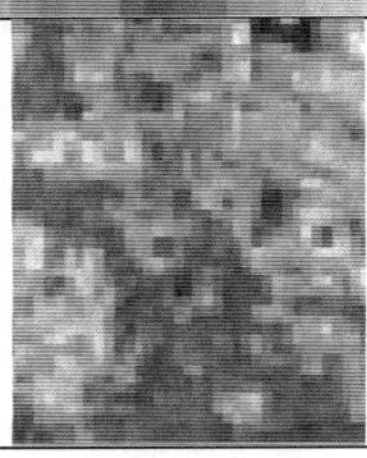	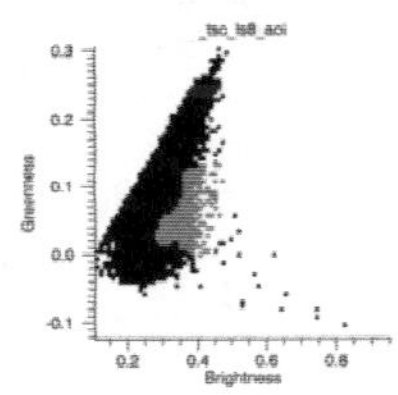	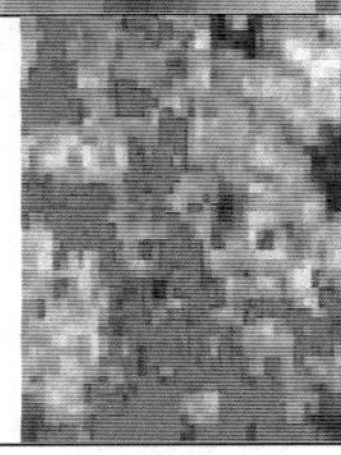

The process in identifiying the object using delineating the scatterplot can be use as the training sample for further classifiction steps. While the classification was conducted using SVM method where

IOP Conf. Series: Earth and Environmental Science **149** (2018) 012041 doi:10.1088/1755-1315/149/1/012041

the area of ex-open mining sand ca. 230.220 Ha in 2009 and increased into 353.92 Ha in 2016, as shown in figure 4.

The DEM created from interferometry of ALOS PALSAR and ALOS PALSAR 2 provide information with vertical accuracy at 70-80 cm and 65-70 cm. The DEM produced by ALOS PALSAR has a prominent of accurate geometric accuration and precision. This because the ALOS PALSAR compose by L band. The L band has better penetration compare to the X and C bands [2,13].

The DEM created using the InSAR method shows detail topographic of the research study, where located at 450 – 1300 meters above sea level. Analysis on overlying the DEM with the information classification of Landsat imagery shows that the mining area in 2009 located at 622.5 - 802.5 meters above sea level and the mining activities continue growth in 2016 to the area that located in higher topographic area, at 825 meters above sea level.

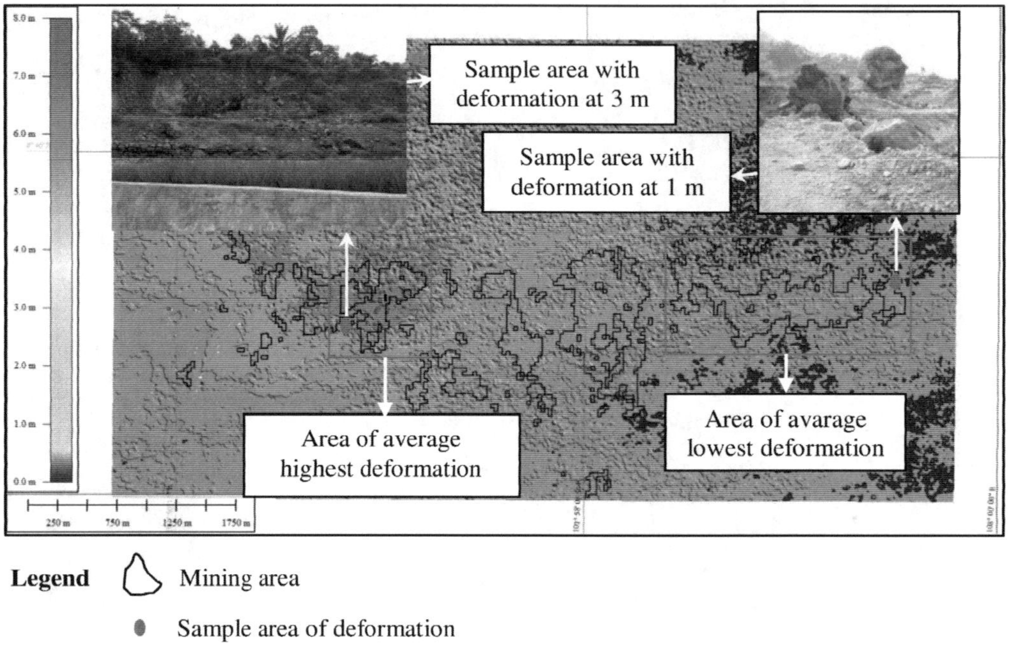

Figure 4. Deformation at mining area.

The DEM from the InSAR method was successfully measured the land deformation in the sand meaning which had decrease 1 – 7 meters within 7 years in the period. This can be seen in the figure 4, where the highest deformation occurred in the western part of ex-mining, while the lowest land deformation occurred in the eastern part of ex-mining area. For more detail, the highest and the lowest land deformation of mining area can be seen in table. 2.

Table 2. The highest and lowest land deformation.

Mining area	Change of elevation
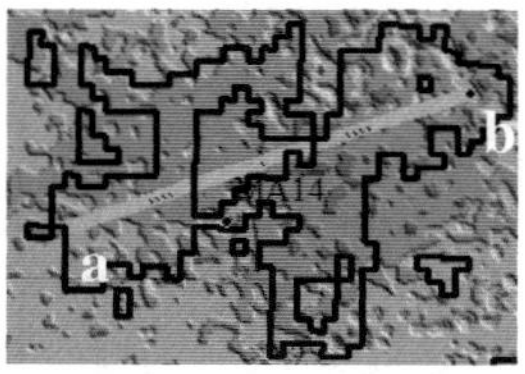 Average change of elevation 5 m	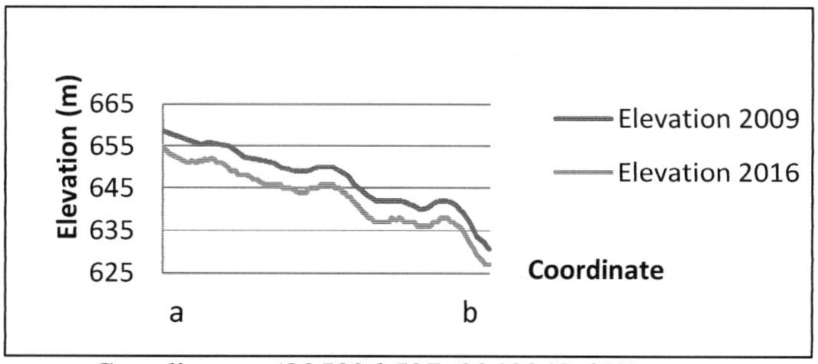 Coordinate a (825826.537; 9248042.659) Coordinate b (826750.905; 9248357.853)
	(a)
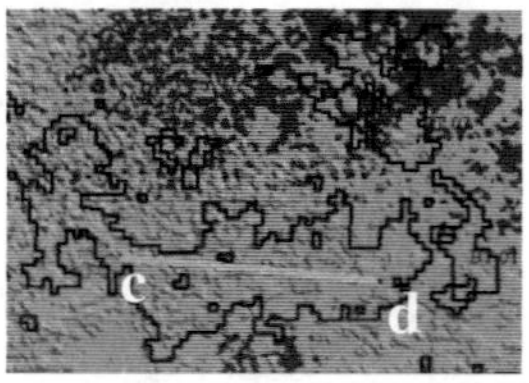 Average change of elevation 1.6 m	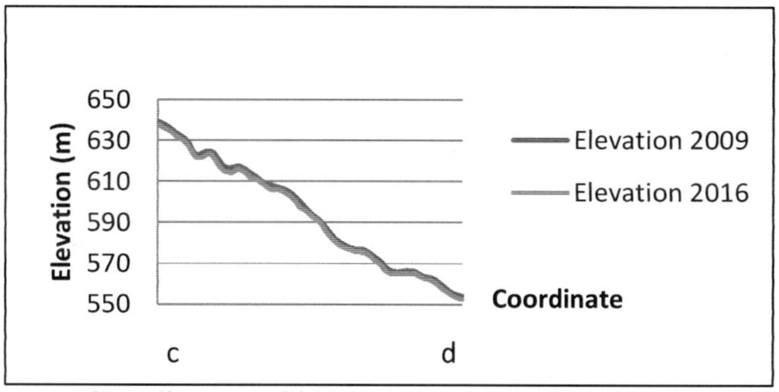 Coordinate c (829331.392; 9248218.101) Coordinate d (830378.574; 9248122.903)
	(b)

The various number of land deformation, where in the eastern part was lower compare to the intesive mining at the western part, which can also be identified in the TCT image of Landsat 5 imagery in 2009 that none of the area was detected and classified as a mining area (figure 3a).

4. Conclusion

TCT method shows an accurate and robust method to detect an accurate location of ex-mining area because its prominent method in differentiating between vegetation and soil. The ex-mining were reached 230.220 Ha in 2009 and reached 353.92 Ha in 2016. The deformation detected from ALOS PALSAR using the InSAR method provide an accurate result. The deformation between 2009-2016 reached 7 meters, while the direction trend for new mining approaching to the higher topography area. Improvement for future research can be done by measuring the actual deformation in field for detail accuracy. This because the current research solely focused on measuring deformation based on satellite imagery and analysis of the landcover changes.

Acknowledgment

The authors would like to thank JAXA for providing the ALOS PALSAR 2 and I would like to thank to Dr. Rahmat Arief for discussing about InSAR.

References

[1] Anonym 2014 *Report on Monitoring and Controlling of Mining Activities of Non-Metal and Rock Mineral Materials* Sumedang (Indonesia: Department of Energy and Land Affairs Sumedang Regency) [In Indonesian]

[2] Chaussard E and Kerosky S 2016 Characterization of Black Sand Mining activities and Their Evironmental Impacts in Philippines Using Remote Sensing *Remote Sensing* **8** 100.

[3] Ross MRV, Brian LM and Emily SB 2016 Deep Impact: Effects of Mountaintop on Surface Topography, Bedrock Structure, and Downstream Water *Environ. Sci. Technol.* **50** (4) 2064–2074 DOI : 10.1021/acs.est.5b04532. ACS Publication

[4] Limpitlaw D 2003 Mapping Mining Waste and Environmental Impact in Zambia With Landsat *Procceedings of the 4th European Congress on Regional Scientific Cartography and Information System*, Bolgna, 17-20 June 2003 **2** 699-671

[5] Alden MG 2009 Remote Sensing Techniques for Monitoring Coal Surface Mining and Reclamation in the Powder River Basin [thesis] College of Art and Science of Ohio University

[6] Townsend PA, Helmers DP, Kingdon CC, McNeil BE, Beurs KMD and Eshleman KN 2009 Changes in Extent of Surface Mining and Reclamation in the Central Appalachians Detected Using a 1976 – 2006 Landsat Time Series *Journal Remote Sensing of Environnment* **113** 67-72

[7] Sutanto A and Tjahjaningsih A 2014 Index of Mined Open Land Using Vegetation Index Differencing Method Case Study: Geumpang Gold Mine, Aceh. Proceedings of the National Seminar on Remote Sensing **2016** [In Indonesian]

[8] Kauth RJ and Thomas GS 1976 The Tasseled Cap - A Graphic Description of the Spectral – Temporal Development of Agricultural Crops as Seen by LANDSAT *Proceedings of the Symposium on Machine Processing of Remotely Sensed Data*, West Lafayette, Indiana, June, 29 – July, 1, 1976 (West Lafayette Indiana, Purdue University) pp. 41-51

[9] Dymond, Caren C, Mladenoff DJ and Radeloff VC 2002 Phenological differences in Tasseled CapIndices Improve Deciduous Forest Classification *Remote Sensing of Environment* **80** 460-472

[10] Sambodo KA and Saleh M 2011 Interferometri SAR Phase Unwrapping Data ALOS PALSAR dengan Menggunakan Algoritma Least-Square Two Dimensional Phase Unwrapping *Proceedings Geomatika SAR Nasional* (GeoSARNas)

[11] Kriswati E 2011 Remote Sensing For Volcano Deformation Monitoring First Series: Interferometric Synthetic-Aperture Radar (InSAR) *Buletin Vulkanologi dan Bencana Geologi* **6** (1) 31-37 [In Indonesian]

[12] Huang C, Wylie B, Yang L, Homer C, Zylstra G 2002 Derivation of Tasseled Cap Transformation Based on Landsat 7 at Satellite Reflectance *International Journal Remote Sensing* **23** (8) 1741-1748

[13] Julzarika A 2016 Rapid and Exact Mapping of Coal Potential in Indonesia With Remote Sensing *Prosiding Seminar Nasional Geomatika* **2016** 273-282 [In Indonesian]

IOP Conf. Series: Earth and Environmental Science **149** (2018) 012042 doi:10.1088/1755-1315/149/1/012042

Carbon stock projection in North Sumatera using multi objective land allocation approach

S N Ichwani[1*], R Wulandari[1], A Ramachandra[2]

[1] PT Hatfield Indonesia, Kusnoto Building, floor 3th, Jl. Ir. H. Juanda No. 18, Bogor
[2] Conservation International Indonesia, Jl. Pejaten Barat No. 16A, Kemang, Jakarta

E-mail : soraya@hatfieldgroup.com

Abstract. Nowadays, GHG emission is a critical issue for environmental management due to the large scale of land cover change, especially forest cover. This study provides a protection development strategy for North Sumatera as one way to manage the area. By using Multi Objective Land Allocation (MOLA), we evaluated two GHG emission scenarios, including a Business As Usual (BAU) scenario and Protection scenario. The result shows that the province will lose the carbon stock up to 24 million tons in the year of 2035 by using a BAU scenario. On the other hand, by implementing the Protection scenario, total carbon stock that is lost in the same period is about 5 millions tons solely. It proves that protection scenario is a good scenario and effective to reduce the carbon loss. Furthermore, this scenario can be an alternative for North Sumatera spatial plan.

1. Introduction

GHG emission issue comes up from environmental destruction impact. It raised year by year mostly caused by deforestation. Therefore, Indonesian Government has a commitment to reduce the emission as much as 26% by 2020 from its own efforts, and 41% from international funding. It was stated by Susilo Bambang Yudhoyono, President of the Republic of Indonesia at that time, in G-20, the international forum for the governments and central bank governors with 20 countries member, in 2009 in Pittsburgh, United States. To support the statement of President, the local government need to ensure that the development on their area will not create a huge impact on the GHG emission. One way to handle this problem is to conduct a study of Strategic Environmental Assessment (SEA).

Strategic Environmental Assessment (SEA) is a series of analysis that are systematic, comprehensive, and participative to ensure that the principles of sustainable development have become the basis and integrated in the development planning within the region. It is expected to reach the prosperity through sustainable development. Moreover, the SEA is mandatory for each region stated in Government Regulation Number 46 Year 2016 about Organizing Procedures of Strategic Environmental Assessment.

The study of carbon stock, which is related to GHG emissions, is part of SEA for North Sumatera Province. It is needed in defining the "alternative" spatial plan for the province. The aim of the study is to calculate the carbon stock in year 2035 that could be "other option" for the spatial land allocation with the highest consideration on the regulation applied in Indonesia and also the sustainability of the landscape in the North Sumatera Province. Moreover, MOLA (Multi Objective Land Allocation) is one approach to process the analysis of carbon stock.

In addition, two selected scenarios will be applied in MOLA process. These two scenarios are Protection scenario and a BAU scenario. The selection between Protection scenario and BAU scenario in order to evaluate GHG emission has different opportunities in analyzing the impact of each scenario. Although, the Protection scenario reduces the possibility of developing land-based economic activity areas compared to BAU scenario, it is projected to be able to reduce the significant carbon loss with increasing protected areas. The objectives of this study is to calculate the carbon stock in year 2035 as the projection by using MOLA approach. Carbon stock projection is needed as part of the SEA and as the alternative of spatial plan in North Sumatera Province. The analysis consists of definition of Protected Area, definition of Cultivation Area, combination of both using MOLA process, and carbon stock calculation.

2. Methods

The Impact analysis is conducted by comparing two scenarios; these are Business as Usual (BAU) Scenario and MOLA Scenario, hereinafter namely Protection Scenario. On BAU scenario, it is assumed that all forested areas fall under Non-Forest Use (*Area Penggunaan Lain*/APL) and Convertible Production Forest Area (*Hutan Produksi Konversi*/HPK) are will be converted to open land in 2035 as a projection. APL and HPK areas used in this scenario are obtained from SK579/Menhut-II/2014 on North Sumatera Forest Area. Next, Protection Scenario follows the same rules as BAU scenario that all forested areas under APL and HPK will be converted to open land area. However, in this scenario, HPK areas are obtained from the MOLA analysis (see Figure 1).

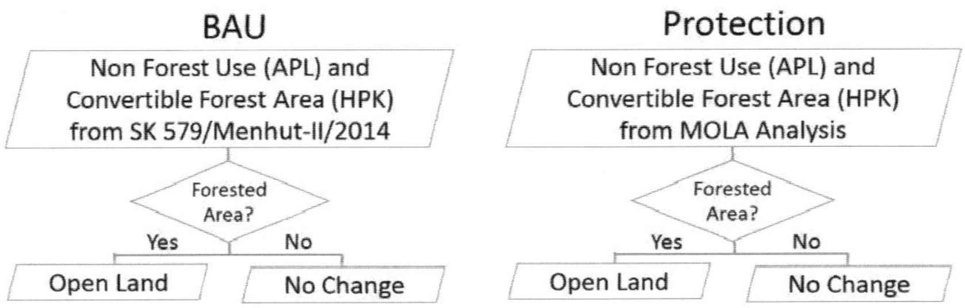

Figure 1 The comparison of BAU scenario and Protection scenario

MOLA (Multi Objective Land Allocation) is an approach that can be used to solve calculation between different uses or transitions to allocate the estimated changes in the simulation step [1]. MOLA could be a solution to the land use conflict of interest. It can allocate a land to a several land use by considering the land suitability, weight priority, and land use space [2]. This analysis has been used in several studies in the world, such as by Eastman JR et al. (1997) [3] and Hajehforooshnia S et al. (2011) [4]. There are several steps to analyze the MOLA, including defining the protected area and cultivation area, selecting the main law and regulation, and overlaying the processed data with land cover. Moreover, the results was calculated by emission factor to find the carbon stock in North Sumatera.

Land cover that is used in this analyses uses data from Ministry of Environmental and Forestry in the year of 2000, 2006, and 2013. It will be a base map of scenario to reflect the future spatial planning.

2.1. *Defining the protected area and cultivation area*

Based on Government of Indonesia Act Number 26 Year 2007 regarding Spatial Planning, it is divided into protected area and cultivation area. To define them, there are law and regulation for each parameter to be used. It is presented in Figure 2. Figure 2 also presents the hierarchy of law and regulation used for MOLA selection.

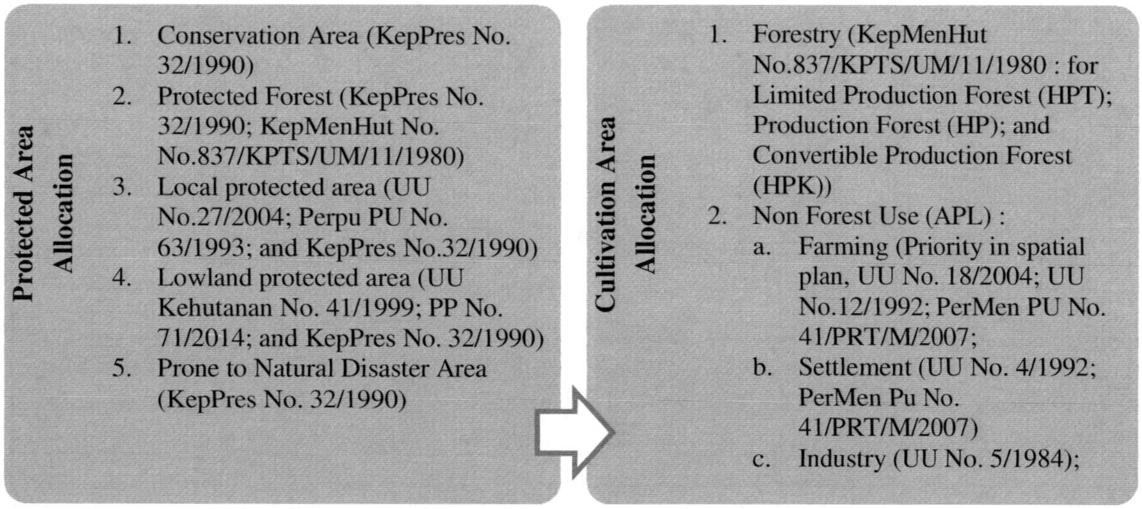

Figure 2 Regulation for defining protected area and cultivation area

Protected area is the area that has been set to protect the environment. In Minister of Public Work Decree Number 15 Year 2009 about Guidance of Province Spatial Planning Arrangement, the protected area are divided into 7 categories including protected forest, lowland protection, local protected area, conservation area, high risk to disaster area, geology protected area, and other protected areas. Nevertheless, this study defines 5 categories only as presented Figure 3. For protected forest, lowland protection, and local protected area, there are some parameters that should be used; while conservation area and high risk to disaster area use the existing data. The data of Conservation area are taken from Ministry of Environmental & Forestry (MOEF) and high risk to disaster area's data are taken from Regional Disaster Management Agency (BPBD).

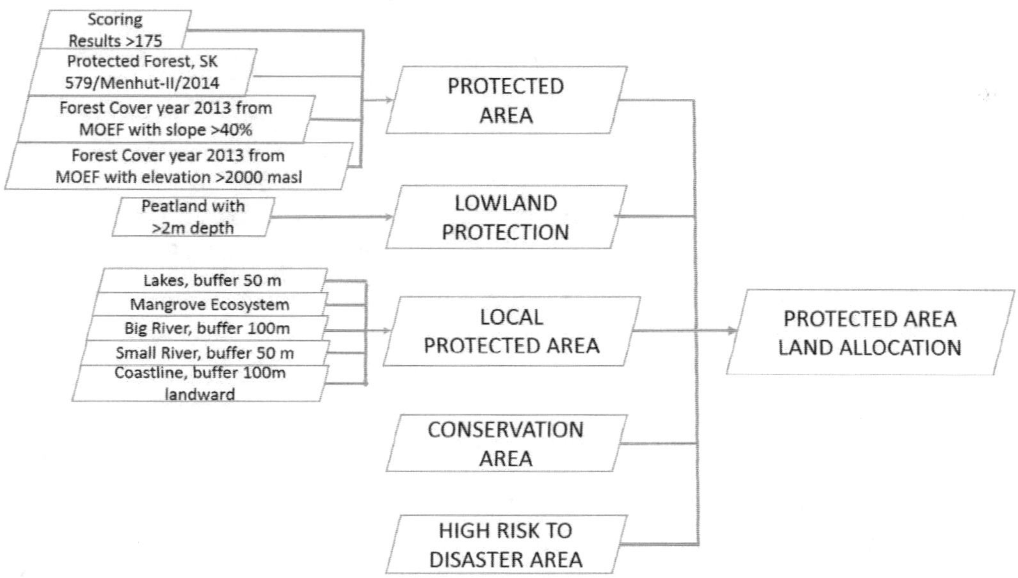

Figure 3 Work flow for protected area land allocation

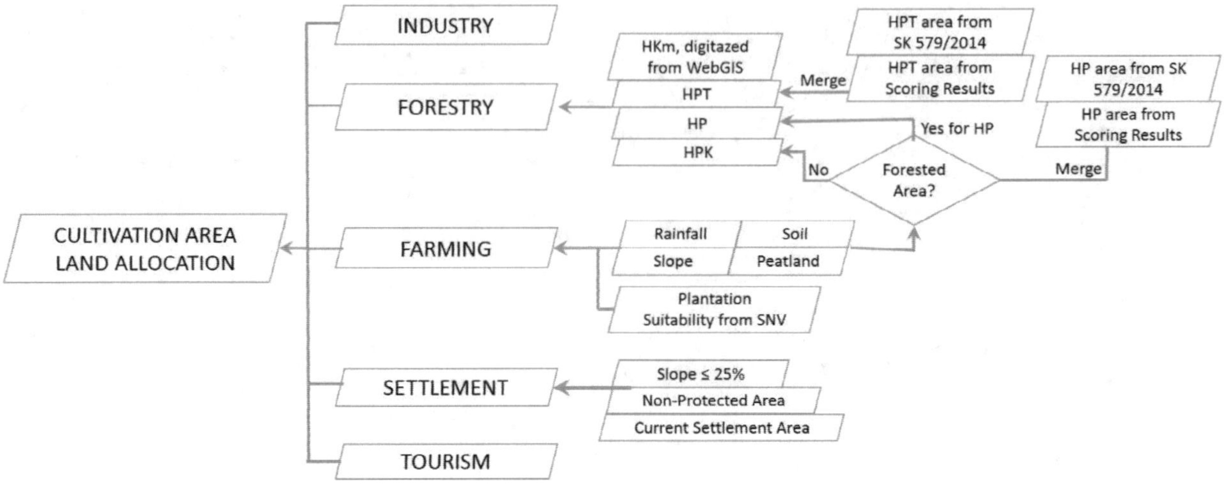

Figure 4 Work flow for cultivation area land allocation

Cultivation area is divided into forestry area, farming area, mining area, fisheries area, industry area, tourism area, settlement area, and other cultivation areas. For this study, 5 categories are defined as mentioned in Figure 4.

After protected area and cultivation area are defined, the data needs to be overlay by land cover in the year of 2000, 2006, and 2013. All forest areas which are selected into BAU Scenario and Protection Scenario will be converted to open land, as predicted in 2035.

2.2. *Calculating the carbon stocks*

The land cover needs to be calculated with emission factor mentioned in **Table 1** to find carbon stocks in North Sumatera Province.

Table 1 Emission factor [5]

No	Land Cover	Emission Factor	No	Land Cover	Emission Factor
1.	Primary Dryland Forest	195	11.	Open Land	2.5
2.	Secondary Dryland Forest	169	12.	Dryland Agriculture	10
3.	Primary Mangrove Forest	170	13.	Mixed Dryland Agriculture	30
4.	Secondary Mangrove Forest	120	14.	Mining	0
5.	Primary Swamp Forest	196	15.	Plantation	63
6.	Secondary Swamp Forest	155	16.	Swamp	0
7.	Plantation Forest	64	17.	Savanna	4
8.	Shrub	30	18.	Paddy Field	2
9	Swamp Shrub	30	19.	Fish Pond	0
10.	Settlement	4	20.	Transmigration	10

3. Results and Discussion

Forest conversion analysis was done based on the assumption that all the forested area under APL and HPK will be cleared. For BAU scenario, APL and HPK area are derived from SK 579/Menhut-II/2014 and the projection year for this analysis is 2035, which is the end period of the Provincial Spatial Plan implementation. The projection of land cover changes for the 2013 to 2035 based on BAU and Protection Scenario are provided in Figure 4.

3.1. BAU Scenario and Protection Scenario

BAU Scenario uses SK 579 that is an existing projection data for North Sumatera Spatial Plan year 2035. The result is shown in Table 2. Based on SK 579, protected area is about 1.7 million hectare. The largest land allocation for protected area is for protected forest. Besides that, cultivation area is about 5.5 million hectares. Plantation is allocated as the largest area for cultivation; that is about 2 million hectares, followed by plantation forest (accumulative number of HPT, HP, and HPK) about 1 million hectares.

Table 2 Protected area land allocation and cultivation land allocation by BAU Scenario

Spatial Plan	Land Allocation		Area (Ha)
Protected Area (SK579)	Protected Forest (HL)		1,206,881
	Natural Reserve Forest (KSA/KPA)		427,009
Cultivation Area (SK579 and North Sumatera Spatial Plan)	Plantation Forest Area	Limited Production Forest (HPT)	641,770
		Production Forest (HP)	704,452
		Convertible Production Forest (HPK)	75,684
	Plantation		2,071,767
	Settlement		209,335
	Wetland Agriculture		558,634
	Dryland Agriculture		1,188,551

The result of Protection Scenario is presented in Table 3. It has been found, first of all, by dividing into protected area and cultivation area. The largest land allocation for protected area is lowland protection area which is 1.2 million hectares. The total of protected area is 2.9 million hectares. The difference with BAU scenario is about 1.5 million hectare. As a consequence, the total of cultivation area is only 4.2 million hectares, where the largest land allocation is still for plantation and followed by plantation forest (accumulative number of HP, HPT, HPK).

Table 3 Protected area land allocation and cultivation land allocation by Protection Scenario

Spatial Plan	Land Allocation		Area (Ha)
Protected Area	Local Protected Area		675,820
	Lowland Protection Area		1,214,392
	Prone to Natural Disaster Area		69,813
	Conservation Area		427,009
	Protected Forest		528,759
Cultivation Area	Plantation Forest Area	Limited Production Forest (HPT)	785,062
		Production Forest (HP)	345,039
		Convertible Production Forest (HPK)	32,684
	Agriculture Area		1,002,702
	Industrial Area		298
	Plantation		2,013,881
	Settlement Area		155,102
	Tourism Area		5

The land allocation of protected area and cultivation area is calculated with land cover in 2000, 2006, and 2013, afterwards. The result is shown in **Figure 5**. Converted area is only for forest area which are located in HPK and other use (APL). The others will be considered not having a land cover change. Based on Figure 5, the graph illustrates that class of forest cover in 2000 to 2013 decreased, Secondary dryland forest and secondary swamp forest are the largest area converted to others class.

In the year of 2035, by using BAU scenario, the conversion from forest area to other use (APL) is larger than that by using Protection scenario. As shown in Figure 5, Protection scenario can reduce the loss of forest. In secondary dryland forest in 2035, it is predicted that the forest loss is only 20 hectare while by BAU scenario its loss becomes nearly 100 thousand hectare.

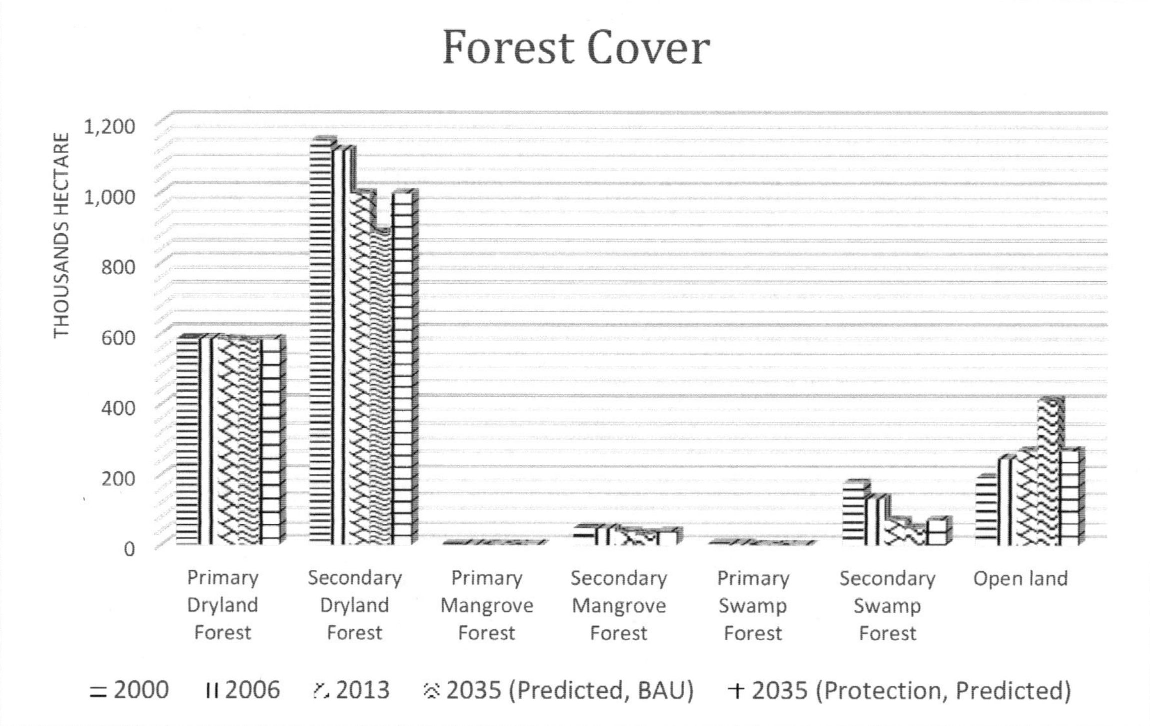

Figure 5 Forest Cover in 2000-2035 (predicted)

Figure 6 illustrates the comparison map of BAU scenario and Protection scenario. The figure shows the largest conversion is located in South West and West of North Sumatera Province. The secondary swamp forest is still exist in South West of province, based on Protection Scenario map, and it has been converted into open land in BAU Scenario map.

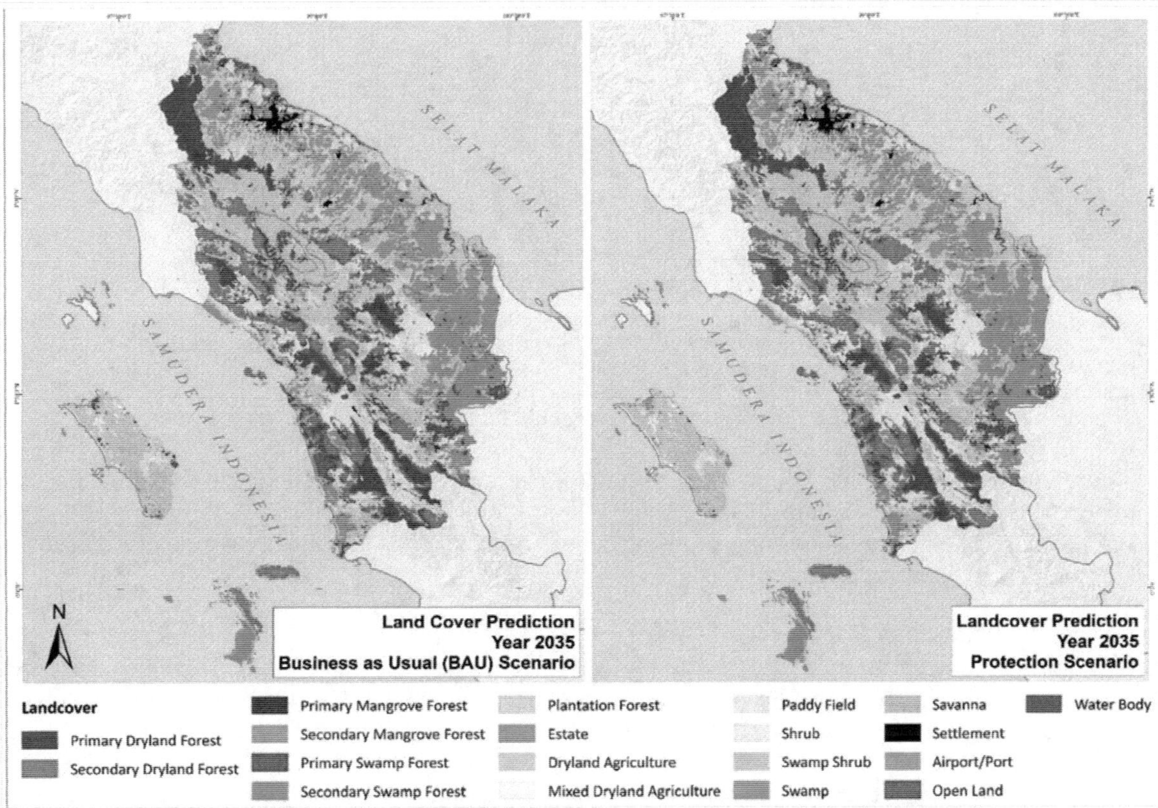

Figure 6 Land cover projection by using BAU Scenario and Protection Scenario

3.2. Carbon Stock

Carbon stock calculation is defined based on the land cover data as "activity layers" and coefficient carbon. Aligned with changes in land cover, carbon stock dynamics also follow the same trends. Historical land cover data were obtained from MoEF year 2000, 2006 and 2013 and projected land cover data were obtained from the land conversion analysis result as explained in the methodology. The result of the carbon stock calculation for BAU scenario and Protection Scenario is shown in Figure 7. Nonetheless, the scenarios is only converted the forested area to open land, so others class will be considered to have the same carbon stock between 2015 and 2035.

Based on **Figure 7**, previous data from year 2000 to 2013 shows that the average decrease of the carbon stock is about 3.5 million ton/year. The largest decrease occurred during the period of 2006 to 2013 when this province experienced a decrease in carbon stock up to 33 million tons. By using BAU scenario, in the projection year, carbon stock will decrease up to 24 million tons. The average of carbon loss is 1 million ton/year from year 2013 to 2035.

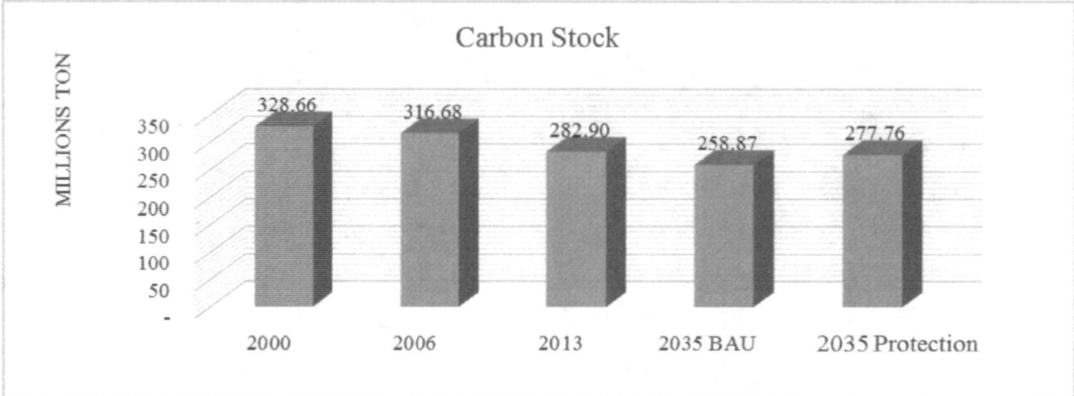

Figure 7 Carbon Stock using BAU and Protection Scenario

Furthermore, Figure 7 also shows the carbon stock using protection scenario. In this scenario, the projection year will have carbon stock for about 277 million tons. It only lost 5 million tons of carbon stock than BAU scenario. The average of losing carbon stock is about 235 thousand tons/year. It proves that protection scenario is effective to reduce the carbon loss.

4. Conclusion and Recommendation

The objectives of this study is to calculate the carbon stock in year 2035 as the projection by using MOLA approach. The result of this study shows that the province will lose the carbon stock up to 1.1 million ton/year from year 2013 to 2035 by using BAU scenario. On the other hand, by implementing the Protection scenario, total carbon stock that is lost at the same period is about 5 million tons. It shows that Protection scenario could effectively prevent the carbon loss rather than BAU scenario which is a current spatial plan in North Sumatera. The Protection Scenario is a good scenario that can be an alternative for North Sumatera spatial plan.

Nevertheless, this study is only a preliminary study. There are many factors that might influence the carbon loss such as economical factor that are not mentioned in this study. Moreover, this study also weighing based on law besides many software, such as Terrset, that can automatically weighing based on priorities in the province. Therefore, the further study is needed in order to prove this argumentation.

References

[1] Camacho O M 2017 Multi-objective Land Allocation (MOLA). In: Camacho Olmedo M., Paegelow M., Mas JF., Escobar F. (eds) Geomatic Approaches for Modeling Land Change Scenarios *Lecture Notes in Geoinformation and Cartography*: Springer, Cham.

[2] Nowar W 2015 *Rencana Penggunaan Lahan untuk Komoditas Unggulan Berbasis Analisis Multi-Objective Land Allocation di Kabupaten Cianjur* Thesis Bogor Agricultural University

[3] Eastman J R, J H, and James T 1997 Multi-criteria and multi-objective Decision Making for Land Allocation using GIS. In: Beinat E., Nijkamp P. (eds) Multicriteria Analysis for Land-Use Management *Environment & Management* **9** 227-251

[4] Hejehforooshnia S, Soffianian A, Mahiny AS, and Fakheran S 2011 Multi Objective Land Allocation (MOLA) for Zoning Ghamishloo Wildlife Sanctuary in Iran *Journal for Nature Conservation* **19** (4) 254-262

[5] Fahmudin A, Santosa I, Dewi S, Setyanto P, Hamrin Y, Wulan Y C, et al. (2013). *Pedoman Teknis Penghitungan Baseline Emisi dan Serapan Gas Rumah Kaca Sektor Berbasis Lahan Buku I – Landasan Ilmiah.* Jakarta: Badan Perencanaan Pembangunan Nasional

IOP Conf. Series: Earth and Environmental Science **149** (2018) 012043 doi:10.1088/1755-1315/149/1/012043

The effect of land cover change to the biomass value in the forest region of West Java province

M I Rahayu*, T Waryono, Rokhmatullah and I P A Shidiq
Department of Geography, Faculty of Mathematics and Natural Sciences, Universty of Indonesia

E-mail : iinsudhie@yahoo.com

Abstract. Due to the issue of climate change as a public concern, information of carbon stock availability play an important role to describe the condition of forest ecosystems in the context of sustainable forest management. This study has the objective to identify land cover change during 2 decades (1996 – 2016) in the forest region and estimate the value of forest carbon stocks in west Java Province using remote sensing imagery. The land cover change information was obtained by visually interpreting the Landsat image, while the estimation of the carbon stock value was performed using the transformation of the NDVI (Normalized Difference Vegetation Index) which extracted from Landsat image. Biomass value is calculated by existing allometric equations. The results of this study shows that the forest area in the forest region of West Java Province have decreased from year to year, and the estimation value of forest carbon stock in the forest region of West Java Province also decreased from year to year.

1. Introduction

Land cover is a fundamental variable that impacts on and links many parts of the human and physical environments [1]. Land cover mapping and monitoring is one of the major application on Earth observing satellite sensor data and is essential for the estimation of land cover change [2].

Climate change is one of global issues that most frequently discussed, and forests play an important role in it. Due to the issue of climate change as a public concern, information of carbon stock availability play an important role to describe the condition of forest ecosystems in the context of sustainable forest management. Forest biomass is an indicator of carbon sequestration, so it can be known how much carbon is lost or accumulated over time [3].

There are several ways to estimate forest biomass, among others sampling by harvesting (destructive), sampling without harvesting (non-destructive), biomass estimation using remote sensing and modeling of biomass estimators [4]. Remote sensing approach can be applied for large extents areas. The advantages of statistical combinations and temporal resolution of remote sensing allows the obtaining of carbon stock information of an area as well as changes over time [5]. Several studies have been done to estimate the value of biomass by using remote sensing data. Roy and Ravan, 1996 suggest approaches for using satellite remote sensing data for regional biomass mapping in Madhav National Park [6], meanwhile, Rakhmawati (2012) found that correlation between NDVI vegetation index and the land cover's biomass was the highest, both of the natural and non natural vegetation covers with R of 60% [7]. Rokhmatullah measured carbon stock based on vegetation (forest) cover derived from remote sensing. The vegetation cover is then converted to carbon by multiplying with biomass carbon conversion factors [8]. This study has the objective to identify land cover change during 2 decades (1996 – 2016) in the forest region and estimate the value of forest carbon stocks in west Java Province using remote sensing imagery.

2. Methods

2.1. Data

The data used in this study is Landsat 5 mosaic year 1996 and 2006, and also Landsat 8 mosaic year 2016. The Landsat data have 30 m resolution and have been corrected either radiometrically or geometrically (Level 1T). The study area is in the forest region of West Java Province (Figure 1). Acoording to UU No.41 Tahun 1999 tentang Kehutanan [9] and PP No.44 Tahun 2004 tentang Perencanaan Kehutanan [10], forest region in West Java Province has 3 main functions as conservation forest, protection forest and production forest.

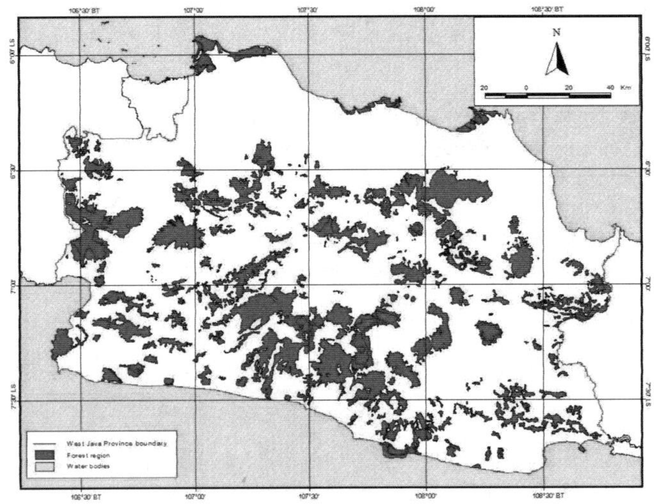

Figure 1. Study area

Field survey were conducted to obtain information needed in estimating biomass value (type of the tree, diameter at breast height, and height of individual tree) and also to validate the result of the land cover classification from Landsat data.

2.2. Method

There are three steps of work in this study, The first step is to create land cover classification from Landsat data year 1996, 2006, and 2016 by visual interpretation, the second step is to calculate vegetation index (NDVI) using band Near Infrared and Visible (Red), and the last step is to estimate biomass value in 1996, 2006, and 2016 (Figure 2)

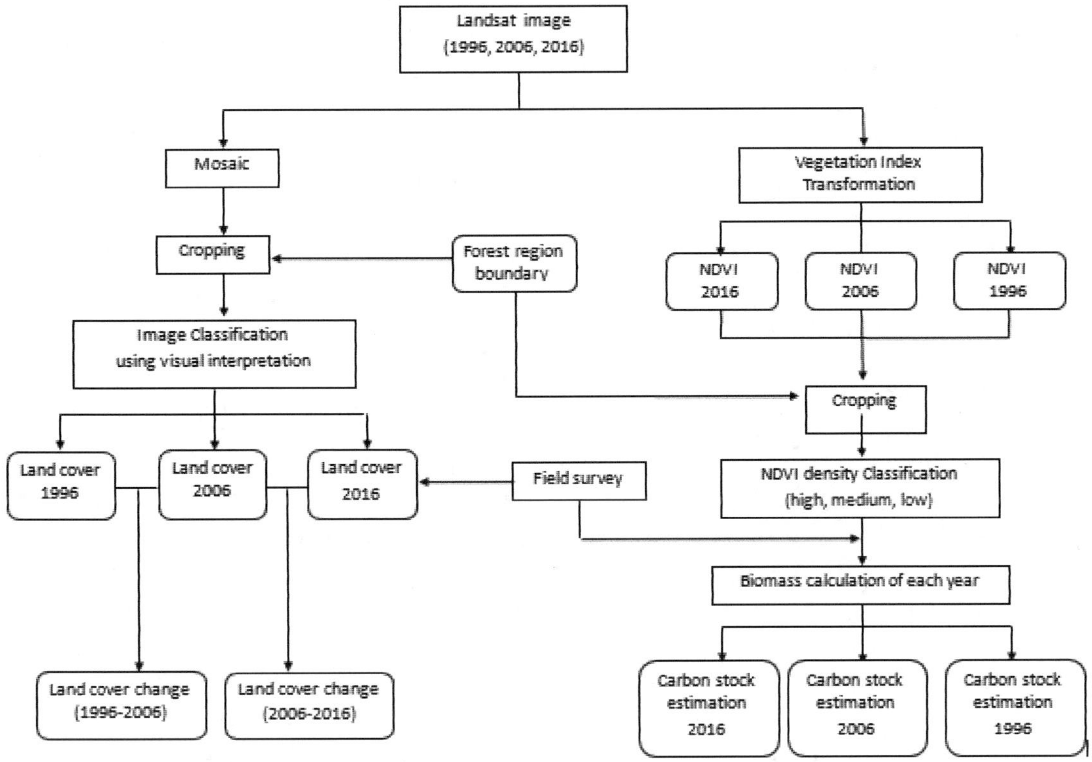

Figure 2. Flowchart of data processing

Land cover classification is done by manually digitizing. Forest classes are separated from other classes including plantation forest. The land cover information was used to calculate land cover change in 1996-2006 and 2006-2016.

NDVI processing is performed on Landsat data. NDVI value is calculated by using reflectant near infrared band and visible band (red). NDVI processing result will be in shapefile format. The shapefile data will be cropped with forest shapefile data so that NDVI in the forest area can be generated. NDVI forest will be grouped into three density classes : high, medium and low density.

The estimation of the biomass at the tree level by using the volume geometric formula is carried out by entering the diameter measurements at breast height, tree height and rod shape figure into the formula [11] :

$$V = 0{,}25\pi \,.\, (D/100)^2 \,.\, H \,.\, F \qquad (1)$$

With V being the volume value of the tree, D is the diameter of the tree at chest height, H is the height of the tree, and F is the correction factor, which is calculated from the ratio of actual stem volume to cylinder volume at the same diameter and height. If barcode number information for a specific type of suspected unavailable, a common stem number value of 0.6 can be used.
The volume value of the tree is then multiplied by the type and BEF value of the tree type.

$$B = \rho \,.\, V \,.\, BEF_{tree} \qquad (2)$$

Estimated biomass is then calculated by summing the individual tree biomass of the constituent tree. The above formula refers to the Guidelines for the Use of Allometric Models for Biomass Assessment and Forest Carbon Stock in Indonesia issued by the Ministry of Forestry and the Environment in 2012[8]. Based on these guidelines, the above formula is the fifth approach used when the allometric model of biomass as well as the allometric model of volume trees developed for a type or type of ecosystem that would otherwise be unavailable, but high data (other than diameter) available from the measurement or inventory of trees.

The 1996 and 2006 biomass values were calculated using the diameter and tree height information in 2016, assuming that NDVI with the same density in 1996, 2006, and 2016 will have the same diameter and tree height information.

3. Result and Discussion

In the period 1996 to 2006 the distribution of unchanged forest land cover is about 167,732.29 Ha (20.85%) of the total forest area (Table 1). While the change of forest cover to non-forest cover is 33923.28 Ha (4.22%), and the change of non-forest cover into forest is 20.53 Ha (0%).

Meanwhile, in the period 2006 to 2016 the distribution of unchanged forest land cover is about 162867.79 Ha (20.25%) of the total forest area. While the change of forest cover to non-forest cover is 4885.02 Ha (0.61%), and the change of non-forest cover into forest is 491.10 Ha (0.06%).

Table 1. Forest Change in the period 1996-2006 and 2006-2016

Change	1996-2006		2006-2016	
	Area (Ha)	%	Area (Ha)	%
Forest - Forest	167732.29	20.85	162867.79	20.25
Forest - Non Forest	33923.28	4.22	4885.02	0.61
Non Forest - Forest	20.53	0.00	481.38	0.06
Non Forest - Non Forest	602725.07	74.93	636166.96	79.09
Total	804401.16	100.00	804401.16	100.00

From 1996 to 2006, the change of forest cover to become non-forest cover was dominated by forest conversion into plantation forests of 29,337.34 Ha (3.65%), whereas changes from forest into other classes were less significant (<1%) (Table 2). While the change from cropland into forest is only 16.33 Ha, and the change from plantation into forest is 4.2 Ha.

From 2006 to 2016, the change of forest cover to become non-forest cover was dominated by forest conversion into cropland of 2482.73 Ha (0.31%), and forest conversion into plantation forests of 1763.07 Ha (0.22%), while the change from non-forest into forest was dominated by conversion from shrubs into forest of 199.27 Ha (0.02%).

IOP Conf. Series: Earth and Environmental Science **149** (2018) 012043 doi:10.1088/1755-1315/149/1/012043

Table 2. Land Cover Change Classes in the period 1996-2006 and 2006-2016

Land Cover Change Classes	1996-2006		2006-2016	
	Area (Ha)	%	Area (Ha)	%
Fixed Classes (F-F, NF-NF)	770457.36	95.78	799034.76	99.33
Forest - Plantation Forest	29337.34	3.65	1763.07	0.22
Forest - Cropland	2564.35	0.32	2482.73	0.31
Forest - Bare Land	884.09	0.11	356.92	0.04
Forest - Plantation	0.00	0.00	117.44	0.01
Forest - Settlement	49.62	0.01	19.21	0.00
Forest - Shrubs	1087.87	0.14	137.55	0.02
Forest - Mining	0.00	0.00	8.10	0.00
Plantation Forest - Forest	0.00	0.00	118.48	0.01
Cropland - Forest	16.33	0.00	119.71	0.01
Bare Land - Forest	0.00	0.00	43.92	0.01
Shrubs - Forest	0.00	0.00	199.27	0.02
Plantation - Forest	4.20	0.00	0.00	0.00
Total	**804401.16**	**100.00**	**804401.16**	**100.00**

Figure 3 and Figure 4 illustrate the spatial distribution of land cover changes for the period 1996-2006 and 2006-2016.

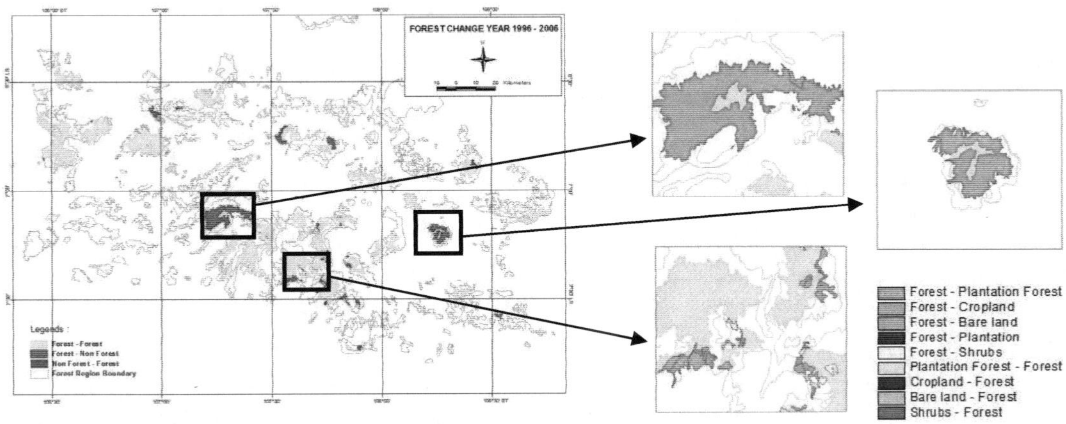

Figure 3. Spatial distribution of land cover changes for the period 1996-2006

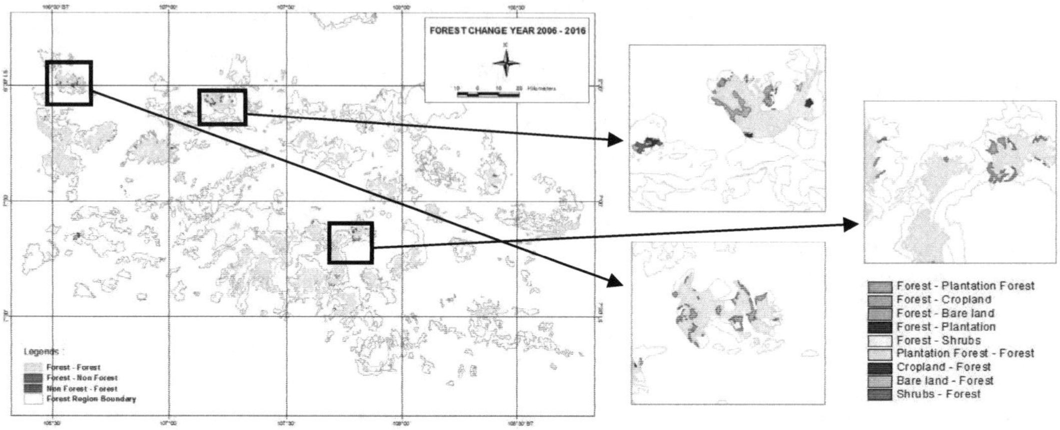

Figure 4. Spatial distribution of land cover changes for the period 2006-2016

Table 3 presents estimation of carbon stock in forest region of West Java province. Total estimation of forest carbon stock in 1996 was 28,154,912.48 ton C/Ha, and in 2006 total forest carbon stock was reduced to 20,677,898.44 ton C/Ha, while in 2016 the total value of forest carbon stock has decreased into 19,925,748.74 ton C/Ha.

Table 3. Estimation of Forest Carbon Stock year 1996, 2006, and 2006

Year	NDVI	Forest	
		Area (Ha)	C (ton C/Ha)
1996	High	159,021.96	25,181,876.62
	Medium	36,609.80	2,729,989.35
	Low	6,023.81	243,046.50
Total		**201,655.57**	**28,154,912.48**
2006	High	100,166.73	15,861,873.65
	Medium	61,044.63	4,552,092.34
	Low	6,541.46	263,932.46
Total		**167,752.82**	**20,677,898.44**
2016	High	98,576.58	15,610,065.90
	Medium	48,069.64	3,584,548.55
	Low	18,120.87	731,134.30
Total		**164,767.09**	**19,925,748.74**

4. Conclusion

During the period 1996 - 2006 the forest area has decreased by 33,923.28 Ha, while the change of non-forest land cover to forest is only 20.53 Ha. Meanwhile, during period 2006 – 2016 the forest area has increased by 491.101 Ha, while the change of forest land cover to non forest is only 4885.021 Ha. However, the forest area in the forest region of West Java Province have decreased from year to year.

The estimation value of forest carbon stock in the forest region of West Java Province also decreased from year to year. In 1996 the value of carbon stocks was 28,154,912.48 ton C/Ha, in 2006 the value of carbon stock was reduced to 20,677,898.44 ton C/Ha, and in 2016 the value of carbon stock became 19,925,748.74 ton C/Ha.

References

[1] Foody 2001 Status of land cover classification accuracy assessment *Remote Sensing of Environment* **80** 185 – 201
[2] Galiano et al 2012 An Assessment of the effectiveness of a random forest classifier for landcover classification *ISPRS Journal of Photogrammetry and Remote Sensing* **67** 93 – 104
[3] Parresol R 1999 Assessing Tree and Stand Biomass: A Review with Examples and Critical Comparisons *Forest Science* **45** 573-593
[4] Dandun S 2009 *Penghitungan Biomasa: Sebuah pengantar untuk studi karbon dan penghitungan karbon* (Bogor: Wetlands International Indonesia Programme)
[5] Wahyuni N I 2012 Integrating of Remote Sensing in Forest Biomass Measurement *Info BPK Manado* **2** (2) 115 – 126
[6] Roy and Ravan 1996 Biomass estimation using satellite remote sensing data – An investigation on possible approaches for natural forest *J.Biosci.* **21** (4) 535-561
[7] Rakhmawati M 2012 Hubungan Biomassa Penutup Lahan Dengan Indeks Vegetasi di Kabupaten Mamuju Utara, Sulawesi Barat *Globe* **14** (2) 157 – 169
[8] Rokhmatullah 2009 Estimation of Carbon Stock Using Remote Sensing : A Case Study of Indonesia *Geospatial World*
[9] Undang-Undang No.41 Tahun 1999 tentang Kehutanan
[10] PP No.44 Tahun 2004 tentang Perencanaan Kehutanan
[11] KLHK 2012 Pedoman Penggunaan Model Allometrik untuk Pendugaan Biomassa dan Stok Karbon Hutan di Indonesia

Identifying the driving forces of urban expansion and its environmental impact in Jakarta-Bandung mega urban region

A E Pravitasari[1,2*], E Rustiadi[1,2], S P Mulya[1,2], Y Setiawan[3], L N Fuadina[1], A Murtadho[1]

[1] Department of Soil Science and Land Resource, Faculty of Agriculture, Bogor Agricultural University, Jl. Meranti, IPB Darmaga Campus, Bogor, West Java, Indonesia, 16680
[2] Center for Regional System, Analysis, Planning and Development (P4W/CRESTPENT), Bogor Agricultural University, Jl. Raya Pajajaran, Bogor, West Java, Indonesia, 16144
[3] Center for Environmental Research, Bogor Agricultural University (PPLH-IPB), Jl. Lingkar Akademik Kampus IPB Darmaga, Bogor, West java, Indonesia 16680

E-mail: andreaemma84@yahoo.com, andreaemma@apps.ipb.ac.id

Abstract. The socio-economic development in Jakarta-Bandung Mega Urban Region (JBMUR) caused the increasing of urban expansion and led to a variety of environmental damage such as uncontrolled land use conversion and raising anthropogenic disaster. The objectives of this study are: (1) to identify the driving forces of urban expansion that occurs on JBMUR and (2) to analyze the environmental quality decline on JBMUR by producing time series spatial distribution map and spatial autocorrelation of floods and landslide as the proxy of anthropogenic disaster. The driving forces of urban expansion in this study were identified by employing Geographically Weighted Regression (GWR) model using 6 (six) independent variables, namely: population density, percentage of agricultural land, distance to the center of capital city/municipality, percentage of household who works in agricultural sector, distance to the provincial road, and distance to the local road. The GWR results showed that local demographic, social and economic factors including distance to the road spatially affect urban expansion in JBMUR. The time series spatial distribution map of floods and landslide event showed the spatial cluster of anthropogenic disaster in some areas. Through Local Moran Index, we found that environmental damage in one location has a significant impact on the condition of its surrounding area.

1. Introduction

Regional development and increasing urban expansion in most developing country including Indonesia affecting a serious problem in environmental condition and decrease carrying capacity of the environment [1]. In Indonesia, two metropolitan areas of Greater Jakarta (Jabodetabek) and Bandung Raya is experiencing the phenomenon of urban expansion. Jabodetabek metropolitan and Bandung Raya became increasingly connected to be a mega-urban area through a corridor called Jakarta-Bandung Mega-Urban Region or JBMUR [2]. The formation of the mega-urban area is caused by a conurbation process that characterized by increased population growth, increased activity in various sectors, and the development of built-up areas such as new city projects and industrial centers which connected through a network of roads or corridors within two metropolitan areas, namely the Bogor-Puncak-Cianjur-Bandung route and the Cipularang Jakarta-Bandung toll road [3]. The conurbation is the process which a metropolitan city experiences an urban physical development that blends with the surrounding cities forming a larger metropolitan area. This larger metropolitan area is an urban area formed by a large city system and its surroundings. Mega urban conurbation process in a globalized world implies in considering it as a complex phenomenon in space and time. It is an extremely large or densely populated urban area which usually consists of several towns merging with the suburbs of a central city. Some important changes introduced in top-down processes of conurbation, mainly through global policies, such as a transportation development, new road or highway construction, new alternative connections among spaces, for instance, that can introduce deep changes in the local structures which need to readapt itself to the new situation and making it possible that emergent processes can arise again but from a different starting point [4].

The increasing population growth of JBMUR especially in urban areas causes the increasing demand for land. Increased demand for such land leads to land conversion, especially the conversion of vegetated lands to built-up area. Uncontrolled land use/land cover changes can cause environmental degradation, which resulting in environmental problems and even anthropogenic disasters such as floods and landslides. The inconsistency between land use and land capacity in relation to environmental carrying capacity may lead to a decrease in environmental quality resulting in anthropogenic disasters [5]. Starting from the beginning of the 1990s, environmental degradation in JBMUR which shown by increasing number of disaster has been a hot issue in the term of sustainable development of mega urban region. The number of floods incidents, landslides and environmental problems that occurred in Indonesia increasingly prompted the importance of the role of disaster risk reduction. According to Yamani et al. [6], land use planning and spatial planning can be used as a tool to reduce disaster risk.

The aims of this study are: (1) identifying the driving forces of urban expansion that occurs in JBMUR; and (2) analyzing distribution patterns of floods and landslides prone areas in JBMUR using time series data and identifying the spatial clustering phenomenon.

Regional development and rapid urban expansion in JBMUR has a significant impact on the local and global environment. Therefore, to prevent uncontrolled rapid urban development in this area which emerging as one of Asian Megacities, must be started from clear understanding of various driving forces that affecting urban expansion [7]. GWR (Geographically Weighted Regression) is an appropriate tool to identify the local spatial dependency of the driving factors which affecting urban expansion. GWR models were used in various studies to identify the spatial variations in the relationship between environmental and socioeconomic indicators, regional development, ecology, natural resourse management and social studies [8-14].

Rapid growth of built-up area as well as uncontrolled land use/land cover change in JBMUR such as converting vegetated lands into built-up area, as well as protected areas into cultivation areas can lead to a decrease in environmental carrying capacity which is the ability of the environment to support human life and other living beings. The decline in environmental carrying capacity can lead to disasters and environmental problems such as floods, landslides, forest fires, land droughts, and increasing critical land. Therefore, it is necessary to conduct such kind of study related to identify the driving forces of land use change especially urban expansion and investigate the distribution patterns of floods and landslides prone areas as the example of environmental impact due to unsustainable patterns of rapid urban development in JBMUR and identify spatial clustering phenomenon on those variables. The results of this study can be used as an evaluation in land use policy and spatial planning (RTRW) and as an effort to prevent and reduce environmental problems such as flood and landslide risk.

2. Method

2.1 Study Area

IOP Conf. Series: Earth and Environmental Science **149** (2018) 012044 doi:10.1088/1755-1315/149/1/012044

This study area is Jakarta-Bandung Mega-Urban Region or JBMUR which located in Java Island and generally covers the administrative area of Greater Jakarta (Jabodetabek Metropolitan Area), Bandung Raya Metropolitan Area, and some areas around the mega-urban corridor. Specifically, the administrative areas in JBMUR consist of 20 regencies/municipalities: five (5) municipalities in DKI Jakarta Province (including East Jakarta, South Jakarta, West Jakarta, North Jakarta, and Central Jakarta Municipality), Tangerang Regency, Tangerang Municipality, South Tangerang Municipality, Bekasi Municipality, Bekasi Regency, Depok Municipality, Bogor Regency, Bogor Municipality, Karawang Regency, Cianjur Regency, Purwakarta Regency, Bandung Barat Regency, Cimahi Municipality, Bandung Municipality and Bandung Regency (figure 1).

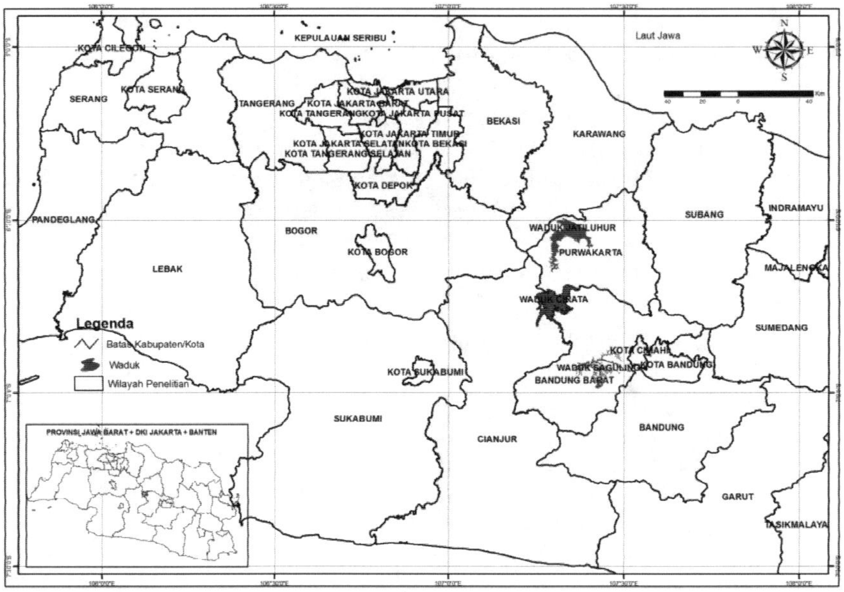

Figure 1. Study Area

2.2 Materials and Methods

The driving forces that affecting urban expansion in JBMUR were identifying using GWR model. GWR is a statistical method to identify local spatial variations. This model works based on the "Tobler's First Law of Geography", which mentioned that everything is usually related with everything else, but those which are near to each other are more related when compared to those that are further away. In GWR model, spatial non-stationarity was assumed and tested. This model addresses the non-stationarity and allows the local spatial variations to vary over space. The result of this analysis is a regression model whose parameter values apply only to each location of observation, and different from other locations. In GWR, we use the weighted matrix element W(i) which the amount depends on the proximity between locations. The weighting function to be used for GWR model in this study is Gaussian Kernel function. The GWR is an expansion of the global regression model. However, unlike global regression that is applied in general at every observation location, GWR produces local model parameter estimators for each observation location using the Weighted Least Square (WLS) method. The formulas and variables used in the model are as follows:

$$Y_j = C_0(u_j, v_j) + \sum_{i=1}^{p} C_i(u_j, v_j) X_{ij} + \varepsilon_j \tag{1}$$

where: Y_j = Dependent variable for observation j; X_{ij} = Independent variable Xi at location *j;* u_j,v_j = Coordinate point for location of observation j; $C_0\ (u_j,v_j)$ = Intercept for observation *j;* $C_i\ (u_j,v_j)$ =

Regression coefficient or local parameter estimate for independent variable X*i* at location *j*. The optimal bandwidth of GWR analysis in this study was determined by minimizing the corrected Akaike Information Criterion (AIC's value) with a correction for finite sample sizes, as described in Fotheringham, Brunsdon, and Chalton [15].

The dependent variable (Y) used in this model is percentage of built-up area (2015); Six (6) independent variables (X) included in this model are: X_1 = population density (person/km^2); X_2 = percentage of agricultural land, excluding ricefield (%); X_3 = distance to the CBD/capital regency (km); X_4 = percentage of household working in agricultural sectors (%); X_5 = distance to the provincial road (km); X_6 = distance to the local road (km).

In this study, we employed Global Moran and Local moran Index (LISA statistics) to analyze distribution patterns of floods and landslides prone areas in JBMUR using time series data and identifying the spatial clustering phenomenon. The formula of Global Moran's Index (I) and Local Moran's Index (I_i) or LISA statistic can be seen as follows:

$$I = \frac{\sum_{i=1}^{n}\sum_{j=1}^{n} W_{ij}(Z_i - \bar{Z})(Z_j - \bar{Z})}{S_Z^{\,2}\sum_{i=1}^{n}\sum_{j=1}^{n} W_{ij}} \quad (2)$$

$$I_i = \frac{\sum_{j=1}^{n} W_{ij}(Z_i - \bar{Z})(Z_j - \bar{Z})}{S_Z^{\,2}\sum_{j=1}^{n} W_{ij}} \quad (3)$$

Where: I= Global Moran's Index; I_i= Local Moran's Index or LISA statistics; Z_i= representing the value of variable Z for point/location i; Z_j= representing the value of variable Z for point/location j (the neighbour); Z = the average value (mean) of Z; W_{ij} = spatial association between i and j (contiquity matrix); representing the proximity of *i*-th village (*desa*) *i*'s and village (*desa*) *j*'s locations; n = the total number of villages (*desa*); S_z^2= the variance of the observed values. Unit of analysis used in this model is village (desa). There are 2955 villages/desa in JBMUR.

3. Results and Discussions

3.1 Identifying Driving Forces which Affecting Urban Expansion in JBMUR

The development of the Jakarta-Bandung corridor has resulted in a massive urban expansion. Rapid urban development in two biggest metropolitan areas (Jabodetabek and Bandung Raya) mainly contributes to urban sprawl and increases population density, as well as encourage urban activities and economic growth in the region and its surrounding area [2]. Based on our previous research, in the period of 1972-2010, urban land area in Jabodetabek increased by about 2,096 km2 because of urbanization and suburbanization's impact [5]. Land use/land cover change continues to increase in JBMUR from 1996 to 2015, especially from vegetated lands such as rice fields and mixed gardens to built-up area. The rapid expansion of built-up area is largely driven by the existence of road networks (especially toll roads) that connect the two metropolitan areas. Areas that have experienced the most widespread land are Bandung, Bogor, Bekasi and Tangerang Regencies. One type of land use that undergoes many changes or conversions into built-up area is rice field.

Based on GWR analysis, the spatial distribution pattern of built-up area of observation and prediction of the model are almost the same (figure 2). From figure 2, it can be seen that the largest percentage of built-up areas which has red colour are mostly located in the center of Jakarta and Bandung metropolitan areas with built-up percentages of over 80%. In the area covered by green colour which closer to the periphery, the percentage of built-up area were lower.

The important analysis results from GWR model can also be known from the value of local R^2. R square (R^2) or coefficient of determination is usually used to calculate the suitability of the model. There is only 1 single value of R^2 in OLS (Ordinary Least Square) model, but in GWR model, each location has their own value of R^2, or often called local R^2. One of the reason using GWR is because local analysis intends to show spatial variability between locations and understand the spatial data more detail. High local R^2 indicates that the local regression model more fits the observations. In figure 3, the value of local R^2 varies for each location with a range between 0.698 and 0.910. The highest local R^2 value is located in the south-eastern part of the study area (especially located in Bandung metropolitan area and some parts of Purwakarta and Cianjur Regencies). Area with a higher local R^2 value is more potential to get the impact of urban expansion, so those areas will easily converted into built-up area compare with the other area with smaller value of local R^2.

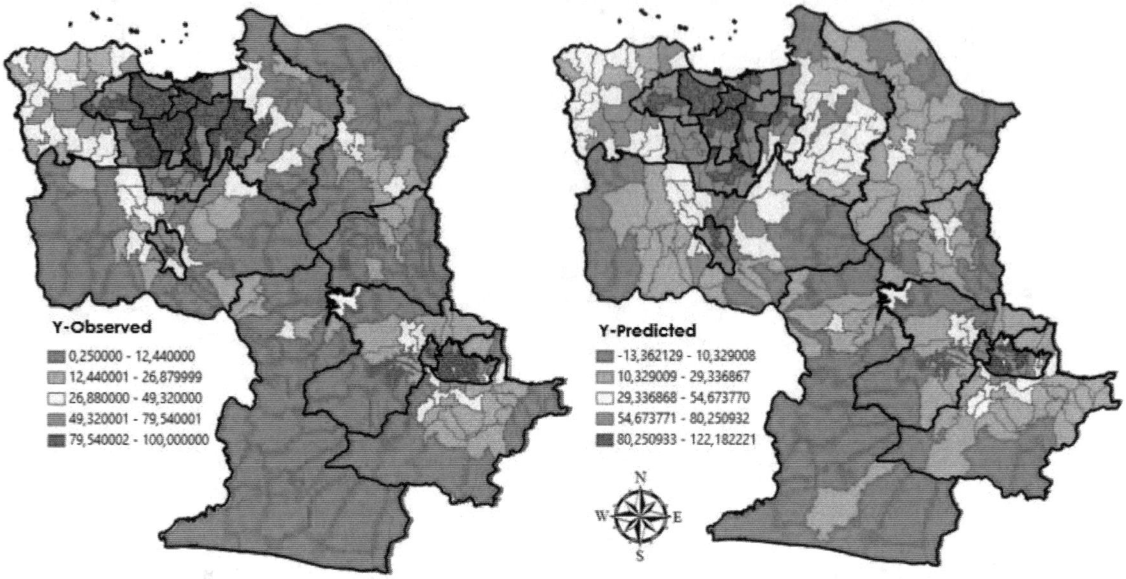

Figure 2. Percentage of built-up area in (1) Observed and (2) Predicted based on GWR Model

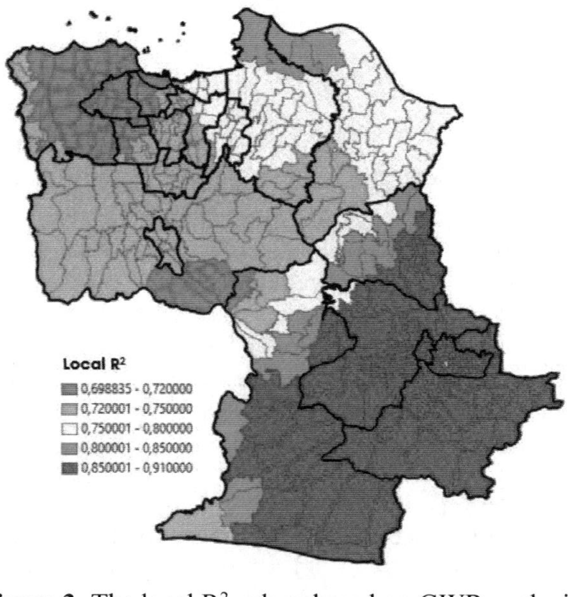

Figure 3. The local R^2 values based on GWR analysis

GWR employs a spatial weighting function with the assumption that everything which located in the near places are more similar than others in distant places (First Law of Geography). figure 4 showed the local variations of parameter estimates (C coefficient) for each independent variable based on GWR results. From figure 4, it can be known the value of local parameter estimates (C) for each independent

IOP Conf. Series: Earth and Environmental Science **149** (2018) 012044 doi:10.1088/1755-1315/149/1/012044

variable (X) used in the model. Based on the results, it seem that the most dominant influence on urban expansion is agricultural activities.

Population density is considered as having relationships with urban expansion [16]. Spatial distribution pattern of the coefficient or parameter estimate (C) for X1 [POPDENS] or population density can be seen in figure 4 (a). POPDENS (X1) has a positive correlation on urban expansion in the whole areas (although the value index are relatively low), which means that increasing population density will increase the percentage of built-up area in that location. An increasing populated region will also potentially lead to increase construction of new settlement location as well as other built-up areas such as new towns, new industrial land, etc [17, 18]. The red colour area in fgure 4 (a) is the location with the biggest influence of population density factor on the increasing of built-up area.

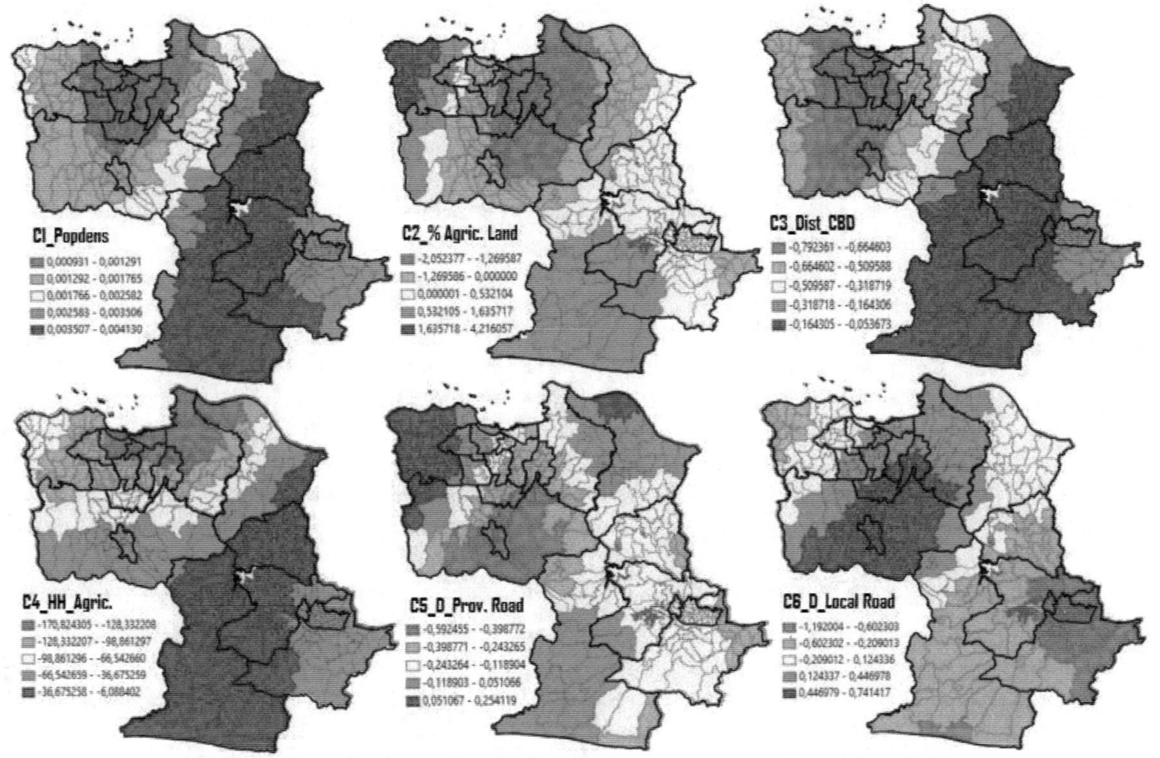

Figure 4. Parameter Estimates (C) for Each Independent Variable

The second and forth independent variables used in the model (X2 and X4) are the percentage of agricultural land area (non-paddy field) [%_AGRIC.LAND] and percentage of the household works in agricultural sector [HH_AGRIC]. Based on the literature review, variable related to agricultural activities were considered as having relationship with urban expansion [19, 20, 21]. The spatial distribution pattern of local parameter estimate values for X2 [%_AGRIC.LAND] can be seen in figure 4 (b). Considering GWR results, the coefficient or local parameter estimate for X2 has a positive to negative value in different location. A positive value of the local parameter estimates means that the effect of X2 is directly proportional to the dependent variable (Y). On the other hand, negative value of local parameter estimates indicating negative relationship between the dependent variable (Y) and X2 [%_AGRIC.LAND]. Locations with negative local parameter estimates values (in figure 4 (b) are light green and dark green colour areas - located in the center of the study area) indicating that at those locations, the reduced percentage of agricultural land (excluding paddy field) will have a significant effect on the increasing of built-up areas. Thus, it can be predicted that the increasing of built-up area in the light green and dark green areas which covering the area of Jakarta, Bogor, Depok, Bekasi and a small area of Tangerang, is the result of land conversion from agricultural land (non-paddy field).

The spatial distribution pattern of local parameter estimate values for X4 [HH_AGRIC] can be seen in figure 4 (d). Considering GWR results, the coefficient or local parameter estimate for X4 has a negative value for the whole areas. This indicates that X4 [HH.AGRIC] has negative relationship with the dependent variable (Y). Similar to the explanation for the variable X2 above, if the area of agricultural land decreases, number of household who works in agricultural sectors will also decrease because of the shrinking of farmland. This shrinking of farmland is caused by land use change or landuse conversion from agricultural to built-up areas.

The third independent variable used in the GWR model (X3) is distance to the CBD (central business district). Distance to the CBD or city center is also considered as one of the driving factors that influence urban expansion [5]. The spatial distribution pattern of local parameter estimate values for X3 [DIST_CBD] can be seen in figure 4 (c). Considering GWR results, the coefficient or local parameter estimate for X3 also has a negative value for the whole areas. This indicates that X3 [DIST_CBD] has negative relationship with the urban expansion. The lands that have been converted to urban expansion are lands which have good accessibility and located close to the central city/CBD. Parameter estimate values for X3 [DIST_CBD] are all negative and range from -0.792 to -0.053. The negative coefficient value indicates that X3 is inversely proportional to the dependent variable [Y]. The lands located close to the CBD will tend to be converted to built-up area.

The last two variables included in GWR model (X5 and X6) are the distance to the provincial road (X5/D_PROV.ROAD) and the distance to the local road (X6/D_LOCAL ROAD). Based on Tuloli [22], Prasetyo, Koestoer, & Waryono [23], and Susanti [24], infrastructure development and accessibility to the roads or highways were considered as driving factors which caused the outspreading of urban expansion. For the distance variable to provincial road (X5), the parameter estimate value ranges from -0.592 to 0.254. As for the distance variable to the local road (X6) the coefficient value ranges from -1.192 to 0.741. Accessibility is a important factor to the development of built-up areas, so [D_PROV.ROAD] and [D_LOCAL ROAD] are included in the GWR model. However, from the GWR results, it can be seen that the magnitude of the influence of the distance variables vary for each location.

3.2 Spatio-temporal distribution patterns and spatial clustering of floods and landslides prone areas in JBMUR

The relationship between urban expansion and environmental deterioration occured in JBMUR was investigated in this study by choosing floods and landslides as representative examples of anthropogenic disaster. Spatial distribution patterns of floods and landslides in JBMUR in the years of 2000, 2006 and 2014 are shown in figure 5 and figure 6. Based on figure 5, number of villages experiencing floods are tend to increase. Blue area shows the location which inundated by floods. Flood incident occurred mostly in the northern part of DKI Jakarta, the northern part of Bekasi Regency, and the northern part of Karawang Regency since the area has a flat topography condition. Floods also tend to occur in the area of Bandung Municipality and the northern part of Bandung Regency because the area has rapid development especially the increasing number of built-up area.

Similar to condition that occurred in flood disaster, landslides in JBMUR also tend to increase from year to year while the spread of landslide events occurred in the southern part of Bogor Regency, Cianjur Regency, West Bandung Regency, and Bandung Regency since those regions have a predominantly steep and hilly topography condition (figure 6). Brown area shows the location which have landslide occurrence. The region has a dominant slope of 41% -60% so vulnerable to the occurrence of landslide disaster.

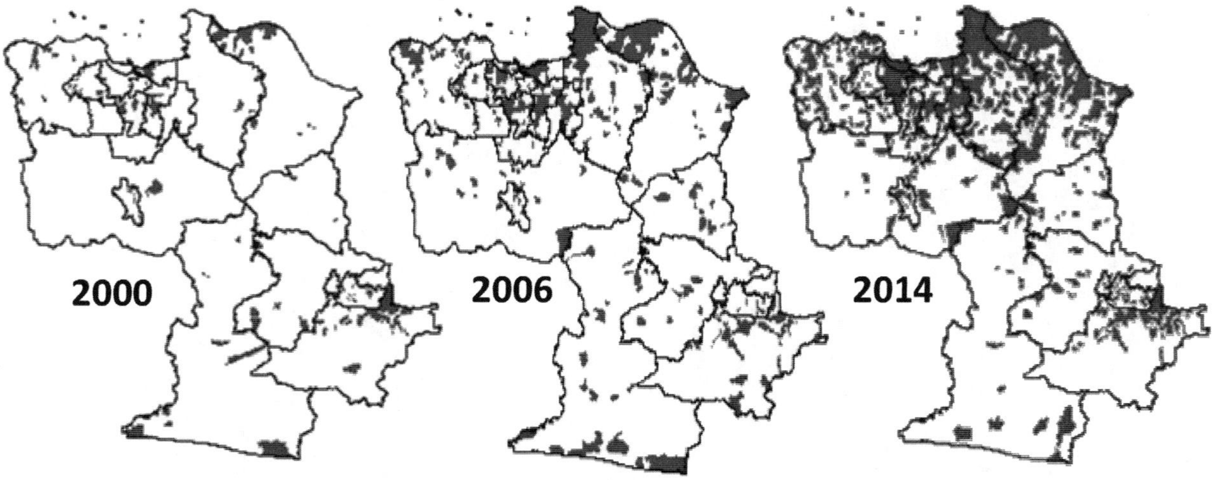

Figure 5. Spatio-temporal distribution patterns of floods in JBMUR

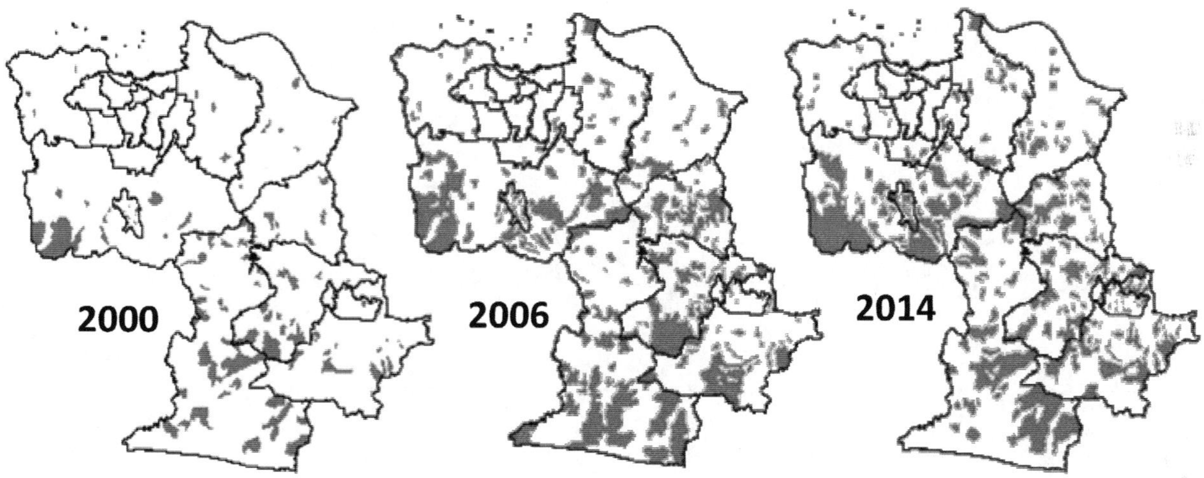

Figure 6. Spatio-temporal distribution patterns of landslides in JBMUR

In this study, we employed spatial autocorrelation analysis using Global Moran Index and Local Moran Index/LISA to identify spatial clusters of flood and landslide prone areas in JBMUR. Flood and landslide's spatial clusters were detected since the spatial autocorrelation or Global Moran Index (Moran's I Index) has a positive value (figure 7). Moran's I Index values for floods and landslides are 0.3 and 0.2, respectively. Those values indicates the spatial clustering phenomenon and through figure 7 it can be seen that the spatial association of floods (based on the number of floods evidence) is higher than landslides, so that the spatial distribution pattern of floods is more clustered than landslide.

Through the LISA cluster map (shown by figure 7), there are 4 categories of spatial association, i.e: (1) high-high/HH (high values surrounding by high values); (2) low-low /LL (low values surrounding by low values); (3) high-low/HL (high values surrounding by low values); or (4) low-high/LH (low values surrounding by high values). The red areas (HH type of spatial association) shows spatial cluster of high vulnerability on floods and landslide prone area while the blue areas (LL type) shows spatial cluster of low vulnerability on floods and landslide prone area. Based on the LISA statistic's result shown in the figure 7, the northern part of Jakarta, Bekasi, Tangerang and Karawang have high vulnerability of floods, while the southern part of Bogor, Cianjur and Purwakarta have high vulnerability of landslides.

IOP Conf. Series: Earth and Environmental Science **149** (2018) 012044 doi:10.1088/1755-1315/149/1/012044

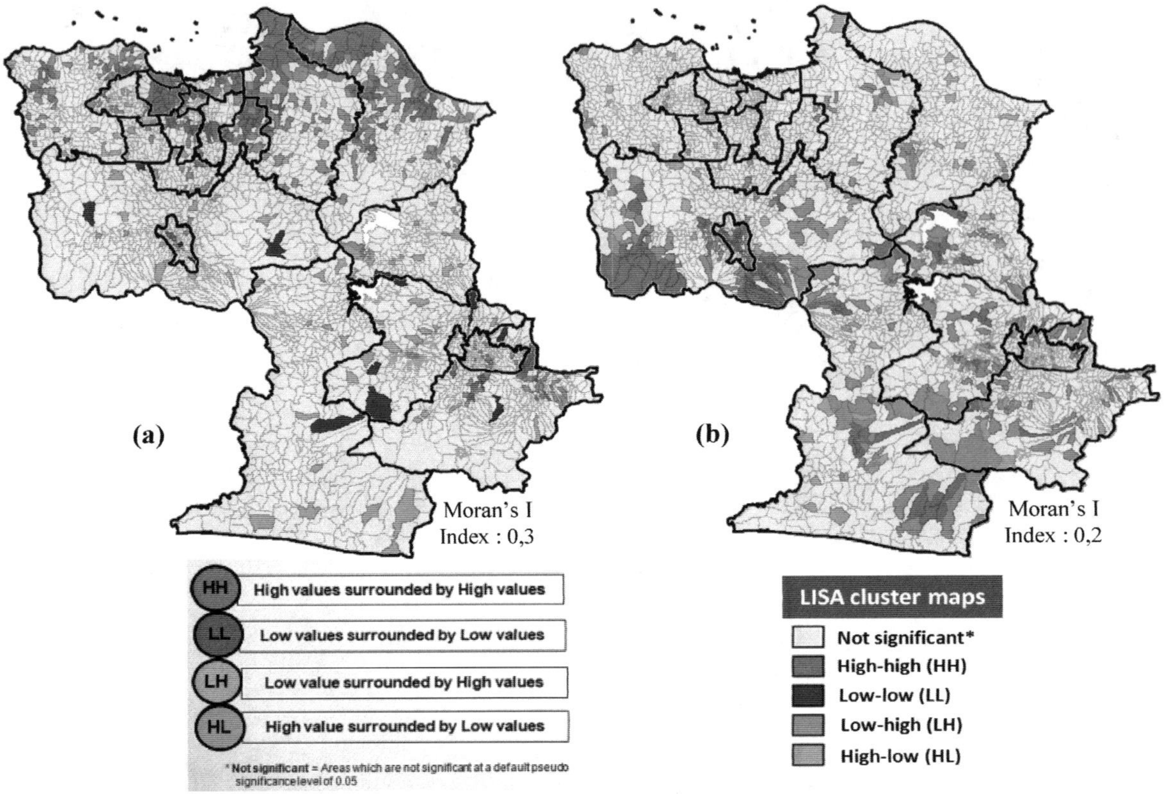

Figure 7. LISA Cluster Map of (a) Floods and (b) Landslides Prone Areas

4. Conclusion

Several kinds of driving forces, namely: local demographic, social, and economic factors including distance to the CBD, provincial and local road spatially affect urban expansion in JBMUR. The timeseries spatial distribution map of floods and landslides showed the spatial cluster of anthropogenic disaster in some areas. Through Local Moran Index, we found that environmental damage such as floods and landslides in one location has spatial association with its surrounding areas. The LISA cluster map from this study could be used as preliminary result to be developed for making floods and landslides zoning map and disaster risk management which useful to minimize environmental deterioration impacts of urban sprawl and uncontrolled urban expansion in megacities/mega urban region.

References

[1] Pravitasari A E, Saizen I, Tsutsumida N, and Rustiadi E 2014 Detection of spatial clusters of flood- and landslide-prone areas using local moran index in Jabodetabek metropolitan area, Indonesia *International Journal of Ecology and Environmental Science* **40** (4) 233-243

[2] Dorodjatoen A M H 2009 The emergence of Jakarta-Bandung mega urban region and it's future challenges *Jurnal Perencanaan Wilayah dan Kota* **20** (1) 15-33

[3] Saifullah K, Barus B, and Rustiadi E 2017 Spatial modelling of land use/cover change (LUCC) in South Tangerang City, Banten *IOP Conf.* **54** (1) 1-11

[4] Rigatti D 2009 Measuring Conurbation. Proceedings of the 7th International Space Syntax Symposium

[5] Rustiadi E, Iman L S, Lufitayanti T, and Pravitasari A E 2013 LUCC Inconsistency analysis to spatial plan and land capability (Case study: Jabodetabek Region) In Y. Himiyama (Ed.),

Toward Sustainable Landuse in Asia (IV) SLUAS Science Report 2013
[6] Yamani A, Rustiadi E, dan Widiatmaka 2015 Evaluasi pola ruang berbasis kerawanan banjir di Kabupaten Pidie *Tata Loka* **17** (3) 130-1
[7] Pravitasari A E, Saizen I, Tsutsumida N, Rustiadi E, and Pribadi D O 2015 Local spatially dependent driving forces of urban expansion in an emerging Asian Megacity: the case of greater Jakarta (Jabodetabek) *The Journal of Sustainable Development* **8** (1) 108-119
[8] Ogneva-Himmelberger, Pearsall, Y H., and Rakshit R 2009 Concrete evidence and geographically weighted regression: A regional analysis of wealth and the land cover in Massachusetts *Applied Geography*
[9] Yu D L 2006 Spatially varying development mechanisms in the Greater Beijing area: A geographically weighted regression investigation *Annals of Regional Science* **40** 173-190
[10] Su S, Xiao R., and Zhang Y 2012 Multi-scale analysis of spatially varying relationships between agricultural landscape patterns and urbanization using geographically weighted regression *Applied Geography* **32** (2) 360-375
[11] Jaimes N B P, Sendra J B, Delgado M G, and Plata R F 2010 Exploring the driving forces behind deforestation in the state of Mexico (Mexico) using geographically weighted regression *Applied Geography* **30** 576-591
[12] Clement F, Orange D, Williams M, Mulley C, and Epprecht M 2009 Drivers of afforestation in Nothern Vietnam: assessing local variations using geographically weighted regression *Applied Geography* **29** (4) 561-576
[13] Malczewski J and Poetz A 2005 Residential burglaries and neighborhood socioeconomic context in London, Ontario: global and local regression analysis *The Professional Geographer* **57** 516-529
[14] Farrow A, Larrea C, Hyman G, and Lema G 2005 Exploring the spatial variation of food poverty in Ecuador *Food Policy* **30** 510-531
[15] Fotheringham A S, Brunsdon C, and Charlton M 2002 *Geographically weighted regression: The analysis of spatially varying relationships* England: John Wiley & Sons
[16] Bilsborrow R E, and Okoth-Ogendo H W O 1992 Population driven changes in land use developing countries *Ambio* **21** 37-45
[17] Hudalah D, and Firman T 2012 Beyond property: Industrial estates and post-suburban transformation in Jakarta Metropolitan Region *Cities* **29** (1) 40-48
[18] Hudalah D, Viantari D, Firman T, and Woltjer J 2013 Industrial land development and manufacturing deconcentration in Greater Jakarta *Urban Geography* **34** (7) 950-971
[19] Van T T 2008 Research on the effect on urban expansion on agricultural land in Ho Chi Minh City by using remote sensing method *VNU Journal of Science Earth Sciences* **24** 104-111
[20] Lopez T, del M, Mitchell A T, and Thomlinson J R 2001 Urban expansion and the loss of prime agricultural lands in Puerto Rico *Ambio* **30** (1) 49-54
[21] Njungbwen E and Njungbwen A 2011 Urban Expansion and Loss of Agricultural Land in Uyo Urban Area: Implications for Agricultural Business *Ethiopian Journal of Environmental Studies and Management* **4** (4)
[22] Tuloli M Y 2012 Pengaruh Faktor Aksesibilitas (Jalan) terhadap Perkembangan Kota Gorontalo *Jurnal Teknik* **10** (1) ISSN: 1693-6191
[23] Prasetyo A, Koestoer R H, and Waryono T 2016 Pola Spasial Penjalaran Perkotaan Bodetabek: Studi Aplikasi Model Shannon's Entropy *Jurnal Pendidikan Geografi* **16** (2) 144-160
[24] Susanti I 2013 Pengaruh Perkembangan Pembangunan Infrastruktur Jalan terhadap Pertumbuhan Pemanfaatan Lahan Kota *Jurnal Rekayasa* **17** (1) 49-57

IOP Conf. Series: Earth and Environmental Science **149** (2018) 012045 doi:10.1088/1755-1315/149/1/012045

Role of cold surge and MJO on rainfall enhancement over indonesia during east asian winter monsoon

R R Fauzi and R Hidayat*

Department of Geophysics and Meteorology, Faculty of Mathematics and Natural Science, Bogor Agricultural University, Bogor, Indonesia

E-mail: rahmath@apps.ipb.ac.id

Abstract. Intensity of precipitation in Indonesia is influenced by convection and propagation of southwest wind. Objective of this study is to analyze the relationship between cold surge and the phenomenon of intra-seasonal climate variability Madden-julian Oscillation (MJO) for affecting precipitation in Indonesia. The data used for identifying the occurrence of cold surge are meridional wind speed data from the ERA-Interim. In addition, this study also used RMM1 and RMM2 index data from Bureau of Meteorology (BOM) for identifying MJO events. The results showed that during East Asian Winter Monsoon (EAWM) in 15 years (2000-2015), there are 362 cold surge events, 186 MJO events, and 113 cold surge events were associated with MJO events. The spread of cold surge can penetrate to equator and brought mass of water vapor that causes dominant precipitation in the Indonesian Sea up to 50-75% from climatological precipitation during EAWM. The MJO convection activity that moves from west to east also increases precipitation, but the distribution of rainfall is wider than cold surge, especially in Eastern Indonesia. MJO and cold surge simultaneously can increase rainfall over 100-150% in any Indonesian region that affected by MJO and cold surge events. The mechanism of heavy rainfall is illustrated by high activity of moisture transport in areas such as Java Sea and coastal areas of Indonesia.

1. Introduction

Indonesia is a country that located in area where crossed by global wind variations. One of the wind patterns that affect climate conditions in this country is the monsoon. Monsoon is a large-scale wind that occurs because the difference between temperature and pressure in the ocean and land. Asian monsoon affect the maximum rainfall in South Asia, especially in Southeast Asia [1]. Meanwhile, one of the extreme weather and climate events in Java and Sumatra were occurred during the influence of monsoon winds [2].

Various occurrences of climate variability often affect the extreme rainfall in some parts of Indonesia. Besides affecting rainfall in annual and inter-annual periods, some climate variability also affects the rainfall in intra-seasonal periods [3]. Extreme rainfall anomaly can be predicted from water vapor that carried away through Intertropical Convergence Zone (ITCZ). Moisture transport defined as a important components to see the potential areas where extreme rain occurs [4]. The amount of water vapor that concentrated in some region causing extreme rain which can be categorized based on threshold and percentile 95% calculation [2]. Some of the intra-seasonal climate variability during boreal winter that can bring water vapor is cold surge and Madden-Julian Oscillation (MJO).

During boreal winter, the flow of surface wind from north that commonly referred as East Asian winter monsoon much observed in East and Southeast Asia [5]. The active winds coincide with the onset of weather extreme where occurred in East Asia that results from interaction between vapor transport and the monsoon. In that condition, often found in the form of a mass flow of cold air which carries more air masses from Siberia. That phenomenon can be called as cold surge [6].

The impact of cold surge not only causes extreme weather where occurred in East Asia, but causes great intensity of rainfall in Southeast Asia, especially in Indonesia [7]. In 2007, there is a strong cold surge occurred five times whose propagation passes through the equator, even the island of Java. The incident became a major factor where Jakarta have a heavy rainfall that causing massive flooding [8]. In addition, some recent study shows that the effect of cold surge takes one day or less to influence the weather of southeast asia region [9]

Further analysis of cold surge with MJO needs to be done to see the influence of intra-seasonal climate variability events against climatological condition in Indonesia. This study was conducted because the previous research is more research about the characteristics of cold surge, while the research emphasizes on influence of cold surge in changing conditions of rainfall patterns in Java Island and the others area are not done yet [2,6,10].

2. Methodology

2.1. Data collection and literature studies

The downloaded data is then selected based on 15°S-25°N and 95°-140°E in this study area. Area studies include Southeast Asia including Indonesia's maritime continent, a small part of the Indian Ocean and Pacific Ocean (see figure 1).

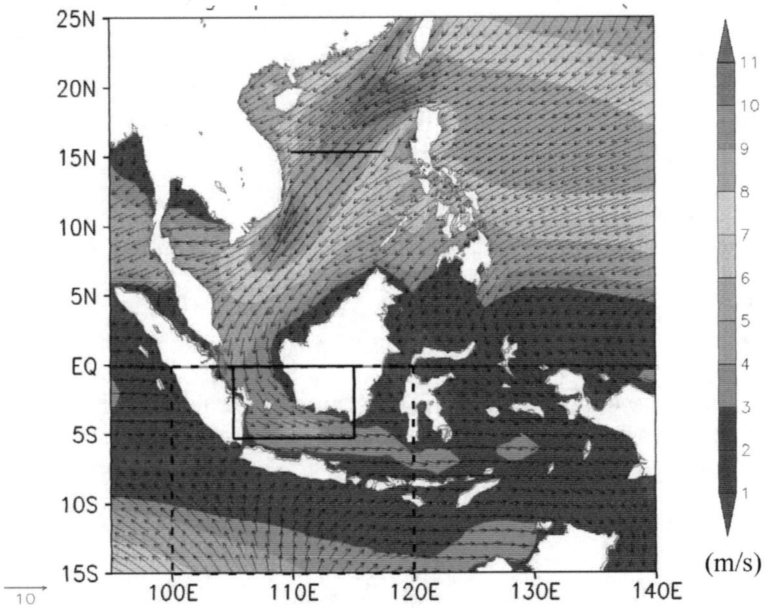

Figure 1. Reseach boundary area and the climatological mean of meredional wind at 925 hPa on EAWM periods.

Wind speed is indicated by shading in map, while the wind direction is indicated by arrows. The fastest meridional wind speed on the East Asia winter monsoon period especially in 2000-2015 located in the South China Sea area, as shown in figure 1. Usually, that area is affected by cold waves formed in October - april with average wind direction coming from NE, NNE, N, NNW and NW [11]. Identifying of cold surge performed using an indication that averaging the value of meridional wind component in the 110°-117.5°E and 15°N region at 925 hPa marked by thick line in figure 1 [12]. Cold surge events can be indicated by the activity of northerly wind speed exceeds 8 m / s and heading south. The analysis period is limited to when the occurrence of EAWM among November-February in 2000-2015 [13]. The thick lines that make up the square plane in figure 1 indicates the area of Cross-Equatorial Northerly

Surge (CENS) events that used as an area of propagation analysis in cold surge event, while the dashed lines that form a square plane signifies the conventional central area of MJO events [10].

2.2. *Preparation of meredional wind data, RMM1, and RMM2 index*
Six-hour data from ECMWF modified into daily averaged with Climate Data Operator (CDO). Daily MJO events affect indonesia's cloud conditions in phases 4 and 5 with the power of amplitude is more than 1, so the selection of phases 4 and 5 is required before composite analysis [2].

2.3. *Plotting the day of cold surge and MJO events in 2000-2015 of the EAWM period*
The result of plotting is a table that contains the date and number of occurrences in each year.

2.4. *Composite analysis of wind and rainfall condition when no cold surge and cold surge events*
This method is done by averaging meridional winds when no cold surge, cold surge, and daily averaged of EAWM periods when has been separated in the previous method. The composite technique is none other than ensemble average, is an arithmetic mean from a data under the same conditions.

2.5. *Calculate moisture transport*
The definition of water vapor transport is a movement of the wind that carries water vapor mass and can be expected from the following equation [4].

$$Q = \frac{1}{g}\int_{pt}^{ps} qV\,dp \qquad (1)$$

Q is moisture transport (kg/m/s), g is earth gravitation (m/s2), q is specific humidity (gr/kg), V is horizontal wind vector (m/s), p is pressure levels with pt and ps is pressure level at 300 hPa and 1000 hPa. The use of data parameters is limited at 300 hPa due to the amount of data lost at 300 hPa upwards that potentially affecting the calculation results of water vapor vertically integration.

2.6. *Composite analysis of rainfall condition when cold surge and MJO events*
This method is the same thing like 5th procedure, but the result is a condition map of rainfall when cold surge and MJO events occurs in millimeters and anomalies units.

2.7. *Wind propagation and its influence analysis when cold surge and MJO events occurs using hovmoller diagram*
Observation of wind movement and rainfall conditions in 105°-115°E With time variations can be seen using hovmoller diagram. The hovmoller diagram is a diagram that can show fluctuations in data values with time and region variations. Latitude variations are limited from 20°N-20°S and its time on 1 November-29 February.

3. Result and discussion

3.1. *Climatological conditions on Indonesian rainfall and wind on East Asian Winter Monsoon (EAWM) periods*
The Asian and Australian monsoons are two global monsoons that mainly affects rainfall in Indonesia. The monsoon is the largest surface wind cycle that has a seasonal pattern in certain areas. One of the important things that should be noted that the beginning of monsoon usually occurs together with high intensity of rainfall in the area that crossed with the monsoon wind pattern [14].

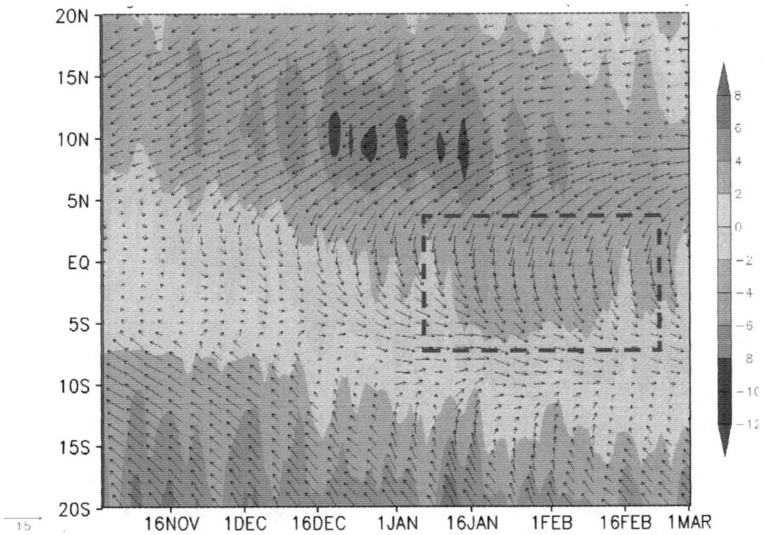

Figure 2. Time-latitude cross-sections of climatological averaged meredional wind in 105-115°BT at 925 hPa on EAWM periods.

Figure 2 shows that the meridional wind on climatological condition across the equator region up to 6°S in mid-January to February with 2-4 m/s. The cold surge event was visible in climatological in December and January that potentially can affect rainfall in Indonesia, especially in Jakarta [8]. Cold surge spread to equator and turn eastward because wind moves from high pressure to low pressure [2]. Deflection is caused by pressure difference that stronger between the eastern and western regions in the equatorial area, compared with the pressure difference between the Continent of Australia and East Asia.

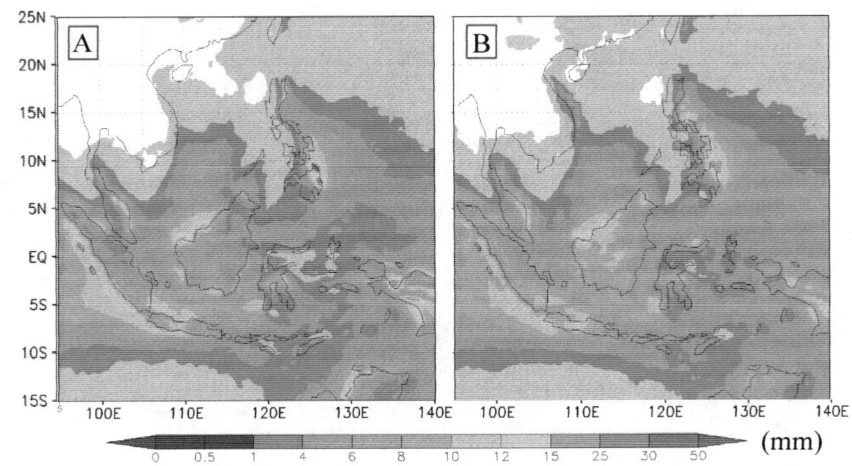

Figure 3. Time-latitude cross-sections of climatological averaged meredional wind in 105-115°BT at 925 hPa on EAWM periods.

This study uses TRMM and GSMaP satellite data to observe the effect of cold surge and MJO against rainfall in Indonesia. The application of TRMM and GSMaP satellite data often used for analyzing the diversity of precipitation in the region that is difficult to observed [10,15]. In figure 3, the difference seen on measurement results of the intensity of rainfall in the oceans and coastal as well as distribution of rainfall in the land. Validation results with station data shows that TRMM rainfall data has correlation value about 0.28 and GSMaP about 0.57. TRMM data visualization tend to display the rainfall value that is too high which compared with observation data, while the GSMaP data capable to displaying the rainfall value that is more precise due to algorithm data organized to further evaluate from performance of several satellites of observation.

3.2. Cold surge propagation in Indonesia Region

Cold Surge (CS) is the spread of cold air mass that moved from East Asia to the Western Pacific Ocean during the Asian winter monsoon [16]. The number of occurrences of cold surge and MJO can be seen on table 1.

Table 1. Number of cold surge and MJO events on EAWM from 2000-2015.

Year	CS	MJO	CS-MJO
2000/01	26	14	7
2001/02	25	14	11
2002/03	14	6	11
2003/04	22	14	10
2004/05	26	8	2
2005/06	28	15	2
2006/07	27	15	4
2007/08	32	10	17
2008/09	26	25	6
2009/10	16	7	6
2010/11	31	9	6
2011/12	20	19	13
2012/13	13	12	4
2013/14	28	3	9
2014/15	28	15	5
Sum	362	186	113

The results show that there are 362 cold surge events, 186 MJO events, and 113 cold surge-MJO events simultaneously, with a total EAWM period in 2000-2015 is 1803 days. The most cold surge events has occur in 2007-2008, MJO in 2008-2009 and the most cold surge-MJO events simultaneously occur in 2007 - 2008.

Cold surge propagation can be seen in figure 4. The negative value in figure 4 defines the north wind, and the positive value defines the south wind. Wind velocity is illustrated with different colors for each interval. Futhermore, meridional wind velocity is represented by color gradation and wind direction represented by arrows with maximum speed is 15 m/s (figure 4). The Domination of north wind caused by different position of the sun in Australia so the pressure gradient in the north-south area became stronger [6].

In figure 4a, the north wind conditions in 2012-2013 from the South China Sea turn eastward at Java Sea with 1-3 m/s. It also happened in 2010-2011 and its climatological condition in 2000-2015. Cold surge is sourced from the South China Sea and can be seen in figure 5c. In climatological condition, cold surge wind pressing the speed of south wind and dominate Java Island especially in Jakarta with 1-6 m/s. Compressive power of cold surge strongly influenced by frequency of its occurence. Cold surge was low in 2012-2013 and less reducing the strength of the north wind (figure 4iii), so the Java Island still crossed by Australian monsoon patterns. The wind pattern will shift up to 15°S when the cold surge is commonlly occurs, like in 2010-2011, with the north wind propagation reach 3 m/s in Java Island. In figure 5c, meridional winds generally reach the island of Sumatra with velocity 1-3 m/s. Cold surge events may increase north wind propagation in the equatorial area as a result of wind propagation towards the south from Siberian-Mongolia High (SMH) [10].

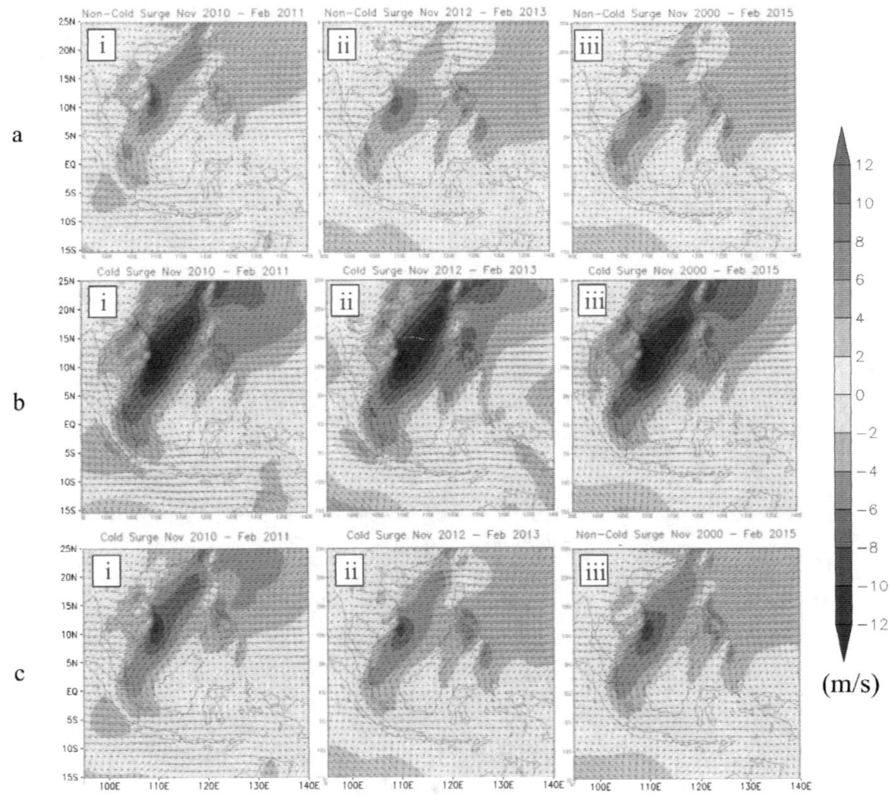

Figure 4. Composite map of average meredional wind on EAWM periods from 2010-2011 (Panel i), 2012-2013 (Panel ii), dan 2000-2015 (Panel iii) when (a) no cold surge, (b) cold surge, and (c) daily averaged of EAWM periods.

3.3. *Influence of MJO on cold surge events in Indonesia*

Defining the MJO phase is performe based on convective region of OLR that greatly affects the water vapor mass and rainfall distribution in the East Indian Ocean, Maritime Continent and West Pacific Ocean [2]. Futhermore, there are 113 total days of MJO and cold surge events simultaneously (table 1).

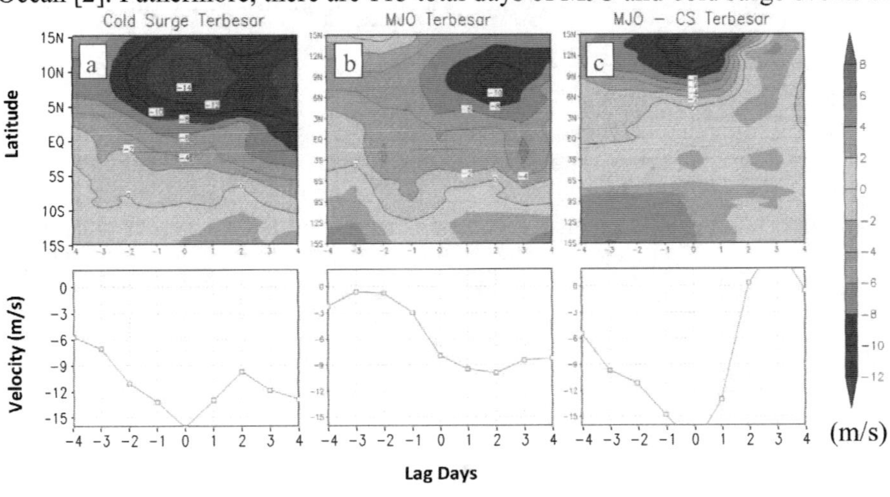

Figure 5. Time-latitude cross-sections of the fastest meredional wind when (a) cold surge, (b) MJO, and (c) cold surge-MJO occurs simultaneously from day -5 to day 5.

Figure 5 shows the effect of MJO oscillation from the west on cold surge propagation to the south at the largest meridional wind speed for each events. The color on hovmoller diagram and the y-axis on the graph below shows wind velocity, with a contour line defines the north wind. The x-axis from figure 5 is the lag day after and the occurrence of intra-seasonal climate variability with the highest meredional wind speed. The rising speed of north wind is seen on five days before the strongest cold surge occurs, while the wind propagation in 10°S requires five days later. Such as Compo *et al.* [17] results, about a strong relationship between cold surge in 15°N, 115°E and convection activities in the southern part of Indonesia.

MJO events with the largest meridional wind causing cold surge pressure stuck by the western wind following the power of convection from MJO in 6°N, so the cold surge reduce the spread of meridional wind. It's influenced by frequency of cold surge itself. When the frequency of cold surge events continually happens, the lag time when wind is going to equator requires 4-6 days after cold surge was detected in Hong Kong [3].

3.4. Characteristics of Indonesian Rainfall when influenced by cold surge and MJO Events

Indonesia has unique patterns of annual and inter-annual climate variability, because there are differences in rainfall patterns in every part of the region [18]. Daily rainfall distribution on a zonal average at 105-115°BT in 2000-2015 shown in figure 6.

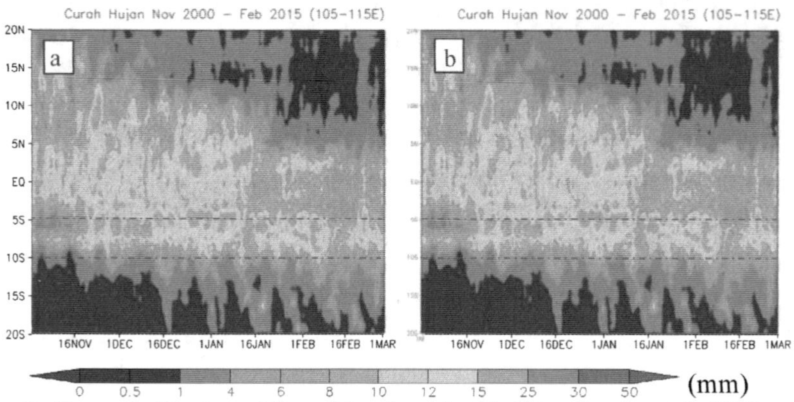

Figure 6. Time-latitude cross-sections of daily rainfall which is averaged on 105-115°E in 2000-2015 displayed by (a) TRMM and (b) GSMaP.

In figure 6, Region of Java Island marked by line breaks at 5-10°S, while at 0-15°N representing the South China Sea region. Java Island has daily rainfall intensity about 10-15 mm. Distribution of 10-25 mm rainfall in the South China Sea varies and have a great intensity from 16 November-16 February then weakened until 1 March. The movement of water vapor horizontally when intra-seasonal phenomena occurs can be seen on figure 7.

Figure 7 showing average mass of water vapor is in 5-20°N. According to Zhou and You [4], the main source of that mass comes from the East China Sea and the Pacific Ocean side. While no cold surge, a low wind speed makes water vapor mass from the East China Sea cannot carried over to Indonesian territory. While cold surge, water vapor mass has been transported from East China Sea and increasing the intensity of water vapor in the area of South China Sea, then the mass spreading to the Java Sea (figure 7b). Futhermore, MJO events can also increasing the speed of water vapor transport in the Philippine Sea and the Java Sea region that come from Indian Ocean (figure 7c). The moisture transport happens because of propagation to the east that will causes rainfall anomaly in some areas of Banda Sea and Java Sea [15]. When both intra-seasonal climate variability occurs simultaneously, rising a mass of large water vapor that concentrate in the South China Sea and Java Sea and also potentionally for produce rain clouds.

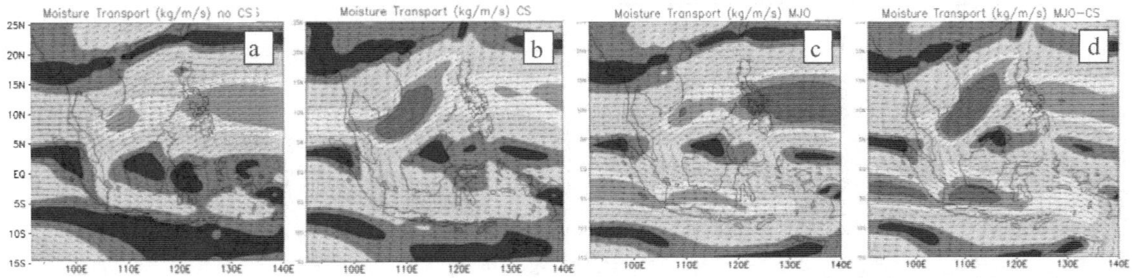

Figure 7. Moisture transport during the occurrence of (a) no cold surge, (b) cold surge, (c) MJO, and (d) cold surge-MJO simultaneously.

The velocity of vertical water vapor mass tend to move fast horizontally in the oceans caused by topographic fluctuations in the Indonesian area that have different geopotential heights [4]. Its raises sharp deflection of water vapor transport in Sunda Strait. Mass movement of moisture in the air have negarife correlation with OLR [2]. The larger water vapor transport then the smaller value of OLR impact on the amount of potential convective cloud formation.

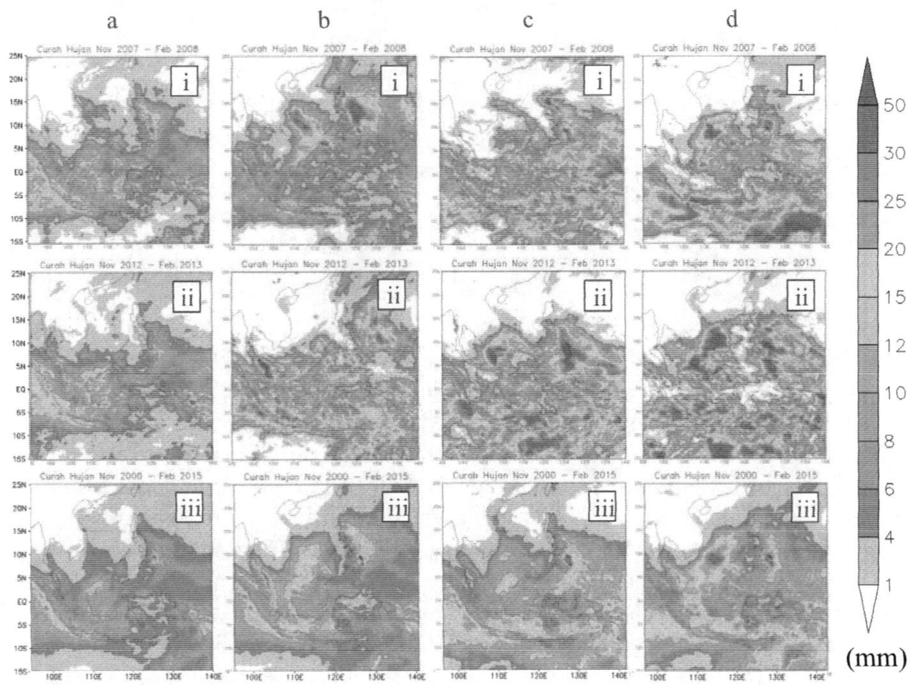

Figure 8. Composite map of daily rainfall (GSMaP) in 2007-2008 (Panel i), 2012-2013 (Panel ii), and 2000-2015 (Panel iii) during (a) no cold surge, (b) cold surge, (c) MJO, and (d) cold surge-MJO simultaneously.

Hattori *et al.* [9] tells that there are 11 cold surge events that greatly affect rainfall enhancement in Indonesia, especially in Java Island. Some of that may spread to Java Island, but its propagation can also be reducting or following the MJO convective power of phases 4 and 5. Cold surge stuck can also be caused by Easterly Wave, Borneo Vortex, and cyclone in the middle of South China Sea [3].

In 2007-2008, the most cold surge and MJO events occur simultaneously. Rainfall intensity in Java Island especially in Jakarta has the intensity 30-50 mm/day when MJO and cold surge occur simultaneously (figure 8). Flood incident in Jakarta caused by the presence of north winds and convective clouds that causing rainfall evenly on the Java Island. Climatologically, rainfall occurs when cold surge has an intensity 10-15 mm/day in Java Sea, South China Sea, Java Island, Sumatra and Kalimantan (figure 9). When MJO occurs, increased rainfall moving up to Eastern Indonesia, but in South China Sea its rainfall intensity not as strong as cold surge events. When both intra-seasonal variability active the instensity of rainfall is up to 15-25 mm/day in Java Sea and South China Sea.

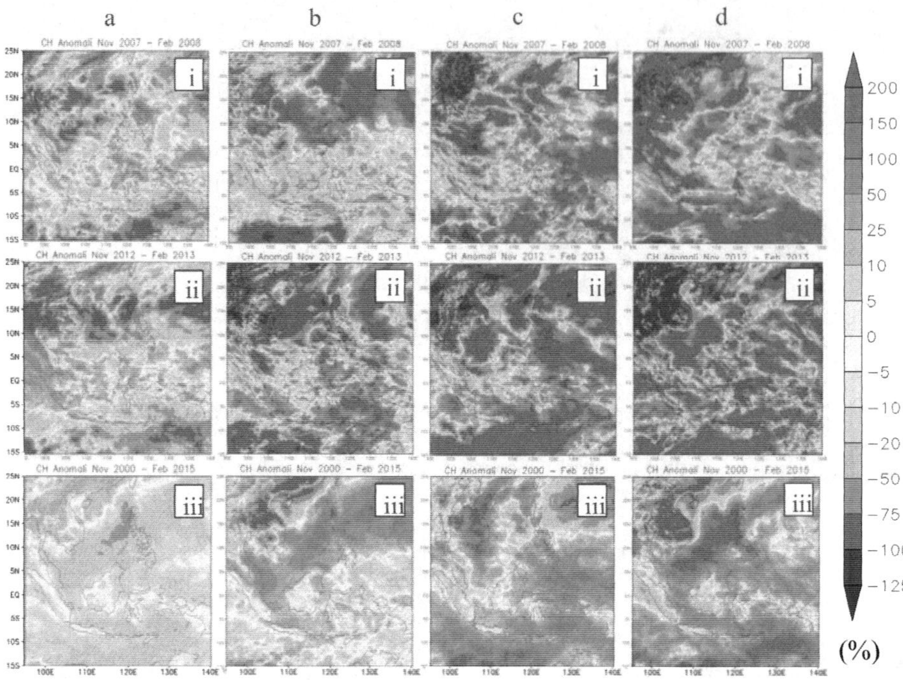

Figure 9. Composite map of daily rainfall anomaly (GSMaP) in 2007-2008 (Panel i), 2012-2013 (Panel ii), and 2000-2015 (Panel iii) during (a) no cold surge, (b) cold surge, (c) MJO, and (d) cold surge-MJO simultaneously.

The anomaly map that shows on figure 9 has a percent unit (%) which indicates increasing and decreasing rainfall climatology on EAWM (figure 3). Decreasing rainfall is represented by negative value and increasing rainfall is represented by positive value. Increased rainfall in 2007-2008 in the South China Sea during cold surge is 150%. But, decreased rainfall is also seen in Indonesian region at 20-75%. While MJO occurs, increased rainfall seen in Eastern Indonesia at 50-150%, East Kalimantan is 10-100% and slightly in Java Island at 5-50%. When MJO and cold surge occur simultaneously can increse rainfall more than 200% in the East Indian Ocean. Futhermore, Increased daily rainfall also seen in the South China Sea, Java Sea, Banda Sea, Central Java, NTT, NTB, and Jakarta at 25-150%.

In 2012-2013, cold surge events can cause positive rainfall anomaly at 10-100% in the Indonesia (figure 9i). The largest increase in rainfall intensity located around Banda Sea and Sunda Strait up to 150%. Decreased rainfall also occurs in Central and West Kalimantan 5-50%. The occurrence of MJO also improves rainfall in area that the conditions is almost same as the effect of cold surge. The greatest increased rainfall is located in the East Indian Ocean region up to 200%. While MJO and cold surge occur simultaneously, rainfall is seen rising in region that affected with MJO and cold surge respectively. However, its large increase is about 25-200%.

Climatologically, cold surge is generally affect on Indonesian Oceans. Cold surge that blows from the South China Sea through Java Sea to Flores Sea in Indonesia. The incidence of rainfall on the track depend on water vapor conditions in the South China Sea. The averaged rainfall increase in the trajectory ranges at 10-50%, including Jakarta. Increased rainfall 100% also seen in the South China Sea and Philippine Sea. The influence of MJO climatologically on West Indian Ocean and Indonesian Ocean with a large increase at 25-150%. While MJO and cold surge occur simultaneously, increased climatological rainfall at 25-150% seen in Indonesian Ocean. However, decreased rainfall usually occurs in mainland area at 5-50%.

IOP Conf. Series: Earth and Environmental Science **149** (2018) 012045 doi:10.1088/1755-1315/149/1/012045

4. Conclusion

In the cold phase, cold surge and MJO events have a relation that can be analyzed with meteorological parameters such as meridional wind, rainfall, and moisture transport. The results show that both phenomena effect on increasing Indonesian rainfall, especially Java Sea. The results showed that during East Asian Winter Monsoon (EAWM) in 15 years (2000-2015), there are 362 cold surge events, 186 MJO events, and 113 cold surge events were associated with MJO events. Moisture transport value shows that East China Sea is the source of extreme rainfall on cold surge events, while MJO has two sources of water vapor transport i.e. Indian Ocean and East China Sea. While cold surge and MJO occurs simultaneously causes the source of water vapor move to Indonesia and increase rainfall up to 100-150% from its climatological conditions.

References

[1] Wang B and Fan Z 1999 Choice of South Asian summer monsoon indices *Bull. Amer. Meteor. Soc* **80** 629-638
[2] Wicaksono GB and Hidayat R 2016 Extreme rainfall in Katulampa associated with the atmospheric circulation *Procedia Environmental Sciences* **33** 155-166
[3] Aldrian E and Susanto RD 2003 Identification of three dominant rainfall regions within Indonesia and their relationship to sea surface temperature *Int. J. Climatol.* **23** 1435-1452
[4] Zhou TJ and Yu RC 2005 Atmospheric water vapor transport associated with typical anomalous summer rainfall patterns in Cina *Journal of Geophysical Research D: Atmospheres* **110** 1-10
[5] Chan JCL and Li C 2004 The East Asia winter monsoon, ed Chang CP (New Jersey: World Scientific Publishing Company)
[6] Aldrian E and Utama GSA 2007 Identification and characteristics of Cold Surge in 1995-2003 *Jurnal Sains Dirgantara* **4** 107-127 [In Indonesian]
[7] Chen TC, Yen MC, Huang WR, and Gallus WA 2002 An East Asian cold surge: case study *Monthly Weather Review* **130** 2271-2290
[8] Wu P, Hara M, Fudeyasu H, Yamanaka MD, Matsumoto J, Syamsudin F, Sulistyowati R, and Djajadihardja YS 2007 The impact of trans- equatorial monsoon flow on the formation of repeated torrential rains over Java Island *SOLA* **3** 93-96
[9] Samah AA, Babu CA, Varikoden H, Jayakrishnan PR, and Hai OS 2016 Thermodynamic and dynamic structure of atmosphere over the east coast of Peninsular Malaysia during the passage of a cold surge *Journal of Atmospheric and Solar-Terrestrial Physics* **146** 37
[10] Hattori M, Mori S, and Matsumoto J 2011 The cross-equatorial northerly surge over the maritime continent and its relationship to precipitation patterns *Journal of the Meteorological Society of Japan* **89A** 27-47
[11] Mo D, Hou Y, Li J, and Liu Y 2016 Study on the storm surges induced by cold waves in the Northern East China Sea *Journal of Marine Systems* **160** 26-39
[12] Chang CP, Harr PA, and Chen HJ 2005 Synoptic disturbances over the equatorial South China Sea and western maritime continent during boreal winter *Monthly Weather Review* **133** 489-503
[13] Ding YH 1994 *Monsoons over China* (London: Kluwer Academic Publisher)
[14] Krishnamurti TN, Stefanova L, and Misra V 2013 *Tropical Meteorology* (Berlin: Springer) p 75
[15] Hidayat R and Kizu S 2010 Influence of the Madden–Julian Oscillation on Indonesian rainfall variability in austral summer *Int. J. Climatol.* **30** 1816-1825
[16] Wang B 2006 *The Asian Monsoon* (Chichester: Praxis Publishing Ltd)
[17] Compo GP, Kiladis GN, and Webster PJ 1999 The horizontal and vertical structure of East Asian winter monsoon pressure surges *Q.J.R. Meteor. Soc.* **125** 29-54
[18] Haylock M and McBride JL 2001 Spatial coherence and predictability of Indonesian wet season rainfall *Journal of Climate* **14** 3882–3887

IOP Conf. Series: Earth and Environmental Science **149** (2018) 012046 doi:10.1088/1755-1315/149/1/012046

Impact of La Niña and La Niña Modoki on Indonesia rainfall variability

R Hidayat, MD Juniarti, U Ma'rufah
Department of Geophysics and Meteorology, Bogor Agricultural University, Bogor, Indonesia

E-mail: rahmath@apps.ipb.ac.id

Abstract. La Niña events are indicated by cooling SST in central and eastern equatorial Pacific. While La Niña Modoki occurrences are indicated by cooling SST in central Pacific and warming SST in western and eastern equatorial Pacific. These two events are influencing rainfall variability in several regions including Indonesia. The objective of this study is to analyse the impact of La Niña and La Niña Modoki on Indonesian rainfall variability. We found the Nino 3.4 index is highly correlated (r = -0.95) with Indonesian rainfall. Positive rainfall anomalies up to 200 mm/month occurred mostly in Indonesian region during La Niña events, but in DJF several areas of Sumatera, Kalimantan and eastern Indonesia tend to have negative rainfall. During La Niña Modoki events, positive rainfall anomaly (up to 50 mm/month) occurred in Sumatera Island, Kalimantan, Java and eastern Indonesia in DJF and up to 175 mm/month occurred only in Java Island in MAM season. La Niña events have strong cooling SST in central and eastern equatorial Pacific (-1.5°C) in DJF. While La Niña Modoki events warming SST occurred in western and eastern equatorial Pacific (0.75°C) and cooling SST in central Pacific (-0.75°C) in DJF and MAM. Walker circulation in La Niña Modoki events (on DJF and MAM) showed strong convergence in eastern Pacific, and weak convergence in western Pacific (Indonesia).

1. Introduction

Indonesia is located between two oceans (Indian and Pacific) and two continents (Asia and Australia). Because of this position, Indonesia has dynamic and complex climate variability. Climate variability that possibly occur in Indonesia, namely La Niña and La Niña Modoki. La Niña refers to ENSO events that cooling SST only occurs in the central and eastern equatorial pacific while La Niña Modoki cooling SST occurs in central equatorial Pacific and warming SST in the eastern and western equatorial Pacific [1].

La Niña events can be seen through a negative Nino 3.4 index and positive Southern Oscillation Index (SOI) whereas La Niña Modoki events can be seen through a negative ENSO Modoki Index (EMI). Nino 3.4 index can be calculated by using Sea Surface Temperature (SST) deviations in the central equatorial Pacific that called the Nino 3.4 region (5 N - 5 S, 120 W - 170 W) [2]. SOI can be calculated by using the difference SST region of Tahiti and Darwin [3]. Variability of SST Nino 3.4 region affects 50% of rainfall variability in Indonesia [4]. ENSO Modoki Index determined through calculation of EMI = [SSTA] A - 0.5 × ([SSTA] B + [SSTA] C) [1]. A region represented by the 165E-140W and 10s-10N, B 110W-70W and 15S-5N, area C 125E-145°E and 10S -20°N [1].

La Niña Modoki is an inter-ENSO variations. It also affected the rainfall variability in some areas [5]. Cai and Cowan [5], examines the impact of La Niña Modoki on Australian autumn rainfall variability. La Niña Modoki lead to rainfall increase extending on northwestern Australia to the northern Murray-Darling Basin, whereas the conventional La Niña lead to increase rainfall occurs only in the Eastern Australia region [5]. Based on the research, we need to understand the impact of La Niña and La Niña Modoki events in Indonesian rainfall variability. This research is expected to provide spatial information about the impact of these events to the Indonesian rainfall variability.

2. Methodology

2.1. Time Series Analysis

Time series analysis used to identify La Niña and La Niña Modoki years. Identification of La Niña based on Nino 3.4 index while La Niña Modoki based on ENSO Modoki Index (EMI). La Niña and La Niña Modoki occurs when Nino 3.4 and EMI show negative value, respectively. Time series data of Nino 3.4 and EMI used in this research are from 1981 to 2010.

2.2. Composite Analysis

Determining seasonal anomalies characteristic of rainfall, Sea Surface Temperature (SST), and wind in Indonesia region during La Niña and La Niña Modoki is done by composite analysis. Rainfall data used for this analysis is CHRPS from 1981 to 2010 with resolution 0.05° x 0.05°, while SST and wind used ECMWF data. Formula used for composite analysis described in equation 1.

$$\text{Composite} = \frac{X_{i1} + X_{i2} + \cdots + X_{in}}{n} \quad (1)$$

Where X_i is rainfall data of given season and n is amount of events.

2.3. Correlation Analysis

Correlation analysis was used to understand the relation between rainfall and two climate variability, La Niña and La Niña Modoki. We used time series and spatial correlation to understand the relation between rainfall and climate variability in long term period and in spatially. Time series correlation between all Indonesia rainfall and climate variability index (Nino 3.4, SOI, and EMI) are done using detrended annual (all month) and seasonal data from 1981 to 2010 through a 13-years sliding window. Spatial correlation calculated using linear cross-correlation at lag 0.

3. Result and Discussion

3.1. Correlation between SOI, Nino 3.4 Index and ENSO Modoki Index with Indonesian Rainfall

The rainfall data used for this correlation is average of spatial rainfall data in all Indonesian region. Correlation between SOI, EMI and Nino 3.4 index with Indonesian rainfall is shown in figure 1. La Niña events were indicated by positive value of SOI and negative value of Nino 3.4 index, while La Niña Modoki events were indicated by negative value of EMI.

Correlation of SOI and Nino 3.4 index with Indonesian rainfall have same fluctuations patterns but have different position. SOI (Nino 3.4) had a positive (negative) correlation in each season. During JJA and SON, both SOI and Nino 3.4 had high correlation reach 0.8 and -0.8, respectively. Whereas, during DJF and MAM, there were some periods that have low correlation. EMI had negative correlation with Indonesian rainfall in all season except MAM. During 1991-1993, EMI had positive correlation with Indonesian rainfall. All the indices had different pattern on DJF and MAM. So for further analysis, we focused only on that season (DJF and MAM).

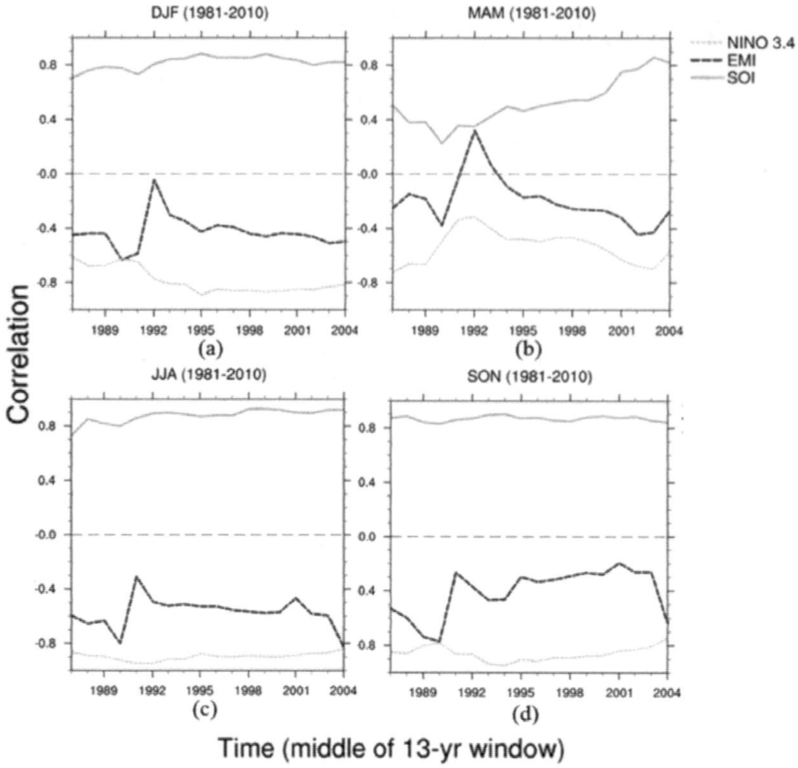

Figure 1. Correlation between SOI, Nino 3.4, and EMI with Indonesian rainfall using detrended trough a 13-years sliding window.

3.2. Spatial Correaltion between Nino 3.4 Index and ENSO Modoki Index with Indonesian Rainfall
Spatial correlation between EMI and Nino 3.4 index with Indonesian rainfall data over 30 years showed different spatial pattern (figure 2). Nino 3.4 had negative correlation with rainfall almost in all Indonesia region except west Kalimantan, Lampung, West Jawa and North Papua (figure 2a). Strong negative correlation (r = -0.9) occurred in East Kalimantan and North Sulawesi and Maluku. According to Aldrian and Susanto (2003), North Sulawesi and Maluku also had strong ENSO influence [6]. On MAM season, the coefficient correlation between Nino 3.4 and rainfall had same pattern with coefficient correlation on DJF, but the correlation was lower.

ENSO Modoki Index (EMI) had a weak negative correlation with Indonesian rainfall on every season (r = -0.3). Spatial extent of region with negative correlation on DJF was wider than MAM. On DJF negative correlation spread on some area such as Sulawesi, North and South Sumatra, East Kalimantan, Sulawesi, Bali, Java, and NTT. On MAM positive correlation was dominated in Kalimantan and Papua.

Generally, Nino 3.4 have stronger influence to rainfall in Indonesia than EMI. It can be seen from the value of coefficient correlation. According to Boer et al. [7], Indonesian rainfall anomalies has a strongest relation with SST anomalies in the Nino 3.4 region (170°-120°BB, 5°LU-5LS) [7]. It is evident that Nino 3.4 has a stronger correlation with Indonesian rainfall than EMI.

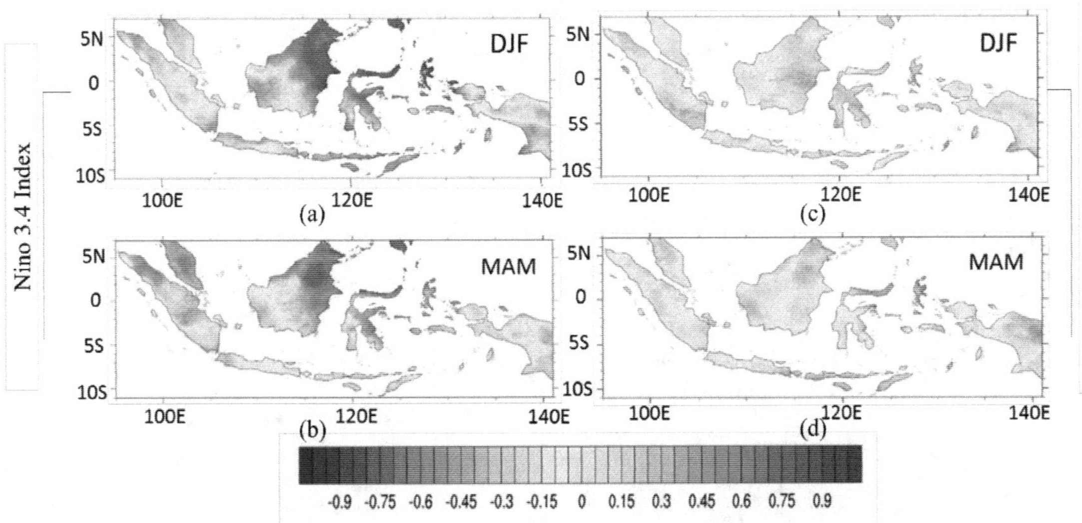

Figure 2. Spatial correlation between Nino 3.4 Index and ENSO Modoki Index with Indonesian rainfall.

3.3. *Identifying La Nina and La Nina Modoki years*

The occurrence of La Niña and La Niña Modoki in past year can be seen from time series data of Nino 3.4 and EMI, respectively. Negative value of Nino 3.4 and EMI indicated that La Niña and La Niña Modoki had occurred. Figure 3 shown La Niña and La Niña Modoki events during 30 years. There were a few times of La Niña and La Niña Modoki events. La Niña occurred in 1985, 1988, 1999, 2007 and 2010 while La Niña Modoki occurred in 1983, 1989, 2000, and 2008. La Niña and La Niña Modoki from 1981 to 2010 occurred five and four time, respectively. The five events of La Niña and and four events of La Niña Modoki in that time will be used to make a composite rainfall anomalies, SST anomalies and Walker circulation anomalies.

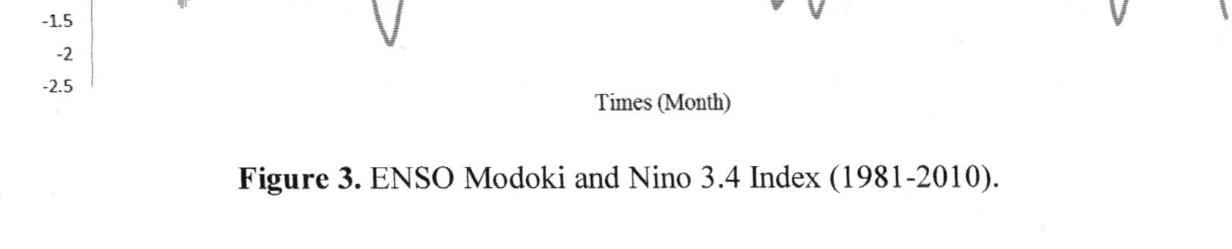

Figure 3. ENSO Modoki and Nino 3.4 Index (1981-2010).

3.4. *Sea Surface Temperature, Zonal and Meridional Wind Anomalies during La Nina and La Nina Modoki*

Indonesia rainfall anomalies are influenced by the SST (sea surface temperature) in the Pacific regions. When La Niña and La Niña Modoki occurred, SST conditions in Pacific has a different pattern than usual. SST conditions can be used as an index to determine the amount of water vapour in the atmosphere [8]. figure 4 shows a composite SST anomalies that is overlaid with the Wind (Zonal and Meridional) at altitude 850 hPa in the Pacific region during the La Niña and La Niña Modoki events. SST anomalies pattern had strong impact on rainfall variability [9]. When SST colder than normal, it can influence the increasing of surface pressure and surface wind divergence. After that it will reduce amount of rainfall [4]. SST Nino 3.4 region affected 50% on rainfall variability in Indonesian [4].

SST anomalies in equatorial pacific during La Niña and La Niña Modoki are different. SST anomalies during La Niña events on DJF is warming up to 1.25°C, while in central and Eastern Pacific is cooling up to -1.5°C. The wind in equatorial moved from east to west. During La Niña Modoki events, western (eastern) equatorial Pacific is warming up to 0.25°C (0.5°C), while in the central equatorial Pacific cooling up to -0.75°C. Generally, wind moved from central to west and east equatorial Pacific.

SST Anomalies during La Niña (Left) and La Niña Modoki (Right)

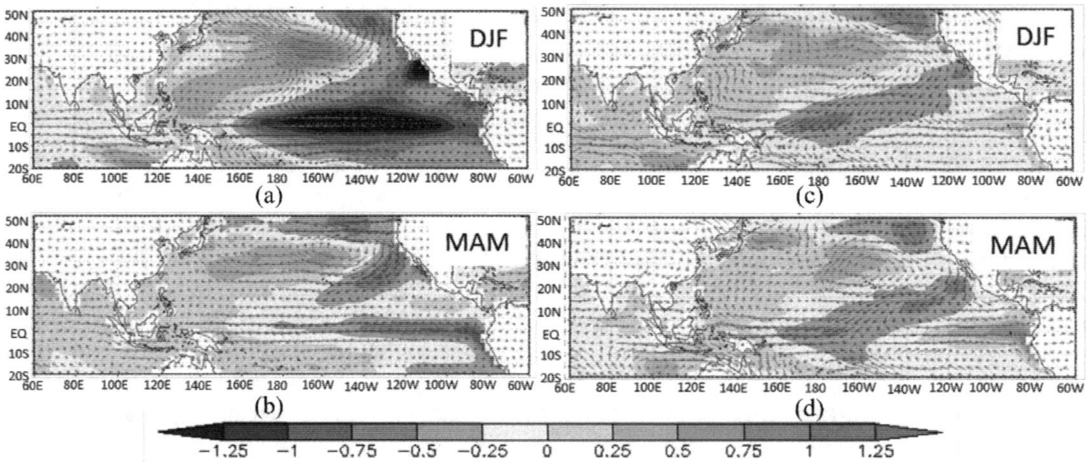

Figure 4. Composite anomalies sea surface temperature (°C), wind (u and v; m/s) level (850 hPa) during La Niña (a-b) and La Niña Modoki (c-d).

SST anomalies in western equatorial Pacific on MAM season during La Niña events (figure 4b) is warming up to 0.25°C while in the central and eastern equatorial Pacific is cooling up to -1°C. Wind movement was centered to the north of eastern Pacific. Conditions of SST anomalies in western equatorial Pacific during La Niña Modoki events on MAM season (figure 4d) is warming up to 0.75°C, in the central equatorial Pacific is cooling up to -0.75°C, and the eastern equatorial Pacific approach the American continent warming again up to 1°C. Wind moved from central to west and east equatorial Pacific.

3.5. Walker Circulation and Zonal Wind during La Niña and La Niña Modoki

Vertical wind profiles or Walker circulation is atmospheric circulation that occurs as a result of the encouragement of sea surface temperature gradient along the equator. The character of this circulation is determined by the interaction between oceans and atmosphere of the tropical areas [10]. Formation of Walker circulation is prevalent in the Pacific region with maximum formation in the upper atmosphere (300-200 hPa) [10]. Strengthening of this circulation are closely related and consistent with cooling and warming SST in the the Pacific [10].

Walker circulation can be analysed through the mean zonal wind (vector) and vertical velocity (contour) at latitude 5°LU-5LS and at the height of 1000-100 hPa. Figure 5 shown a composite Walker circulation anomalies during La Niña and La Niña Modoki events. Negative value of vertical velocity indicated the convection while positive value of vertical velocity indicated the subsidence. Walker circulation conditions during La Niña on DJF season (figure 5a) shown that many convection occurred in parts of Indonesia, while the Pacific region has a significant subsidence. During La Niña Modoki, convection occurs in Indonesia and eastern Pacific, while the subsidence occurs in the central Pacific region (figure 5c).

Walker circulation during La Niña events on MAM season (figure 5b) showed that Indonesia has a convection but not very strong, while central Pacific region has low subsidence. During La Niña Modoki events on MAM season (figure 5d), there was convection in eastern and western Pacific. The convection in eastern was stronger than in western.

Walker Circullation Anomalies during La Niña (Left) and La Niña Modoki (Right)

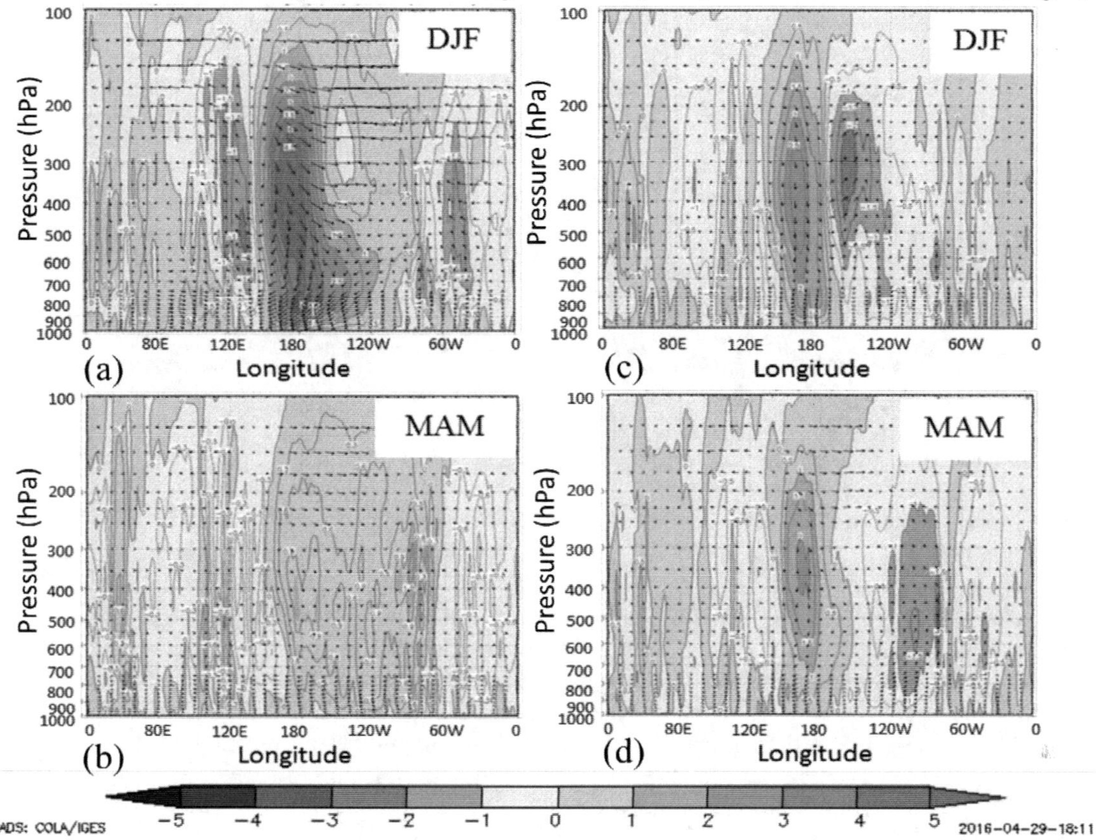

Figure 5. Composite seasonal Walker circulation anomalies using zonal wind (u; m/s) and vertical velocity (ω; 10-2Pa/s) level 1000-100 hPa during La Niña (a-b) and La Niña Modoki (c-d).

3.6. Rainfall Anomalies During La Nina and La Nina Modoki

La Niña and La Niña Modoki caused different impact to rainfall anomalies in Indonesia. Rainfall anomalies indicate standard deviation of rainfall that occurred during La Niña and La Niña Modoki events. The monthly spatial data of Indonesian rainfall are used to composite analysis. Based on the result on figure 6, the range of rainfall anomalies was high that was around -250 mm/month to 250 mm/month.

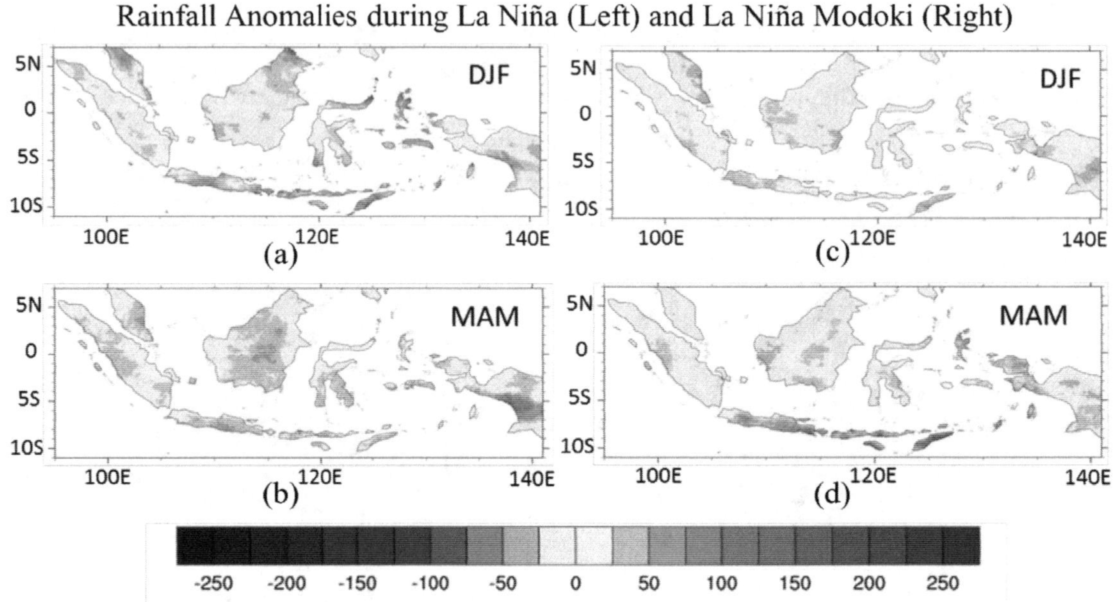

Figure 6. Composite Indonesian rainfall anomalies La Nina (a-b) and La Nina Modoki (c-d).

Positive rainfall anomalies during La Niña events on DJF season (figure 6a) is up to 150 mm/month that occurred in Sulawesi, Maluku, and southern part of Indonesia, while negative rainfall anomalies up to -75 mm/month occurred in some part of Sumatera Island, Kalimantan and eastern Indonesia. During La Niña Modoki (figure 6c) positive rainfall anomalies up to 50 mm/month occurred in part of Sumatera Island, Java, Kalimantan and eastern Indonesia. While negative rainfall anomalies up to -75 mm/month occurred in parts of Borneo, eastern Indonesia and Sulawesi Island.

La Niña events on MAM season (figure 6b) affected positive rainfall anomalies up to 175 mm/month that occurred in all of Indonesia regions. Condition of rainfall anomalies during La Niña Modoki events (figure 6d) on MAM season showed that Java Island has a strongest positive anomaly up to 175 mm/month. La Niña Modoki also causing negative rainfall anomalies up to -75 mm/month in part of Sumatera Island and eastern Indonesia.

4. Conclusion

Nino 3.4 and EMI have negative correlation in every season except on MAM, while SOI has positive correlation in all season except on MAM. Spatially, Nino 3.4 index has strong negative correlation (r = -0.95) with Indonesian rainfall, while EMI have weak correlation (r = -0.3).

SST anomalies during La Niña events indicate that cooling SST occurred in the central and eastern equatorial Pacific up to -1.5°C in every season. In La Niña Modoki events, warming SST occurred in the western and eastern equatorial Pacific up to 0.75°C and cooling in the central equatorial Pacific up to -0.75°C in both seasons. Walker circulation anomalies that occurred during La Niña events on DJF season have strong convection in the western equatorial Pacific, but during MAM season experience weak convection. La Niña Modoki caused strong convection in the eastern Pacific and weak convection in the western Pacific (Indonesia) during DJF and MAM season. La Niña caused positive rainfall anomalies in Indonesia up to 200 mm/month on DJF and MAM, but on DJF season some areas such as Sumatera Island, Kalimantan and eastern Indonesia had negative rainfall anomalies. La Niña Modoki during DJF caused weak positive rainfall anomalies (50 mm/month) in some are of Sumatera Island, Borneo, Java and eastern Indonesia, and caused a strong positive anomaly up to 175 mm/month only in Java Island during MAM season.

References

[1] Ashok K, Behera S K, Rao S A, Weng H, Yamagata T. 2007. El Niño Modoki and its possible teleconnection *Journal of Geophysical Research* **112** 1-27
[2] Bunge L, Clarke A J 2009 A Verified Estimation of the El Niño Index Niño-3.4 since 1877 *American Meteorological Society* **22** 3979-3992
[3] Trenberth K E 2001 *Encyclopaedia of Ocean Sciences (Second Edition)* (Academic Press) pp 228-240
[4] Hendon H H 2003 Indonesia rainfall variability: impacts of ENSO and Local Air-Sea Interaction *Journal of Climate* **16** 1775-1790
[5] Cai W and Cowan T 2009 La Niña Modoki impacts Australia autumn rainfall variability *Geophysical Research Letters* **36** 1-4
[6] Aldrian E, Susanto R D 2003 Identification of three dominant rainfall regions within Indonesia and their relationship to sea surface temperature *International Journal Climatol* **23** 1435-1452
[7] Boer R, Notodipuro K A, Las I 2007 Prediction of daily raindall characteristic from monthly climate indicate *Jurnal Agromet Indonesia* **21** 12-20
[8] Syaifullah D 2010. Kajian sea surface temperature (SST), southern oscillation index (SOI) dan dipole mode pada kegiatan penerapan teknologi modifikasi cuacadi propinsi riau dan sumatera barat juli – agustus 2009 *Jurnal Sains & Teknologi Modifikasi Cuaca.* **11** 1-7
[9] Estiningtyas W, Ramadhani F, Aldrian E 2007 Analisis korelasi curah hujan dan SPL SPL wilayah Indonesia, serta implikasinya untuk prakiraan curah hujan (studi kasus Kabupaten Cilacap) *Jurnal Agromet Indonesia* **21** 46-60
[10] Lau K M, Yang S 2002 Walker circulation *Journal Elsevier Science* **6** 1- 6

Fine-tuning satellite-based rainfall estimates

Hastuadi Harsa[1], Agus Buono[2], Rahmat Hidayat[3], Jaumil Achyar[4], Sri Noviati[1], Roni Kurniawan[1] and Alfan S Praja[1]

[1] Research and Development Center, Indonesia Meteorology Climatology and Geophysics Agency, Jl. Angkasa I No. 2, Kemayoran, Jakarta 10720, Indonesia
[2] Computer Science Department, Mathematics and Natural Science Faculty - Bogor Agricultural University, Bogor, Indonesia
[3] Geophysics and Meteorology Department, Mathematics and Natural Science Faculty - Bogor Agricultural University, Bogor, Indonesia
[4] Database Center, Indonesia Meteorology Climatology and Geophysics Agency, Jl. Angkasa I No. 2, Kemayoran, Jakarta 10720, Indonesia

E-mail: hastuadi@gmail.com

Abstract. Rainfall datasets are available from various sources, including satellite estimates and ground observation. The locations of ground observation scatter sparsely. Therefore, the use of satellite estimates is advantageous, because satellite estimates can provide data on places where the ground observations do not present. However, in general, the satellite estimates data contain bias, since they are product of algorithms that transform the sensors response into rainfall values. Another cause may come from the number of ground observations used by the algorithms as the reference in determining the rainfall values. This paper describe the application of bias correction method to modify the satellite-based dataset by adding a number of ground observation locations that have not been used before by the algorithm. The bias correction was performed by utilizing Quantile Mapping procedure between ground observation data and satellite estimates data. Since Quantile Mapping required mean and standard deviation of both the reference and the being-corrected data, thus the Inverse Distance Weighting scheme was applied beforehand to the mean and standard deviation of the observation data in order to provide a spatial composition of them, which were originally scattered. Therefore, it was possible to provide a reference data point at the same location with that of the satellite estimates. The results show that the new dataset have statistically better representation of the rainfall values recorded by the ground observation than the previous dataset.

1. Background

Rainfall measurement is carried out on the ground by the observation sites. The measured values are able to represent the rain events. However, the observations are scattered in non-homogeneous distance among others. On the other hand, satellite-based rainfall observations are able to provide data spatially, i.e. in homogeneous distance between their adjacent data locations. Therefore, the satellite-based data can be used as alternative sources where the observation sites do not present.

The data obtained from the satellite estimates are not true rainfall intensity. They are products of transformation algorithms applied to the reflectance received by the satellite sensors into rainfall intensity value. As the products of algorithms, they may contain intrinsic biases. There have been numerous researches regarding the comparison between data produced by satellite estimates and ground

IOP Conf. Series: Earth and Environmental Science **149** (2018) 012047 doi:10.1088/1755-1315/149/1/012047

observations. These researches identified bias of the satellite estimates to the ground observations data, so that the bias of satellite products can be suppressed. Toté [1] evaluated the satellite rainfall product for drought and flood monitoring in Mozambique. Vernimmen [2] studied three satellite estimates product in Indonesia and compared them with ground station data. The results were able to reduce the error in the satellites estimates. As-Syakur [3] compared daily, monthly, and seasonal satellite rain rates products with that of rain gauges in Bali.

CHIRPS (Climate Hazard Infra-Red Precipitation with Stations), a relatively new rainfall estimates dataset, is available at Climate Hazard Group (CHG) [4]. CHIRPS is derived from several data sources. It has 0.05° spatial resolution (~5 kilometers) and is delivered in daily, pentadal, and monthly forms. The CHIRPS dataset coverage area spans from 50°S to 50°N for all longitudes. Its temporal availability starts from 1981 to near-present. CHIRPS incorporates data blending procedure between ground measurements and the satellite-only rainfall estimates, another product of CHG: CHIRP (Climate Hazard Infra-Red Precipitation).

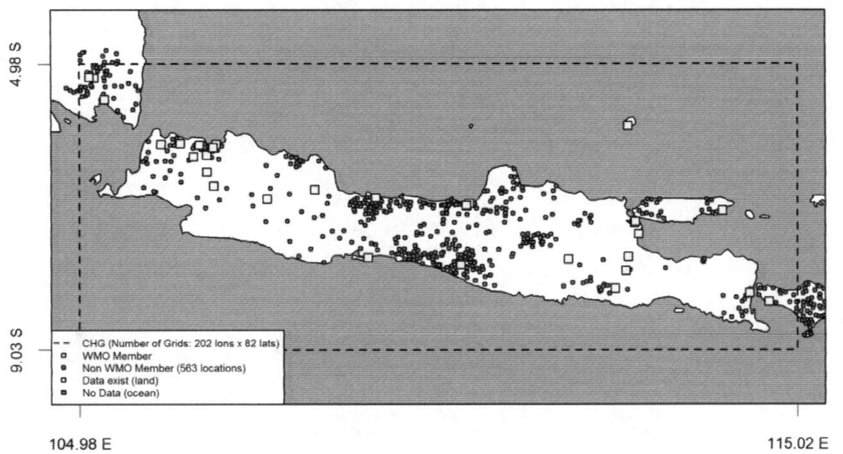

Figure 1. The study area: Java Island.

2. Data and study area

In this research, the data were located in Java Island. Java has the densest population in Indonesia and also contributes significant agricultural harvests to the national Gross Domestic Product. Rice, corn, and sugar are the examples of common food crops in Java. A better understanding of rainfall characteristics may increase the success of the farmers to supply the demand of these agricultural products.

Figure 1 shows the study area. The area spans from 104.975°E to 115.025°E and from 9.025°S to 4.975°S, as specified by dashed bounding box. The spatial resolution of CHIRPS data is 0.05° or approximately 5 kilometers. By that resolution, there are 202 longitude units and 82 units of latitude, i.e. 16564 data locations (represented in grid points). CHIRPS data are available on the land only, so that the number of grid points having numerical data is 5177. The rest of the grid points do not contain data as they are located on the ocean. The available data locations is indicated by the white area on the map in figure 1, while the ocean is indicated by the grey area.

The observation data were available from Indonesia Meteorology Climatology and Geophysics Agency (BMKG). There are two types of BMKG ground observation data: the open-access data and the reserved data. The BMKG open-access data locations are shown by the yellow rectangle while the red circle show the BMKG reserved data locations. There were 29 locations represented the open-access data and 563 locations represented the reserved data. The BMKG open-access data locations are World Meteorology Organization (WMO) members. CHG also collects the observation data from those open-access locations to perform the data blending process between the ground-based observation and the satellite-only data. Both the observation data and the CHIRPS data collected for this research are in the length of 12 years: from January 1st, 2005 until December 31st, 2016. The data are in daily temporal resolution. These data denote the rainfall intensity values and have the units of millimeter per day.

3. Methods

This research aims to modify the CHIRPS data using additional ground observation data. It is expected that this procedure is able to make CHIRPS data characterize the ground observation data. In order to accomplish this objective, there are two methods utilized: Quantile Mapping (QM) and Inverse Distance Weighting (IDW).

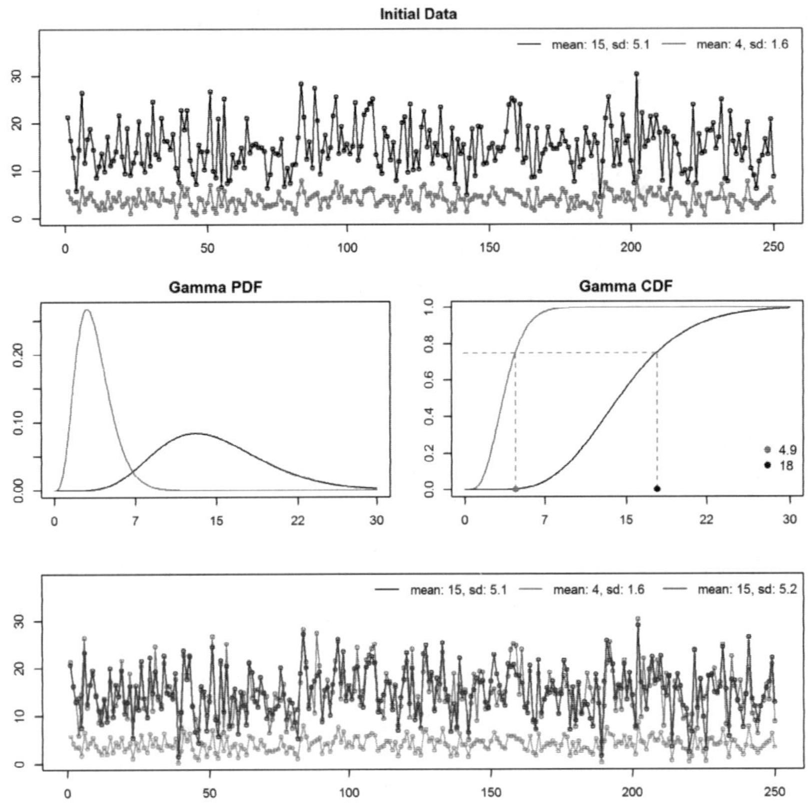

Figure 2. Quantile mapping illustration.

3.1. *Quantile mapping*

QM is a non-parametric bias correction (BC) method [5]. This method transform all value in a dataset onto new values. The transformed values construct a new dataset. The values in the new dataset can eventually have the occurrence probability as desired. This process requires a determination of new probability function to map a value having a particular probability occurrence from previous dataset.

Piani [6] demonstrated the use of QM to correct the bias of global simulated daily precipitation and temperature. The illustration of QM is given by figure 2. Suppose that there are two datasets as plotted at the top row, the data in red will be adjusted to be closer to the data in black. By calculating the mean and standard deviation of each dataset, then the Probability Density Function (PDF) of the values occurred in each dataset can be obtained as shown by the left chart at middle row. Note that in generating the PDFs here, the values are presumed to be in Gamma distributed data, as the rainfall values are. Nevertheless, another distribution function may be used regarding to the nature of the data. After the PDFs are available, the Cumulative Distribution Function (CDF) of each PDF can be calculated, as shown by the right chart at middle row. The transformation of a value (4.9 for example as shown in the

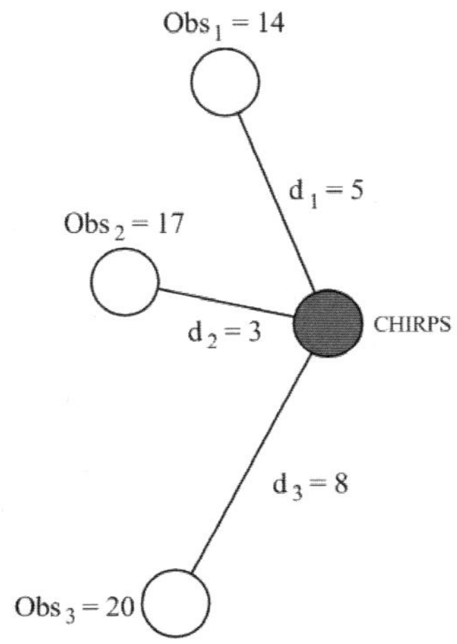

$$W = (d)^{-2}$$
$$y = (\sum W_i \cdot x_i) \cdot (\sum W_i)^{-1}$$
$$w \in \{1/5^2, 1/3^2, 1/8^2\}$$
$$y_{CHIRPS} = (14/5^2 + 17/3^2 + 20/8^2)/(1/5^2 + 1/3^2 + 1/8^2)$$

Figure 3. Inverse distance weighting illustration.

chart) is performed by looking up its location at the CDF of the data being corrected (around 0.7 at the red CDF curve) and looking up the values from the chosen CDF (black curve) with the same location of CDF (also 0.7). The location 0.7 belongs to the value 18 in black CDF. Therefore, every time value 4.9 occurred in the red dataset, it is replaced by the value 18. These steps are completed for every value found in the data being corrected and yield a new dataset as plotted by the blue lines at the bottom figure.

The blue data, derived from the previous data in red, have more similar mean and standard deviation with the reference data in black. The blue dataset has similar oscillation with that of the red dataset. The blue dataset is still a representation of the red dataset, but it is now shifted up and stretched.

To accommodate the non-stationary characteristic of rainfall data, the time series of every CHIRPS grid and the ground observation locations were grouped in month (January to December). The mean and standard deviation were calculated for each month. Hence, there were 24 features calculated for every CHIRPS grid and every ground observation location. This process was applied to every CHIRPS grid and ground observation locations. The monthly mean and standard deviation of every CHIRPS grid were then used to construct the monthly PDF and CDF of their corresponding grid.

3.2. Inverse Distance Weighting

After the monthly CDFs of every CHIRPS data were available, the next process was determining the reference CDFs. Each CHIRPS grid must have its own reference CDF. The reference CDF was constructed from the interpolation of mean and standard deviation of the ground observation locations. The interpolation procedure was necessary because the position of CHIRPS data locations did not exactly match the location of the ground observation. The number of ground observation locations included in interpolation process for one CHIRPS grid was determined to be three locations. The reference CDFs was provided for each CHIRPS grid.

The interpolation method used in this research was Inverse Distance weighting (IDW), as highlighted by Beek [7] and Sluiter [8] regarding its usage in meteorological data. Chen [9] also used IDW to estimates the spatial rainfall distribution. In this paper, IDW was applied to interpolate the mean and standard deviation of three nearest ground observation location to one CHIRPS grid. The power parameter for distance was set to 2. An illustration of IDW is shown by figure 3. The overall procedures in this research were as follows:

- Extract monthly mean and monthly standard deviation for each observation and for each grid point of estimates (yields 12 mean and 12 standard deviation values) from daily data.
- Generate monthly PDF of all CHIRPS grid.
- Generate monthly CDF from the monthly PDF of all CHIRPS grid.

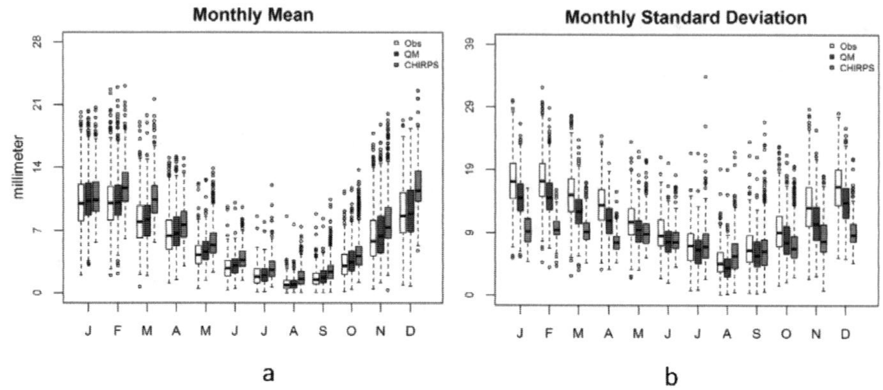

Figure 4. Boxplot of ground observation locations and their nearest CHIRPS grid: a. monthly mean, b. monthly standard deviation.

- Provide interpolated mean and standard deviation using IDW from three nearest ground observation for a CHIRPS grid.
- Generate monthly reference PDF of all CHIRPS grid.
- Generate monthly CDF from all monthly reference PDF.
- Transform the value of CHIRPS using QM based on its monthly CDF and its monthly reference CDF.

4. Results and Discussion

The default condition of the data from the ground observation and CHIRPS is shown by the boxplots in figure 4. Both datasets have the same monthly rainfall pattern, i.e. monsoon, which constitute the 'U' letter in their mean values. The estimates (red) have higher rainfall mean than the observation (white), but lower at the standard deviation. IDW adjusted the mean and standard deviation of estimates so that they are closer to the observation as shown by the blue boxplot in figure 3 as well. Note that the boxplot in figure 4 is derived from the condition of all ground observation locations and their nearest CHRPS grid. The example of spatial monthly mean condition is shown by figure 5. Only the mean condition of

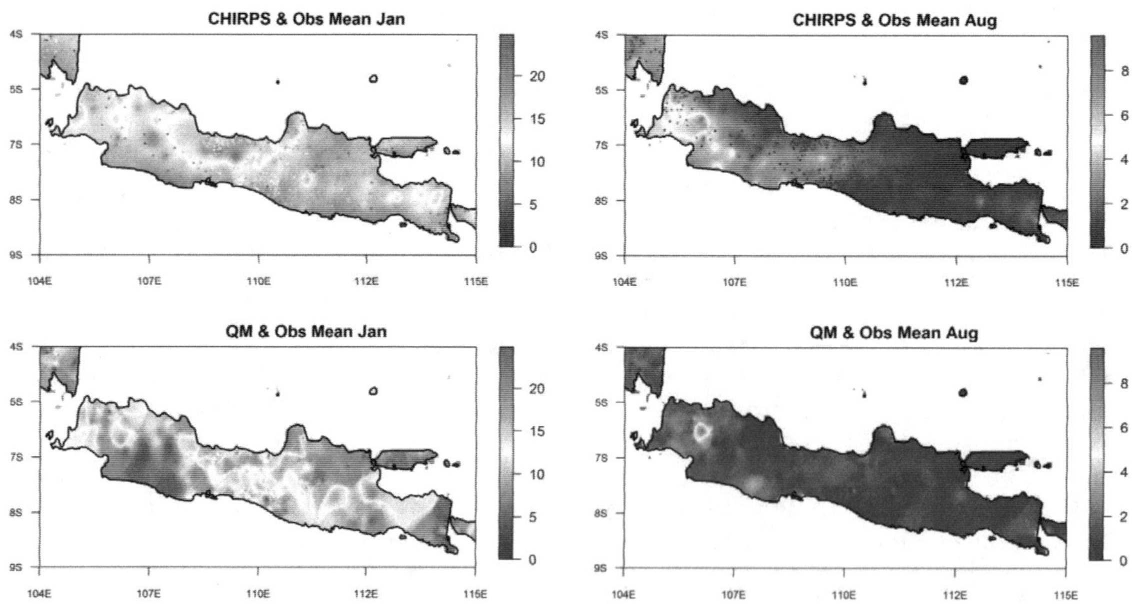

Figure 5. Spatial condition of CHIRPS mean and its correction.

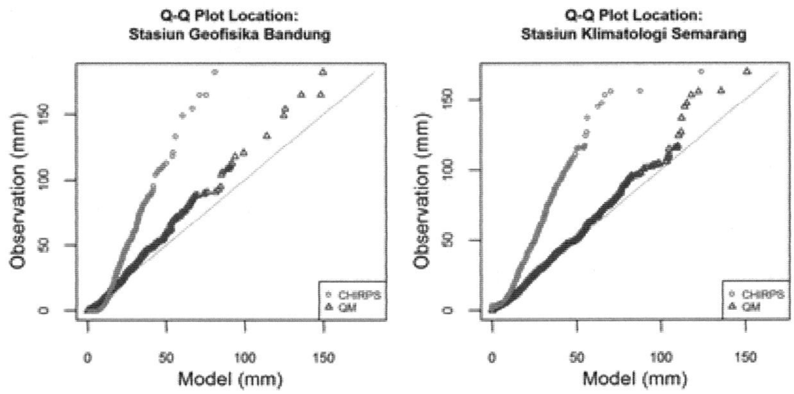

Figure 6. Quantile-Quantile plot of four examples of CHIRPS versus ground observation (red) and corrected CHIRPS versus ground observation (blue).

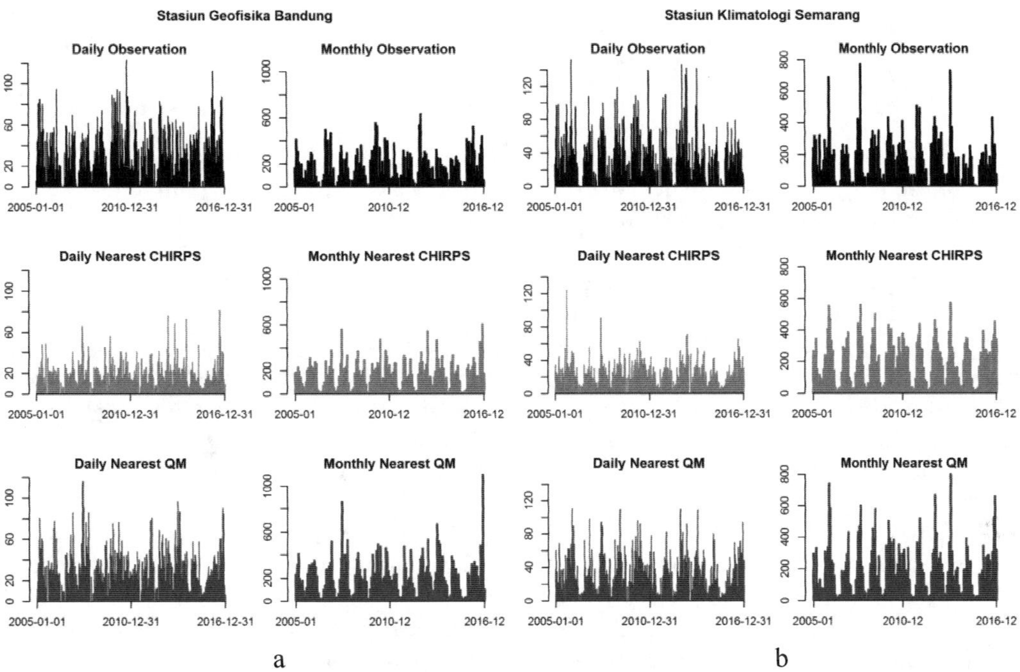

Figure 7. Time series of ground observations (black), CHIRPS (red), and the new dataset (blue) of two selected locations: a. Bandung Geophysics Station, b. Semarang Climatology Station.

the wettest and driest month (January and August) are displayed. The dots represent the ground observation locations. They are drawn in front of the CHIRPS grids. Therefore, if they still appear, there are still differences between the CHIRPS grids and the ground observations. It is found that CHIRPS still has many different values with the ground observations. The result of bias correction has modified the mean of CHIRRS to be closer to the ground observation as indicated by less visible dots appearing. The modification also affected the standard deviation.

Figure 6 shows the Quantile-Quantile (QQ) plot of CHIRPS versus the ground observation it is found that the data are distributed in Gamma as indicated by the denser distribution near zero value and getting sparse as the values getting higher. The bias corrected of CHIRPS lies nearer the perfect slope than the CHIRPS original data. The time series example of the new dataset which is derived from CHIRPS is shown by figure 7. The time series charts are taken from the same locations of figure 6. The time series

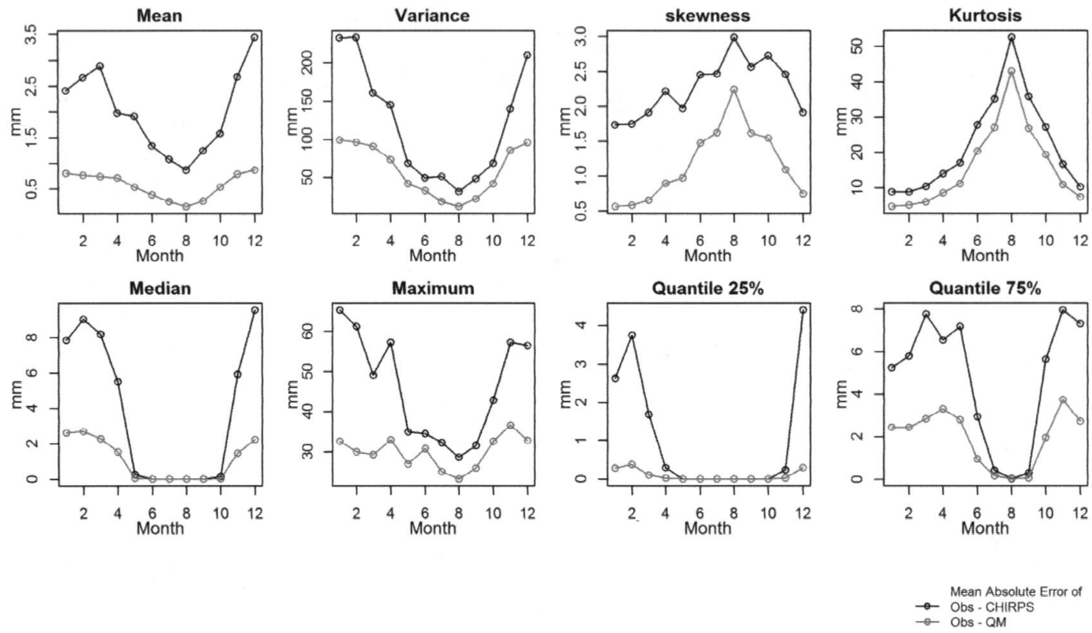

Figure 8. Monthly mean of statistical difference between each ground observation location and its nearest grid of CHIRPS (black) and the new dataset (red).

indicate that the new dataset (blue) has bigger range than CHIRPS (red). Some statistical calculations of the CHIRPS and the new dataset compared to the ground observation dataset is shown by figure 8. The monthly mean of statistical parameters calculation was performed on every locations of the ground observations, CHIRPS, and the new dataset. To compare the results, the monthly mean of statistical parameters of CHIRPS and the new dataset were each subtracted from the monthly mean of statistical parameters of ground observations. Note that the locations included in figure 8 are all ground observation locations and their nearest grid of CHIRPS and the new dataset.

5. Conclusion

This research has produced an alternative dataset of daily rainfall intensity derived from the satellite based combined with ground station (CHIRPS) products. The new data were formed by modifying the probability of the values obtained from CHIRPS using Quantile Mapping (QM). To establish the QM process, a parameter estimation procedure was first conducted to construct new spatial Probability Density Function (PDF) and Cumulative Density Functions (CDF) subsequently over grids of satellite estimates using Inverse Distance Weighting (IDW). IDW provided the spatial interpolation of the CDF parameters obtained from the observation as the reference, which was located sparsely. The results show that IDW-QM succeeds in pulling the satellite estimates closer to the observation.

References

[1] Toté C, Patricio D, Boogaard H, Wijngaart RVD, Tarnavsky E and Funk C 2015 Evaluation of satellite rainfall estimates for drought and flood monitoring in Mozambique *Remote Sensing* **7** 1758-1776

[2] Vernimmen RRE., Hooijer A, Mamenun and Aldrian E 2012 Evaluation and bias correction of satellite rainfall data for drought monitoring in Indonesia *Hydrology and Earth System Sciences* **16** 133–146

[3] As-Syakur A R, Tanaka T, Prasetia R, Swardika I K and Kasa I W 2011 Comparison of TRMM multisatellite precipitation analysis (TMPA) products and daily-monthly gauge data over Bali *International Journal of Remote Sensing* **32** (24) 8969-8982

[4] Funk C, Peterson P, Landsfeld M, Pedreros D, Verdin J, Shukla S, Husak G, Rowland J, Harrison L, Hoell A and Michaelsen J 2015 The Climate Hazards Infrared Precipitation with Stations—A New Environmental Record for Monitoring Extremes *Scientific Data* **2**

[5] Valdés-Pineda R, Demaría E M C, Valdés J B, Wi S and Serrat-Capdevil A 2016 Bias correction of daily satellite-based rainfall estimates for hydrologic forecasting in the Upper Zambezi, Africa *Hydrology and Earth System Sciences*

[6] Piani C, Haerter JO and Coppola E 2010 Statistical bias correction for daily precipitation in regional climate models over Europe *Theoretical and Applied Climatology* **99** 187–192

[7] Beek EG 1991 Spatial interpolation of daily meteorological data; theoretical evaluation of available techniques. *DLO The Winand Staring Centre* Report 53.1

[8] Sluiter R 2009 Interpolation methods for climate data: literature review *Roy. Netherlands Meteorological Inst. Tech. Rep.*

[9] Chen F W and Liu C W 2012 Estimation of the spatial rainfall distribution using inverse distance weighting (IDW) in the middle of Taiwan *Paddy Water Environ* **10** 209

The influence of surface roughness and turbulence on heat fluxes from an oil palm plantation in Jambi, Indonesia

Tania June[1], Ana Meijide[2,3], Christian Stiegler[2], Alan Purba Kusuma[1], Alexander Knohl[2]

[1]Agrometeorology Division, Department of Geophysics and Meteorology, Faculty of Mathematics and Natural Sciences, Bogor Agricultural University (IPB), Darmaga campus, telp/facs:+62 251 8623850, Indonesia
[2]University of Goettingen, Faculty of Forest Sciences, Bioclimatology, Büsgenweg 2, 37077 Göttingen, Germany
[3]University of Granada, Department of Ecology, Avda. Fuente Nueva s/n 18071, Granada, Spain

Email : tania.june@yahoo.com

Abstract. Oil palm plantations are expanding vastly in Jambi, resulted in altered surface roughness and turbulence characteristics, which may influence exchange of heat and mass. Micrometeorological measurements above oil palm canopy were conducted for the period 2013–2015. The oil palms were 12.5 years old, canopy height 13 meters and 1.5 years old canopy height 2.5 m. We analyzed the influence of surface roughness and turbulence strenght on heat (sensible and latent) fluxes by investigating the profiles and gradient of wind speed, and temperature, surface roughness (roughness length, z_o, and zero plane displacement, d), and friction velocity u*. Fluxes of heat were calculated using profile similarity methods taking into account atmospheric stability calculated using Richardson number Ri and the generalized stability factor ζ. We found that roughness parameters (z_o, d, and u*) directly affect turbulence in oil palm canopy and hence heat fluxes; they are affected by canopy height, wind speed and atmospheric stability. There is a negative trend of *d* towards air temperature above the oil palm canopy, indicating the effect of plant volume and height in lowering air temperature. We propose studying the relation between zero plane displacement *d* with a remote sensing vegetation index for scaling up this point based analysis.

1. Introduction

Current land transformation in Jambi resulted in a rapid expansion of oil palm plantations. Oil palm (Elaeis guineensis) is one of the crops which has high economic value as global market demands for its products are very high. Currently, along with Malaysia, Indonesia produces more than 80% of palm oil to meet global market demands [1,2]. With increasing global need and price, the demand for oil palm is projected to increase sharply, leading to further land use changes towards oil palm plantations.

Transformation from forest to oil palm plantation will result in changing surface biophysical parameters Biophysical surface characteristics determine microclimate variability which influences many ecological processes at the land surface as well as results in feedback to the atmosphere [3]. In other words, there is an interrelationship between atmosphere and the vegetation covering the ground surface [4]. Interest in understanding this vegetation-atmosphere feedbacks increases continuously, however, direct measurements of micrometeorological parameters in oil palm plantations are still scarce.

Biophysical parameters of the land surface include surface roughness (roughness length, z_o, zero plane displacement, d), friction velocity u* (indicating strength of turbulence), sensible and latent heat (water) fluxes and CO_2 fluxes. By analysing wind speed profiles together with air temperature above the related surface, we can obtain the three parameters z_o, d, and u* [5-7]. These roughness parameters are useful for describing surface-atmosphere exchange of heat, water vapor, and greenhouse gases such as CO_2. The sink strength of the land surface for momentum [8] determines the dynamics of the boundary layer, especially for characterizing the surface boundary layer. Surface roughness determines the vertical wind speed and air temperature profile, mesoscale and global momentum exchange [9], land-atmosphere energy exchange [10], and turbulent air mixing beneath, within, and above its roughness elements.

In this paper we focus on the relationship of surface roughness with air temperature and sensible heat flux. Our objectives are (a) to estimate key surface biophysical parameters such as surface roughness (roughness length, z_o, zero plane displacement, d), friction velocity u* (indicating strength of turbulence), (b) to calculate sensible and latent heat fluxes, and (c) to investigate their interactions.

2. Material and method

2.1. Research Sites

The research sites are a young oil palm plantation at Pompa Air Village and a mature oil palm planation at PTPN VI Jambi, Indonesia (Figure 1), where micrometeorological towers were installed. The age of the oil palms plantations were 1 and 12 years, with canopy height of 2.5 m and 13 m respectively. The tower in Pompa Air is located at S 01°50' 7.6", E 103°17' 44.2" altitude 75 m a.s.l (-1.83545 °S, 103.29562 °E) and the tower in PTPN VI is located at S01°41'35.052", E 103°23'29.040", altitude 76 m a.s.l. (-1.69307 °S 103.3914 °E) (Figure 2 and 3). Soil types around both towers are mineral. The plantation at PTPN VI has an area of 2025 ha.

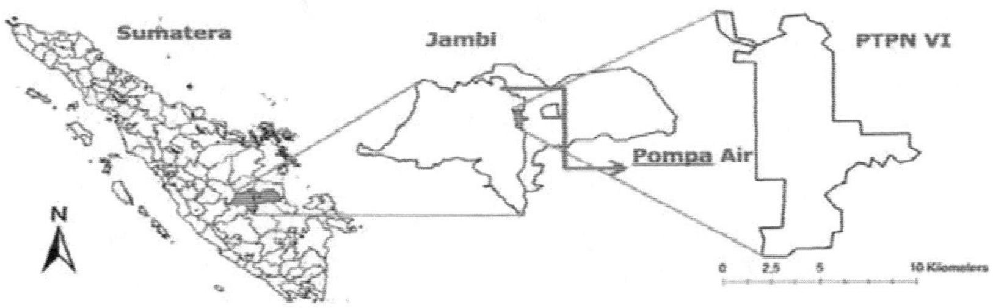

Figure 1. Research sites, Pompa Air Village and PTPN VI oil palm plantation, Jambi.

IOP Conf. Series: Earth and Environmental Science **149** (2018) 012048 doi:10.1088/1755-1315/149/1/012048

Figure 2. Research site in Pompa Air Village with micrometeorology tower installed on a 2.5 years old oil palm.

Figure 3. Research in PTPN VI with micrometeorology tower installed within 13 years old oil palm.

2.2. Instrumentation and Data Analysis

Micrometeorological instruments were installed on a tower at several heights which include air temperature sensors above canopies; atmospheric pressure sensors, humidity sensors, wind direction and wind speed sensors. Above the 1 years-aged plantation, four cup anemometers were installed at 2.4 m, 3.15 m, 4.14 m, and 5.08 m above the ground, whereas in the mature 12 years old plantation, cup anemometers were installed at 13, 15.4 and 18.5 m above the ground. The sensors for temperature and relative humidity were installed at the same height as the anemometers. Data analysis were conducted for the period 2013-2015. The data were recorded every 10 minutes. The meteorological tower is part of the CRC 990 EFForTS project, a collaboration between the University of Göttingen, Bogor Agricultural University, University of Jambi and the University of Tadulako. Details on the tower instrumentation are found in [11].

2.2.1 Roughness characteristics

Three roughness parameters *i.e.* zero-plane displacement (*d*), friction velocity (u*), and roughness length (z_0) were derived from the logarithmic wind profile equation under neutral conditions [12,13]:

$$n(z-d) = \frac{k}{u^*}u(z) + lnz_0 \qquad (1)$$

where k is von Karman's constant (0.40) and z denotes measurement height (m). Zero-plane displacement were estimated based on [14].

2.2.2 Sensible and latent heat flux

Under neutral condition, sensible heat flux Q_H (W m-2) was approximated by equation:

$$Q_H = \rho c_p K_H \frac{dT}{dz} \qquad (2)$$

where c_p is specific heat at constant pressure (1004.2 J kg^{-1} K^{-1}), K_H is eddy diffusivity of heat, and dT/dz is vertical gradient of air temperature. The equation of Q_H under non-neutral condition was corrected as followed:

$$Q_H = \rho c_p k^2 \frac{(u_2-u_1)(\theta_2-\theta_1)}{\left[ln\left(\frac{z_2-d}{z_1-d}\right)\right]^2 \varphi_m \varphi_s} \quad (3)$$

where φ_s (the dimensionless gradient of θ)) and φ_m (the dimensionless wind shear) were calculated using the following equations:

$$\varphi_s = (1\text{-}15\,\zeta)^{-1/2}, \text{ and } \varphi_m = (1\text{-}15\,\zeta)^{-1/4} \qquad \text{if } \zeta < 0 \quad (4)$$

$$\varphi_s = 1+5\,\zeta, \text{ and } \varphi_m = 1+5\,\zeta \qquad \text{if } \zeta \geq 0 \quad (5)$$

ζ denoting Monin-Obukhov stability parameter, and is related to Richardson Number, Ri [15,7]:

$$\zeta = Ri \qquad \text{if } Ri < 0 \quad (5)$$

$$\zeta = Ri/(1\text{-}5\,Ri) \qquad \text{if } 0 \leq Ri \leq 0.1 \quad (6)$$

$$\zeta = 0.2 \qquad \text{if } Ri > 0.1 \quad (7)$$

$$Ri = \frac{g\left(\frac{\partial\theta}{\partial z}\right)}{\theta_a\left(\frac{\partial u}{\partial z}\right)^2} \quad (8)$$

$Ri < -0.01$ refers to unstable condition, $-0.01 < Ri < 0.01$neutral condition, and $Ri > 0.01$ stable condition. Equation (8) shows that Ri is classified based on the ratio of gradient of potential air temperature (θ) and gradient of horizontal wind velocity, g is acceleration due to gravity, and θ_a denotes average potential air temperature (K).

2.2.3 Turbulence and Turbulent Kinetic Energy (TKE)

TKE were calculated for conditions of unstable atmospheric stratification during day-time as TKE production is less important during night-time conditions due to predominantly stable atmospheric stratification [16,7] as followed:

$$TKE = 0.5[\sigma_u{}^2 + \sigma_v{}^2 + \sigma_w{}^2] \quad (9)$$

where $\sigma_u{}^2$, $\sigma_v{}^2$, and $\sigma_w{}^2$ are the variances of the longitudinal, lateral, and vertical wind velocity components. These three variances are defined as the square of standard deviations of wind velocities given by equations 10,11 and 12 [16]:

$$\sigma_u = 0.11 w_B (z/z_i)^{1/3}[1 - 0.7(z/z_i)] \quad (10)$$

$$\sigma_v = 0.08 w_B [0.5 + 0.4(1 - z/z_i)^2] \quad (11)$$

$$\sigma_w = 0.11 w_B (z/z_i)^{1/3}[1 - 0.8(z/z_i)] \quad (12)$$

Those equations are functions of buoyancy velocity scale (w_B) and mixing layer depth (z_i). w_B can be determined by using equation 13 and 14:

$$w_B = \left(\frac{1}{0.08}\right) w^* \quad (13)$$

$$w^* = \left[\frac{|g| z_i}{T_v}\left(\frac{Q_H}{c_p \rho}\right)\right] \quad (14)$$

where w^* is the Deardorff velocity, T_v is virtual air temperature (K), and Q_H denotes sensible heat flux (Wm^{-2}).

3. Results and discussion

3.1. Roughness characteristics

Zero plane displacement (*d*) increased with age of the plantation with an average value of 1.87 m and 11.13 m for the young and mature oil palm plantation, respectively (Figure 4) revealing that d value depends on both vegetation height and density. Beside of height and density of the standing vegetation, [17,18] clarified that the mechanical sturdiness of vegetation stem also defines the displacement height where the wind speed is close to zero. Roughness length (z_0) as well as friction velocity (u*) follow similar pattern , where the highest z_0 and u* were found in the mature compared to the young plantation. They were 0.4 m, 0.22 m s-1 and 0.2 m, 0.08m s-1 respectively (Figure 4, Figure 5). The value of d tends to be more persistent in comparison with z_0 and u*. The values of z_0 dropped with increasing wind velocity. The values of z_0 varied largely during calm wind conditions, particularly in the 12 years-old plantation. Friction velocity (u^*) was strongly positively related to wind speed.

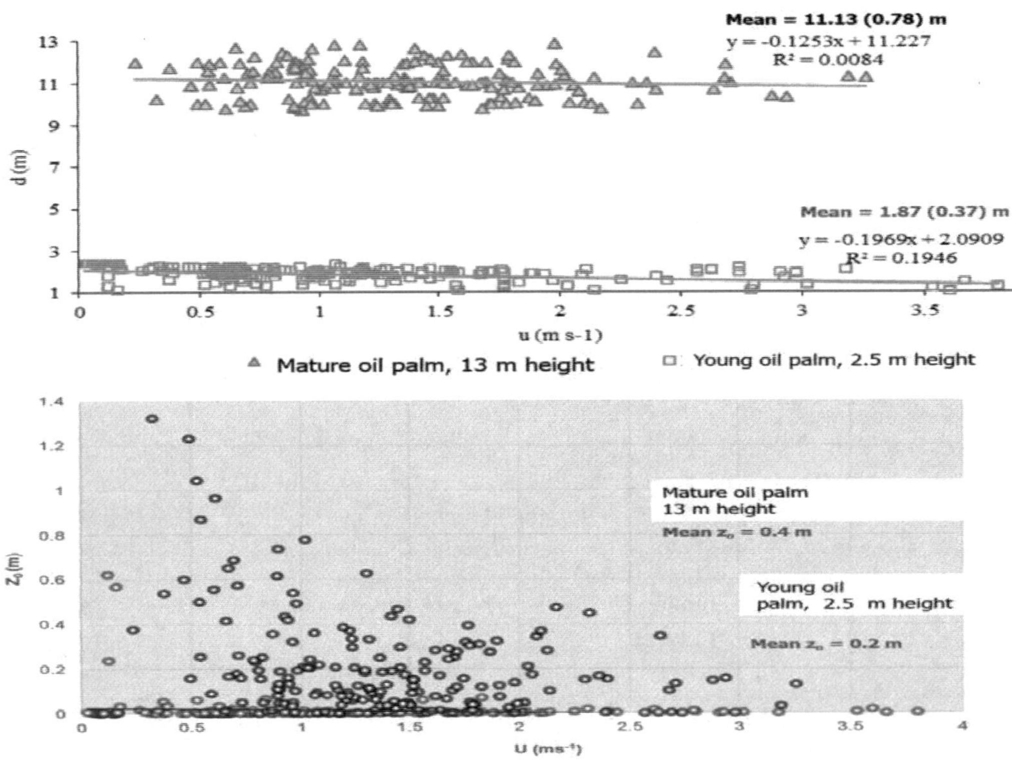

Figure 4. Zero plane displacement (d) and roughness length (z_o) of the young and mature oil palm plantation.

The value of d describes the height inside the canopy where wind speed starts to be extinguished due to the friction imposed by the plant canopy. The value of z_0 shows the position where momentum is well absorbed by the roughness element. Unlike these two roughness parameters, which characterize the unique dimension of the surface roughness, u* explains a tangential velocity of air parcel movement primarily due to mechanical turbulence. That is why u* is usually determined under neutral condition, when buoyancy effects are insignificant in the absence of intensive surface heating. Thereby u* can completely depict the roughness of the underlying surface. According to [8], the 12 years-old oil palm plantation can be classified as a very rough surface (terrain category: 3, class: 6) which is the same as orchards and bushes, whereas the 1 years old oil palm plantation tended to be classified between a smooth surface (terrain category: 1, class 2) and an open surface (terrain category: 2, class 3). The roughness length of a vegetated surface can be a function of wind speed [19, 6]. Strong gusts of wind can bend the stems of plants parallel to the wind direction which results in lower z_0. Unlike flexible stems which are easily bent by gust of wind, the sturdy stems of oil palms with a high density will have a much higher roughness length due to a reduced bending effect. The sparsely vegetated surface of the 1 years old oil palm plantation obviously had a lower roughness length compared to the 12 years old plantation, since it did not have a large absorption area for momentum. In the 1 year old plantation, where plant spacing is of the order of the canopy height or larger (about 8 m distance between palms) a larger part of the momentum is absorbed by the ground surface. Therefore, the 1 year-old plantation had very small roughness length compared to the 12 year old plantation. The higher values of u* in the 12 year-old plantation compared to the 1year-old planation indicate that greater shear stress was generated in the older plantation and triggered bigger turbulence production. The reason for the greater shear stress is the higher surface roughness in the 12 year-old plantation. Shear stress also increases with increasing wind speed, thus explaining the increase in u* with wind speed. (Figure 5).

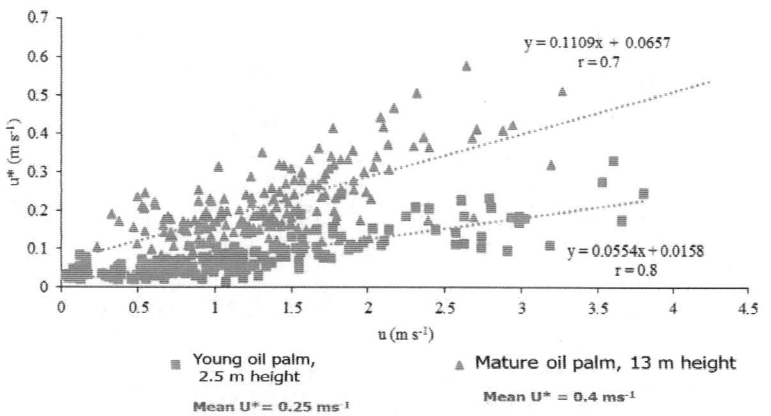

Figure 5. Friction velocity (u*) versus horizontal wind speed (u) at the young and mature oil palm plantations.

Environmental problems such as soil erosion either by wind gust or run off are expected to occur more often when an oil palm plantation is still young, short, sparse and relatively flexible in terms of bending moment. However, this problems will be gradually reduced by the growth of the plants (the increase of roughness characteristics). At the optimum growth phase when the oil palm canopies will fully cover the ground, z_0 and d are expected to reach maximum values, and may act as windbreaker reducing wind erosion.

3.2. Turbulence and Heat Fluxes (Sensible and Latent heat fluxes)

Sensible and latent heat fluxes as well as u* are positively correlated with turbulent kinetic energy (TKE, Figure 6). During the day, incoming short wave radiation is absorbed by the surface and warms the atmosphere upwards, the potential air temperature gradient becomes negative (unstable atmospheric condition) as the sensible heat flux transfers heat from warmer surface to the cooler atmosphere above. On the contrary, at night, when the ground is usually cooler than the air, the potential air temperature gradient becomes positive (stable atmospheric condition) and the flux direction changes from the atmosphere to the surface. In case of neutral conditions, the vertical gradient of potential air temperature is close to zero, meaning no or negligible sensible heat transfer, neither from the surface to the atmosphere or in opposite direction. The sensible heat flux was larger in the young compared to the mature plantation. One of the main drivers of heat flux variation is turbulence production. If we talk about turbulence, we usually relate it to atmospheric stability which is represented by the Richardson number Ri. A negative Ri refers to unstable condition, this is when the turbulences are intensively generated, whereas a positive Ri refers to stable condition when the turbulence is dampened. Rougher surface such in mature plantation usually has greater turbulence production due to bigger shear stress (can be shown by the higher u* in Figure 5).

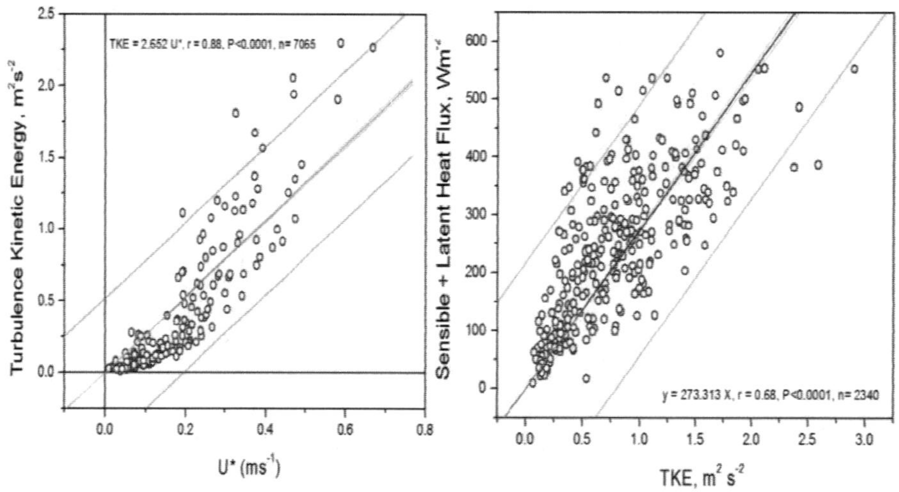

Figure 6. Influence of TKE on heat fluxes, both sensible and latent heat fluxes (right). TKE was significantly correlated with u* (left).

The existence of turbulence (TKE) advances the flux transfer process (Figure 5). Various factors have simultaneously a role in the heat flux process, such as surface roughness, air temperature gradient, turbulence generation, and solar radiation (peak time and maximum surface heating during day time). Turbulence indicates an irregular behavior of wind speed deviating from its mean velocity profile. Generally, turbulent motion is generated by two main factors, i.e. buoyancy production and wind friction between air movements and the rough surface [7]. Turbulence reaches a maximum when surface heating is at maximum under unstable condition resulting in high buoyant productions as well as when wind shears near the surface is high. We found that TKE above the mature plantation was greater than TKE above the young plantation. Under unstable conditions, beside of buoyancy production and wind shears, the tendency of TKE to increase or decrease can be affected by transport of TKE advected by mean wind speed from one location to another. TKE also can be transported vertically, where the excess of TKE near the surface will tend to be moved up to a higher location by turbulent motions [16].

3.3. Turbulence and air temperature

One of the driving factors determining the vertical air temperature profile in vegetated areas is the heating process by solar radiation. Similarly to leaf area index, the zero-plane displacement height, d, which is a function of canopy height and density) can be a good parameter characterizing the effect of plants on microclimate. [20] stated that the increase of d was proportional to the increase of the Normalized Density Vegetation Index generated from remote sensing data. During day time, a sparse canopy (young plantation) absorbs less solar radiation and thus more radiation reaches the ground compared to the dense canopy of the mature plantation. As a consequence, the surface heats stronger generating more sensible heat. The sensible heat will create buoyancy effects stimulating turbulent mixing forcing hot air parcels to move upwards and thus rapidly increase air temperature above the plantation canopies. Therefore, the mean daily cycle of air temperature in the young plantation was higher than in the mature site (Figure 7). Although turbulence generated above the mature plantation was greater than above the young plantation, the heating process within this denser canopy takes longer than the ground heating process, therefore the air temperature above the young plantation increased faster than the mature site. Additionally, the higher turbulence of the "rougher" mature plantation, was allocated for latent heat fluxes (resulting from the larger water consumption of the mature oil palms), which lowered the surface and hence air temperature of the mature plantation.

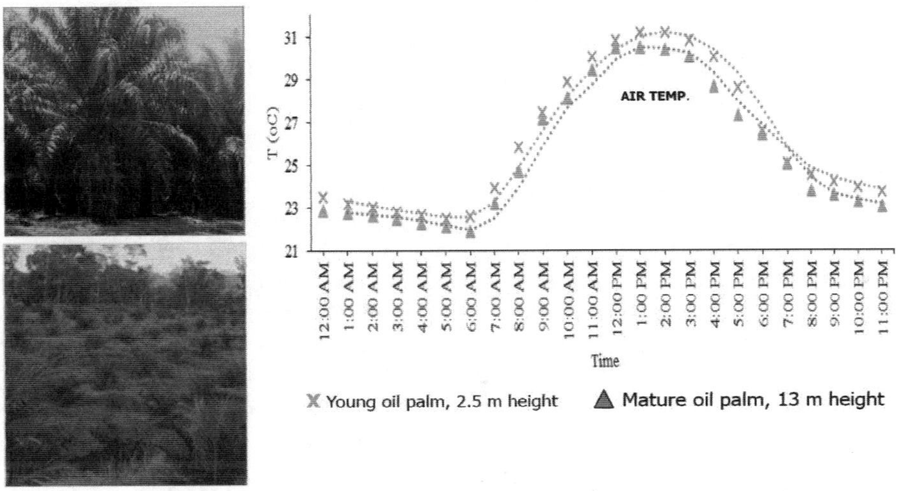

Figure 7. Mean daily cycle of air temperature at the young and mature oil palm plantations.

Similar to [21], the mean daily maximum air temperature had a strong negative relationship with roughness expressed as *d*, while its mean daily minimum did not (Figure 8). This indicates the interplay of surface roughness, water consumption and air temperature dynamics.

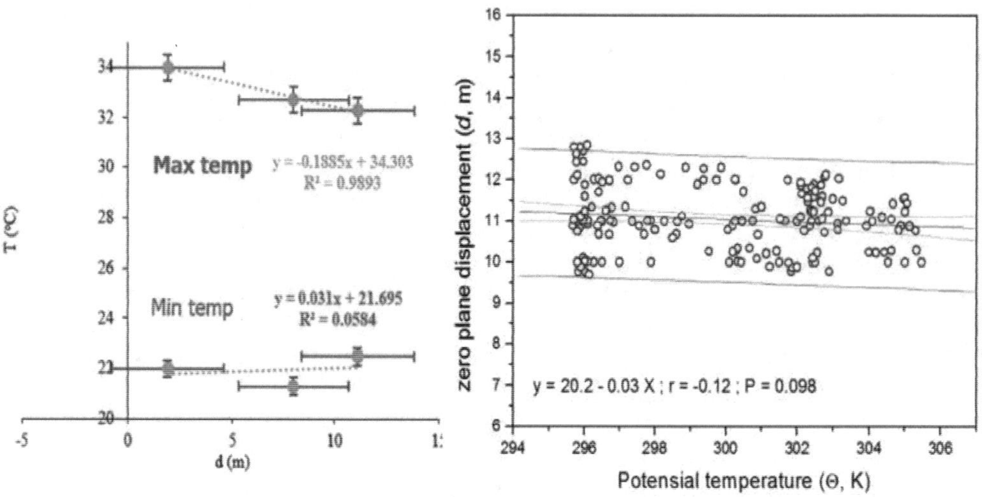

Figure 8. Increasing roughness characteristic (d) results in decreasing maximum air temperature, but had no significant effect on minimum air temperature (left). Weak negative correlation between d with potential temperature (right).

4. Conclusion

Surface roughness influences the dynamics of the surface boundary layer, including vertical wind speed and temperature profile, momentum flux, sensible and latent heat fluxes, and turbulence. Parameters of surface roughness which include zero-plane displacement (d), friction velocity (u*), and roughness length (z_0) increased with plant height (and age) as shown by the comparison between a 1 and 12 years old plantation. The 12 years old plantation was 'rougher' than the 2.5 years old plantation generating bigger shear stress indicated by higher u*. Proportional to the shear generated, both turbulence intensities and TKE increases towards maximum value with increasing surface heating, when the mixing processes were more effective. Surface roughness (z_o and d) affects the transferred sensible and latent heat by influencing vertical mixing of air, through u* and TKE, in the surface boundary layer, thereby contribute to regulating surrounding air temperature.

Young plantations showed higher air temperature and had higher sensible heat fluxes (despite lower roughness characteristics), while mature plantations invest more energy towards latent heat fluxes and have lower surface temperature. Thereby, evaporative cooling is the main determinant of regulating temperature in oil palm as for most vegetated surfaces when water is available [22]. There is a decreasing trend of maximum air temperature with increasing surface roughness (d, z_o) above the oil palm canopy, indicating the effect of increasing plant volume in reducing air temperature. This is also supported by Landsat derived parameter NDVI [20,23]. We therefore recommend the use of remote sensing (and LIDAR) products for scaling up to larger area.

Research on surface roughness, turbulence, sensible and latent heat fluxes on oil palm is very limited. Therefore it would add valuable information needed to support the analysis of land use change effects on energy, mass fluxes and micro/regional scale climate [24].

Acknowledgements
This study was financed by the Deutsche Forschungsgemeinschaft (DFG) in the framework of the collaborative German-Indonesian research project CRC990 (subprojects A03), a research collaboration between University of Gottingen and Bogor Agricultural University, as well as the Universities of Jambi and Tadulako. The authors would like to thank Edgar Tunsch, Dietmar Fellert, and Malte Puhan as well as our field assistants from the University of Jambi. We also thank PTPN VI and the plantation owner at Pompa Air for allowing us to conduct our research at their plantation.

References
[1] Koh, L P and Wilcove D S 2007 Cashing in palm oil for conservation. Nature. **448** 7157 993-4
[2] Fitzherbert E B, Struebig M J, Morel A, Danielsen F, Brühl C A, Donald P F, Phalan, B 2008 Trends *Ecol E* **23** 10 538-45 doi: 10.1016/j.tree.2008.06.012
[3] McPherson R A 2007 A review of vegetation-atmosphere interactions and their influences on mesoscale phenomena. *Prog. Phys. Geogr.* **31** 261–285
[4] Richardson A D, Keenan T F, Migliavacca M, Ryu Y, Sonnentag O, Toorney M 2013 Climate change, phenology, and phenological control of vegetation feedbacks to the climate system. *Agricultural and Forest Meteorology* **169** 156– 173
[5] Tian X, Li Z Y, van der Tol C, Su Z, Li X, He Q S, Bao Y F, E X Chen, Li L H 2011. Estimating zero-plane displacement height and aerodynamic roughness length using synthesis of LiDAR and SPOT-5 data. *Remote Sensing of Environment* **115** 2330-2341
[6] Zhibaol D, S Gao, Fryrear D W 2001 Drag coefficients, roughness length and zero-plane displacement height as disturbed by artificial standing vegetation. *Journal of Arid Environments* **49** 485-505 doi: 10.1006/jare.2001.0807
[7] Arya A P 2001 *Introduction to Micrometeorology* Academic Press, INC
[8] Wierringa, J 1993 Representative roughness parameters for homogeneous terrain. *Boundary Layer Meteorology* **63** 323-363
[9] Raupach M R 1994 Simplified expressions for vegetation roughness length and zero-plane displacement as functions of canopy height and LAI. *Boundary Layer Meteorology* **71** 211-216
[10] Yang R, Friedl M A 2003 Determination of roughness lengths for heat and momentum over boreal forests. *Boundary Layer Meteorology* **107** 581-603
[11] Meijide A, Röll A, Fan Y, Herbst M, Niu F, Tiedemann F, June T, Rauf A, Hölscher D, Knohl A 2017 Controls of water and energy fluxes in oil palm plantations:Environmental variables and oil palm age, *Agricultural and Forest Meteorology* **239** 71-85
[12] Oke T R 1978 *Boundary Layer Climates*. London: Methuen & Co Ltd
[13] Kaimal, J C, Finnigan, J J 1994 *Atmospheric boundary layer flows: Their Structure and Measurement* Oxford University Press
[14] Riou C H 1984 Simplified calculation of the zero-plane displacement from wind speed profiles. *J. Hydrol.* **69** 351-357
[15] Businger J A., Wyngaard J C, Izumi Y, Bradley E F 1971. Flux-profile relationships in the atmospheric surface layer. *J. Atmos. Sci.* **28** 181-189
[16] Stull R 2000. Meteorology for Scientist and Engineers 2nd Edition. USA: Brooks/Cole Thomson Learning
[17] Chang J 1968. *Climate and Agriculture*. An ecological Survey. Aldine, Chicago, USA
[18] June T 1987 *Medan Angin pada Pertanaman Kacang Kedelai (Glycine max (L.) Merr) dengan Arah Baris Berbeda*. Institut Pertanian Bogor Bogor: Indonesia
[19] Sutton O G 1953 *Micrometeorology* New York: Mc Graw-Hill

[20] Ramli, N I, Idris Ali, Syamsyul M, M Saad H, Majid TA 2009. Estimation of the roughness length (zo) in Malaysia using satellite image. The Seventh Asia-Pacific Conference on Wind Engineering, November 8-12, 2009, Taipei, Taiwan

[21] Hardwick S R., Toumi R, Pfeifer M, Turner E C, Nilus R, Ewers R M 2015. The relationship between LAI and microclimate in tropical forest and oil palm plantation: Forest disturbance drives changes in microclimate. *Agricultural and Forest Meteorology* **201** 187–195

[22] Sabajo C R, le Maire G, June T, Meijide A, Roupsard O, Knohl A 2017: Expansion of oil palm and other cash crops causes an increase of the land surface temperature in the Jambi province in Indonesia, Biogeosciences **14** 4619-4635 doi.org/10.5194/bg-14-4619-2017

[23] Zhao K, Popescu S 2009 Lidar-based mapping of leaf area index and its use for validating GLOBCARBON satellite LAI product in a temperate forest of the southern USA. *Remote Sensing of Environment* **113** 1628–1645

[24] Pitman A J 2003. The evolution of, and revolution in, land surface schemes designed for climate models. *Int. J. Climatol.* **23** 479–510

Fishery Household (FH) database and cultivation areas in Indramayu regency to develop Shrimp and Milkfish farming based on GIS

Rita Rostika, Noir P. Purba, Iskandar, Lantun Paradhita Dewanti, and Izza Mahdiana Apriliani
Faculty of Fisheries and Marine Sciences, Padjadjaran University

E-mail: ritarostika_unpad@yahoo.com

Abstract. With a coastline length of 114 km, the utilization of the coastline areas is potential for especially shrimp and milkfish cultivations, which have a higher economic value. However, appropriate development strategies are highly required. The purpose of this research is to examine the existing conditions and organize integrated information for fishery household as well as the areas utilized for shrimp and milkfish cultivations, especially in coastal areas. The methods used include an analysis from Landsat 8 imaginary, field survey, and interviews with various sources. This research conducted in June-November 2015. Data from remote sensing were digitized and utilized as mark point to survey area of cultivation. The results show that the distributions of RTP data cover *Windu* and *Vaname* shrimp polycultures by 1,727 RTP, milkfish by 1,551 RTP, *Vaname* shrimp by 2,953 RTP, and *Windu* Shrimp by 88 RTP. The area, which may be utilized as ponds, is 9,854.1 ha. The area for milkfish ponds is 12,065.08 ha, while the digitation result is 10,801.92 ha. It shows that Indramayu coastal areas still have numerous and various potentials, which may be utilized for shrimp and milkfish cultivations.

1. Introduction

Fish becomes one of the world's strategic commodities expected to be greatly increased as the future demand increases [1,2] especially in Indonesia and local regency such as Indramayu. Freshwater aquaculture in Indonesia started with the stocking of common carp in backyard ponds in West Java and subsequently expanded to other parts of Java, Sumatra and Sulawesi in the early twentieth century [3]. Indramayu has a great potential fishery which is supported by its natural factors, with its coastline length of 114 km, both in cultivation and capture fishery. Cultivation is considered as the main economic source in coastal areas of Indramayu Regency. Aquaculture includes fish farming in freshwater, brackish and marine aquaculture in Indramayu is now sufficiently developed, it is characterized by the activity of aquaculture that have been spread in 31 districts in Indramayu [4].

The utilization of coastal areas for cultivation is highly potential to provide great opportunities for the communities in obtaining various benefits to improve their life welfare. One of them is shrimp and milkfish cultivation which has high economic value, yet requires the appropriate strategies for its development [5]. The cultivation of milkfish (*C. chanos*) in Indonesia has a very good prospect. In 2008 the production of milkfish cultivation (*C. chanos*) was 422,086 tons higher than the total production in the Philippines at 349,432 tons. The production of milkfish (*C. chanos*) in Indonesia increased again in 2012 at 482,930 tons [6].

Aquaculture has a quite significant trend increase [7]. This condition can provide competitive advantage between fish farmers in Indramayu and farmers from other regions [8]. One of them is milkfish and shrimp commodities, which ranks the first in the fishery production in Indramayu [9]. In addition, the pond areas are extended from 13,700 to 22,000 in 2014 since aquaculture system is easier to be developed and periodically monitored. However, there are still various challenges and obstacles faced in the management of aquaculture fisheries in Indramayu Regency.

The potencies of Indonesia to be the source of fish in the world are supported by the fact that its natural sources of ocean has not been optimally explored, land to be used as both production and hatcheries is available, and climate to enhance the optimal growth of fish farming throughout the year is suitable [10]. One of the challenges faced by Indramayu Regency is due to the development of cultivation areas, which should be optimally utilized. Information on the utilization of coastal areas is urgently required, particularly on the potential of coastal areas, which are integrated with the ocean. The integrated information system, which is considered as the rapid and simple data storing, processing, as well as delivery from various sectors is in the form of Geographic Information System (GIS). Utilization of Geographic Information Systems (GIS) promising resource management and modeling primarily quantitative models become easier and simpler. GIS is an efficient and effective way to determine the characteristics of an area of land and development potential [11,12]. GIS may be integrated with the Remote Sensing Technology, which has various advantages in providing the multi-temporal spatial data that the integration may have broader coverage, ability to reach the remote areas, and early information regarding to the organization of fishery household database in the coastal areas of Indramayu Regency, such as milkfish and shrimp cultivation.

Inventory and development are required in order to improve the fishery production and to support the government programs [13,14]. This research aims at identifying the existing shrimp and milkfish cultivation areas and potential which recently have become the cultivation development areas and Household Fishery (RTP) in Indramayu Regency.

2. Method

The data collection was conducted through a delayed-mode image, field survey, and interviews with various informants. The primary data were obtained from fish and shrimp cultivators through interviews supported with a list of structured questions (questionnaires) to obtain the required information. Meanwhile, the secondary data were obtained from the related institutions, in this case, from the Fisheries Office, the Regional Development Planning Agency (*Badan Perencanaan Pembangunan Daerah/Bappeda*), and Central Bureau of Statistics (*Badan Pusat Statistik/BPS*) of Indramayu Regency. The sub-districts under study are Cantigi, Indramayu, Kandanghaur, karangampel, Krangkeng, Lohbener, Losarang, and Pasekan. The first is data processing conducted with google earth (delayed data SPOT 5 and Landsat 8 TM), and further processed to determine the cultivation locations. Due to the interpretation results, the validation points are determined and then field survey was conducted using GPS handheld Garmin. In addition, field documentation and interviews with informants were eventually performed.

3. Result and Discussion

In general, all coastal districts Indramayu potential for the development of pond farming. Currently, there are business activities of shrimp and shrimp pond farming, vaname shrimp and milkfish. Traditional systems managed by the community as a monoculture and polyculture. The condition of fishpond location in Indramayu Regency is the high of pond embankment of 0.75-1 m and the embankment of the pond 0.5-1 m, the area of the pond varies 1.5-5 ha / plot with 1 door, the water depth of the pond 0, 3-0.5 m with an average of 0.4 m [15].

Determination of suitable location for the aquaculture activities should be done carefully by considering the condition of land and water (biological, physical and ecological), socio-economic condition and infrastructure support. Placement of suitable aquaculture activities will impact on the reduction of negative influence on the environment, to minimize conflicts of interests among land users, and to maximize the production of the farm produced [16] and the cultivation of shrimp and fish in Indramayu majority still run traditionally. Cultivators are still less responsive to technology [4].

3.1. Shrimp

Windu shrimp is one of the superior commodities in Indramayu Regency because it has a relatively high selling value with bigger shrimp size. The total area for shrimp cultivation in Indramayu Regency has reached 8,594.60 ha. Tiger shrimp production target of increasing the average is 10% per year, while shrimp vaname 17% per year [17]. The sub-districts which have the largest shrimp ponds in Indramayu Regency are Cantigi, Losarang and Pasekan. Sub-district Pasekan with major commodity shrimp, designated as a development area Minapolitan Cultivation in Indramayu in 2012 [8].

Fattening segmentation is the most preferable segment selected by the Windu shrimp cultivators because the cultivation is relatively simple and appropriate with the regional pond water characteristics in Indramayu Regency. This commodity is cultivated in monoculture while the cultivation scale for Vaname shrimp is included into the simple cultivation scale. Until now, the sub-districts which cultivate Windu shrimp include Pasekan, Sindang and Patrol.

Table 1. The comparison of cultivation areas for Windu shrimp ponds.

No.	District	Land Potential (Ha)*	Result of Dgitation Area (Ha)
1	Cantigi	2,171.22	2,775.00
2	Pasekan	1,507.00	1,375.00
3	Losarang	2,017.00	2,391.20
4	Indramayu	531.18	736.00
5	Sindang	716.00	698.00
6	Krangkeng	352.00	339.30
7	Arahan	714.06	755.00
8	Kandanghaur	207.00	230.00
9	Lohbener	259.70	104.50
10	Balongan	74.96	51.90
11	Juntinyuat	19.50	3.73
12	Karangampel	22.00	5.98
13	Patrol	2.00	2.17
	Total	8,594.60	9,467.78

Table 1 above shows the comparison between the potential cultivation areas for windu shrimp [9] with the digitization result on pond areas in Indramayu sub-district. Simultaneously, there is an addition potential pond area by 8,594.60 hectares. However, based on the digitization results on windu pond area is equal to 9467.78 hectares. The addition of shrimp cultivation pond areas is predicted taking place due to the addition of Fishery Household (FH) in the related sub-district or the existence of other cultivators who switch into cultivating Windu shrimp. The greatest pond area increase is experienced by Cantigi sub-district. Most pond area increase takes place because the related sub-district has the Fishery Household (FH) for milkfish cultivation made utilizing poly-culture system with Windu shrimp that the number of Fishery Households (FH) increases, from 476 to 707.

IOP Conf. Series: Earth and Environmental Science **149** (2018) 012049 doi:10.1088/1755-1315/149/1/012049

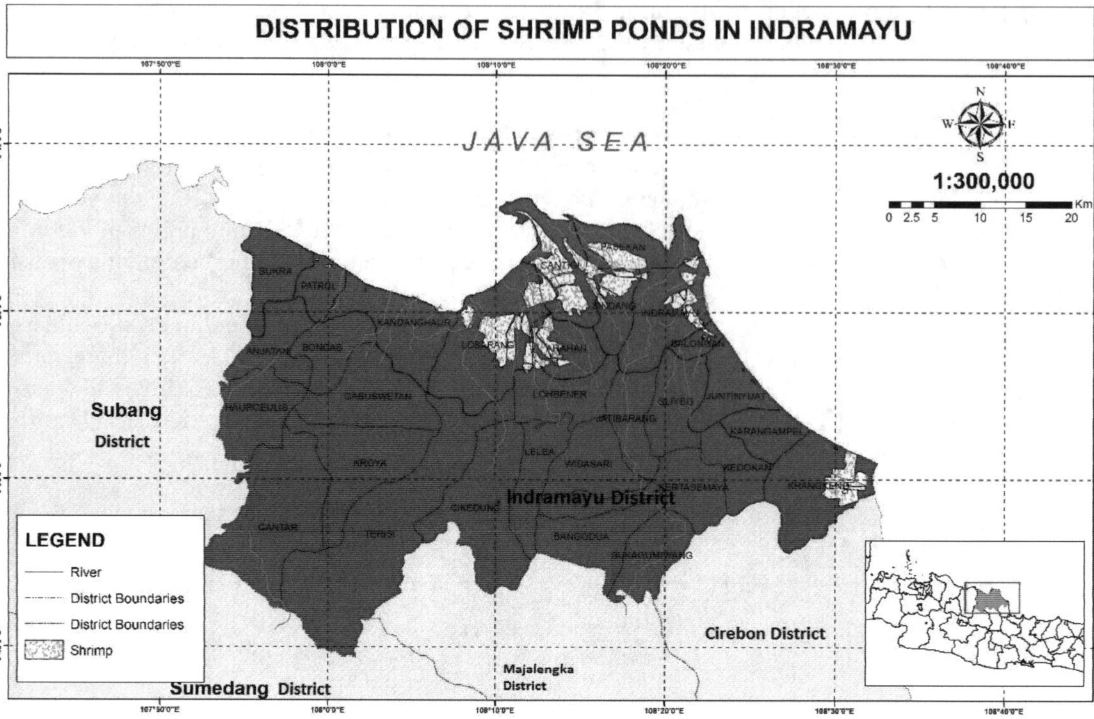

Figure 1. Distribution of Shrimp ponds.

The decreasing number of pond area takes place in most sub-districts, covering Pasekan, Sindang, Krangkeng, Lohbener, Balongan, Juntinyuat, and Karangampel (Figure 1) due to the dry season that the water sources are limited. In order to maintain the production, the cultivators made a poly-culture between Windu shrimp and milkfish to maximize the potential of pond areas and water resources. Thus, the number of Fishery Households for milkfish cultivation is more than that for Windu shrimp cultivation.

3.2. Milkfish

The milkfish cultivation area in Indramayu Regency has reached 12,063.08 hectares which is wider regarding to its good market share, cultivation which is relatively easier, and appropriateness to the pond water characteristics in Indramayu Regency when compared to the shrimp. The existing milkfish cultivation in this region belongs to the fattening segmentation with simple cultivation scale. The milkfish cultivation areas are spread in Krangkeng, Juntinyuat, Balongan, Sindang, Sukra, Patrol, Cantigi, Arahan, Losarang, and Indramayu sub-districts.

Based on table 2 which is regarding to the comparison of milkfish cultivation area, it shows that the potential of the whole pond area is 12,065.08 hectares and the result of digitization on milkfish cultivation area is 10,801.92 Ha. The difference of milkfish pond area between its potential and digitization is 1,263.16 Ha. This huge difference shows that almost all sub-districts have experienced pond area decrease, including Cantigi, Pasekan, Losarang, Indramayu, Sindang, Krangkeng, Arahan and Balongan. In addition, there is a seeding segmentation conducted in Pasekan sub-district by only 0.65 hectares.

The largest pond area decrease is in Cantigi sub-district due to its transition from monoculture to poly-culture cultivation. The milkfish Fishery Households make the poly-culture between milkfish and shrimp that the number of poly-culture fishery households is 707, while the milkfish fishery households is only 435. Similarly, the other sub-districts also experience pond area decrease. Meanwhile, the milkfish cultivation area experiences increase, such as in Kandanghaur, Juntinyuat, Sukra, and Patrol sub-districts with the greatest milkfish pond area addition is in Kandanghaur sub-district (Figure 2). It is proven that the number of fishery households for milkfish cultivation is more dominating than that for the shrimp which is only equal to 176 fishery households.

Table 2. The Comparison of Milkfish Cultivation Area.

No	Distirct	Land Potential (Ha)	Result of Digitation Area (Ha)
1	Cantigi	4,092.00	3,505.00
2	Pasekan	3,223.00	2,892.00
3	Losarang	2,519.00	2,320.00
4	Indramayu	977.79	908.50
5	Sindang	444.00	377.00
6	Krangkeng	12.00	11.92
7	Arahan	309.84	274.00
8	Kandanghaur	346.00	454.30
9	Balongan	123.50	49.60
10	Juntinyuat	4.00	6.00
11	Sukra	11.00	11.50
12	Patrol	2.00	2.10
	Total	12,065.08	10,801.92

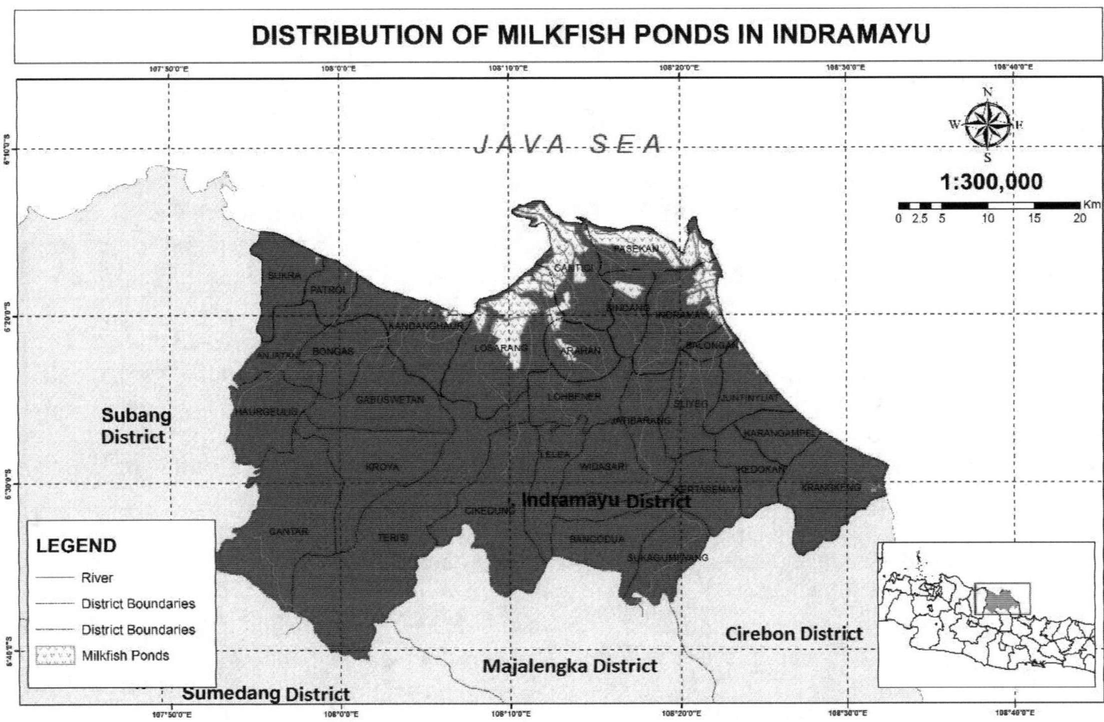

Figure 2. Distribution of Milkfish Ponds.

4. Conclusion
The results show that the data distribution of fishery households consists of 1727 fishery households for Windu shrimp polyculture, 1,551 milkfish fishery households, and 88 Windu shrimp fishery households. The pond area which may be utilized is 9,854.1 hectares. The milkfish pond area is 12,065.08 hectares, while the digitization result is 10,801.92 hectares.

Acknowledgments
Graceful thanks to the Maritime and Fisheries Affairs (Diskanla-Indramayu) for providing the required data. We also supported by our survey team: Dwima Derry, Fiqi Fadilah, Cuncun, Darmadi, and Fenta.

References
[1] Akbar J 2014 *Potential and challenge of cultivation the swamp fishgroup (black fish and white fish) in South Kalimantan* (Bandung: Unpad Press) p 233
[2] Rimmer, Sugama M K, Rakhmawati D, Rofiq R and Habgood R 2013 A review and SWOT analysis of aquaculture development in Indonesia *J. Reviews in Aquaculture* **5** 1-25
[3] Budiman A 2007 *Freshwater fish seed resources in Indonesia. In: Bondad-Reantaso MG (ed.) Assessment of Freshwater Fish Seed Resources for Sustainable Aquaculture* (Rome: Food and Agriculture Organization of the United Nations) p 341
[4] Suryanegara E and Hikmah 2012 The Relationship between Shrimp Tiger Budidaya (Panaeus monodon) with Milkfish (chanos chanos) in Indramayu District West Java *J. Bull. Sosek Kelautan dan Perikanan* **7** 35-40
[5] Tan E O, de Guzman D L, Darvin L C and Balgos M C 1984 *Milkfish research in the Philippines. In Juario J V, Ferraris R P, Benitez L V (Eds.) Advances in milkfish biology and culture, Iloilo City* (Philippines: Island Pub. House in association with the Aquaculture Department, Southeast Asian Fisheries Development Center and the International Development Research Centre) pp 171-181
[6] Huniyah A, Alamsyah M A and Pursetyo K T 2015 Financial Analysis of Fish Enlargement Milkfish (Chanos chanos) in Pond with Traditional Farming and Monoculture-Polyculture System in Mulyorejo Surabaya East Java *J. Ilmiah Perikanan dan Kelautan* **7** 169-176
[7] Tran N R U, Chan C, Phillips M, Mohan C, Henriksson P, Koeshendrajana S, Suri S and Hall S 2017 Indonesian aquaculture futures: An analysis of fish supply and demand in Indonesia to 2030 and role of aquaculture using the AsiaFish model *J. Marine Policy journal* **79** 25-32
[8] Sari S W, Nurmalina R and Setiawan B 2014 Efficency Permorfance of Fish Supply Chain in Indramayu District, West Java *J. of Management and Agribusiness* **11** 12-23
[9] Marine and Fishery Agency [DKP] 2014 *Statistic of Fishery Report in Indramayu District* (Jakarta: Statisctic of Fish Culture)
[10] Suharno I, Susilowati and Firmansyah 2017 Management of the traditional milkfish culture in Indonesia: an approach using technical efficiency of the stochastic frontier production *J. Bioflux* **10** 578-586
[11] Fauzi, Yulian, Susiloboko and Mayasari Z M 2009 Analysis of Land Suitability of Coastal Area of Bengkul Province Trough Spatial Model Design and Geographic Information System *J. Forum Geografi* **23** 101-111
[12] Ross L G, Handisyde and Nimmo D C 2009 Spatial decision support in aquaculture: the role of geographical information systems and remote sensing *J. New Tech. in Aquaculture* pp 707-749
[13] Yi D, Reardon T and Stringer R 2016 Shrimp aquaculture technology change in Indonesia: Are small farmers included? *J. Aquaculture* 1-10
[14] Herianto A S, Wastutinigsih S P, Foster D, Rimmer M and Callinan R 2010 Agricultural and fisheries extension in Indonesia – origins, transitions and current challenges *J. Extension Farming Systems* **6** 23-31

[15] Utojo A, Mustafa and Suhaimi R 2013 Characterization of Aquaculture Management Area in Gresik East Java *J. Riset Akuakultur* **8** 311-324

[16] Suplicy F M, Vianna L F N, Rupp G S, Novaes A L T, Garbossa L H P, Souza R V, Guzenski J, Costa S W, Silva F M and Santos A A 2017 Planing and Management For Sustainable Coastal Aquaculture Development in Santa Catarina State, south Brazil *J. Reviews in Aquaculture* **9** 107-124

[17] Wati, L A, Wen I, Chang and Mustadjab M M 2013 Competitiveness of Indonesian Shrimp Compare with Thailand Shrimp in Export Market *J. Wacana* **16** 24-31

Effects of oceanographic factors on spatial distribution of Whale Shark in Cendrawasih Bay National Park, West Papua

Maulida Ranintyari[1], Sunarto[2], Mega L Syamsuddin[2], Sri Astuty[2]
[1]Bachelor Degree, Marine Science Program, Faculty of Fisheries and Marine Science, Padjadjaran University, Jl. Raya Bandung-Sumedang Km. 21, Jatinangor, West Java 45363, Indonesia
[2]Department of Marine Science, Padjadjaran University, Jl. Raya Bandung-Sumedang Km. 21, Jatinangor, West Java 45363, Indonesia

E-mail: nairnamaulida@gmail.com

Abstract. Whale sharks are a leading species in Cendrawasih Bay due to its benign nature and its regular appearance. Recently, whale sharks are vulnerable to scarcity and even extinction. One of the efforts to maintain the existence of the whale shark population is by knowing its spatial distribution. This study aims to analyze how the oceanographic factors affect the spatial distribution of whale sharks in Cendrawasih Bay National Park. The method used in this research is descriptive with the quantitative approach using *the Generalized Additive Model* (GAM) analysis. The data consisted of the whale shark monitoring data in TNTC taken by WWF-Indonesia, and image data of sea surface temperature (SST) and chlorophyll-a concentration of Aqua-MODIS, and also sea surface current from Aviso. Analyses were conducted for the period of January 2012 until March 2015. The GAM result indicated that sea surface current was better than the other environment (SST and chlorophyll-a concentration) as an oceanographic predictor of whale shark appearance. High probabilities of the whale shark's to appear on the surface were observed in sea surface current velocities between 0.30-0.60 m/s, for SST ranged from 30.50-31.80 °C, and for chlorophyll-a concentration ranged from 0.20-0.40 mg/m³.

1. Introduction

Whale sharks are one of the most common marine species found in TNTC waters [1]. The whale shark (*Rhincodon typus*) is the largest fish in the world, it has slowly grown and slow process of genital maturity, the number of tillers that it produces relatively little and live longer. These characteristics make the whale shark range scarcity and even extinction if exploitation uncontrollably [2].

Currently the whale shark population has declined globally, and has made the whale shark into the vulnerable status of IUCN Red List. In addition, whale sharks also enter into Appendix II of CITES and Annex I issued by UNCLOS. In Indonesia, the Ministry of Marine Affairs and Fisheries of the Republic of Indonesia has established the status of the protection of whale sharks through the Decree of the Minister of Marine Affairs and Fisheries no. 18 of 2013.

Knowledge of whale sharks is very important, it is intended for the sustainability of living whale sharks. However, in Indonesia, research about whale sharks is relatively rare, especially whale sharks in TNTC. So still needed a study of whale sharks to add information and efforts in preserving the populations about this largest fish in their natural habitat. The purpose of this study was to analyze the effect of whale shark habitat seen from oceanographic factors in its spatial distribution.

Habitat whale sharks do not escape from the dynamic nature of oceanography. The occurrence of changes in a habitat will definitely affect the lives of biota living in the habitat. So we need a calculation that can calculate and analyze the relationship between variables in a non-linear manner. Analysis techniques that will be used in this research is using a statistic model of GAM (Generalized Addictive Model). GAM statistical modeling is a new modeling method that has strong strengths for non-linear approaches, so GAM can be conducted to support ecological-related research. Valavanis et al. (2008), the Generalized Additive Model (GAM) is one of the most common nonlinear statistical models used in the analysis of fisheries data [3].

2. Study area, material, and method

2.1. Study area

This study was conducted in August-December 2016 with study area is the waters of Teluk Cendrawasih that enter into the Cendrawasih Bay National Park area (figure 1). Geographically, Cendrawasih Bay National Park is located at coordinates 1°43'-3°22' S and 134°06'-135°10' E. Data processing is carried out in the Computer Laboratory of Fisheries and Marine Sciences Faculty, Padjadjaran University.

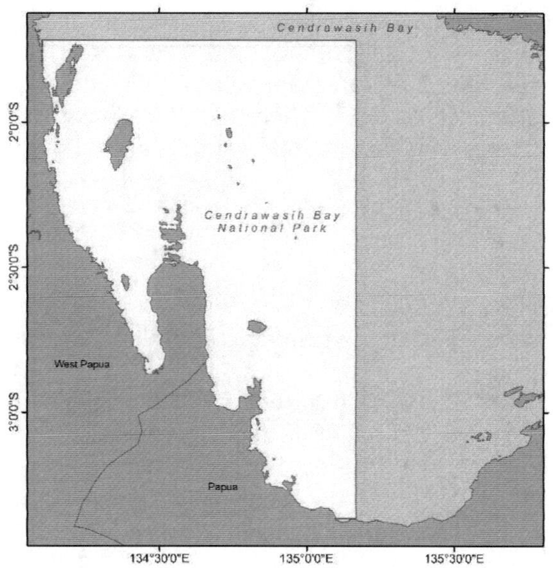

Figure 1. Study Area.

2.2. Material

Software used in this research shown in table 1.

Table 1. Software.

No.	Software	Utility
1.	ArcGis	Create a study area map Create merging of distribution of whale shark with oceanographic factors
2.	R	GAM analysis
3.	SeaDas	Data extracting from. TIFF format
4.	Ms. Excel	Data control process

A material used in this research shown in table 2.

Table 2. Material.

No.	Dataset (2012-2015)	Resolution		Source
		Temporal	Spatial	
1.	Sea Surface Temperature (SST)	2012-2015	4 Km	www.oceancolor.gsfc.nasa.gov
2.	Chlorophyll-a	2012-2015	4 Km	www.oceancolor.gsfc.nasa.gov
3.	Sea Surface Current	2012-2015	Wind Direction	www.aviso.altimetry.fr
4.	Point of whale shark's appearance	2012-2015	Point of occurrence	WWF Indonesia TNTC

2.3. *Generalized Additive Model (GAM)*
In general, the nature of oceanography and whale sharks that are dynamic, difficult to control and unpredictable, make the necessary calculations that can know the correlation between the two. Statistical calculations that have a non-linear approach can be the solution. This research data will be processed using statistical models of GAM (Generalized Additive Model). GAM static modeling calculations use the mgcv package contained in software R, using the following formula:

$$g(u_i) = \alpha_0 + S_1(X_{1i}) + S_2(X_{2i}) + S_3(X_{3i})$$

Where:
G: The link of function
A: The expected value of the dependent variable (Whale Sharks's Appearance)
α_0: The model constant
$\chi_{1,2,3}$: Parameter values (Sea surface current-SST-Chlorophyll-a)
Sn: A smoothing function for each of the model covariates

We used the GAM model to determine the relation between parameters of oceanography to the distribution of whale sharks in TNTC. GAM modelling is done by using Gaussian distribution and identity link function [3]. GAM model formation starts with every single explanatory variable, followed by a combination of two and three explanatory variables.

2.4. *Comparative analysis*
The results of the calculations using statistical modelling were analysed descriptively to get a descriptive comparison between oceanographic factors to the spatial distribution of whale sharks. Comparative descriptive analysis was also conducted to determine which oceanographic factors have a greater influence by comparing the oceanographic factors to the spatial whale shark distribution.

3. Results and discussion

3.1. *Whale sharks distribution*
The data used to determine the spatial and temporal distribution of whale sharks in TNTC were collected by fisherman and TPHP WWF Indonesia TNTC from 2012 to 2015. On the table 3, the number of observations made each month was not the same means observation was not done every day. This causes the number of occurrences of whale sharks recorded each month to differ so that the number of whale sharks from 2012 to 2014 was much different. The difference in the number of whale sharks can also be caused by the behavior of whale sharks.

IOP Conf. Series: Earth and Environmental Science **149** (2018) 012050 doi:10.1088/1755-1315/149/1/012050

Table 3. Whale Shark's Occurrence in TNTC.

Month	2012			2013			2014			2015		
	D	A	D/A	D	A	D/A	D	A	D/A	D	A	D/A
Jan	31	610	20	18	115	6	9	24	3	14	39	3
Feb	29	580	20	28	396	14	16	54	3	11	23	2
Mar	31	853	28	31	248	8	16	27	2	5	17	3
Apr	30	544	18	30	156	5	17	7	0	-	-	
May	23	258	11	30	156	5	19	48	3	-	-	
June	27	91	3	17	99	6	19	0	0	-	-	
July	18	35	2	19	81	4	21	0	0	-	-	
Aug	8	14	2	17	55	3	21	0	0	-	-	
Sept	21	44	2	26	100	4	16	0	0	-	-	
Oct	19	104	5	10	30	3	10	167	17	-	-	
Nov	8	15	2	4	27	7	18	57	3	-	-	
Dec	11	15	1	5	21	4	9	33	4	-	-	
Total	**256**	**3163**		**235**	**1484**		**191**	**417**		**30**	**79**	

D: number of observations (days); A: number of occurrences (whale sharks); D/A: average daily

The appearance of whale sharks to sea level was related to the behaviour of whale sharks. Whale sharks will swim and occurrence on the sea surface to search for food, but the whale shark will also dive into the water column to follow where the food goes. Generally, whale sharks spend time in shallow waters less than 50 m or near the surface because whale sharks are plankton-eating sharks and are animal filter feeders. According to Brunnschweiler *et al*. (2009), the tendency of whale shark to dive in the deep waters allegedly to follow the movement of food or to detect the condition of a water [2].

The number of occurrences of zero (0) in a given month could be said that whale sharks at the same time are swimming in the water column, so monitors did not see whale sharks appearing on the surface. The absence of whale shark's recording may be due to non-monitoring at any given time. Many factors that cause monitoring could not be done. In addition to knowing the number of whale sharks occurring over a period of time, the observations made in the TNTC region also showed the point of whale sharks. At one point the emergence of whale sharks, not only shows an individual that appears, but can represent more than one individual that appears (figure 2). This showed that whale sharks aggregate or stay in the same place within a few days.

The presence of differences in the point of occurrence could be due to the whale shark was an active swimmer. This was evidenced by the telemetry study by Stewart (2014), that there are six individual whale sharks attached to satellite markers showing movement in and out of TNTC [4].

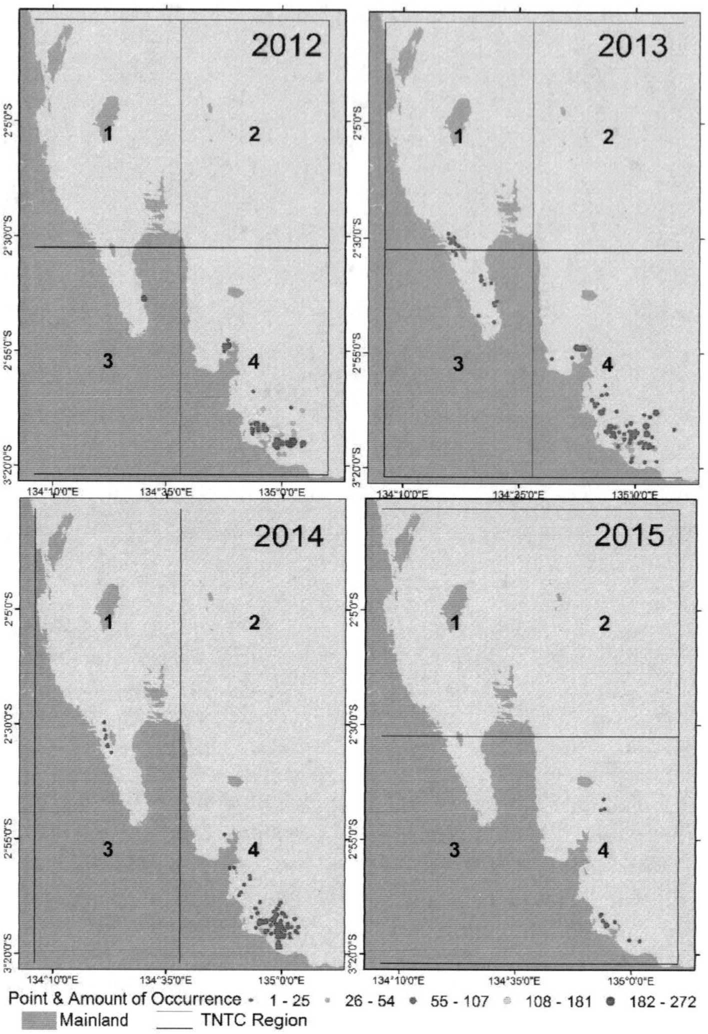

Figure 2. Point of Whale Shark's Occurrence.

3.2. GAM analysis

GAM model forming techniques began with one explanatory variable with a combination of two, and three explanatory variables or vice versa. In the model calculation results, there was a value of deviance and value of Akaike's Information Criterion (AIC). These two values indicate the level of accuracy of the explanatory variables in explaining the variation of response variables in each GAM model equation. The greater the value of deviance and the smaller the AIC value means the higher the accuracy of the GAM model in explaining the variation of response variables [3,5].

The GAM model equation by combining the three explanatory variables has the highest deviance value and the lowest AIC value. This shows that the combination of GAM models with the three explanatory variables has a high degree of accuracy. The significance level of each explanatory variable is grouped by Verzani (2005). The results of significance obtained in each explanatory variable are extremely significant and highly significant (table 3).

Table 4. GAM Results.

Model	Var.	Dev.	*P-Value*	AIC
s(SST) +s(Chl-a) +s(Current)	Current	3.14	0.000096***	0.0000000862
	SST	2.69	0.000354***	0.000000356
	Chl-a	1.85	0.001563**	0.00025
Deviance (%)			6.16	
*** *Extremely Significant*			** *Highly Significant*	

At the time of formation of GAM model equations, there was a smoothing curve function used to model the relationship between response variables and their explanatory variables. The calculation result of smoothing curve estimation on each explanatory variable can be shown in graphical form using *the plot command* in program Rb (table 4).

The estimation graph of the smoothing curve consists of the x and y axis. The x-axis shows the values of the explanatory variables of the oceanographic parameter, and the y-axis shows the result of smoothing against the corresponding response variable value. The black lines on the graph show the results of GAM calculations visualized in graphical form, while the dashed black lines represent the 95% confidence interval. The red horizontal line represents the zero limit of data, where the data below the zero limit can be interpreted as the absence of correlation between the response variable and the explanatory variable or it can be said that the range of values is not the optimum value of the whale shark to surface (figure 3).

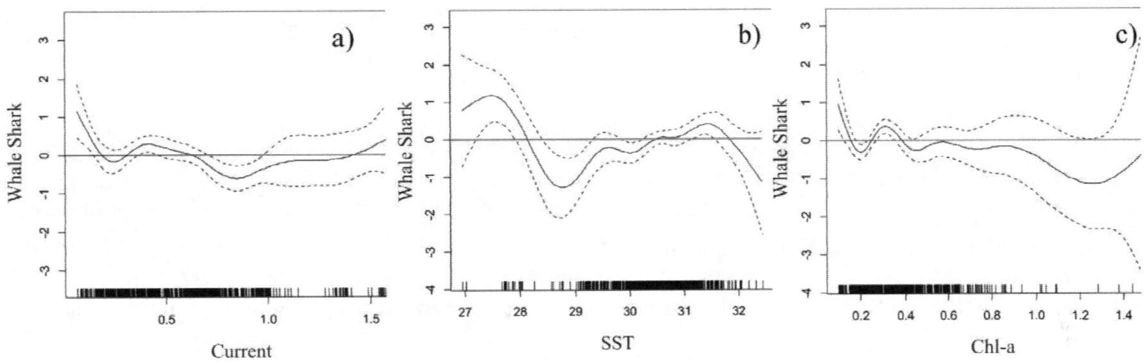

Figure 3. Effect of the Three Oceanographic Variables.

Whale sharks are plankton-eating animals. They may follow prey, such as various plankton, which are associated with high chlorophyll-a concentration [6]. Besides phytoplankton, there are many categories of zooplankton and small fishes and squids, which are also prey of the whale shark, and quantifying the prey in several areas where sharks pass through is needed [6].

The variable of chlorophyll-a concentration showed the lowest significance value when compared with other explanatory variables. Although whale sharks target many plankton as food, but whale sharks are more targeted zooplankton as its prey, so the analysis results showed smaller numbers of significance. In addition, the chlorophyll-a pigment is only owned by phytoplankton as the primary producer, so there was no direct contact between phytoplankton and the whale shark.

The occurrence of whale sharks in coastal waters is believed to coincide with productivity events that provide an ample supply of zooplantonic/larval food [7]. Our results strongly suggest that whale sharks target zooplankton as their primary food source in Bahia de Los Angeles and forage using several strategies that appear to be linked to prey abundance and composition [8].

In contrast to chlorophyll-a, the SST variable yields a very strong significance value, but is not as strong as a sea surface current variable. SST is believed to affect marine life, as well as the distribution of whale sharks. The cause of SST is not the most influential variable first, perhaps because whale sharks can tolerate a wide range of temperatures.

Whale shark dive to over 979.5 m, and can tolerate a temperature range 26.4 ^{0}C [9]. According to Colman (1997), a fairly large distribution of whale sharks can also show that whale sharks are tolerant of temperature, although whale sharks still favour warm temperatures [4]. The distribution of whale sharks is found in warm tropical and subtropical waters (18-30 ^{o}C) between 30° N and 30° S, except in the Mediterranean Sea [4].

Compared with SST and chlorophyll-a, surface current variables have the greatest effect on the distribution of whale sharks in TNTC. Tracking data show that the direction of travel in three sharks was influenced by prevailing geo-strophic currents [10]. The dominant whale shark's behaviour was near the surface of the sea, allegedly used by whale sharks to reach high-productivity areas. However, in the previous discussion, it appears that the movement of whale sharks was not affected by the speed and direction of surface currents. It was important to note that the whale sharks' data used were not data from whale shark tagging, so the movement of individual whale sharks was unknown.

Overall, the results of GAM analysis used to determine the magnitude of the influence of the three explanatory variables on response variables can only explain 6.16%. This value shows a small value. The low deviance value of the resulting GAM equation is thought to be due to the small number of datasets used. Although the data set used amounted to 1887, but the amount of time used was inconsistent in each year. In addition, the coordinates used in the dataset are coordinates that are almost always the same at each time, not the coordinates that cover the entire TNTC region.

Based on GAM plots, it could be seen the range of values of the explanatory variables that had a positive influence on the response variable. This positive influence showed the range of explanatory variable values favored by whale sharks to rise to the surface of the ocean. Positive effects for surface current variables ranged from 0.30 to 0.60 m / s. Positive effects of SST variables ranged from 30.50 to 31.80 °C, whereas for the variable chlorophyll-a concentration ranged from 0.20-0.40 mg / m^3.

3.3. *Prediction of whale sharks in TNTC*

Prediction of whale shark distribution in TNTC was spatially conducted to determine the number of individual whale sharks likely to surface or visit one of the areas in TNTC. This prediction calculation used the *predict. gam* function in the mgcv package that is inputted and calculated using the software R. The result of *predict. gam* function calculation is the number of individuals who have the opportunity to appear at the same point with the inputted data. The predictions map presented only consists of two months, where the month is the highest average and the lowest average appearance of whale sharks for four years.

In March, predictions show that about 2-4 individuals of whale sharks appear in all areas of TNTC. In March 2012, almost all areas of TNTC are predicted to be visited by 3-4 individuals of whale sharks. Didn't much different from March 2012, March 2013-2015 is predicted to be visited by 2-3 individuals whale sharks. In each year, whale sharks are predicted to often visit area 4 or in areas close to the mainland.

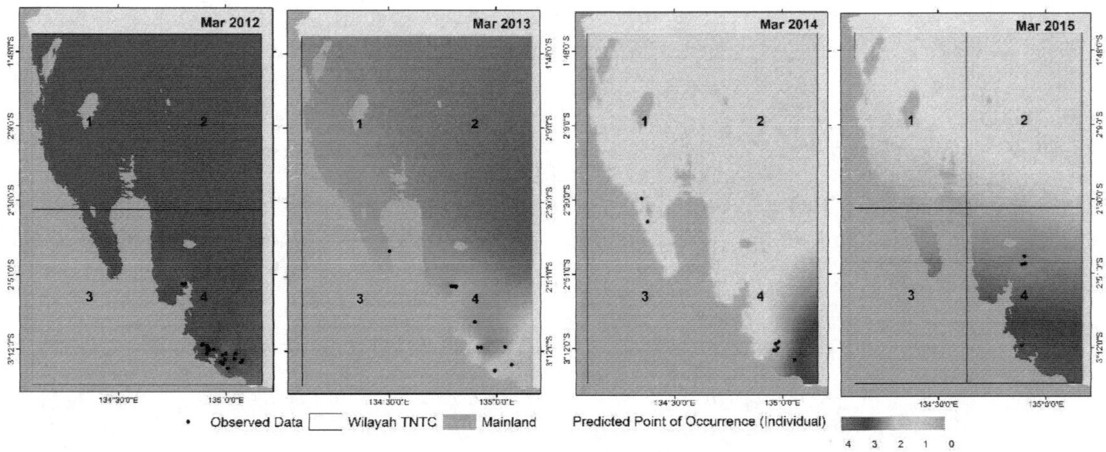

Figure 4. Prediction of Whale Shark's Occurrence on March.

Did not much different from March, August prediction shows 2-3 individuals whale sharks will appear in region 4. In August 2012, it's predicted that about 3 individual whale sharks will visit region 4, whereas in other regions only predicted about 2 individuals. In contrast to August 2012, August 2013 predicted whale sharks will visit the entire region of TNTC with the same number of individuals, which are about 2-3 individuals (figure 4).

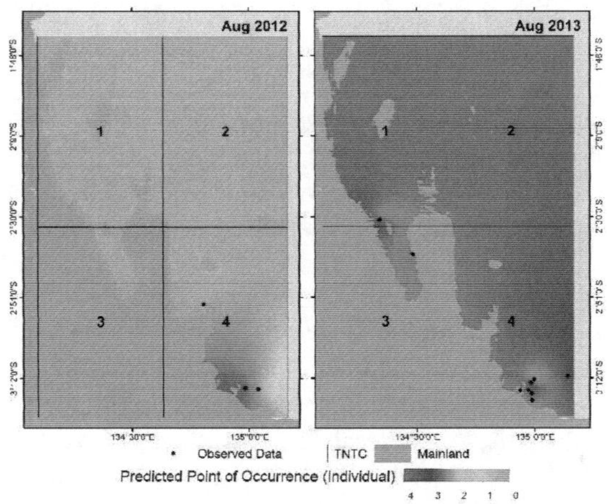

Figure 5. Prediction of Whale Shark's Occurrence on August.

This prediction results are similar to the results of monitoring conducted by TPHP WWF Indonesia TNTC. Whale sharks would still always visit region 4 as its favourite territory when compared to other regions. However, there was a little difference in March 2013, whereby whale sharks predicted dominant visiting in areas close to the high seas compared to areas close to the mainland (figure 5).

The differences could be attributed to the location of the monitoring and behaviour of the whale shark itself. Monitoring locations didn't cover the entire territory of TNTC made the lack of data in certain locations. The behaviour of whale sharks that are active but dominant swimmers make use of ocean surface currents, making the reason strong enough if whale sharks are constantly migrating and migrating for food or for finding new habitat locations.

Predicted results for both months showed similar results with monitoring results. The predominant individual is predicted to appear or visit on region 4 or in areas close to the mainland. This is related to the depth of the waters adjacent to the different land with deep water depths with the land. Whale sharks

IOP Conf. Series: Earth and Environmental Science **149** (2018) 012050 doi:10.1088/1755-1315/149/1/012050

are known to favour shallow waters, which in the shallow waters of whale sharks can find their prey with ease, as well as warmer temperatures in the waters near land than those farther offshore.

4. Conclusion

Based on the results of the study and discussion, it can be concluded that the sea surface current has the high significant value on the emergence of whale shark Rhincodon typus in TNTC region, compared to sea surface temperature (SPL) and chlorophyll-a concentration. The range of oceanographic factors favoured by the whale shark to appear on the sea surface, respectively, is the velocity of sea surface current between 0.30-0.60 m/s; SPL between 30.50-31.80°C, and chlorophyll-a concentration between 0.20-0.40 mg/m^3.

References

[1] Stewart B S 2011 *Workshop and Monitoring Training for Whale Sharks in Cendrawasih Bay National Park, West Papua 2-7 May 2011 Nabire, Papua* (San Diego: Hubbs-Seaworld Research Technical Report 2011) 375

[2] Direktorat KKJI 2015 *Pedoman Umum Monitoring Hiu Paus di Indonesia* (Jakarta: Direktorat KKJI Ditjen KP3K KKP RI) 8

[3] Wibawa T A 2011 Pemanfaatan Data Satelit Oseanografi untuk Prediksi Daerah Potensial Penangkapan Tuna Mata Besar (Thunnus obesus) di Samudra Hindia Selatan Jawa-Bali *J. Segara* **7** 1-12

[4] Tania C and Noor B A 2014 *Pemantauan Hiu Paus di Taman Nasional Teluk Cendrawasih* (Wasior: WWF Indonesia)

[5] Zuur A F, Ieno E N, Walker N J, Saveliev A A and Smith G M 2009 *Mixed Effects Models and Extensions in Ecology with R* (New York: Spinger)

[6] Hsu H H, Joung S J, Liao Y Y, Liu K M 2007 Satellite Tracking of Juvenile Whale Sharks, Rhincodon typus, in the Northwestern Pacific *Fisheries Research*

[7] Sleeman J C, Meekan M G, Fitzpatrick B J, Steinberg C R, Ancel R and Bradshaw C J 2010 Oceanographic and Atmospheric Phenomena Influence The Abundance of Whale Sharks at Ningaloo Reef, Western Australia *Journal of Experimental Marine Biology and Ecology* **382** 77-81

[8] Nelson J D and Eckert S A 2007 Foraging Ecology of Whale Sharks (Rhincodon typus) within Bahia de Los Angeles, Baja California Norte, Mexico *Fisheries Research* **84** 47-64

[9] Graham R T, Roberts C M and Smart J C 2006 Diving Behavior of Whale Shark in Relation to a Predictable Food Pulse *Journal of The Royal Society Interface* **3** 109-116

[10] Rowat D and Gore M 2007 Regional Scale Horizontal and Local Scale Vertical Movements of Whale Sharks in The Indian Ocean Off Seychelles *Fisheries Research* 32-40

IOP Conf. Series: Earth and Environmental Science **149** (2018) 012051 doi:10.1088/1755-1315/149/1/012051

Phytoplankton community in lake Ebony, Pantai Indah Kapuk, North Jakarta

N T M Pratiwi, I P Ayu, S Hariyadi, D Mulyawati, A Iswantari
Department of Aquatic Resources Management, Faculty of Fisheries and Marine Science, Bogor Agricultural University, Indonesia

Email: niken_tmpratiwi@yahoo.com

Abstract. Lake Ebony is an ornamental lake in coastal area of North Jakarta, located at 6°6'18"S-6°6'35"S and 106°44'39"E-106°44'56"E. Phytoplankton community in Lake Ebony lives in high organic materials received from domestic waste. A spatio-temporal observation at five sites was carried out to understand the spatial distribution of phytoplankton at each group of time of observation and the succession of phytoplankton. Spatial analysis was carried out to map the distribution pattern of plankton,using ArcGIS 10.1 with IDW (Inverse Distance Weighted) interpolation method. Spatial clustering was determined by Canberra Index. The succession of phytoplankton was shown by graph of Frontier succession models, SDI (rate of succession), and SIMI. There were two clustered groups of site. Based on graph of Frontier succession, phytoplankton in Lake Ebony was at Stage 2 and 3 with the rate of succession ranged from 0.008 to 0.003, and value of SIMI ranged from 0.68 to 0.97. There was different spatial distribution pattern of phytoplankton in three groups of observation time, with low rate of succession.

1. Introduction

Lake Ebony is one of ornamental lake in residential place of Bukit Golf Mediterania, Pantai Indah Kapuk (PIK), North Jakarta (6°6.572'S and 106°44.755'E). The lake is laid in northern part of PIK, facing Java Sea. This shallow waters roles as water pool and part of polder system, and completed with sewage water treatment plant (STP) for residential wastewater came through it and surrounded by a supporting channel as addition drainage to throw water directly to the sea. The channel receives water from STP and residence ditches. Unfortunately, the overflow from the channel, with high organic content, enters to the lake.

High organic content of wastewater leads to increasing of nutrients for phytoplankton growth. The organic materials are decomposed by heterotrophic bacteria becoming inorganic nutrients compounds or bioavailable nutrients. Those are nutrients that can be directly absorbed by phytoplankton.

The performance of phytoplankton growth could change along with the variation of water quality [1-4] since of its species specific requirement. The change of nutrients concentration will effect on the quantity or density of phytoplankton. Furthermore, the change of nutrients composition, especially N:P ratio will effect on the quality or species composition of phytoplankton.

Hence, the composition, density, and the biomass of phytoplankton varied spatially and temporally or succession [5,6]. Therefore, succession shows the influence of water quality towards phytoplankton community [7]. Then, the succession is the change of species relative abundance of a phytoplankton community [5] or the change of dominance species, the biomass, and the density of phytoplankton [2].

2. Material and Methods

2.1. Data collection

The study was carried out in one year with monthly observation, from April 2014 to March 2015. The water samples for phytoplankton and water quality, including nutrient analysis were taken from five sampling sites of Lake Ebony, Bukit Golf Mediterania, Pantai Indah Kapuk (PIK), North Jakarta (Figure 1). One year observation, based on [5], is adequate to study temporal variation of phytoplankton.

The principles of sampling technique, measurement, and water analysis were following [8]. The water quality parameters comprised of temperature, transparency, conductivity, pH, salinity, DO, ammonium, nitrite, nitrate, and orthophosphate. Meanwhile, the identification of phytoplankton was based on [9-13].

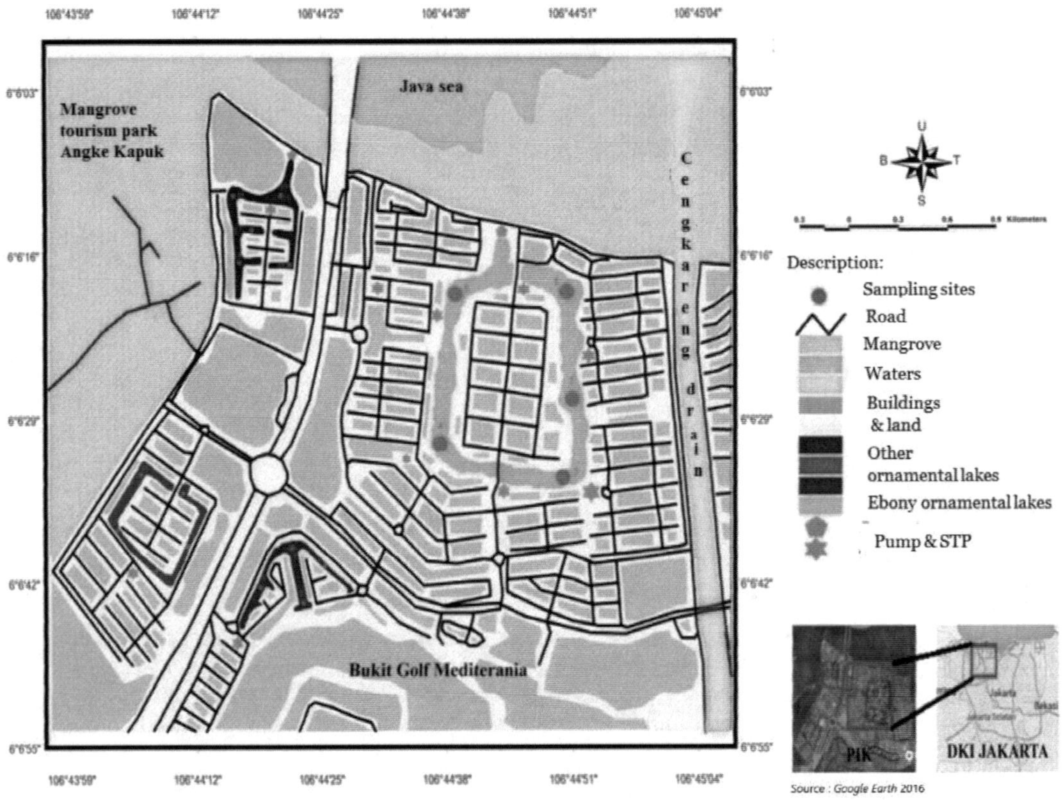

Figure 1. Location of sampling site in Lake Ebony.

2.2. Data analysis

The community structure of phytoplankton was determined by the calculation of several diversity indices, such as Shannon-Wiener diversity index, Evenness index, and Simpson dominance index [14]. Furthermore, the density of phytoplankton and water quality were analyzed spatially and temporally. The spatial analysis of phytoplankton was meant to study the distribution of species density; and the temporal analysis was meant to study the phytoplankton succession.

There was a pre-analysis to determine group of observation time, using Stander's Similarity Index (SIMI) [15]. For each group of observation time, there was analysis of phytoplankton distribution. There was also a spatial analysis to determine group of sampling sites using Canberra Similarity Index [16] based on water quality data.

Another spatial analysis was estimating phytoplankton distribution using ArcGIS 10.1 software with interpolation method. Interpolation method is a mathematical function to estimate values of locations with lack of data. Contextually, in ArcGIS spatial analysis, this method is used to give a prediction of each raster pixel values based on limited data of sampling sites.

The interpolation method in this study is IDW (Inverse Distance Weighted). This method uses an assumption that the result of interpolation is more similar to the nearest sampling site data than the furthest. According to [17], IDW method has higher accuracy than other methods, such as kriging or spline method.

There were also other analyses to understand the succession of phytoplankton in Lake Ebony. The analyses comprise of plot of Frontier's graph models [18] and the rate of succession, using Summed Difference Index or SDI [5].

3. Results

3.1. Phytoplankton composition and density

Figure 2 and 3 illustrates one year variation of phytoplankton densities and number of species in Lake Ebony, spatially and temporally. The phytoplankton density of Site 4 is very much lower than that of other sites, while the number of species of each site was relatively similar. In temporal, monthly condition of those two biological parameters relatively varied.

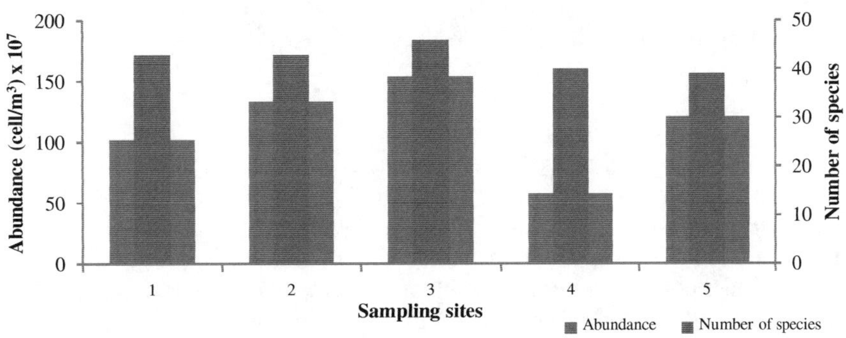

Figure 2. Spatial density and number of species of phytoplankton in Lake Ebony.

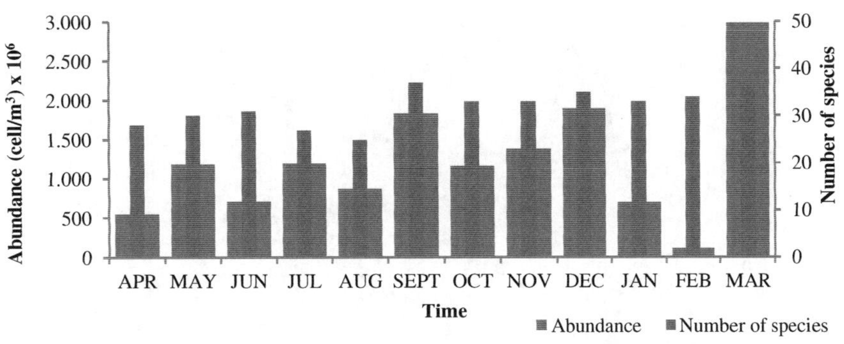

Figure 3. Temporal density and number of species of phytoplankton in Lake Ebony.

There were found five class of phytoplankton with 48 species in total; seven species belong to Cyanophyceae, 14 species of Bacillariophyceae, 20 species of Chlorophyceae, 4 species of Euglenophyceae, and 3 species of Dinophyceae. The monthly composition of phytoplankton density and number of speciesvaried (Figure 4a and 4b). The density of phytoplankton was mainly dominated by

Chlorophyceae, but the highest number of species was Cyanophyceae, followed by Chlorophyceae. The community structure of phytoplankton was characterized as freshwater influenced by saline water.

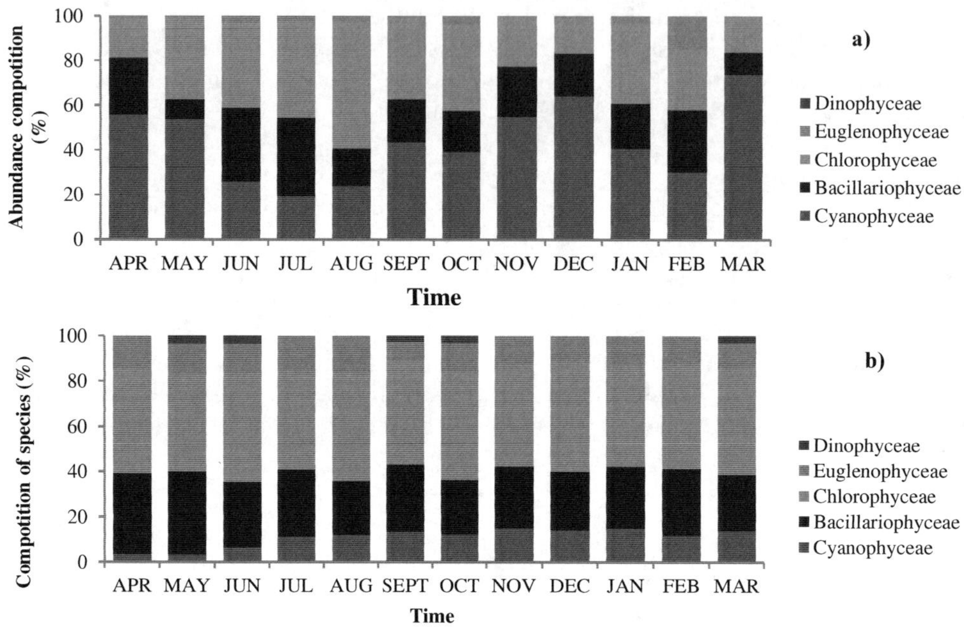

Figure 4. Composition of phytoplankton; a) density and b) number of species.

3.2. Nutrients and water quality

Nutrients condition in Lake Ebony and water quality varied temporally (Figure 5 and Table 1). Quite similar pattern occurred between ammonia and ammonium, and between nitrate and nitrite. The phosphorous also fluctuated, but in different level of concentration. The lowest turbidity occurred in the least depth of waters. However, the light transparency was relatively low.

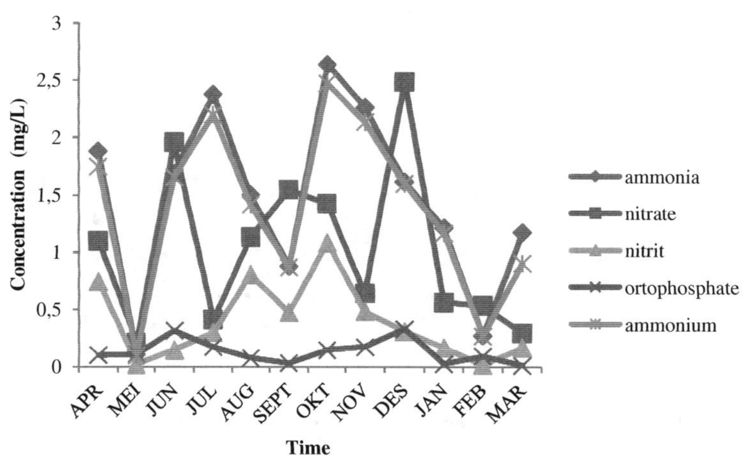

Figure 5. Nutrients concentration of Lake Ebony.

Table 1. Water quality of Lake Ebony.

Parameters	Temperature	DO	Depth	Transparency	Turbidity	Conductivity	pH	Salinity
Time	°C	mg/L	cm	Cm	NTU	µmHos/cm	-	PSU
Apr	30±0.4	6±1.2	99±2.1	25±5	48±20.0	3808±216	8±0.0	1±0.9
May	31±0.3	4±1.5	98±1.7	27±7	84±69.7	3567±121	8±0.5	1±0.6
Jun	30±0.4	4±1.6	101±1.9	30±3	89±154.0	2734±262	7±0.7	1±0.4
Jul	30±0.2	4±1.3	96±1.2	23±6	138±134.5	4201±196	8±0.3	2±0.0
Aug	30±0.4	6±2.0	95±1.8	27±5	47±13.6	3403±281	8±0.3	1±0.9
Sept	30±0.3	5±0.7	108±1.4	28±2	113±54.4	4620±370	7±0.4	2±0.4
Okt	30±0.4	5±2.8	105±1.3	15±2	57±6.6	5289±142	8±0.0	2±0.6
Nov	31±0.4	5±1.2	74±2.5	14±3	26±11.1	5140±806	8±0.1	2±0.0
Dec	30±0.6	5±0.9	92±2.3	24±2	36±5.7	3709±539	7±0.3	2±0.0
Jan	29±0.3	4±0.9	81±1.3	25±5	42±9.7	4439±375	7±0.8	2±0.4
Feb	26±0.4	4±0.3	97±0.8	25±4	56±27.3	1231±296	7±0.4	1±0.0
Mar	31±0.7	9±4.2	97±1.8	21±4	75±30.9	3581±638	8±0.7	1±0.9

3.3. Spatial distribution of phytoplankton

The density and composition of phytoplankton varied, both spatial and temporal. As a whole, there were 48 species of phytoplankton that were found in Lake Ebony. The most abundant species was *Merismopedia* (Cyanophyceae), followed by *Cyclotella* (Bacillariophyceae), and *Scenedesmus* (Chlorophyceae). In group of class, Cyanophyceae and Chlorophyceae dominated.

In order to get the description of spatial distribution of phytoplankton, there was a pre-analysis of temporal cluster, resulting in three groups of time (Figure 6), depicting . Then for those three groups there were three illustrations of phytoplankton distribution in Lake Ebony waters (Figure 7a, 7b, and 7c).

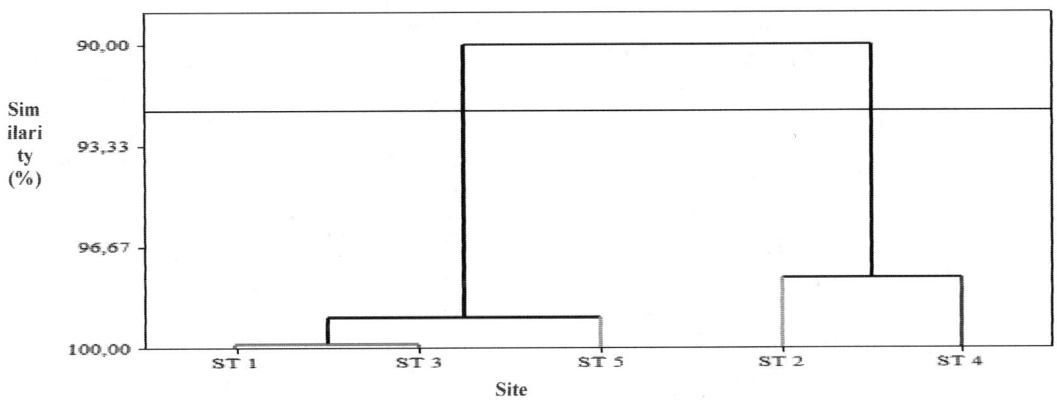

Figure 6. Temporal clustering of Lake Ebony.

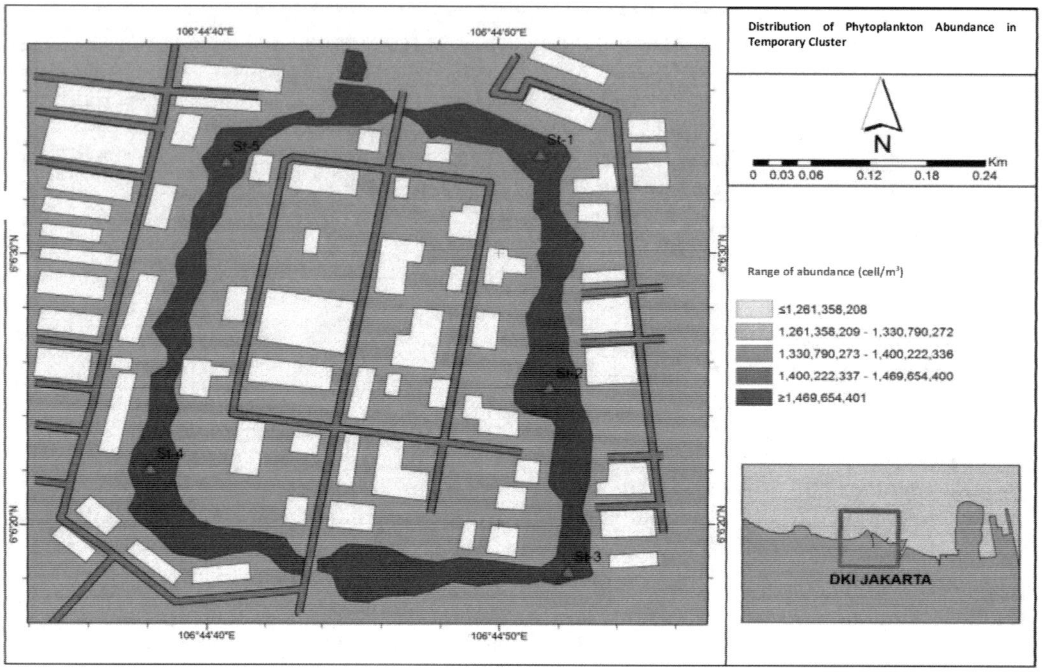

(a)

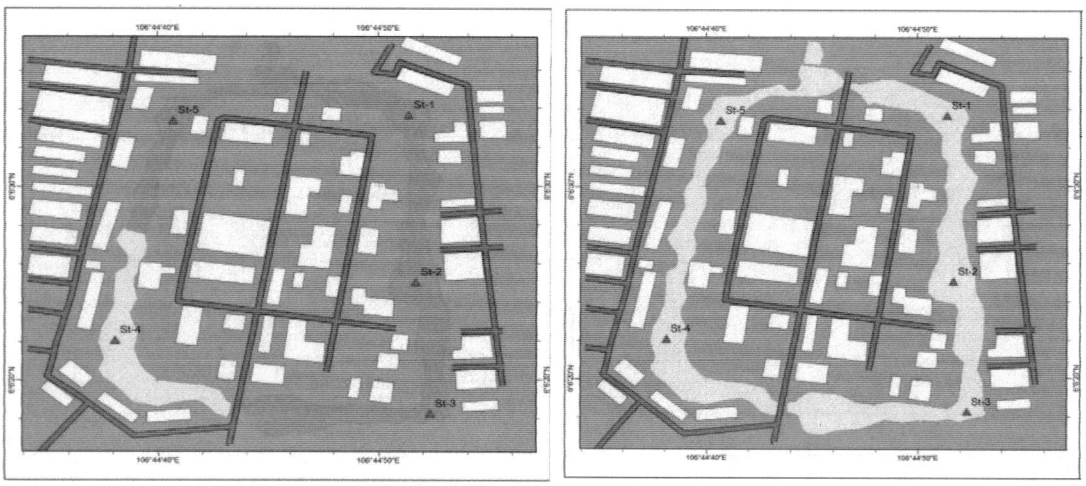

Figure 7. Phytoplankton distribution of Lake Ebony at a) Time 1, b) Time 2, and c)Time 3.

3.4. Succession pattern of phytoplankton

The varied density and composition of phytoplankton indicated that there was a succession phenomenon. In order to understand the phenomenon, preliminary analysis was conducted, to get an illustration onspatial zonation (group of sampling sites) of Lake Ebony first. The result shows that there were two groups (Zone 1 and Zone 2) of sampling sites (Figure 8).

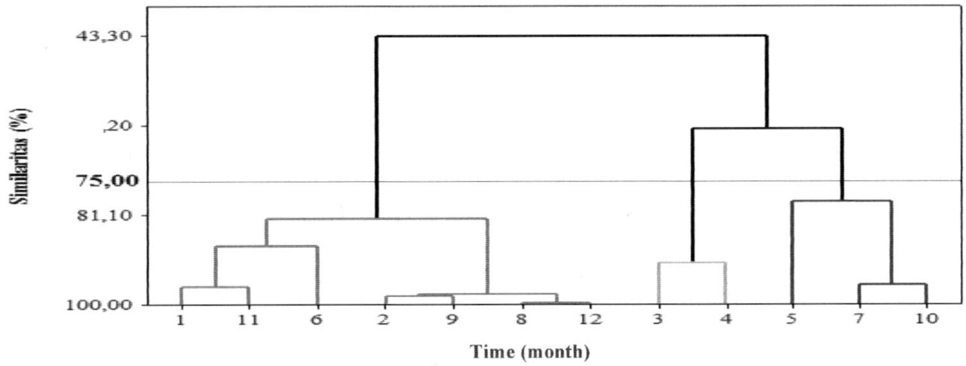

Figure 8. Group of sampling sites of Lake Ebony.

Furthermore, the analyses of succession wereperformedfor each of those groups. The first analysis was rank-frequence Frontier Graph, as the illustration of succession pattern resultingin two stadia of succession, Stadium 2 for Zone 1 and Stadium 3 for Zone 2 (Figure 9).

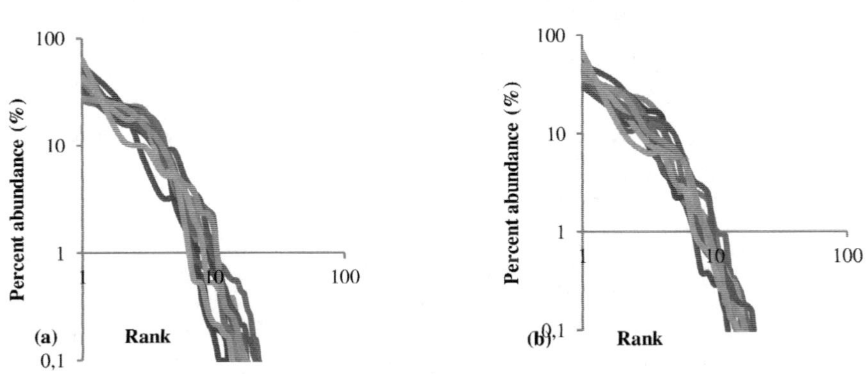

Figure 9. Succession pattern on rank-frequence plot: (a) Stadium 2; (b) Stadium 3.

The rate of succession and the inter-time difference of species for the two zones were illustrated in Figure 10a and 10b. It is shown that the rate of phytoplankton succession of Zone 2 was higher than of Zone 1. Furthermore, the relatively less inter-time difference of species occurred in Zone 1.

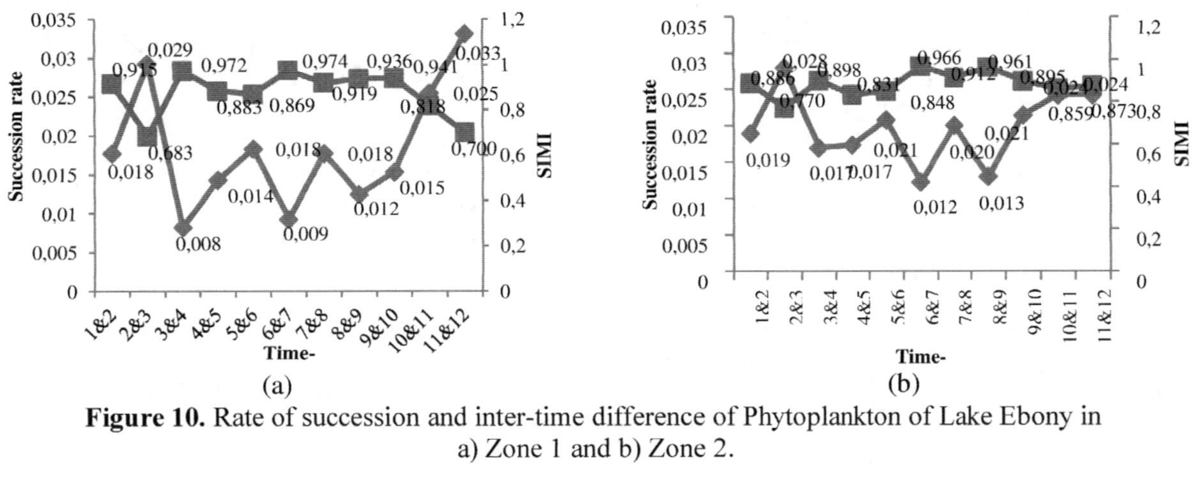

Figure 10. Rate of succession and inter-time difference of Phytoplankton of Lake Ebony in a) Zone 1 and b) Zone 2.

3.5. Community structure of phytoplankton

Diversity Index usually used to describe the community structure of phytoplankton. The most common diversity indexes are Shannon-Wiener diversity index (H'), Evenness index (E) and Simpson dominance index (C). The value of those indexes for phytoplankton in Lake Ebony is laid in Table 2. Nodominance species occurred.The level of evenness was low to moderate, and low diversity.

Lake Ebony receives overflow from supporting channel with high content of organic materials that will be transformed into inorganic nutrients in decomposition process. The nutrients will be utilized by phytoplankton.

High density and number of species at most of sampling sites indicates that the nutrients have high concentration. High density of Cyanophyceae also indicates high content of organic materials. In this condition, Cyanophyceae appears in high content of organic materials and high concentration of nutrients or eutrophic condition. This condition is supported by [19-22] which mentioned that Cyanophyceae predominant in eutrophic waters.

The spatial distribution shows different pattern in time. Furthermore, the density was higher in the middle of the time of observation. It is presumed that it was correlated to weather or season. In detail, each of time periods has different profile of phytoplankton distribution among sites or sampling sites. At the Time 1 and 2, the phytoplankton was distributed evenly in all sampling sites. Meanwhile, at the Time 3, the phytoplankton was less concentrated in Southwest part of the waters. This was presumed that the wind direction influences the position of concentrated population of the community. In this case, the wind blew to the East part of waters.

Table 2. Diversity Index of phytoplankton (a) Zone 1 and (b) Zone 2 in Lake Ebony.

Time (a)	**APR**	**MAY**	**JUN**	**JUL**	**AUG**	**SEP**	**OCT**	**NOV**	**DEC**	**JAN**	**FEB**	**MAR**
H'	1.32	1.47	1.73	1.71	1.98	1.88	2.03	1.65	1.76	2.08	2.05	1.31
E	0.40	0.48	0.55	0.55	0.62	0.52	0.59	0.47	0.51	0.60	0.58	0.38
C	0.39	0.33	0.23	0.23	0.18	0.22	0.18	0.29	0.26	0.18	0.18	0.43
Time (b)	**APR**	**MAY**	**JUN**	**JUL**	**AUG**	**SEP**	**OCT**	**NOV**	**DEC**	**JAN**	**FEB**	**MAR**
H'	1.36	1.47	1.82	1.76	1.92	1.77	1.96	1.44	1.70	2.07	1.84	1.27
E	0.46	0.46	0.55	0.57	0.63	0.53	0.58	0.44	0.49	0.62	0.55	0.37
C	0.36	0.37	0.21	0.22	0.20	0.24	0.20	0.40	0.30	0.17	0.22	0.47

The community structure has synchronization with the succession stadium. The stadium 2 and 3 were characterized by less dominance of species [18], but the phytoplankton community was relatively stable. It means that the composition of the community will not change rapidly with the change of water quality or other environmental conditions.

Community structure in Stadium 2 shows a stable condition, lowering diversity, with relatively high productivity [18]. The well-developed phytoplankton in Lake Ebony were Cyanophyceae (*Microcystis* sp.,*Merismopedia* sp., *Anabaena* sp.), Bacillariophyceae (*Melosira* sp., *Stephanodiscus* sp., *Cyclotella* sp.), Chlorophyceae (*Scenedesmus* sp., *Pediastrum* sp., *Chlamydomonas* sp.). Those species are highly tolerated to high concentration of organic materials [23].

With Stadium 2 and 3, the phytoplankton community shows low rate of succession, hence there was a relatively similar species inter-time of observation. The density was not significantly different, and the predominance species was not change inter-time of comparison.

4. Conclusion

There were 48 species of phytoplankton in Lake Ebony from five classes, Cyanophyceae, Bacillariophyceae, Chlorophyceae, Eulenophyceae, and Dinophyceae. The most abundant species was from Cyanophyceae as indicator of eutrophic waters. There was different spatial distribution pattern of phytoplankton in three groups of observation time, with low rate of succession.

References

[1] Tilman D, Kilham S S, Kilham P 1982 Annu. Rev. Ecol. *Evol. Syst.* **13** 349-72
[2] Wetzel RG 2001 *Limnology: Lake and River Ecosystems* 3rd edition London: Academic Press
[3] Phlips E J, Badylak S, Grosskopf T 2002 *Estuar Coast. Shelf Sci.* **55** 385–402
[4] Khan TA 2003 *Limnologica* **33** 327-39
[5] William M, Lewis JR 1978 *Am. Nat.* **112** 401-14
[6] Ariyadej C, Reungchai T, Pimpan T, Saowapa A 2004 *Songklanakarin J. Sci. Technol.* **26** 595-607
[7] Sommer U Z, Maciej G, Winfried L, Annie D 1986 *Arch Hydrobiol* **106** 433-72
[8] APHA 2012 *Standard Methods for Examination of Water and Wastewater* 22nd ed Washington
[9] Edmonson W T 196 *Fresh Water Biology*2nd ed Washington: John Wiley & Sons Inc
[10] Prescott G W 1970 *The Freshwater Algae* Dubuque Iowa: Brown Company Publishers
[11] Belcher H, Erica S 1978 *A Beginner's Guide to Freshwater Algae* London: Her Majesty's Stationery Office
[12] Mizuno T 1979 *Illustrations of The Freshwater Plankton of Japan*, revised ed Osaka: Hoikusha Publishing Co
[13] Yamaji IE 1979 *Illustrasion of The Marine Plankton of Japan* Osaka: Hoikusha Publishing Co.
[14] Krebs C J 1972 *Ecology: The Scientific Study of the Interactions that Determine the Distribution and Abundance of Organisms* New York: Harper & Row Publisher
[15] Stander J M 1970 *Diversity and Similarity of Benthic Fauna off Oregon* Thesis Corvallis: Oregon State University
[16] Legendre L, Legendre P 1983 *Numerical Ecology* Amsterdam: Elsevier Scientific Publishing Company
[17] Pramono G 2008 *J. Forum Geografi* **22** 145-58
[18] Frontier S 1985 *Diversity And Structure in Aquatic Ecosystems* London: Aberdeen University Press
[19] Rawson D S 1956 *Limnol. Oceanogr.* **1** 18-25
[20] Henderson-Seller B, Markland H R 1987 *Decaying Lakes—The Origins and Control of Cultural Eutrophication* New York: John Wiley&Sons
[21] Mur L R, Skulberg O M, Utkilen H 1999 *Cyanobacteria in the Environment: Toxic Cyanobacteria in Water-A Guide to their Public Health Consequences, Monitoring and Management* London: WHO
[22] Robarts R D, Tamar Z *Journal of Marine and Freshwater Research* **21** 391-99
[23] Mason C F 1993 *Biology of Freshwater Pollution* 2nd ed New York: John Wiley &Sons Inc

Mapping shallow waters habitats using OBIA by applying several approaches of depth invariant index in North Kepulauan Seribu

V P Siregar*, S B Agus, T Subarno, N W Prabowo

Department of Marine Science and Technology, Bogor Agricultural University, Kampus IPB Darmaga, Bogor 16680, West Java, Indonesia

E-mail: vingar56@yahoo.com

Abstract. The availability of satellite imagery with a variety of spatial resolution, both free access and commercial become as an option in utilizing the remote sensing technology. Variability of the water column is one of the factors affecting the interpretation results when mapping marine shallow waters. This study aimed to evaluate the influence of water column correction (depth-invariant index) on the accuracy of shallow water habitat classification results using OBIA. This study was conducted in North of Kepulauan Seribu, precisely in Harapan Island and its surrounding areas. Habitat class schemes were based on field observations, which were then used to build habitat classes on satellite imagery. The water column correction was applied to the three pairs of SPOT-7 multispectral bands, which were subsequently used in object-based classification. Satellite image classification was performed with four different approaches, namely (i) using DII transformed bands with single pair band input (B1B2), (ii) multi pairs bands (B1B2, B1B3, and B2B3), (iii) combination of multi pairs band and initial bands, and (iv) only using initial bands. The accuracy test results of the four inputs show the values of Overall Accuracy and Kappa Statistics, respectively 55.84 and 0.48; 68.53 and 0.64; 78.68 and 0.76; 77.66 and 0.74. It shows that the best results when using DII and initial band combination for shallow water benthic classification in this study site.

1. Introduction

The benefits of remote sensing technology in providing spatial information is undoubted, given the few advantages it offers. Remote sensing has the ability to record information from a large area with a relatively short time so that can save costs. In coastal and marine areas, remote sensing has been widely-used, for example, to extract information from shallow waters. Shallow waters in the context of remote sensing are the water column where light can penetrate to the bottom, generally at a maximum depth of 25 m [1].

Coral reefs and seagrasses are the main components of shallow water ecosystems. In addition to the two major ecosystems, the complexity of shallow water habitat plays an important role in providing ecological roles in the marine environment. Detailed spatial information on the composition of shallow water habitat covers is certainly required in planning and managing coastal and marine areas. However, the limitations of available information and the uncertainty of the methods and schemes used in mapping shallow water ecosystems are factors that still need attention. The uncertainty of the method and the classification scheme may be caused by several things, including the accuracy resulting from satellite image extraction.

The water column is one of the contributing factors that determine the accuracy of satellite image classification results [1]. The attenuation of energy propagates in the water column thus affecting the information received by the satellite sensor. On the other hand, an area and others have diverse characteristics that have different attenuation coefficients. This is certainly difficult when it will apply a similar model in various waters environment. Correction of the water column is sometimes necessary to improve the accuracy of extraction of shallow shallow water information. One method of water column correction is known and widely applied in shallow water mapping, the Depth Invariant Index (DII) introduced by Lyzenga [2]. The advantage of such method is not necessarily in situ measurement to obtain attenuation coefficient. By using a combination of band pairs, we can estimate the value of the ratio of the attenuation coefficient of the two bands (k_i/k_j).

Generally, the classification approach of remote sensing imagery can be grouped into two, pixel-based and object-based. Pixel-based classification is a technique that has been commonly known, so it can be categorized as a conventional technique. In recent years, object-based classification is quite popular and widely used in remote sensing data processing. Anggoro, et al. [3] for example, using OBIA to map the geomorphological zone of shallow waters in Pari Island. OBIA techniques are also widely used to map shallow water habitats as well as coral reefs [4-6].

The application of remote sensing for mapping shallow water habitats, water column correction such as DII transformation is commonly used in pixel-based classification. The using of water column correction on OBIA is still limited to the literature that discusses the method. Among the studies that applying water column correction on object-based classification was Wahidin, et al. [6] which obtains 56 - 73% classification accuracy from Landsat image for mapping coral reef ecosystem in Morotai Island with 7 classes.

The purpose of this study was to determine the ability of object-based classification on SPOT-7 imagery to map shallow water habitats in the northern of Kepulauan Seribu by using several scenarios. The scenarios are (i) using DII transformed bands with single pair band input, (ii) multi pairs bands, (iii) combination of multi pairs band and initial bands, and (iv) only using initial bands.

2. Method

2.1. Study Site

This research was conducted in North of Kepulauan Seribu, there are five small islands included in research location, that are Harapan Island, Kelapa Island, Kelapa Dua Island, Kaliageh Island, and some areas of Panjang Island. Harapan and Kelapa are two islands that side by side and connected each other. Administratively, the research location is included in the North Kepulauan Seribu sub-district, Kepulauan Seribu Regency - Jakarta. Based on geographical position, the study site extends between 106°33'9.3"E and 5°38'35"S – 106°35'44"E and 5°39'58.39"S. Three of the five islands in this study site are inhabited islands consisting of 2 villages, namely Harapan Island and Kelapa Island. Figure 1 shows the location of the study sites.

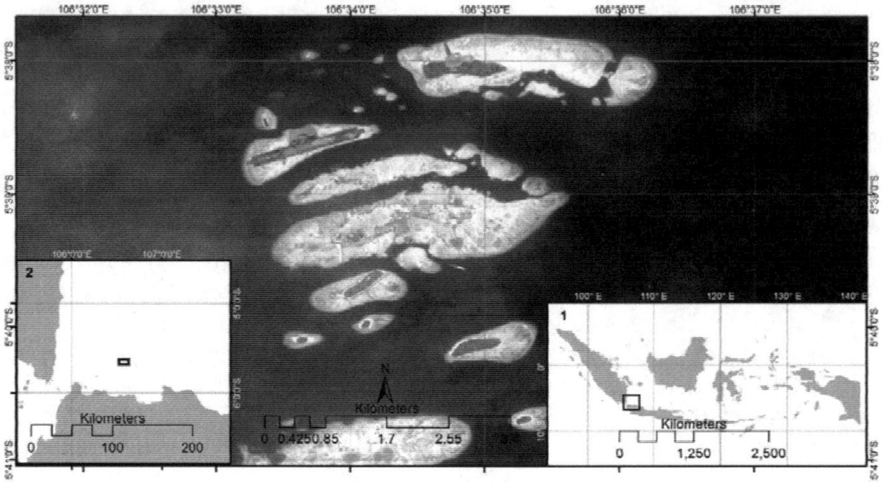

Figure 1. Study site located in the north of Kepulauan Seribu, Indonesia.

2.2. *Data Collection*

The main data used was a SPOT-7 multispectral image, consisting of four multispectral bands, blue, green, red and NIR bands. The imagery was acquired on 12 of June 2016. The image has spatial resolution on multispectral bands of 6m, and 1.5m on panchromatic bands (Table 1). In addition, field data were used as training and validation of image classification results. Field data were obtained from a field survey conducted in June and November 2016.

Table 1. The characteristics of SPOT-7 imagery

Band	Wavelength	Spatial Resolution
Band 1 (Blue)	485 nm	6 m
Band 2 (Green)	560 nm	6 m
Band 3 (Red)	660 nm	6 m
Band 4 (NIR)	825 nm	6 m
Panchromatic	597,5 nm	1,5 m

Field data were collected using systematic random sampling method, with the distance between sampling points about 20m. The determination of the observation location to represent the study area was undertaken during the designing of the field survey, by visual interpretation of the WorldView-2 imagery. We did not use SPOT-7 for designing field survey because in that time the imagery not available yet. Data collection considers the diversity of shallow water benthic covers and the spatial resolution of satellite imagery used in the classification. Field data collection was assisted by using a square transect and GPS Trimble GeoXT 2008. Each data point was equipped with transect photographs using an underwater camera to facilitate post-processing and categorization of each data set.

2.3. *Data Analysis*

Digital image processing was done by applying pre-processing and analysis through classification. Digital image processing began by applying some basic corrections, then classifications were performed in two levels, level 1 to separate land, shallow and deep waters, then level 2 to classify shallow waters into several habitat classes. Generally, the processes of satellite image processing performed is presented in Figure 2 below.

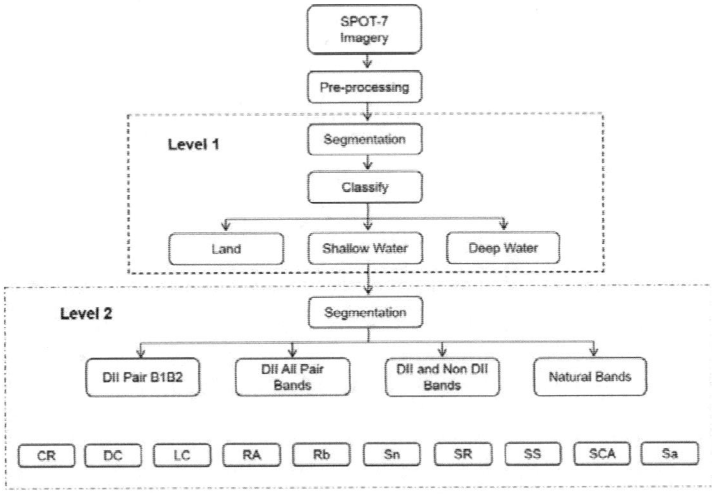

Figure 2. Hierarchy of classification using OBIA in this study

2.3.1. Pre-processing

Pre-processing was done on field data and satellite image data. Field observation data were then grouped into the habitat classes. The field data categorization then obtained the result of habitat class definition as many as 10 class. The habitat classes are used, i.e coral with rubble (CR) live coral (LC), dead coral (DC), rubble with algae (RA), rubble (Rb), sand (Sn), sand with rubble (SR), sand with seagrass (SS), sand coral and algae (SCA), and seagrass (Sa). Initial correction applied on SPOT-7 image was atmospheric correction using FLAASH (Fast Line-of-Sight Atmospheric Analysis of Spectral Hypercubes) method, geometric correction, then image cropping to focus on the study area.

Water column correction used was the method of Depth Invariant Index (DII) developed by Lyzenga [2]. The method has been widely known and widely used in shallow water mapping. The DII equation is written as:

$$DII = ln(L_i) - \left[\left(\frac{k_i}{k_j}\right)\right].\ln(L_j) \quad (1)$$

DII = Depth invariant index; Li = the radiance of band i; Lj = radiance of band j; and k_i/k_j = the ratio of the attenuation coefficients of band i and band j. The values of k_i/k_j were obtained from sample extraction of imagery from the substrate i.e. B1B2 0.5837, B1B3 0.2384, and B2B3 0.5029.

2.3.2. Image Classification

As shown in Figure 2, OBIA classification of SPOT image was performed with two levels. Level 1 was for land and deep water masking then takes only shallow waters for classification in level 2. In level 1, the classification was done by applying the threshold values of some transformations for 3 classes. The threshold values used are, land B1/B4<3; shallow waters B1/B3<1.3, and deep water B1/B3>=1.3. Classification at level 2 applied several approaches to obtain the most accurate classification results and to determine the effect of water column correction (DII) on the classification results. The input method used in level 2 i.e (i) using one pair band of DII; (ii), a combination of 3 pairs bands of DII; (iii) combination of 3 pairs bands of DII and initial bands; and (iv) using natural bands without DII transformation.

The method (i), the selection of DII band pairs based on the regression parameters of natural logarithm values of the two bands used, from 3 pairs of DII bands on SPOT-7 image, i.e pairs of B1B2, B1B3, and B2B3, used in the experiment (i) was B1B2. As for the method (ii) used the three pairs of DII band combinations. The relationship between DII band pairs is shown in Figure 3.

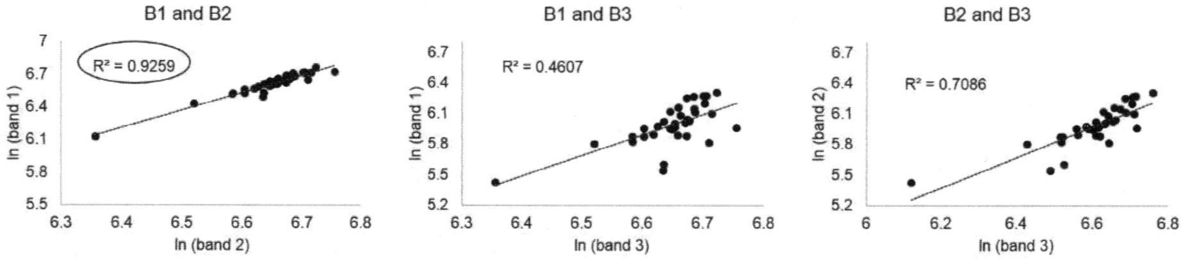

Figure 3. The relation between band pairs of DII from SPOT-7 imagery

Segmenting or grouping pixels into objects is one of the steps in object-based classification. The segmentation was applied multiresolution segmentation algorithm. The size of the object formed from the segmentation result is represented by the scale parameter of the object used in the segmentation process. Determination of parameter scale greatly affects the accuracy of classification [7], so that value is of concern in the classification of satellite imagery. In this study, after experimenting with some value of object scale, finally, we used the value of 40 for level 1 and 10 for level 2. In Table 2 can be seen the number and size of objects formed from the results of segmentation both in 2 levels.

Table 2. Scale parameters, number of objects, and object sizes from 2 levels segmentation

Level	Scale Segmentation	Number of objects	Area (m^2)		
			Min	Max	Mean
L1	40	532	341.14	460,033.86	22,964.39
L2	10	3764	42.64	7,419.66	1,119.00

2.3.3. Accuracy Assessment

The accuracy assessment was done to obtain the accuracy level of thematic map resulting from the four approaches used in classification. This accuracy assessment was done by constructing a confusion matrix, three general parameters obtained from the matrix, i.e overall accuracy (OA), producer's accuracy (PA), and user's accuracy (UA). OA is expressed as the ratio of the amount of data used as a validation sample and is well explained against the number of classes defined. UA is translated as the average probability (%) an object will be correctly classified and on average showing how well each class in the field has been classified. UA is the average probability (%) of an object of a classified image, actually representing the classes in the field [1]. The Kappa statistic value was also calculated from each classification result to compare the approach that gave better classification the benthic habitat.

$$Kappa\ (K) = \frac{N \sum_{j=1}^{r} - \sum_{i=1}^{r}(x_i + .x_{+i})}{N^2 - \sum_{i=1}^{r}(x_j + .x_{+i})} \quad (2)$$

IOP Conf. Series: Earth and Environmental Science **149** (2018) 012052 doi:10.1088/1755-1315/149/1/012052

3. Results and Discussion

Spatial distribution of classification results from shallow water habitat shows different pattern and extent of each method applied. The classification using only one pair of DII bands, the coral reef class shows significant differences from the other 3 methods. Figure 5 (c) shows the LC class was found only in some spots with a smaller area than other methods. Similarly, for seagrass class (Ss), both in methods (i) and (ii) indicate patterns of distribution that show significant differences. The seagrass class on the two initial approach classification seems to be spreading to the zone of reef slope.

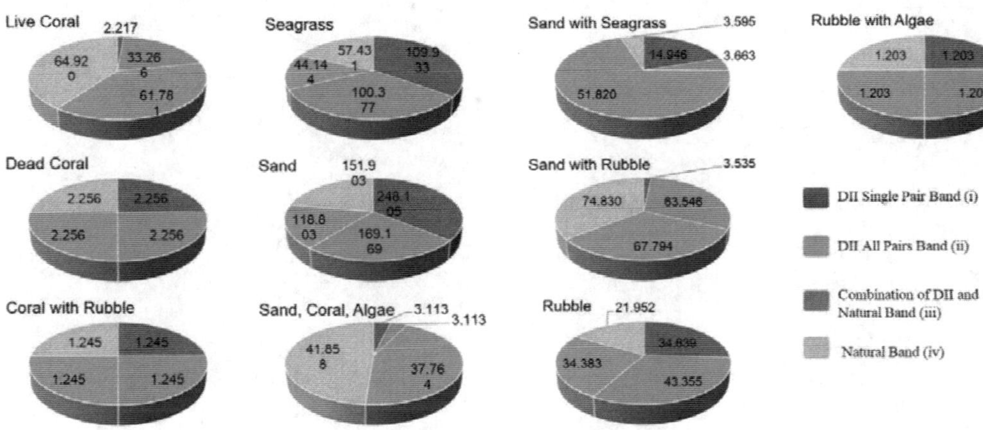

Figure 4. Area (ha) of habitat classes form different method used in classification

The area of classification result from the benthic habitat as shown in Figure 4 appears to vary according to the approach methods used. Habitat classes show significant differences in LC, SCA, SS, and SR. In the use of the first method (i), obtained LC, and SR classes with the smallest than 3 others methods, while SCA and SS obtained the smallest distribution when applying method (ii). From Figure 4 it is also known that the SS class was obtained by the largest area on the use of the method (iii). To determine the most optimal method of mapping the habitat class in this study area, it can be assessed based on the accuracy values presented in table 3 and table 4.

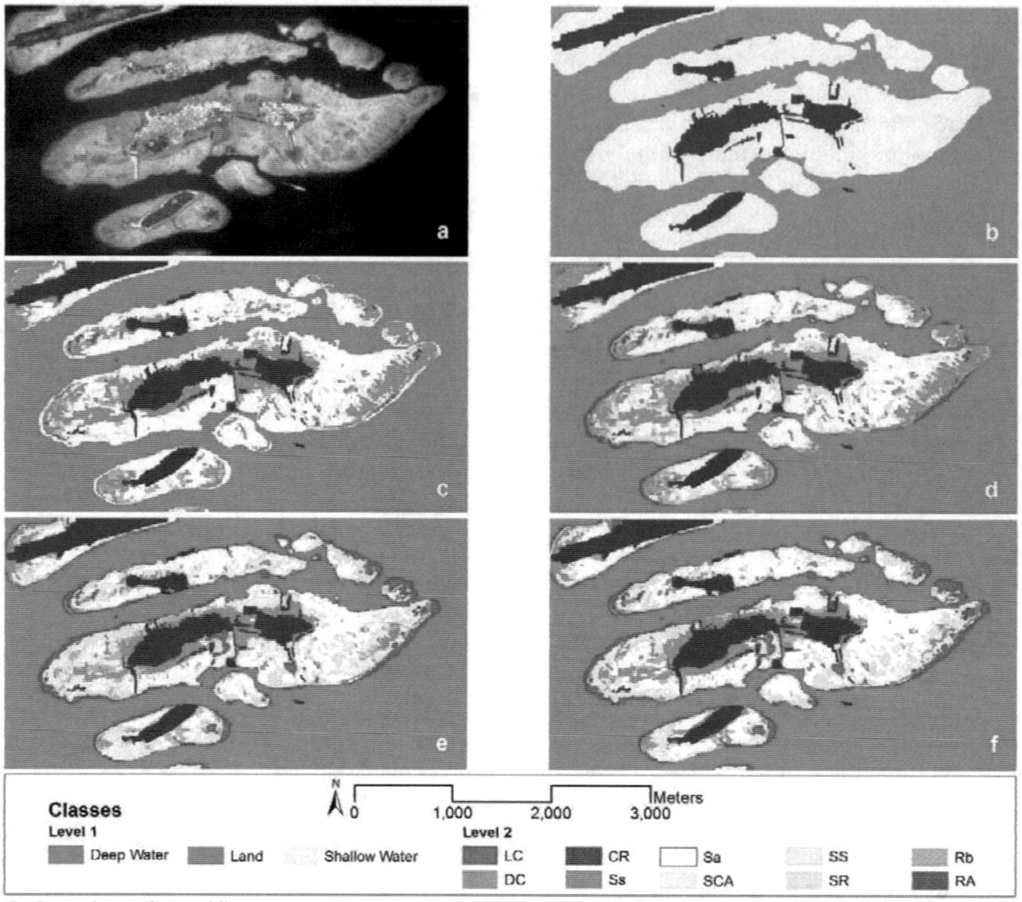

Figure 5. Overview of classification results by applying OBIA, (a) RGB composite of SPOT-7; (b) Classification in level 1; (c) Classification with DII single pair band; (d) Classification of DII all pairs bands (e) Combination of DII all pairs bands and initial bands; and (f) Classification using initial bands

The overall accuracy and Kappa statistics from the four methods applied, obtained the accuracy range 55.84 – 78.68 (%) for OA and 0.48 – 0.74 for Kappa values (Table 3). Table 3 shows the OA and Kappa statistics of each method applied, where the bold value indicates the method with the best accuracy of the four approach methods used. With the number of habitat class schemes of 10 classes, the OA value of 78.68% is considered to be passable to map shallow water habitats at the study site. Previously in the same location, research was done to map habitat with 11 classes and obtained OA value of 76.8% [8]. Another study conducted in the Kepulauan Seribu to map the benthic habitat with 7 classes using WorldView-2 imagery, obtained the accuracy of 72.38 - 84.29% with the use of DII and PCA transformations PCA [9].

Table 3. The overall accuracy (OA) and Kappa Statistic calculated form each classification approach

Method	OA (%)	Kappa
(i) DII Pair B1 and B2	55.84	0.48
(ii) DII All Pair Bands	68.53	0.64
(iii) Combination DII and non DII Band	**78.68**	**0.76**
(iv) Non-DII (Initial bands)	77.66	0.74

The UA and PA values in detail refer to the method used are presented in Table 4. Both UA and PA, the lowest values are found when using the single pair bands, the lowest UA on the Ss class, while the lowest PA found on CR class. The highest PA value on RA class for method (iv), while on the other three methods was highest in class Sa.

Table 4. User's accuracy (UA) and producer's accuracy (PA) values from each method used

Classes \ Methods	DII B1 and B2 (i)		DII All Pair Bands (ii)		Combination (iii)		Non DII (iv)	
	UA	PA	UA	PA	UA	PA	UA	PA
Live Coral (LC)	80.00	22.22	91.67	61.11	83.33	83.33	83.33	83.33
Dead Coral (DC)	100.00	33.33	100.00	47.62	100.00	61.90	100.00	66.67
Coral with Rubble (CR)	100.00	21.43	90.00	64.29	84.62	78.57	100.00	78.57
Seagrass (Ss)	34.69	77.27	44.74	77.27	81.82	81.82	73.91	77.27
Sand (Sa)	53.62	**84.09**	72.55	**84.09**	84.44	**86.36**	80.85	86.36
Sand, Coral and Algae (SCA)	100.00	30.00	100.00	50.00	60.00	60.00	58.33	70.00
Sand with Seagrass (SS)	56.00	58.33	69.57	66.67	71.43	83.33	70.83	70.83
Sand with Rubble (SR)	72.73	44.44	50.00	61.11	60.87	77.78	56.52	72.22
Rubble (Rb)	61.90	72.22	73.68	77.78	73.68	77.78	82.35	77.78
Rubble with Algae (RA)	100.00	50.00	71.43	62.50	100.00	75.00	87.50	**87.50**

The application of DII in classification by using only one pair of bands (i) obtained the classification results with low PA values for the class of LC, DC, CR, and SCA which only ranged from 21 - 30%. While 3 other methods gave better value, especially on method (iii) and (iv).

Based on PA values in Table 4, it shows that methods (iii) and (iv) consistently provide better classification results for the four methods used. The differences methods (iii) and (iv) are found in the SS and RA classes, where (iii) gave better results for the SS class, while in the RA class the method (iv) gave the better result in recognizing the class. Another situation was found that, using the combination DII bands and initial bands (iii) able to explain seagrass and sand mixed with seagrass (Ss and SS) with better results.

From the experimental results in this study by involving some difference of inputs in OBIA, it was generally known that the use of single pair band of DII was not able to give a better result. Similarly, the combination of some DII bands (ii) does not give better results when compared to DII band combined with initial bands (iii), or by using only initial bands (iv). The use of methods (i) and (ii) reduce the area and distribution of the reef classes, i.e. LC, DC, and RC, those classes are more prevalent on the reef slopes. It was be expected because of the depth factor, the conditions in the field showed the distribution of reef class in the deeper areas than the seagrass class. When applied water column correction and only involve single pair band or multi pairs band of DII in classification, reef class reduced to sand and seagrass. Unlike the case when it initial bands including as part of the classification inputs both in (iii) and (iv) methods, it shows differences in classification results, especially on reef classes. On mapping benthic covers, some researchers that applied water column correction i.e. [10] for reefs mapping, [11] mapping seagrass using Sentinel-2 imagery, and [12] that using geomorphological zonation and water column correction approach.

4. Conclusion

Classification with including DII band combine with the initial band (iii) gives classification result with higher accuracy (OA 78.68% and Kappa 0.76). While the lowest accuracy was obtained when applying method (i) which only reached 55.84% for OA and 0.48 for Kappa Statistic. Based on the results obtained in this study site, the method (iii) was well used when mapping the seagrass, but to map the coral reefs

IOP Conf. Series: Earth and Environmental Science **149** (2018) 012052 doi:10.1088/1755-1315/149/1/012052

should be obtained better results when applying the method (iv). It seems to be better to use water column correction when mapping seagrass, but sometimes it does not be needed when mapping reefs.

References

[1] Green EP, Mumby PJ, Edwards AJ, and Clark CD 2000 *Remote Sensing Handbook for Tropical Coastal Management Coastal Management Sourcebooks 3*

[2] Lyzenga DR 1981 Remote sensing of bottom reflectance and water attenuation parameters in shallow water using aircraft and Landsat data *International Journal of Remote Sensing* **2** 71-82

[3] Anggoro A, Siregar VP, and Agus SB 2015 Geomorphic zones mapping of coral reef ecosystem with OBIA method, case study in Pari Island *Jurnal Penginderaan Jauh* **12** 1-12

[4] Roelfsema C, Phinn S, Jupiter S, Comley J, and Albert S 2013 Mapping coral reefs at reef to reef-system scales, 10s–1000s km2, using object-based image analysis *International Journal of Remote Sensing,* **34** 6367-6388

[5] Zhang C, Selch D, Xie Z, Roberts C, Cooper H, and Chen G 2013 Object-based benthic habitat mapping in the Florida Keys from hyperspectral imagery *Estuarine, Coastal and Shelf Science* **134** 88-97

[6] Wahidin N, Siregar VP, Nababan B, Jaya I, and Wouthuyzen S 2015 Object-based image analysis for coral reef benthic habitat mapping with several classification algorithms *Procedia Environmental Sciences* **24** 222-227

[7] Jhonnerie R, Siregar VP, Nababan B, Prasetyo LB, and Wouthuyzen S 2015 Random forest classification for mangrove land cover mapping using Landsat 5 TM and ALOS PALSAR imageries *Procedia Environmental Sciences* **24** 215-221

[8] Siregar VP, Agus SB, Subarno T, and Prabowo NW 2016 Mapping of shallow water benthic habitat on reef flat zone using object based and pixel based image classification in Harapan-Kelapa Islands, Kepulauan Seribu," presented at the The 3rd International Symposium on LAPAN-IPB Satellite for Food Security and Environmental Monitoring, Bogor, Indonesia, 2016

[9] Manuputty A, Gaol JL, Agus SB, and Nurjaya IW 2017 The utilization of Depth Invariant Index and Principle Component Analysis for mapping seagrass ecosystem of Kotok Island and Karang Bongkok, Indonesia *IOP Conference Series: Earth and Environmental Science* 012083

[10] Yasir Haya LOM and Fujii M 2017 Mapping the change of coral reefs using remote sensing and in situ measurements: a case study in Pangkajene and Kepulauan Regency, Spermonde Archipelago, Indonesia *Journal of Oceanography*

[11] Traganos D and Reinartz P 2017 Mapping Mediterranean seagrasses with Sentinel-2 imagery *Mar Pollut Bull*

[12] Selamat MB, Jaya I, Siregar VP, and Hestirianoto T 2012 Geomorphology Zonation and Column Correction for Bottom Substrat Mapping using Quickbird Image *Jurnal Teknologi Perikanan dan Kelautan* **3**

Water mass characteristic in the outflow region of the Indonesian throughflow during and post 2016 negative Indian ocean dipole event

A Bayhaqi[1*], I Iskandar[2], D Surinati[1], A S Budiman[1], A K Wardhana[1], Dirhamsyah[1], D Yuan[3] and D O Lestari[4]

[1]Physical Oceanography and Climate Laboratory, Research Center for Oceanography, Indonesia Institute of Sciences (LIPI), Jakarta, Indonesia
[2]Department of Physics, Faculty of Mathematic and Natural Sciences, Sriwijaya University, Palembang, South Sumatera, Indonesia
[3]Institute of Oceanology, Chinese Academy of Sciences (IOCAS), Qingdao, People Republic of China
[4]Graduate School of Environmental Sciences, Sriwijaya University, Palembang, South Sumatera, Indonesia

E-mail: ahmad.bayhaqi@lipi.go.id

Abstract. Strong El Niño and positive Indian Ocean Dipole (pIOD) events in 2015/2016 followed by relatively strong negative Indian Ocean Dipole (nIOD) and weak La Niña in 2016 events have affected hydrography conditions in the Indonesian Throughflow (ITF) region. Two research cruises were conducted using RV Baruna Jaya VIII in August and November 2016. These cruises aim to evaluate possible impact of those two climate mode events on the water mass characteristic in the outflow region of the ITF. Hydrographic data from those two cruises were combined with the sea surface temperature (SST) from the Advanced Very High Resolution Radiometer (AVHRR) and surface wind data from the European Centre for Medium-Range Weather Forecasts (ECMWF). The results showed that in the 2016 anomaly year, the cooler sea surface temperature was observed during the negative IOD (nIOD) event while the warmer temperature was found in the post of nIOD event. The observed water mass characteristics in the outflow region of the ITF revealed that the upper layer was dominated by the Indian Ocean water mass, while the Pacific Ocean water mass was observed in the deeper layer. The observed current data across the Sumba Strait showed that the South Java Coastal Current (SJCC) was observed in the upper layer, propagating eastward toward the Savu Sea. A few days later, the observed currents in the upper layer of the Ombai Strait revealed the ITF flow towards the Indian Ocean. Meanwhile, the lower layer showed an eastward flow towards the Ombai Strait.

1. Introduction

El Niño-Southern Oscillation (ENSO) is a naturally climate mode phenomenon involving ocean temperature fluctuation and the changes in the atmosphere circulation. This phenomenon has a great impact for the life such as flood and drought. Since 1950, 23 El Nino events have been recorded and the most severe El Nino event was recorded in 1997/98. However, recent El Nino event in 2015/2016 was reported as the strongest event since 1997/98 [1]. In addition, this strong El Nino event coincidentally followed by the positive Indian Ocean Dipole (pIOD) event. During pIOD event, negative Sea Surface Height Anomaly (SSHA) could trigger upwelling anomaly in the western tropical

Pacific via Indonesian seas [2]. As an impact, sea surface temperature (SST) in south of Java was cooling during May through July as a sign of pIOD event [3].

The tropical Indonesian seas, an inter-oceanic exchange of Pacific-Indian Ocean, hold an important role to modulate the global climate through the Indonesian Throughflow (ITF), part of the global thermohaline circulation. It is clear that the ITF brings freshwater budget into the Indian Ocean along its pathways within the Indonesian archipelago via western (Makassar strait) and eastern entrance (Maluku or Halmahera Sea) [4,5,6]. The North and South Pacific water mass are the sources of the ITF water mass [7,8]. The northern water sources consist of the North Pacific Subtropical Water (NPSW) and the North Pacific Intermediate Water (NPIW) [9]. These water masses mostly exit through the Ombai Strait and the Timor Passage while the rest flows through the Lombok strait [6,10].

The outflow region of the ITF is a cross road of water masses from the Pacific Ocean, the internal Indonesian Seas and the Indian Ocean. About 15 Sv water masses flow through this region annually [10,11]. Previous study has examined the water characteristic and its variability in the ITF outflow region [12]. The authors suggested that salinity maximum of the North Pacific Water is stronger during the transition between two monsoons and there was a phase lag between the Lombok Strait and Ombai Strait as well as the Timor Passage around one and five months. Moreover, the relationship between ITF flow and the IOD event was showed the increase of ITF transport in the upper layer of outflow strait during the strong IOD positive in late 2006 while as the opposite condition in the IOD negative 2004, that upper flow showed the decrease value [10]. Thus, in this study, we will evaluate not only the the flow but also the water mass characteristic in the outflow region of the ITF during an extreme climate event in 2016 with the newer observation data than the previous study.

2. Method

2.1 Observational Data

Observations were conducted on August and November 2016 using Conductivity Temperature Depth (CTD) SBE and Shipboard Acoustic Doppler Current Profiler (SADCP) mounted at the RV Baruna Jaya VIII. The SADCP recorded ocean data along the cruise track. During August 2016, the Widya Nusantara Expedition (EWIN) cruise recorded hydrography data at 14 CTD stations. Meanwhile, the joint research cruise between Research Center for Oceanography, Indonesian Institute of Sciences (RCO LIPI) and Institute of Oceanology, Chinese Academy of Sciences (IOCAS) recorded hydrography data at 13 CTD stations during November 2016. CTD was casted around 50-meter depth from the bottom for EWIN cruise while it was around 1000-meter depth for RCO LIPI – IOCAS joint cruise with acquisition rate of 24 Hz. It means that the CTD can record 24 data pulse in 1 second. The CTD stations are located on the outflow of ITF such as ombai strait and timor passage following the method from the previous study [13]. This figure 1 presents the stations for the CTD measurements in August and November.

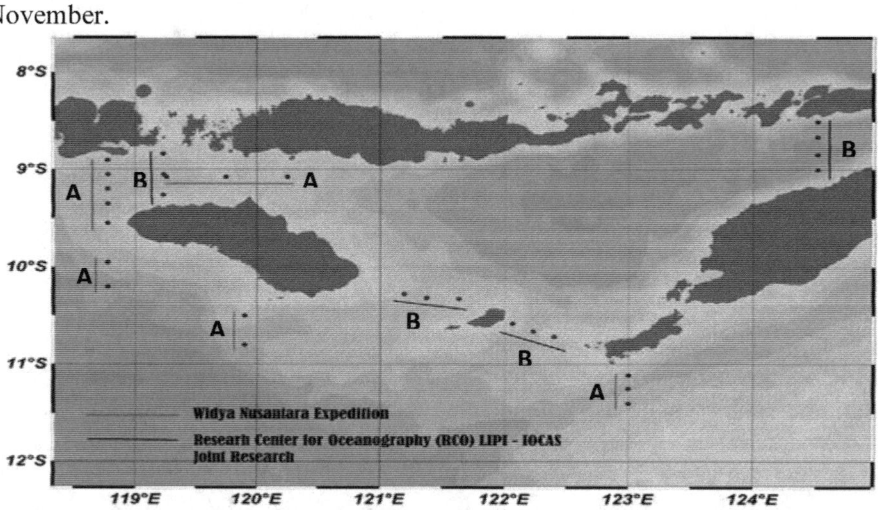

Figure 1. The stations for CTD measurements. The red lines (A) indicate EWIN stations while the blue lines (B) indicate RCO LIPI-IOCAS stations

2.2. Satellite Data
Monthly sea surface temperature (SST) data were obtained from the National Oceanographic and Atmospheric Information (NOAA) Optimum Interpolation (OI) SSST with a horizontal resolution of 0.25° for a period of January 2000 to December 2016. Meanwhile the European Centre for Medium-Range Weather Forecast (ECMWF) ERA Interim monthly data sets are retrieved from January 1979 to December 2016 having resolution of 0.5°. Note that the monthly climatology fields were calculated from the time series over the period of January 2001 until December 2015. The anomaly field are, then, defined as the deviation from the mean climatology. In this study, we calculate the DMI from the different value of sea surface temperature anomaly between the western pole (10°S to 10°N and 50°E to 70°E) and the eastern pole (0°S to 10°S and 90°E to 110°E) following [14].

3. Result and Discussion
3.1. Climate mode events 2015-2016
Figure 2 shows the time series of Nino 3.4 index and Dipole Mode Index (DMI) Respectively. Note that the Nino 3.4 index is defined as averaged SST anomaly in the central equatorial Pacific Ocean over the box 170°W - 120°W, 5° - 5°N [13,15]. Meanwhile, the DMI is the east-west temperature gradient across the tropical Indian Ocean between the western equatorial Indian Ocean (50°E – 70°E and 10°S – 10°N) and the southeastern equatorial Indian Ocean (90°E - 110°E and 10°S – 0°N) [14].

It is shown that during 2015, a strong El Niño event took place in the tropical Pacific (figure 2a), accompanied by a positive Indian Ocean Dipole (IOD) in the tropical Indian Ocean (figure 2b). Evolution of the El Niño event was started in boreal spring 2015, peaked in boreal winter 2015 and terminated in boreal spring 2016 [16]. Meanwhile, the positive IOD event was lasting very short: started in June, peaked in September and terminated in November 2015.

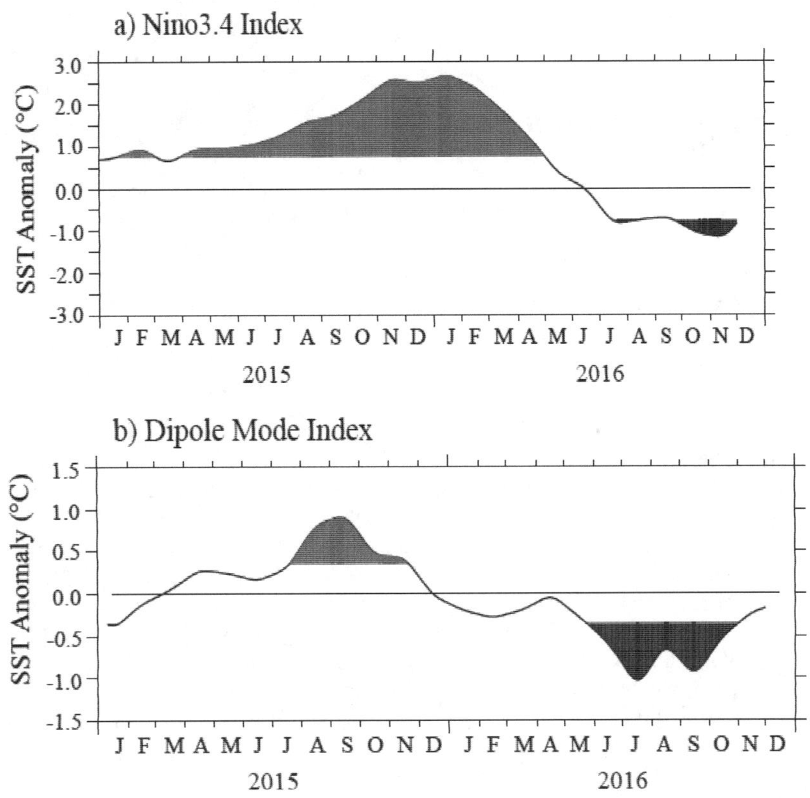

Figure 2. Time series of (a) Niño 3.4 index and (b) Dipole Mode Index (DMI) from January 2015 until December 2016. The red shaded-curves indicate the El Niño or positive Indian Ocean Dipole (IOD) events while the blue shaded-curves indicate La Niña or negative IOD events.

The situation was reverse in 2016. Weak La Niña event was observed at the end of 2016 (figure 2a) while relatively strong negative IOD event took place in the tropical Indian Ocean from May until November 2016 (figure 2b). Note that the EWIN and joint research observations were conducted during and post negative IOD event and during the initial phase of weak La Niña event in August and November 2016. Figure 3 shows the spatial-temporal evolutions of the SST and wind over the eastern tropical Indian Ocean and within the Indonesian seas. It is clearly shown that the EWIN observation was conducted during the peak phase of negative IOD event which is indicated by warm SST anomaly in the eastern tropical Indian Ocean associated with westerly wind anomaly (figure 3b). On the other hand, the joint research observation was conducted after termination of the negative IOD event and during the initial phase of weak La Niña. Therefore, the SST anomaly in the study area was relatively normal and the wind was south-easterly wind anomaly (figure 3e).

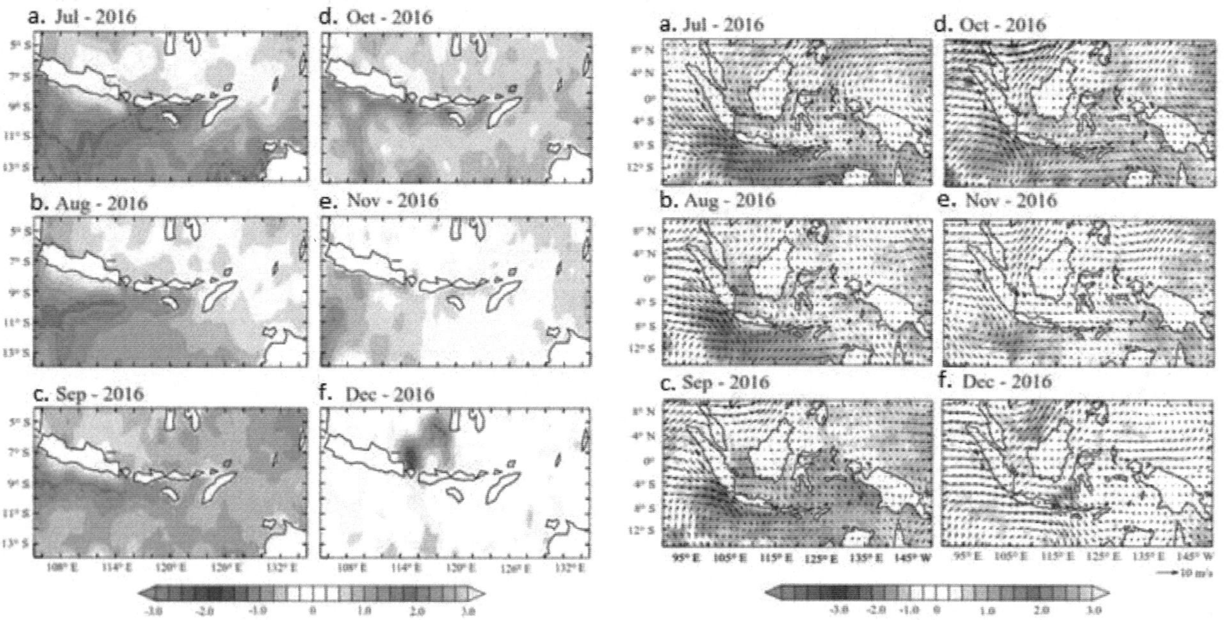

Figure 3. Monthly spatial-temporal evolution of the sea surface temperature (SST) and wind anomalies in the eastern tropical Indian Ocean and within Indonesian seas during July – December 2016

3.2. Water mass characteristics

Figure 4 shows the T-S diagram of the water mass observed during both observational cruises in August and November 2016. There were six kinds of water masses observed in the research area. In the upper layer (0-500 m) we found the Bay of Bengal Water (BBW) from the Indian Ocean, Indian Equatorial Water (IEW), Indonesian Upper Water (IUW) and Western North Pacific Central Water (WNPCW). Meanwhile, in intermediate layer between 500-1500 m, the Antarctic Intermediate Water (AIW) and the Indonesia Intermediate Water (IIW) were identified. It is also shown that the water mass has divided the water column into three regions. Firstly, the surface layer (<50 m) was dominated by water mass from the Indian Ocean, secondly several water mass resources from the Indonesian water and the Pacific Ocean trigger mixing within 100-500 m depth and lastly the Antarctic and the Indonesian water are dominant resources in the intermediate layer.

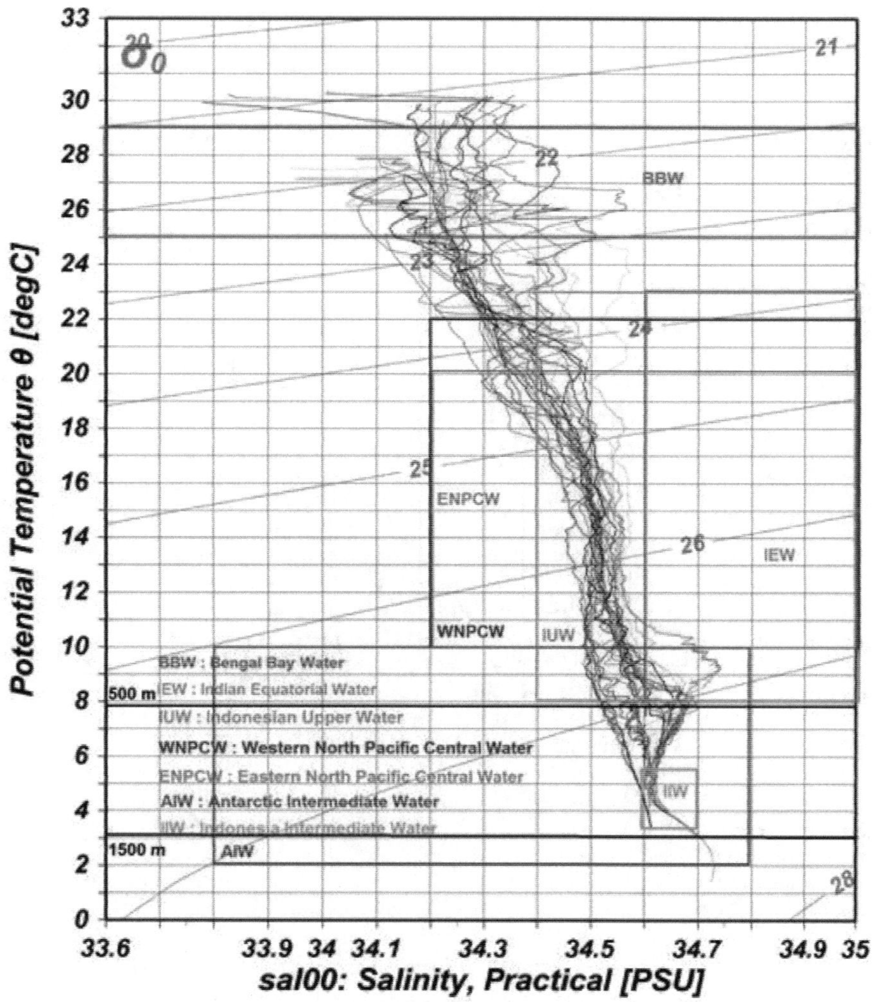

Figure 4. The T-S diagram for all CTD stations from August and November 2016 measurements.

3.3 Spatial Distribution of Observed water mass

Figure 5 shows the observed water mass in the northern parts of the ITF outflow region while figure 6 presents the observed water mass in the southern parts of ITF outflow region. In the western Savu sea, we found the BBW and IEW water masses (figure 5, *left panel*). Fresh water mass was observed in the upper layer within the Savu sea (figure 5, *upper panel*). On other hand, in the Ombai strait, the Pacific water mass in the upper layer (figure 5, *right panel*). In the southern parts of ITF outflow region, the observed water masses were saltier than those observed in the northern parts (figure 6). In the western part, water masses from the equatorial Indian (IEW) and Bay of Bengal Water (BBW) are found (figure 6, left panel). On other hand, the water masses observed in the Timor Passage and in the southern part of Savu sea were dominated by water massed from the Pacific Ocean (figure 6, lower and right panels).

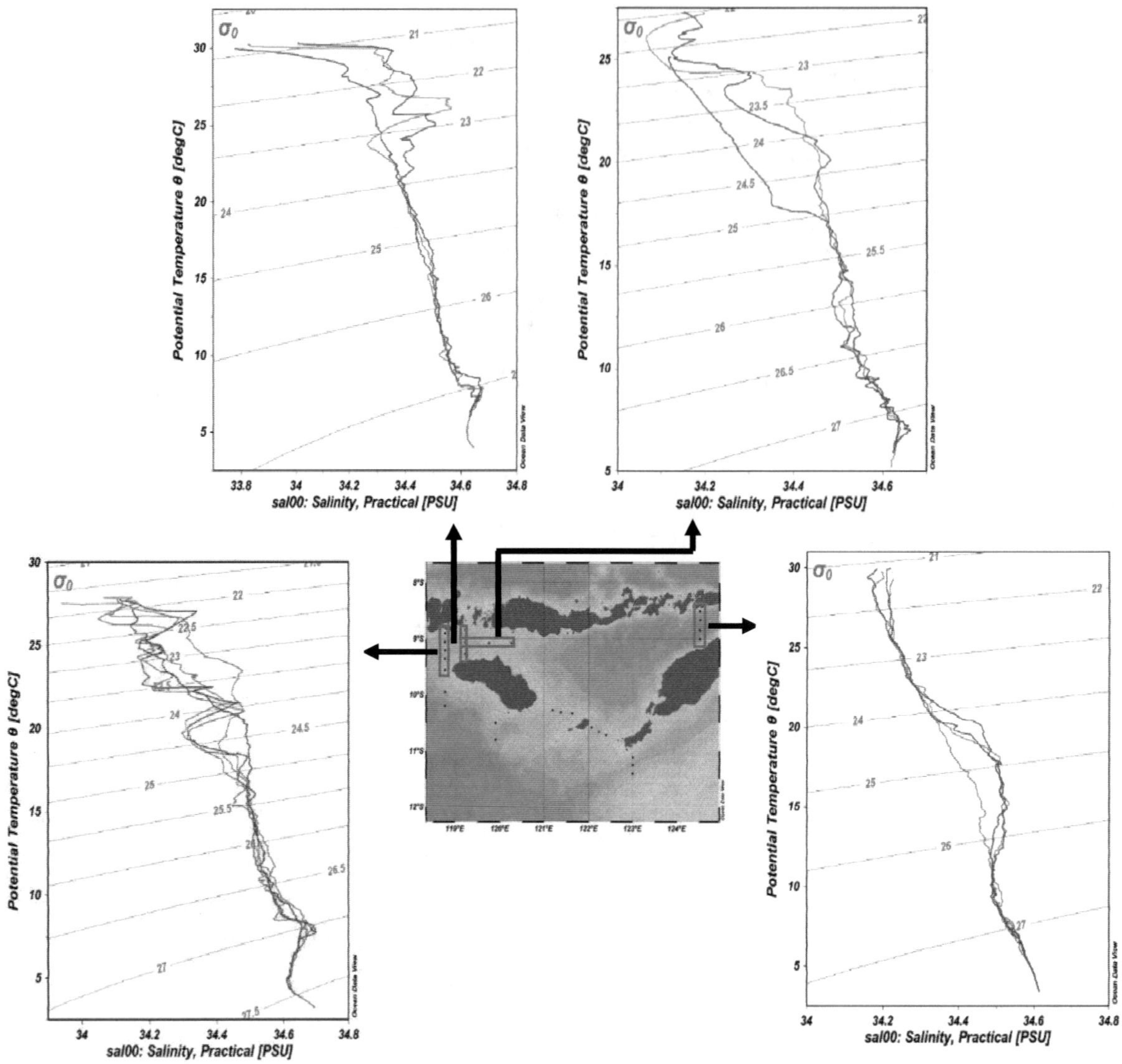

Figure 5. The T-S diagram of observed water mass in the northern area of ITF outflow region

The observed water in this area has unique characteristics. The maximum salinity value ~34.60 psu indicating the North Pacific Subtropical Water (NPSW) water mass be inivisible. Vertical mixing in this region is estimated affecting the water mass. Thus, it has the same characteristic with Banda and Timor sea water mass which is marked by the maximum salinity ~34.50 and the same profile until the bottom layer. The difference value 0.1 – 0.2 psu between them is resulted from the intrusion of Indian Central Water which has high salinity. This water mass is one of the primary sources for the ventilation of permanent thermocline in the Indian ocean. It is formed by the subduction in the region of negative wind stress curl and moves eastward with the south Indian current [17]. This water mass result also showed that there is the Indian ocean water trace which enters the Sumba strait. On the other hand, figure 6 showed the distribution of water mass in the southern region of ITF outflow. That figure illustrated the southern region is slightly more influenced by Indonesian Intermediate Water (IIW) which characterised by the range of temperature value from 3.5°C – 5.5°C and the value of salinity 34.6 psu – 34.7 psu

Related to the sea surface temperature (SST), both figure 5 and 6 provided the different value in temperature during two cruises. The maximum value of SST in the EWIN cruise and Joint research cruise is 27.8°C and 30°C respectively. The monsoonal condition potentially contributes in making that difference since August is still in the southeast monsoon where in this period can trigger the upwelling phenomenon and reduce the sea temperature. In the opposite way, the temperature can be warmer when the northwest monsoon. That condition indicates that during 2016 anomaly year, the cooling temperature is followed by nIOD event coincidentally and the temperature will increase after the nIOD phase in this region.

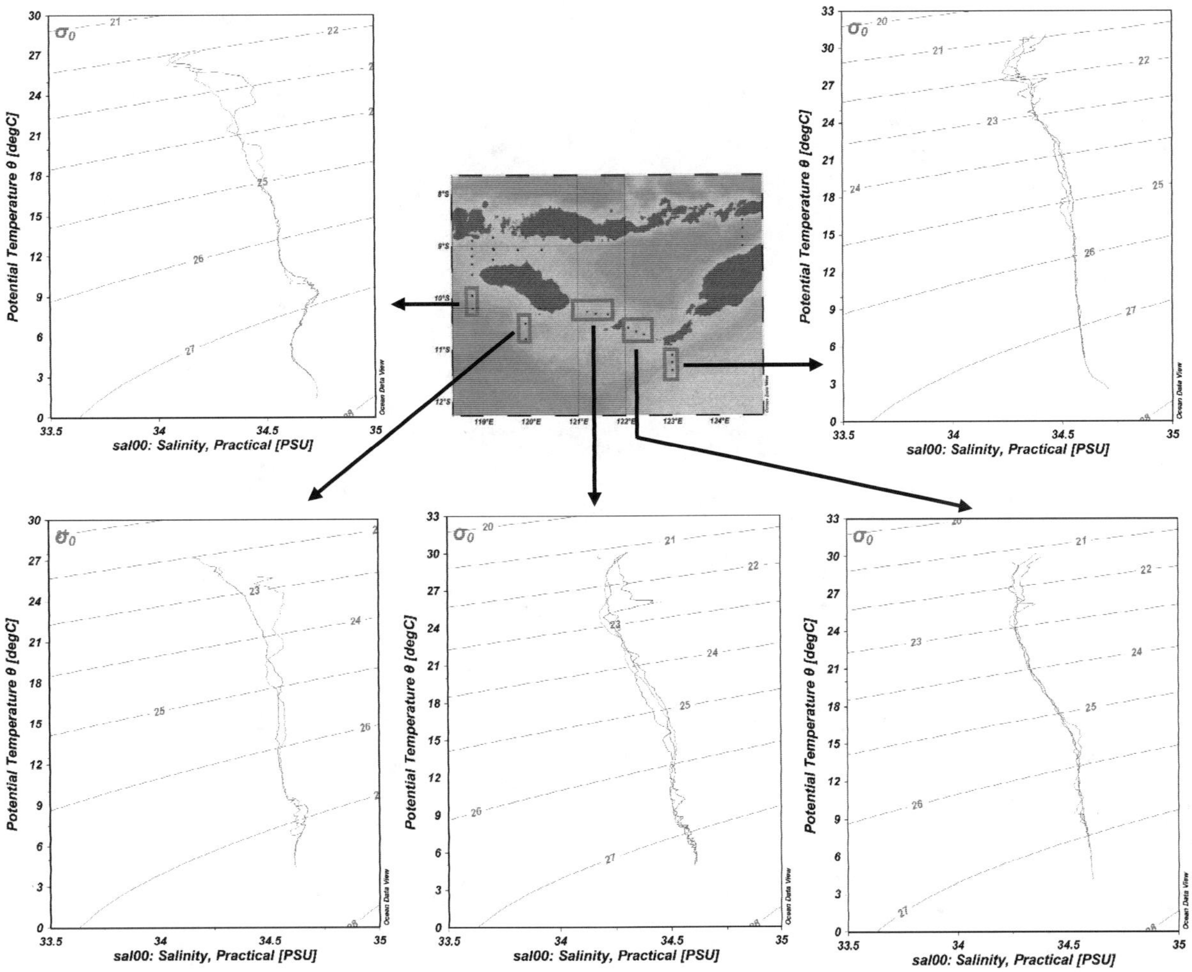

Figure 6. The T-S diagram of observed water mass in the southern area of ITF outflow region

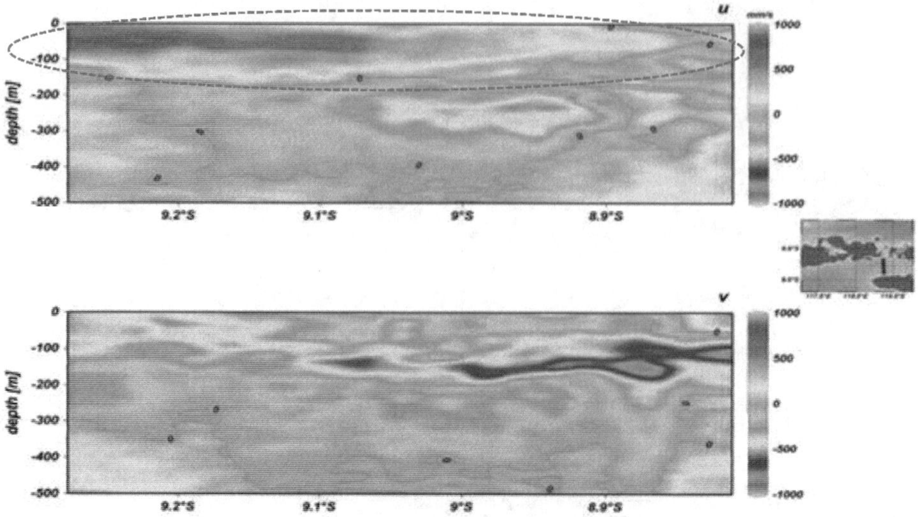

(a)

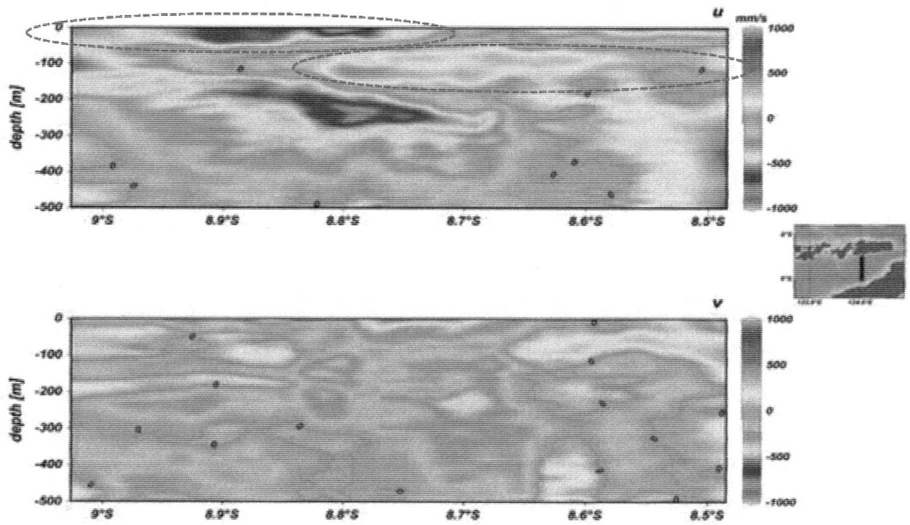

(b)

Figure 7. Vertical section of Zonal (*upper panel*) and meridional (*lower panel*) current in (a) Sumba Strait (b) Ombai Strait

To evaluate a possible dynamic underlying the spatial distribution of water masses during the observational periods, we analyse the observed currents in two straits, namely Sumba and Ombai Strait (figure 7). It is shown that the penetration of the Indian Ocean water into the ITF outflow straits was associated with strong eastward zonal current observed in Sumba Strait (figure 7a, *upper panel*). This eastward zonal current is possibly propagation of South Java Coastal Current (SJCC). The Pacific and Indian Ocean water masses observed in the southern part of ITF outflow area associated with strong eastward current in between two westward ITF cores (figure 7b, *upper panel*). This subsurface eastward current is known as South Java Coastal Under Current (SJCUC). These results fit with [18] where observation measurement from 2004-2006 showed the trace of SJC moving to the east into the Savu sea. From this condition, the flow in the Sumba strait confirms that there is a contribution from the Indian Ocean water mass through the SJC.

4. Conclusion

Warm SST anomaly in the ITF outflow region was detected during the negative IOD phase. As the opposite, SST anomaly was cooling down after the stoppage of negative phase. The cooler sea sea surface temperature was observed during the EWIN cruise in southeast monsoon (nIOD event 2016) while warmer temperature was found during joint rsearch cruise RCO LIPI-IOCAS in north monsoon (post nIOD event 2016) indicating that there is a monsoonal contribution in this region. The upper layer of western ITF outflow region was dominated by the Indian Ocean water mass induced by South Java Coastal Current while the deeper layer was dominated by Pacific origin water masses induced by ITF. Meanwhile, in the eastern ITF outflow region, Pacific Ocean water mas via ITF was the biggest resource both in upper and deeper layer.

Acknowledgements

This study was supported by the Joint Research Project between Research Center for Oceanography, Indonesian Institute of Sciences (RCO LIPI) and Institute of Oceanology, Chinese Academy of Sciences (IOCAS). The second author is supported by Ministry of Research, Technology and Higher Education through *Hibah Berbasis Kompetensi 2017.*

References

[1] Varotsos C A, Tzanis C G and Salis N V 2016 On Progress of the 2015-2016 El Nino Event *Atmos.Chem.Phys* **16** 2007-2016

[2] Zhou Q, Duan W, Mu M and Feng R 2015 Influence of Positive and Negative Indian Ocean Dipoles on ENSO via the Indonesian Throughflow: Results from Sensitivity Experiments *Advances in Atmospheric Sciences* **32** 783-793

[3] Delman A S, Sprintall J, McClean J L and Talley L D 2016 Anomalous Java Cooling at the Initiation of Positive Indian Ocean Dipole Events *J Geophysical Res Oceans* **121** 5805-5824

[4] Mayer B and Damm P E 2012 The Makassar Strait Troughflow and its Jet *J Geophys.Res* **105** 11243-11258

[5] Susanto R D, Ffield A, Gordon A L and Adi T R 2012 Variability of Indonesian Throughflow within Makassar Strait, 2004-2009 *J Geophys.Res* **117** C09013.

[6] Iskandar I, Mardiansyah W, Setiabudidaya D, Poerwono P and Syamsuddin F 2015 *Interannual variation of the Indonesian throughflow in the Timor Passage as revealed in SODA: 1958* – 2008 AIP Conference Proceeding **1677** 2-15

[7] Gordon A L and Fine R A 1996 Pathways of Water between the Pacific and Indian Oceans in the Indonesian seas *Nature* **379** 146-149

[8] Ilahude A G and Gordon A L 1996 Thermocline Stratification within the Indonesian Seas. *J Geophys Res* **101** 12401-12409

[9] Gordon A L, Susanto R D, Ffield A, Huber B A, Pranowo W and Wirasantosa S 2008 Makassar Strait Throughflow, 2004-2006 *Geophysical Letter* **35** L24605

[10] Sprintall J, Wijffels S E, Molcard L and Jaya I 2009 Direct Estimates of the Indonesian Throughflow entering the Indian Ocean: 2004-2006 *J Geophys Res* **114** C07001

[11] Iskandar I, Masumoto Y, Mizuno K, Sasaki H, Affandi A K, Setiabudidaya D and Syamsuddin F 2014 Coherent intraseasonal oceanic variations in the eastern equatorial Indian Ocean and in the Lombok and Ombai Straits from observations and a high-resolution OGCM *J Geophy Res* **119** 615-630

[12] Atmadipoera A, Molcard R, Madec G, Wijffels S, Sprintall J, Koch-Larroy A, Jaya I and Supangat A 2009 Characteristic and Variability of the Indonesian Throughflow water at the Outflow Straits *Deep Sea Research I* **56** 1942-1954

[13] Trenberth K E, Caron J M, Stepaniak D P and Worley S 2002 The evolution of ENSO and global

atmospheric surface temperatures *J. Geophys. Res* **107** 4065
[14] Trenberth K E and Stepaniak D P 2001 Indices of El Niño evolution *J. Clim* **14** 1697– 1701
[15] Saji N H, Goswami B N, Vinayachandran P N and Yamagata T 1999 A Dipole Mode in the Tropical Indian Ocean *Nature* **401** 360-363
[16] Iskandar I, Utari P A, Lestari D O, Qurnia W S, Setiabudidaya D, Khakim M Y N, Yustian I, and Dahlan Z 2017 *Evolution of 2015/2016 El Niño and its impact on Indonesia* AIP Conference Proceeding **1857**
[17] Kartensen J and Tomzcak M 1997 Ventilation processes and water mass ages in the thermocline of the southeast Indian Ocean *Geophysical Research Letter* **24**(22) 2777-2780
[18] Sprintall J, Wijffels S, Molcard R, Jaya I 2010 Direct Evidence of the South Java Current System in Ombai Strait *Dynam Atmos Oceans* **50** 140-156

IOP Conf. Series: Earth and Environmental Science **149** (2018) 012054 doi:10.1088/1755-1315/149/1/012054

Response of near-surface currents in the Indian Ocean to the anomalous atmospheric condition in 2015

P A Utari[1], M Y Nurkhakim[2], D Setiabudidaya[2], I Iskandar[2*]
[1] Program Studi Ilmu Lingkungan, Program Pascasarjana, Universitas Sriwijaya, Palembang, Sumatera Selatan, Indonesia
[2] Jurusan Fisika, Fakultas Matematika dan Ilmu Pengetahuan Alam, Universitas Sriwijaya, Palembang, Sumatera Selatan, Indonesia

Email: iskhaq@mipa.unsri.ac.id

Abstract. Anomalous ocean-atmosphere conditions were detected in the tropical Indian Ocean during boreal spring to boreal winter 2015. It was suggested that the anomalous conditions were characteristics of the positive Indian Ocean Dipole (pIOD) event. The purpose of this investigation was to investigate the response of near-surface currents in the tropical Indian Ocean to the anomalous atmospheric condition in 2015. Near-surface current from OSCAR (Ocean Surface Current Analyses Real Time) reanalysis data combined with the sea surface temperature (SST) data from OISST – NOAA, sea surface height (SSH) and surface winds from the ECMWF were used in this investigation. The analysis showed that the evolution of 2015 pIOD started in June/July, peaked in the September and terminated in late November 2015. Correlated with the evolution of the pIOD, easterly winds anomalies were detected along the equator. As the oceanic response to these easterly wind anomalies, the surface currents anomalously westward during the peak of the pIOD. It was interesting to note that the evolution of 2015 pIOD event was closely related to the ocean wave dynamics as revealed by the SSH data. Downwelling westward propagating Rossby waves were detected in the southwestern tropical Indian Ocean. Once reached the western boundary of the Indian Ocean, they were redirected back into interior Indian Ocean and propagating eastward as the downwelling Kelvin waves.

1. Introduction

The prior investigations have advocated that the sea surface temperature (SST) variability in the Indian Ocean is strongly tight to the ocean-atmosphere interaction and also gives contribution to the large scale atmospheric circulation [1,2]. The anomalous of warmer (colder) SST in the western (eastern) tropical Indian Ocean indicates the positive Indian Ocean Dipole (pIOD). Correlated with this SST anomaly, easterly wind anomalies were detected along the equatorial Indian Ocean and southeasterly anomalies were detected along the southern coast of Sumatra and Java [3–5]. The result of the upper ocean to the easterly wind anomalies along the equator, eastward upwelling Kelvin waves were resulted along the equator. Meanwhile, westward downwelling Rossby waves were resulted in the off equatorial region [6–8]. The intension of the IOD event is signified by the SST gradient between the western equatorial Indian Ocean (50°E - 70°E, 10°S - 10°N) and the eastern Indian Ocean (90°E - 110°E, 10°S - 10°N) [9–11].

Previous study has demonstrated that the air-sea flux is a major forcing for the outgrowth of the IOD event [12]. In addition, the pIOD event has caused hot and dry condition over Sumatera and Java [12,13]. One of the previous studies investigated the backdrop of warming and cooling SST's correlated with

IOP Conf. Series: Earth and Environmental Science **149** (2018) 012054 doi:10.1088/1755-1315/149/1/012054

ENSO and generated by the interactions internal of Ocean Atmosphere to the Indian Ocean [13]. The same studies disclosed the connection which was ensued between ENSO and Indian Ocean warming during summer [8]. While the Coastal wind and Ekman pumping possibly maintained by causing the blooms . Due to these reasons, the pattern of warming and cooling Indian Ocean became the investigation highlight to observe on the response of near-surface ocean currents to the wind field anomalies during 2015 over the tropical Indian Ocean during 2015.

2. Data and Method

In this investigation, daily surface wind data were regained from the European Centre for Medium-Range Weather Forecasts (ECMWF) which is acquire at *http://www.apps.ecmwf.int/* [15]. Data period of 1 January 2006 until 31 December 2015 with horizontal resolution of 0.25° × 0.25° were used in this observation. The daily sea surface height (SSH) data were also acquired from the ECMWF. The data were available for the period of 1 January 2006 until 15 December 2015 with horizontal resolution of 0.25°× 0.25°.

The data of daily SST were acquired from OI-SST analysis (Optimum Interpolation-Sea Surface Temperature. This dataset was created by merging all available inspections from diverse platforms (satellites, ships, buoys). The SST data were used in this investigation with the spatial resolution of 0.25°× 0.25° from 1 January 2006 until 31 December 2015.

Near-surface current data were acquired from the Ocean Surface Current Analyses Real-Time (OSCAR). The data were available from 21 October 1992 until present with the spatial resolution of 1° × 1°. These data represent near-surface currents at 15 m depth and on the global grid every 5 days and they were available at *http://apdrc.soest.hawaii.edu*. In order to evaluate the intensity of the 2015 pIOD event, the Dipole Mode Index (DMI) from the Ocean Observations Panel for Climate (OOPC) was used in this investigation and it was available at *http://stateoftheocean.osmc.noaa.gov*. The data was available from 4 November 1981 until 5 April 2017. Note that the DMI was defined as the SST gradient between the western tropical Indian Ocean (50°E - 70°E, 10°S - 10°N) and the eastern tropical Indian Ocean Indian Ocean (90°E - 110°E, 10°S – equator) [14].

To define the pIOD and nIOD, we defined the result of SST gradient between the eastern and the western Indian ocean to find the significant value of DMI. These values were categorized as pIOD event which had value more than +1.5 standard deviation. On the other side, the nIOD event was classified by -1.5 standard deviation [16]. From these standard deviation, the anomalous conditions of each variable can be determined by observing the evolution of pIOD during 2015. The climatology of all fields was estimated on the time sequence from January 2006 until December 2014. Then, the anomalous fields were specified as the deviation from each climatological field. The wind data, currents, and SSH anomalies were processed by a low-pass filter with cut-off period of 15 day as suggested in the previous investigation [17].

3. Results and Discussion

3.1. The evolution of 2015 pIOD event

In this section, we evaluated the evolution of pIOD event in 2015 by analyzing the time series of the DMI (figure 1). It was signified that the 2015 pIOD event was started in boreal summer (August) 2015. Interestingly, the DMI showed three DMI peaks, with weakening signals in between indicating intraseasonal disturbances. The first peak was detected in early September, the second peak was found in the mid-October, while the last peak was detected in the mid-November. The pIOD event was terminated in early boreal winter (late November/early December).

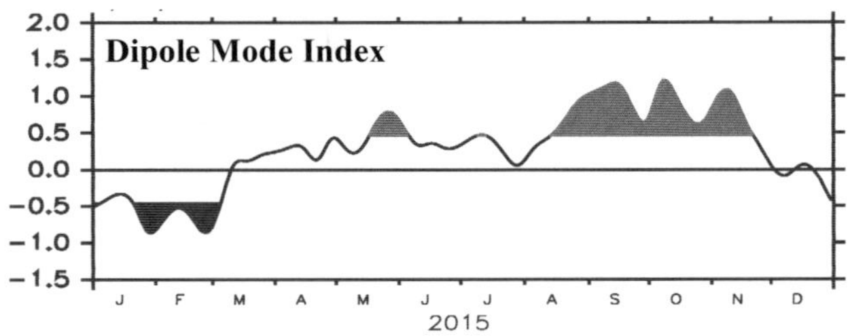

Figure 1. The Dipole Mode Index during January – December 2015

To see the mean fields of atmospheric circulation in the normal conditions, the sea surface temperature and surface wind became the variable which can be represented the atmospheric circulation. The monthly averaged climatological of sea surface temperature and surface wind was provided in figure 3 during July – December. While south-easterlies was ensued during boreal summer (July – August 2015), Boreal fall (September – November 2015) was dominated by north-westerlies and strengthened till the earliest boreal winter (December 2015). The evolutions of surface wind influenced the variability sea surface temperature in the topical Indian Ocean. Cold SST dominated in the western tropical Indian Ocean and the northern tropical Indian Ocean while the warm SST experienced in the eastern tropical Indian Ocean and the equatorial tropical Indian Ocean. The warm SST was strengthened by south-easterlies from July to October 2015.

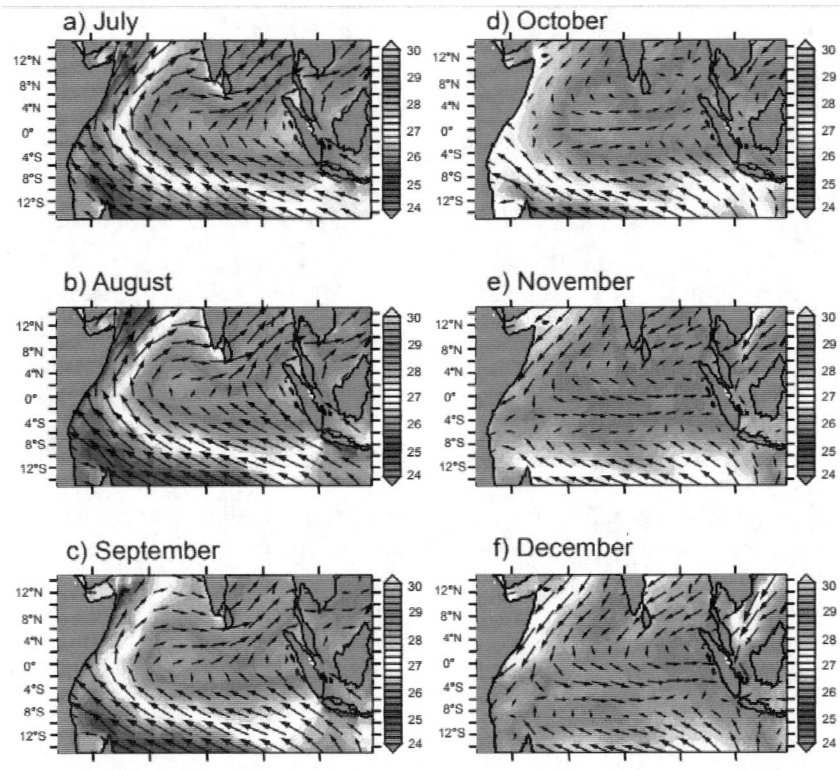

Figure 2. Monthly average of the sea surface temperature (*shaded* in °C) and the surface wind (*vector* in m/s) in the tropical Indian Ocean during July – December 2006 - 2015.

In order to evaluate the spatio-temporal evolution of the 2015 pIOD event, we then analyzed the monthly variation of the SST and wind field anomalies over the tropical Indian Ocean (figure 2). The evolution of the surface wind anomalies correlated with the 2015 pIOD event was started in July as shown figure 2a. In July, the equatorial region and off Sumatera occurred anomalous easterly winds and southeasterly winds. These easterly and southeasterly wind anomalies were correlated with the cold SST anomaly developed off South Java and warm SST anomaly occupied the western and central tropical Indian Ocean. As the anomalous winds were strengthened, the SST dipole pattern was also developed, though it did not show an extreme cold SST anomaly in the eastern tropical Indian Ocean (figure 2b-d). The pattern of SST was weakened as the wind anomaly was also weakened in November (figure 2e). Then, the SST dipole pattern was terminated and the tropical Indian Ocean experienced basin-wide warming in December (figure 2f).

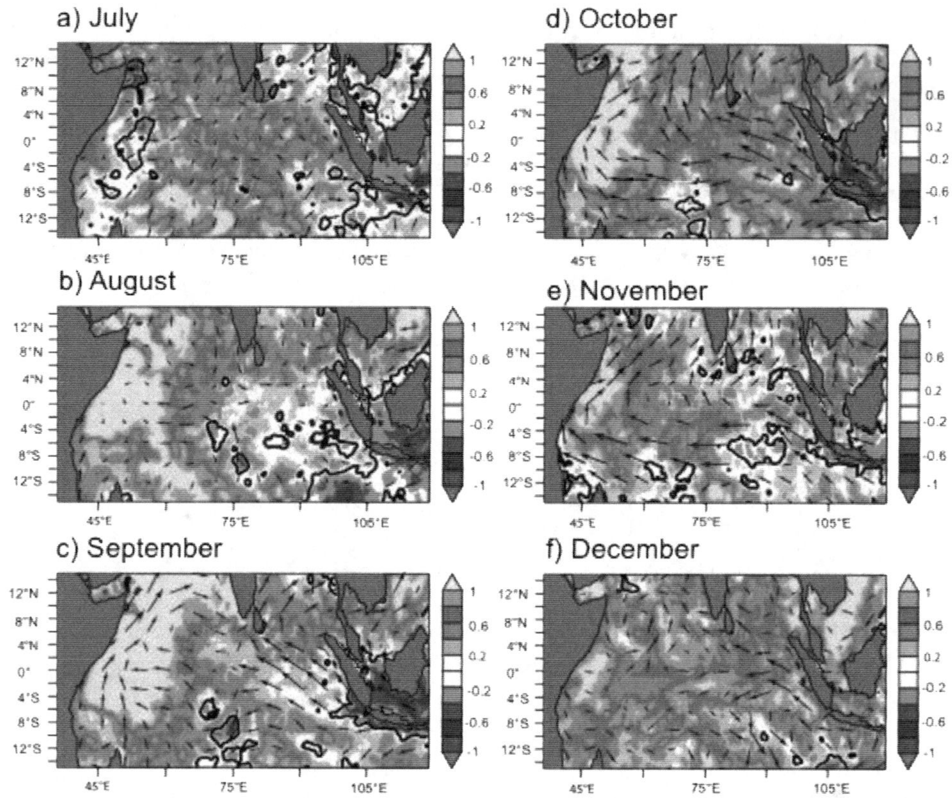

Figure 3. Anomalous evolutions of the sea surface temperature (*shaded* in °C) and the surface wind (*vector* in m/s) in the tropical Indian Ocean during July – December 2015.

3.2. Response of the near-surface currents to anomalous wind fields

In this observation, we investigated the response of near-surface currents to the anomalous wind fields over the equatorial Indian Ocean during 2015. The time-longitude diagrams of the anomalous near-surface current, SSH and surface wind was signified along the equatorial Indian Ocean in figure 3, respectively. Figure 3a showed that along the equatorial Indian Ocean indicated relatively weak easterly wind anomalies in July. From figure 3b weak westward currents along the equator was forced by those weak easterly wind anomalies.

The easterly wind anomalies in the eastern basin was strengthened from August to September 2015 (figure 3a). As a response, the eastern half of the equatorial Indian Ocean occurred relatively strong westward currents during this period (figure 3b). However, we was not detected a clear eastward propagating upwelling Kelvin waves (negative SSH anomaly) along the equator during this period (figure 3c). Instead, the SSH data revealed clear westward propagating downwelling Rossby waves (positive SSH anomaly) in the central equatorial Indian Ocean during this period (figure 3c). This downwelling Rossby wave was resulted as a response of the SSH in the off equatorial region to the easterly wind anomalies along the equator. Previous study has suggested that downwelling Rossby wave off south equator play an important role in initiating the development of pIOD event by inducing warm SST in the western pole of the pIOD.

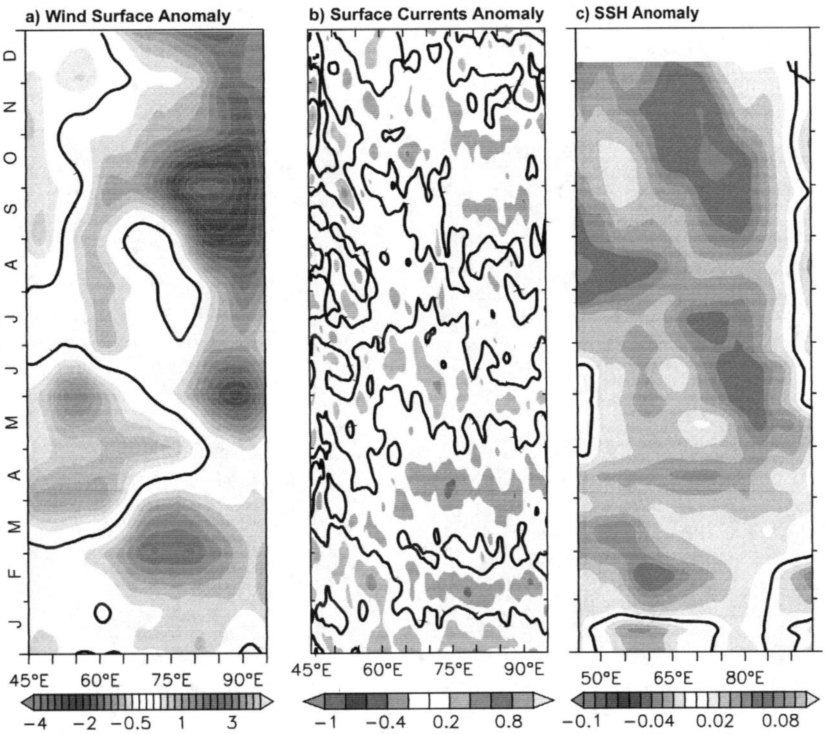

Figure 4. Hovmuller diagram of the a) surface wind anomaly (m/s), b) surface currents anomaly (m/s), and c) sea surface height anomaly (m) along the equator during January – December 2015.

To see the mean fields of ocean circulation in the normal conditions, the sea surface height (SSH) and surface current became the variable which can be represented the ocean circulation. Figure 3 showed the climatology of SSH and surface current during July – December. From figure 5, the western tropical Indian Ocean experienced the Negative SSH and the positive SSH was ensued in the eastern tropical Indian Ocean. However, the equatorial Indian Ocean did not show the significant indication of upwelling and downwelling signals during July 2015. In the western tropical Indian Ocean experienced downwelling Rossby waves during July till December and was strengthened in October. On the other side, the upwelling Kelvin waves occurred in the eastern tropical Indian Ocean and was intensified in October. In the southern of Java, the positive SSH occurred from July – October and the peaked signal took place in the September. The dynamic of surface current dominated westward current during July – September and the eastward current controlled during October – December which was illustrated by vector in figure 5.

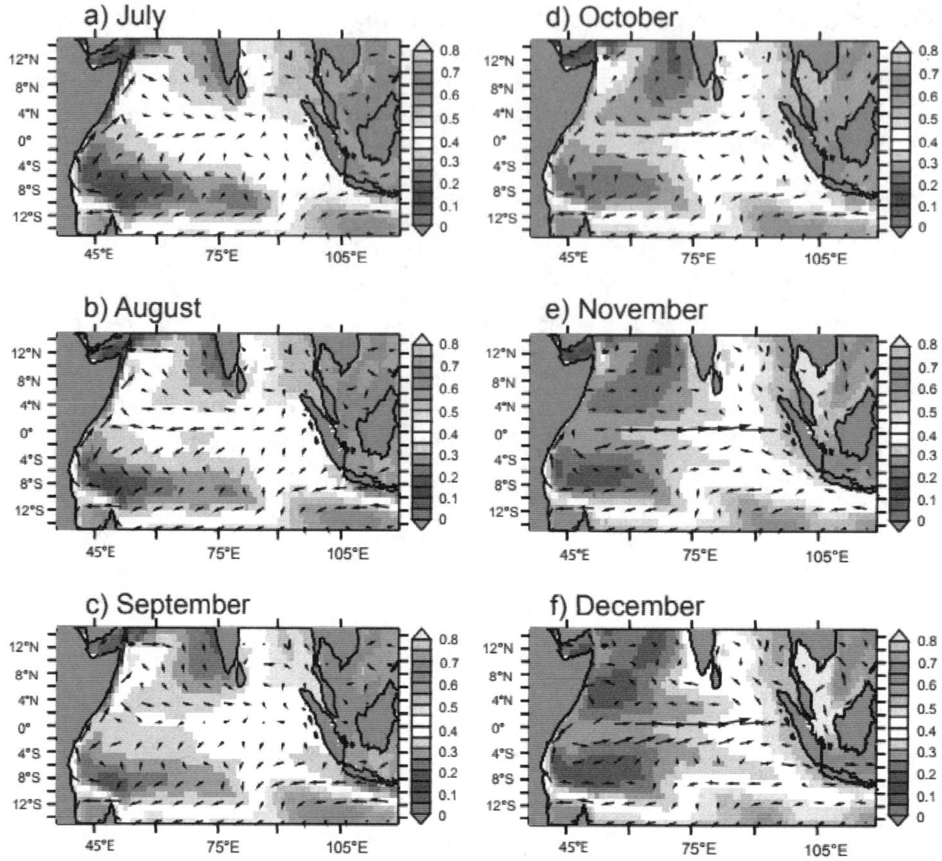

Figure 5. Monthly average of the sea surface height (*shaded* in meter) and the surface current (*vector* in m/s) in the tropical Indian Ocean during July – December 2006 - 2015.

To have a better view of spatio-temporal evolution of the near surface current and SSH to the wind field anomaly, the monthly evolution of the near-surface current and the SSH in the tropical Indian Ocean was specified in figure 3. As expected, there was no clear sign of the upwelling Kelvin waves along the equatorial Indian Ocean in response to the easterly wind anomalies (figure 6a-b). However, westward propagating downwelling Rossby waves were clearly detected from the development phase until the termination phase of the pIOD event, in particular in the southern hemisphere. Once the downwelling Rossby waves reached the western boundary of the Indian Ocean, part of the energy was redirected back as eastward propagating downwelling Kelvin waves (figures 6c-d). In the eastern equatorial Indian Ocean off west Sumatra coast, however, we detected upwelling signals (negative SSH anomalies) during the peak phase of the pIOD event starting from September (figure 6c). This negative SSH signal was strengthened in October and propagated northward and southward (figures 6d). In contrast to the SST pattern (figure 4), the SSH has signified a clear dipole pattern during the peak phase of the pIOD event in September-October 2015 with negative (positive) SSH anomalies occupied the eastern (western) tropical Indian Ocean (figures 6c-d).

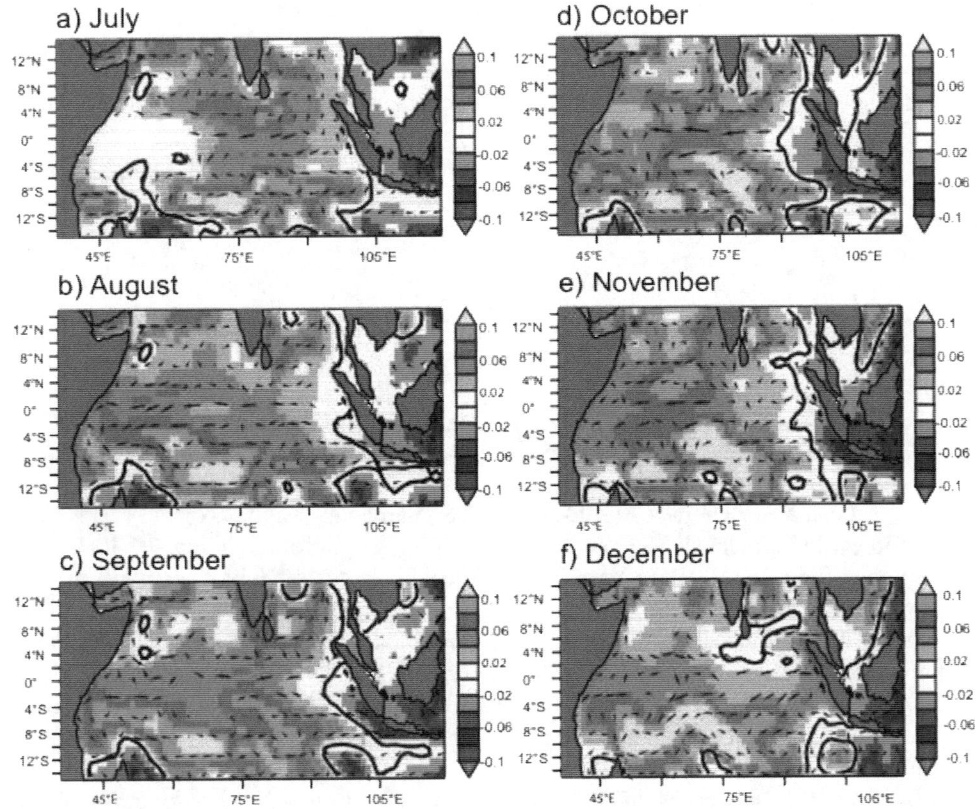

Figure 6. Anomalous evolutions of the sea surface height (*shaded* in meter) and the surface current (*vector* in m/s) in the tropical Indian Ocean during July – December 2015.

Meanwhile, the response of near-surface currents to the anomalous winds along the equator was obvious. Anomalous westward currents were detected along the equator during the peak phase of the pIOD in from August to October 2015 (figures 6b-d). These anomalous westwards currents were terminated in November as the easterly wind anomalies were weakened (figure 6e). The near-surface currents were reverse their direction in December although in the eastern equatorial Indian Ocean still occurred the easterly wind anomalies (figure 6f). These observations were the same analysis which was done by Nababan (2009) that there was the anomalous of surface current in the equator Indian Ocean during October – December. The northern-southeastern of tropical Indian Ocean took place the reverse direction of surface current.

4. Conclusion

A short-lived pIOD event took place in the tropical Indian Ocean, which was developed in August. The highest intensity occured in September/October and ceased in late-November/early December. Correlated with this pIOD event, strong easterly wind anomalies along the equatorial Indian Ocean perceived strong easterly wind anomalies during the peak phase of the event. The effect of easterly wind anomalies to the oceanic dynamic, along the equatorial Indian Ocean detected strong westward current anomalies.

Interestingly, the SSH pattern did not show a clear pattern of the upwelling Kelvin waves as a response of the surface ocean to the wind fields. However, in off equatorial Indian Ocean, in particular in the southern hemisphere, robust westward spreading downwelling Rossby waves were detected from the development phase until the cessation phase of the pIOD event. These westward downwelling Rossby waves induced warm SST to the western pole of the pIOD in agreement with previous investigation [16]. Once reached the western boundary of the Indian Ocean, these downwelling Rossby waves were delineated back into the interior Indian Ocean as eastward propagating downwelling Kelvin waves.

Acknowledgement
The first author was supported by the PMDSU Scholarship from the Ministry of Research, Technology and Higher Education, Indonesia. Gratitude was also conveyed by the author to the anonymous reviewers for their valuable comments. The last auhtor was sponsored by the Ministry of Research, Technology and Higher Education through the *Penelitian Berbasis Kompetensi 2017*.

References

[1] Schott F A, Xie S-P and McCreary Jr J P 2009 Indian Ocean Circulation and Climate Variability *Rev. Geophys.* **47** 1–46
[2] Cherchi A, Gualdi S, Behera S, Luo J J, Masson S, Yamagata T and Navarra A 2007 The influence of tropical Indian Ocean SST on the Indian summer monsoon *J. Clim.* **20** 3083–105
[3] Ajayamohan R S, Rao S A and Yamagata T 2008 Influence of Indian ocean dipole on poleward propagation of boreal summer intraseasonal oscillations *J. Clim.* **21** 5437–54
[4] Dwivedi S 2012 Forecasting the peak anomalies of dominant intrinsic modes of Indian Ocean Dipole Deep. *Res. Part I Oceanogr. Res. Pap.* **70** 73–82
[5] Jarvis C, Darbyshire R, Eckard R, Goodwin I and Barlow E 2018 Influence of El Niño-Southern Oscillation and the Indian Ocean Dipole on winegrape maturity in Australia *Agric. For. Meteorol.* **248** 502–10
[6] Nagura M and McPhaden M J 2010 Dynamics of zonal current variations associated with the Indian Ocean dipole *J. Geophys. Res. Ocean.* **115** 1–12
[7] Deser C, Alexander M A, Xie S-P and Phillips A S 2010 Sea Surface Temperature Variability: Patterns and Mechanisms *Annu. Rev. Mar. Sci.* **2** 115–43
[8] B A, Cherchi A and Ratna S B 2017 ENSO and the recent warming of the Indian Ocean *Int. J. Climatol.* **1**
[9] Saji N H, Goswami B N, Vinayachandran P N and Yamagata T 1999 A dipole mode in the tropical Indian Ocean *Nature* **401** 360–3
[10] Ashok K, Guan Z, Saji N H and Yamagata T 2004 Individual and Combined Influences of ENSO and the Indian Ocean Dipole on the Indian Summer Monsoon *Am. Meterological Soc.* **17** 3141–55
[11] Saji N H and Yamagata T 2003 Possible impacts of Indian Ocean Dipole mode events on global climate *Clim. Res.* **25** 151–69
[12] Li T, Wang B, Chang C-P and Zhang Y 2003 A Theory for the Indian Ocean Dipole-Zonal Mode* *J. Atmos. Sci.* **60** 2119–35
[13] Murtugudde R, McCreary J P and Busalacchi A J 2000 Oceanic processes associated with anomalous events in the Indian Ocean with relevance to 1997–1998 *J. Geophys. Res.* **105** 3295
[14] Nababan B 2009 Unusual upwelling evidence along easten part of equator in Indian Ocean during 1997-1998 El-Niño event *J. Kelaut. Nas.* **4** 16–31
[15] Gupta N and Bhaskaran P K 2017 Inter-dependency of wave parameters and directional analysis of ocean wind-wave climate for the Indian Ocean *Int. J. Climatol.* **37** 3036–43
[16] Sun S, Fang Y, Feng L and Tana 2014 Influence of the Indian Ocean Dipole on the Indian Ocean Meridional Heat Transport *J. Mar. Syst.* **134** 81–8
[17] Kanzow T, Cunningham S a, Rayner D, Hirschi J J-M, Johns W E, Baringer M O, Bryden H L, Beal L M, Meinen C S and Marotzke J 2007 Observed flow compensation associated with the MOC at 26.5 degrees N in the Atlantic. *Science* **317** 938–41

IOP Conf. Series: Earth and Environmental Science **149** (2018) 012055 doi:10.1088/1755-1315/149/1/012055

Indonesia sea surface temperature from TRMM Microwave Imaging (TMI) sensor

Y Marini *, **K T Setiawan**

Remote Sensing Application Center of LAPAN, Jl. Kalisari No. 8 Pekayon, Pasar Rebo, Jakarta Timur - 16418

E-mail: yennie.marini@lapan.go.id

Abstract. We analysis the Tropical Rainfall Measuring Mission's (TRMM) Microwave Imager (TMI) data to monitor the sea surface temperature (SST) of Indonesia waters for a decade of 2005-2014. The TMI SST data shows the seasonal and interannual SST in Indonesian waters. In general, the SST average was highest in March-May period with SST average was 29.4°C, and the lowest was in June – August period with the SST average was 28.5°C. The monthly SST average fluctuation of Indonesian waters for 10 years tends to increase. The lowest SST average of Indonesia occurred in August 2006 with the SST average was 27.6° C, while the maximum occurred in May 2014 with the monthly SST average temperature was 29.9 ° C.

1. Introduction

Sea surface temperature is one of the parameters that determine the waters quality, and directly affect the life of marine organisms as well. Temperature changes will affect the metabolism, reproduction, and distribution of fish in the sea [1]. Sea surface temperature (SST) variations experienced from time to time in accordance with the natural conditions that affect the waters. These changes occur on a daily, seasonal, interannual and long-term (decadal). Monitoring long-term changes in the ocean temperature and circulation is important for understanding both the causes and results of climate change, since the oceans provide a net sink for both anthropogenic CO2 emissions and the additional heat trapped by changes in the atmospheric composition, as well as providing an important SST-dependent feedback mechanism via evaporation [2].

Currently SST measurements have been simplified with the presence of remote sensing technology that can reach synoptic sea-level areas to detect dynamic sea-level physical changes when compared to in situ field observations. the National Oceanic and Atmospheric Administration's geostationary satellite (NOAA) was the first satellite to measure sea surface parameters using infrared sensors in the 1970s [3]. Infrared sensors provide measurements with relatively high spatial resolution, but unable penetrate clouds and aerosols, this is the limitation of using IR sensor measurements of all weather conditions.

This is different when using a microwave sensor because thermal radiation at microwave frequencies is less to be affected by cloud cover and water vapor contamination as well [4-5]. Those advantages make the microwave SST data widely used for ocean research and SSTs play an important role in Earth science research [6-11]. Satellite microwave radiometry able to measure of SST in all-weather condition except area of active rain, sun-glitter and close to the shoreline due to side-lobe contamination. Side-lobe contamination caused ± 50 km from shore have no SST value [12].

TRMM Microwave Imager (TMI) is one of five instruments carried by Tropical Rain Measuring Mission (TRMM) satellite. TRMM is one passive microwave satellite which was launched in 1997 by NASA (National Aeronautics and Space Administration) and NASDA (National Space Development Agency of Japan), but on April 8th, 2015 the satellite was shut down, however we still can access the TMI data from Remote Sensing System website. Remote Sensing System (RSS) is established by Frank J. Wentz in 1974. It is a scientific research company with specializing in satellite microwave remote sensing on earth. RSS started to process the TRMM data on December 7th, 1997 until April 8th, 2015.

Microwaves penetrate clouds with little attenuation, giving an uninterrupted view of the ocean surface. The comparisons TMI SST with ocean buoys show a root mean square difference of about 0.6°C, which is partly due to the satellite-buoy spatial-temporal sampling mismatch and the difference between the ocean skin temperature and bulk temperature [4]. Gross [13] compared 4 years of TMI SSTs by the Tropical Atmosphere Ocean/Triangle Trans-Ocean Buoy Network (TAO/TRITON), and Pilot Research Moored Array in Tropical Atlantic (PIRATA) moored buoy in situ SSTs, and the result shown that TMI SSTs have a mean bias of -0.07°C and a standard deviation of 0.57°C. A study of daily SST availability from an IR polar orbiter, IR geostationary, and TMI gave SST coverage of 48%, 56%, and 78%, respectively [14], and in this study, we will demonstrate the ability of TMI STT data that can provide free cloud cover SST information to describe the sea surface temperature in Indonesia waters for a decade of 2005-2014.

2. Method

The research was done in the Indonesian water with boundary coordinate 10°N – 16°S and 92° –143°E. Data used in this study were monthly TMI SSTs data from 2005 – 2014. TMI data was downloaded from Remote Sensing Systems (RSS) Website (http://data.remss.com/tmi/).

The RSS is provided TMI SST data as daily, 3-day, weekly, and monthly data. Daily data consists two data coverage data which are ascending orbit segment (daytime) and descending orbit segments (night time). 3-day data is the average of 3 days ending on and including file date. For example, 3-day data dated April 3, 2014, then the data is the average of data dated April 1, 2, and April 3, 2014. Weekly data is an average of 7 days, starting from Sunday through Saturday. For example, the date of weekly SPL data is April 12, 2014, then the data is the average of the daily data of April 6, 2014 (Sunday) to April 12, 2014 (Saturday), while the monthly data is the average of all data within the calendar month.

Any binary data is available at http://data.remss.com/tmi, consists of fourteen grid for daily data and six grids for average data (3-day, weekly and monthly data) sized 0.25 X 0.25 degrees from (1440,720) byte maps.

The data values fall between 0 and 255. Specific values have been reserved:

0 – 250 = valid geophysical data
251 = missing SST or wind speed due to rain, or missing water vapor due to heavy rain
252 =*not used in this data set*
253 = TMI observations exist, but are bad
254 = no TMI observations
255 = land mass

The data values between 0 and 250 need to be scaled to obtain meaningful geophysical data. To scale the data, multiply by the scale factors with the equation below:

$$\mathbf{(SST * 0.15) - 3} \quad (1)$$

The valid data range is -3°C to 34.5°C.

The TMI binary data can be extracted using mathlab, fortran, idl or phyton software. After the data are extracted, then the data are converted into a data format that can be read by image processing software such as ER Mapper, Envi, etc. for further analysis. The flowchart of TMI SSTs data processing presented on figure 1.

IOP Conf. Series: Earth and Environmental Science **149** (2018) 012055 doi:10.1088/1755-1315/149/1/012055

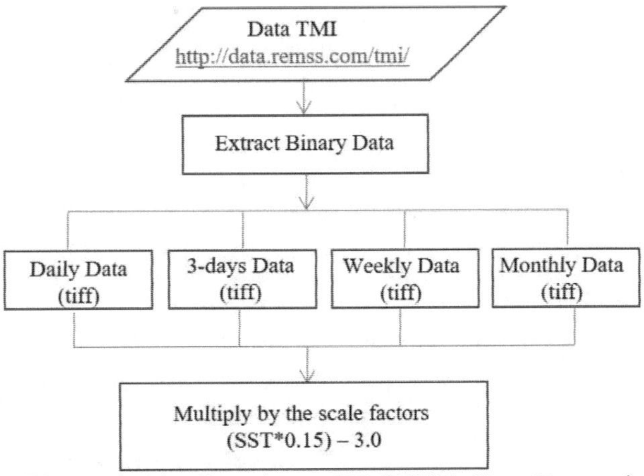

Figure 1. The flowchart of TMI SSTs Data Processing

3. Result and Discussion

The extraction result of TMI SST binary data is shown in figure 2. The SST value is still in the range of 0-255, in order to produce the SST value in °C, the TMI SST binary data should be multiplied by the scale factor shown in equation 1, SST in °C is shown in figure 3. Figure 4 shows the monthly TMI SSTs data along 2005 and 2014.

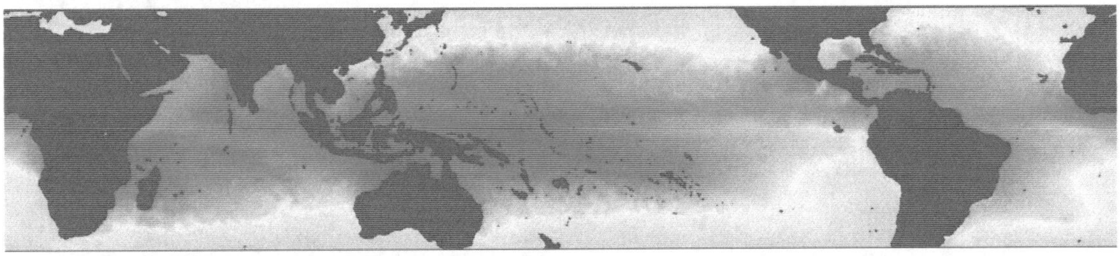

Figure 2. December 2014 monthly TMI SST distribution on scale 0 - 255

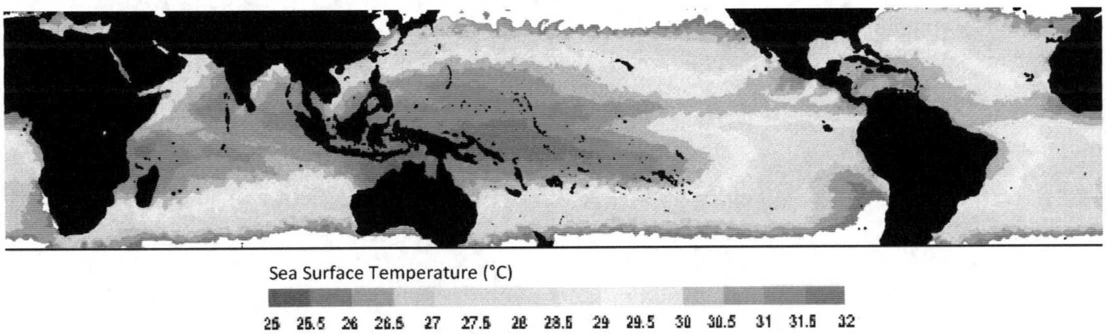

Figure 3. December 2014 monthly TMI SST distribution on celsius degree (°C)

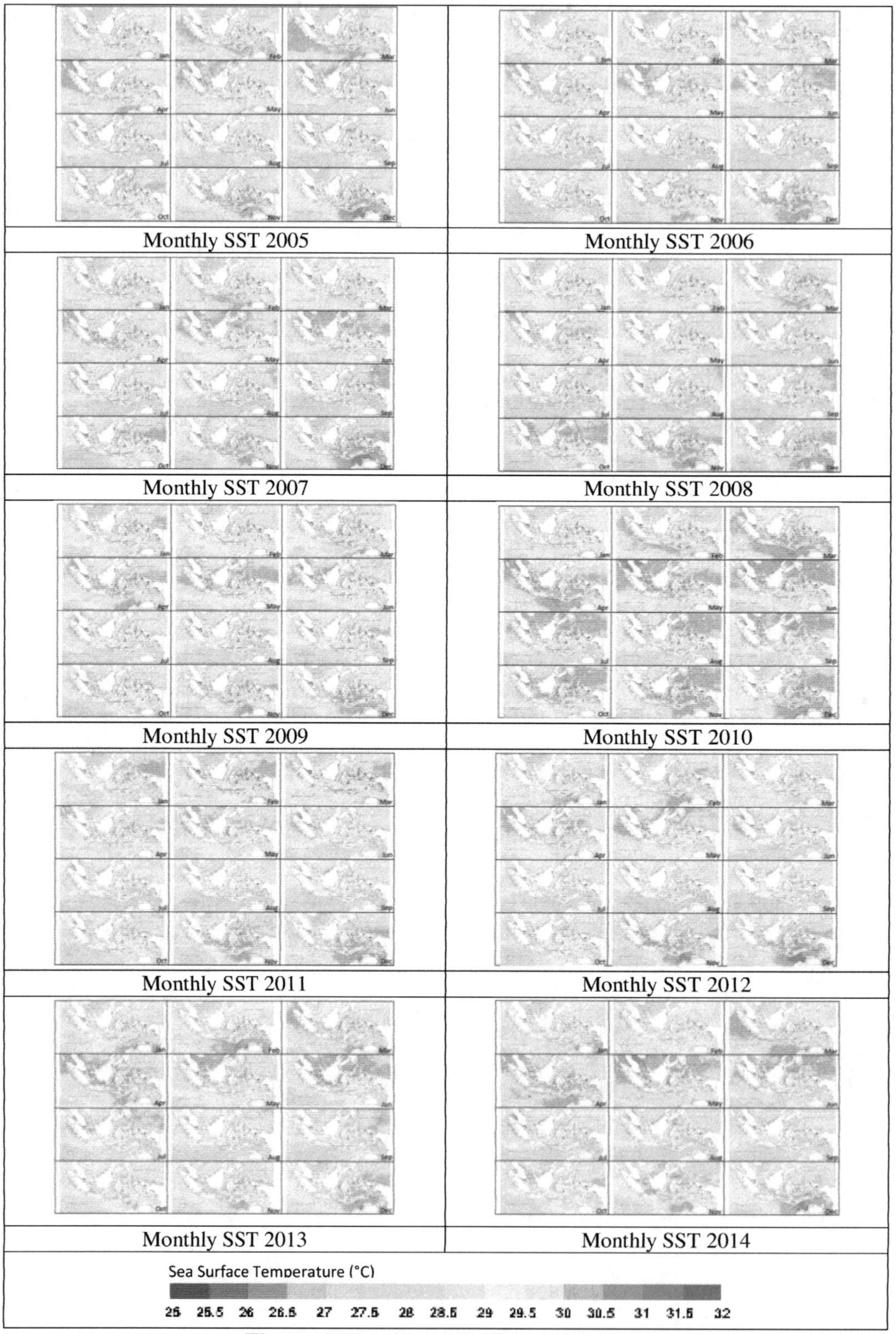

Figure 4. Monthly TMI SSTs 2005-2014

The annual average SST images displayed in figure 5. It was derived from calculating the monthly average SST data for a year. In general, the SST distribution in Indonesian waters is similar, northern parts of Indonesian waters have a higher SST distribution compared to the southern part.

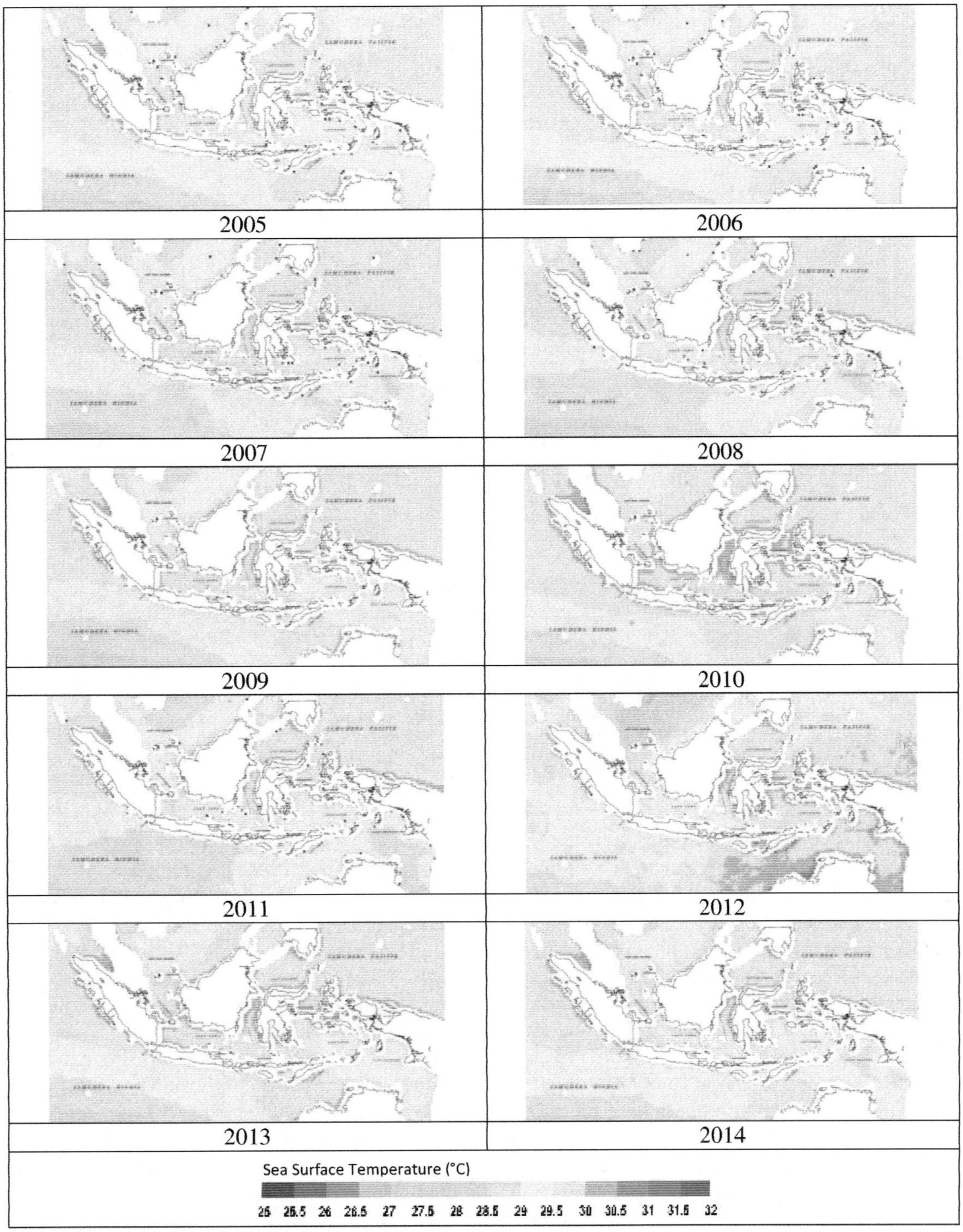

Figure 5. Annual average of TMI SSTs 2005-2014

SST fluctuation from TMI data from 2005 to 2014 is presented as a graphic in figure 6. From the graphic we can see, generally, the SST fluctuations in Indonesian waters is sinusoidal form, where the highest peak usually occurs in May and lowest in August. This is in line with the previous research [15]. Eningtyas [15] performed an analysis of the rainfall and sea surface temperature correlation in Indonesia using NOAA data from 1986 to 2006. The results show SST was relatively higher compared to other months starting from January and reaching its peak in May. The low water temperatures in August due to the influence of upwelling zone south of Java and over the Arafura Shelf. The cool waters appear to be carried into the eastern Java Sea, then flow northwestward and enter the South China Sea through Karimata Strait [16].

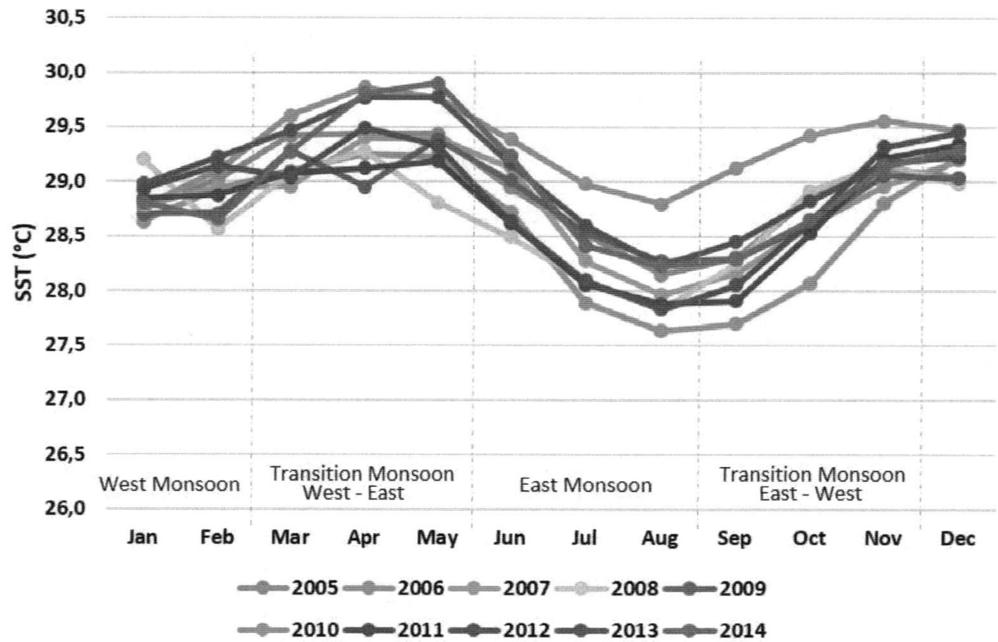

Figure 6. Graph of temporal variability monthly SST average in Indonesian waters

Indonesian SST characteristics during 10 years based on TMI data ranged from 27.6° to 29.6°C, with the dominant temperature ranging from 28.0 – 29.0°C. Indonesia's sea surface temperature varies seasonally. Generally, the SST average was highest in March-May period with SST average was 29.4°C, and the lowest was in June – August with the SST average was 28.5°C. Another study showed the same result [17]. Habibie [17] studied the trend of SST changes in Indonesia by using the Integrated Global Ocean Services System (IGOSS) SST data from 1982 – 2009. The result showed that in March–May period generally was the warmest SST in Indonesia compared to other periods. March-May period is the transitional monsoon of the West monsoon to the East monsoon, it is said that during the transition monsoon the distribution of SPL in the territory of Indonesia is almost evenly distributed, there is no significant difference between the northern and southern hemispheres, in contrast to the conditions during the monsoon period in June - August which coincides with the period of the east monsoon or the Australian monsoon, the southern hemispheres much cooler than northern hemispheres.

Temporarily, the variation of sea surface temperature in the period 2005 - 2014 can be seen in figure 7. The monthly SST average fluctuation of Indonesian waters for 10 years tends to increase. The lowest SST average of Indonesia occurred in August 2006 with the SST average was 27.6° C, while the maximum occurred in May 2014 with the monthly SST average temperature was 29.9 ° C.

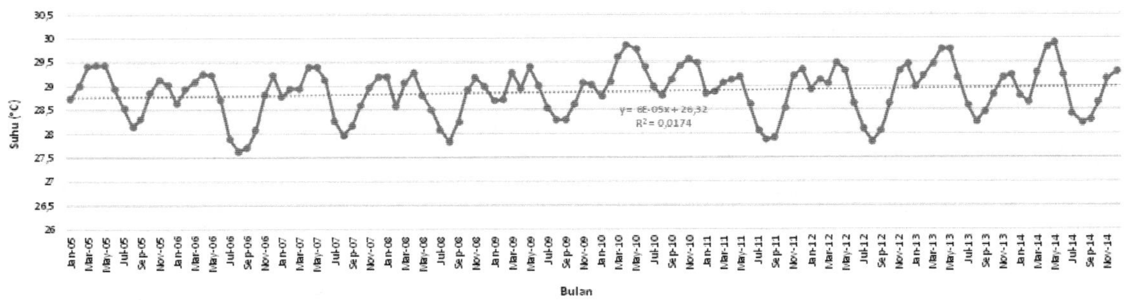

Figure 7. Graph of temporal variability monthly SST average in Indonesian waters during 2005-2014

The annual SST variation of Indonesian waters during 2005-2014 based on TMI data were ranging from 1.1°-1.7°C, the lowest annual SST variation occurred in 2009 and 2010 at 1.1°C while the highest occurred in 2012 and 2014 at 1.7°C. This corresponds to [18] statement, that the annual variation of Indonesian sea surface temperature is about 2° C. This condition also corresponds to the statement of [19] that the condition of the tropical sea surface layer is warm and the annual temperature variation is small and the daily temperature variation is not too high compared to subtropics and polar. The average annual temperature variation of tropical waters is less than 2 ° C in the equatorial area. The graphic annual variation and average of Indonesia's SST during the 2005-2014 period can be seen in figure 8.

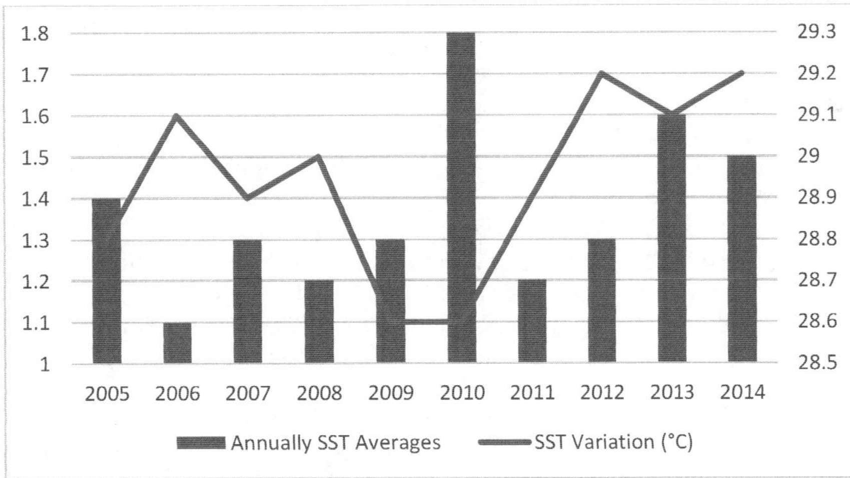

Figure 8. Annual variation and average TMI SSTs 2005 – 2014

4. Conclusion

This research showed that TMI data can describe the SST variability of Indonesian water monthly, seasonal, and interannual. Knowing the variability of the SST helps us to understand the anomaly of SST that occurred in Indonesian waters. Indonesian SST characteristics during 10 years based on TMI data ranged from 27.6° to 29.6°C, with the dominant temperature ranging from 28.0 – 29.0°C.

During the interval 2005-2014 the monthly SST average fluctuation of Indonesian waters for 10 years tends to increase, the lowest annual average of SST in Indonesian waters occurred in 2006 which is 28.6°C and the highest is in 2010 with the average temperature is 29.3°C. Generally, the SST average was highest in March-May period with SST average was 29.4°C, and the lowest was in June – August with the SST average was 28.5°C, with he annual SST variation of Indonesian waters during 2005-2014 based on TMI data were ranging from 1.1°-1.7°C.

5. Acknowledgement

The authors are grateful to Dr. Rokhis Khomarudin, Director of Remote Sensing Applications Centre and Syarif Budhiman, M.Sc., Head of Program and Facilities, Remote Sensing Applications Centre of Indonesian National Institute of Aeronautics and Space (LAPAN) for valuable support to this research. TMI data were produced by Remote Sensing Systems and sponsored by the NASA Earth Sciences Program. Data are available at www.remss.com/missions/tmi. Special thanks to NASA Goddard Space Flight Centre for providing access to TMI monthly data files.

References

[1] Nybakken J W 1988 *Marine biology: An ecological approach*
[2] Gentemann C L 2004 In situ validation of Tropical Rainfall Measuring Mission microwave sea surface temperatures *Journal of Geophysical Research*
[3] McClain E P, Pichel W G and Walton C C 1985 Comparative performance of AVHRR-based multichannel sea surface temperatures *Journal of Geophysical Research* **90** 11587–11601
[4] Donlon C J, Gentemann C L, and Wentz F J 2001 Measuring surface temperature with microwave sensors *Backscatter* **12(2)** 37–39
[5] Gentemann C L, Minnett P J, Le Borgne P, and Merchant C J 2008 Multi-satellite measurements of large diurnal warming events *Journal of Geophysical Research Letters* **35** (L22602) 1-6
[6] Chelton D B and Wentz F J 2005 Global microwave satellite observations of sea surface temperature for numerical weather prediction and climate research *Journal of Climate* **86** (80) 1097–1115
[7] Chelton D B, Wentz F J, Gentemann C L, Szoeke R A D, and Schlax M G 2000 Satellite microwave SST observations of transequatorial tropical instability waves *Journal of Geophysical Research Letters* **27**(9) 1239–1242
[8] Chelton D B, Esbensen S K, Schlax M G, Thum N, Freilich M H, Wentz F J, Gentemann C L, McPhaden M J, and Schopf P S 2001 Observations of coupling between surface wind stress and sea surface temperature in the eastern tropical Pacific *Journal of Climate* **14** (7) 1479–1498
[9] Stammer D, Wentz F J, and Gentemann C L 2003 Validation of microwave sea surface temperature measurements for climate purposes *Journal of Climate* **16** (1) 73–87
[10] Wentz F J, Gentemann C L, Smith D K, and Chelton D B 2000 Satellite measurements of sea surface temperature through clouds *Science* **288** (5467) 847–850
[11] Gentemann C L, Wentz F J, Carls A M, and Deborah K S 2004 In situ validation of tropical rainfall measuring mission microwave sea surface temperatures *Journal of Geophysical Research* **109** (C04021) 1-9
[12] Emery, William J, Brank P, Funk A, and Boning C 2006 A comparison of sea surface temperatures from microwave remote sensing of the Labrador Sea with in situ measurements and model simulations *Journal of Geophysical Research* **111** (C12013)
[13] Gross M G 1990 *Oceanography A View of Earth. Prentice Hall. Englewood Cliffs, New Jersey.*
[14] Guan L, and Kawamura H 2003 SST availabilities of satellite infrared and microwave measurements *Journal of Oceanography* **59** 201–209
[15] Eningtyas W, Ramadhani F, and Aldrian E 2007 Correlation analysis of rainfall and Indonesia sea surface temperature and its implications on rainfall prediction: Cilacap case study Journal of Agromet Indonesia **21** (2) 46-60
[16] Qu T, Du Y, Strachan J, Meyer G S, and Slingo J 2005 Sea surface temperature and its variability in The Indonesian region *Journal of Oceanography* **18** (4) 51-61
[17] Habibie M N, Nuraini T A 2014 The characteristic and trend of sea surface temperature over Indonesia in 1982-2009 Jurnal Meteorologi dan Geofisika **15** (1) 37-49
[18] Reynolds R W, Gentemann C L, and Wentz F J, 2004 Impact of TRMM SSTs on a climate-scale SST analysis Journal of Climate **17** (8) 2938–2952
[19] Wyrtki K 1961 *Physical Oceanography of The Southeast Asean Waters. University of California, La Jolla. California.*

IOP Conf. Series: Earth and Environmental Science **149** (2018) 012056 doi:10.1088/1755-1315/149/1/012056

Diversity and distribution of Dorippid Crabs (Brachyura: Dorippidae) in East Coast of Lampung, Indonesia

Zairion, AA Hakim, A Mashar, A Fahrudin, L Adrianto, B Widigdo, Y Wardiatno
Department of Aquatic Resources Management, Faculty of Fisheries and Marine Sciences, Bogor Agricultural University (IPB), Bogor 16680, Indonesia

E-mail: zairion.zai@gmail.com

Abstract. Dorippid crabs (Brachyura: Dorippidae) commonly live in sand or mud substrate at shallow coastal water. These crabs are a macro-benthos, mostly found in tropic, have no economic value, but have high abundance in blue swimming crab habitat and lack information in Indonesia. This study aims to determine diversity and species distribution of dorippid crabs in East Coast of Lampung, Indonesia. This study was conducted on March-April 2017 in seven locations along coastal area (i.e. Labuhan Maringgai, Kuala Penet, Sekopong, Wako, Seputih, Sungai Burung, and Kuala Teladas). The specimen was collected from the blue swimming crab gill-net fishery as by-catch. Each location found by this family was noted at coordinate point at satellite imagery map and processed using ArcGIS. Two species have been found, there are *Dorippe quadridens* and *Dorippoides facchino*. Both species were found at every location, except *D. quadridens* at Sungai Burung as well as *D. facchino* in Sekopong. There are 74% among the sample of fishermen caught *D. quadridens* in those areas with an average number of 36 individuals and weight of 959 grams. While, 40% of fishermen caught *D. facchino* with an average number of 17 individuals and weight 152 grams.

1. Introduction

Research on the taxonomic diversity of decapod crustaceans has increased considerably during the last decades [1], whereas the diversity, distribution and abundance of marine crabs is one among the issues. Spatially or temporally distribution of a species is their distribution across its habitats, from home range to utilization distribution, and its implications for its life history [2]. An others consideration are related to fisheries, such as the unintentionally caught species or bycatch related to ecological impact of fishing.

The coastal waters of Eastern Lampung is one of distribution area of the blue swimming crab (*Portunus pelagicus*) in Indonesia, which is part of coastal waters and shallow seas of Southeast and Eastern Asia as main distribution area of this species [3]. This area is potential for blue swimming crab wild catch in Indonesia and the species is one of important fishery resources that exploited by small-scale fishery, mostly caught with bottom gill-nets and has been exploited more than two decades [4-7]. Bottom gill nets were operated as a static fishing gear in the bottom of the waters. The use of net caused abundant non target species or bycatch. In Thailand, local fishermen used a stationary gear traditionally. They believed that this gear has not much effect to bycatch, however, they still get non-target species or bycatch The bycatch consists of other crustaceans, molluscs, echinoderms, fish and soft coral [8]. According to [9], the bycatch information in the blue swimming crab (BSC) fishery is limited. A further case to consider is unintentionally caught species, so-called "bycatch", with a portion being marine crabs. Most of them are non-economic species and are usually discarded. Bycatch has been the world's most important fisheries issue since the 1990s. As a result of bycatch issues, many species of marine crab are currently threatened. Unfortunately, there has been a limited study on the species of diversity and ecology of marine crabs for conservation purposes.

Crustaceans might the largest proportion of bycatch from the blue swimming crab net fishery and dorippid crabs (Brachyura: Family Dorippidae) were included [8], but it was excluded collapsible crab trap fishery [9]. Research on the family Dorippidae is more about the existence of species, such as species records in Madagascar [10], Australia [11,12], Thailand [13], New Caledonia, Indonesia and Philippines [14]. However, there is limited research and information on dorippid crabs in Indonesian waters, including their occurrence as bycatch in certain fishery or their biological and ecological aspects.

The IUCN status of dorippid crabs is Not Evaluated for most of them. This status reflected the species that have very little information or research that has been done in the world. Researches on biological aspects have not been done in Indonesia, even very little research in other countries. For performing of resources management, basic information was required such as biological information [15]. In addition, the East Lampung coastal waters is one of blue swimming crab fisheries (BSC) center in Indonesia. One of blue swimming crabs main food was crustacea [16,17]. As an example, 78.43% of total food was crustacea for the adult crabs in India [18]. The condition assumed that dorippid crab might one of BSC main food sources or associated species, because they were very abundant. The terms have no economic value, but the species have a large ecological role so it was important to examine. This study aims to determine the diversity and distribution of dorippid crabs in East Coast of Lampung, Indonesia.

2. Method

2.1. Data Collection
This study was conducted on March-April 2017 in the East Coast of Lampung, Indonesia. There were seven locations in coastal area from south to north of the East Coast of Lampung have selected to be sampling station, which was represent a coastal water habitat related to coastal morphology, vegetation coverage based on satellite imagery in google map. Those sampling site were Labuhan Maringgai, Kuala Penet, Sekopong, Wako, Seputih, Sungai Burung, and Kuala Teladas (figure 1). Dorippid crabs are among bycatch from the blue swimming crab bottom gill-net fishery. One fisherman has a set of bottom gill-net. Some fishermen who were fishing in certain fishing ground around sampling site have chosen purposively to be unit of sample. All catches from blue swimming crab net of fisherman sample were taken and separated between target and non-target catches. Non-targeted species were grouped into crustacean, fishes, molluscs, and others. All non-targeted species were identified, separated by genus or species, calculated in number, and weighed in total from each species. Weight was measured by digital scale to the nearest 1 gram. Identification of species from the Dorippidae family is based on the identification book [19].

2.2. Data Analysis
All non-targeted species were measured their proportion in each group and analyses diversity of Dorippid among sampling site. Each location found by this family was noted at coordinate point and processed using ArcGIS 10.2. The spatial distribution of abundance and biomass of Dorippid crabs was measured based on total number and weight of certain species at standardized net length and visualized using ODV 4. Ocean Data View 4 (ODV 4) is a computer software program that serves to display the exploration results of oceanography and geo-reference view, as well as sequence of data (grid data) interactively. One of the function is interpolation of data points into grid with weighted average gridding method.

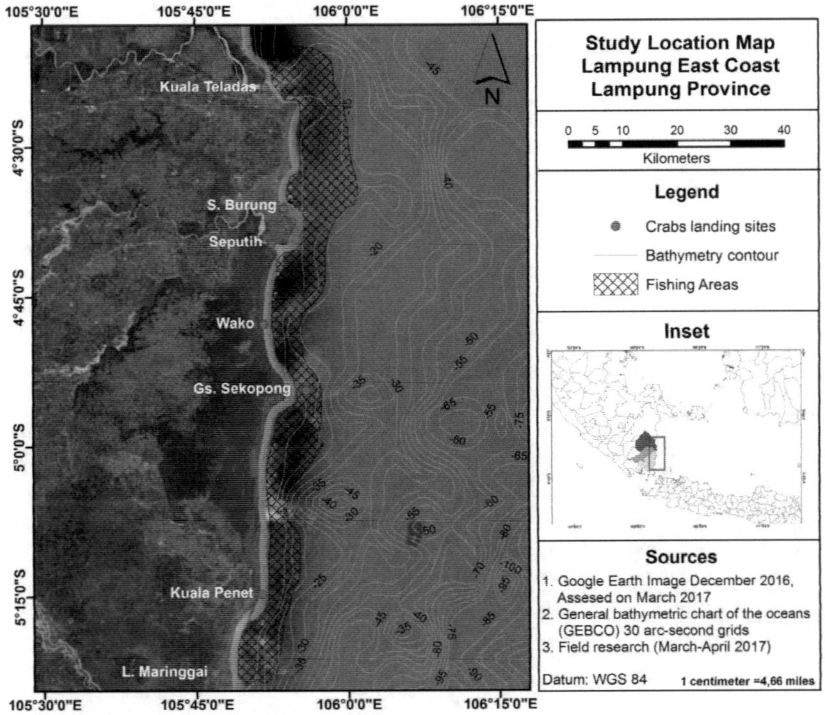

Figure 1. Study area of dorippid crab and sampling sites in East Coast of Lampung, Indonesia.

3. Result and Discussion

3.1 Non-Targeted Catches

Data retrieval had been done for two months and had received non-targeted catch data. The proportions of non-targeted catches based on the weight of each group were varies in each location (table 1). Crustaceans were dominant catches in volume at Kuala Penet, Wako, and Sungai Burung, whiles molluscs in Sekopong, Seputih, and Kuala Teladas. There was only one location (Labuhan Maringgai) found that the volume of non-target catches dominated by fishes. The average proportion of each group at all sites was also dominated by crustaceans, followed by molluscs, other species and fishes. Crustaceans occupied 41%, whiles molluscs 29%, other species 16%, and fish at least 14% (figure 2). This condition might influenced by diversity of habitat along coastal waters at eastern Lampung related to coastal coverage area, input from land trough the river and oceanographic condition. It might also affect by gill-net mesh size used whereas the fisher in Labuhan Maringgai use larger net mesh-size (4-4.5 inch) than the other areas (3.5-4 inch).

Blue swimming crab caught by using bottom gill nets where operated as a static fishing gear in the bottom of the waters. Mollusks and most of crustaceans groups attached on aquatic substrates and sessile organism, so the two most dominant groups were caught. Several species of mollusks attach to shallow marine substrate, gastropod more commonly found on sand substrate, sand-bottom bottom and reed zones [20]. In addition, other groups also many caught from starfish and soft coral (sponge). Several species of coral sponges were found in shallow waters with mud substrate [21].

Table 1. Proportion (%) of each non-targeted catches group from 7 locations.

Group	Location						
	Labuhan Maringgai	Kuala Penet	Sekopong	Wako	Seputih	Sungai Burung	Kuala Teladas
Crustaceans	27	75	29	44	31	50	21
Fish	34	5	4	11	25	2	17
Molluscs	18	7	36	39	36	40	36
Others	21	13	31	6	8	8	26

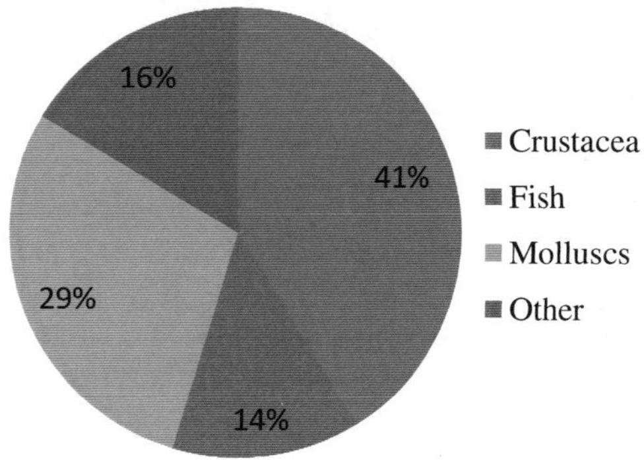

Figure 2. The average proportion of each group of non-targeted catches at all locations.

3.2 Diversity

One of the dominant crustaceans is dorippid crabs from the familly of Dorippidae. The specimens were called as "trash" by fisherman because the species was no economic value and broke the net. Two species have been found, there were *Dorippe quadridens* (Fabricius, 1793) and *Dorippoides facchino* (Herbst, 1785), which indicate the diversity of Dorippidae species in East Coast of Lampung. Dorippid crabs are commonly live in sand or mud substrate at shallow coastal water. These crabs are a macro-benthos, mostly found in tropic, and have no economic value and consumption. Familly Dorippidae consists of nine genera, two of which are genus *Dorippe* and *Dorippoides*. There are seven species of genus *Dorippe* [11] and 2 species of genus *Dorippoides* [19] are known from the Indo-West Pacific region.

The results of morphological identification were the same as those described by [19]. Description of *Dorippe quadridens* is carapace strongly sculptured, granulated, bearing pubescence and grooves evident (figure 3a). Tubercles usually well indicated and rather high. Merus covered all over with dense short pubescence. Abdomen with 3 distinct teeth on second somite. Lower orbital margin with 5-6 spines. The fourth and fifth legs are covered with dense short pubescence and end with hooks. While description of *D. quadridens* was carapace convex laterally and posteriorly, more flattened in middle and anteriorly (figure 3b). Front consisting of 2 distinct, triangular teeth, directed forward, or of blunter teeth, directed outward. Chelipeds distinct unequal, right larger, left as in female. Second and third pereiopods slender.

Both species were also found at every location, but *D. quadridens* was not found at Sungai Burung as well as *D. facchino* in Sekopong (table 2). Generally, there were seventy four percent (50/68) of fishermen caught *D. quadridens,* while almost forty percent (27/68) of fishermen caught *D. facchino* in all area. The largest number of sample was 19 fishermen at Kuala Penet, while the lowest was four fishermen at Sungai Burung. Those samples were representative among the blue swimming crab fishers who have fishing ground in adjacent of sampling areas. However, the highest percentage of fisher caught *D. facchino* and *D. quadridens* occurred at Wako and Seputih. It's seemed that those two species distributed widely around those two areas with the high probability captured by the fishermen. On the other hand, the sea surface sediment at those two areas seems visually as muddy substrate as one of Dorippid main habitat [11].

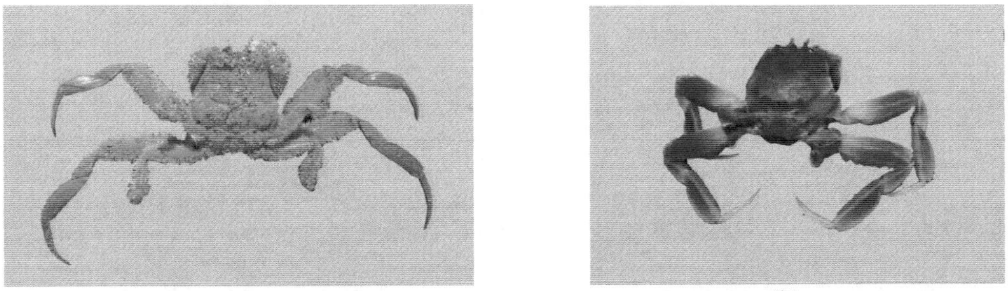

(a) (b)

Figure 3. (a) *Dorippe quadridens* and (b) *Dorippoides facchino* from East Coast of Lampung.

Table 2. Number of fishermen who get *Dorippe quadridens* and *Dorippoides facchino*.

Location	Number of fisherman		
	Sample	Caught *D. quadridens*	Caught *D. facchino*
Labuhan Maringgai	9	3 (33%)	2 (22%)
Kuala Penet	19	14 (74%)	5 (26%
Sekopong	12	11 (92%)	0 (0%)
Wako	9	9 (100%)	8 (89%)
Seputih	6	6 (100%)	5 (83%)
Sungai Burung	4	0 (0%)	3 (75%)
Kuala Teladas	9	7 (78%)	4 (44%)
	68	50 (74%)	27 (40%)

The *D. quadrides* found in sublittoral, shallow water, substrate shelly sand or mud to sand bottoms and rocks [12, 19]. In addition, this species also reported in muddy-sand bottoms on the back reef lagoon, without sea-grass meadows [22]. This species was caught in depths of 7 to 39 m [12], varies from 1 to 73 m and most commonly occurs in 1 to 30 m and ever found 1 individual in 415 m [19]. In this study we found *D. quadrides* at a depth of 2 to 9 m with a substrate dominated by small grain size, and categorized as silt and silty loam sediment.

3.3 Spatial Distribution

The spatial distribution was conducted by weighted average gridding method. The method was extrapolating of closest value and weighting from distance and value of each coordinate. The spatial distribution of the abundance of *D. quadrides* and *D. facchino* was shown in figure 4. Based on the distribution of individual abundance, the most abundance of two species was found at the Kuala Penet location. It is shown with a reddish orange color. The lowest abundance at Labuhan Maringgai and Kuala Teladas location was shown in blue for *D. quadrides*, and at Kuala Teladas location was shown in purple for *D. facchino*. The average number of *D. quadridens* and *D. facchino* were 36 and 17 individuals.

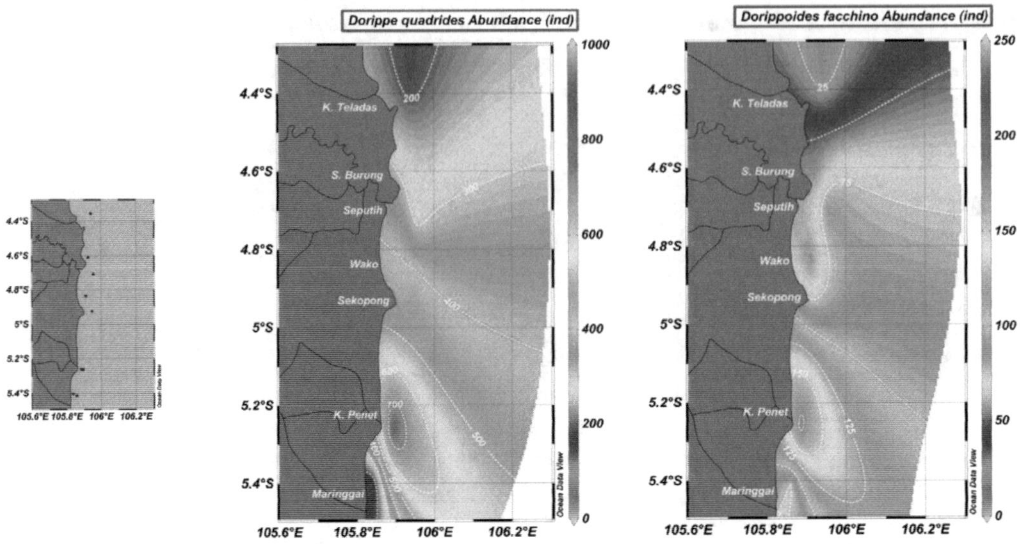

Figure 4. The abundance distribution of *Dorippe quadridens* and *Dorippoides facchino* from East Coast of Lampung.

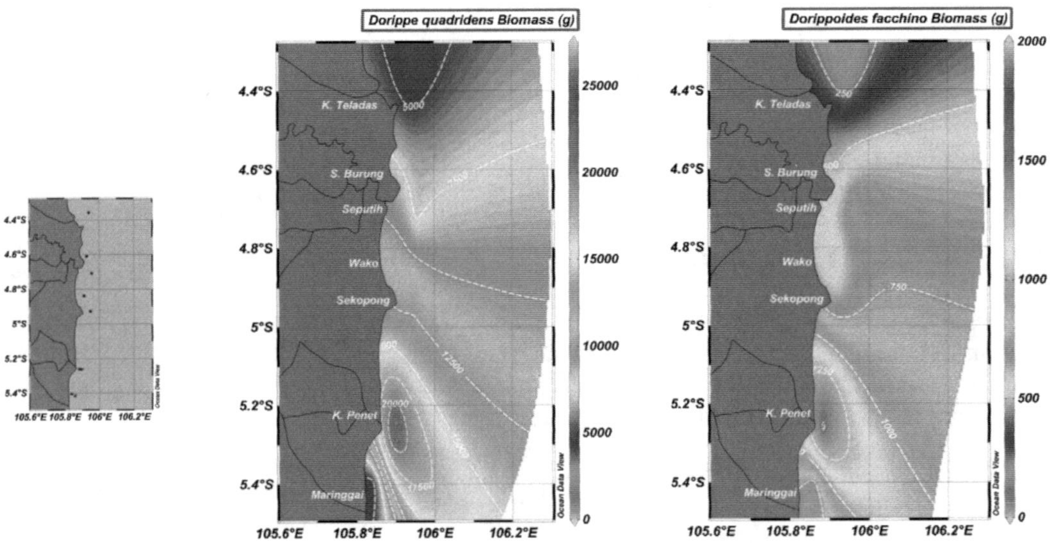

Figure 5. The biomass distribution of *Dorippe quadridens* and *Dorippoides facchino* from East Coast of Lampung.

The biomass distribution of *D. quadrides* and *D. facchino* was shown in figure 5. The highest biomass of two species was found at the Kuala Penet location. It was shown with a reddish orange color. The lowest abundance at Labuhan Maringgai and Kuala Teladas location was shown in blue for *D. quadrides*, and at Kuala Teladas location was shown in purple for *D. facchino*. The average estimated biomass of *D. quadridens* and *D. facchino* were 959 and 152 grams.

The highest abundance and estimated biomass of those two dorippid species was not coincide with the probability of captured by the fisher. This condition might be related to the behavior of those two founded dorippid species, such as schooling.

D. quadrides has a wide distribution within the Indo-West Pacific region, extending from the Suez Canal, the Red Sea and southeastern Africa to Hong Kong, the Philippines, Indonesia, and Australia [19]. This species was one of the most widely distributed species of Dorippinae and the greatest number of specimens and has the widest geographical range [19].

4. Conclusion

Two species were found at every location as represent on diversity of dorippid crabs, except *D. quadridens* at Sungai Burung as well as *D. facchino* in Sekopong. The highest probability of captured to those two species occurred in Wako and Seputih, but the highest abundance and biomass found at Kuala Penet, and might related to the schooling or patchy behavior of the species. This study provided biological information that can be used as a basis for utilization and management of blue swimming crab and dorippid crab resources. In addition, further studies should be conducted on whether there was symbiosis, competition, or predators-prey relationship between dorippid crabs and blue swimming crab as one of importantt fishery resources in Indonesia.

Acknowledgments

This research was funded by the Government of Indonesia through Directorate General of Strengthening Researches and Development, Ministry of Researches, Technology, and Higher Education. The Authors wish to thank to Muhamad Radifa for helping satellite imagery and GIS analysis.

References

[1] Ng P K L, Guinot D and Davie P J F 2008 Systema Brachyurorum: Part I. An annotated checklist of extant brachyuran crabs of the world *The Raffles. Bull. Zool.* **17** 1

[2] Armstrong J D, Kemp P S, Kennedy G J A, Ladle M and Milner N J 2003 Habitat requirements of Atlantic salmon and brown trout in rivers and streams *Fish. Res.* **62** 143

[3] Lai J C Y, Ng P K L and Davie P J F 2010 A revision of the *Portunus pelagicus* (Linnaeus, 1758) species complex (Crustacea: Brachyura: Portunidae), with the recognition of four species *The. Raffles. Bull. Zool.* **58** 199

[4] Zairion, Boer M, Wardiatno Y and Fahrudin A 2014a Composition and size of the blue swimming crab (*Portunus pelagicus*) caught at several bathymetric stratifications in East Lampung waters *J. Lit. Perikan. Ind.* **20** 199

[5] Zairion, Wardiatno Y, Fahrudin A and Boer M 2014b Spatial temporal distribution of *Portunus pelagicus* breeding population in East Lampung coastal waters *Bawal* **6** 95

[6] Zairion, Wardiatno Y, Boer M and Fahrudin A 2015a Reproductive biology of the blue swimming Crab *Portunus pelagicus* (Brachyura: Portunidae) in East Lampung Waters, Indonesia: Fecundity and reproductive potential *Trop. Life. Sci. Res.* **26** 67

[7] Zairion, Wardiatno Y and Fahrudin A 2015b Sexual maturity, reproductive pattern and spawning female population of the blue swimming crab, *Portunus pelagicus* (Brachyura: Portunidae) in East Lampung coastal waters, Indonesia *Indian. J. Sci. Technol.* **8** 596

[8] Fazrul H, Hajisamae S, Ikhwanuddin M and Pradit S 2015 Assessing impact of crab gill net fishery to bycatch population in the Lower Gulf of Thailand *Turkish. J. Fish. Aquat. Sci.* **15** 761

[9] Kunsook C and Dumrongrojwatthana P 2017 Species diversity and abundance of marine crabs (Portunidae: Decapoda) from a Collapsible Crab Trap Fishery at Kung Krabaen Bay, Chanthaburi Province, Thailand *Trop. Life. Sci. Res.* **28** 45

[10] Chen H L 1987 Dorippidae (Crustacea Decapoda Brachyura) collected in Madagascar waters *Bull. Mus. natn. Hist. nat. Paris.* A **3** 677

[11] Manning R B 1993 Two New Dorippid Crabs from Australia (Crustacea: Decapoda: Dorippidae *Rec. Aust. Mus.* **1** 1
[12] Thoma B T 2007 Notes on crabs of the families Goneplacidae and Dorippidae (Decapoda: Brachyura) from the Dampier Archipelago, Western Australia *Rec. West. Aust. Mus.* **73** 299
[13] Ng P K L and Davie P J F 2002 A Checklist of the Brachyuran Crabs of Phuket and Western Thailand *Phuket. Mar. Biol. Center. Sp. Pub.* **23** 369
[14] Chen H L 1993 Crustacea Decapoda: Dorippidae of New Caledonia, Indonesia and the Philippines. *Resultats des Campagnes MUSORSTOM* (vol 10) ed A Crosnier (Paris: Mem Mus natn Hist nat) pp 315-345
[15] Widigdo B, Rukisah, Laga A, Hakim A A and Wardiatno Y 2017 Carapace length-weight and width-weight relationships of *Scylla serrata* in Bulungan District, North Kalimantan, Indonesia *Biodiversitas.* **18** 1316
[16] Zainal K A Y 2013 Natural food and feeding of the commercial blue swimmer crab, *Portunus pelagicus* (Linnaeus, 1758) along the coastal waters of the Kingdom of Bahrain *J. Ass. Arab Univ. for Basic and Appl. Sci.* **13** 1
[17] Kunsook C, Gajaseni N and Paphavasit N 2014 The feeding ecology of the blue swimming crab, *Portunus pelagicus* (Linnaeus, 1758) at Kung Krabaen Bay, Chanthaburi Province, Thailand *Trop. Life. Sci. Res.* **25** 13
[18] Josileen J 2011 Food and feeding of the blue swimmer crab, *Portunus pelagicus* (Linnaeus, 1758) (Decapoda, Brachyura) along the coast of Mandapam, Tamil Nadu, India *Crustaceana.* **84** 1169
[19] Holthuis L B and Manning R B 1990 *Researches on Crustacea, Special Number 3* (Tokyo: The Carcinological Society of Japan) p 151
[20] Filippenko D 2011 Fauna of gastropod molluscs in the Curonian Lagoon littoral biotopes (Baltic Sea, Kaliningrad region, Russia) *Malacologica. Bohhemoslovaca.* **10** 79
[21] Longakit M B A, Sotto F B and Kelly M 2015 The shallow water marine sponges (Porifera) of Cebu, Philippines *Sci. Diliman.* **17** 52
[22] Thomassin B A 1974 *Soft bottom carcinological fauna sensu lata on Tuléar Coral Reef Complexes (S.W. Madagascar): distribution, importance, roles played in trophic food-chains and in bottom deposits* vol 1 (Brisbane: Great Barrier Reef Committee) pp 297-320

IOP Conf. Series: Earth and Environmental Science **149** (2018) 012057 doi:10.1088/1755-1315/149/1/012057

Waveform identification and retracking analyses of Jason-2 altimeter satellite data for improving sea surface height estimation in Southern Java Island Waters and Java Sea, Indonesia

Bisman Nababan[1], Muhammad R. Hakim[1], and James P. Panjaitan[1]
[1] Department of Marine Science and Technology, Faculty of Fisheries and Marine Sciences, Bogor Agricultural University, Bogor, Indonesia

Email: bisman@ipb.ac.id

Abstract. Indonesian waters containing many small islands and shallow waters leads to a less accurate of sea surface height (SSH) estimation from satellite altimetry. Little efforts are also given for the validation of SSH estimation from the satellite in Indonesian waters. The purpose of this research was to identify and retrack waveforms of Jason-2 altimeter satellite data in southern Java island waters and Java Sea using several retrackers and performed improvement percentage analyses for new SSH estimation. The study used data of the Sensor Geophysical Data Record type D (SGDR-D) of Jason-2 satellite altimeter of the year 2010 in the southern Java island waters and 2012-2014 in Java Sea. Waveform retracking analyses were conducted using several retrackers (Offset Center of Gravity, Ice, Threshold, and Improved Threshold) and examined using a world reference undulation geoid of EGM08 and Oceanic retracker. Result showed that shape and pattern of waveforms were varied in all passes, seasons, and locations specifically along the coastal regions. In general, non-Brownish and complex waveforms were identified along coastal region specifically within the distance of 0-10 km from the shoreline. In contrary, generally Brownish waveforms were found in offshore. However, Brownish waveform can also be found within coastal region and non-Brownish waveforms within offshore region. The results were also showed that the four retrackers produced a better SSH estimation in coastal region. However, there was no dominant retracker to improve the accuracy of the SSH estimate.

1. Introduction

Thousands of small islands in Indonesian waters are particularly vulnerable to sea level rise due to global warming such as melting ice sheets and glaciers in Antarctica and Greenland, and as a result of sea water expansion [1-4]. For example, the rate of sea-level rise in Indonesian waters was about 5.82 mm/yr or about twice of the world sea level rise rate [5, 6]. Therefore, the dynamics of Indonesian sea level rise should be monitored using synoptic and near-real time data that can be obtained from altimeter data.

Altimeter satellite data have been successfully used to accurately estimate sea surface height especially for offshore waters [7-15]. For offshore waters, modern microwave altimeter satellites can measure instant sea surface height with an accuracy of about 4.1 cm [16]. However, altimeter satellites still face problems for estimating sea surface height in coastal waters or shallow waters. Altimeter satellite data in coastal areas or shallow waters may be contaminated by complex topographic surfaces and coastlines of radar echoes, strong marine dynamics including tides, and land locations close to the sea that make it impossible to make accurate geophysical corrections such as correction wet troposphere, tidal correction, and atmospheric correction [17-19]. The terrestrial effects can induce many additional peaks in the waveform of distorted Brown's theoretical model (waveform) as a surface wave signal (radius ~ 7 km in 1Hz sampling) as it approaches or leaves the coastline [16].

Previous study using altimetry Jason-2 satellite data of 2013 in southern Java island waters and based on waveform identification, about 69% (the highest number) of non-Brown-waveform was found within 0-10 km of the coastline. Meanwhile, within the distance of 10-50 km and 50-100 km from the shoreline, the numbers of non-Brown waveforms were 5% and 3%, respectively [20]. Using similar data of 2012-2014 along central-eastern part of Java Sea, [21] also found that majority non-Brown waveform was found near to the coast and sometime also found in offshore.

Based on waveform retracking analyses conducted along central-eastern part of Java Sea of 2012-2014, the waveform retracking analyses were able to improve the accuracy of SSH estimation about 29.7% in the north coast and 56.4% in the south coast of total non-Brown-waveform in each region. However, there was no specific retracker producing dominant improvement percentage (IMP) of sea surface height (SSH) estimation [21].

The results of previous studies showed that the shape and pattern of waveform altimetry Jason-2 satellite data in the waters of southern Java island and central-eastern part of Java Sea varied based on location and season and did not have a definite pattern, especially in coastal waters. In general, the non-brownish waveform was found in coastal waters particularly near the coast and brownish waveform in offshore waters. However, non-brownish waveforms were sometimes found in offshore waters. Therefore, the waveform identification of the altimetry Jason-2 satellite data is very important to be conducted for all passes and the waters of southern Java island and the Java Sea. The previous results of retracking analysis were also showed that there were no single (dominant) retracker to produce the best improvement percentage (IMP) of sea surface height (SSH) estimates in areas with non-brownish waveform. However, some retrackers may provide improvements to SSH estimates in areas that were previously invalid (non-brownish waveform) by the standard (oceanic) retracker. Therefore, further retracker analyses are very essential to be conducted for all Indonesian waters to obtain optimal results in improving the accuracy of SSH estimates from altimetry satellite data. Increasing the accuracy of SSH estimation results from satellite data through the effort of waveform identification and retracking analyses are very important considering that Indonesian waters containing lots of shallow/coastal waters and are susceptible to less accurate estimates of SSH. The objectives of this study were to identify waveform and perform retracking analyses for altimetry satellite Jason-2 data to improve accuracy of SSH estimation especially along the coastal region.

2. Methods

2.1. Study Areas and Data

The study areas were the waters of southern Java island and Java Sea with limit of about 2.5°-7.5°S and 106°-116°E (yellow rectangle in figure 1).

We used altimetry Jason-2 satellite data of D-type sensor geophysical data record (SGDR) of 2010 for the region of southern Java island waters (Indian Ocean off Java island) and 2012-2014 for the region of Java Sea (figure 1). The data were obtained from the "NOAA's comprehensive large array-data stewardship system" (www.class.ncdc.noaa.gov) data website, AVISO (http://www.aviso.oceanobs.com), and the global geoid undulation model of EGM2008 version WGS 84 obtained from "NGA: National Geospatial-Intelligence Agency" data network (http://earth-info.nga.mil/GandG/wgs84/gravitymod/egm2008/egm08_wgs84.html).

The ascending pass (a pass of the satellite crossover moving from the southern part of the earth to the northern part of the earth (latitude -66.15° to +66.15°) and given an odd number) of 229, 051, 121, and 203 were used in this study. The descending pass (a pass of the satellite crossover moving from the northern part of the earth to the southern part of the earth (latitude +66.15° to -66.15°) and given an even number) of 242, 604, 140, and 216 were also used in this study (figure 1). One cycle is the time required by the satellite to cross the same pass which takes about 9.9156 days (~10 days) with orbital accuracy ± 1 km [22]. In this study, the retracking waveform analyses was performed on the above eight Jason-2 satellite passes.

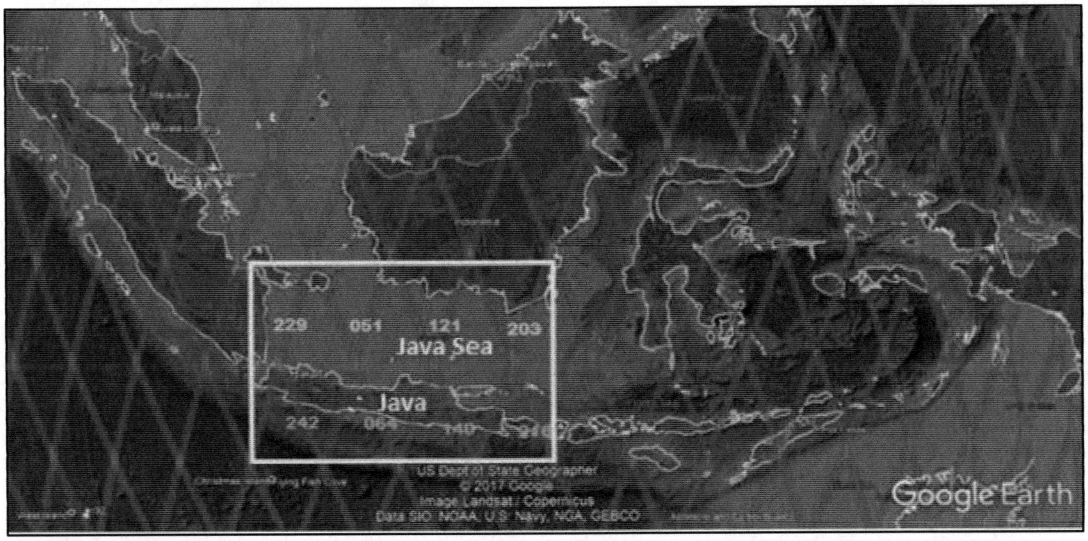

Figure 1. Altimetry Jason-2 satellite passes in Indonesian waters. The yellow colored box was the location of the identification and retracking waveform analyses of altimetry Jason-2 satellite data in this study.

2.2. Waveform Identification

The altimeter Jason-2 satellite measures the altitude of the sea level by transmitting microwaves to the sea surface and then measuring the time difference during the wave reception and reflection. The following was SSH calculations according to [23]:

$$SSH = Altitude - Range \quad (1)$$

where: SSH = sea surface height (m), Altitude = satellite elevation of the ellipsoid reference (m), Range = satellite altitude of sea level (m). The equations used to obtain the Range value are shown in the following equaton [24]:

$$Range = c\,(tr - te)\,/\,2 \quad (2)$$

where: Range = satellite height from sea surface (m), c = speed of light (299792458 m/s), te = time when satellite transmits microwaves to sea surface (s), tr = time when satellite receives wave reflected from sea surface (s).

The raw data of altimetry Jason-2 satellite is in the form of wave (power in y-axis and gate number in x-axis) and it is called a waveform. The center point at the leading edge of a waveform corresponds to the range at the nadir point of the sea surface [25-28]. On the offshore region where generally found the Brownish-waveform, the midpoint of the leading edge waveform lies on the 32nd gate so that the midpoint of the on-board leading edge on the Jason-2 Satellite is set at the 32nd gate [25]. However, in shallow coastal waters due to its waveform shape which normally does not follow Brownish-waveform model, it causes the midpoint position of the leading edge to vary the extent of the sea surface physical condition so that it does not correspond to the value of the leading edge position assigned to the satellite. This causes the range calculation of SSH to produce a less accurate value [25-

28]. Gate is the time of observation when the reflection wave occurs. The Jason-2 satellite has 104 gates where the gate spacing is 3.125 ns (nano second) so that the time span required to record a single data waveform is 325 ns [29].

In this study, we used in-house built Matlab program to perform waveform identification processes. The main idea of this proces was to identify each waveform either belong to the form of Brownish or Non-Brownish waveform.

2.3. Waveform Retracking Analyses

Retracking waveform is a method for finding the right tracking gate cutting the midpoint of the leading edge to get the actual range value [27]. There are several methods of retracking waveform (retracker) used in this research i.e., ocean, ice, OCOG, threshold (10%, 20%, and 50%), and improved threshold (10%, 20%, and 50%). The ocean retacker method (standard/default retracker) is an on-board retracker in the SGDR-D Jason-2 data so that the range value of the retracker does not need to be counted again [25]. The ocean retracker was developed using a Brown-waveform model [30]. The ice retracker was originally developed to observe the polar ice heights which developed based on the 30% threshold retracker using the average power quantity of the waveform previously calculated by the OCOG method [31]. OCOG retracker was developed by [32] where the purpose of this method is to obtain the center point of the wave-form. The formula and explanations in the calculation of the retracker threshold method (10%, 20%, and 50%), as well as the improved threshold (10%, 20%, and 50%) are presented in [13] and [28].

2.4. Estimating Sea Surface Height (SSH)

According to [16] and [24] the result of the retracking waveform analysis is the value of the range correction calculated by putting aside the gate value from the midpoint position of the leading edge of the retracking waveform result with the midpoint position of the on-board leading edge (equation 3). The retracking range is obtained from the addition of Jason-2 Satellite on-board range with correction range (equation 4):

$$dr = c \times \Delta Ga \times (Gr - G0) / 2 \quad (3)$$

$$Rr = R + dr \quad (4)$$

where: dr = satellite altitude correction with retacked sea level (m), c = speed of light (299792458 m/s), ΔGa = time interval for one gate Jason-2 satellite (3.125 ns), Gr = gate observation at midpoint leading edge of retracking, Go = gate on-board observation of Jason-2 Satellite (32), R = range of satellite measurement (m), Rr=range of retracking results (m).

The next calculation is to add a correction of geophysical and atmospheric disturbance to the retracking range (Rr) to get the corrected range (Rcorr), as shown in equation 5 [33]:

$$R_{corr} = R_r - \Delta h_{dry} - \Delta h_{wet} - \Delta h_{iono} - \Delta h_{ssb} - h_{tides} - h_{atm} \quad (5)$$

where: R_{corr} = corrected range (m), R_r = range of retracking (m), Δh_{dry} = dry troposphere correction (m), Δh_{wet} = wet tropospheric correction (m), Δh_{iono} = ionospheric correction (m), Δh_{ssb} = surface shape correction sea (m), h_{tides} = tidal correction (m), h_{atm} = dynamic atmospheric correction (m).

The last step to get the SSH value of retracking waveform was by using equation from Yang *et al.* (2008) as shown in Equation 6. The Jason-2 satellite had an altitude value or satellite elevation of an ellipsoid reference of 1336 km in the equatorial region [22].

$$SSH_{retracking} = Altitude - R_{corr} \quad (6)$$

where: $SSH_{retracking}$ = retracked sea surface height (m), Altitude = satellite heights of the elipsoid reference (m), R_{corr} = corrected range (m).

In this study, we used in-house built Matlab program to perform retracking analyses for altimetry Jason-2 satellite data and produce SSH estimation specifically for coastal waters region.

2.5. Improvement Percentage Determination and SSH Retracking Validation

The performance of the retrackers can be calculated using the Improvement Percentage (IMP) formula based on [28] and developed by [25] so that the ability of each retracker can be calculated by making the ocean retacker (standard/on-board retracker) as a benchmark for improving the ability. Improvement Percentage (IMP) formula of retracked SSH versus standard (on-board) estimated SSH is shown in equation 7.

$$IMP = (\sigma_{ocean}-\sigma_{retracker})/\sigma_{ocean} \times 100\% \quad (7)$$

where: IMP = percentage improvement (%), σ_{ocean} = standard deviation of on-board SSH divide by geoid undulation (m), $\sigma_{retracker}$ = standard deviation of SSH retracking with geoid undulation (m).

The percentage value of the IMP of each retracker shows how much improvement of the retracker produce against the on-board retracker so the greater the value of the IMP the better the ability of the retracker and vice versa. Validation of retracked SSH was compared to its profile against the geoid undulation profile (EGM08). If the profile of retracked SSH did not resemble the geoid undulation profile then the retracked SSH was considered to be not valid values even if it had been able to correct the errors of SSH on-board [25-27, 34-35].

3. Result and Discussion

3.1. Waveform Identification

Using in-house built Matlab program, we performed identification on shape and pattern of waveform of Jason-2 altimeter satellite data of the year of 2010 in the southern Java island waters and during 2012-2014 in Java Sea region. In previous study, Nababan et. al. reported that the brownish waveform was very rare in coastal areas at a distance of 0-10 km from the coastline [20]. Generally, the brownish waveform was found in offshore waters at a distance of 50-100 km from the coastline. Given the non-uniformity pattern and shape of the waveforms based on the track and distance from the coastline or based on the seasons, therefore, identification of the shape and pattern of the Jason-2 altimeter satellite data for the year of 2010 in the southern Java island waters was continued to be performed in this study. All ascending (229, 051, 121, and 203) and descending (242, 604, 140, and 216) passes were analysed in this study.

The results showed that generally non-brownish waveform was found in coastal waters, especially near the coast. However, the shape, pattern, and frequency of occurrence in each pass did not show the same pattern (figure 2). In pass 051, the non-brownish waveform was found only in coastal waters about 0-5 km from the coastline and the rest was seen as brownish waveform. In pass 64, a non-brownish waveform was encountered in coastal waters at a distance of 0-10 km from coastline, whereas in pass 127 some non-brownish waveform were encountered at a distance of 0-5 km and 17-23 km from coastline. Likewise on pass 140, the non-brownish waveform group was encountered in two locations, namely at a distance of 0-8 km and 17-19 km. In pass 203 and 242, non-brownish waveforms were formed at a distance of 0-15 km from the coastline. In pass 229, the non-brownish waveforms were also found near to coast at about 0-3 km from the coastline (figure 2). In general, the non-brownish waveforms commonly encountered in coastal waters near the shoreline because radar waves from altimeter satellites were affected by shoreline or land. The foot print of radar waves from altimeter satellites reaching the sea surface was also strongly influenced by the condition of the surrounding waters to a distance of about 2-5 km while the distance between these two satellite data points was about 300 m [36-37].

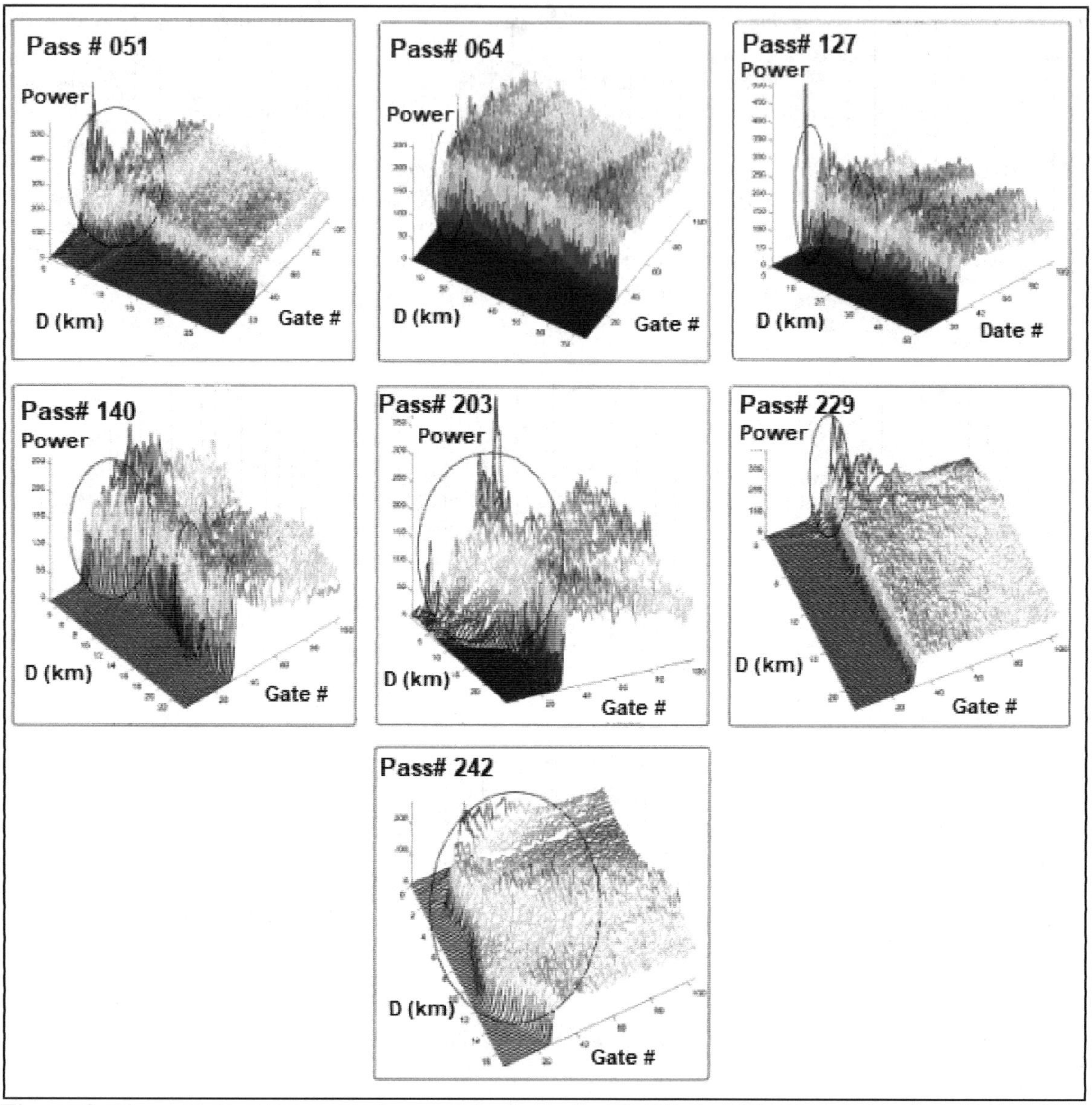

Figure 2. Shape and patterns waveform of the Jason-2 altimeter satellite data in southern Java island waters of the year 2010 along the coastal waters. The waveforms circled in black were non-brownish waveform. Note: D = distance from shore.

The identification of the shape and pattern waveform of the altimetry Jason-2 satellite data for offshore water region was also conducted in this study with the distance from the coastline limited to about 100 km from the coastline in the southern Java island waters. We limited to this distance because in general the waveform of altimetry Jason-2 satellite data along the offshore region (at the depths >1000 m) was generally in the form of Brownish waveform [7-15].

In general, the waveform patterns in offshore waters were found as brownish waveform, however, sometimes non-brownish waveforms were also found in offshore waters (eq. pass 64, 140, and 203) (figure 3). The black-circled waveforms were non-brownish waveforms because the middle edge of the leading edge waveform was not around the 32nd pre-given gate. The Jason-2 satellite had a pre-given gate on the 32nd gate [25]. By using the ocean retracker, the waveform circled in black would result in a less accurate SSH estimation or invalid value (NaN).

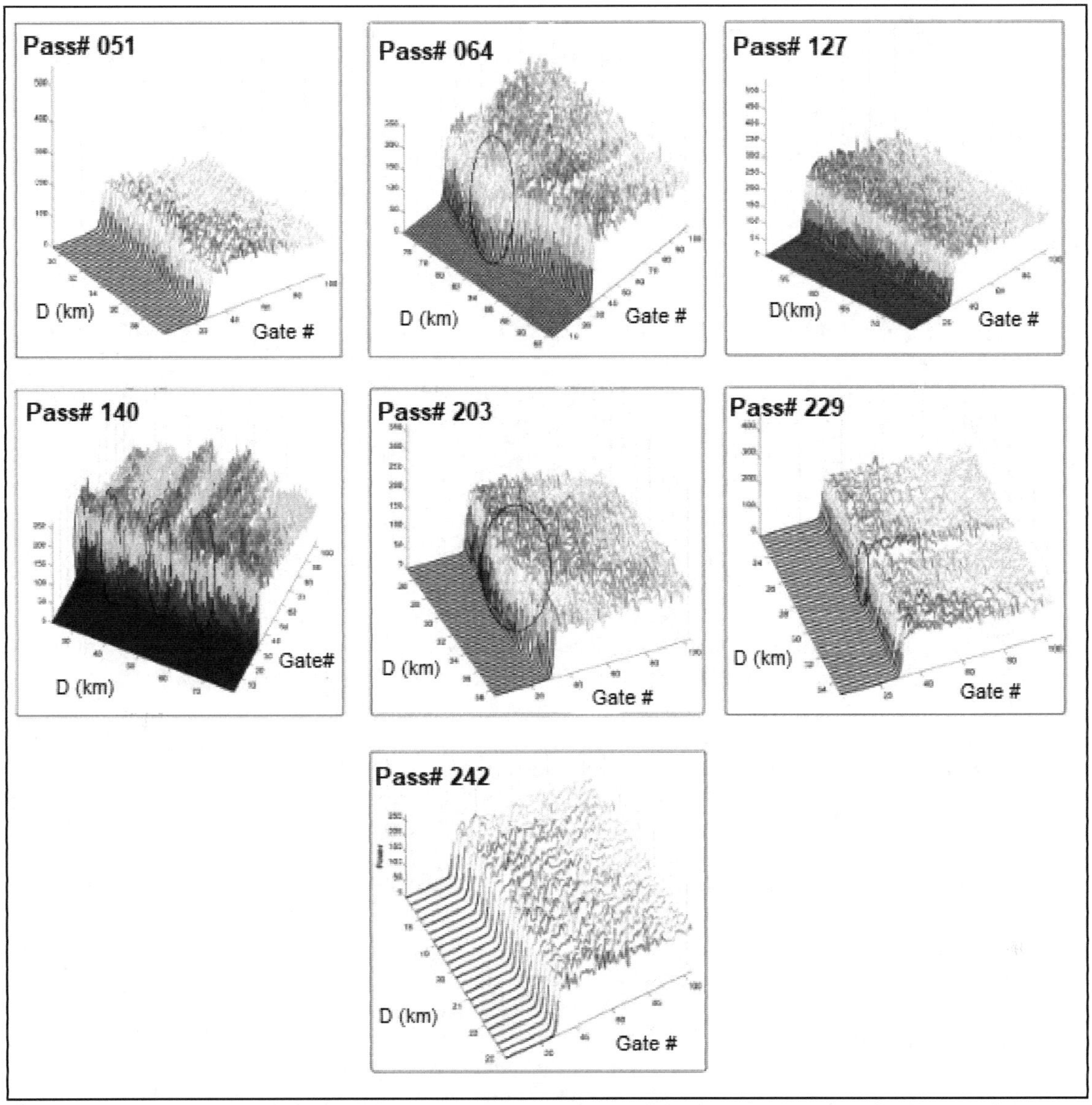

Figure 3. The waveform shape and pattern of the Jason-2 altimeter satellite data for the southern Java island waters of the year 2010 along the offshore waters. The waveforms circled in black were non-brownish waveforms. D=distance from shore.

For Java Sea, waveform identification showed that generally non-brownish waveforms were found along coastal areas and brownish waveforms along the offshore region with average of 94.4%. The percentage of non-bownish waveform for all passes during the period of 2012-2014 was found about 5.6% (table 1). The highest percentage of non-brownish waveforms was encountered in the pass number 203 with a value of 6.9%, while the lowest percentage of non-brownish waveform was found in pass number 140 of 4.9% (table 1). The waveform pattern in the Java Sea generally had a similar pattern with the southern waters of Java island specifically along the coastal areas which predominantly non-brownish waveforms. Non-brownish waveforms in coastal areas were caused by reflected signals from the mainland, the bias of the sea state, and the complex shoreline [20, 27].

Table 1. Average percentage of non-brownish waveforms of all Jason-2 altimeter satellite waveform data of 2012 - 2014 on each pass in the Java Sea.

Pass number	Pass length (km)	Total waveform	Number brownish waveform	Number non-brownish waveform	Non-brownish waveform (%)
051	426.3	1,537	1,449	88	5.7
064	405.7	1,463	1,391	72	4.9
127	386.0	1,392	1,315	77	5.5
140	423.7	1,528	1,452	76	5.0
203	352.4	1,271	1,183	88	6.9
216	391.0	1,410	1,334	76	5.4
229	356.1	1,286	1,214	72	5.6
242	133.5	482	454	28	5.8
Average	133.5	1,296.13	1,224	72.13	5.6

3.2. Waveform Retracking Analyses for Improving SSH Estimation Accuracy

Waveform retracking analyses for Jason-2 altimeter satellite data of 2010 for the southern Java island waters and for the period of 2012-2014 for the Java Sea were conducted using retracker method like OCOG, Threshold 10%, Threshold 20%, Threshold 50%, Improved Threshold 10%, Improved Threshold 20%, Improved Threshold 50%, and Ice. After performing retracking waveform analyses, the sea surface height (SSH) estimate was also calculated based on each retracker.

For non-brownish waveform conditions, the standard ocean retracker generally produce inaccurate or invalid (NaN) SSH estimate values. To improve this inaccurate or invalid SSH estimate values, therefore, retracking was done using another retracker such as OCOG, Threshold 10%, Threshold 20%, Threshold 50%, Improved Threshold 10%, Improved Threshold 20%, Imrpoved Threshold 50%, and Ice. The success of each retracker to produce a better SSH estimate value was determined by calculating the value of improvement percentage (IMP). The IMP was calculated from the percentage between the difference of standard deviation of the on-board SSH (standard ocean retracker) with the geoid and standard deviation of the other SSH retracker with the geoid divided by the standard deviation of the standard on-board retracker. Figure 4 showed an example of retracker analyses for SSH estimation in the southern coastal waters of Java island for pass number 242. In the figure 4, it showed that at a distance of 0-7 km from the coastline, the standard oceanic retracker did not succeed in providing SSH estimation value (invalid SSH estimation or produce NaN value). This meant that the waveforms in this area were identified as non-brownish waveforms so that the standard ocean retracker produced an invalid value (NaN) for SSH estimation. However, the 20% threshold retracker was succesfully provide valid SSH estimation values up to a distance of about 1.8-7 km from the coastline where the standard ocean retracker gave the invalid (NaN) SSH values. OCOG retracker was also able to generate SSH estimation value at a distance of 1.8-7 km from the coastline but the value was considered less accurate because it was too much different from the comparator value of SSH based on the EGM08 model.

The results of the waveform retracking analyses in the Java Sea using Jason-2 altimeter satellite data in the period 2012-2014, it showed an improvement of SSH estimate values in the range 31.9-62.8% (table 2). Unlike in the southern Java island waters which the improvement can reach of 95.24%, in the Java Sea the maximum percentage of improvement was only of 62.8%. This was because the Java Sea was relatively shallower (<200 m) and flanked by two coastlines with some small islands within the Java Sea. These conditions were more likely to have more complex waveform shapes and patterns compared to the shape and pattern of waveforms in the deeper water. The roughness of the sea surface may also give noise to the waveform and causes the waveform to be more difficult to analyze [27]. Each retracker had specific advantages and disadvantages with its area of study so that each retracker performs differently depending on the characteristics of its waters [16, 26]. It seems that Threshold 10% retracker provide better percentage improvement for SSH estimate in Java Sea (tabel 2).

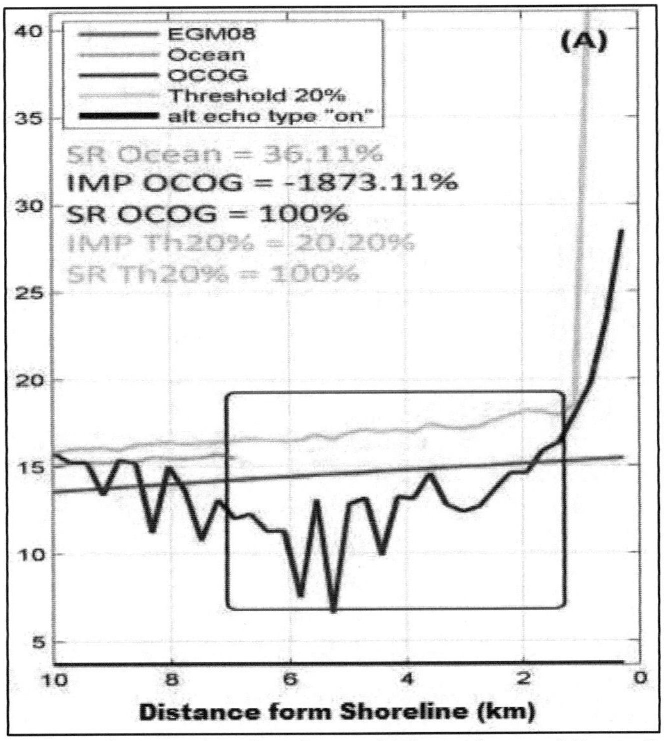

Figure 4. An example of retracking analyses of Jason-2 satellite altimeter data pass number 242 in the southern Java island waters. In the black box, SSH estimates by 20% Threshold retracker produced IMP of 20.20%.

Table 2. Best retracker and percentage improvement of SSH estimate for Jason-2 altimeter satellite data in Java Sea waters.

Pass number	Pass length (km)	Best retracker	Improvement Percentage (IMP)
051	426.3	Threshold 10%	61.0
064	405.7	Threshold 10%	61.5
127	386.0	Ice	56.7
140	423.7	Threshold 10%	57.5
203	352.4	Threshold 50%	31.9
216	391.0	Ice	62.8
229	356.1	Improved Threshold 20%	58.4
242	133.5	Threshold 10%	55.6

The results of the retracker analyses for 2010 Jason-2 altimeter satellite data in the southern Java island waters showed various improvements to the SSH estimation particularly in coastal/shallow

IOP Conf. Series: Earth and Environmental Science **149** (2018) 012057 doi:10.1088/1755-1315/149/1/012057

waters compared to the standard ocean retracker. The best results were generated by the 50% Threshold retracker which provided a 95.24% improvement in coastal waters (0-10 km distance) on pass 242 in the southern Java island waters. While a small percentage of the improvements was generated by retracker of 50% Improved Threshold which results in an improvement of 8.51% at a distance of 0-10 km from the coastline (pass number 127, table 3). However, the results of this analyses showed that no dominant retracker producing the best SSH estimate. For that, we need to perform retracking analyses for all Jason-2 altimeter satellite data specifically in shallow waters.

Table 3. Best retrackers and percentage improvement of SSH estimate for Jason-2 altimeter satellite data in southern Java island waters.

Pass number	Distance from coastline (km)	Best retracker	% Improvement (IMP)
051	0-10	Improved Threshold 20%	67.04
	10-50	Threshold 10%	41.26
	50-100	Ice	14.96
064	0-10	Ice	11.33
	10-50	Improved Threshold 10%	49.16
	50-100	Threshold 20%	36.60
127	0-10	Improved Threshold 50%	8.51
	10-50	Threshold	19.0
	50-100	Improved Threshold 20%	60.85
140	0-10	Improved Threshold 10%	86.34
	10-50	Improved Threshold 10%	83.32
	50-100	Threshold 20%	38.73
203	0-10	Threshold 20%	76.16
	10-50	Improved Threshold 20%	54.99
	50-100	Threshold 20%	85.04
229	0-10	Threshold 50%	59.95
	10-50	Ice	44.11
	50-100	Threshold	24.88
242	0-10	Treshold 50%	95.24
	10-50	Ice	12.92
	50-100	Improved Threshold 20%	49.21

4. Conclusion

In general, non-Brownish and complex waveforms were identified along coastal region specifically within the distance of 0-10 km from the shoreline. In contrary, generally Brownish waveforms were found in offshore. However, Brownish waveform can also be found within coastal region and non-Brownish waveforms within offshore region. Therefore, the identification of the waveform for all passes of Jason-2 altimeter satellite data in Indonesian waters remains necessary. The number of non-brownish waveforms in the Java Sea was found more in the northern part of the Java Sea (coast around southern Borneo island) than in the southern part of Java Sea (north coast of Java island). This relates to the more complex topography of the coastline in the northern part of the Java Sea (the southern coastline of the Borneo island) compared to the southern part of the Java Sea (the northern coastline of Java Island).

The waveform retracking analyses of Jason-2 altimeter satellite data were successfully improve the accuracy of sea level estimation (SSH) in the southern Java island waters and Java Sea particularly in coastal region. However, there was no dominant retracking method to improve the accuracy of the SSH estimate. These results indicated that it is necessary to perform retracking waveform analyses for all Indonesian waters specifically along the coastal or shallow waters.

Acknowledgments
This research was supported by the Directorate Research and Social Services, Directorate General of Research Strengthening and Development, Ministry of Research, Technology and Higher Education, Republic of Indonesia with grant No. 011/SP2H/LT/DRPM/IV/ 2017 dated, 20 April 2017 and addendum contract No. 011/SP2H/LT/DRPM/VIII/2017 dated 21 August 2017. Special thanks goes to AVISO and NOAA which provided Jason-2 altimeter satellite data for this research. We would also like to acknowledge reviewers who provided some comments and inputs to improve this paper.

5. References

[1] Hansen J, Sato M, Hearty P, Ruedy R, Kelley M, Masson-Delmotte V, Russell G, Tselioudis G, Cao J, Rignot E, Velicogna I, Kandiano E, von Schuckmann K, Kharecha P, Legrande A N, Bauer M, and Lo KW 2015 Ice melt, sea level rise and superstorms: evidence from paleoclimate data, climate modeling, and modern observations that 2°C global warming is highly dangerous *Atmospheric Chemistry and Physics* **15** pp 20059–20179

[2] Levermann A, Peter U, Clark, Marzeion B, Milne G A, Pollard D, Radic V, and A Robinson 2013 The multimillennial sea-level commitment of global warming *PNAS* **110** pp 13745–13750 PMC 3752235 Freely accessible PMID 23858443

[3] Velicogna I 2009 Increasing rates of ice mass loss from the Greenland and Antarctic ice sheets revealed by GRACE *Geophysical Research Letters*, **36** (19) Bibcode: 2009GeoRL. 3619503V

[4] IPCC 2007 *Climate Change 2007*: Impacts, Adaptation and Vulnerability. Contribution of Working Group II to the Fourth Assessment Report of the Intergovernmental Panel on Climate Change eds Parry, Martin L, Canziani, Osvaldo F, Palutikof, Jean P, van der Linden, Paul J, and Hanson, Clair E 1000 pp (Cambridge, United Kingdom: Cambridge University Press)

[5] Nababan B, Hadianti S, and Natih N M N 2015 Dynamic of sea level anomaly of Indonesian waters (in Bahasa, abstract in English) *J. Ilmu dan Teknologi Kelautan Tropis* **7** (1) pp 259-272

[6] Nababan B, and Gaol J L 2013 Reconstruction of sea surface height of Indonesian coastal regions through retracking analyses of altimeter satellite data for monitoring sea level rise and detection of potential fishing ground (in Bahasa) Final report year-1: variability of sea level rise of Indonesian waters and retracking analyses of South Java island waters 111pp (Bogor, Indonesia: Bogor Agricultural University)

[7] Wunsch C 2006 Altimetry: Past, Present, and Future *Proc. of 15 years of progress in radar altimetry* (Venice, Italy, 13-18 March 2006) (ESA Special Publication SP-614 ESA Publications Division, ESTEC 2200 AG Noordwijk, The Netherlands)

[8] Evans D L, Alpers, Cazenave W, Elachi A, Farr C, and T et. al. 2005 Seasat--A 25-year legacy of success *Remote Sensing of Environment* **94** pp 384-404

[9] Leben R R, and Powell B S 2003 Accuracy assessment of Jason-1 and TOPEX/POSEIDON along-track sea surface slope *Marine Geodesy* **26** (3-4) pp 355-366

[10] Leben R R, Born G H, and Engebreth B R 2002 Operational altimeter data processing for mesoscale monitoring *Marine Geodesy* **25** 3-18

[11] Carton J A and Chao Y 1999 Caribbean sea eddies inferred from TOPEX/ POSEIDON altimetry and a 1/6° Atlantic Ocean model simulation *J. of Geophys. Res.* **104** (C4) pp 7743-7752

[12] Legresy B, and Remy F 1997 Surface characteristics of the Antarctic ice sheet and altimetric observations *J. of Claciology* **43** (114) pp 197-206

[13] Davis C H 1996 A robust threshold retracking algorithm for extracting ice-sheet surface elevation from satellite radar altimeters *IGARSS* **3** pp 1783-1787

[14] Leben R and Born G H 1993 Tracking loop current eddies with satellite altimetry *Advances in Space Research* **13** (11) pp 325–333

[15] Martin TV, Zwally H J, Brenner A C, and Bindschadler R A 1983 Analysis and retracking of continental ice sheet radar altimeter waveform *J. Geophys. Res.* **88** pp 1608-1616

[16] Yang L, Lin M, Bai Y, and Pan D 2008 Retracking Jason1 altimeter waveform over China Coastal Zone *Proc. of SPIE* vol **7154** 71540K-10

[17] Deng X, Featherstone W, Hwang C, and et. al. 2002 Estimation of contamination of ERS-2 and Poseidon satellite radar altimetry close to the coasts of Australia *Mar. Geod.* **25** (1) pp 249-271

[18] Brooks R L, Lockwood D W, Lee J E, Hancock D W III, and Hayne G S 1998 Land effects on TOPEX radar altimeter measurements in Pacific Rim coastal zones ed Brown, R.A. Remote sensing of the Pacific by satellite (Glebe, Australia: *Earth, Ocean, and Space Pty.*) pp 175-198

[19] Hancock D W III, and Lockwood D W 1990 Effects of island in the geosat footprint *J. Geophys. Res.* **95** (C3) pp 2849-2855

[20] Nababan B, Hakim R, Adrian D, dan Gaol J L 2014 Waveform identification of altimeter satellite data in coastal and deepwater waters of the southern Java island waters (in Bahasa, abstract in English) *J. Ilmu dan Teknologi Kelautan Tropis* **6** (2) pp 445-459

[21] Hakim M R, Nababan B, dan Panjaitan J P 2015 Increasing the accuracy of sea surface height estimates through Jason-2 satellite retracking waveform analyses in the Java Sea (in Bahasa, abstract in English) *J. Ilmu dan Teknologi Kelautan Tropis* **7** (2) pp 771-790

[22] Dumont J P, Rosmorduc V, Picot N, Bronner E, Desai S, Bonekamp H, Figa J, Lillibridge J, and Scharoo R 2011 *OSTM/Jason-2 Products Hand-boo*k Available online http://www.aviso.altimetry.fr/fileadmin/documents/data/tools/hdbk_j2.pdf (accessed on 20 July 2017)

[23] Chelton D B, Ries J C, Haines B J, Fu L, and Callahan P 2001 Satellite Altimetry eds Fu L and Cazenave A *Satellite altimetry and earth sciences* A handbook of techniques and applications 1-132pp (San Diego: Academic Pres, International Geo-physics Series vol 69)

[24] Anzenhofer M, Shum C K, and Rentsh M 1999 Coastal Altimetry and Applications Tech. Rep. No. 464 36p (The Ohio State University, Columbus, USA: Geodetic Science and Surveying)

[25] Lee H, Shum C K, Emery W, Calmant S, Deng X, Kuo C Y, Roesler C, and Yi Y 2010 Validation of Jason-2 altimeter data by waveform retracking over California coastal ocean *Marine Geodesy* **33** (S1) pp 304-316

[26] Guo J Y, Gao Y G, Hwang C, and Sun J L 2010 A multi-subwaveform parametric retracker of the radar satellite altimetric waveform and recovery of gravity anomalies over coastal oceans *Science China – Earth Sciences* **53 pp** 610-616

[27] Deng X, and Featherstone W E 2006 A coastal retracking system for satellite radar altimeter waveforms: applica-tion to ERS-2 around Australia *J. of Geophys. Res.* **111**

[28] Hwang C, Guo J Y, Deng X, Hsu H Y, and Liu Y T 2006 Coastal gravity anomalies from retracking Geosat/GM altimetry: improvement, limitation and the role of airbone gravity data *J. Geod.* **80** pp 204-216

[29] Gommenginger C, Thibaut P, Fenoglio-Marc L, Quartly G, Deng X, Gomez-Enri J, Challenor P, and Gao Y G 2011 Retracking altimeter waveforms near the coasts eds Vignudelli S,

Kostianoy A G, Cipollini P, and Benveniste J *Coastal Altimetry* 61-101pp (Berlin Heidelberg: Springer-Verlag)

[30] Brown G S 1977 The average impulse response of a rough surface and its applications IEEE Trans. *Antennas Propag*. **25** pp 67-74

[31] Bamber J L 1994 Ice sheet altimeter processing scheme *International J. of Remote Sensing* **15** (4) pp 925-938

[32] Wingham D J, Rapley C G and Griffiths H 1986 New technique in satellite altimeter tracking systems *IGASS'86 Symposium Digest* **1** (ESA Publications Division, Noordwijk) pp 185-190

[33] Andersen O B and Scharroo R 2011 Range and geophysical corrections in coastal regions: and implications for mean sea surface determination eds Vignudelli S, Kostianoy A G, Cipollini P, and Benveniste J *Coastal Altimetry* pp 104-142 (Berlin Heidelberg: Springer-Verlag)

[34] Bao L, Lu Y, and Wang Y 2008 Improved retracking algorithm for oceanic altimeter waveforms *Progress in Natural Science* **19** pp 195-203

[35] Fenoglio-Marc L, Fehlau M, Ferri L, Becker M, Gao Y, and Vignudelli S 2010 Coastal sea surface heights from improved altimeter data in the Mediterranean Sea ed Mertikas S.P. *IAG Symposia* **135** (Gravity, Geoid and Earth Observation, Springer Verlag) pp 253-262

[36] Idris N H, Deng X, and Idris N H 2017 Validation of sea levels from coastal altimetry waveform retracking expert system: a case study around the Prince William Sound in Alaska *IOP Conf. Series: Journal of Physics: Conf. Series* **852** (2017) 012029

[37] Smith W H F and Scharroo R 2015 Waveform aliasing in satellite radar altimetry *IEEE Transactions on Geoscience and Remote Sensing* **53** (4) pp 1671-1682

IOP Conf. Series: Earth and Environmental Science **149** (2018) 012058 doi:10.1088/1755-1315/149/1/012058

Application of rain scanner SANTANU and transportable weather radar in analyze of Mesoscale Convective System (MCS) events over Bandung, West Java

G A Nugroho, T Sinatra, Trismidianto and I Fathrio
Center of Atmospheric Science and Technology, National Institute of Aeronautics and Space, Bandung, Indonesia

E-mail address: ginaldi.ari@lapan.go.id

Abstract. Simultaneous observation of transportable weather radar LAPAN-GMR25SP and rain-scanner SANTANU were conducted in Bandung and vicinity. The objective is to observe and analyse the weather condition in this area during rainy and transition season from March until April 2017. From the observation result reported some heavy rainfall with hail and strong winds occurred on March 17th and April 19th 2017. This events were lasted within 1 to 2 hours damaged some properties and trees in Bandung. Mesoscale convective system (MCS) are assumed to be the cause of this heavy rainfall. From two radar data analysis showed a more local convective activity in around 11.00 until 13.00 LT. This local convective activity are showed from the SANTANU observation supported by the VSECT and CMAX of the Transportable radar data that signify the convective activity within those area. MCS activity were observed one hour after that. This event are confirm by the classification of convective-stratiform echoes from radar data and also from the high convective index from T_{bb} Himawari 8 sateliite data. The different MCS activity from this two case study is that April 19 have much more MCS activity than in March 17, 2017.

1. Introduction

Mesoscale convective system (MCS) are defined as an organized convective components that have a longer duration than the individual convective elements and also have the largest area than the convective storms [1]. In the tropics, MCS play different part in the mass transport and heating process [2-3], where made an impact on the global general circulation [4-5] .

MCS in a meteorological radar point of view is a precipitation that consist of two type of precipitation (convective and stratiform precipitation) [6], with a minimum of 30% are associated with convective rainfall [7]. Characteristic between convective and stratiform precipitation based on meteorological radar detection showed different pattern of gradient reflectivity, where convective precipitation have a sharp peaks gradient of reflectivity resemble its core while stratiform precipitation have a weak gradient reflectifivity [8]. Spatial distribution of the reflectivity pixel from meteorological radar data were also used to distinguish between convective, stratiform and also mixed precipitation [9].

Simultaneous observation of two x band radar system, transportable weather radar LAPAN-GMR25SP and rain-scanner SANTANU, were conducted during rainy season and transition season from March until April 2017. The observation area were focused in the Bandung Basin in West Java Indonesia. During this period, there are reported of some heavy rainfall with hail and strong winds occurred at Bandung and vicinity on March 17th and April 19th 2017. The heavy rainfall are lasted within 1 to 2 hours damaged some properties and trees in Bandung. These two type of x band radar system are complement to each other. The transportable weather radar is a doppler x band radar that could generate doppler velocity and spectrum width beside reflectivity that could not generated in SANTANU. But since this doppler radar needs to conduct a volume scan, its temporal resolution are less than SANTANU.

This paper are discussed about the application of this two type of X-band radar during the heavy rainfall in two case study. Different with other observation [10] , in this paper two type of X-band radar data are used to investigate not only the spatial distribution but also the vertical structure of the precipitation [11] and also to analyze if there is any connection between the heavy rainfall with the MCS activity. The MCS indicator are based from the radar data and Tbb Himawari satelitte data.

2. Data and method

Transportable weather radar is a cost effective x band radar manufactured by GAMIC mbH with the type of GMWR-25-SP. This radar is a single polarization doppler weather radar mounted on a vehicle for conducting weather observation in different location site [12]. This radar were operated by LAPAN for research activities mainly on atmospheric research. A volume scan strategy using 20 elevation scan of altitude were applied. The volume data can reach from ground surface until up to 15 km of detection, with spatial and temporal resolution of 250 x 250 meter in every 10 minutes. Three main product from this radar were reflectivity, doppler velocity, and spectrum width, with other derivative product such as CAPPI, CMAX, VSECT, etc [13]. CAPPI (Constant Altitude Plan Position Indicator) represent the radar data in horizontal cross section at selected height above sea mean level. CMAX (column maximum) represent the vertical column maximum value. VSECT (Vertical Section) represent the vertical cross section through the volume radar data between two coordinate (start and stop coordinate).

SANTANU is a x band rain scanner based on marine radar development. This radar have a wide vertical beam that could scan a very large volume of precipitation. A clutter map were applied to extract the ground clutter in order to achieve the precipitation data [14]. Spatial and temporal resolution are 240x240 meter in every 2 minutes. High temporal resolution can be utilized to analyse the precipitation movement and also the rate of growth/decay area of precipitation [15].

Three X band radar are used in this research. Two rain scanner radar type and one transportable Doppler radar type are used to scan around Bandung area. Figure 1 is the scheme of the location of the X-band radar, with coloured circle represent the coverage area. A and B are the two X band rain scanner SANTANU. Each of the SANTANU radar are complementing each other due to limited coverage area and also very sensitive to attenuation. Meanwhile C is the X band Transportable Doppler radar.

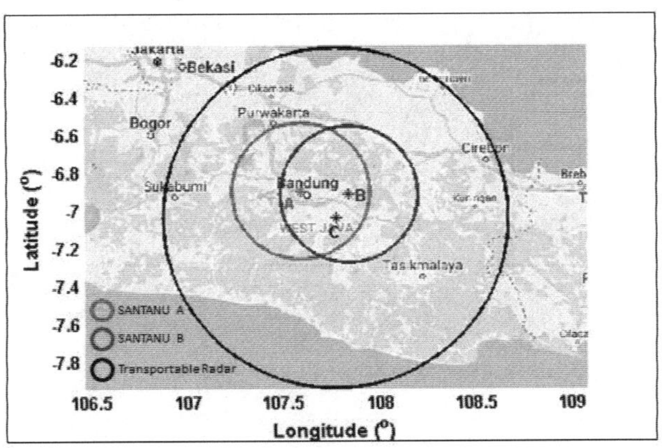

Figure 1. Coverage area of SANTANU A, B, and Transportable Weather Radar.

Observation period were conducted from March 2017 until End of April 2017. During the period of March 17 and April 19 2017 there are a heavy storm that suspected from the generation of an MCS phenomena. In order to prove this, several data were used such as VSECT and CMAX of reflectivity and doppler velocity from transportable radar to analyse the convective process within those heavy storm. In order to distinguish between convective and stratiform precipitation, Steiner method were applied to the CAPPI 3 km data. Meanwhile PPI of reflectivity from SANTANU were used to confirm the location of the heavy storm followed with heavy precipitation.

Steiner method are able to distinguish between convective and stratiform precipitation from the reflectivity radar data. This method are based from the intensity of the reflectivity and also the sharpest maxima from the spatial pattern of the radar reflectivity [16]. Steiner method in this paper are only applied to the Transportable Radar since the SANTANU have less range of reflectivity value than the Transportable radar. Steiner method applied to the CAPPI 3 km product since this product have less ground clutter and noise but still represent the echo activity in the mid-level of the precipitation [17]. Satelitte data from Himawari were also used to confirm the MCS activity during the observation period. MCS is indicated by the largest area and the black body temperature (hereafter Tbb) value [18].

3. Result and discussion

3.1. Case study in March 17, 2017

Two case study were presented in this paper. In March 17. Precipitation were occurred from 11.00 LT until 20.00 LT to 21.00 LT. This precipitation were spread almost the entire Bandung area based on SANTANU observation (figure 2). SANTANU A and B are detected the early stage of precipitation in the middle of Bandung city area which started around 11.00 LT. In thirty minutes this object then grow into a wider area, with the SANTANU B showed a unique form similar with a v shape. In the next thirty minutes the object lost the v shape consistency and change its form until disappeared in 12.48 LT. There are time with less precipitation from 13.00 until 13.30, but after that several precipitation were occurred. Long span of precipitation area were observed between 16.00 – 16.30 LT in the north side of Bandung with reflectivity from 15 – 28 dbz. Large area of precipitation were observed between 18.00 – 20.00 LT in the east and in the south east of Bandung ranging from 12- 30 dbz.

Figure 3 is the CMAX product from the Transportable radar in March 17 2017 from 10.50 LT until 11.30 LT with no data after that due to electrical failure. This figure not only confirm the early stage of the convective precipitation but also more detailed information about the forming of the v-shape area. From the maximum value of the CMAX in 11.30 LT there are differences between the two part of the V-shape (left flank and right flank), with the right flank possess higher reflectivity than the other flank. Using a vertical point of view from the VSECT data showed more detail of the vertical structure of this right flank. Figure 4 is the VSECT in 11.30 LT data of the v-shape side (right flank). This VSECT showed towering reflectivity from near surface until 12 km. Based on the higher reflectivity, there are three consecutive convective cell. First cell is located in the altitude between 620 meter until 7 km, stretched horizontally within 6 km. Second cell is a possess smaller reflectivity with altitude from 2 km until 4 km. Meanwhile the third cell showed a vertical convective reflectivity with altitude in 620 meter until 3.5 km.

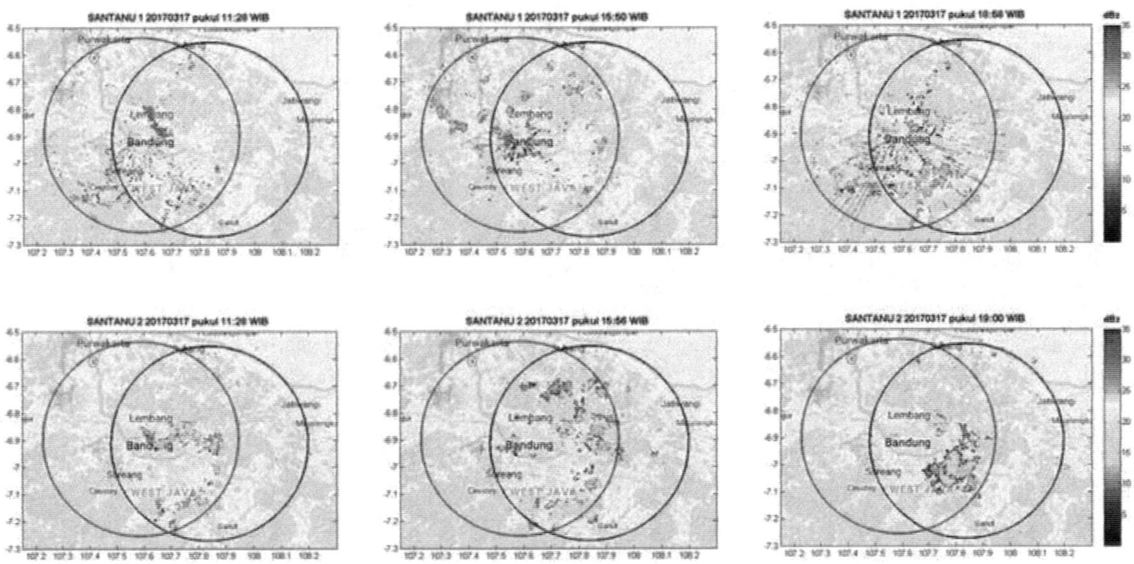

Figure 2. SANTANU in March 17.

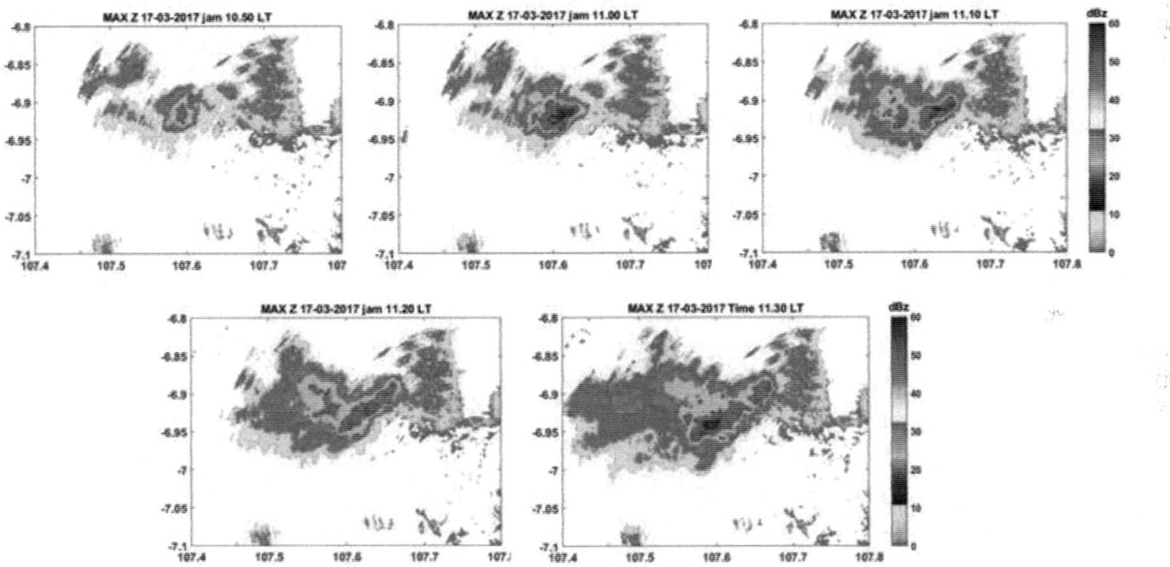

Figure 3. CMAX in March 17.

Based on the steiner method on the v shape (figure 5) object contain 2 type of precipitation (convective and stratiform). First stage from this figure showed two early convective in 10.50 LT and combined into one single convective cell in 11.00 LT, with the later process of convective type are located inside the V-shape area with the stratiform around those area from 11.10 until 11.30 LT.

IOP Conf. Series: Earth and Environmental Science **149** (2018) 012058 doi:10.1088/1755-1315/149/1/012058

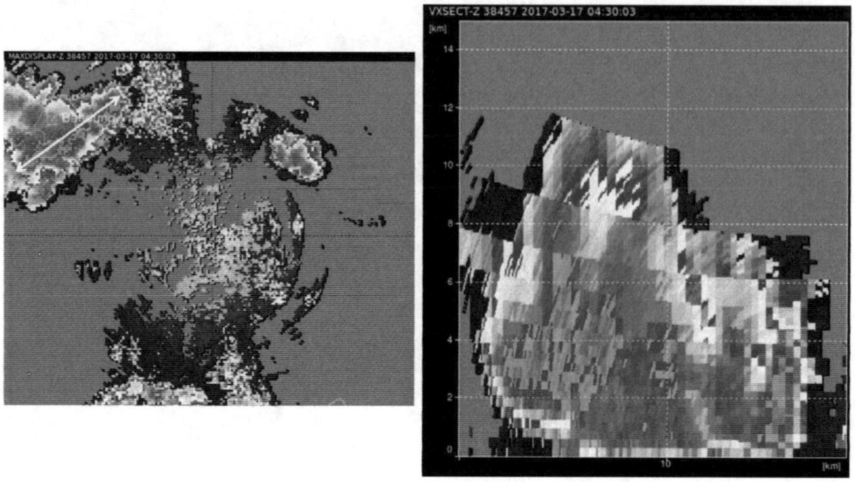

Figure 4. VSECT in March 17.

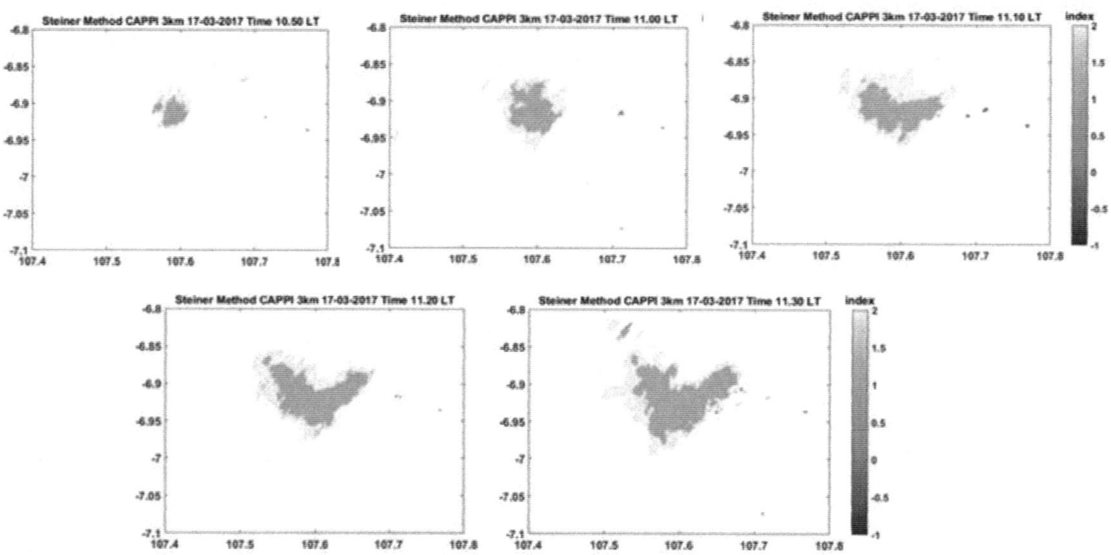

Figure 5. Steiner method in March 17.

Based on the Tbb from Himawari data (figure 6) showed MCS development occurs started from 13 LT until 21 LT. Cloud activity was still low, and then developed into a mesoscale convective system several hours after 11 LT The Tbb activity were captured above Bandung area at 15 LT until 16 LT

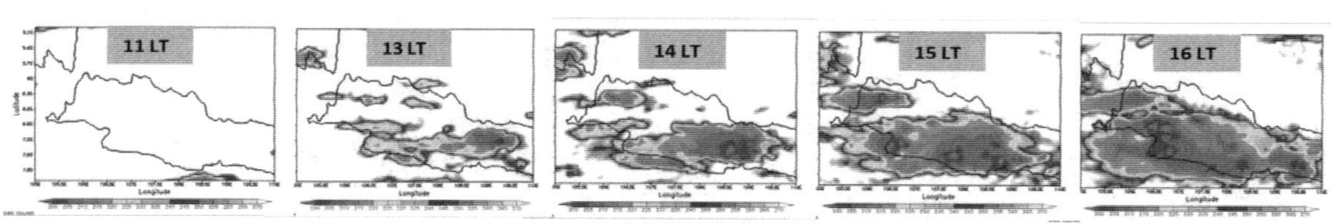

Figure 6. Tbb Himawari on March 17.

3.2. Case study in April 19, 2017

In April 19, SANTANU were detecting precipitation from 11.00 LT until 19.00 to 20.00 LT (figure 7). Various precipitation object were observed during those period. Heavy rainfall are occur in the middle of Bandung area from 13.20 until 14.40 LT. Twenty minutes later heavy precipitation were also observed in almost every coverage area of the SANTANU A and B.

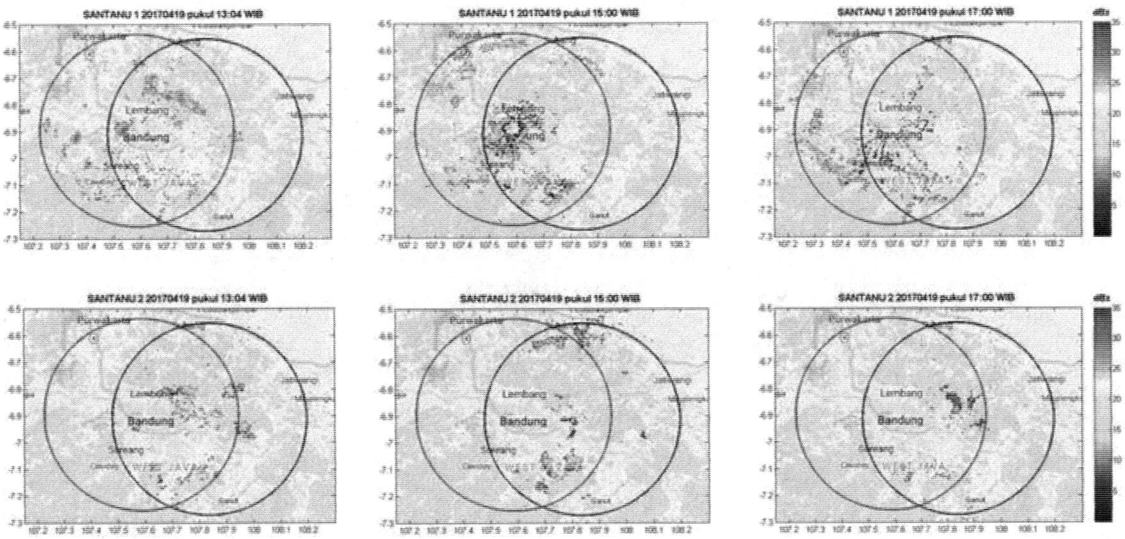

Figure 7. SANTANU in April 19.

Figure 8 is the CMAX from the heavy rainfall in the middle of Bandung city. Convective cell were located in Bandung city on 13.20 LT. This cell then grew until 13.50 LT and decrease until 14.10 LT. Based on data sample in 13.30 LT, VSECT data observed (figure 9) showed having maximum values at 1-2 km height and 4-6 km height, with the cloud top height reached about 12 km. After this event from 14.20 until 18.00 LT, CMAX observed more reflectivity value with the maxima in around 15.00 LT until 17.00 LT. In 16.30 LT detected a widespread reflectivity area reaching until 40 - 50 km.

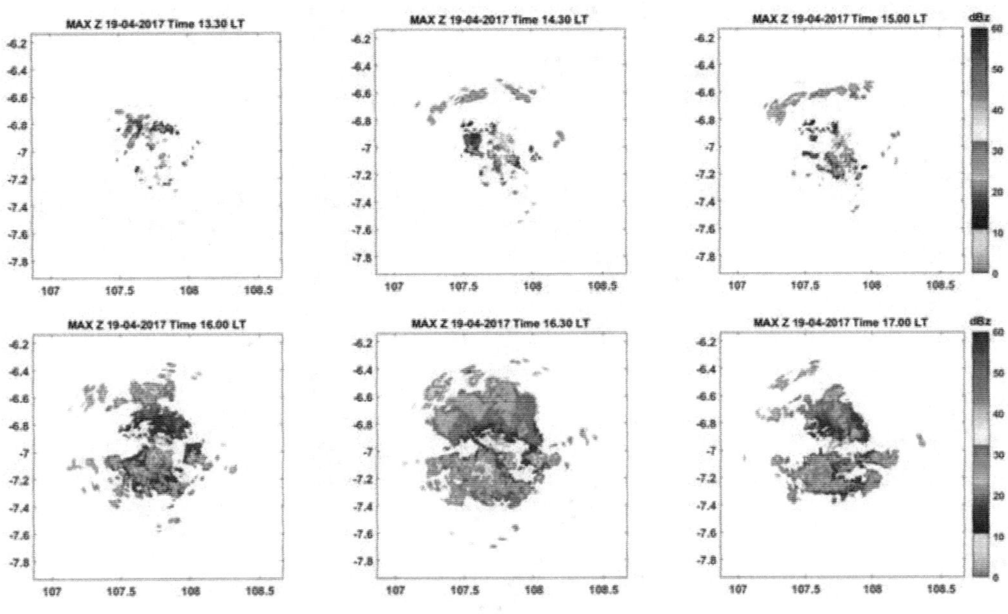

Figure 8. CMAX in April 19.

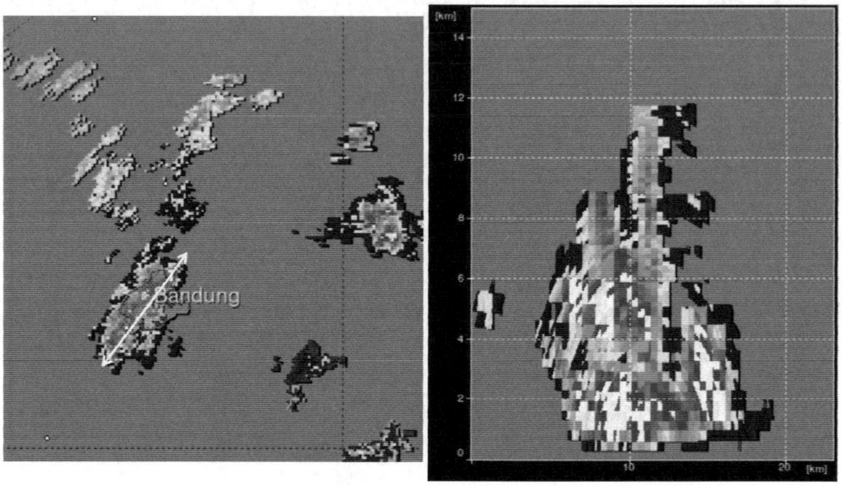

Figure 9. VSECT in April 19.

From steiner method showed a combination between convective and stratiform in case study on April 19 (figure 10). In 16.30 LT with maximum area are also having a combination between convective and stratiform with the convective index are having more than 50% of the area.

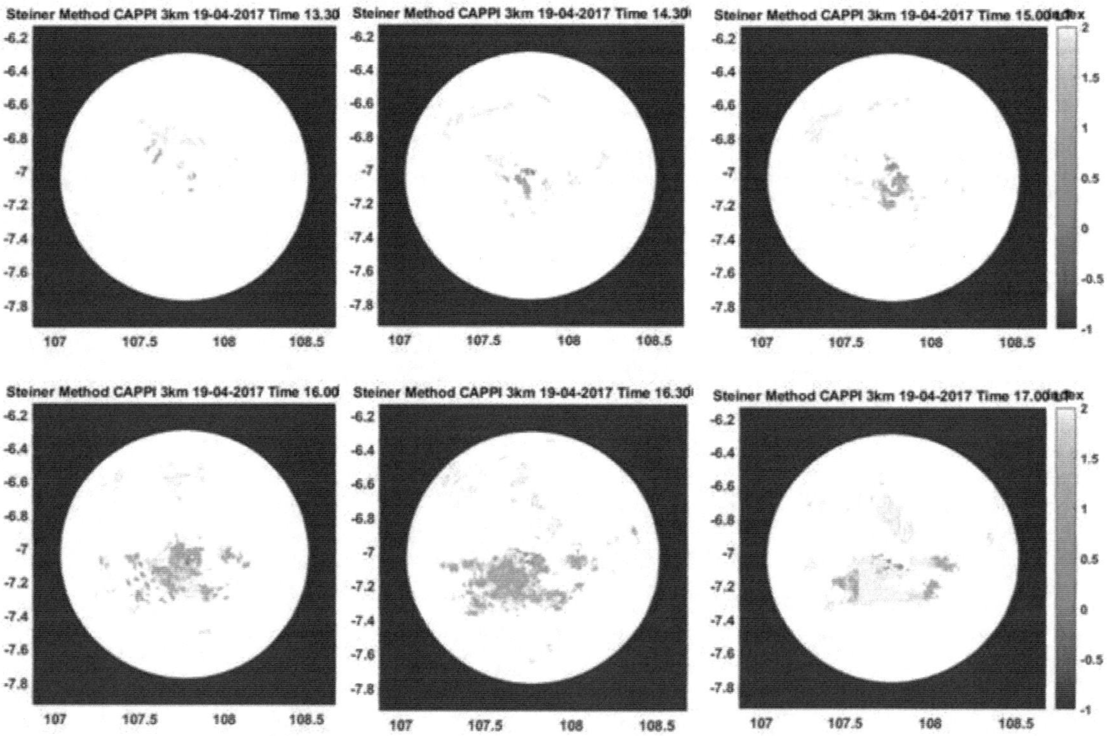

Figure 10. Steiner method in April 19.

In contrast with case study before, the Tbb from Himawari data showed a larger scale of MCS development. This occur from 14 LT until 18 LT. There are no data available in 17 LT.

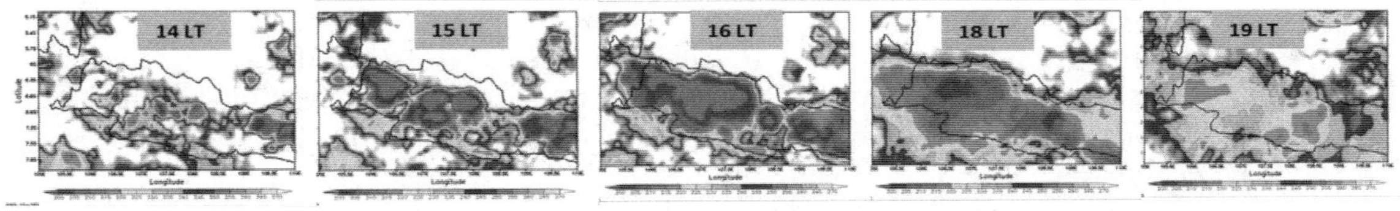

Figure 11. Tbb Himawari on April 19.

4. Conclusion

The heavy rainfall on March 17 and April 19 that occur in Bandung city around 11.00 until 13.00 LT are based from a local convective cell in the middle of Bandung city based on two type of x-band radar. This local convective activity are showed from the SANTANU observation supported by the VSECT and CMAX of the Transportable radar data that signify the convective activity within those area. A unique reflectivity forming a v-shape area are observed from this x-band radar.

After this heavy rainfall, there are MCS activity occur around Bandung area. This event are confirm by the classification of convective-stratiform echoes from radar data and also from the Tbb Himawari 8 sateliite data. The different MCS activity from this two case study is that April 19 have much more MCS activity than in March 17, 2017. The maximum of MCS activity in April 19 on 16.30 LT were having a combination of convective and stratiform with an area reached up to 40 - 50 km. More observation are needed to observe the MCS based on combination of X-band radar and satellite data.

Acknowledgement

This research were supported by the Center of Atmospheric Science and Technology, National Institute of Aeronautics and Space - PSTA LAPAN and also from Indonesian Center of Excellence Research Fund. The authors thank the Rancakasumba electric substation for support and providing the radar location site.

References

[1] Maddox R A 1980 Mesoscale convective complexes *Bull. Amer. Meteor. Soc.* **61** 1374-1387
[2] Houze R 1989 Observed structure of mesoscale convective systems and implications for large-scale heating *Quart. J. Roy.Meteor. Soc.*115,425–461
[3] Johnson R H, Rickenbach T M, Rutledge S A, Ciesielski P E and Schubert W H 1999 Trimodal characteristics of tropical convection *J. Climate*. **12**, 2397–2418
[4] Hartmann D L, Hendon H H and Houze Jr. R A 1984 Some implications of the mesoscale circulations in tropical cloud clusters for large-scale dynamics and climate *J. Atmos. Sci.* **41** 113–121
[5] Schumacher C, Houze Jr. R A and Kraucunas I 2004 The tropical dynamical response to latent heating estimates derived from the TRMM Precipitation Radar *J. Atmos. Sci.* **61** 1341–1358
[6] Houze R 1993 Cloud Dynamics, *International Geophysic series*, Academic Press, **53**
[7] Rigo T and Llasat M C 2004 A Methodology for the Classification of Convective Structures using Meteorological Radar: Applica-tion to Heavy Rainfall Events on the Mediterranean Coast of the Iberian Peninsula *Nat. Hazards Earth Syst. Sci.* **4** 59–68
[8] Steiner M, Houze Jr. R A and Yuter S E 1995 Climatological characterization of three-dimensional storm structure from operational radar and rain gauge data. *J. Appl. Meteor.* **34,** 1978–2007
[9] Nanding N, Rico-Ramirez M A and Han D 2015 Comparison of different radar-raingauge

rainfall merging techniques *Journal of Hydroinformatics* **17** 422-445
[10] Rigo T and Llasat M C 2005 Radar analysis of the life cycle of Mesoscale Convective Systems during the 10 June 2000 event *Natural Hazards and Earth System Sciences* **5** 959–970
[11] Davini P, Bechini R, Cremonini R and Cassardo C 2012 Radar-Based Analysis of Convective Storms over Northwestern Italy *Atmosphere* **3** 33-58
[12] Sinatra T, Noersomadi, Nugroho G A and Trilaksono N J 2014 Characterizing the diurnal cycle of convective activity over the south of west java using doppler x-band radar observation *The 4th international symposium for sustainable humanosphere (ISSH)-a forum of the humanosphere science school (HSS)* **2014** ISSN:2088-9127
[13] Ahmed S M 2014 GAMIC meteor product description manual, Aachen, Edition 2 Rev 25, GAMIC mbH.
[14] Nugroho G A, Munir M M and Khairurrijal 2015 Computer-Based Marine Automatic Radar for Rain Detection *Applied Mechanics and Materials* **771** (2015) pp 9-12
[15] Nugroho G A, Sinatra T, Awaludin A, Halimurrahman and Handiana D 2016 Hail Detection By a Low Cost Local Weather Radar Operated For Disaster Early Warning System *Proceeding of The 7 Indonesian Japan Joint Scientific Symposium* **2016**
[16] Steiner M R and Houze Jr. R A 1997 Sensitivity of the estimated monthly convective rain fraction to the choice of Z–R relation *J. Appl. Meteor.* **36** 452–462
[17] Ahmed S M 2012 ENIGMA IV user manual, Aachen, Edition 2 Rev 19, GAMIC mbH.
[18] Trismidianto, Yulihastin E, Satyawardhana H and Ishida S 2017 A composite analysis of the Mesoscale Convective Complexes (MCCs) development over the Central Kalimantan and its relation with the propagation of the rainfall systems *IOP Conf. Series: Earth and Environmental Science* **54** (2017)
[19] Houze Jr. R A 2004 Mesoscale convective systems *Rev. Geophys.* 42-43
[20] Leary C A and Houze Jr. R A 1979 The structure and evolution of convection in a tropical cloud cluster *J. Atmos. Sci.* **36** 437–457

Design and Implementation of Effective Electrical Power System for Surya Satellite-1

A H Sulistya[1], W Hasbi[2], R Muhida[1,3]

[1] Surya University, Campus Hub (Unity Tower Building) Jl. Boulevard Gading Serpong kav. M5 no 21, Summarecon Serpong, Tangerang 15810; Banten – Indonesia
[2] LAPAN Satellite Technology Center, Jl. Cagak Satelit KM 04 Rancabungur, Bogor 16310; Jawa Barat – Indonesia
[3] Bandar Lampung University, Jl. Zainal Abidin Pagar Alam No.26, Labuhan Ratu, Kedaton, Kota Bandar Lampung 35142, Lampung – Indonesia

E-mail: afiqhs@gmail.com

Abstract. Surya Satellite-1 is a nanosatellite developed by students of Surya University. The subject of this paper is the design and implementation of effective electrical power system for Surya Satellite 1. The electrical power system role is to supply other systems of the satellite with appropriate electrical power. First, the requirements of the electrical power system are defined. The architecture of the electrical power system is then designed to build the prototype. The orbit simulation is calculated to predict the power production. When prototype test and simulation data is gained, we make an operation scenario to keep the produced power and the consumed power in balance. The design of the modules of the electrical power system is carried out with triple junction solar cells, lithium ion batteries, maximum power point trackers, charging controllers, power distributions, and protection systems. Finally, the prototypes of the electrical power system are presented.

1. Introduction

Nanosatellite has emerged as a popular platform among academic and industrial space communities. They need less cost and time to develop than traditional and sophisticated satellites but can perform space technology demonstrations. The miniaturization of satellite technology allows the increase of interest among space communities to develop small spacecraft missions based on CubeSat standards [1]. The major challenge of designing nanosatellite are the limitation in volume, mass, and power available.

Surya Satellite-1, that commonly abbreviated as SS-1, is the first nanosatellite developed by undergraduate students and to be launched by Indonesia. This 1-unit CubeSat carried automated package reporting system (APRS) communication mission run on amateur radio bands. The use of APRS SS-1 are transmitting and repeating telemetry data from satellite and remote area to the ground station and amateur radio users.

This paper describes the electrical power system (EPS) built for SS-1 to provide the power to conduct the mission, how the design, the power produced, and the operation usage [2]. The EPS harvests energy form sun by solar panels, stores it in the batteries, and distributes to the subsystems specifically and safely. The challenge of the EPS design is to provide reliable power with the strict limitation of mass and volume available in 1U that is around 1.3 kg and 1 liter for the rest of the subsystem. Then the

IOP Conf. Series: Earth and Environmental Science **149** (2018) 012059 doi:10.1088/1755-1315/149/1/012059

amount of energy harvested is limited and so the storage size. The additional challenge is the lack of attitude control so SS-1 has a spinning rate that limit the power harvested from its body mounted solar panels. The satellite operation must consider the energy balance since it has eclipse phase and the dynamic demands of power.

Because the commercial CubeSat modules are yet rare and costly, the subsystem modules of SS-1 are mainly self-developed. The goal of developing the EPS is to demonstrate the reliable system but cost-effective constructed from commercially of the shelf (COTS) components to operate at the required conditions.

2. Requirements of the electrical power system

2.1. Space Environment
SS-1 is expected to orbit at low earth orbit (LEO) with the altitude at 400 km and inclination at 51.6 degree. In this environment the orbit duration is around 90 minutes with the 1/3 is in eclipse and 2/3 of it is in daylight. Then the temperature range within the orbit is around -20°C to 60°C. During the daylight, the solar panels harvest energy to be used by the subsystems and stored the excess. As the hot environment, the components selected must be survive and operate normally. During the eclipse, the temperature will be cold and there is no power harvested from solar panels. Lithium ion battery is the component that could not operate in the cold environment so that the battery temperature need to be maintained above 0°C by active thermal control.

Deployed from the ISS, the expected lifetime of the CubeSat is predicted to be 16 months according to the simulation with Jacchia Roberts atmospheric model. Then the batteries depth of discharge (DOD) need to be maintained to operate at 10% of the actual capacity to attain charge-discharge cycle till the end of life (EOL). Then the sizing of the battery capacity can be calculated.

2.2. Functional
The EPS should effectively be designed for the operational need of the other subsystems. EPS duties are to generate power from sun illumination by solar panels, harvest energy efficiently by maximum power point tracker (MPPT), store the energy for eclipse and peak demand use, distribute the power effectively to the loads, and provide protection systems for the components.

The subsystems are mainly operating at 5.0V but the communication module is operating between 3.5V to 4.5V. Then the power sources are designed to operate at the required bus voltage. The 5V bus regulated by a boost converter. And the communication module uses unregulated bus directly from the batteries. That is why the lithium ion battery is chosen because the nominal operating voltage is between 3.6V to 4.1V so there is no dissipated power from the dc-dc converter.

The solar panels are also designed to be suitable for the power demand. Each side of the body is installed two cells of triple-junction GaAs with 1.2W 2.4V connected in parallel. Then the power is harvested with MPPT that works as boost converter, so the voltage is high enough be used by charging controller to charge the lithium ion batteries.

The EPS should be equipped with protection system from reverse current, over-charge and deep-discharge, overcurrent, over and under temperature [3].

2.3. Power budgets
EPS responsible to provide the required electric power to entire the CubeSat system. There are the power required by the subsystems.

Table 1. Power Budgets of SS-1.

Subsystem	Peak Power (mW)	Standby Power (mW)	Power Bus
COM VHF	3000	240	Unregulated
Payload VHF	400	250	5V
COM UHF	3000	240	Unregulated
Payload UHF	400	250	5V
Heater	3000	0	Unregulated
Data Handling		40	5V

3. Design of electrical power system architecture

The EPS architecture design consist of three separate functional blocks: energy harvesting, energy storage, and power distribution. The voltage on the main power switch (MPS) is unregulated, normally 3.6 to 4.2 V, staying on the same voltage level as the battery cells. This architecture was chosen to minimize the need for voltage conversion steps for battery charging and discharging, therefore increasing efficiency. This architecture is also popular among other CubeSats with MPPT system, because it is very efficient. This configuration also allows battery cells to work as voltage stabilizers, reducing noise on the main power bus and therefore in the whole system.

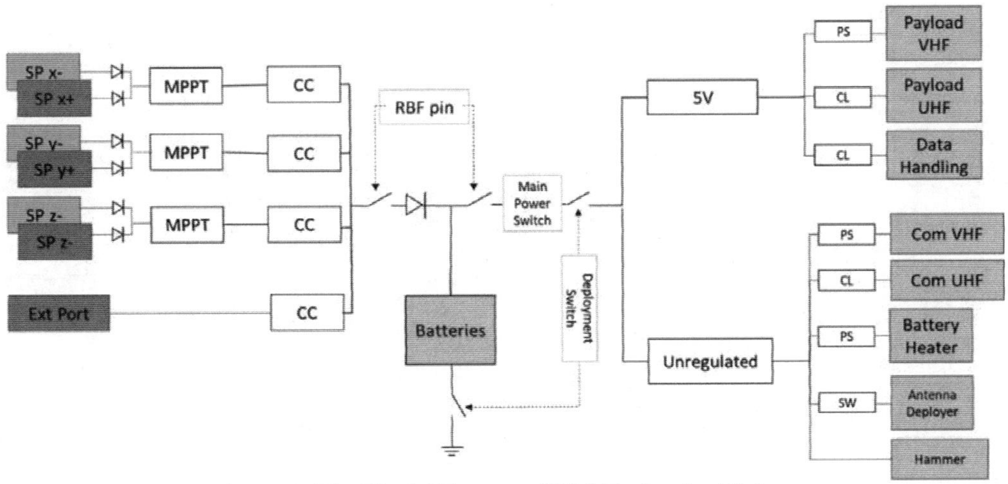

Figure 1. The Block Diagram of EPS Design for SS-1.

3.1. *Energy Source (Solar cells)*

The energy harvesting subsystem of the EPS collects energy from 12 solar cells. Cells are placed pairwise on all of the 6 sides of the satellite. In this configuration only three sides of the satellite can be in direct sunlight at any given time, meaning that only the cells on those sides produce power.

Table 2. The specified design of solar panels.

Side	X+	X-	Y+
Area	2 x (4cm x 8cm)	2 x (4cm x 8cm)	2 x (4cm x 6cm)
Pmax (W)	2.432	2.432	2.071
Layout	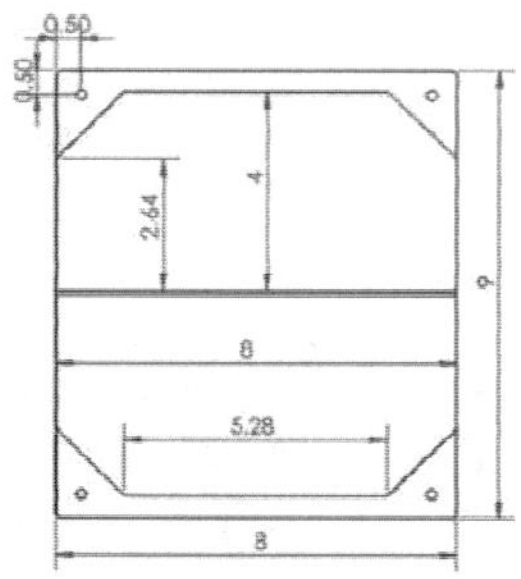	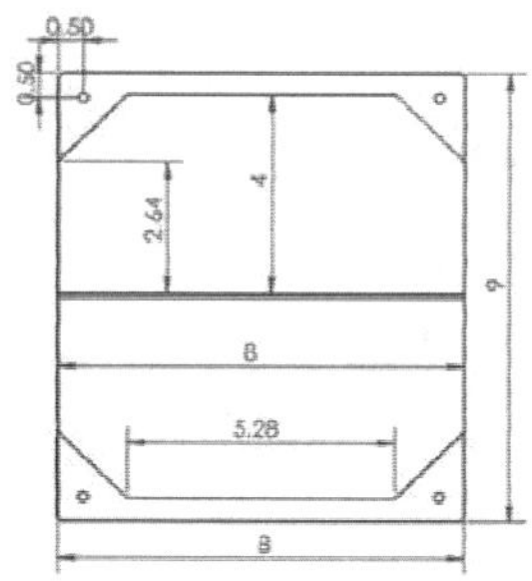	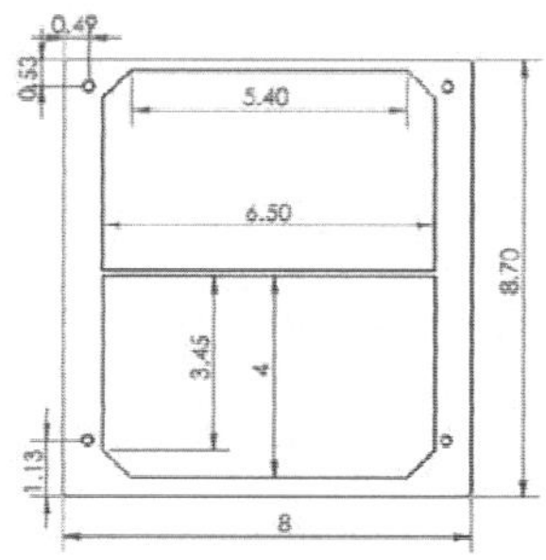

Side	Z+	Z-	Y-
Area	2 x (4cm x 8cm)	2 x (4cm x 8cm)	2 x (3cm x 4cm)
Pmax (W)	2.432	2.432	1.257
Layout	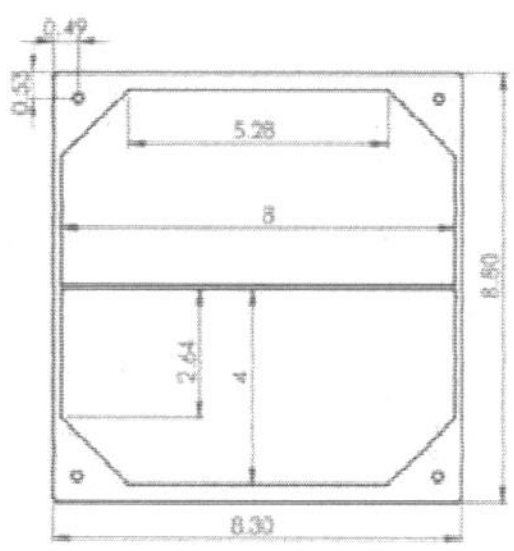	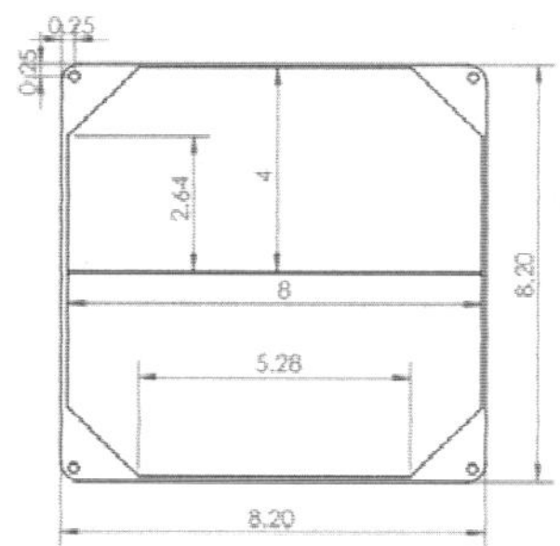	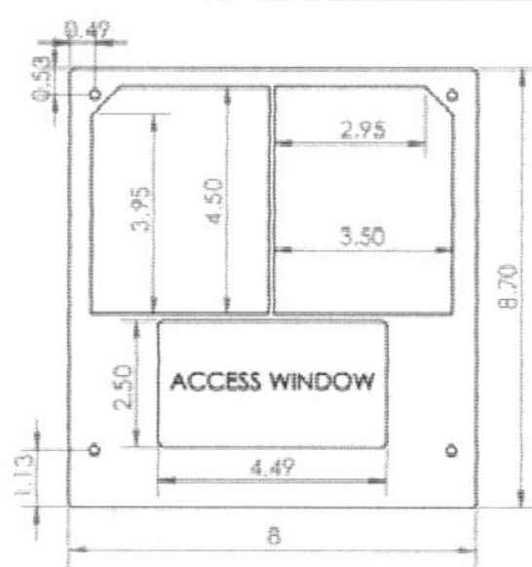

3.2. Power Harvesting (Maximum power point trackers)

The most effective way of collecting energy from the solar cells is by using maximum power point tracking. Due to the placement of the solar cells, the satellite needs to track only three power points at any given moment in time and therefore three separate MPPT systems were planned for the EPS. In the final design, SPV1040 maximum power point tracking chips from STMicroelectronics were used. These chips were originally meant for battery charging and regulate their output voltage to match the effective input resistance of consumers with the optimal consumer resistance, corresponding to the maximum power point of the solar cells, making them ideal for the chosen power topology. The chips only work in boost mode, meaning that the solar cells had to be placed in parallel [4], which in turn creates relatively large currents and low voltages, making voltage drops on components an issue. To get around this problem, an ideal diode system was used, consisting of MOSFETs driven by LTC4415 ideal diode controllers from Linear Technology.

3.3. Energy Storage (Lithium-Ion Batteries)
For energy storage, lithium ion batteries were chosen because the higher energy density among the other type of batteries [5] [6]. NCR18650B cylindrical lithium ion battery cells were used with nominal capacity of 3200 mAh. There are three cells of batteries connected in parallel therefore the capacity is summed but the voltage remains the same as single cell. This simplified the bus voltage and charging system because the parallel connected batteries can balance each other without any balancer circuits needed.

3.4. Charger controllers
To protect battery cells and to provide suitable charging requirements, EPS is equipped with independent battery charger controller circuits, which allow to protect the cells from overvoltage, overcurrent, and charging outside the safe temperatures. The system was implemented through L6924D charger controller that allow input power to be turned off when the battery is fully charged. As an extra layer of protection, the temperature the batteries cells are constantly monitored. In case of battery temperature exceeding 60°C or dropped below 0°C charging will be turned off.

3.5. Power distribution
The power distribution subsystem uses the energy, harvested from the solar cells and stored in the battery cells, to power all of the subsystems. To do this, switching regulators and control circuits were used. For powering the other subsystems per requirements, it was decided to create two voltage buses within the EPS those are 5V bus and unregulated bus. The 5V bus is regulated by boost converter and the unregulated bus is directly taken from the battery voltage.

3.6. Protections
Safety requirements are needed to be implemented on EPS before and while in operation use. Before the operational use, there are two levels of protection to prevent the satellite turned on until deployed into orbit. The first level is the remove before flight (RBF) pin that disconnect the power harvester, storage, and subsystems until been pulled out by astronaut after the satellite has placed in JSSOD [7]. The second level is the deployment switch that disconnect the loads from the power source at both vcc and gnd.

While the satellites are operational, the loads need to be protected from fault such as in rush and over-current by current limiter. For additional protection, there is a hammer system that periodically reboot the entire system in case there are any error or lagging on microcontroller devices that can't be solved from the ground.

4. Power Production

4.1. Power harvested
Power production came from the harvested energy from the solar panels. As the satellite is spinning and has the daylight-eclipse cycle, then the power produced of the panels in an orbit be like the following graph that came from the orbital simulation. The simulation parameters set at solar irradiance of 1366 W/m^2 under AM-0 condition at the beginning of life (BOL) that the solar cells efficiency remain 30%.

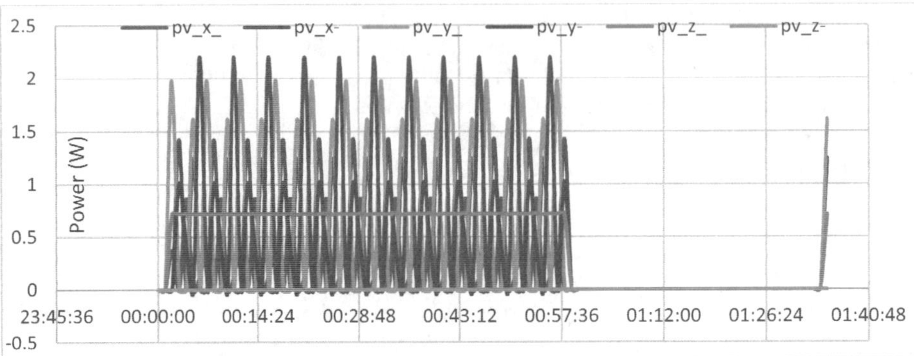

Figure 2. The solar power production in one orbit.

By calculating the sort of data repetition and considering the degradation of solar cells, then the power produced at EOL is summarized as the table below. The average power produced by solar cells during daylight is 2.844 W.

Table 3. Solar power production at end of life.

Average	Energy (J)	Duration (s)	Power (W)
Orbit Cycle	9609.13	5490	1.750
Daylight	9544.38	3356	**2.844**
Eclipse	0	2134	0

4.2. Efficiencies

The distributed power to the subsystems were dissipated during the conversion. Because the power converters have efficiencies when performing their duties. The EPS efficiency is the product of each converter efficiencies.

Table 4. Efficiencies of power converters.

Converter	Efficiency (%)
MPPT	0.95
Charger	0.95
Boost	0.90
EPS	0.81

Based on the average power production and the EPS efficiency, the calculated power distributed to the subsystem are the product of both value, that is 2.310 W.

5. Operation Strategy

5.1. Energy balance

The average power of the communication and payload subsystem that conduct the mission plan have calculated to be 1.47 W. Then the energy balance can be explained by the following table. The balance has deficit of 1101.5 J of energy in every orbit.

Table 5. Power balance in one orbit cycle.

Load	Demand			Supply			Balance of energy (J)
	Average Power (W)	Duration (s)	Energy (J)	Average Power (W)	Duration (s)	Energy (J)	
Com + Payload	1.47	5490.00	8042.85	2.31	3356.00	7752.42	-290.43
Heater	3.00	200.00	600.00				-600.00
Microcontroller	0.04	200.00	8.00				-8.00
Hammer	0.04	5490.00	203.13				-203.13
Total			8853.98			7752.42	-1101.56

5.2. Scenario

The deficit of energy balance can be overcome by the operation strategy. The simplest strategy is to hibernate the satellite periodically to perform exclusive charging. The sum of deficit orbits can be overcome by once hibernation as the simulation table below.

Table 6. Simulation of deficits and hibernation orbits.

Number of orbit	Deficit of Energy (J)	Charged Energy (J)	Energy balance (J)
1	-1101.56	7752.42	6650.87
2	-2203.11	7752.42	5549.31
3	-3304.67	7752.42	4447.76
4	-4406.22	7752.42	3346.20
5	-5507.78	7752.42	2244.65
6	-6609.33	7752.42	1143.09
7	-7710.89	7752.42	41.53
8	-8812.45	7752.42	-1060.02
maximum number of orbit before hibernation orbit			**7**

The satellite should be hibernated once in every seven orbits to keep the balance between energy produced and demanded.

6. Conclusion

The requirements and design of the electrical power system for the Surya Satellite-1 has been implemented. The EPS can supply reliable power production for the satellite with module built from commercially of the shelf components. The design has satisfied the requirements of environment, functional, and power demand. The solar power can be harvested by 95% efficiency and the total efficiency of module is 81%. To achieve the balance between supply and demand the operation strategy is implemented.

Acknowledgements

The authors are grateful to the Head of Satellite Technology Center - LAPAN and all the employees who have provided guidance and facilities during the design and construction of electrical power system for Surya Satellite-1. The author is also grateful to the team members of Surya Satellite-1 project who have given advices and helps to support the smoothness of this work.

References

[1] A. Poghosyan and A. Golkar, "CubeSat evolution: Analyzing CubeSat capabilities for conducting science," *Progress in Aerospace Sciences,* pp. 59-83, 2017.

[2] Wertz, Space Mission Analysis and Design, California: Microcosm Press; Kluwer Academic Publishers, 1999.

[3] P. Thirion, "Design and Implementation of On-board Electrical Power Supply of Student Nanosatellite OUFTI-1 of University of Liège.," University of Liège, Liège, 2005.

[4] M. Pajusalu, E. Ilbis, T. Ilves, M. Veske, J. Kalde, H. Lillm, R. Rantsus, M. Pelakauskas, A. Leitu, K. Voormansik, V. Allik, S. Lätt, J. Envall and a. M. Noorm, "Design and pre-flight testing of the electrical power system for the," in *Proceedings of the Estonian Academy of Sciences*, Tartu, 2014.

[5] M. Robyn, L. Thaller and D. Scott, "Nanosatellite Power System Considerations," NASA Scientific and Technical Information Program, Houston, 1995.

[6] B. T. Scheidegger, "Performance Characterization of High Energy," NASA Scientific and Technical Information, Ohio, 2010.

[7] JAXA, *Small Satellite Deployment Interface Control Document,* JEM Payload Accommodation Handbook, 2013.

IOP Conf. Series: Earth and Environmental Science **149** (2018) 012060 doi:10.1088/1755-1315/149/1/012060

Band co-registration modeling of LAPAN-A3/IPB multispectral imager based on satellite attitude

P R Hakim, A H Syafrudin, S Utama and A P S Jayani

Satellite Technology Center, National Institute of Aeronautics and Space (LAPAN), Rancabungur, Bogor 16310, Indonesia

E-mail: patriarachmanhakim@yahoo.com

Abstract. One of significant geometric distortion on images of LAPAN-A3/IPB multispectral imager is co-registration error between each color channel detector. Band co-registration distortion usually can be corrected by using several approaches, which are manual method, image matching algorithm, or sensor modeling and calibration approach. This paper develops another approach to minimize band co-registration distortion on LAPAN-A3/IPB multispectral image by using supervised modeling of image matching with respect to satellite attitude. Modeling results show that band co-registration error in across-track axis is strongly influenced by yaw angle, while error in along-track axis is fairly influenced by both pitch and roll angle. Accuracy of the models obtained is pretty good, which lies between 1-3 pixels error for each axis of each pair of band co-registration. This mean that the model can be used to correct the distorted images without the need of slower image matching algorithm, nor the laborious effort needed in manual approach and sensor calibration. Since the calculation can be executed in order of seconds, this approach can be used in real time quick-look image processing in ground station or even in satellite on-board image processing.

1. Introduction

LAPAN-A3/IPB satellite is third Indonesian micro-satellite with earth observation and remote sensing missions, with 4-bands multispectral pushbroom imager, which are blue, green, red and near-infrared (NIR), as main payload. The images produced have 16 bit radiometric resolution with 15 meter ground sampling resolution and 120 km swath-width [1]. Same as other multispectral satellite images, raw images captured from LAPAN-A3/IPB satellite multispectral imager suffer from both geometric and radiometric distortion which should be systematically corrected. There are several distortions which commonly occur on satellite multispectral images, most importantly are radiometric vignetting distortion [2] and geometric distortion due to satellite movement instability while imaging [3][4]. Another significant geometric distortion which also occurs on LAPAN-A3/IPB multispectral images is band co-registration distortion, which is caused by position and orientation differences of band detectors respect to lens center [5]. The distortion causes an object in four channel images are not align each other, therefore the composite image produced seems being blurred since same object in different channel image have different position and orientation within the image.

Band co-registration can be corrected using several methods, which are manual approach, rigorous sensor modeling and image matching method. Every approach has its own different advantages and disadvantages. Manual approach is done by producing many ground control points (GCP) for each channel image, which is referenced to referenced image. The referenced image could be a validated referenced map or one of the multispectral channel images which is chosen as reference. This manual approach could produce an accurate result if done very carefully by the expert, but it needs laborious efforts and time consuming. Another approach to correct band co-registration distortion is rigorous sensor modeling which will directly calculate the distortion parameters based on known and validated satellite and imager parameters, both intrinsic and extrinsic ones [6]. However, this approach needs accurate satellite and sensor geometry to produce an accurate result.

Currently, band co-registration distortions occur on LAPAN-A3/IPB multispectral images are corrected using image matching approach, in which NIR, green and blue channel image is referenced to red channel image. The approach divides each channel image into many sub-images and then utilizes Canny edge detection algorithm to pre-process the image before applying cross-correlation based image matching [7][8][9]. This method can be executed autonomously without knowing any satellite and imager characteristics, and based on numerous LAPAN-A3/IPB images which have been processed, this approach is able to produce quite accurate result. However the algorithm still needs time to process the images, which is around 4-5 minutes in average for one image. While the time needed is still considered acceptable in most application, it clearly cannot be used in real time satellite image processing, for example in quick-look application on real time acquisition mode.

This research aims to develop a simple approach which is able to calculate band co-registration that occurs on LAPAN-A3/IPB multispectral images, based on satellite attitude data provided by star tracker sensor (STS). In this approach, the influence of various satellite attitudes when imaging to band co-registration error produced by image matching algorithm is deeply investigated based on numerous actual LAPAN-A3/IPB multispectral images which are captured under various attitudes. The method works entirely based on image matching approach which is used on current setup, however the manner in which the calculation is done is significantly simplified. The main idea of this approach is to find the relationship between satellite attitude while imaging and band co-registration distortion characteristics produced by image matching calculation on LAPAN-A3 multispectral images. Once the band co-registration models have been established, they can be used to calculate band co-registration parameters in no time with similar accuracy compare to image matching approach used in standard processing.

2. Methodology

Based on rigorous sensor modeling theory, band co-registration distortion is influenced by satellite attitude while imaging. The calculation of band co-registration distortion on more than 150 images of LAPAN-A3/IPB imager by using image matching approach, which is composed of Canny edge detection and Fast Fourier Transform (FFT), also shows this characteristics. Nadir observation tends to produce small band co-registration distortion while off-nadir observation often produces significant distortion. The model of this relationship however, is still unknown, therefore needs to be determined for each pair of band co-registration.

As with any other modeling, this research starts with tabulating all variables which will be modeled, both for dependent and independent variables. In this modeling, dependent variables consist of band co-registration parameters of each pair of channel images. These parameters essentially determine translational difference of two channel images both in vertical and horizontal direction. These dependent variables are produced by image matching algorithm which is used in current setup of LAPAN-A3/IPB multispectral image pre-processing. In the other hands, independent variables in this modeling consist of satellite attitude of yaw, pitch and roll angle when the imager capturing the image. These satellite attitudes when imaging are derived from start tracker sensor (STS) data and by using SGP-4 orbit model [10].

After all dependent and independent variables are gathered, modeling of band co-registration distortion for each band with respect to satellite attitude can be done by using least-square regression method. Since the number of observation data is much higher than model parameters which will be determined, pseudo-inverse approach will be used in this regression process [11]. There are six models that will be developed, consisting two models for each pair of band co-registration. Then the accuracy of each models produced is evaluated. Finally, actual image correction using the resulted band co-registration models is done to validate the results visually. Figure 1 shows general flowchart that is used in this research.

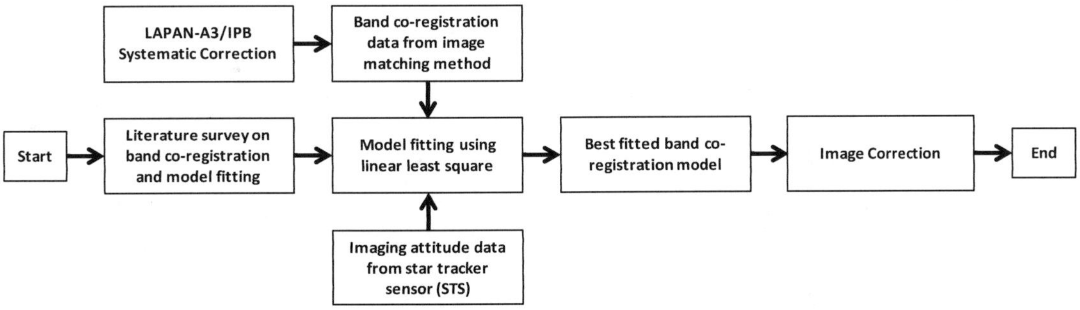

Figure 1. General flowchart used in this research.

3. Modeling Result and Analysis

As explained earlier, the modeling process is based on image pre-processing that has been done on numerous actual LAPAN-A3/IPB multispectral images. Figure 2 shows several examples of band co-registration distortion on parts of LAPAN-A3/IPB multispectral images in NRG composite format, which clearly showing that each color-channel is not aligned to each other.

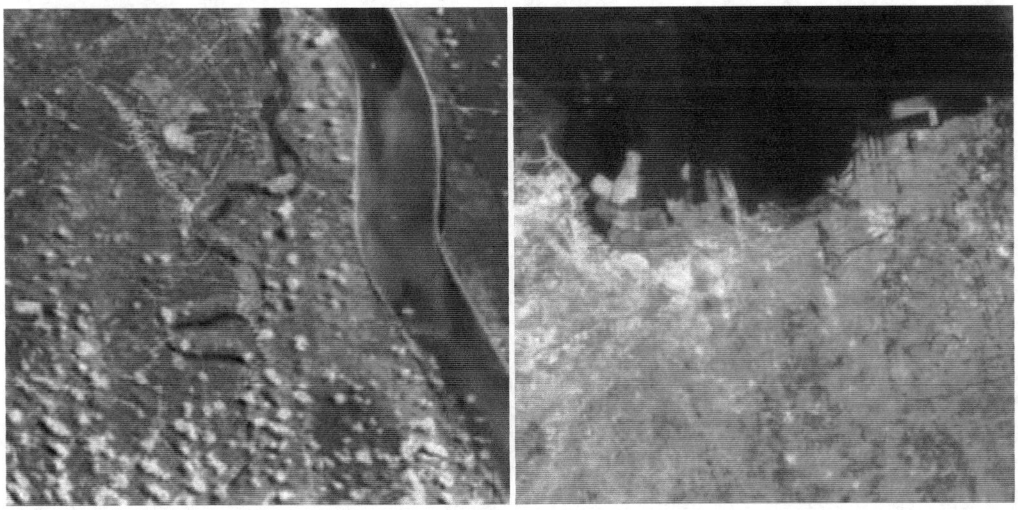

Figure 2. Examples of band co-registration distortion on LAPAN-A3/IPB multispectral images.

It can be seen on these two example raw images that co-registration distortion occurs on many objects across the image, such as cloud, river, road and other human-made structure. The distortion appears on every LAPAN-A3/IPB multispectral images, however the magnitude of co-registration which occurs between images varies depending on satellite attitude when the imager capturing the image on each particular day.

3.1. Observation Data

Table 1 shows example of several data which relate band co-registration distortion produced by image matching algorithm and satellite attitude while imaging, only for NIR to red pair of co-registration. These sampled data are obtained from images taken by LAPAN-A3/IPB multispectral imager during

August 2017. Band co-registration values on the table represent both horizontal and vertical pixel differences between two channel images, which further divided into three sampled region of image, which are left and right side as well as center region of the image. Meanwhile, satellite attitude values on the table represent average satellite attitude while imaging in each yaw, pitch and roll axis.

Table 1. Example of observation data used in the modeling.

Image Data		Satellite Attitude			NIR-Red Co-registration Values					
					Horizontal			Vertical		
Date	Target	Yaw	Pitch	Roll	Left	Mid	Right	Left	Mid	Right
20170801	Muntok-Lampung	4	-7.4	-1.9	-49	0.3	33	-71	-81	-113
20170802	Ciledug-Kalipucang	5.1	-1.5	-1.5	-48	-0.1	32	-69	-79	-112
20170803	Sipitang-Tinotuhan Kalteng	6.7	-2	-2.4	-51	-3.4	27	-70	-79	-111
20170809	J P Coen Peak-Kimaam Papua	5.3	-5.5	-2.7	-51	-2.7	28	-70	-80	-113
20170810	Waikabubak NTT	10	-6.6	-7	-55	-6.2	23	-70	-81	-116
20170811	Maluku Utara	10	-9.4	-6.5	-55	-7.3	23	-70	-81	-116
20170812	Pangkalpinang-Ujungkulon	4.8	-3.6	-11	-48	0.3	32	-69	-80	-115
20170813	Nangatayap Kalbar	9.3	-5.8	-1.8	-54	-6.4	23	-70	-79	-111
20170815	Lombang-Somba Sulbar	0.5	-12	0.5	-46	4.4	40	-72	-81	-114
20170816	Salongo-Banggai Sulawesi	2.3	-3.9	-3.2	-46	3.4	38	-69	-78	-110
20170817	Kepri-Jambi	5.6	1.3	-0.3	-49	-0.1	32	-68	-77	-110
20170819	Brunei-Nangaipinoh Kalbar	0.2	13	-2.7	-42	5	36	-71	-82	-116
20170820	Senggi-Okaba Papua	2.5	15	1.7	-44	2.7	32	-74	-84	-117
20170823	Merak	0.1	0.5	1.5	-42	6.3	40	-69	-78	-109
20170824	Ambarawa-Purworejo	-0.7	-0.7	2.5	-43	5.9	40	-69	-78	-108
20170825	Kaltim-Kalsel	-1.4	-3.2	1.4	-43	5.6	40	-69	-78	-109
20170826	Palu-Polewali	1.7	-4	2.5	-43	5.6	39	-69	-78	-108
20170827	Manado-P Masoni Sulawesi	-2.1	-3.9	5.8	-40	9.1	45	-70	-78	-109
20170828	Palembang-Martapura	0.7	-7.6	-8.6	-41	7.7	44	-70	-80	-114
20170830	Tuban-Blitar	-0	1.3	0.1	-42	6.1	40	-69	-78	-109

Based on attitude filtered data used in this modeling, Figure 3 shows example of strong relationship between satellite attitudes in yaw axis to NIR-red band co-registration values in horizontal direction, while the values of band co-registration in vertical axis has fairly good relationship with pitch and roll angle. The co-registration values in Figure 3 are the values obtained from center part of the image.

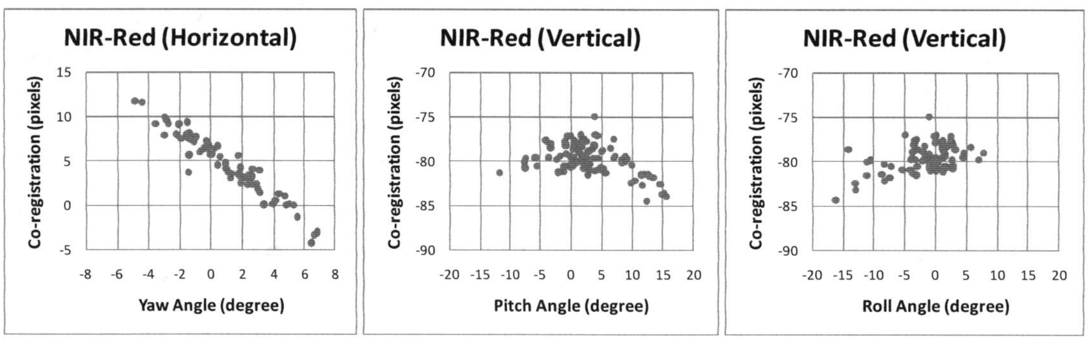

Figure 3. Strong relationship between satellite attitudes to NIR-red band co-registration values.

3.2. Linear Least Square Regression

As can be seen from Figure 3, yaw angle has a strong influence to horizontal co-registration value, while pitch and roll angle has fairly influence to vertical co-registration value. However, type and magnitude of the relationship still need to be determined. Therefore, modeling band co-registration as

IOP Conf. Series: Earth and Environmental Science **149** (2018) 012060 doi:10.1088/1755-1315/149/1/012060

a function of satellite attitude is conducted by using linear least square approach. The model of each band co-registration in both directions could be linear or nonlinear, first order or second order, as well as could be sinusoidal or exponential. Initially, trial and error approach is used to determine the best approximation for each co-registration model, where all potential candidate models are evaluated. Model with the smallest error is chosen as the best model for that particular co-registration model. Table 2 shows sample of list of models which are evaluated for NIR-Red band co-registration in vertical axis, producing second order cosine model with pitch and roll couplings as the best model.

Table 2. The best model determination of vertical NIR-Red co-registration.

No	Vertical NIR-Red Co-registration Model	Model Error (Pixels)			
		Left	Mid	Right	Mean
1	$K_1\sin(Yaw) + K_2$	1,75	2,25	4,08	2,69
2	$K_1\cos(Yaw) + K_2$	1,77	2,12	3,79	2,56
3	$K_1\cos(Pitch) + K_2$	1,32	1,74	3,52	2,19
4	$K_1\cos^2(Pitch) + K_2\cos(Pitch) + K_3$	1,17	1,52	3,12	1,94
5	$K_1\cos(Pitch)\cos(Roll) + K_2$	1,36	1,39	2,42	1,73
6	$K_1\cos^2(Pitch)\cos^2(Roll) + K_2\cos(Pitch)\cos(Roll) + K_3$	1,21	1,19	2,15	1,51

3.3. Regression Result and Analysis

The same approach is used to find the best model of all pair of band co-registration in both horizontal and vertical direction. Table 3 shows the best model obtained for each band co-registration. It can be seen that all band co-registration in particular direction has the same type of model, although the magnitude might not be the same.

Table 3. The best model of each pair of band co-registration.

Band	Co-registration Model	Model Error (Pixels)			
		Left	Mid	Right	Mean
N2R (H)	$K_1\sin(Yaw) + K_2$	1,85	1,22	3,08	2,05
N2R (V)	$K_1\cos^2(Pitch)\cos^2(Roll) + K_2\cos(Pitch)\cos(Roll) + K_3$	1,21	1,19	2,15	1,51
G2R (H)	$K_1\sin(Yaw) + K_2$	1,56	1,33	1,97	1,62
G2R (V)	$K_1\cos^2(Pitch)\cos^2(Roll) + K_2\cos(Pitch)\cos(Roll) + K_3$	1,48	1,35	2,38	1,74
B2G (H)	$K_1\sin(Yaw) + K_2$	0,37	0,30	0,46	0,38
B2G (V)	$K_1\cos^2(Pitch)\cos^2(Roll) + K_2\cos(Pitch)\cos(Roll) + K_3$	0,42	0,15	1,72	0,76

Based on regression results obtained, it can be concluded that band co-registration value in horizontal axis is strongly influenced by yaw satellite angle, which can be modeled as first order of sine function. Meanwhile band co-registration value in vertical axis is influenced by both pitch and roll satellite angle, which can be modeled as second order of couplings cosine function.

These obtained models have an average 1-2 pixels relative accuracy compared to current image matching approach. Meanwhile, absolute accuracy of band co-registration correction currently employed on LAPAN-A3/IPB multispectral images by using image matching approach is around 2-3 pixels error [7]. Therefore, the overall accuracy of the proposed band co-registration in this research by using attitude modeling approach is around 3-5 pixels error. However, accuracy of the model obtained seems to be good only for near nadir satellite attitude, since most of the images used in this research were taken when the satellite not far away from nadir. Therefore, the band co-registration model of each band should be updated by using every new image captured in the future, to cover wider range of satellite attitude while imaging.

4. Image Correction Analysis

The band co-registration models that have been obtained are then used to produce band co-registered composite image of actual LAPAN-A3/IPB multispectral images. The models directly replaces image matching algorithm used in standard LAPAN-A3/IPB systematic image processing algorithm. Figure 4 shows image correction results in NRG composite format, showing that the models successfully correct band co-registration distortion occurs on raw satellite image.

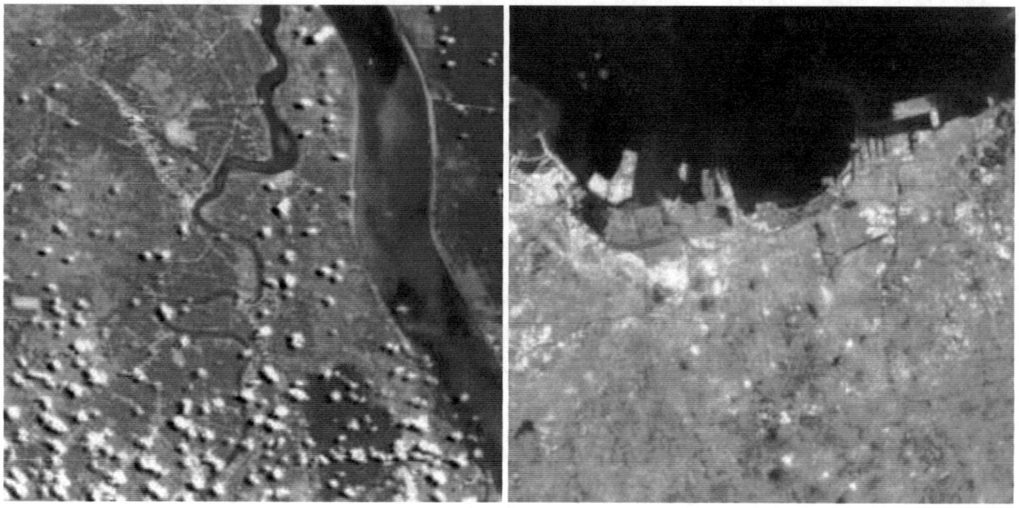

Figure 4. Systematic image correction of LAPAN-A3/IPB multispectral images.

From Figure 4, it can be seen that every object in each corrected channel image has aligned to each other, which can be seen clearly on the shape and color of clouds and river. The images have also been geo-referenced to referenced map as well as radiometric corrected for vignetting effect using pre-flight radiometric calibration data.

While the accuracy produced seems not the ideal performance to achieve, the proposed band co-registration algorithm can be executed in no time. For typical LAPAN-A3/IPB images, the algorithm can be executed averagely in 5-10 seconds, compare to 4-5 minutes needed when executing current image matching algorithm. The trade-off between band co-registration accuracy and processing time needed by the algorithm should be treated carefully, maintaining good accuracy with fast computation. Based on numerous image processed, more complex model such as non-linear sinusoidal model should always been chosen since it produces better accuracy with similar computation time.

5. Conclusions

Band co-registration of LAPAN-A3/IPB multispectral images can be corrected by using simple modeling approach, based on satellite attitude data which provided by star tracker sensor. The band co-registration distortion in horizontal axis can be modeled as yaw angle function with first order sine model, while distortion in vertical axis can be modeled as pitch and roll coupling function with second order cosine model. Accuracy of the proposed band co-registration models are around 2 pixels error compare to results produced by image matching algorithm. Meanwhile, the processing time of the proposed band co-registration algorithm is much faster compare to the other methods, allowing real time implementation. Further research should be conducted about the possibility of satellite attitude determination based on calculated co-registration values from image matching algorithm, thus greatly improves systematic image correction performance in case of the absence of satellite attitude data provided by star sensor tracker.

Acknowledgments
The authors would like to thank Director of Satellite Technology Center (LAPAN) and Chief Engineer of LAPAN-A3/IPB satellite for their support so that this research can be successfully completed.

References

[1] Hasbi W and Suhermanto 2013 Development of LAPAN-A3/IPB satellite an experimental remote sensing microsatellite *34th Asian Conference on Remote Sensing* Bali
[2] Hakim P R, Syafrudin A H and Salaswati S 2015 Analysis of radiometric calibration on matrix imager of LAPAN-A3 satellite payload *IEEE International Conference on Aerospace Electronics and Remote Sensing Technology* Bali
[3] Hakim P R, Hasbi W and Syafrudin A H 2014 ADCS requirements of LAPAN-A3 satellite based on image geometry analysis *IEEE International Conference on Aerospace Electronics and Remote Sensing Technology* Yogyakarta pp. 142-146
[4] Perrier R, Arnaud E, Sturm P and Ortner M 2010 Estimating satellite attitude from pushbroom sensors *23th IEEE Conference on Computer Vision and Pattern Recognition* San Francisco pp. 591-598
[5] Tahir A M, Hakim P R and Syafrudin A H 2016 Image-focusing quality improvement on LAPAN-A3 satellite multispectral imager *Journal of Aerospace Technology* **14(1)** pp. 37-50
[6] Boukerch I, Hadeid M, Mahmoudi R, Takarli B and Hasni K 2014 Geometry based co-registration of ALSAT-2A panchromatic and multispectral images *Journal of Indian Society of Remote Sensing*
[7] Hakim P R, Syafrudin A H, Wahyudiono A and Utama S 2017 Autonomous band co-registration of LAPAN-A3 multispectral imager using edge detection and fast-fourier transform *unpublished*
[8] Tan W J and Kwoh L K 2012 Improvement of micro-satellite multispectral pushbroom sensor band co-registration: an XSAT case study *The 33rd Asian Conference on Remote Sensing* Pattaya
[9] Dell'Endice F, Nieke J, Schlapfer D and Itten K I 2007 Scene-based method for spatial miss-registration detection in hyperspectral imagery *Applied Optics* **46(15)** pp. 2803-2816
[10] Riesing K 2015 Orbit determination from two line element sets of ISS-deployed CubeSats *29th Annual AIAA/USU Conference on Small Satellites* pp. 1-9
[11] Ben-Israel A and Greville T N E 2003 *Generalized Inverses: Theory and Application 2nd edition* (New York: Springer)

IOP Conf. Series: Earth and Environmental Science **149** (2018) 012061 doi:10.1088/1755-1315/149/1/012061

Acquisition performance of LAPAN-A3/IPB multispectral imager in real-time mode of operation

P R Hakim, R Permala and A P S Jayani

Satellite Technology Center, National Institute of Aeronautics and Space (LAPAN)

E-mail: patriarachmanhakim@yahoo.com

Abstract. LAPAN-A3/IPB satellite was launched in June 2016 and its multispectral imager has been producing Indonesian coverage images. In order to improve its support for remote sensing application, the imager should produce images with high quality and quantity. To improve the quantity of LAPAN-A3/IPB multispectral image captured, image acquisition could be executed in real-time mode from LAPAN ground station in Bogor when the satellite passes west Indonesia region. This research analyses the performance of LAPAN-A3/IPB multispectral imager acquisition in real-time mode, in terms of image quality and quantity, under assumption of several on-board and ground segment limitations. Results show that with real-time operation mode, LAPAN-A3/IPB multispectral imager could produce twice as much as image coverage compare to recorded mode. However, the images produced in real-time mode will have slightly degraded quality due to image compression process involved. Based on several analyses that have been done in this research, it is recommended to use real-time acquisition mode whenever it possible, unless for some circumstances that strictly not allow any quality degradation of the images produced.

1. Introduction

LAPAN-A3/IPB satellite is an experimental remote sensing satellite which brings a multispectral pushbroom imager as main payload. The imager has four color channels of blue, green, red, and near-infrared (NIR) with 15 meters ground sampling resolution and swath-width of 120 km [1]. The imager has been continuously producing images which cover all Indonesia regions with 5 GB of daily image produced. Currently, LAPAN-A3/IPB multispectral image acquisition is done by using recorded mode where the satellite records the image captured in on-board memory and then downloads the images later in the day from polar ground station. Limitation of satellite on-board RAM memory buffer storage limits the imager to capture the image for only around two minutes duration for single acquisition pass. Furthermore, limitation of number of the payload data downloading in polar ground station, which is once a day, also limits the size of images that can be produced daily which is around 5 GB a day. With current setup, LAPAN-A3/IPB multispectral imager produces two images each day with averagely 700x120 km coverage area per image with revisit time of 21 days [2].

To build up national capability and independency of foreign technologies, in 2018 LAPAN will have an X-band antenna installed in Bogor ground station, which will directly replace polar ground station for downloading the payload data. In addition of this daily payload downloading, there will be an option for LAPAN-A3 multispectral imager to capture western parts of Indonesian regions in real-time acquisition mode, thus the imager will be able to observe full area within its ground-track path inside LAPAN ground station coverage. However, in real-time mode acquisition, the image captured must be compressed on-board before transmitted to ground station since the image data rate is beyond satellite transmission rate of 105 Mbps [3]. LAPAN-A3/IPB multispectral images can be compressed by using lossless method of Differential Pulse Code Modulation (DPCM) [4][5] or lossy method of Fast Fourier Transform (FFT) [6][7].

This research will discuss about performance of LAPAN-A3/IPB multispectral imager in real-time acquisition mode of operation. The goal of this research is to determine the image coverage and image quality produced by the imager when the real-time mode is used. Image coverage analyses consist of comparison between daily Indonesia images produced by current recorded mode and the proposed real-time mode of acquisition, considering satellite on-board limitations as well as satellite transmission capability. Image compression performance which will be used in real-time operation mode will also analyzed deeply since it plays important role in both image coverage and image quality analysis. The research will utilize numerous actual LAPAN-A3/IPB multispectral images that have been captured up to recently. With careful setup, the trade-off between the coverage area and image quality can be optimized to produce optimal LAPAN-A3 multispectral imager real-time acquisition performance.

2. Methodology

This research aims to evaluate performance of real-time operation mode of LAPAN-A3 multispectral imager, in terms of image coverage and image quality. This research consists of several steps, starting by calculating average daily image coverage produced in recorded acquisition mode by using current acquisition setup. Then the average of daily image coverage produced in real-time acquisition mode supported by the addition of LAPAN X-band receiver in Bogor is calculated. The quality of image produced in real-time acquisition mode, which uses either lossless or lossy image compression, is analyzed based on numerous actual LAPAN-A3/IPB multispectral images that have been captured. Finally, the trade-off between the quantity and quality of image produced in real-time acquisition mode compare to current setup is evaluated. Figure 1 shows general flowchart used in this research.

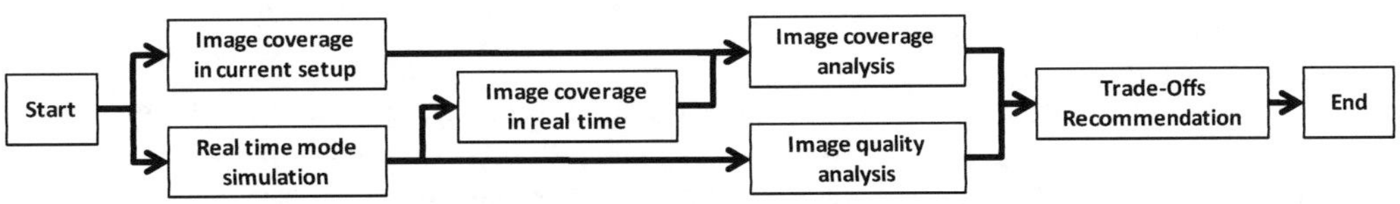

Figure 1. General flowchart of this research.

The calculation of daily image coverage of LAPAN-A3/IPB multispectral imager of current setup is done based on numerous images that have been captured. In the other hands, the calculation of daily image coverage of LAPAN-A3/IPB multispectral imager in real-time mode needs the calculation of data rate of the imager, image compression ratio produced, as well as estimated of the length of satellite pass where the imager would capture the image. Table 1 shows LAPAN-A3/IPB multispectral imager and satellite operation parameter which are used in image coverage analyses.

Table 1. LAPAN-A3/IPB multispectral imager observation parameters.

Imager Data Rate		Transmission Parameters	
Parameter	Value	Parameter	Value
Color channels	4 bands	RAM capacity	4 GB
Array of detectors	8000 pixels	Transmission data rate	105 Mbps
Radiometric resolution	16 bit	Daily polar transmission	~10 minutes
Repetition period	1,9 ms	Daily Bogor transmission	~15 minutes
Ground sampling distance	15 meter	Lossless compression	DPCM 10 bit
CCSDS format	+20%	Lossy compression	FFT 12 bit

IOP Conf. Series: Earth and Environmental Science **149** (2018) 012061 doi:10.1088/1755-1315/149/1/012061

3. Image Coverage Analysis

Image coverage area by LAPAN-A3/IPB multispectral imager will be analyzed based on experience of numerous acquisitions that have been done since the satellite launch in 2016. One of limitations that limit LAPAN-A3/IPB multispectral image coverage area in current setup is 4 GB of RAM which causes the imager could only capture maximum of 4 GB image data for single pass. Another limitation is in current setup, downlink transmission is done only once a day, limiting roughly 5,6 GB daily image data that can be downloaded from satellite. Figure 2 shows the mosaic of more than 200 images captured by the imager since the satellite launch, although effectively the imager started continuously to capture the data since January 2017. The images have been systematically corrected, including radiometric correction [8][9] and geometric correction [10][11]. It can be seen that, in many cases, the imager could only capture small portion of the longest possible path in the satellite pass. Even the longest image captured often could only cover half of its potential area that should be covered.

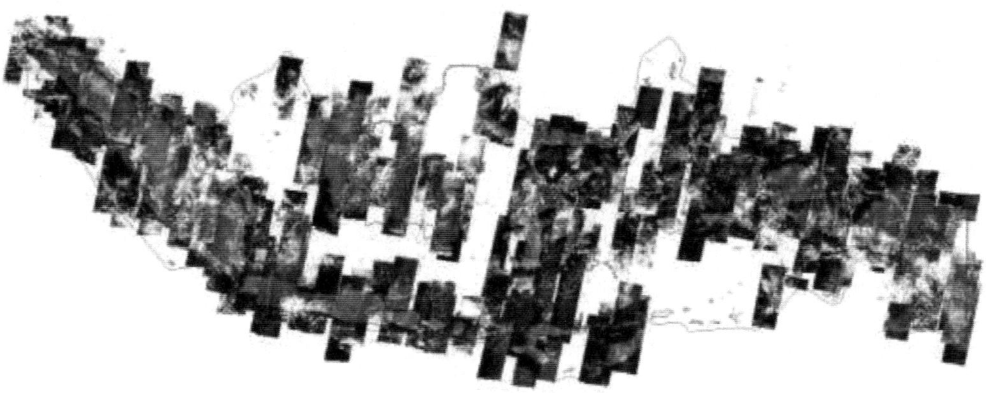

Figure 2. Mosaic of LAPAN-A3/IPB multispectral images on Indonesia region.

With the addition of X-band receiver in Bogor ground station, there are two major fields of improvement that could be exploited, which are the ability to download more payload data and also the ability of LAPAN-A3/IPB multispectral imager to cover much larger Indonesia area by using real-time acquisition mode.

3.1. Data Rate of LAPAN-A3/IPB Multispectral Imager

Since multispectral imager of LAPAN-A3/IPB satellite has four color channels, which are blue, green, red and near-infrared (NIR), with 8000 effective pixels for each channel and 16 bit radiometric resolution for each pixel, the imager will produce 4x8000x16=0,512 Mbit data for every repetition period. With current setup of 1,9 ms repetition interval, then the imager will produce 269,47 Mbps data rate. Considering LAPAN-A3/IPB satellite uses CCSDS packet data format for its payloads, the actual multispectral imager data rate would be around 323,37 Mbps.

However, LAPAN-A3/IPB satellite downlink transmission capacity is only 105 Mbps. Therefore, to be able to transmit multispectral imager data in real-time mode, imager data rate should be lowered from 323 Mbps to fit the downlink transmission capacity of 105 Mbps. Data rate reduction ratio of 3,1 or higher is needed to achieve multispectral imager real-time acquisition. The data rate reduction can be accomplished by several ways, including least significant bit (LSB) truncation, image dimension reduction, or the most efficient ways, by using image compression techniques.

3.2. Actual LAPAN-A3/IPB Multispectral Image Compression Ratio

LAPAN-A3/IPB satellite employs both lossless and lossy compression algorithm to reduce imager data rate to fit 105 Mbps satellite transmission rate. In lossless compression method, the image is down-sampled from 16 bit into 10 bit, and then lossless algorithm of DPCM is used. Meanwhile, in lossy compression method, the image is down-sampled from 16 bit to 12 bit, and then lossy algorithm of FFT is used. Table 2 shows compression ratio produced for several LAPAN-A3/IPB multispectral images by using both lossless and lossy algorithm. The images were taken from March to May 2017 by observing area relatively near to Bogor ground station, to simulate the real-time acquisition mode which can only happens when the satellite observing area covered by Bogor ground station. Parameter of K=512 in lossy algorithm is called scaled-factor, one of important parameter that can be adjusted in lossy compression algorithm. Value of K=512 is the lowest possible value that can be set, producing the least compression ratio with maintaining the best decompressed image quality.

Table 2. Compression ratio of LAPAN-A3/IPB multispectral images.

Source Image		Image Compression Ratio									
		Lossless Mode				Lossy Mode K=512				Average	
Date	Target	B	R	N	G	B	R	N	G	Lossless	Lossy
13/03/17	Jakarta	2.58	2.51	2.53	2.51	3.68	3.04	3.27	3.04	2.53	3.26
24/03/17	Rengasdengklok	2.63	2.56	2.58	2.58	3.95	3.28	3.68	3.26	2.59	3.54
04/04/17	Pamanukan	2.61	2.56	2.57	2.57	3.80	3.21	3.46	3.18	2.58	3.41
09/04/17	Ujungkulon	2.59	2.59	2.56	2.59	3.54	3.03	3.15	3.01	2.58	3.18
15/04/17	Losarang	2.57	2.55	2.57	2.56	3.25	2.74	3.16	2.77	2.56	2.98
20/04/17	Banten	2.63	2.58	2.60	2.59	3.98	3.24	3.61	3.32	2.60	3.54
26/04/17	Cirebon	2.56	2.53	2.55	2.53	3.37	2.79	3.10	2.84	2.55	3.02
01/05/17	Kepulauan Seribu	2.62	2.57	2.59	2.59	3.93	3.23	3.71	3.26	2.59	3.54
07/05/17	Cirebon	2.57	2.58	2.57	2.58	3.38	2.95	3.09	2.92	2.57	3.09
12/05/17	Sindangbarang	2.59	2.56	2.57	2.56	3.65	3.07	3.44	3.09	2.57	3.31
Average		2.60	2.56	2.57	2.57	3.65	3.06	3.37	3.07	2.57	3.29

From Table 2, it can be seen that lossless algorithm produce 2,57 compression ratios, while lossy algorithm with K=512 produce 3,29 compression ratios. The requirement of data reduction ratio of 3,1 or higher clearly fulfilled by this lossy compression, and since setting value of K=512 is enough to achieve 3,1 compression ratios, it is not recommended to increase the value of K since higher value of K will lower image quality with unnecessary increasing compression ratio. In the other hand however, lossless compression ratio still need further reduction, since it only produces 2,57 compression ratios, lacking around 1,2 reduction ratios to fulfill the requirement of 3,1 compression ratio. This problem can be solved by reducing multispectral image dimension that will be transmitted. In LAPAN-A3/IPB lossless mode, there is parameter that can be adjusted to control the swath-width of image that will be transmitted. For example, to achieve the additional 1,2 reduction ratio, the parameter can be set so that 1000 pixels on each image side will be discarded, meaning that only 6000 pixels in the middle that will be transmitted in this lossless real-time acquisition mode.

3.3. Image Coverage Analysis under Several Compression Modes

Based on above image compression ratio calculation, it can be said that LAPAN-A3/IPB multispectral imager real-time acquisition can be realized either by using lossless method with 2000 pixels swath-width reduction or by using lossy method with full image swath-width. This section will analyze the comparison of daily image coverage area produced by LAPAN-A3/IPB multispectral imager when either lossless and lossy images is used in real-time acquisition of western parts of Indonesia region, compare to image coverage area produced by recorded mode only, without real-time acquisition of western parts of Indonesia region.

Before calculating the influence of real-time acquisition mode to daily image coverage area, the image coverage area produced by recorded-only observation will be analyzed. Since Bogor ground station will have new X-band receiver antenna, LAPAN-A3 satellite would be able to download its payload data more frequent than before, which currently only able to download once a day in polar ground station. Currently, LAPAN-A3 satellite is able to download roughly 5,632 GB multispectral imager data from polar ground station, or about 1383 km length by 120 km swath-width coverage area. This image coverage area usually consists of same portion of eastern parts and western parts of Indonesian region.

By addition of X-band receiver antenna in Bogor ground station, LAPAN-A3 satellite would be able to increase its download capacity. With rough calculation, two satellite passes every night in Bogor ground station would produce at least 8 GB of multispectral imager data that can be downloaded daily. Even without further downloading payload data in polar ground station, Bogor ground station alone can support a full RAM-limited observation of 1006 km length for both eastern and western parts of Indonesia region, for total of 2012 km length by 120 km swath-width coverage area. The addition of polar ground station for receiving LAPAN-A3/IPB data to Bogor ground station will not add the daily image coverage of Indonesia area produced, since RAM memory in PDHS LAPAN-A3/IPB satellite could only allow the imager to capture 4 GB of image in one single pass (1006 km length). However, polar ground station downloading capability could increase the overall satellite mission flexibility and can overcome the problems when the transmission failure occurs in Bogor ground station.

Although without real-time acquisition mode it seems that LAPAN-A3/IPB multispectral imager could produce vast coverage area, the 1006 km length coverage for each satellite pass is not even close to observe full Indonesia region in some particular satellite path. The longest path that the satellite will observe Indonesia region in one single pass is around 1800 km length, which happens when the satellite passes through North Sulawesi province all the way to East Nusa Tenggara province. In this particular pass, LAPAN-A3/IPB multispectral imager needs two-passes to observe this full 1800 km path, which can only realized by waiting for 21 days of imager revisit time. Figure 3 shows illustration of the length of satellite passes when observing Indonesia [12]. As we can see, half of the path has more than 1006 km length, which indicates that many area in Indonesia region cannot be observed properly every 21 days of imager revisit time.

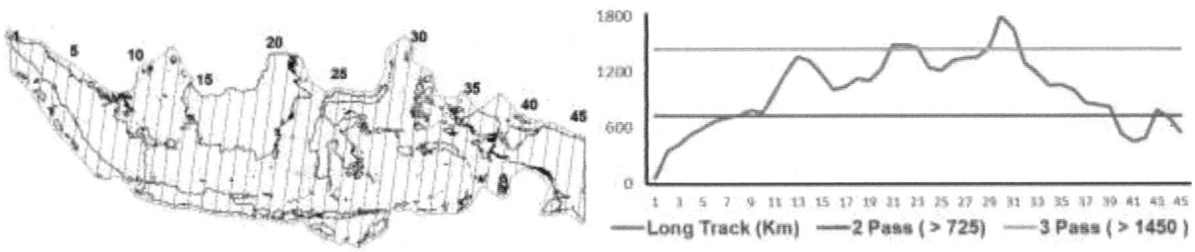

Figure 3. The length of LAPAN-A3 satellite passes when observing Indonesia. [12]

With real-time acquisition mode when the satellite observing western parts of Indonesia region, this problem can be partially solved. For any satellite passes that covered by Bogor ground station, it can be guaranteed that the imager could observe and transmit all of the area it passes. Real-time acquisition does not need the use of on-board memory storage, so the imager could produces image lines infinitely to be transmitted. As long as the ground station receiver is able to receive the data transmitted by the satellite, which is constrained roughly by satellite line of sight from the ground station, the real-time acquisition mode could potentially produce more than 6 GB of compressed images or equivalently around 4000 km by 120 km coverage area, assuming 8 minutes effective ground station contact. This memory-less mode of acquisition of western part of Indonesia allows acquisition of eastern part of Indonesia, which must be executed in recorded mode, for the RAM memory buffer limit of 1006 km by 120 km coverage area.

This calculation of course, depends on minimum elevation angle needed for the satellite to be able to transmit the payload data properly. From polar ground station experiences, 15 degrees elevation angle is enough to transmit the data properly. Figure 4 shows Bogor ground station coverage area, which covers all western parts of Indonesia up to Sulawesi and Timor islands. It can be said that only Maluku islands and Papua province that cannot be observed by using real-time acquisition mode. The calculation will be slightly different, if for example, the minimum elevation angle is somewhat larger than 20 degree which can be caused by significant frequency-interferences from ground station neighborhood or there are obstacles such as mountain though it is very unlikely.

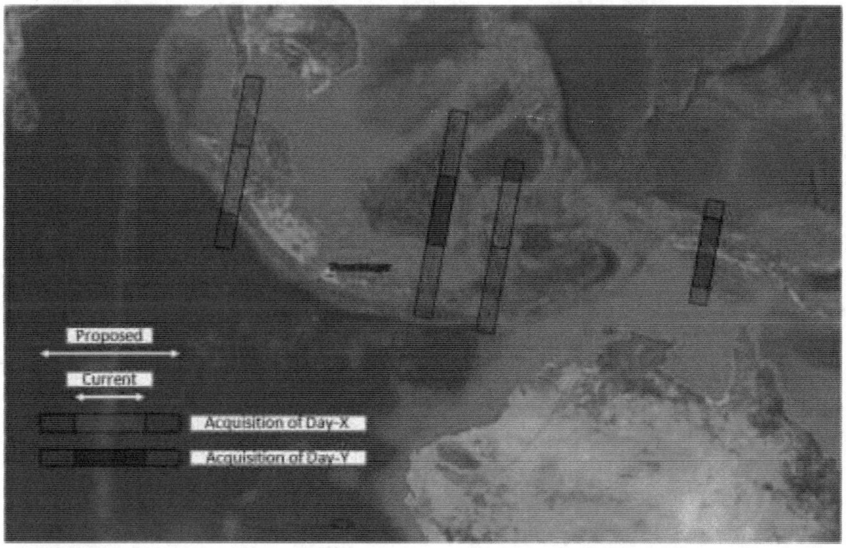

Figure 4. Illustration of daily LAPAN-A3/IPB multispectral observation.

Based on the estimated length of each satellite passes on figure 3 and illustration of daily LAPAN-A3/IPB multispectral imager observation on figure 4, table 3 summarizes image coverage produced by LAPAN-A3/IPB multispectral imager when real-time acquisition mode is used when observing western parts of Indonesia. The table also compares image coverage produced to image coverage of recorded-only observation, considering the utilization of Bogor and polar ground station to download the payload data.

Table 3. Comparison of image coverage area of LAPAN-A3/IPB multispectral imager.

Operation Mode	Daily Image Coverage (km)			Daily Area Covered (ha)	
	Eastern Indonesia	Western Indonesia*	Others	Indonesia**	Others
Current Setup: Recorded, Polar	692x120	692x120	0	16,60 ha	0
Recorded Mode, Bogor	1006x120	1006x120	0	24,14 ha	0
Lossless Mode, Bogor	1006x120	4000x90	1006x120	48,07 ha	12,07 ha
Lossy Mode, Bogor	1006x120	4000x120	1006x120	60,07 ha	12,07 ha
Recorded Mode, Bogor-Polar	1006x120	1006x120	1383x120	24,14 ha	16,60 ha
Lossless Mode, Bogor-Polar	1006x120	4000x90	2389x120	48,07 ha	28,67 ha
Lossy Mode, Bogor-Polar	1006x120	4000x120	2389x120	60,07 ha	28,67 ha

* Area within satellite pass covered by Bogor ground station, including South China Sea and Indian Ocean

** Including countries around Indonesia such as Singapore, Malaysia, Brunei and Thailand

From table 3 above, it can be concluded that in terms of image coverage point of view, real-time acquisition should always be used when LAPAN-A3/IPB multispectral imager observe western parts of

Indonesia region. In addition of capability in observing all Indonesia regions properly every 21 days of imager revisit time, the real-time acquisition mode also allows the imager to capture another part region of the world, including parts of South-East Asia countries or even Europe, Africa and America continent for another satellite missions.

4. Image Quality Analysis

In real-time acquisition mode, LAPAN-A3/IPB multispectral image is compressed to accommodate satellite downlink transmission rate of 105 Mbps by using either lossless or lossy algorithm. The quality of the decompressed images produced will be degraded, which is caused either by digital number down-sampled for both algorithm, or reduced swath-width in lossless algorithm, or direct quality degradation of lossy compression process. The down-sampling of image digital number from the original 16 bit into 12 bit for lossy or 10 bit for lossless would not cause serious effect since the effective radiometric resolution of any imager system would be less than the maximum bit, due to the presence of noise. In the other hand, the reduced swath-width in lossless algorithm will cause the image produced losses perhaps 25 percent its swath-width, but this can be well overshadowed by the much higher coverage improvement in along-track direction produced by real-time acquisition of LAPAN-A3 multispectral imager.

4.1. Lossless Compression Image

Normally, lossless compression algorithm will produce decompressed image without any image quality degradation. However, digital number down-sampling from 16 bit to 10 bit in LAPAN-A3 multispectral image lossless compression causes the decompressed image produced suffers from slight degradation in image quality. Based on the imager radiometric calibration, standard deviation of each pixel readings is around 0,5 percent of its nominal value. Table 4 shows examples of digital number value readings from several sampled images of the imager when observing uniform light source of integrating sphere. It can be assumed that for the actual digital number value of 10000 for example, the pixel would produce digital number readings ranging between digital numbers of 9950 to 10050. This characteristic exceeds the digital number down-sampling of 6 bits, therefore the effect of this down-sampling is theoretically not significant.

Table 4. Standard deviation of digital number readings of imager pixels.

Calibration Images	Nominal Value of Pixel Digital Number				Standard Deviation of Pixel Readings			
	G	N	R	B	G	N	R	B
Image-1 (5111)	11034	15375	15585	2893	59	77	75	42
Image-2 (5311)	11837	17467	17461	3199	61	82	80	43
Image-3 (5193)	28233	35557	35630	8334	159	199	194	118
Image-4 (1002)	21451	36177	34393	7327	98	139	132	66
Image-5 (5321)	27796	42417	41843	8061	147	196	190	103
Image-6 (5421)	28637	44754	44052	8289	151	202	196	105
Image-7 (1012)	25706	56518	52810	7727	85	129	124	51

Figure 5 shows decompressed image of LAPAN-A3/IPB image by using both lossless and lossy compression algorithm. The image was taken on May 1, 2017, capturing Indonesia capital province of Jakarta with Thousands Islands in the north. The image is visualized on NRG composite format, which shows vegetation land cover in red color and non-vegetation land cover in green color. The image has also been corrected systematically after decompression process, including radiometric vignetting correction and geometric band co-registration correction.

4.2. Lossy Compression Image

As explained earlier, in LAPAN-A3/IPB lossy compression algorithm, there are several parameters that can be adjusted to produce optimal performance between compression ratio and decompressed image quality. One of these parameters is compression scaled-factor which can be set to either of 512, 1024, 2048 or 4096. The higher scaled-factor will produce higher compression ratio but with lower decompressed image quality. Figure 5 shows results example of decompressed images of LAPAN-A3/IPB image by using scaled-factor of 512 and 4096. It can be seen that by using lossy compression method with any scaled-factor, there is no significant degradation of image quality that can be seen visually. However, for scaled factor of 4096, the more detailed decompressed image will show noticeable image degradation in several parts of the image.

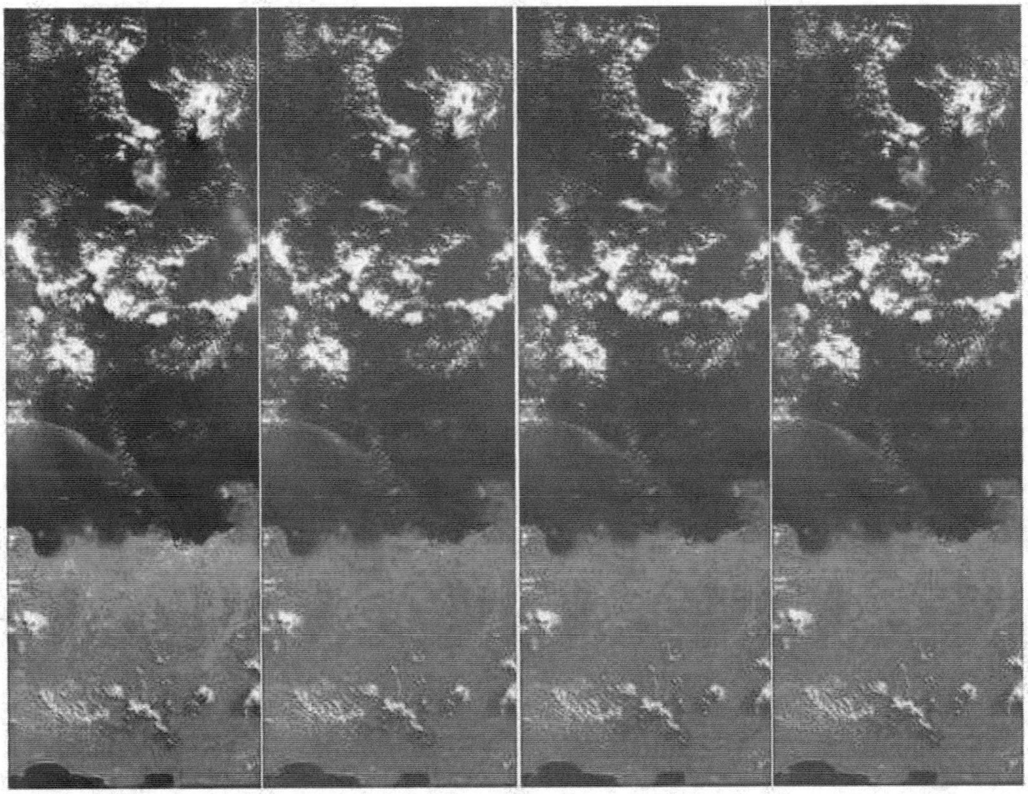

Figure 5. Decompressed images of LAPAN-A3/IPB image; from left to right: original image, lossless image, lossy image with K=512, and lossy image with K=4096.

4.3. Trade-Offs Analysis between Image Coverage and Quality

Based on Figure 5, it can be seen that there is no significant image quality degradation resulted from image compression process, both for lossless and lossy compression with K=512. Therefore, LAPAN-A3/IPB multispectral imager should use real-time operation mode whenever possible to maximize image coverage area produced. Based on several analyses about the trade-offs between image coverage and image quality produced by real-time mode operation of LAPAN-A3/IPB multispectral imager, it can be concluded that in circumstances that need the best images possible, image acquisition must be done in recorded mode as current setup. For circumstances that need collection of vast coverage area in short period of time, real-time acquisition mode with lossy algorithm could be used. Meanwhile for circumstances in between which needs fast coverage with moderate quality, real-time acquisition mode with lossless algorithm could be used.

IOP Conf. Series: Earth and Environmental Science **149** (2018) 012061 doi:10.1088/1755-1315/149/1/012061

5. Conclusions

By using real-time acquisition mode, LAPAN-A3 multispectral imager could produce images at least twice as many as images produced by using recorded acquisition mode, thus greatly improves the ability of the imager to provide more coverage of Indonesia area in shorter period of time. The quality of images captured in real-time acquisition mode is slightly degraded due to image compression used to accommodate 105 Mbps downlink transmission rate. When slight image quality degradation is allowed, it is recommended to always use real-time acquisition mode whenever it possible. Further research of finding optimal Huffman table that will be used for image compression should be conducted to increase the quality of image which is transmitted from the LAPAN-A3/IPB satellite in real-time acquisition mode, both for lossless or lossy compression scenario.

Acknowledgments

The authors would like to thank Director of Satellite Technology Center (LAPAN) and Chief Engineer of LAPAN-A3/IPB satellite for their support so that this research can be successfully completed.

References

[1] Hasbi W and Suhermanto 2013 Development of LAPAN-A3/IPB satellite an experimental remote sensing microsatellite *34th Asian Conference on Remote Sensing* Bali

[2] Hakim P R and Utama S 2017 Imaging frequency improvement of LAPAN-A3/IPB satellite multispectral imager using off-nadir observation maneuver *SIPTEKGAN XXI* Serpong

[3] Nasser E N, Suhermanto and Hasbi W 2013 Design, implementation and test of payload data handling LAPAN-A3/IPB satellite *34th Asian Conference on Remote Sensing* Bali

[4] Hakim P R and Permala R 2017 Analysis of LAPAN-IPB image lossless compression using differential pulse code modulation and huffman coding *IOP Conference Series: Earth and Environmental Science* **54** 012096

[5] Anitha S 2015 Lossless image compression and decompression using Huffman coding *International Research Journal of Engineering and Technology* Vol. **2**

[6] Hakim P R and Permala R 2015 Compression algorithm performance analysis of LAPAN-A3 satellite image using fast-fourier transform *Advance in Science and Technology of Indonesian Aircraft, Rocket and Satellite* pp. 97-108

[7] Sahnoun K and Benabadji N 2014 Satellite image compression algorithm based on FFT *The International Journal of Multimedia and Its Application* Vol. **6(1)**

[8] Hakim P R, Syafrudin A H, Hasbi W and Salaswati S 2016 Dark-image modeling analysis of LAPAN-A3 satellite imager *International Seminar on Aerospace Science and Technology IV* Lombok pp. 58-66

[9] Zhang Y and Chen S 2001 Preflight calibration of HY-1 satellite 4-band CCD camera *22nd Asian Conference on Remote Sensing* Singapore

[10] Hakim P R, Rahman A, Suhermanto and Rachim E 2012 Pushbroom imager data systematic geometric correction model using colinear projection method *Journal of Aerospace Technology* **10(2)** pp. 121-132

[11] Cramer M, Stallmann D and Haala N 2000 Direct georeferencing using GPS/inertial exterior orientations for photogrammetric applications *IAPRS XXXIII* Amsterdam

[12] Syafrudin A H, Hakim P R, Najati N and Hasbi W 2017 Mapping strategy for large area using optical payload on microsatellite Bogor *unpublished*

IOP Conf. Series: Earth and Environmental Science **149** (2018) 012062 doi:10.1088/1755-1315/149/1/012062

Momentum Biased Performance of LAPAN-A3 Satellite for Multispectral Pushbroom Imager Operation

S Utama, M A Saifudin and M Mukhayadi

Satellite Technology Center (Pusteksat) Indonesian National Institute of Aeronautics and Space (LAPAN) Bogor, Indonesia
Jl. Cagak Satelit km 0.4 Rancabungur, Kabupaten Bogor, Indonesia

E-mail: utamasatriya@gmail.com

Abstract. One of LAPAN-A3 satellite's missions is for earth observation using multispectral pushbroom imager. This mission needs a stable and accurate attitude to capture any area of interest. To accomplish the mission LAPAN-A3 satellite use momentum biased attitude control. The satellite use magnetic torquers to control the angular momentum vector and a reaction wheel to spin the satellite to remain nadir pointing. When the satellite spinning there are nutation and precession occurred. This paper observes attitude accuracy and stability due to nutation and precession of the satellite's momentum and the effect to pushbroom imager operation. Nutation observed with 0.28° amplitude and 73 seconds period, and precession observed with 1° amplitude and 92 minutes period. This nutation and precession profile will lead to 2.6° attitude accuracy and maximum movement on ground track 0.055 km/s in along track direction and 0.259 km/s (0.026°/s) in cross track direction. Both of attitude accuracy and movement are outperformed the limitation. However due to satellite movement in orbit there is still blurring effect on the imager.

1. Introduction

LAPAN-A3 is one of Indonesian satellites that made by LAPAN. LAPAN-A3 already launched at 2016. LAPAN-A3 has 97° inclination orbit with 508 km altitude. The primary mission of the satellite is earth observation using multispectral pushbroom imager [1]. For this mission, LAPAN-A3 use four band pushbroom imager with swath 120 km and GSD 15 m at altitude of 508 km.

To support the mission LAPAN-A3 is equipped with some sensor and actuator for attitude determination and control system. For attitude determination LAPAN-A3 have two star sensors, six coarse sun sensors, one horizon sensor and three fiber optic laser gyros. In order to control the attitude LAPAN-A3 employs four reaction wheels and three magnetic torquers. Figure 1 shows some of LAPAN-A3's sensor in satellite body axes.

The mission operation of LAPAN-A3 use momentum bias attitude control that is well implemented in the previous LAPAN satellite i.e. LAPAN-TUBSAT [2]. For a microsatellite class, the momentum bias attitude control strategy has been proved to support the high resolution video surveillance and imaging mission using CCD matrix camera[3][4]. This method implements a momentum exchange technique by utilizing a device that can rotate in the spacecraft such as a momentum or reaction wheel. By using this method, there is an angular momentum vector aligned to perpendicular of the orbit plane. Magnetic torquers are used to maintain the angular momentum vector to the desired direction and magnitude. Here the reaction wheel on *y* axis, which is defined as pitch axis and perpendicular to the optical axis of pushbroom imager, has to absorb the angular momentum. It means the spacecraft only need one reaction wheel in that axis to keep the satellite in nadir pointing using certain pitch rate. However, due to external torque such as earth magnetic field and gravity gradient, the angular momentum vector will always change. The movement of angular momentum vector can be a short period of nutation, or longer period of precession. While the precession can only be handled by the right compensation setting of magnetic torquer, the nutation damping is simply managed by utilizing of the second wheel that perpendicular to the pitch axis. Although it will not perfectly eliminate the nutation, the utilization of the reaction wheel will reduce most of it.

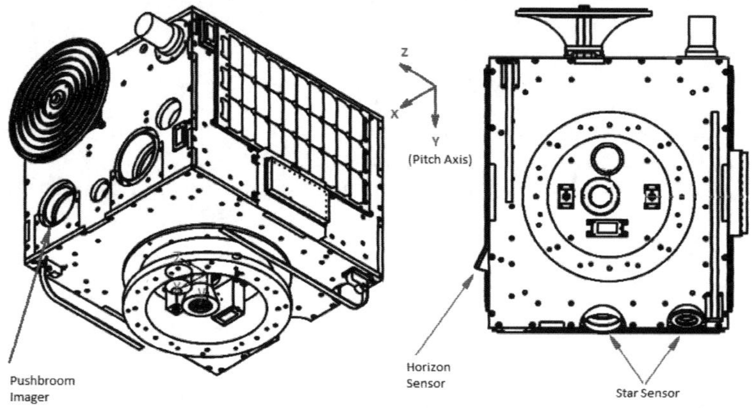

Figure 1. Position of Star Sensor and Pushbrooom Imager in LAPAN-A3 Body Axes.

There are already some researches about effect of satellite attitude movement to imaging operation such as on ALOS [5], ZY-3 [6], Alsat-1B [7], Yaogan-26 [8]. Some method to reduce the attitude movement also already studied and proposed [9][10]. Some attitude requirement for LAPAN-A3's multispectral pushbroom imager has already calculated [11]. The research is also study about how stable the attitude for this imager. The aim of this paper is to observe attitude accuracy and stability due to nutation and precession of LAPAN-A3.

2. Methodology

To measure the performance of LAPAN-A3 momentum biased attitude control for pushbroom imager operation, two steps have been established. The first step is angular momentum vector movement observation and the second step is to calculate the effect of angular momentum vector movement to multispectral pushbroom imager operation.

To observe angular momentum vector movement, star sensor will collect the spacecraft orientation data that transformed into direction of $-y$ spacecraft body axis which is parallel with angular momentum vector direction. The additional gyros reading are also taken as comparison to the star sensor data.

The movement of angular momentum vector will directly influence the satellite attitude and affect multispectral pushbroom imaging. Regarding operation of pushbroom imager, two parameters of attitude should be observed, they are attitude accuracy and attitude stability. While attitude accuracy is defined as how far the attitude different from target, attitude stability means how fast the attitude change. When taking the images, attitude accuracy will affect image accuracy and attitude stability will affect quality of image.

Some condition will be simulated to see the effect of angular momentum movement. The first condition is perfectly nadir, without any angular momentum movement. Second condition is nadir pointing with angular momentum movement. Third condition is off nadir operation. All calculations are done for 5 minutes imager operation over Indonesian region.

3. Nutation and Precession Observation

Nutation is rotational motion for which the instantaneous rotation axis is not aligned with the principal axis [12]. Since star sensor is the most accurate attitude sensor[13][14], the observation of nutation and angular momentum precession use star sensor in inertial frame that expressed in right ascension and declination coordinate. The angular momentum vector has been set perpendicular to the orbit plane that move about 0.9887 °/day due to sunsynchronous precession and expressed by right ascension and declination of angular momentum ($R_{\vec{A}}$ and $D_{\vec{A}}$). Then the observation will find how far the right ascension and declination of *-y* spacecraft body axis (R_S and D_S) from the angular momentum. The different between spacecraft attitude to its target of angular momentum is shown as follow

$$\Delta = R_{\vec{A}} - R_S$$
$$\Delta = D_{\vec{A}} - D_S$$

Data from star sensor is taken every 270 ms. For this research, star sensor data is collected for one orbit period. Due to strong sunlight when crossing equator on daylight, star sensor may not read any star to calculate satellite attitude. From Figure 2 can be seen a region where there are no star sensor data, marked with red box. When the data taken on November 2017, the blind spot of star sensor is happens when declination is in the lowest position.

From Figure 2 can be easily seen short period and long period oscillation. Beside these two oscillations there is also a rather small ripple that may come from sensor accuracy or noise. This ripple can easily be smoothed by simple moving average.

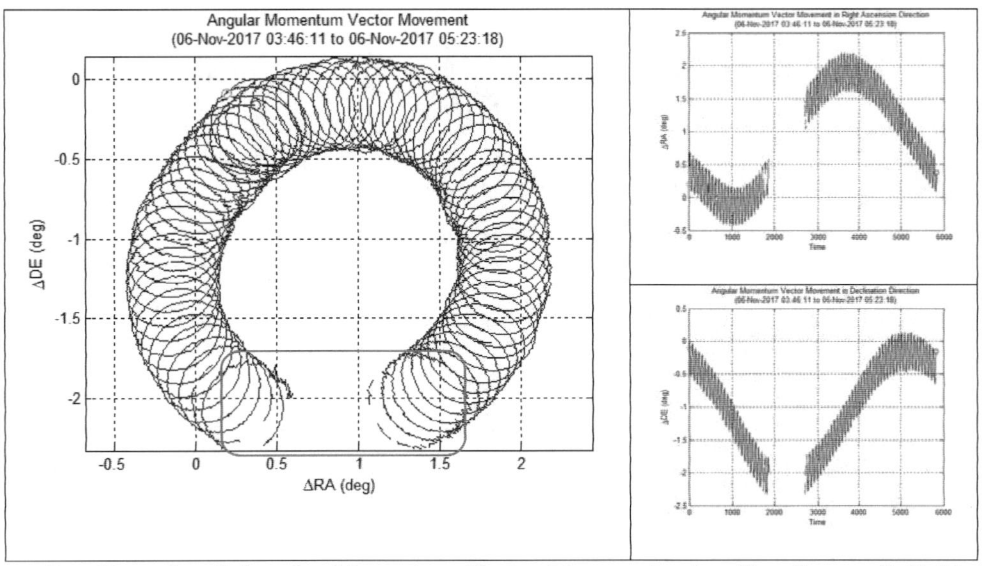

Figure 2. Example of Angular Momentum Vector Movement using Star Sensor, (left) Plot of Declination and Right Ascension Movement Shows Some Smaller Cones in One Bigger Cone,(top right) Plot of Right Ascension to Time Observation, (bottom right) Plot of Declination to Time Observation.

After smoothing of the data, the next step is to find amplitude and period for short period oscillation, which is called nutation. This is done by measure the value and time of maximum local and minimum local. By using these values, amplitude and period for every half oscillation is obtained. Figure 3 shows an example of plot of every half oscillation amplitude and period for one observation. From Figure 3 can be seen amplitude and period value are spreading with majority of data are gathered in smaller region. Centroid of the data is calculated to find one value of amplitude and one value of period for one observation.

Table 1 shows amplitude and period of nutation from some star sensor observation. The observation is taken for the magnitude of angular momentum around 0.53 kg-m^2/s or equivalent with 5500 rpm of LAPAN-A3's reaction wheel. The observation obtained nutation's amplitude of 0.2783° in right ascension and 0.2714° in declination with period 74.38 seconds in right ascension and 73.5 seconds in declination.

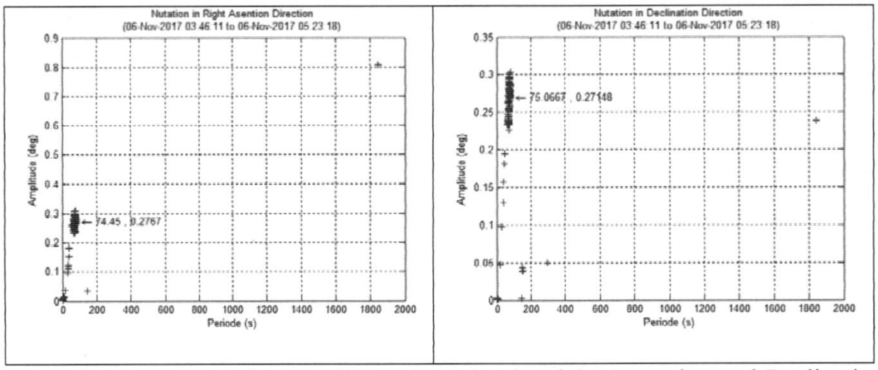

Figure 3. Amplitude and Period of Half Wave Nutation in Right Ascension and Declination in One Observation using Star Sensor.

Table 1. Amplitude and Period of Nutation from Observation using Star Sensor.

Time		Nutation				Angular Momentum	
		RA		DE			
Start	Stop	Amplitude (deg)	Period (s)	Amplitude (deg)	Period (s)	(rpm)	(kg-m^2/s)
2017/10/31 03:21:38	2017/10/31 04:58:45	0.2807	74.33	0.2642	67.20	5474	0.52314
2017/11/01 03:04:52	2017/11/01 04:41:57	0.2812	75.20	0.2762	76.20	5508	0.52638
2017/11/05 05:33:10	2017/11/05 07:10:18	0.2781	72.07	0.2725	73.40	5577	0.53298
2017/11/06 03:46:11	2017/11/06 05:23:18	0.2767	74.45	0.2715	75.07	5591	0.53432
2017/11/11 03:51:26	2017/11/11 05:28:50	0.2749	74.33	0.2740	73.73	5586	0.53384
2017/11/12 03:33:43	2017/11/12 05:10:53	0.2785	75.87	0.2700	75.40	5575	0.53279
Mean		0.2783	74.38	0.2714	73.50		

The next step is to identify the longer period oscillation called precession by averaging the nutation circle.

Table **2** show the amplitude and period of precession from some star sensor observations. There is difference between right ascension and declination period that happened because of star sensor blind spot. Period that calculated from right ascension is more representative than declination because in minimum region of declination there are no star sensor data. From these observations are obtained precession's amplitude is 1.02° and precession period is 5542.87 s.

Table 2. Amplitude and Period of Precession from Observation using Star Sensor.

Time		Precession						Ang. Momentum	
		RA		DE		Center			
Start	Stop	Amplitude (deg)	Period (s)	Amplitude (deg)	Period (s)	RA (deg)	DE (deg)	(rpm)	(kg-m ^{2}/s)
2017/10/31 03:21:38	2017/10/31 04:58:45	1.0776	6395.00	0.9964	6390.60	0.29	0.44	5474	0.52314
2017/11/01 03:04:52	2017/11/01 04:41:57	1.0297	6177.80	0.9856	4840.00	-0.53	-1.02	5508	0.52638
2017/11/05 05:33:10	2017/11/05 07:10:18	1.0106	5409.40	0.9720	6132.40	1.12	-0.85	5577	0.53298
2017/11/06 03:46:11	2017/11/06 05:23:18	1.0251	5083.60	0.9351	6534.20	0.88	-1.07	5591	0.53432
2017/11/11 03:51:26	2017/11/11 05:28:50	1.0046	5254.80	1.0185	6562.20	2.59	0.27	5586	0.53384
2017/11/12 03:33:43	2017/11/12 05:10:53	0.9848	4936.60	1.0106	6400.00	1.66	1.12	5575	0.53279
Mean		1.0221	5542.87	0.9864	6143.23				

Gyros are also used to observe the angular momentum vector movement as comparison for star sensor observation. Nutation observation using gyros is done by observe the movement angle in x and z axis. Figure 4 shows example of nutation for observation in October 31st 2017 between 05:57:10 until 06:03:00. Gyro data is taken every 10 seconds for around 4-5 minutes. Table 3 shows some nutation data. It can be seen that both amplitude and period are depending on magnitude of angular momentum vector. The bigger magnitude of angular momentum vector will make nutation smaller and faster. In normal operation LAPAN-A3 has angular momentum around 5500 rpm. By applying linear regression to data in Table 3 for magnitude of angular momentum 5500 rpm is obtained nutation with period 74.4 s and amplitude 0.25° in x direction and period 74.5 s and amplitude 0.28°in z direction. These values are not far from star sensor observation (0.28° in right ascension and 0.27° in declination with period 74.4 seconds in right ascension and 73.5 seconds in declination).

Table 3. Amplitude and Period of Nutation from Observation using Gyros.

Start Time	Angular Momentum		Angle X		Angle Z	
	(rpm)	(kg-m ^{2}/s)	Amplitude (deg)	Period (s)	Amplitude (deg)	Period (s)
2017/10/20 07:31:50	3264.70	0.31	0.43	124.29	0.47	124.29
2017/10/20 10:41:10	3256.69	0.31	0.44	124.62	0.48	123.33
2017/10/25 09:13:40	4410.19	0.42	0.32	93.00	0.35	92.38
2017/10/25 12:21:30	5147.04	0.49	0.26	78.62	0.28	78.62
2017/10/26 07:21:50	5340.27	0.51	0.26	77.04	0.29	77.04
2017/10/26 12:04:50	5428.56	0.52	0.26	75.83	0.29	75.65
2017/10/26 13:37:10	5458.74	0.52	0.25	75.26	0.29	75.14
2017/10/27 05:28:30	5656.69	0.54	0.25	72.73	0.28	72.73
2017/10/27 07:04:10	5689.23	0.54	0.25	71.85	0.28	71.85
2017/10/31 05:57:10	5530.74	0.53	0.25	73.33	0.28	75.00
2017/11/01 10:18:20	5617.72	0.54	0.25	73.33	0.28	73.14
2017/11/01 13:29:00	5644.84	0.54	0.25	73.08	0.28	73.08

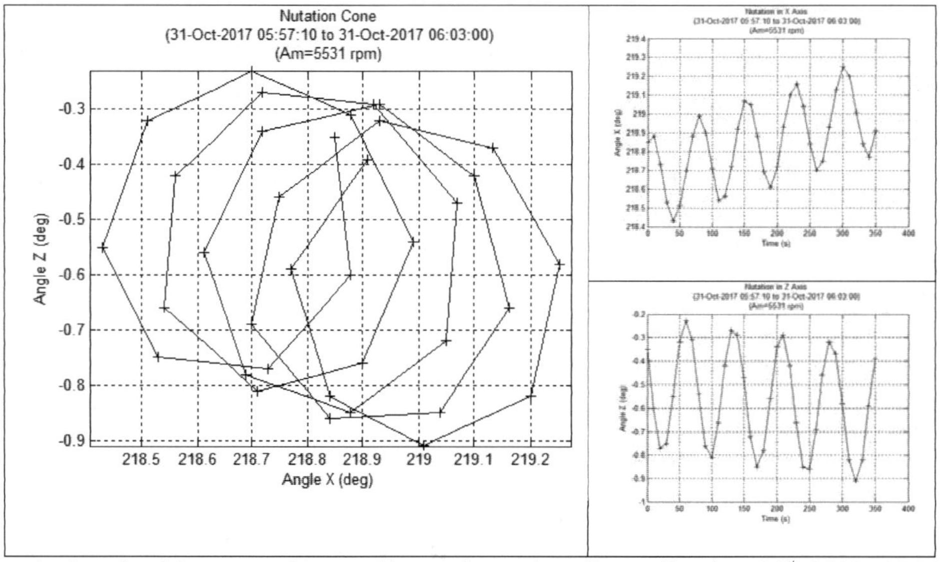

Figure 4. Angular Momentum Vector Observation using Gyros (October 31st 2017, 05:57:10 - 06:03:00)(left) Plot of Angle Movement in *z* and *x* Axis Shows Some Nutation Cones, (top right) Plot of Angle *x* to Time Observation, (bottom right) Plot of Angle *z* to Time Observation

For precession observation longer data is needed. For longer time observation, frequency of data taken from Gyros need to be reduced. **Figure 5** shows example for long observation using Gyros data that taken every 80 seconds for 82 minutes. It's rather difficult to see the nutation from that graph because the sampling time is too big. Angle *x* tends to increase and angle z tends to decrease with oscillation compared to star sensor observation in **Figure 2** show that the results are very difference. This may happen because Gyros has bias error. In LAPAN-A3, the Gyros has bias error 4°/hour [15].

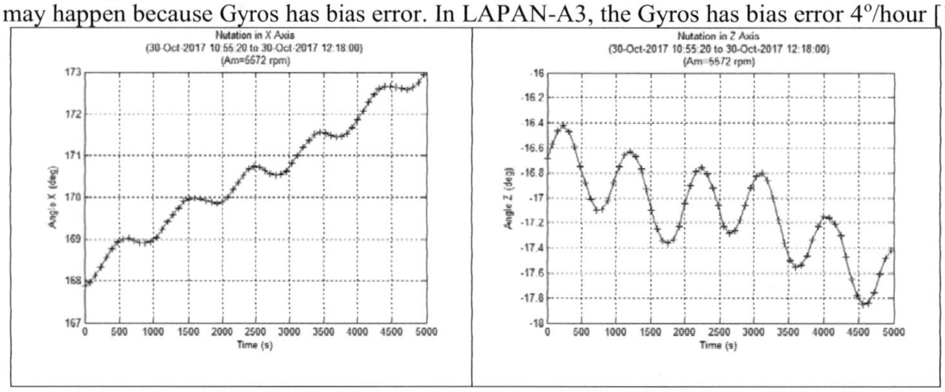

Figure 5. Example of Angular Momentum Vector Movement Long Observation using Gyros, (left) Plot of Angle X to Time Observation, (right) Plot of Angle Z to Time Observation.

4. Effect on Multispectral Pushbroom Imager

When observe momentum bias it's easier to use inertial frame, but when it come to imager operation it's easier to use satellite body frame or horizon local frame. Unfortunately, transformation from inertial frame to body frame or horizon local frame is not very direct, there are coupling between the axes. Movement in one direction of inertial frame may affect two directions in body frame or horizon local frame. For this reason, attitude accuracy and stability calculation must be on satellite body frame or distance in ground (horizon local frame).

Attitude accuracy is easily only twice of amplitude of precession and nutation. So, from star sensor observation, for around 5500 rpm angular momentum, the attitude accuracy is 2.6°, almost the same with DLR-TUBSAT satellite [16] which has 3° attitude accuracy. Using 508 km LAPAN-A3's attitude, this attitude accuracy equivalent to 23.08 km in perfectly nadir pointing. Compared to swath of pushbroom imager that is 100 km, this attitude accuracy is acceptable.

For attitude stability analysis, there are three condition calculated; perfectly nadir without nutation and precession, nadir with precession and nutation, off nadir operation.These calculations done for 5 minutes simulated time when crossing Indonesia. Movement of area taken in ground is observed to see blurring effect. **Table 4** shows movement limitation from [5] and corrected limitation due to updated orbit.

Table 4 Image Blurring Criteria for LAPAN-A3's Multispectral Pushbroom Imager

	Orbit Altitude (km)	GSD (m)	Exposure Time (ms)	Maximum Movement (km/s)
[5]	600	18	1.25	7.20
Updated	508	15	1.80	4.17

4.1. Perfectly Nadir Operation

Figure 6 shows perfectly nadir operation. In this ideal condition, there are no nutation and precession. Satellite ground track is coincided with center of imager (both showed in redline). Green line show area that covered by the imager. Since altitude of LAPAN-A3 is 508 km, the ground speed is 7.61 km/s, compared to Table 4 this speed already exceeded the limitation that will lead to 1 pixel blurring in along track direction. This blurring is due to combination of satellite movement in orbit and high exposure time not due to attitude stability.

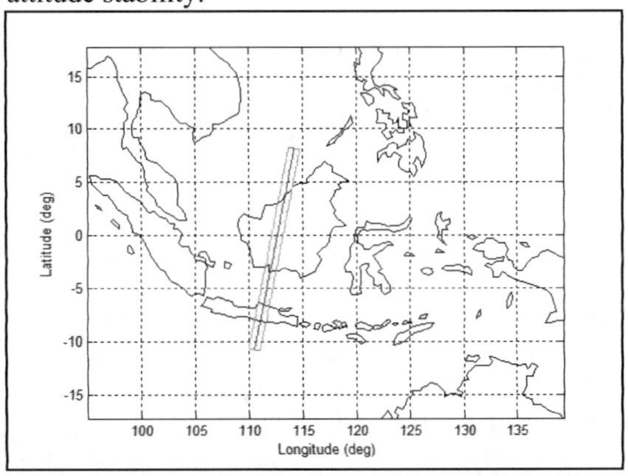

Figure 6. Perfectly Nadir Pointing for LAPAN-A3's Multispectral Pushbroom Imager for 5 Minutes Operation over Indonesia.

4.2. Nadir with Precession and Nutation

In this calculation, precession and nutation are included. Precession and nutation are modeled with amplitude and period using data that observed from star sensor. Figure 7 shows result for this calculation. Left side image shows area of image taken by the imager. From this image can be seen that center of imager are oscillating along ground track. Right side image shows plot of euler angle and center image movement to cross track and along track direction. Maximum movement in cross track direction is 0.233km/s, and in along track direction is 0.055km/s. Movement in along track direction is not significant compared to satellite ground speed and still make 1 pixel blurring. In cross track direction there is no blurring because the movement is better than the limitation.

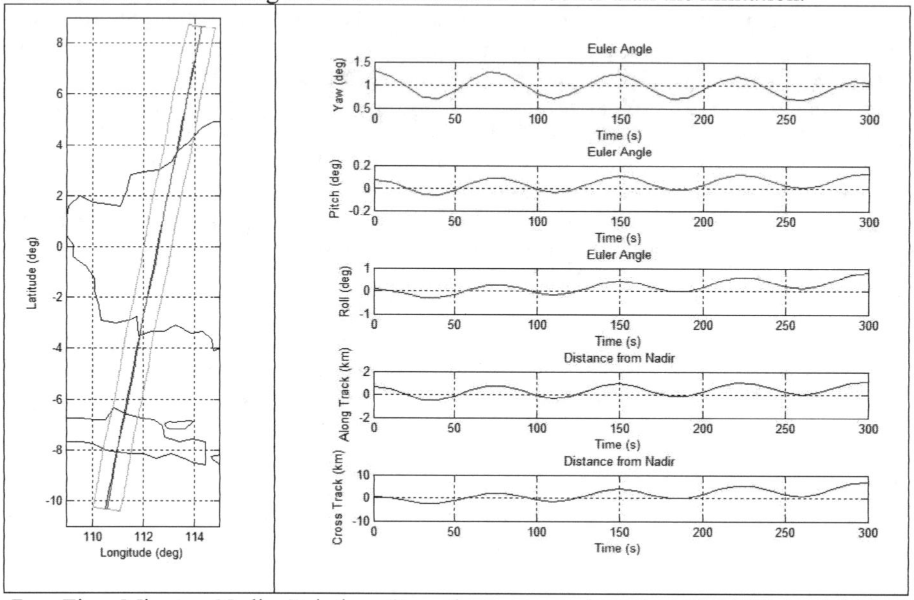

Figure 7. Five Minutes Nadir Pointing Operation over Indonesia with Effect of Nutation and Precession, (left) Ground Plot, blue line show satellite ground track, red show center of imager, green show area covered by imager, (right) Movement in satellite body frame (yaw, pitch, roll) and in local horizon frame (cross track and along track).

4.3. Off Nadir Operation

To increase frequency of taking image in certain area, off nadir operation is needed. Off nadir operation until 20° in roll is still recommended [17]. This part will calculate 20° roll operation. This roll angle is achieved by move the angular momentum vector 20° in right ascension direction. This method can be applied because target area is near equator so moving angular momentum vector in right ascension will directly affect satellite roll angle.

Figure 8 shows result for this off nadir operation. This method cannot achieve 20° roll angle, cross track distance on ground that achieved is around 180 km. There is not only rolling that happened but also considerable pitching and yawing in the satellite. Maximum movement in cross track direction is 0.2594km/s and in along track direction is 0.0385km/s. These movements are almost the same with in nadir operation, there will be blurring in along track direction and no blurring in cross track direction.

By these calculation can be seen that momentum bias of LAPAN-A3 for both nadir and off nadir operation (until 20°) is stable enough to avoid adding blurring image. Image blurring than happened is due to satellite movement in orbit that can be avoided by setting faster exposure time. Figure 9 shows image collected by LAPAN-A3 imager until October 2017. It shows a lot off region already covered by the imager. Most of them are taken in nadir pointing.

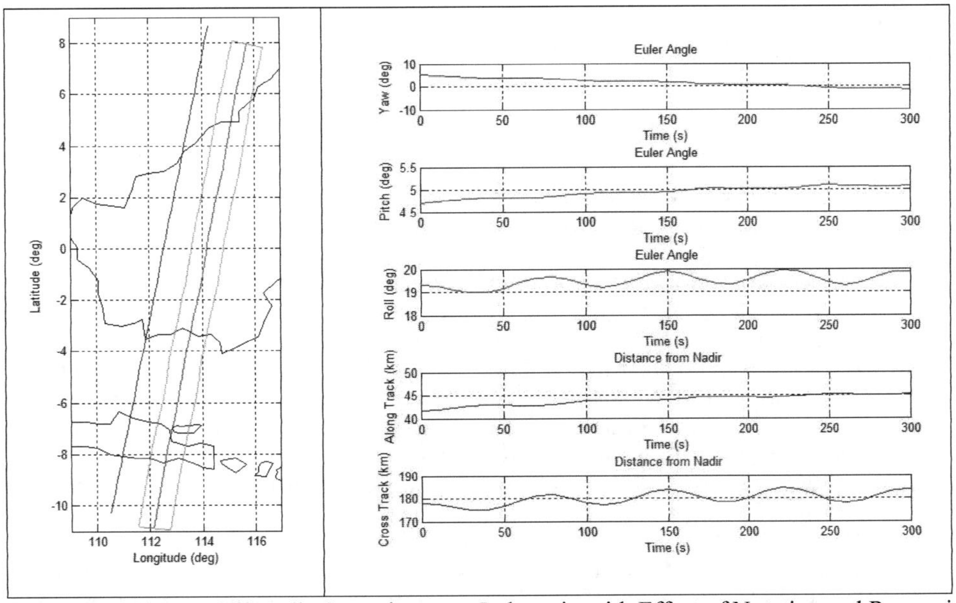

Figure 8. Five Minutes Off Nadir Operation over Indonesia with Effect of Nutation and Precession, (left) Ground Plot, blue line show satellite ground track, red show center of imager, green show area covered by imager, (right) Movement in satellite body frame (yaw, pitch, roll) and in local horizon frame (along track and cross track).

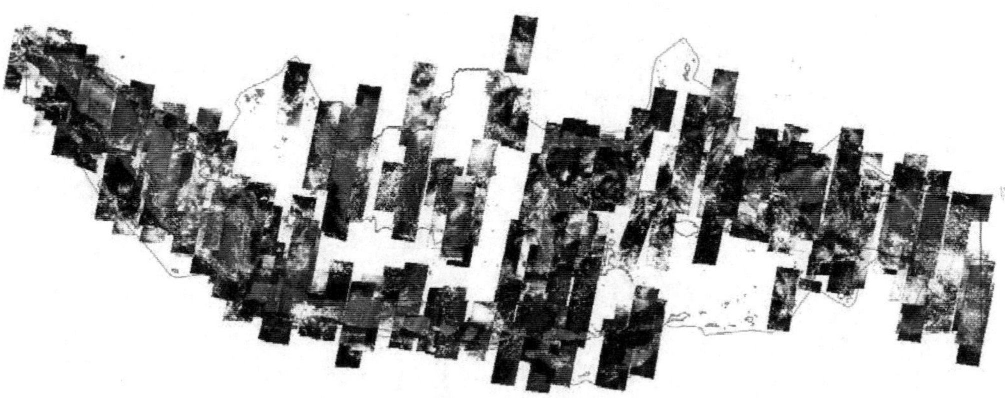

Figure 9 Indonesia Region Covered by LAPAN-A3's Imager until October 2017

5. Conclusion

Momentum bias that applied to LAPAN-A3 satellite has nutation with 0.28° amplitude, 73 seconds period and precession with 1 deg amplitude, 92 minutes period for 5500 rpm (0.53 kg-m^2/s). This nutation and precession lead to 2.6° attitude accuracy that means every target still in imager coverage although not always in center of imager. In terms of attitude stability, precession and nutation lead to maximum 0.055 km/s in along track direction and 0.259 km/s (equivalent with 0.026° deg/s) in cross track direction. These values are outperforming the requirement. The blurring effect in along track direction is happened but due to satellite movement in orbit and high exposure time, not due to attitude stability. It can be concluded that momentum bias of LAPAN-A3 has a good performance for multispectral pushbroom imager operation.

Acknowledgments
The authors would like to thank Prof. Udo Renner for thesupervition and LAPAN satellite operator for their support.

References

[1] Hasbi W and Suhermanto 2013 Development of LAPAN-A3 / IPB Satellite an Experimental Remote Sensing Microsatellite *34th Asian Conf. Remote Sens. 2013, ACRS 2013* 8

[2] Mukhayadi M, Madina R and Renner U 2014 Attitude control of bias momentum micro satellite using magnetic and gravity gradient torque *Proceeding - ICARES 2014 2014 IEEE Int. Conf. Aerosp. Electron. Remote Sens. Technol.* 132–6

[3] Steckling M, Renner U and Röser H-P 1996 DLR-TUBSAT, qualification of high precision attitude control in orbit *Acta Astronaut.* **39** 951–60

[4] Roemer S 2003 Flight Experience with the Micro Satellite Maroc-Tubsat *54th International Astronautical Congress of the International Astronautical Federation, the International Academy of Astronautics, and the International Institute of Space Law* (Reston, Virigina: American Institute of Aeronautics and Astronautics) pp 3–5

[5] Takaku J and Tadono T 2010 High resolution DSM generation from ALOS PRISM - Processing status and influence of attitude fluctuation - *Int. Geosci. Remote Sens. Symp.* 4228–31

[6] Tong X, Li L, Liu S, Xu Y, Ye Z, Jin Y, Wang F and Xie H 2015 Detection and estimation of ZY-3 three-line array image distortions caused by attitude oscillation *ISPRS J. Photogramm. Remote Sens.* **101** 291–309

[7] Serief C 2017 Estimate of the effect of micro-vibration on the performance of the Algerian satellite (Alsat-1B) imager *Opt. Laser Technol.* **96** 147–52

[8] Wang M, Fan C, Pan J, Jin S and Chang X 2017 Image jitter detection and compensation using a high-frequency angular displacement method for Yaogan-26 remote sensing satellite *ISPRS J. Photogramm. Remote Sens.* **130** 32–43

[9] Lyu W, Dai S, Dong Y, Shen Y, Song X and Wang T 2017 Attitude motion compensation for imager on Fengyun-4 geostationary meteorological satellite *Acta Astronaut.* **138** 290–4

[10] Janschek K, Harnisch B, Tchernykh V and Dyblenko S 2017 SmartScan: a robust pushbroom imaging concept for moderate spacecraft attitude stability *International Conference on Space Optics — ICSO 2006* vol 2006(SPIE)p 90

[11] Hakim P R, Hasbi W and Syafrudin A H 2014 ADCS Requirements of Lapan-A3 Satellite Based on Image Geometry Analysis *IEEE Int. Conf. Aerosp. Electron. Remote Sens. Technol.* 142–6

[12] Burke C A 2010 *Nutation in the spinning SPHERES spacecraft and fluid slosh* (Massachusetts Institute of Technology)

[13] Saifudin M A, Amin D El and Huzain M F 2016 Development of Experimental Star Sensor for LAPAN-A3 / IPB Microsatellite *Adv. Aerosp. Sci. Technol. Indones.*

[14] Saifudin M A and Triharjanto R H 2010 Algoritma Pengenalan Pola Bintang untuk Deteksi Posisi Bintang pada Star Sensor Satelit LAPAN *J. Teknol. Dirgant.* **8** 36–42

[15] Anon 2016 *LAPAN-A3/IPB Microsatellite for Remote Sensing Experiment Detail Design* (Bogor)

[16] Roemer S and Renner U 2003 Flight experiences with DLR-TUBSAT *Acta Astronautica* vol 52pp 733–7

[17] Hakim P R and Utama S 2017 Imaging Frequency Improvement of LAPAN-A3/IPB Satellite Multispectral Imager using Off-nadir Observation Maneuver *SIPTEKGAN XXI*

Requirements and design structure for Surya Satellite-1

H Steven[1], and M F Huzain[2]
[1]Engineering Physics Department, Surya University, Tangerang, Banten, 15810 Indonesia
[2]Senior Engineer, National Institute of Aeronautics and Space, Centre of Satellite Technology, Bogor, West Java, 16310 Indonesia

E-mail : hery.mindarno14@student.surya.ac.id

Abstract. Currently, there are various references on the manufacture of nanosatellite specifications weighing 1KG - 10KG.The Surya Satellite-1 is the first nanosatellite made by universities in Indonesia. The Surya Satellite-1 team gets a launch offer from Japan Aerospace Exploration Agency (JAXA) and, all the nanosatellites manufacturer racers at ICD (Interface Control Document) obtained from JAXA. The formation of the Satellite-1 Surya framework is also based on the provisions of JAXA. The various specifications and requirements specified by the JAXA space agency consisting of specific specifications such as the mass of nanosatellite 1U (10cm x 10cm x 11.65cm) size of at least 0.13KG and a maximum of 1.33KG, with the determination of a gravity point not exceeding 2 cm from the nanosatellite geometry center point. In the case of preventing solar radiation in space, there is a requirement that the structure of satellite structures on hard black anodization should be more than 10 meters in the surface of the satellite structure. In terms of detail, the satellite structure is a black hard anodized aluminum after its manufacturing process derived from the MIL-A-8625 document, type 3.

1. Introduction

In many developed countries, Nanosatellite design by students has become very common, but in Indonesia, this is a remarkable achievement. Surya Satellite-1 is a Surya University student project that works with National Institute of Aeronautics and Space (LAPAN) and Amateur Radio Organization of Indonesia (ORARI). The aim of this project is to become a pioneer in making the first cubesat in Indonesia and to be the inspiration of other students to design their own cubesat that meets the criteria and space agency requirements. The Surya Satellite-1 mission is an Automatic Packet Reporting System (APRS) communication that orbits in Low Earth Orbit (LEO). The Surya Satellite-1 mass is 1.15 kg which is planned to be launched with Japan Aerospace Exploration Agency (JAXA) J-SSOD (JEM-Small Satellite Orbital Deployer) module. So all the requirements and design of Surya Satellite-1 refer to the standards contained in JEM Payload Accommodation Book Vol.8 Small Satellite Development, Interface Control Document, JAXA [1,2].

Like satellite in general, the role of satellite structure is very important, causing the research and development of Surya Satellite-1 structure to be very important as well. Therefore, this paper will explain the design, analysis, manufacture, and testing of Computer Aided Design (CAD), Engineering Model (EM) model, to Surya Satellite-1 Flight Model (FM) [2,3,5]. Satellite structures must be able to withstand the burden and force provided by the rocket when launched into space. Structures should be designed with careful calculations and careful considerations, both from their own mechanical systems and every subsystem contained in the satellites. The inspiration in the Cubesat structure design process can be obtained from the design developed by the spacecraft company, which is tailored to the needs and requirements of the launcher, due to the lack of practical experience in designing the satellite while the small error in the design of satellite structures can cause fatal damage on the satellite as a whole as well as on each component of the satellite subsystem [4,6]. The purpose of this research is to design an optimum Surya Satellite-1 structure in accordance with the satellite mission and in accordance with the requirements of the launcher.

In making the Surya Satellite-1 structure, the Research Team refers to the Interface Control Document (ICD) provided by the JAXA Space Agency. The construction of the Surya Satellite-1 Structure is 1U in shape, which means it has a volume of 100 mm x 100 mm x 113.5 mm with a structure weight of approximately 554 grams. The material selection process for the Surya Satellite-1 structure not only takes into account the weight factor, but also the thermal power factor of conductivity, manufacturing process and cost. The design process of the Surya Satellite-1 CAD model is done by using Solidwork 2015 software by considering several factors. In addition to the design drawings, the design of this CAD model can be used for Finite Element Method (FEM), Buckling, and Vibration testing in ANSYS 2014 software [15,16]. The test results using this software will serve as an initial prediction of the strength of satellite structures prior to laboratory scale testing which should be done as validation.

2. Surya Satellite -1

2.1. Standart Cubesat Requirement
The design of the Surya Satellite-1 structure design refers to the ICD provided by the JAXA to any cube maker that will launch its satellites using JAXA's launch vehicle. The launcher provided various requirements regarding the cubesat structure design specifications to be launched [9,11,13]. The requirements consist of Mass Requirement, Dimension Requirement, and Material Structure Requirement [7,8,10,12,14].

2.1.1 Mass Requirement
1. Mass Total Satellite size 1U minimum 0.13 kg and maximum 1.33 kg.
2. Center of gravity of all satellites cannot be more than 2 cm from satellite geometry center.

2.1.2 Dimension Requirement
1. 1U magnification size of 100mm + -0.1mm for X and Y coordinates, for Z coordinates of 113.5mm + -0.1mm.
2. Minimum Rails width should be 8.5mm.
3. The outer surface of Cubesat shall not touch the inside of the J-SSOD module, other than the rails portion of cubesat.
4. No component exceeds the total area of the cubesate of 6.5 mm.
5. The creation of the cubesat accesss window must be in the coordinates of -Y or + X.
6. For cubesat size 1U and 2U there shall be a hole for separation spring provided by the space agency, for 3U cubesat size is not required.

2.1.3 Material Structure Requirement

1. All materials and components must be installed during launching, ejection, and cubesat operation.
2. Aluminum 7075 and 6061 materials should be used for the manufacture of main structures and Cubesat rails.
3. The cubesat surface and rails shall be anodized with a minimum thickness of 10 microns meters, in accordance with the requirements of MIL-A-8625, type 3 document.
4. Toxic and hazardous materials should not be used in the manufacture of cubesat.

2.2. Methodology of Surya Satellite - 1 Structure

In essence, the construction of the Surya Satellite-1 structure is based on satellite and ICD missions provided by the launcher. Preparation of satellite structure design requires several steps in the engineering process, among others: initial design, material selection, analysis using simulation method, discussion of simulation results, to obtain Engineering data from the simulation process [17-19]. The simulation results are then analyzed and interpreted based on ICD. For more details, the following schematic diagram of the design of Surya Satellite-1 structure design:

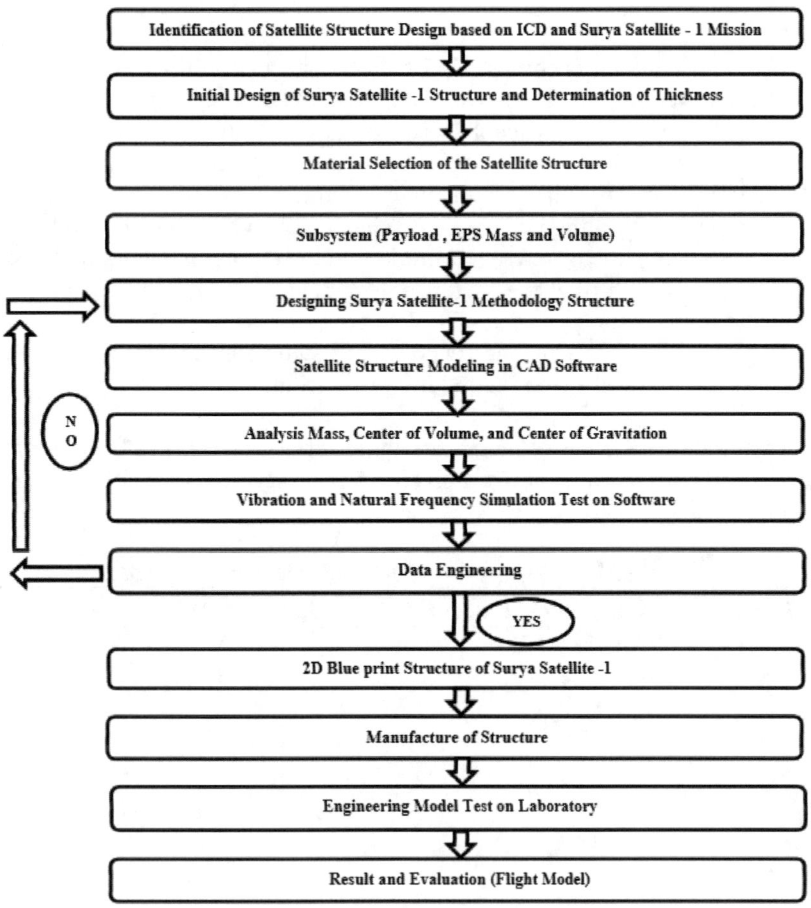

Figure 1. Methodology of Surya Satellite-1 structure making.

2.3. Component of Surya Satellite -1

The 1U Surya Satellite -1 array consists of various subsystems inside which are arranged in 6 layers of Printed Circuit Board (PCB). To find out the mass detail of each subsystem are found in Table 1.

Table 1. Component of Surya Satellite-1.

No.	Subsystem	Mass (gram)
1	Structure	554
2	Deploy Antenna System	21
3	Module Radio	45
4	Payload	40
5	On Board Data Handling	42
6	Battery	174
7	Electronic Power System	53
8	Solar Cell	189

3. Selection Material of Surya Satellite –1

3.1. Structure Material

The selection of the next material becomes one of the most important elements in making the Surya Satellite-1 structure. In order to minimize failure / damage, in addition to being based on ICD of JAXA, the selection of materials must also be in accordance with the recommendations that have been published by space agencies such as NASA, ESA or space agencies, particularly regarding the strength of the material and its outgassing properties in vacuum chamber and others [20,21]. Particularly in the creation of a 1U satellite structure a small design change alone can affect subsystems in satellites. Not only is the weight factor considered, but many other parameters such as thermal conductivity, manufacturing process, and cost [22,23]. Based on these factors selected material used in the manufacture of Surya Satellite -1 structure is Aluminum type 7075-T6. This type of material is in addition to use for satellite skeletons, many also used for other aerospace needs such as commercial aircraft frameworks. For more details, the materials and composition contained in the aluminum 7075 type can be seen in Table 2 and Table 3.

From Parties JAXA allows 2 types of aluminum that is 7075 and 6061 in the manufacture of cubesat structure. After doing various research and search some reference got comparison of characteristics and criteria contained in Table 4.

Table 2. Component of material Aluminum 7075.

Component	Al	Cr	Cu	Fe	Mg	Mn	Si	Ti	Zn
Wt. (%)	87.1 - 91.4	0.18 - 0.28	1.2 - 2	Max 0.5	2.1 - 2.9	Max 0.3	Max 0.4	Max 0.2	5.1 - 6.1

Table 3. Material properties of Aluminum 7075.

Material Properties			
Physical Properties		**Electrical Properties**	
Density (gr/cc)	2.81	Electrical Resistivity (ohm.cm)	5.15 e-006
Material Properties		**Thermal Properties**	
Hardness Brinel (Hb 500)	150	Specific Heat Capacity (J/g – °C)	0.96
Ultimate Tensile Strenght (MPa)	572	Thermal Conductivity (W/m-K)	130
Tensile yield Strenght (MPa)	503	Melting Points (°C)	477-635
Elongation at break (%)	11	Solidus (°C)	477
Modulus of Elasticity (GPa)	71.7	Liquidus(°C)	635
Poisson's Ratio	0.33	**Processing Properties**	
Fatigue Strength (MPa)	159	Annealing Temperature (°C)	413
Machinability (%)	70	Solution Temperature (°C)	466-482
Shear Modulus (GPa)	26.9	Aging Temperature (°C)	121
Shear Strength (MPa)	331		

Table 4. Comparison of Alumunium 7075 dan 6061.

Criteria	**Percentage**	**Al 7075**		**Al 6061**	
		Score	**Value**	**Score**	**Value**
Stiffness to Mass Ratio	40	10	400	9	360
Manufacture Time	15	10	150	10	150
Uniformity	25	10	250	10	250
Durability	20	10	200	10	200
Total			1000		960
Priority		1		2	

3.2. CAD Modelling

The purpose of this research is to design an optimized Surya Satellite-1structure in accordance with the satellite mission and in accordance with the requirements of the launcher. Things to consider in designing satellite structures are the size of the rails contained in the corners of the structure, the installation of solar panels, and the six sides of the satellite structure that must conform to the "Cube Sat Design Specification". The selection of Aluminum 7075 material as a Surya Satellite-1 structure material has become the best choice in accordance with the given requirements. In addition, the satellite structure must be adaptive so that if during installation and testing stages, hindrance occurs in subsystems or other disturbed parts, there is no need to redesign the entire satellite structure [24-27]. From the various requirements, the main model of Surya Satellite -1 is designed, which also consists of 6 sides, including Antenna Deployment System section. These 6 sections will be combined using the M3 and M2 size bolts.

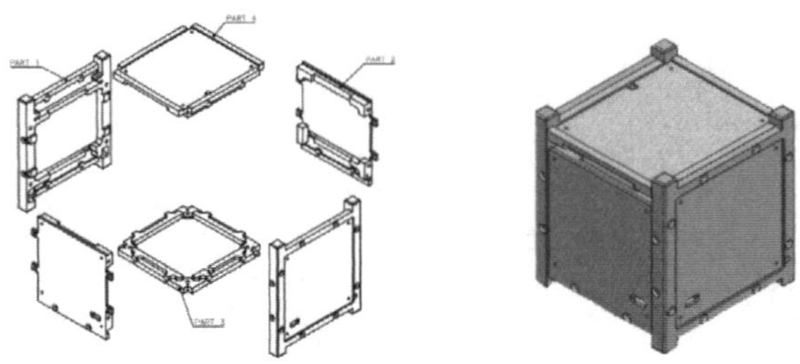

Figure 2. The unification mechanism of Surya Satellite Structure-1.

3.2.1 Antenna Deployment System

The Surya Satellite-1 Antenna Deployment Mechanism uses a nylon string disconnect system by NiCr wire which is powered by certain voltages and currents. The NiCr wire will heat up in a few moments after being set by the timer to disconnect the nylon string. After the nylon is broken then the retainer (blue colour) made of Teflon PTFE is no longer stuck by nylon. Antenna VHF and UHF will certainly force it out of the module, so the holder will be pushed out. Equally important is the satellite antenna deployment module installed 2 switches on each antenna (1 main switch and 1 backup switch) so that there are a total of 4 switches that detect the position of the antenna VHF and UHF is still in folded or has been unfolded [28,29].

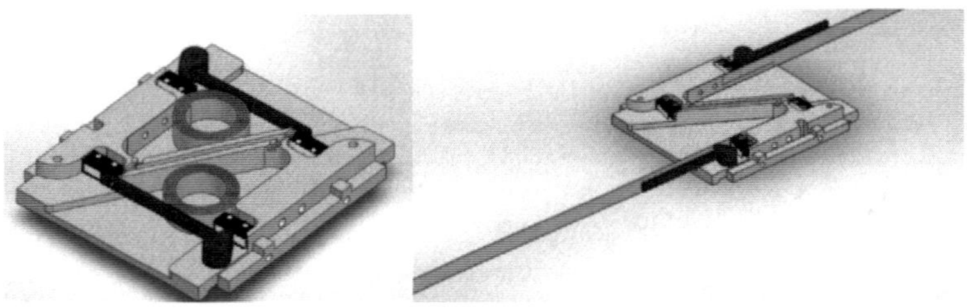

Figure 3. Antenna deployment system.

3.2.2 Surya Satellite-1 CAD Modelling

The Surya Satellite structure -1 is designed to meet the requirements of "Cubesat Standard Design". The structure weighs 554 grams to carry the main load, arranged like a rack with 6 layers of PCB with different functions (see figure 4 left). PCBs are stacked to be reinforced with structures on the Z axis. The connection between PCBs is used spacer made of stainless steel as amplifier between PCB and upper (Z +) and bottom (Z -) structures [30-32]. For the exterior of the structure are made space used for solar panels on the six sides of the satellite (see figure 4 right). Design of the overall design of the Surya Satellite-1 structure is done on Solidwork 2015 software.

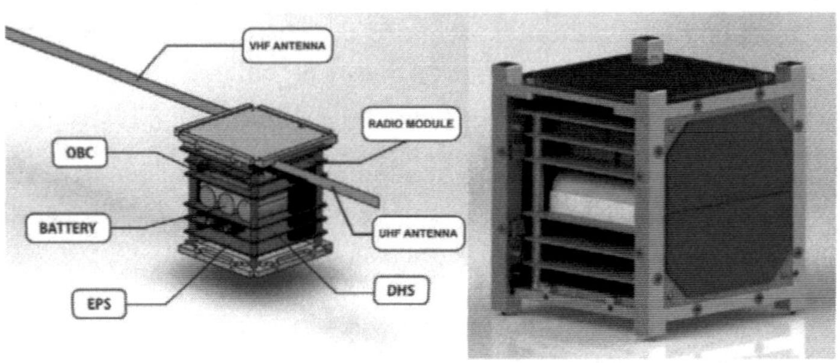

Figure 4. Surya Satellite-1 3D CAD Modelling.

4. Structural Analysis and Results Summary

Analysis of the main satellite structure strength was done by simulation and FEM analysis on ANSYS 14 software [33-34]. Various experiments were conducted to obtain static load values, natural frequencies, and strength limits of satellite structures [35-39]. The analysis of the satellite structure was initiated by modelling 3D in CAD software. Then, simulation of natural frequency and static is performed [41-45]. The reference value / limit value of this simulation is adjusted to the choice of JAXA launcher vehicle offered on the ICD of the launcher rocket [45-49].

4.1 Static Analysis

The first step in performing static analysis on ANSYS software is to process the meshing on the satellite structure model. The meshing process can be done by joining the segments of the image so that the 3D image can be used as a model for simulation in ANSYS software 14. In static experiments analysis of satellite structure of 1U Cubesat satellite yields the total deformation value shown in the figure. The analysis shows the total deformation value of 0.0001242 mm and this value is very small when compared to the thickness of the Surya Satellite-1 structure. The total value of Von Misses pressure is 0.45076 MPa This value is much smaller than the strength of Yield Strength material Aluminum 7075 of 331 Mpa (figure 5)

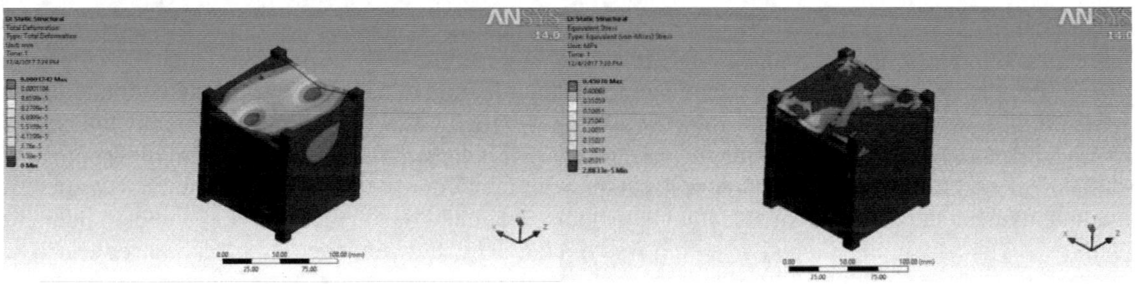

Figure 5. Maximum deformation and maximum von mises stress.

4.2 Frequency of Analysis

The satellite structure must be designed to with stand shocks during the launch process. To prevent resonance at launching, the experiment is performed in several modes. All modes in the natural frequency analysis are in the table 5 and the results of this analysis are in the figure 6.

Table 5. Natural frequencies

S .No	Mode	Natural Frequency (Hz)	S .No	Mode	Natural Frequency (Hz)
1	1	3199.1	9	9	5080.5
2	2	3231.7	10	10	5116.3
3	3	3484.1	11	11	5653.5
4	4	3858.1	12	12	5953.1
5	5	4584.2	13	13	6210.9
6	6	4698.7	14	14	6277.6
7	7	4808.5	15	15	6541.7
8	8	5016.9			

Natural Frequency at mode 1 and mode 15 with the total deformation at first mode and 15th mode of 2011.7 mm and 2743.7mm was shown in figure 6.

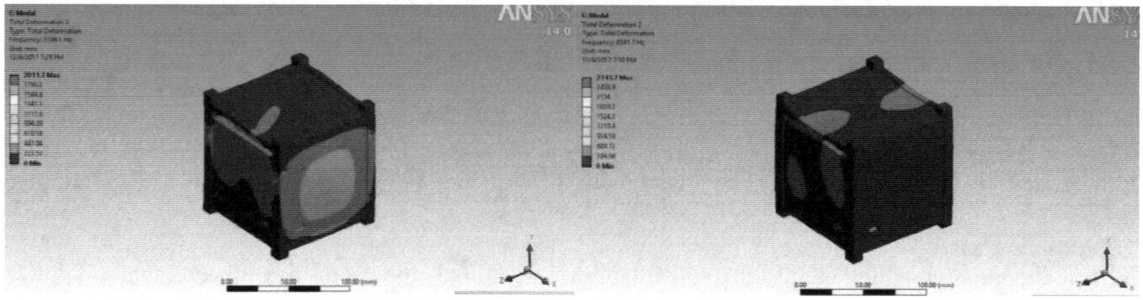

Figure 6. Deformation at mode 1 and mode 15

5. Conclusion

The cubesat Surya Satellite-1 structure design has a mass of 480 grams with a volume of 100mm × 100mm × 113.5mm (h) using Aluminum material 7075-T6. Material selection not only takes into account the weight factor, but many other parameters such as thermal conductivity, manufacture process, and cost. 3D CAD modelling is done on Solidworks 2015 software and simulated in software ANSYS 14. Simulation is done with the aim to get maximum deformation value; maximum von misses pressure, and natural frequency mode of Surya Satellite-1 structure.

The simulation results show the von mises pressure value of 0.45076 MPa which is smaller than the yield strength of aluminum 7075 of 331MPa. While the total deformation value of 0.0001242 mm is still very small when compared to the thickness of Surya Satellite-1 structure. This simulation is done by FEM method and refers to the largest static quasi value. From the results of this simulation we can conclude that the designed Surya Satellite- 1 structure has met the requirements of "Cubesat Standard Design" in accordance with what is written on JAXA ICD. The simulation result would be tested on vibration test by laboratory scale.

Acknowledgment

The authors are grateful to the Head of Satellite Technology Centre, Head of Program and Facilities of Satellite Technology Centre, all researchers / engineers and employees of Satellite Technology Centre - LAPAN who have provided guidance and advice during the construction of Surya Satelit-1 structure. The author is also grateful for the facilities provided by Satellite Technology Centre to support the smoothness of this project.

References

[1] Japan Aerospace Exploration Agency (JAXA) 2013 *JEM Payload Accommodation Book Vol.8 Small*
[2] NanoRacks 2013 *NanoRacks CubeSat Deployer (NRCSD) Interface Control Document LLC 0.36*
[3] Melahat C, Aykut Ç, Kaya, İnalhan 2011 Design and Analysis of an Innovative Modular Cubesat Structure for ITU-pSAT II *Journal of IEEE* **11** 494–499
[4] Fiona S 2012 Preliminary System Development and Detailed Structural Design and Analysis for the CanX-7 Nanosatellite *A thesis submitted in conformity with the requirements for the degree of Master of Applied Science Graduate Department of Institute for Aerospace Studies University of Toronto*
[5] Arunkumar R 2015 Design and Development of UWE-4 Integration of Electric Propulsion Units, Structural Analysis and Orbital Heating *Analysis Master Thesis in Space Science and Technology University of Würzburg*
[6] Kendyl P and Julia S 2002 CubeSat Specification Document Structures *ICE Cube Final Specification Document, Cornell University*
[7] Simon L, Amy H, Armen T, Wenscel L, Riki M 2009 Cubesat Design Spesification **12** 7-15

[8] CubeSat Design Specification 2014 The CubeSat Program *California Polytechnic State University* **13**
[9] CalPoly 2014 *CubeSat Design Specification* (California: California Polytechnic State University)
[10] CalPoly-P-POD-Rev.E 2014 Poly Picosatellite Orbital Deployed User Guide (California, California Polytechnic State University)
[11] Heidt H, Puig-Suari J, Moore AS, Nakasuka S, and Twiggs R J 2001 CubeSat: A New Generation of Picosatellite *14th Annual AIAA/USU Conference on Small Satellites, Logan, UT*
[12] Wells J 2002 *Canada's Smallest Satellite: The Canadian* (Canada: University of Toronto Institute for Aerospace Studies)
[13] Diana A, Benjamin J, and Guillaume R 2006 *Design of a Swiss Cube Structure and Configuration* (Switzerland: Goldstem)
[14] Braegen E, Hayward D, Hynd G, and Thomas A 2007 Design, Build and *Launch of a Small Satellite Based on CubeSat Designs* (Australia: The University of Adelaide)
[15] Huzain MF, Trihajanto RH 2013 Pengukuran Karakteristik Dinamika Struktur Satelit LAPAN-ORARI/A2 *Jurnal Teknologi Dirgantara* **11** 107-116
[16] Huzain MF 2013 Analisis Dinamik Struktur Satelit LAPAN-ORARI/A2 Menggunakan Simulasi Numerik Finite Element Method *Teknologi Pengamatan dan Informasi Kedirgantaraan* **51**
[17] Budiantoro PA, Huzain MF 2013 Perancangan Struktur Satelit LAPAN-ORARI/A2 untuk Menghasilkan Mass Properties Yang Optimal *Pengembangan Teknologi Satelit di Indonesia* **21**
[18] Triharjanto RH 2007 Desain dan Pengukuran Struktur Satelit Mikro LAPAN – TUBSAT
[19] Triharjanto RH, Huzain MF 2011 Pengembangan Model Dinamika Struktur Satelit Mikro di LAPAN
[20] Armittede, Scott E 2014 *Mechanical and embedded systems design for Generic Nanosatellite Bus spacecraft* (Canada, University of Totonto)
[21] NASA-STD-6016 2008 *Standard Materials and Processes Requirements for Spacecraft, in NASA Technical* (Washington DC: Standard National Aeronautics and Space Administration) p 67
[22] Chiranjeeve HR, Kalaichelvan K, Rajadurai A 2014 Design and vibration analysis of a 2U-Cubesat structure Using AA 6061 for AUNSAT-II IOSR *Journal of Mechanical and Civil Engineering* 61-68
[23] Julie F 2015 *Stress simulation of the SEAM CubeSat structure during launch* (Swedia: KTH Royal Institute of Technology School of Engineering Sciences)
[24] Yevgeniy B 2017 2U Cubesat Structural Design and Integration *A Thesis Presented to the Faculty of the College of Science Morehead State University*
[25] Herrera JEA, Ferrer-Pérez JA, Colín A and Bermúdez B, Reyes 2016 CubeSat System Structural Design *67th International Astronautical Congress (IAC)*
[26] Yoshiki S, Shinichi N, Kenji H, Chisato K, Kanichi K and Takanori O 2009 Structure and thermal control of panel extension satellite (PETSAT) *Journal of Acta Astronautica* **65** 958–996
[27] Arjomandi, Maziar, Brad G 2010 *Build and Launch of a Small Satellite Based on CubeSat Standards* (Australia: Adelaide University)
[28] David F, Bill H 2014 Antenna Design Challenges for New-Generation Nano Satellites *Microwave Journal* **65** 24-34
[29] Farr D, Henderson B 2014 Antenna Design Challenges for New-Generation Nano Satellites
[30] Oliva, Andrew et al. 2011 *Design and Analysis of Subsystems for a CubeSat Mission MQP Report NAG-1102* (Worcester: Worcester Polytechnic Institute)
[31] Dopart, C et al. 2012 *Design and Analysis for a CubeSat Mission MQP Report NAG1102* (Worcester: Worcester Polytechnic Institute)
[32] Pumpkin Inc. 2012 *3D Models of the CubeSat Kit Retrieved 5-9*
[33] Grindrod JL 2016 CXBN-2 Vibration Test Report *Morehead State University, Space Science Center. Document number: 161118-CXBN2-VIBTR.*

[34] Quiroz-Garfias C, Silva-Navarro G, and Rodriguez-Cortes H 2007 Finite Element Analysis and Design of a CubeSat Class Picosatellite Structure *Paper presented at the Electrical and Electronics Engineering ICEEE 2007*

[35] S. Mauthe 2012 AISSat-2 Vibration Test Plan Tech. *Rep. SFL-AIS-SYS-TP003. UTIAS Space Flight Laboratory*

[36] S Libin and F Hui 2002 Static and Modal Analysis of Telescope Frame in Satellite (Beijing: Tsinghua University and University of Science and Technology)

[37] M Sedighi and Mohammadi B 2003 On the Static and Dynamic Analysis of Small Satellite (Mesbah) *IROST, Tehran, Iran*

[38] Israr A 2014 Vibration and Modal Analysis of Low Earth Orbit Satellite *Shock and Vibration* **2014** 1-8

[39] Joe B, Michael, Kaitlyn K, Emie M, Sam N, Alex O, Tim S and Andrew W 2012 *Mechanical, Power and Thermal Subsystem Design for a CubeSat Mission* (Massachusetts: Worchester Polytechnic Institute)

[40] Afendi M, Amin NAM, Abdul MMS 2014 Stress and Thermal Analysis of Cubesat Structure *Conference Paper in Applied Mechanics and Materials*

[41] Barkanov E 2001 *Introduction to the Finite Element Method, Institute of Materials and Structures* (Baltics: Riga Technical University)

[42] Ontac S¸ Dag S, and Gokler MI 2007 Structural finite element analysis of stiffened and honeycomb panels of the RASAT satellite *3rd International Conference on Recent Advances in Space Technologies (RAST '07)* pp. 171–175

[43] Kudzanai S and Tawanda M 2017 Finite element analysis of a cubesat *International Symposium on Industrial Engineering and Operations Management (IEOM) Bristol, UK*

[44] Saurabh and Yudhvir Yadav 2016 Literature Review on Finite Element Method. International *Journal of Enhanced Research in Science, Technology & Engineering* **5** 267-269.

[45] Kelly SG 2012 Mechanical vibrations: theory and applications SI (America: Cengage Learning)

[46] Cihan M 2008 *A Methodology for the Structural Analysis* (Turkey: Cubesat Department of Space Engineering, Istanbul Technical University)

[47] Asif I 2014 Vibration and Modal Analysis of Low Earth Orbit Satellite *Shock and Vibration* **2014**

[48] Jayaram S, McQuilling M, Condoor S 2012 PRO/MECHANICA-based structural and random vibration analysis of picosatellite structure *Int. J. Computer Aided Engineering and Technology* **4** 90-100

[49] Randy T, Alex Y 2009 *Material Selection and Finite Element Analysis for Structural Support of UCI CubeSat Spacecraft* (Irvine: University of California)

IOP Conf. Series: Earth and Environmental Science **149** (2018) 012064 doi:10.1088/1755-1315/149/1/012064

An Application of spectral analysis method investigating the anomalous change in the QBO for period of 1997 to 2017, especially at 50 hPa related to the ozone's concentration

Eddy Hermawan
Center for Atmospheric Science and Technology of National Institute of Aeronautics and Space Republic of Indonesia

E-mail: eddy_lapan@yahoo.com

Abstract. This study is mainly concerned an application of the spectral tehnique analysis in investigating the anomalous change of the Quasi-Biennial Oscillation (QBO) over the equatorial region, especially over the Singapore related to the ozone's concentration at about 50 hPa. This is important since we need to know that characteristics of ozone concentration at that layer due to the ozone depletion caused by the effect of the global warming. Please note here, ozone is already stated as one of the most important parameter needed to protect our earth from Ultra Violet (UV) light. No much study of ozone's protection, especially near the equatorial region. In this preliminary results, we show the characteristics of QBO taken from the Singapore 's QBO data analysis. The QBO has represented as the downward propagating of zonal wind variation. Then, we investigate impact of QBO on the Ozone's Concentration anomaly at stratospheric layer. We found the pre-dominant peak oscillation of QBO at around 28 month. We found also an a good agreement (pattern) between QBO and Ozone's Concentration anomaly, especially at 50 hPa layer with a correlation coefficien of about 0.5. By this formula, we can estimate the next behavior of ozone's concentration anomaly taken from the QBO index that updated regularly.

1. Introduction

The quasi-biennial oscillation (QBO) or alternation of zonal wind (from easterly to westerly and inversely) in the equatorial stratosphere in the layer from about 100 hPa (16 km) to about 3 hPa (40 km) is known for more than half a century. A review of the current understanding of the QBO can be found in Baldwin et al. [1]. Zonal wind QBO is a prominent feature of the equatorial stratosphere, and being an important component of atmospheric general circulation causes the interannual variability of many parameters and processes in Earth's atmosphere.

Interannual variability of equatorial total ozone (TOZ) is dominated by the quasi-biennial oscillations that distinctly appear after removing the annual cycle from the original data. The QBO signal varies between ±10 Dobson units (DU), approximately ±4% of the background equatorial TOZ [1,2]. Longitudinal structure of the ozone QBO demonstrates zonally uniform phase changes (Shiotani, M) [3]. A large number of studies have shown that equatorial total ozone variation is characterized by strong coupling with the QBO of the equatorial zonal wind in the lower stratosphere. High correlation ($r > 0.7$) between total ozone and zonal wind in tropical zone has been obtained using the equatorial zonal wind at 30 and/or 50 h.

IOP Conf. Series: Earth and Environmental Science **149** (2018) 012064 doi:10.1088/1755-1315/149/1/012064

There is an interaction that occurs between the troposphere and stratosphere layer known as Stratosphere and Troposphere Exchance (STE). One is the vertical dynamics of Ozone concentration anomalies (O3) especially that occuring at the stratosphere layer. Many factors influence it, one of them is the dynamics or behavior of the direction and speed of zonal wind (East-West) in the stratosphere layer known as *quasi-biennial oscillation* (QBO). In this present study we mainly concerned to investigate the characteristics of QBO taken from the Singapore's QBO data analysis that taken from this web address (http://www.geo.fu-berlin.de/en/met/ag/strat/produkte/qbo/index.html), and to investigate impact of QBO on the Ozone's Concentration at stratospheric layer, especially at about 50 hPa.

2. The theory of QBO

The study of the stratospheric region at the equator is dominated by the phenomenon of downward propagation of westerly and easterly that firstly is discovered by Ebdon [4] and Reed et al. [5] which came to be known as *quasi-biennial oscillation* or QBO. QBO is a westernized oscillation of winds and clouds that propagate downward in the equatorial stratosphere-bottom region with an average period of 28 months [6], where the QBO lead phase has a stronger maximum amplitude than the western phase and the QBO amplitude is of a nature symmetry of the equator following the Gaussian distribution [6,7]. The mechanism of formation of QBO is proposed by Holton and Lindzen [5], where QBO is the result of interaction between the mean flow with the equatorial wave equation mode. Holton and Lindzen [8] were the first to model QBO based on the planetary equatorial wave propagation, Kelvin waves and Rosby-gravity waves.

Holton [9] states that QBO occurs only in the equatorial region because *Kelvin waves* and *Rosby-gravity waves* only form in the region of about 12^0N - 12^0S. The area provides considerable energy derived from the release of latent heat in large-scale convective cloud clouds in the tropics to form planeter-scale waves and push them down to the lower stratosphere. Therefore QBO is not formed in high latitudes because the resulting energy is not strong enough to push the equatorial wave into the stratosphere, although in high latitudes there is cloud formation and front region formation.

In addition to discussing the phenomenon of zonal wind oscillations, QBO also has an influence on chemical and physical processes in the lower stratosphere. Holton [9] has proven that QBO can satisfy thermal wind equations if the vertical shifts of zonal winds (wind shear) are associated with the temperature in the stratosphere. Meanwhile Baldwin et al. [1] show that to analyze QBO can use temperature and ozone data. Although QBO occurs only in the stratosphere-below the equator region, but QBO can affect the stratosphere region globally [10]. Because QBO affects the process of distribution and transport of substances in the stratosphere and may be a factor that causes ozone depletion.

Previous study of QBO has been done by Pascoe et al. [6] using ERA-40 data. But in this study not only QBO quantifying by using different methods, but also linking QBO with chemical and physical processes that occur in the lower stratosphere. It is hoped that by analyzing the characteristics of QBO and its relation to chemical and physical processes in the lower stratosphere can provide a better understanding of the phenomenon of QBO. In addition, it is expected to open ideas and insights in order to continue the study or research on broader QBO.

3. Data and Method

The main data of his study is taken from the Singapore's QBO data analysis (http://www.geo.fu-berlin.de/en/met/ag/strat/produkte/qbo/index.html). The availability data is started from Januari 1979 until now which devided from 1000 to 1 hPa [11]. While, the monthly of Ozone Total Coloum is taken from AIRS satellite of Giovanni NASA (https://giovanni.gsfc.nasa.gov/giovanni/ for period of January 1997 to June 2017. In this study, we apply the spectral technique analysis that represented as the *Power Spectral Density* (PSD) and Wavelet. Wavelet analysis known as an effective tool to analysed power variations in time series. Time series were decomposed into time frequency space, then will determine variability dominant modes and variation modes in time as well. The wavelet transform has

IOP Conf. Series: Earth and Environmental Science **149** (2018) 012064 doi:10.1088/1755-1315/149/1/012064

been used in previous studies, including tropical convection [12], the El Niño–Southern Oscillation (ENSO) [13,14], atmospheric cold fronts [14], central England temperature [15], the dispersion of ocean waves [16], wave growth and breaking [17], and coherent structures in turbulent flows [18]. In this present study we estimate the power spectral density using FFT (Fast Fourier Transform) that appeared at Matlab software. Detailed this technique can be seen at this web site (https://www.mathworks.com/help/signal/ug/power-spectral-density-estimates-using-fft.html).

4. Results and Discussion

Figure 1 is showing the PSD the monthly of QBO data for period of January 1997 to June 2017 which starting from 100 to 10 hPa's layer for every 10 hPa in altitude height. From this figure, we can see the pre-dominant peak oscillation an about 28 month. In the same figure, we found also the second pre-dominant oscillation an about 21 month, but the value smaller than 28 month. Since our analysis is focussed on 50 hPa layer, we make the contour plot of that QBO data as showing at figure 2.

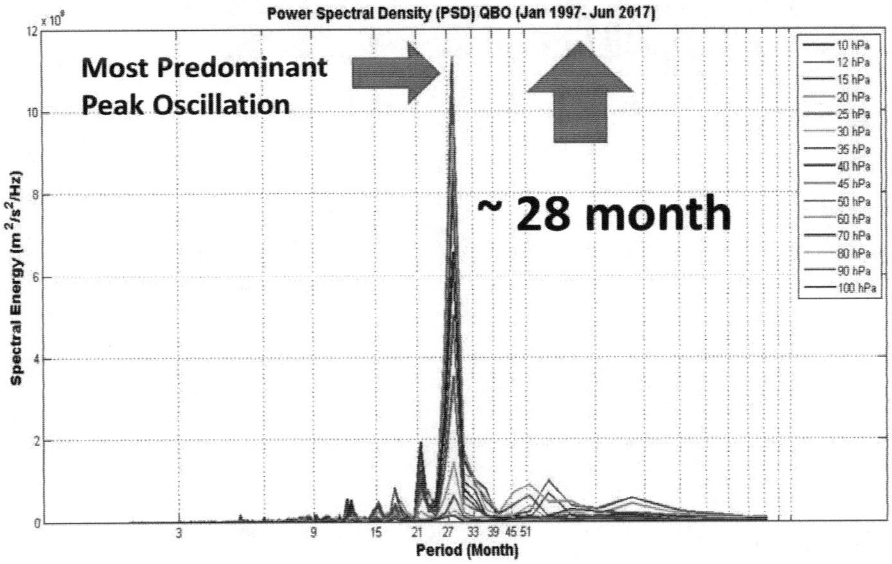

Figure 1. The Power Spectral Density (PSD) of QBO (Jan 1997-Jun 2017).

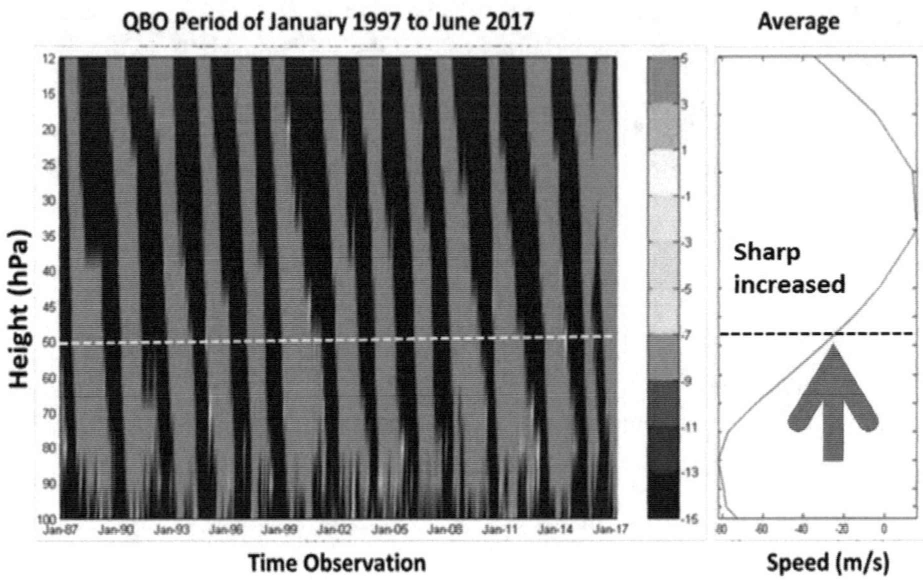

Figure 2. The time-height section of QBO signal for period of January 1997 to June 2017.

From figure 2 above, we can see the sharp increased of QBO signal at around 50 hPa with the wind speed averaged is about 20 m/s. To investigate when that strong signal is occurred, we applied the wavelet technique analysis. Then, we obtain the pre-dominant peak oscillation of 50 hPa is still 28 month and occurring at around 2003 as showing at figure 3.

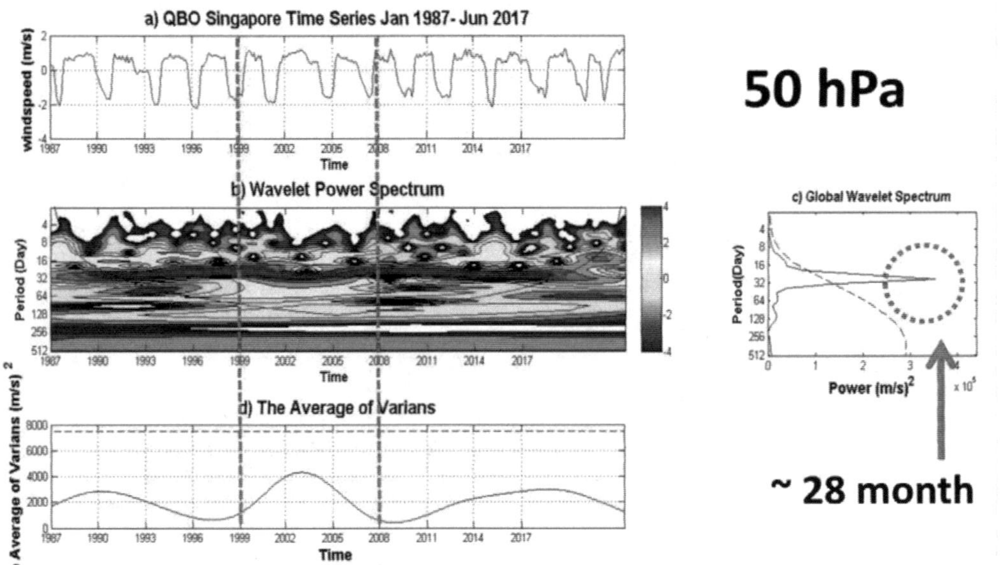

Figure 3. As the same as Figure 1, but for wavelet technique analysis.

Basically, we expand our analysis using the longer data started from 1987 to 2017 by comparing with other layers. They are 10, 20, 30, 50, and 70 hPa, respectively as showing at figure 4. From this figure, we can see the consistency of 50 hPa and other layers.

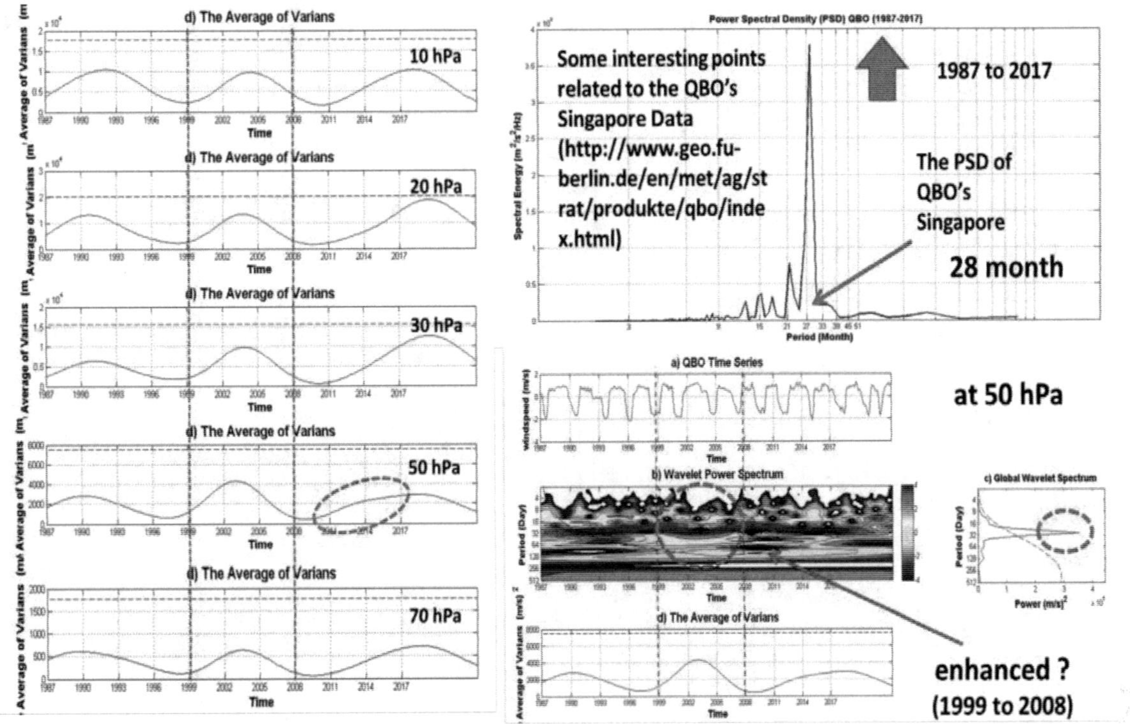

Figure 4. As the same as Fig. 1, but for longer data 1987 to 2017.

Then, finally we concentrate to analyze the comparison between QBO and ozone anomalies over Kototabang, West Sumatera, Indonesia started from September 1, 2002 to April 4, 2017 as showing at figure 5. From this figure, we can see an a good pattern both parameters.

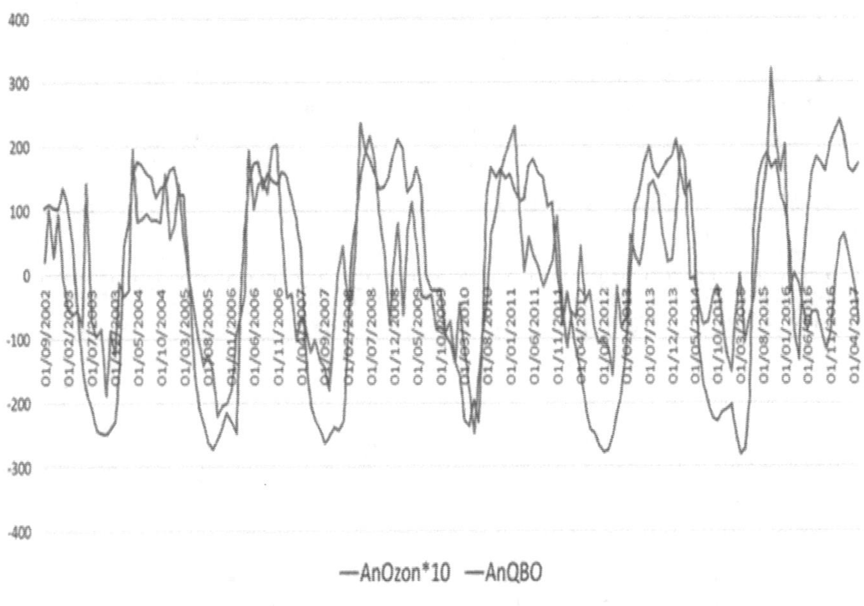

Figure 5. The time series ozone anomaly and QBO started from September 1, 2002 to April 4, 2017.

Please note here, this results has already confirmed with Gabis [19] and Fadnavis and Beig [20] when they investigated the ozone behaviour at equatorial stratosphere. Finally, we are showing the scatter plot between ozone anomaly and QBO using Fig. 4 data as showing at figure 6. Although, the coeficien correlation is not too big, but it is enough, since the value is close to 0.5.

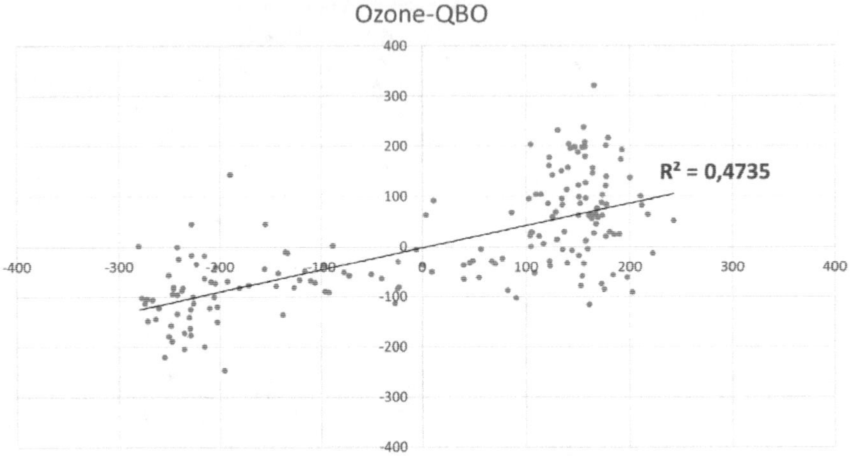

Figure 6. The scatter plot between ozone anomaly and QBO.

5. Conclusion

Ozone (O_3) is one of the most important gases at the stratosphere that can be protected the UV light from the sun. Recently, the behavior of ozone concentration especially at the equatorial region is very important to be investigated due to the ozone depletion at this region. Since ozone concentration is related well to the wind speed and direction at the stratosphere, in this present study we investigate the Quasy Biennial Oscillation (QBO). We used the QBO data taken from Singapore that has longer the QBO data. We applied the spectral technique that represented by Power Spectral Density (PSD) and wavelet, respectively. We found the pre-dominant peak oscillation of QBO at around 28 month and located around 50 hPa. Then, we compared with the ozone concentration anomaly for period of September 1, 2002 to April 4, 2017. We found an a good pattern (agreement) between two parameter that represented by coeficien correlation (R^2) of about 0.5. Both figures look similar each other. By this reason, we can mention that we can estimate the next of ozone concentration by using the QBO data that producted by the Singapore regularly.

6. Future Works

Since, we need to know the next QBO anomaly and their impact on the global and regional climate, especially over the Maritime Continent, we need to make prediction, such as using the ARIMA technique.

Acknowledgments

The author wish to thank to head of Center for Atmospheric Science and Technology of National Institute of Aeronautics and Space Republic of Indonesia who has already supported this study.

References

[1] Baldwin M P et al. 2001 The Quasi-biennial Oscillation *Rev. Geophys.* **39** 179-229
[2] Bowman 1989 Global patterns of quasi-biennial oscillation in total ozone *J. Atmos. Sci.* **46** 3328-3343

[3] Shiotani M 1992 Annual quasi-biennial and El Niño-Southern Oscillation (ENSO) time-scale variations in equatorial total ozone *J. Geophys. Res.* **97** 7625-7633
[4] Ebdon R A 1960 Notes on the Wind Flow at 50 mb in Tropical and Subtropical Regions in January 1957 and in 1958 *Q. J. R Meteorol. Soc.* **86** 540-542
[5] Reed R G, Campbell W J, Rasmussen L A, Rogers D G 1961 Evidence of downward-propagating annual wind reversal in the equatorial stratosphere *Journal of Geophysical Research* **66** 813-818
[6] Pascoe C L *et al.* 2005 The Quassi-biennial Oscillation: Analysis Using ERA-40 data *J. Geophys. Res.* **110** 1-13
[7] Holton J R, Lindzen R S 1972 An updated theory for the quasi-biennial cycle of the tropical stratosphere *American Meteorological Society* **29** 1076-1080
[8] Holton J R 2004 An Introduction to Dynamics Meteorology 4th Edition (Burlington: Elsevier)
[9] Holton J R, Tan H C 1980 The influence of the equatorial Quasi-Biennial Oscillation on the global atmospheric circulation at 50mb *Journal of Atmospheric Science* **37** 2200-2208
[10] Dee D P *et al.* 2011 The ERA-Interim Reanalysis: Configuration and Performance of the Data Assimilation System *Q. J. R. Meteorol. Soc.* **137** 553-597
[11] Weng H, Lau K M 1994 Wavelets, period doubling, and time-frequency localization with application to organization of convection over the tropical western Pacific *J. Atmos. Sci.* **51** 2523–2541
[12] Gu D, Philander S G H 1995 Secular changes of annual and interannual variability in the Tropics during the past century *J. Climate* **8** 864–876
[13] Wang B, Wang Y 1996 Temporal structure of the Southern Oscillation as revealed by waveform and wavelet analysis *J. Climate* **9** 1586–1598
[14] Gamage N, Blumen W 1993 Comparative analysis of lowlevel cold fronts: Wavelet, Fourier, and empirical orthogonal function decompositions *Mon. Wea. Rev.* **121** 2867–2878
[15] Meyers S D B, Kelly G, O'Brien J J 1993 An introduction to wavelet analysis in oceanography and meteorology: With application to the dispersion of Yanai waves *Mon. Wea. Rev.* **121** 2858–2866
[16] Liu P C 1994 Wavelet spectrum analysis and ocean wind waves *Wavelets Analysis and Its Applications* **4** 151 166
[17] Farge M 1992 Wavelet transforms and their applications to turbulence *Annu. Rev. Fluid Mech.* **24** 395–457
[18] Foufoula-Georgiou E Kumar P 1995 *Wavelets in Geophysics* Academic Press p 373.
[19] Gabis I P 2014 Quasi-biennial oscillation (QBO) of tropical total ozone under *Advances in Space Research* **54** 2499
[20] Fadnavis S and Beig G 2009 Quasi-biennial oscillation in ozone and temperature over tropics *JASTP* **71** 257-263

Study of 3D bathymetry modelling using LAPAN Surveillance Unmanned Aerial Vehicle 02 (LSU-02) photo data with stereo photogrammetry technique, Wawaran Beach, Pacitan, East Java, Indonesia

N M Sari, J T Nugroho, G A Chulafak and D Kushardono

Remote Sensing Applications Center, Indonesian National Institute of Aeronautics and Space (LAPAN)

E-mail: nurwita.mustika@lapan.go.id

Abstract. Coastal is an ecosystem that has unique object and phenomenon. The potential of the aerial photo data with very high spatial resolution covering coastal area is extensive. One of the aerial photo data can be used is LAPAN Surveillance UAV 02 (LSU-02) photo data which is acquired in 2016 with a spatial resolution reaching 10cm. This research aims to create an initial bathymetry model with stereo photogrammetry technique using LSU-02 data. In this research the bathymetry model was made by constructing 3D model with stereo photogrammetry technique that utilizes the dense point cloud created from overlapping of those photos. The result shows that the 3D bathymetry model can be built with stereo photogrammetry technique. It can be seen from the surface and bathymetry transect profile.

1. Introduction

The Aerial remote sensing using airborne becomes one of remote sensing techniques that can be used to obtain remote sensing data with very high spatial resolution. Aerial remote sensing becomes an alternative of satellite remote sensing to overcome the existing problems related to cloud cover, expensive satellite operational costs, until acquisition period. By using the aircraft, remote sensing data acquisition can be more easily, efficiently and flexibly in expected high spatial resolution related to adjustable flying height [1,2]. Several types of airbornes used in the acquisition of aerial photograph data can be either manned or unmanned aircraft. In a paper review by [3] described how UAS (Unmanned Aerial Systems) develops in photogrammetry and remote sensing both from the airborne and sensor.

Other remote sensing for coastal and small island study is important to help better coastal management including monitoring and identification of changes or dynamics of coastal and marine environments [4]. Several remote sensing studies for coastal and small islands application have been conducted, for mapping shallow water habitats by considering basic topography of waters; Analysis of growth rate, productivity rate, productivity of seagrass enhalus acoroides; seagrass mapping to species level based on spectral library in situ measurement using the SAM algorithm (spectral angle mapper) then perform the accuracy assessment [5-7]. Meanwhile another study was conducted to determine the value and variability of remote sensing reflectance (R) from sea levels in different seasons and different types of seawater [8]. Remote sensing with UAV data has been widely used for coastal area monitoring even the quantitative measurement of dune and coastal surface that can be conducted immediately after storm, which was previously difficult [9].

Bathymetry depicting seafloor topography has been studied using remote sensing data by utilizing existing bands as performed in the bathymetric mapping accuracy testing between single band and ratios was also performed in [10], it indicated that the ratio bands were able to normalize the variation of underwater coverage, it was shown with lower SE than single band.

The method that will be used in this research is bathymetry modeling using stereo technique which utilizing photographs to create a 3D model. UAV utilization has been used in surface modeling by using overlap points from overlaid photographs, until the quality assessment of photogrammetric UAV products with DSM accuracy [11]. In small and medium scale, method for obtaining high-resolution data for landform modelling is Structure from Motion (SfM) technique by using overlaid point cloud on images as the basis for building their 3D models [12]. Photogrammetry technique by utilizing photo sequences have been applied in various studies, including for measuring the construction of underground coal mines [13]. Further research to create a three-dimensional model with photogrammetry was made in [14] to build a three-dimensional model of urban areas by utilizing a combination of oblique and nadir photogrammetry that produced texture information for subsequent three-dimensional model. The creation of a three-dimensional surf zone waves model with stereo techniques has also been performed [15]. In another study surface modeling is done with Structure-from-Motion (SfM) Multi-view Stereo (MVS) method of photogrammetry which obtain accuracy up to 2 mm [16]. Micro-bathymetry measurement with underwater stereo camera has been done and obtained precision to the millimeter level [17]. Another study of shallow water bottoms measurement made with UAV data has also been performed [18].

In previous research, the method that has been used in various studies to build a three-dimensional model using photo sequences, has not been attempted on remote sensing aerial data acquired with an unmanned aerial vehicle where the spatial resolution reaches more than 10cm in particular for bathymetry modeling. This study aims to create an initial model bathymetry with stereo photogrammetry technique using LSU-02 data.

2. Materials and methods

2.1. Materials

The aerial photography data used in this initial three-dimensional bathymetry study was acquired using LAPAN Surveillance Unmanned Aerial Vehicle 02 (LSU-02) on April 7, 2016 at 15.11 WIB. The camera used is the Sony Alpha A6000 with a pixel size of 6000 x 4000 pixels with spatial resolution reaches 10cm. The areas included in the study area are Wawaran Beach, south coast of Pacitan Regency, East Java (figure 1).

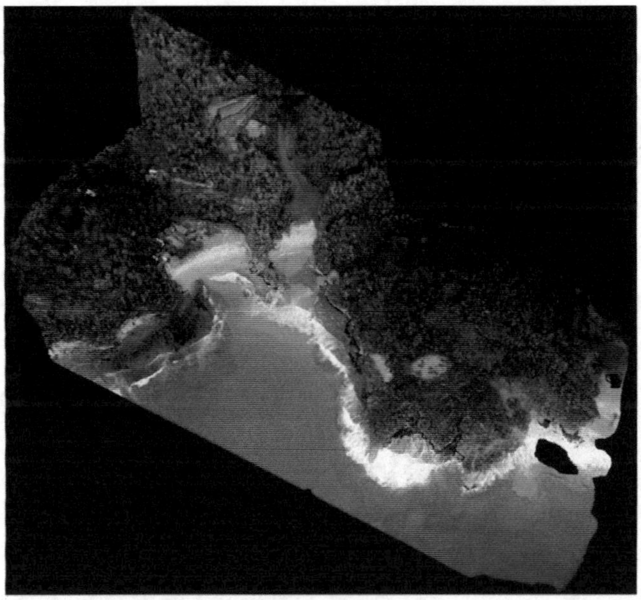

Figure 1. Orthomosaic image of sudy area.

IOP Conf. Series: Earth and Environmental Science **149** (2018) 012065 doi:10.1088/1755-1315/149/1/012065

2.2. Methods

In this research, study of bathymetry modelling uses a stereo photogrammetry technique consist of overlaid point cloud of photographs that will be the mosaic, which then constructed to be a three-dimensional model of the point cloud. Figure 2 shows the flowchart of bathymetry modeling using LSU-02.

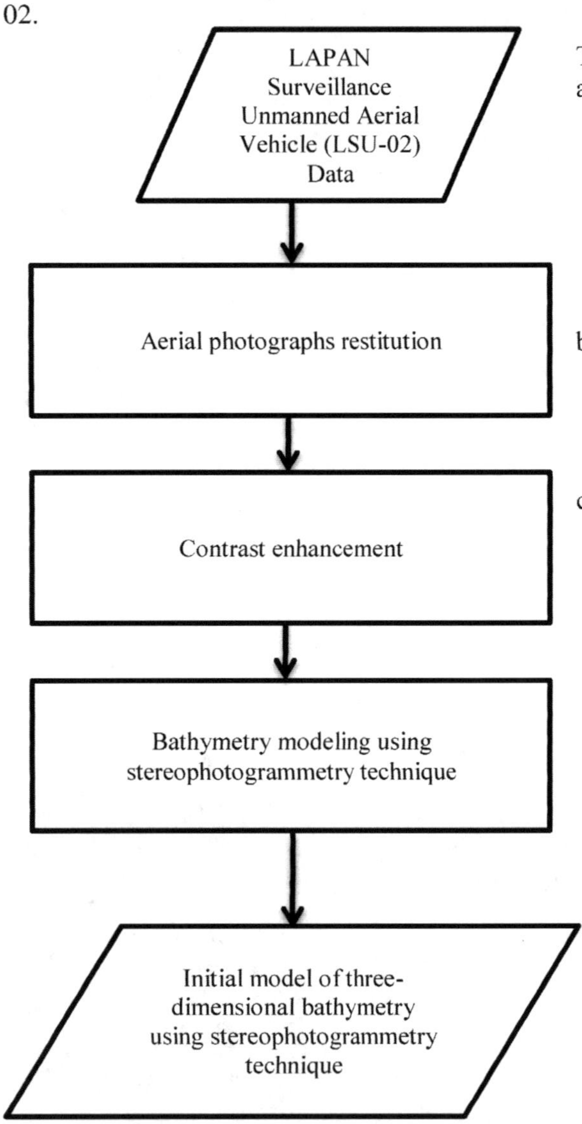

Figure 2. Flowchart of bathymetry modeling using LSU-02.

The stages of data processing is,

a. Restitution of stereo aerial photographs
The initial stage of data processing is restitution or reconstruction of stereo aerial photographs that will be constructed to become three-dimensional model from two-dimensional data or overlaid photographs. These overlaid photographs generate orthomosaic image data of study area.

b. Contrast enhancement
Further process after generating orthomosaic image is contrast enhancement that aims to clarify the color gradation of bathymetry study area.

c. Bathymetry modeling with stereo photogrammetry
Stages conducted in this modeling are as in [19] where the dense point cloud is built in several stages, named search for points on all scales and locations, determining the exact location of the point, and determination of point direction.

- Search for points on all scales and locations,
  In this stage the searching process of points on all scales and locations of the study area is applied with the principle of Difference of Gaussian (DoG). This formula has been developed in the theory of edge detection [20].

$$G(x, y, \sigma) = \frac{1}{2\pi\sigma^2} e^{x^2+y^2 / 2\sigma^2} \quad (1)$$

Where
x,y = position
σ = scale

- Determining the exact location of the point,
  Determination of exact point location is conducted by measuring stability and interpolation with Taylor expansion as in [19]. Taylor expansion is written in the formula.

$$D(X) = D + \frac{\partial D^T}{\partial X} x + \frac{1^2}{2} X^T + \frac{1}{2} X^T \frac{\partial D^T}{\partial X} X \tag{2}$$

Where
x= (x,y)= offset of the point

- Determination of point direction
  At this stage, each key point is determined in a direction based on the direction of the slope. This is the assignment stage of orientation as has been done in [19] to place the direction.

$$m(x, y) = \sqrt{((L(x+1, y) - L(x,1, y))^2} \tag{3}$$

$$\theta(x, y) = \arctan \frac{L(x,1, y) - L(x,1, y)}{L(x, y-1) - L(x, y-1)} \tag{4}$$

Where
L (x, y) = sample image
m = gradient magnitude
θ= direction

When the points used to create a three-dimensional model (point cloud) has been obtained then the next stage is the finalization of 3D model bathymetry and analysis.

3. Result and discussion

3.1. Contrast enhancement of orthomosaic images

A total of eight photos were used in this study to become an orthomosaic image (figure 2) and experienced several processes to be a 3D model. The result shows that the coverage area including land and water in the coastal area. The existing land covers in land area are settlement, forest and moor. Deep blue color is a natural character of pure water color where the particles, one of them is sand, can affect the absorption which, of course affect the natural color of seawater [21]. Thus shallow seawater that has suspended particles and closer to the seafloor with certain materials will result the lighter color of the seawater than the deep seawater. Before the contrast enhancement process, the orthomosaic image is still the image with natural color, seen how the color of water has a gradation of light blue in the shallow area and darker blue in the deeper seawater. In sea water area, the contrast enhancement process is conducted to reinforce the color gradation in seawater.

Contrast enhancement was performed in the orthomosaic image of the study area as shown in figure 3 where the color used is false color, the water color is brown gradation and vegetation color is green. Through the enhancement result, the color of vegetation becomes darker. From the result it is known that the water color gradation from light to dark in the area of sea water means that the depth is vary. In figure 3, the color of the darker water is as the distance from the land area further. It indicates that the depth of the seawater increases.

IOP Conf. Series: Earth and Environmental Science **149** (2018) 012065 doi:10.1088/1755-1315/149/1/012065

3.2. Digital elevation model

The digital elevation model (DEM) of the study area is made by converting the mosaic result into two dimensional view, with color gradation that represents the elevation of the area. The DEM here was produced through the stereo photogrammetry technique (figure 4).

For DEM accuracy in land area, DEM accuracy assessment for DEM generated from UAV data at a small area and capable of achieving very high accuracy has conducted in [22]. This shows that the acquisition of DEM from the UAV is able to show the morphological condition in the field. The color gradation that appears is yellow to red and green to blue which means that the yellow color is the highest point while the dark blue is the deepest point. This is confirmed by the transect crossing from the yellow point to the dark blue point is from highest to lowest point.

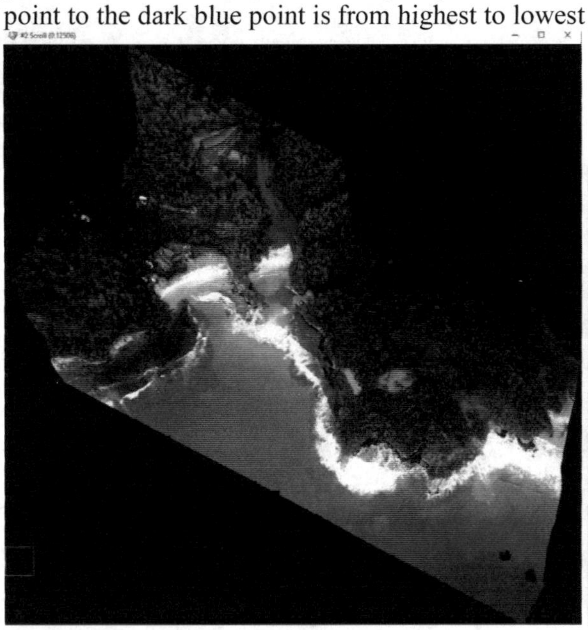

Figure 3. Result of contrast enhancement of orthomosaic image.

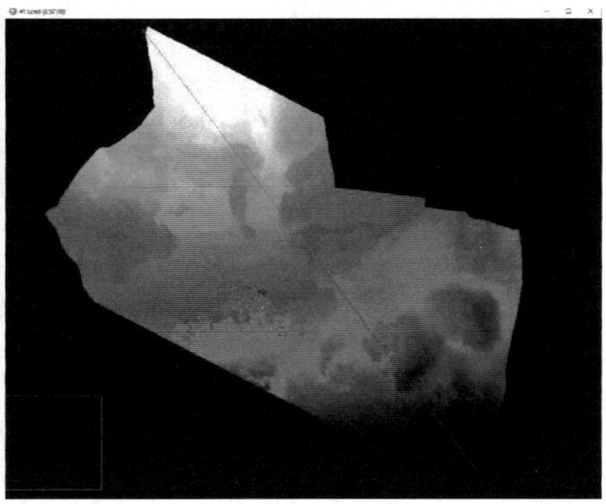

Figure 4. Digital elevation model of the study area.

3.3. Profile of surface and bathymetry

Figure 5 shows how the profile of the surface and bathymetry in the study area. The transect line is red as shown in figure 4. In figure 5 the blue line shows how the relief on the surface and under the sea. Horizontally, it represents the distance along the transect while vertically shown profile from the surface to the bottom of the sea water. The transect starts from the surface with hill morphology that is the highest point with bright yellow color in figure 4, and the transect terminates at the deepest point with dark blue color of sea water in figure 4. Based on y-axis it is known that the range of vertical value is the highest value reaching more than 200 while the lowest value reaching -100. This value is the measurement in the digital number resulting from the overlapping of LSU-02 photo data.

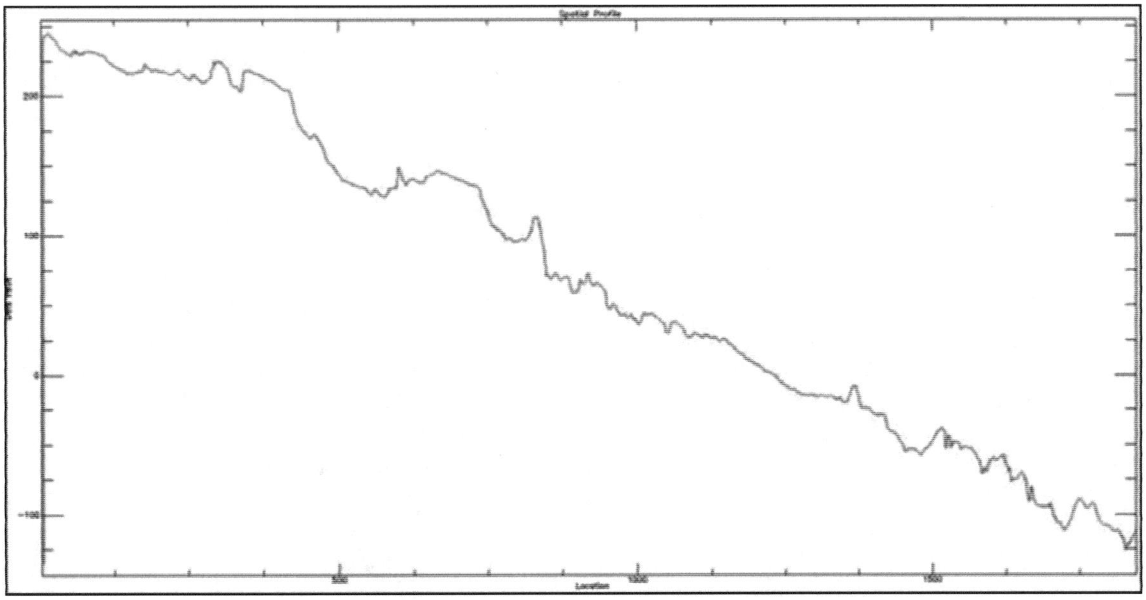

Figure 5. Profile of surface and bathymetry.

3.4. Perspective of 3D surface and bathymetry

Figure 6 shows perspective of 3D surface and bathymetry of the study area, both land and sea water area. From the figure, it can be seen how the appearance of land and sea on the coast in the study area. The land part of the study area appears to have higher areas on the ridge or hill and undulating area. Land covers in this section are vegetation consists of forest and moor and settlement on the coastal plain. Steep slopes on certain part of the coast are also seen in the picture.

The color of seawater visible in the study area is graded from light to dark which indicates depth difference. The gradation, that is, in the area close to the land has lighter water color which indicates shallow water. Meanwhile in the region that is farther from the land is deeper which is shown by the darker color of seawater.

This condition is clearly represented by a 3D model built from overlapping points that form the dense point cloud. The 3D model shows that in areas close to the land, the condition is shallower than area further away from the land.

Figure 6. Perspective of 3D surface and bathymetry.

4. Conclusion
The proposed method of this research to build bathymetry model with stereo photogrammetry technique is capable of displaying the three-dimensional model of sea water depth in Wawaran Beach, Pacitan.

Acknowledgments
We would like to deliver our gratitude to Drs. Gunawan Setyo Prabowo, M.T., Head of Aeronautics Technology Center, Prof. Dr. Ir. Vincentius P. Siregar, DEA. as our resource person.

References
[1] Sofiati R 2011 Teknologi pesawat tanpa awak untuk pemetaan dan pemantauan tanaman dan lahan pertanian *Informatika Pertanian* **20** 2 pp 58 - 64
[2] Kushardono D 2014 Teknologi akuisisi data pesawat tanpa awak dan pemanfaatannya untuk mendukung produksi informasi penginderaan jauh *Inderaja* **V** 7
[3] Colomina and Molina 2014 Unmanned aerial systems for photogrammetry and remote sensing: A review *ISPRS Journal of Photogrammetry and Remote Sensing* **92** pp 79–97
[4] Syah A F 2010 Penginderaan jauh dan aplikasinya di wilayah pesisir dan lautan *Jurnal Kelautan* **3** 1 pp 18-28
[5] Setyawan I E, Siregar V P, Pramono G H and Yuwono D M 2014 Pemetaan profil habitat dasar perairan dangkal berdasarkan bentuk topografi: studi kasus pulau panggang, kepulauan seribu Jakarta *Majalah Ilmiah Globe* **16** 2 pp 125-132
[6] Rustam A, Bengen D G, Arifin Z, Gaol J L and Arhatin R E 2013 Growth rate and productivity dynamics of enhalus acoroides leaves at the seagrass ecosystem in pari islands based on in situ and alos satellite data *International Journal of Remote Sensing and Earth Sciences* **10** 1 pp 37-46
[7] Aziizah N N, Siregar V P and Agus S B 2016 Penerapan algoritma spectral angle mapper (sam) untuk klasifikasi lamun menggunakan citra satelit Worldview-2 *Jurnal Penginderaan Jauh dan Pengolahan Data Citra Digital* **13** 2 pp 61-72
[8] Nababan B, Wirapramana A A G and Arhatin R E 2013 Spektral remote sensing reflektansi permukaan air laut *Jurnal Ilmu dan Teknologi Kelautan Tropis* **5** 1 pp 69-84
[9] Turner I L, Harley M D and Drummond C D 2016 UAVs for coastal surveying *Coastal Engineering* **114** pp 19–24
[10] Wicaksono P 2015 Perbandingan akurasi metode band tunggal dan band rasio untuk pemetaan batimetri pada laut dangkal optis *Simposium Nasional Sains Geoinformasi IV* pp 802-810
[11] Ruzgienė B, Berteška T, Gečyte ., Jakubauskienė E and Aksamitauskas V Č 2015 The surface modelling based on UAV Photogrammetry and qualitative estimation *Measurement* **73** pp 619–627
[12] Westoby M J, Brasington J, Glasser N F, Hambrey M J and Reynolds J M 2012 Structure-from-Motion photogrammetry: A low-cost, effective tool or geoscience applications *Geomorphology* **179** pp 300–314
[13] Slaker B A and Mohamed K M 2017 A practical application of photogrammetry to performing rib characterization measurements in an underground coal mine using a DSLR camera *International Journal of Mining Science and Technology* **27** pp 83–90
[14] Yalcin G and Selcuk O 2015 3D city modelling with oblique photogrammetry method *Procedia Technology* **19** pp 424-431
[15] de Vries, S, Hill D F, de Schipper M A and Stive M J F 2011 Remote sensing of surf zone waves using stereo imaging *Coastal Engineering* **58** pp 239–250
[16] Ferreira E, Chandler J, Wackrow R and Shiono K 2017 Automated extraction of free surface topography using SfM-MVS photogrammetry *Flow Measurement and Instrumentation* **54** pp 243-249
[17] Schmidt V E and Rzhanov Y 2012 Measurement of micro-bathymetry with a GOPRO underwater stereo camera pair *IEEE Xplore Document Oceans* doi: 10.1109/OCEANS.2012.6404786

[18] Partama I G, Kanno A, Akamatsu Y, Inui R, Goto M and Sekine M 2017 A simple and empirical refraction correction method for uav-based shallow-water photogrammetry I*nternational Journal of Environmental, Chemical, Ecological, Geological and Geophysical Engineering* **11** 4

[19] Tao W, Lei Y and Mooney P 2011 Dense point cloud extraction from uav captured images in forest area *Proceedings of IEEE International Conference on Spatial Data Mining and Geographical Knowledge Services*

[20] Marr D and Hildreth E 1980 Theory of Edge Detection *Proceedings of the Royal Society of London. Series B, Biological Sciences* **207**(1167) pp 187-217

[21] Morel A and Prieur L 1977 Analysis of variations in ocean color *Lymnology and Oceanography* 22 4 pp 709-722

[22] Uysal M, Toprak A S and Polat N 2015 DEM generation with UAV photogrammetry and accuracy analysis in Sahitler Hill. *Measurement* **73** pp 539–543

IOP Conf. Series: Earth and Environmental Science **149** (2018) 012066 doi:10.1088/1755-1315/149/1/012066

Comparison Spatial Pattern of Land Surface Temperature with Mono Window Algorithm and Split Window Algorithm: A Case Study in South Tangerang, Indonesia

Tasya Bunai[1], Rokhmatuloh[2], Adi Wibowo[3]

Department of Geography, Faculty of Mathematcs and Natural Science, Universitas Indonesia, Kampus UI Depok, Depok 16424, Indonesia

E-mail: rokhmatuloh.ssi@sci.ui.ac.id

Abstract. In this paper, two methods to retrieve the Land Surface Temperature (LST) from thermal infrared data supplied by band 10 and 11 of the Thermal Infrared Sensor (TIRS) onboard the Landsat 8 is compared. The first is mono window algorithm developed by Qin et al. and the second is split window algorithm by Rozenstein et al. The purpose of this study is to perform the spatial distribution of land surface temperature, as well as to determine more accurate algorithm for retrieving land surface temperature by calculated root mean square error (RMSE). Finally, we present comparison the spatial distribution of land surface temperature by both of algorithm, and more accurate algorithm is split window algorithm refers to the root mean square error (RMSE) is 7.69° C.

1. Introduction

The land surface temperature is an essential parameter in the physical process and the interaction between the land and the atmosphere [1]. Land surface temperature data can provide information on spatial and temporal variations of the Earth's surface in various applications on a global scale [2]. It has been widely used in many studies such as estimation of energy inventories, moisture monitoring and evapotranspiration [3-14], climate change, urban heat island [15-27], and environmental studies requiring land surface temperature as the basic input.

Increasing the number of land surface temperature used in various fields of study has implications in developing of remote sensing techcnique to retrieving the land surface temperature value [28]. In recent decades, satellite sensors such as the Moderate-resolution Imaging Spectroradiometer (MODIS) [29] and the National Oceanic and Atmospheric Administration - Advanced Very High-Resolution Radiometer (NOAA-AVHRR) provide twice thermal data in a day using two Long Wave Infra Red (LWIR). Landsat 5 Thematic Mapper (TM) [30] and Landsat 7 Enhanced Thematic Mapper Plus (ETM +) have 16-day repeat cycle and provide thermal data with higher image rsolution using infrared band. Then Landsat 8 Thermal Infrared Sensor (TIRS) is the latest generation of landsat series that has two infrared bands with the same spatial and temporal resolution as the previous Landsat [31]. Since remote sensing satellites can provide thermal data with short time intervals, it makes remote sensing an important tool in monitoring land surface temperature [32].

Along with development of remote sensing techniques for retrieving land surface temperature, algorithm, which is used to obtain land surface temperature values, is also developed. Radiative transfer equation (RTE) was the first algorithm used [33]. The next mono-window algorithm was developed by Qin et al (2001) [34] and single channel by Jiminez-Munoz and Sobrino (2003) [35]. Furthermore, split windows [36], and multi-angle algorithms [37] are the advanced methods used. The method which is mostly used by many researchers is the NASA methodology contained in the Landsat science data users handbook in each series of landsat [38-40]. In this book, the value of land surface temperature starts from the conversion of the digital number (DN) to spectral radian and then obtained the brightness temperature (K) [41]. The advantage of NASA methodology is easy and quick to implement but, it can not describe actual field conditions because, it only refers to the theory of blackbody [42].

Meanwhile, to obtain the land surface temperature values from satellite data well there are three main variables that must be considered and corrected: atmosphere, angular and emissivity [43]. Therefore, the author used the mono window and split window algorithm in the derivation of land surface temperature in South Tangerang City because, they used two of the three variables (atmosphere and emissivity). The purpose of this paper is to perform the spatial distribution of land surface tempertaure using two algorithms, as well as to determine more accurate algorithm for retrieving land surface temperature value.

2. Material and Method

2.1. Study Area

South Tangerang (106°38’00”E/06°13’30”S to 106°47’00”E/06°22’30”S) is used as a study area, it is located in the eastern of Banten Province with an area of 147.19 km² or 1.63% of the total area of Banten Province. While administratively, South Tangerang consists of seven districts.

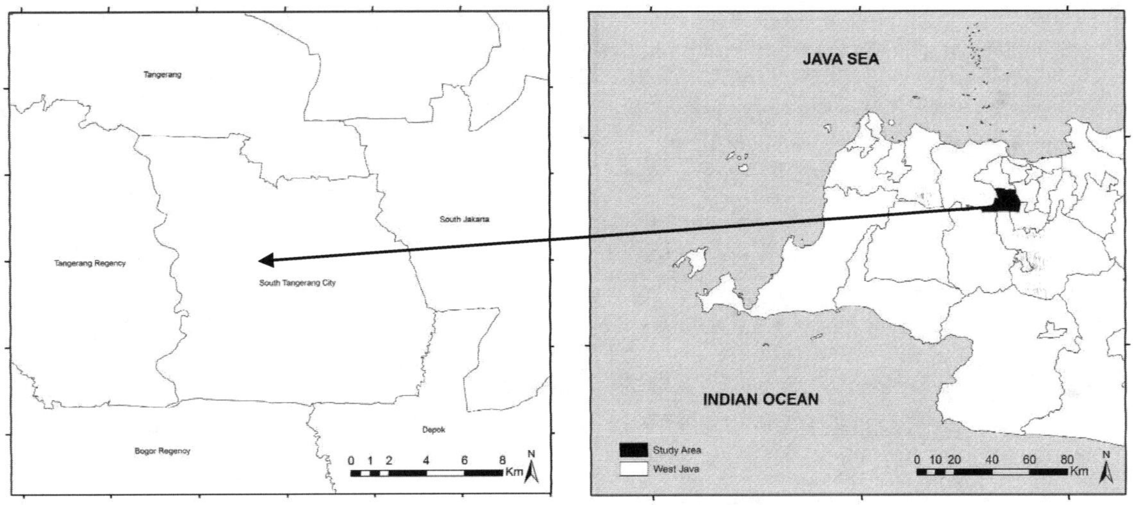

Figure 1. Map of South Tangerang City in Banten Province, Indonesia.

2.2. Data

The land surface temperature data obtained from Landsat 8 OLI-TIRS path 122 row 64, recording date March 29 2017 at 09.20 for South Tangerang Region, and land surface temperature data based on field measurement used infrared thermomether. The training sample taken as many as 30 points and is used for calculating root meat square error as fitting parameter.

2.3. *Mono Window Algorithm*

Mono-window algortihm developed by Qin et al. (2015) [44], the formulation of LST can be expressed as follows.

$$T_s = [a_{10}(1 - C_{10} - D_{10}) + (b_{10}(1 - C_{10} - D_{10}) + C_{10} + D_{10})T_{10} - D_{10}T_a]/C_{10} \quad (1)$$

$$C_{10} = \tau_{10}\varepsilon_{10} \quad (2)$$

$$D_{10} = (1 - \tau_{10})[1 + (1 - \varepsilon_{10})\tau_{10}] \quad (3)$$

Where a_{10} = -62.7182 and b_{10}= 0.4339 are model constants, when the temperature varies from 0 to 50 ° C; ε_{10} is the emissivity for Landsat 8 OLI-TIRS, it is obtained from the NDVI (Sobrino, Jimenez-Munoz, & Paolini, 2004); τ_{10} is the atmospheric transmittance for Landsat 8 OLI-TIRS; Is the effective mean atmospheric temperature for tropical model, where near air surface temperature in which has been taken from meteorological stations for the acquisition date of Landsat; $T_{10}$and is the brightness temperature of TIRS band 10.

2.4. *Split Window Algorithm*

Split-window algorithm was developed by Rozenstein et al. (2014) [45]. This algorithm uses atmospheric transmittanceand emissivity as inputs. The land surface temperature was estimated using the following equation.

$$T_s = A_0 + A_1 * T_{10} - A_2 * T_{11} \quad (4)$$

Where T_s is the value of LST in celcius degree (°C), T_{10} and T_{11} are the brightness temeperatures obtained of band 10 and 11, respectively, and A_0, A_1, and A_2 are coefficients determined by the emmissivity and atmospheric transmittance of both TIRS bands :

$$C_{10} = \varepsilon\,\tau_{10} \quad (5)$$

$$C_{11} = \varepsilon\,\tau_{11} \quad (6)$$

$$D_{10} = (1 - \tau_{10})\big(1 + (1 - \varepsilon\,\tau_{10})\big) \quad (7)$$

$$D_{11} = (1 - \tau_{11})\big(1 + (1 - \varepsilon\,\tau_{11})\big) \quad (8)$$

$$E_0 = D_{11}C_{10} - D_{10}C_{11} \quad (9)$$

$$E_1 = D_{11}(1 - C_{10} - D_{10})/E_0 \quad (10)$$

$$E_2 = D_{10}(1 - C_{11} - D_{11})/E_0 \quad (11)$$

$$A = D_{10}/E_0 \quad (12)$$

$$A_0 = E_1 * a_{10}/E_2 * a_{11} \quad (13)$$

$$A_1 = 1 + A + E_1 * b_{10} \quad (14)$$

$$A_2 = A + E_2 * b_{11} \quad (15)$$

2.5. *Root Mean Square Error*

Root Mean Square Error (RSME) was considered as fitting parameters between predicted and observation data. Where $Y_{observed}$ shows the land surface temperature measured in situ and $Y_{predicted}$ shows the of land surface temperature from Landsat image processing.

$$\mathrm{RMSE} = \sqrt{\frac{1}{n}\sum_{i=1}^{n}(y_{observed} - y_{predicted})^2} \quad (16)$$

3. Results and Discussion

The result shows, both of the land surface temperatures distribution have got similar pattern. Figure 2 shows the LST by mono window algorithm value has got range 19,0 - 31,0 °C, and Figure 3 LST value by split window algorithm has got range 29,9 - 51,8 °C. Higher land surface temperature value for both of algorithms are dominated in southern part of South Tangerang.

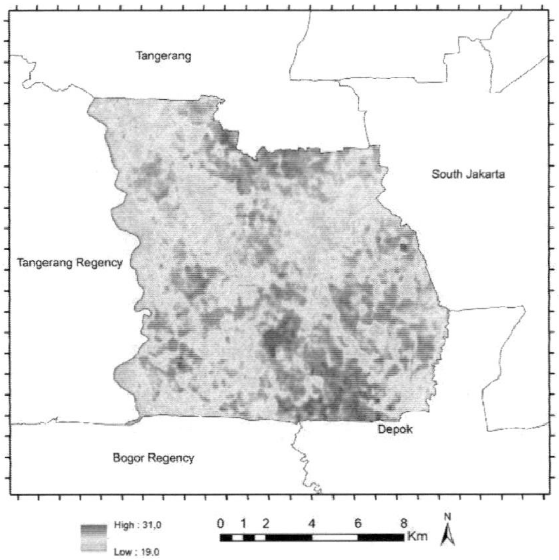

Figure 2. Land surface temperature by mono window algorithm.

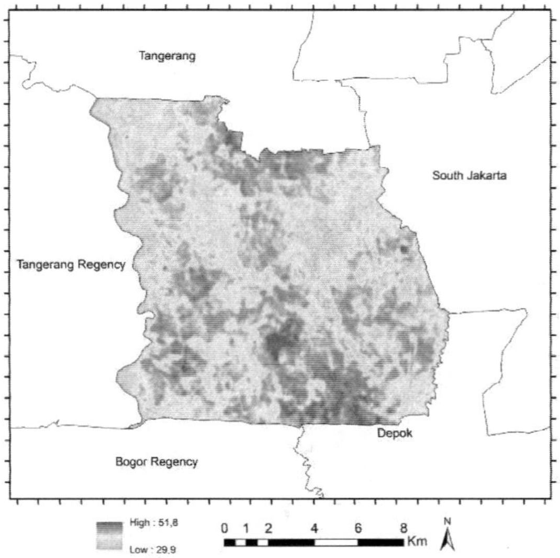

Figure 3. Land surface temperature by split window algorithm.

The way to find a more accurate method of land surface temperature value using RMSE (Root Mean Squared Error) as fitiing parameter. Based on RMSE calculation of land surface temperature with mono window algorithm is 11.42 °C and for Split Window is 7.69°C. This result shows that Split Window algorithm is more accurate than mono window algorithm because it has got smaller error value.

Table 1. Root mean square error for both algorithms.

LST Algorithm	RMSE °C
Mono Window Algorithm	11.42
Split Window Algorithm	7.69

4. Conclusion

The mono window algorithm and split window algorithm were applied to map the land surface temperature distribution in South Tangerang . Based on the maps of land surface temperature distribution for both algorithms, it is clear that that they have got similar pattern. Higher land surface temperature values dominated in the southern part of South Tangerang, and best result is obtained with split window algorithm with an RMSE of 7.69°C.

Aknowledgments

This study was conducted in Department Geography, Faculty of Mathematics and Natural Science Universitas Indonesia. We wish to thank Mr. Rokhmatuloh, and Mr. Adi Wibowo who have supported in this study.

References

[1] Wang S, Ding H, Liang H, and Ma Q 2016.*Resources Conservation and Recycling* **128** 526-534.
[2] Kerr Y H, Lagourade J P, and Imbernon J 1992 *Remote Sensing of Environment* **33** 197-209.
[3] Allen R G, Tasumi M, and Trezza R 2007 *Irrigation and Drainage Engineering* **133** 380-394.
[4] Zhao W, Hu Z, Li S, Guo Q, Liu Z, Zhang L, 2017 *Science of The Total Environment* **599-600** 891-898.
[5] Elmes A, Rogan J, Williams C, Ratick A, Nowak D, Martin D, 2017 *ISPRS Journal of Photogrammetry and Remote Sensing* **128** 338-353.
[6] Li G, Zhang F, Jing Y, Liu Y, Sun G, 2017 *Science of The Total Environment* **596-597** 256-265.
[7] Qiu G Y, Zou Z, Li Z, Li H, Guo Q, Yan C, Tan S, 2017 *Habitat International* **68** 30-42.
[8] Mathew A, Khandelwal S, Kaul N, 2018 *Energy and Buildings* **159** 271-295.
[9] Ziaul S, Pal S, 2018 *Urban Climate* **24** 34-50.
[10] Poon P K, Kinoshita A M, 2018 *Journal of Hydrology* **559** 71-83.
[11] Mean F H, Yusop, Yusof F, 2018 *Atmospheric Research* **201** 102-115.
[12] Jepsen S M, Harmn T C, Ficklin D L, Molotoch N P, Guan B, 2018 *Journal of Hydrology* **556** 645-659.
[13] Romaguera M, Vaughan R G, Ettema J, Izquierdo-Verdiguier E, Hecker CA, Van Der Mer F D, 2018 *Remote Sensing of Environment* **204** 534-552.
[14] Urqueta H, Jodar J, Herrera C, Wilke H G, Medina A, Urrutia J, Custodio E, Rodriguez J, 2018 *Science of The Total Environment* **612** 1234-1248.
[15] Bechtel B, Hoshyaripour G, and Zaksek K 2012 *Remote Sensing* **4** 3184-3200.
[16] Hereher M E, 2017 *Journal of African Earth Science* **126** 75-83.
[17] Zhao Z Q, He B J, Li G L, Wang H B, Darko A, 2017 *Energy Building* **155** 282-295.
[18] Zhang X, Estoque R C, Murayama Y, 2017 *Sustainable Cities and Society* **32** 557-568.
[19] Chen Y, Wang X, Jiang B, Wen Z, Yang N, Li L *Landscape and Urban Plnning* **162** 68-79.
[20] Sheng L, Tang X, You H, Gu Q, Hu H, 2017 *Ecological Indicators* **72** 738-746.
[21] Cui Y, Yang D, Hong T, Ma Jingin, 2017 *Energy* **130** 286-297.
[22] Estoque R C, Murayama Y, 2017 ISPRS Journal of Photogrammetry and Remote Sensing **133** 18-29.
[23] Bevilacqua P, mazzeo D, Bruno R, Arcuri N, 2017 *Energy and Buildings* **150** 318-327.
[24] Hwang R L. Linn C Y, Huang K T, 2017 *Energy and Buildings* **152** 804-812.
[25] Kotharkar R, bagade A, 2018 *Landscape and Urban Planning* **169** 92-104.
[26] He B J, 2018 *Urban Climate* **24** 26-33.
[27] Wang S, Ma Q, Ding H, Liang H, 2018 *Resources Conservation and Recycling* **128** 526-534.
[28] Li Z L, Ren H, Sobrino J A, Tang B H, Trigo I F, Wan Z, . . . Yan G 2013 *Remote Sensing environment* **131** 14-37.
[29] Duan S B, Li Z L, Cheng J, Leng P, 2017 ISPRS Journal of Photogrammetry and Remote Sensing **126** 1-10.
[30] Chatterjee A S, Singh N, Thapa S, Sharma D, Kumar D, 2017 International Journal of Applied Earth Observation and Geoinformation **58** 264-277.
[31] Sekertekin A, Kotuglu S H, and Kaya S 2016 *Environment Monitoring Assess* **188** 30.
[32] Zhou J, Zhang X, Zhang W, and Zhang H 2014 *Remote Sensing* **6** 5344-5367.
[33] Berk A, Bernstein L S, and Robertson D C 1989 MODTRAN : A Moderate Resolution Model for LOWTRAN 7. Tech. Rep. GL-TR-89-0122 Air Force Geophysics Laboratory Hanscom Air Force Base Bedford MA.
[34] Qin Z, Berliner P, and Karnieli A 2001 *Remote Sensing* **22** 3719-3746.
[35] Jiminez-Munoz J C, and Sobrino J A 2003 *Journal of Geophysical Research* **108** 4688.
[36] Sobrino J A, Jiminez-Munoz J C, and Paolini L 2003 *Remote Sensing of Environtment* **90** 434-440.
[37] Dash P, Gotsche F M, Olesen F S, and Fisher H 2002 *International Journal of Remote Sensing*

23, 2563-2594.
[38] Chen X, Li P, Yin Z, and Zhao H 2005*Remote Sensing of Environment* **104** 133-146.
[39] Singh K B, Grover A, and Zhan J 2014 *Energies* **7** 1811-1828.
[40] Wibowo A, and Rustanto A 2013 *Indonesian Journal of Geography* **45** 101-11.5
[41] Ardiansyah. (2015). Pengolahan Citra Penginderaan Jauh Menggunakan ENVI 5.1 dan ENVI LIDAR (Teori dan Praktik). Jakarta Selatan: PT. LABSIG INDERAJA ISLIM.
[42] Meneses S F 2015 Thermal Remote Sensing at Lyte Geothermal Production Field Using Mono Window Algorithm. *Proceeding World Geothermal Congress.* Melbourne.
[43] Franca G B, and Cracknell A P 1994 *International Journal of Remote Sensing* **15** 1695-1712.
[44] Qin Z, Wang F,Karnieli A, Song C, Tu L, and Zhao S 2015 *Remote Sensing* **7** 4269-4289.
[45] Roznstein O, Derimian Y, Karnieli A, and Qin Z 2014 *Sensors* **14** 5768- 5780.

IOP Conf. Series: Earth and Environmental Science **149** (2018) 012067 doi:10.1088/1755-1315/149/1/012067

Experimental test for receiving X-Band data LAPAN-A3 Satellite with 5.4m antenna diameter

Sonny Dwi Harsono* and Wahyudi Hasbi

Satellite Technology Center, Indonesian National Institute of Aeronautics and Space (LAPAN), Jl Cagak Satelit Km.04 Rancabungur, Bogor 16310

E-mail: sonny.harsono@lapan.go.id

Abstract. LAPAN-A3 / LAPAN-IPB Satellite launched on June 22, 2016 (03:56 UTC) as an experimental micro-satellite for remote sensing and monitoring of maritime traffic. The Satellite was launched as a secondary payload on ISRO Cartosat-2C as its main payload, the launch carried out at SDSC (Satish Dhawan Space Centre) in India using PSLV-C34 rocket launcher. The Satellite was in orbit polar sun-synchronous with a height of 505 km above sea level. It has an inclination angle of 97 degrees and heavy satellite 115 kg, with this orbit, the satellite will pass through Ground station 4 times (2 times during the day and two times at night) with a duration of the track at the time of the pass about 10-15 minutes. The Satellite payload carried 4 bands Line Scan Cameras and Digital Imager (SpaceCam). For main mission is the earth observation for food vegetables And as additional mission is carrying AIS (Automatic Identification System) receiver to monitor maritime traffic in the region of the poles, then Star Sensor made by LAPAN for qualifying room, then for scientific contained magnetometer sensor for monitoring the Earth's Magnetic field. The purpose of this scientific paper is to test the reception of data payloads of the LAPAN-A3 satellite on X-Band frequency of 8.2 GHz using a 5.4 M solid antenna Ground Stations LAPAN in Pare-Pare. The purpose of this experiment will tell us with 5.4 meter of diameter solid antenna is capable or not enough for HDRM receiver to lock a signal and produce the data output, and how this result if compare with 11 meter of diameter antenna in Splitzberg Groundstation in Norway.

1. Introduction

The first phase of LAPAN micro satellite program was started by developing LAPAN-TUBSAT. This satellite is still operational in orbit since January 2007. Gaining experience in developing, launching, and operating micro-satellite as well as in developing and operating satellite's ground station, LAPAN continue in developing its second satellite, named LAPAN-A2 also known as LAPAN-ORARI. The mission for LAPAN-A2/ORARI is Earth observation using RGB digital camera, maritime traffic monitoring, and amateur radio communication in near equatorial orbit. And these day LAPAN have LAPAN-A3/IPB satellite which become the first Indonesian Remote Sensing Experimental Satellite. In figure 1, it shows the orbital characteristic of LAPAN-A2/ORARI and LAPAN-A3/IPB satellite [1]. LAPAN-A3 / LAPAN-IPB Satellite also known as LISAT (LAPAN IPB Satellite), is a program of cooperation in terms of microsatellite remote sensing between Space agency (National Institute of Aeronautics and Space of Indonesia) in Jakarta and Bogor Agriculture University (Institut Pertanian Bogor) located in Bogor, Indonesia. The purpose of this mission is to monitor the food sources in Indonesia and to provide environmental monitoring. The real purpose of the satellite LAPAN-A3/ LAPAN-IPB is to provide actual data, frequent and accurate to observe and predict the condition of the Indonesian archipelago. Republic of Indonesia with an estimated population of 258 million is the largest archipelago (island countries) in the world with more than 17,500 islands. Two third of country territory is covered by the sea and 1/3 of the country most of the land is covered by forests and agriculture. In this cooperation project, LAPAN responsible for the design and development of microsatellites, while IPB is responsible for algorithm development and application datasets. This satellite one of main program of PUSTEKSAT (Pusat Teknologi Satelit) programs from National Institute of Aeronautics and Space (LAPAN) Indonesia, this satellite designed for specific experimental mission such as remote sensing, marine and fisheries research, land cover and ship monitoring using Automatic Identification System (AIS), that were needed to observe of Indonesian territory. In order to simulate LAPAN-A3/IPB satellite for line imager data acquisition is required high data rate modulator and demodulator (HDRM) [2]. This satellite will have the main mission as an imager satellite with 4 bands multi spectral imager using optical line scan camera to monitor the land cover, cultivated area and sea, and also ship monitoring for additional mission. Because of this satellite have a polar orbit path in 24 hour is only 4 times passed across the groundstation with 2 in the morning and 2 in the night. LAPAN-A3 Satellite used 8.2GHz X-Band frequency for main payload whereas early LAPAN Satellites use S-Band frequency, one facilities for X-Band that LAPAN have is 5.4m antenna in Pare-Pare which as usually used for receive Modis data.

This research will show an experimental result for receiving X-Band data from LAPAN-A3 Satellite by using 5.4 meter diameter antenna and also compared to 11 meter diameter antenna that has used so far in Splitzberg Norway.

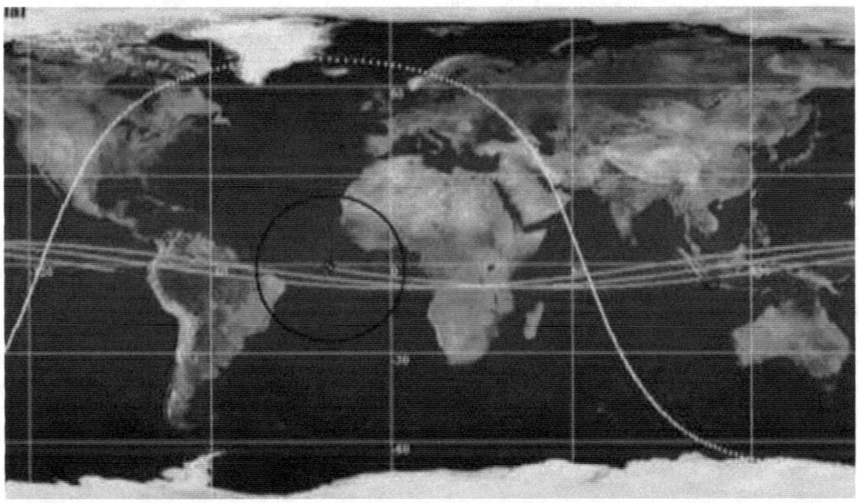

Figure 1. LAPAN-A2 and LAPAN-A3 Orbit (A2 Blue, A3 Yellow)[1].

2. Methodology

Figure 2 describes how to receive a data from LAPAN-A3 Satellite, from X-Band antenna with diameter of 5.4 meter and LHCP polarization Feed go through Low Noise Amplifier (LNA). This frequency is converted from 8.2 GHz to 720 MHz using X-Band Down Converter. This is due to the Demodulator HDRM is only support 720 MHz or 1200 MHz (Mix) as its input, and the data will be stored in RAW Data model in a Hard Drive HDRM itself.

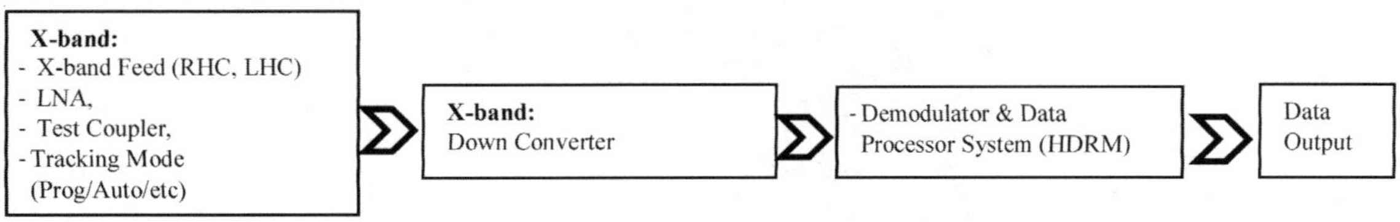

Figure 2. Flowchart System Test Receive X-Band LAPAN-A3 Satellite [10].

Modulation is a technique to carry digital data over analog waveform core of signal modulation is to put the message and place it on a carrier signal for transmission. While Demodulation is the reverse process, i.e. to get the message signal by splitting the carrier signals. To facilitate the operation of the transmission signal frequency used is generally a high frequency for reasons easy to be separated. LAPAN-A3 / LAPAN-IPB Satellite is very similar to a satellite in particular LDCM satellite LANDSAT-8, so it is identical to the modulation scheme. The modulation scheme of Landsat Satellite can be seen in figure 3.

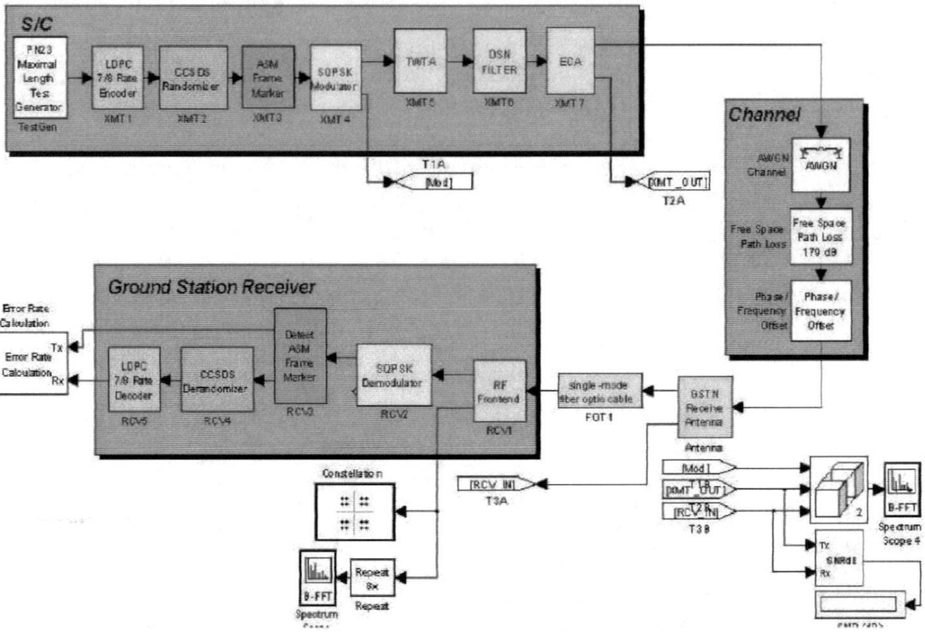

Figure 3. System flow data reception LDCM [2].

High Data Rate Modem is a multi-receiver satellite data, data processing, and system simulation system that supports CCSDS (Consultative Committee for Space Data Systems) with CDL standards, DVB - S2, and the spectrum of a direct connection. This HDRM demod perform data reception, entering, processing, distribution, and archiving functions at speeds up to 1.7 Gbps. It is most commonly used in ground stations, but is also ideal for applications including integration and test a satellite or aircraft data acquisition.

HDRM demod (figure 4) is implemented as a digital, software radio. can be configured to accommodate multiple modulation schemes, data rates, coding algorithms, and data formats. It provides demodulation filters that can be tuned to match the characteristics of the satellite transmitters, and can improve communication performance. Linear wideband system allows adaptive filtering of the received signal based on the symbol rate and Doppler. These characteristics also ensures compatibility with complex modulation and coding which is suitable to be developed for future missions.

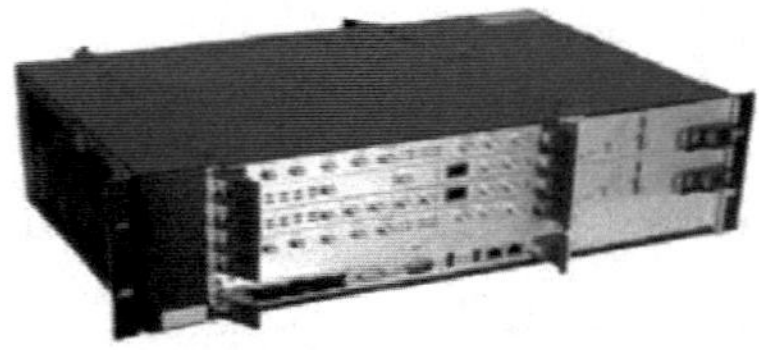

Figure 4. Demodulator HDRM LAPAN-A3/LAPAN-IPB [2].

HDRM used is highly modular system that can be configured. HDRM It has a high level of full modulation and demodulation, the ability to use multiple commands modulation and Forward Error Correction algorithms to support a diverse set of waveforms. There are lots of applications and scenarios that allow configuration to meet the requirements of a variety of different missions. Different applications can be achieved through configuration software and a wide selection of modular hardware. Figure 5 illustrates the modular architecture.

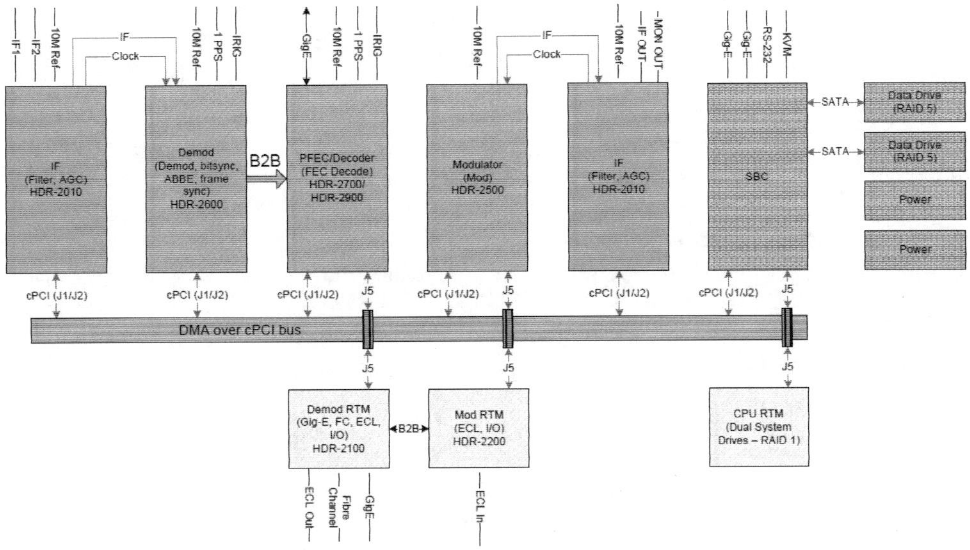

Figure 5. Architecture scheme HDRM [8].

The purpose of this test was to try receiving X-Band signal and data from LAPAN-A3 / LAPAN-IPB Satellite, the data received in real time from satellite to 5.4-m antenna diameter in Pare-Pare Ground station.

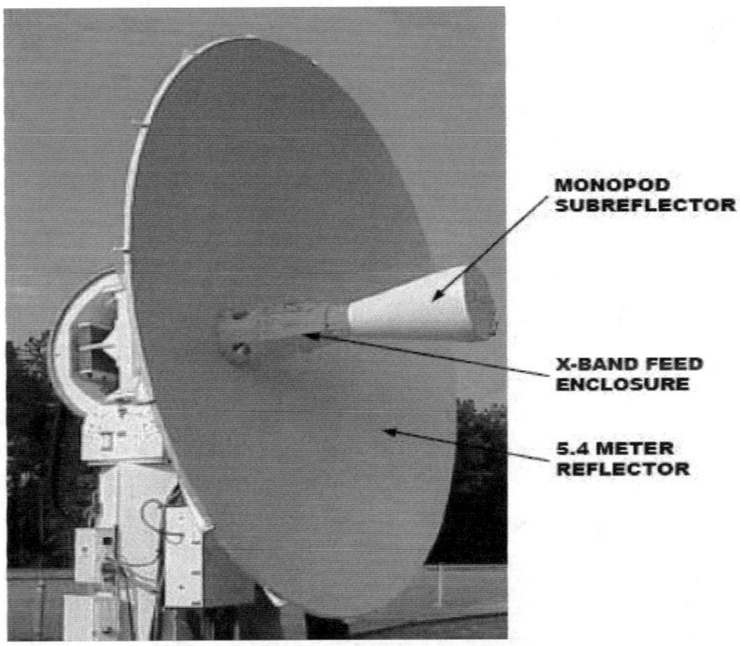

Figure 6. Antenna Subsystem ViaSat 5.4-m [3].

The antenna consists of a 5.4-meter dual shaped reflector and Cassegrain feed. Both the high efficiency X-band feed and the sub reflector are enclosed in one monopod type structure which maintains the accuracy of the alignment. This monopod structure is heated and pressured to prevent ice formation and any negative performance effect due to weather (figure 6).

This system features a high performance autotracking X-band feed. The Y- over X-axis pedestal configuration is mounted on a rigid base extension suitable for ground or rooftop installation. The dual shaped optics use a monopod feed/subreflector to optimize efficiency. The result is superior G/T performance.

In telecommunication, free-space path loss (FSPL) is the loss in signal strength of an electromagnetic wave that would result from a line-of-sight path through free space (usually air), with no obstacles nearby to cause reflection or diffraction. It is defined in "Standard Definitions of Terms for Antennas", as "The loss between two isotropic radiators in free space, expressed as a power ratio". Usually it is expressed in dB, The FSPL is rarely used standalone, but rather as a part of the Friis transmission equation, which includes the gain of antennas.

Free-space path loss is proportional to the square of the distance between the transmitter and receiver, and proportional to the square of the frequency of the radio signal.

The equation for FSPL is

FSPL = (4πd/λ) 〚^2〛 (1)

= (4πdf/c) 〚^2〛

where:

λ = is the signal wavelength (in meters),
f = is the signal frequency (in hertz),
d = is the distance from the transmitter (in meters),
c = is the speed of light in a vacuum, 2.99792458 × 108 meters per second.

This equation is only accurate in the far field where spherical spreading can be assumed; it does not hold close to the transmitter.

A convenient way to express FSPL is in terms of dB

$$\begin{aligned}\mathbf{FSPL(dB)} &= 10\log_{10}\left(\left(\frac{4\pi df}{c}\right)^2\right) \\ &= 20\log_{10}\left(\frac{4\pi df}{c}\right) \\ &= 20\log_{10}(d) + 20\log_{10}(f) + 20\log_{10}\left(\frac{4\pi}{c}\right) \\ &= 20\log_{10}(d) + 20\log_{10}(f) - 147.55\end{aligned} \quad (2)$$

The size of antenna is determines by link budget calculations, the estimation calculation is used for antenna development, link availability, power requirements, bit error rate, as well as the overall customer satisfaction with the satellite service. This link budget is a method for assessing the power received and the noise ratio in link radio wave. The calculations must be calculated for an individual transponder, and must be recalculated for each of the individual links. Link budgets are usually calculated for a worst-case scenario, the one in which the link will have the lowest C/N ratio or lowest tolerable availability [4]. Image of LAPAN-A3/LAPAN-IPB Satellite, its specification, its satellite link, satellite link parameter, and system scheme module grooves on Demod HDRM can be seen in figure 7, table 1, figure 8, table 2, and figure 9 respectively.

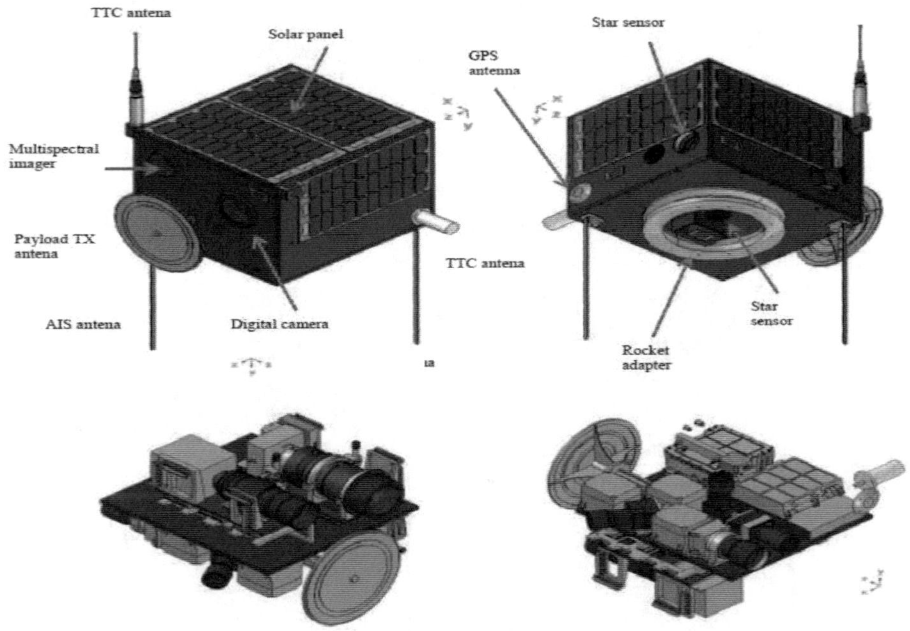

Figure 7. LAPAN-A3/LAPAN-IPB Satellite [5].

Table 1. LAPAN-A3 Satellite specification.

Mission	Payload	Spectral Resolution	Spacial Resolution	Orbit	Communication	Total Weight
Design, Integration, Test and Operation is down in Indonesia	Push broom 4 band multispectral imager Experiment with 18m resolution and 100 Km swath.	0.45 - 0.52 Blue	18 m resolution and 100 Km swath	Polar sun synchronousat 97° inclination	Payload : Tx X-Band (8200 MHz, BW 168 MHz)	80 - 100 Kg
Earth observation with 4 band multispectral imager for land use classification and environment observations	Digital space camera with 4 m resolution and 8 Km swath.	0.52 - 0.60 Green	4 m resolution and 8 Km swath		TTC : UHF	
Supporting global maritime awareness by the reception of AIS signal from ship	Automatic Indentification System (AIS) receiver ex. AIS Sat (Norwegian Satellite)	0.63 - 0.69 Red 0.76 - .090 NIR				

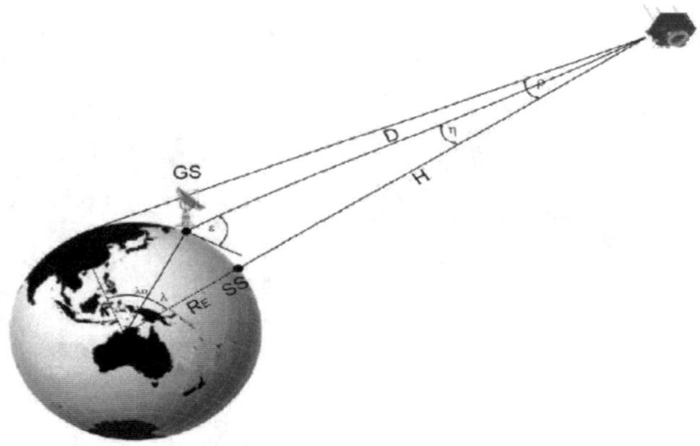

Figure 8. LAPAN-A3/LAPAN-IPB Satellite Link [6].

Table 2. LAPAN-A3 Satellite link parameters

No.	Link Parameters	Value	Unit
1	Altitude	505	Km
2	Slant Range (5 deg elev)	2448	Km
3	Transmission Power TX (5 Watt)	7	dBW
4	Frequency Operation	8200	MHz
5	Waveguide / cable loss (satellite)	1	dB
6	Antenna gain (satellite)	5	dBi
7	EIRP (satellite)	11	dBW
8	Free space loss	178.5	dB
9	Atmosphere Absorption Loss	0.1	dB
10	Antenna Gain (Groundstation)	52.1	dBi
11	Received waveguide / Cable Loss (Groundstation)	1	dB
12	Received Carrier Power (Groundstation)	-35.4	dBm
13	G/T (Groundstation)	29.5	dB/K
14	Boltzmann's constant	-228.6	dBW/Hz/K
15	Data Bandwidth (168 MHz)	82.3	dB Hz
16	Data Rate (105 Mbps)	81.8	dB
17	Carrier to Noise ratio (C/N)	8.3	dB
18	Eb/No	10.3	dB

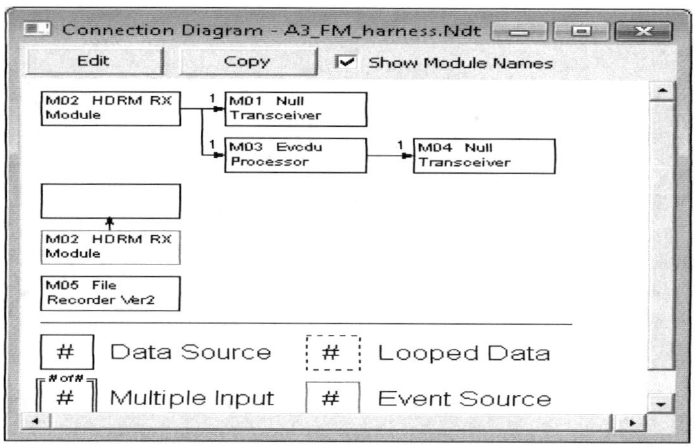

Figure 9. LAPAN-A3 system scheme module grooves on Demod HDRM [9].

IOP Conf. Series: Earth and Environmental Science **149** (2018) 012067 doi:10.1088/1755-1315/149/1/012067

3. Results and analysis

For the configuration at 5.4m antenna the first things should do is to add LAPAN-A3 Satellite schedule for tracking activity, because the satellite in polar orbit which is moving from the North Pole to South Pole of the Earth depending the day-night time when the satellite is passing above Groundstation. For configuration editor program for input of RF setting for LAPAN-A3 Satellite which as center frequency is 8200 MHz (8.2GHz) and polarization is Left Hand Circular Polarization (LHCP).

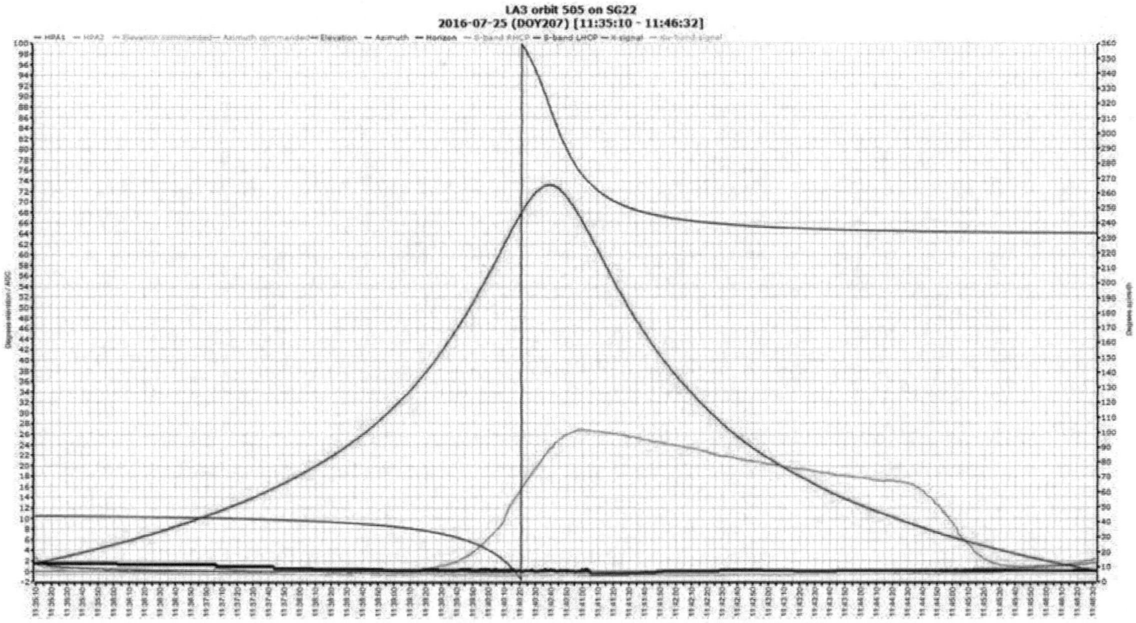

Figure 10. X-Band signal LAPAN-A3 Satellite acceptable received.

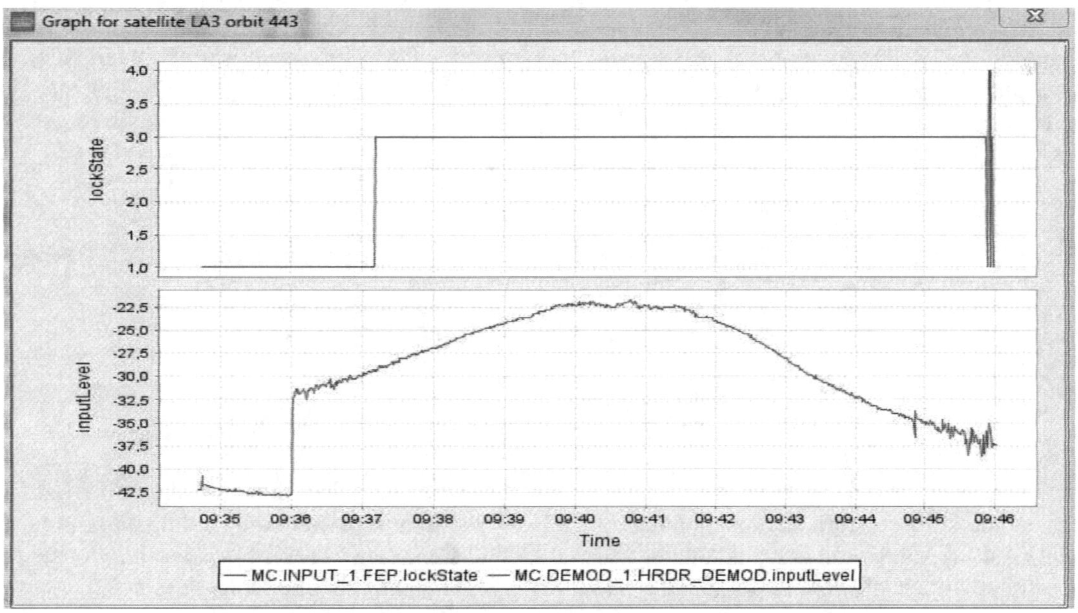

Figure 11. Locked signal X-Band LAPAN-A3 Satellite.

IOP Conf. Series: Earth and Environmental Science **149** (2018) 012067 doi:10.1088/1755-1315/149/1/012067

The result of receiving X-Band Signal of LAPAN-A3 Satellite for acceptable condition is shown in figure 10 and figure 11. Which as in figure 10 is a data for a movement of antenna azimuth and elevation when tracking a satellite and also a data when a signal X-Band is receive in receiver or spectrum analyzer. Figure 11 describes for signal lock on receiver, from the data we can see for locking signal condition is start from around -30 dBm and lost at -35 dBm, this data is get when using a 11 meter of diameter in Splitzberg Groundstation in Norway.

For this experiment we used a 5.4 meter of diameter for receive a signal X-Band from LAPAN-A3 Satellite, for controlling an antenna movement is using a program at Antenna Control Unit. This program can track a satellite position and directed an antenna pointing to that position, this means that a azimuth and elevation of antenna is a same position with LAPAN-A3 Satellite.

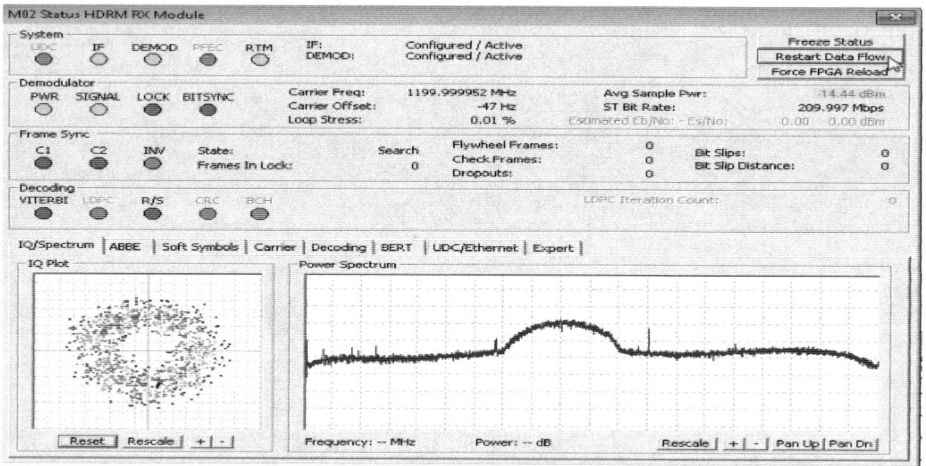

Figure 12. Received signal X-Band LAPAN-A3 Satellite in HDRM.

The result for this experiment with using 5.4-meter diameter is shown on figure 12 and figure 13. Which as in figure 12 is signal receive in HDRM receiver, the status for receiving signal is the HDRM can detect a signal from a satellite (this show on indicator signal in demodulator turn on) this result is not enough for HDRM to demodulate a signal and to decode. The same result is also shown when used a spectrum analyzer for see a signal strength, the signal receive is below from -35 dBm this meaning a signal receive with 5.4 meter diameter is not enough for HDRM receiver to lock a signal.

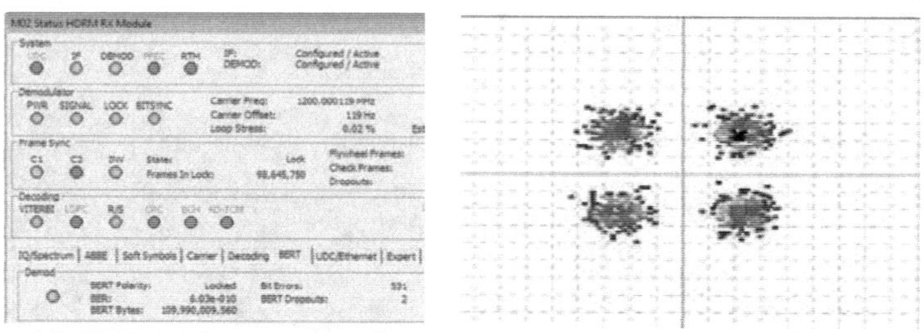

Figure 13. HDRM Status Signal Condition For Demodulate X-Band Signal.

For condition how to HDRM receiver can modulate and to decode a signal is shown in figure 13, When indicator is on (turn green) this mean a HDRM can demodulate and decoding a signal and its shown a QPSK signal condition is how to be. The QPSK signal is in locking condition is located to 4 position in diagram whereas in not locking condition a QPSK signal is in circular shape in diagram.

4. Conclusion

Quadrature-phase shift keying (QPSK) is a form of phase-shift keying in which four different phase angles are used [7]. In QPSK, the four corners are usually separated by 90° spacing [8]. LAPAN-A3 Satellite is using QPSK for modulate a signal, and for receive this a Groundstation used a HDRM receiver. At the end, we compare a receiving signal X-Band of Satellite LAPAN-A3 with 11 meter diameter antenna in Splitzberg Norway and antenna 5.4 meter of diameter in LAPAN Pare-Pare Groundstation. The best result is get from antenna 11 meter, the signal more than enough to reached minimum signal for demodulate data, but for 5.4 meter antenna a signal can be receive but not enough for HDRM receiver to demodulate it, that's why in this experiment with using 5.4 meter antenna the result is cannot be used.

Acknowledgement

The author would like to say Thank you to Mr. Abdul Karim, MT, as Head of Satellite Technology (PUSTEKSAT) LAPAN, on direction so that scientific papers can be resolved properly. My Thanks also go to Mr. Munawar as Head of LAPAN PARE-PARE and all Engineer, and all my friends at Dissemination and Program Division in PUSTEKSAT who have helped and facilitated this research, and R & D.

References

[1] Hasbi W and Suhermanto 2013 Development of LAPAN-A3/IPB satellite an experimental remote sensing microsatellite **34** 1508-1515

[2] Harsono SD and Suhata 2015 Analysis of Transmission System and Receiver Satellite Data LDCM based High Data Rate Modulator *Aeronautics and Space Technology, Research and Engineering of UAV, Rocket, and Satellite* 140–149

[3] Dankberg MD 2017 5.4 Meter Fixed X-Band Tracking System: Operation and Maintenance. Carlsbad (California: ViaSat)

[4] Atayero AA, Luka MK, and Alatishe AA 2011 Satellite Link Design: A Tutorial *Inter. Jour. of Elect. and Comp. Sci.* **11**(4)

[5] Judianto CT and Nasser EN 2015 The Analysis of LAPAN-A3/IPB Satellite Image Data Simulation Using High Data Rate Modem. *Procedia Environ. Sci.* **24** 285–296

[6] Judianto CT and Wurianto E 2014 Analisis Potensi Gangguan Interferensi Microwave Link Terhadap Operasi Satelit Lapan-A3 *Jurnal Teknologi Dirgantara* **12**(1) 59-72

[7] Asif R and Haithem M 2017 10 Gbit/s mode-multiplexed QPSK transmission using MDM-to-MFDM based single coherent receiver for intra- and inter data center networking *Opt. Commun.* **391** 106–110

[8] Branning T 2010 High Data Rate Modem: Operations and Maintenance Manual. Chantilly (Virginia: Avtec Systems, Inc.,)

[9] Branning T 2011 Kratos Company September 2011 Programmable Telemetry Processor : User Guide. Chantilly (Virginia: Avtec Systems, Inc.,)

[10] Branning T 2013 HDR-8SE-64R10: Configuration & Installation Document. Chantilly (Virginia: Avtec Systems, Inc.,)

Pre-Flight Radiometric Model of Linear Imager on LAPAN-IPB Satellite

A Hadi Syafrudin*, Sartika Salaswati, Wahyudi Hasbi

Satellite Technology Center, Indonesian National Institute of Aeronautics and Space (LAPAN), Jl Cagak Satelit Km.04 Rancabungur, Bogor 16310

E-mail: ahaddi@yahoo.com

Abstract. LAPAN-IPB Satellite is Microsatellite class with mission of remote sensing experiment. This satellite carrying Multispectral Line Imager for captured of radiometric reflectance value from earth to space. Radiometric quality of image is important factor to classification object on remote sensing process. Before satellite launch in orbit or pre-flight, Line Imager have been tested by Monochromator and integrating sphere to get spectral and every pixel radiometric response characteristic. Pre-flight test data with variety setting of line imager instrument used to see correlation radiance input and digital number of images output. Output input correlation is described by the radiance conversion model with imager setting and radiometric characteristics. Modelling process from hardware level until normalize radiance formula are presented and discussed in this paper.

1. Introduction

Multispectral Imager is one main payload of LAPAN-IPB Satellite. This Imager Consist of 4 Channel are Blue, Green, Red, and Near Infrared. Multispectral imager on satellite show in figure 1. Imager like standard Camera can produce image with Digital Number (DN) representation of light Intensity of object in some spectral Response. Radiometric model is formula to predict radiance input value on entered in imager surface. This Model will convert each DN to radiance unit. Normally Imager have setting to enhanced of DN images for suitable image looking with high dynamic range. With Limited mechanical and electronic of Lens and CCD sensor cause different response each pixel. Need Radiometric Model to convert DN to Radiance with specific imager setting and Algorithm correction to create DN images which uniform radiance representation. Process Photon energy to DN is depends of Hardware level on CCD Sensor and Analog Digital Converter (ADC).

Figure 1. Multispectral imager on LAPAN-IPB Satellite.

IOP Conf. Series: Earth and Environmental Science **149** (2018) 012068 doi:10.1088/1755-1315/149/1/012068

Satellite images have own calibration and have correlation with calibration in ground or pre-flight. Possible different factor is vacuum and temperature which can't simulation in ground. The pre-flight data can be references for orbit calibration and estimation error caused and study for next calibration project. The accuracy of the pre-launch calibration is estimated to approximately 8 percent [1].

One important for radiometric quality is uniformity each pixel which describes reflectance from earth surface. First to correction is uniformity each pixel from light coming in lens surface or call path radiance before error cause external factor like atmospheric, sun angle, and terrain or variation of topography. Image after correction generally representation of dynamic range of uniform responsivity pixel of object after remove two noise parameter. Radiometric characterization of the sensor (actually whole camera) are linearity, DSNU (Dark Signal Uniformity) and PRNU (Photo Response Non Uniformity) [2].

Each imager have own structure mechanical, electronic, and methods for capture of images. Imager design decision have limitation for device used, budget, material, assembly, integration, and test for this component. The knowledge of models of these errors allows for developing some methods of reducing some of the noise and for removing some artefacts from the images [3]. Radiometric Correction algorithm is representation of hardware flow and imager setting when images captured.

2. Methodology

Hardware level is represent each parameter setting which effect DN value from ADC digitation. Model define by conversion of hardware flow to mathematic representation. Methods for radiometric calibration typically involve up to four steps [4].

1. The camera response is corrected to be linear.
2. The camera's black response is subtracted.
3. Corrections are applied for spatial variations in the effective aperture of the lens and the sensor sensitivity. This correction yields a relative radiometric calibration.
4. Converts the relative calibration to absolute calibration by applying an appropriate scaling factor.

All each parameter have Constanta based on measurement statistic. To define Parameter characteristic use two measurement, first is spectral measurement by Monochromator and second is radiometric measurement by integrating sphere with some variation setting. In the present study a monochromator is used for precise measurement of the spectral sensitivity of the camera [5]. Spectral measurement show in figure 2. Monochromator will be set to 5-10 nm from 400nm – 1100 and parallel record on four Band.

Figure 2. Test with monochromator.

First measurement for Full width at half maximum (FWHM) of four channel of camera. This useful to calculate to define part of spectral power base on FWHM of integrating sphere which already known. Second measurement to get correlation between power input and DN of images with Setting Parameter. These methods included the prelaunch calibration using spherical integrating sources [6]. Test with integrating sphere with output 14 inch describes in figure 3. Nine lamp on integrating sphere can switch on-off and one can smooth tuning with aperture control.

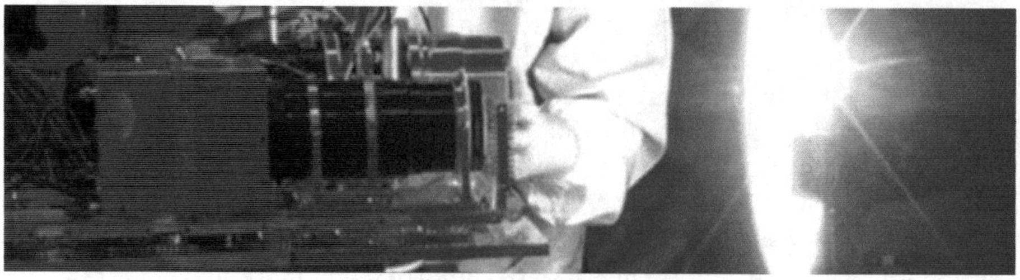

Figure 3. Test with integrating sphere.

3. Result and discussion

3.1. Hardware level

Hardware level of multispectral imager show in figure 4. Photon light energy will be charge photodiode array and store to CCD Cells during exposure time and conversion analogue voltage. This signal will transmit to ADC with clock transfer video signal with correlation double sampling. Signal will be add voltage from offset Digital to Analog Converter (DAC) and multiple by Power Gain Amplifier (PGA) and quantification 16 Bit ADC with range input 0-4 volt. Offset and Gain decided to control signal not saturation or Cut off.

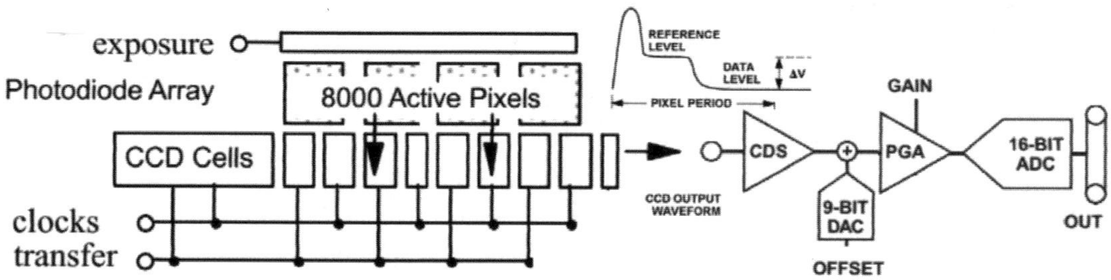

Figure 4. Hardware level diagram of multispectral imager.

On this diagram have four setting parameter will cause increase or decrease of DN value with same input. Another control like repetition line rate is not have directly effect to converting from DN to radiance. Repetition rate cause percentage of overlap each line to make image size geometry between along track in two dimensional XY. Setting value will adjust after images operate in orbit, it possible have a lot combination to get better images. For this reason all value will convert to output CCD voltage before convert to DN which uniform radiance representative. This methods to remove of signal conditioning setting.

3.2. Spectral and radiometric

On figure 5 FWHM spectral define and imager input from integrating sphere calculated beside on this FWHM. Output intensity integrating sphere not same in Multispectral channel because lamp halogen spectral characteristic. FWHM and bandwidth value describes in table 1 column number 2 and 3. Every pixel has illuminated 99.2% uniform in surface lens imager with fix power depend of lamp sphere on.

In figure 6 describes voltage to radiance. Constanta DN to radiance conversion get from DN of images after subtract by dark image and input power sphere comparison. Average voltage to radiance value conversion describes in table 1 in last column. Constanta DN to Radiance is correlation with responsivity pixel in line imager. In edge DN have more radiance representation than centre pixel because vignette effect. It is typically caused by sensor components (e.g. barrel) occluding light from the detector plane at wide angles [7].

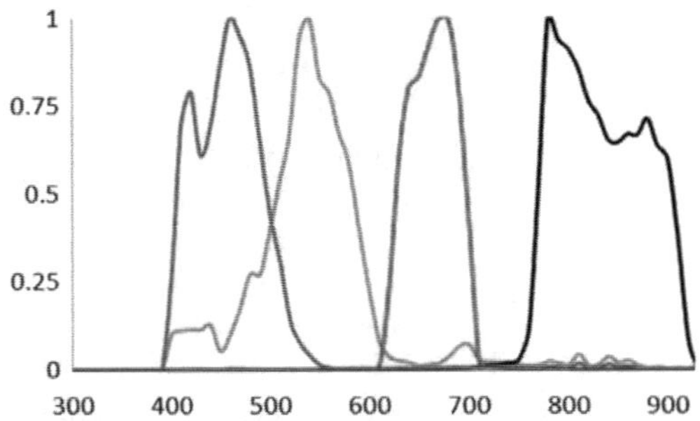

Figure 5. Normalize spectral value multispectral imager.

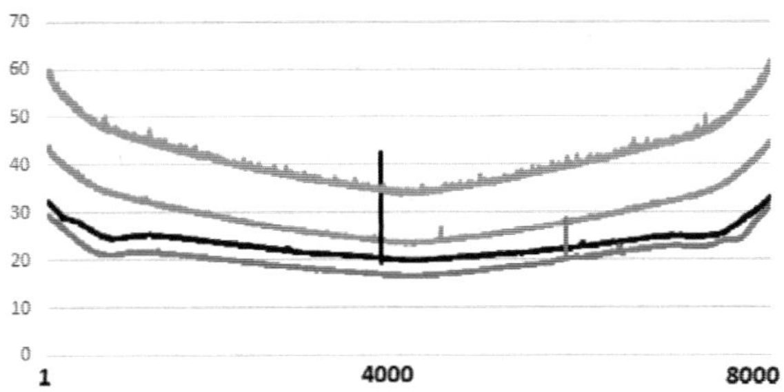

Figure 6. Voltage to radiance value Detector Index each band.

Table 1. FWHM and average radiance voltage to radiance Constanta.

Band	FWHM	Bandwidth	Radiance (mW/cm2-sr-um)
Blue	0.410 - 0.490	0.080	41.76
Green	0.510 - 0.580	0.070	29.69
Red	0.630 - 0.700	0.070	20.45
NIR	0.770 - 0.900	0.130	23.43

3.3. Setting characteristic

Model created use voltage CCD output prediction this means precise value between nominal setting and real value is important. Exposure can set with precision 10 Nano Second. Gain have 64 setting with decibel gain, and offset will increase plus or minus 1.18 millivolt. Dark current have increase versus long time of imager ON. The setting characteristic describes in figure 7. On this result generally is same with datasheet but have small response different in high setting value.

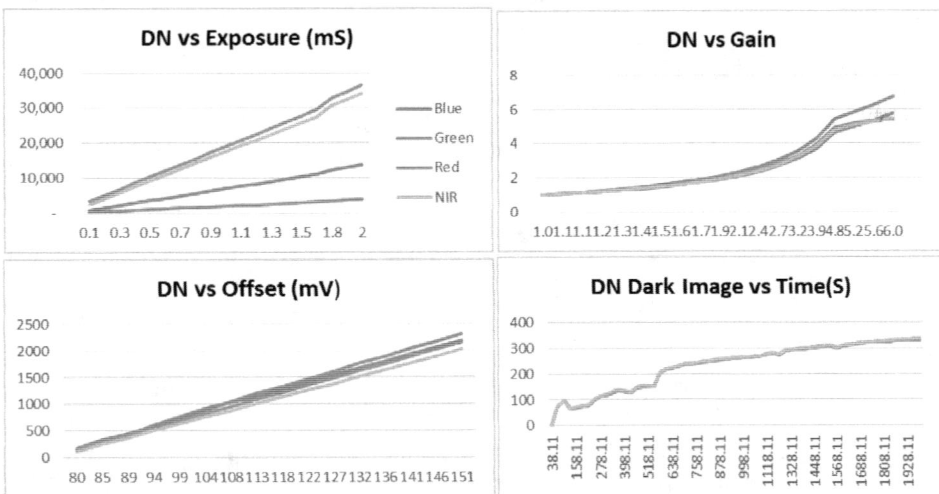

Figure 7. Setting Characteristic of average DN each channel with variety exposure, variety gain, and variety offset. Also response dark current vs time.

Table 2 describes 10 measurement procedure with variety setting and two combination value of input integrating sphere. Input each band in one procedure is different because spectral output have unique every wavelength. Value PRNU is high every procedure measurement and consistent around less than

30%. Nominal response dark DN and flat test describes in figure 8 and 9. Dark test use close lens with dark object and flat test use integrating sphere.

Table 2. Result of PRNU and DSNU calculation before correction. Measurement with input different radiance (mW/cm2-sr-μm) from integrating sphere and different setting from multispectral imager.

	1	2	3	4	5	6	7	8	9	10
Eksp (mS)	1.0	1.0	1.0	1.0	1.5	1.5	1.5	1.0	1.0	1.5
Gain	1.6	1.6	2.9	2.9	1.6	2.9	2.9	1.6	1.6	1.6
Offset(mV)	75.2	150.5	75.2	150.5	75.2	75.2	150.5	75.2	150.5	75.2
B Inp (rad)	2.8	2.8	2.8	2.8	2.8	2.8	2.8	5.6	5.6	5.6
B DSNU (%)	1.6	0.8	1.6	0.8	1.9	1.9	1.0	1.9	0.9	2.3
B PRNU (%)	27.7	27.5	28.1	27.8	27.7	27.2	27.7	27.4	27.3	27.3
G Inp (rad)	7.3	7.3	7.3	7.3	7.3	7.3	7.3	14.5	14.5	14.5
G DSNU (%)	0.6	0.4	0.7	0.5	0.8	0.9	0.6	0.8	0.6	0.9
G PRNU (%)	29.6	29.6	30.0	29.8	29.7	29.5	29.6	29.6	29.5	29.3
R Inp(rad)	12.8	12.8	12.8	12.8	12.8	12.8	12.8	25.2	25.2	25.2
R DSNU (%)	2.9	1.5	3.1	1.5	3.2	3.3	1.6	3.3	1.6	3.5
R PRNU %	29.2	29.3	29.3	29.2	29.2	29.2	29.2	29.2	29.2	28.9
N Inp (rad)	16.0	16.0	16.0	16.0	16.0	16.0	16.0	31.6	31.6	31.6
N DSNU (%)	1.5	0.6	1.4	0.6	1.8	1.5	0.6	1.68	0.6	1.9
N PRNU (%)	26.8	26.8	26.9	26.8	26.6	26.6	26.6	26.94	26.9	25.5

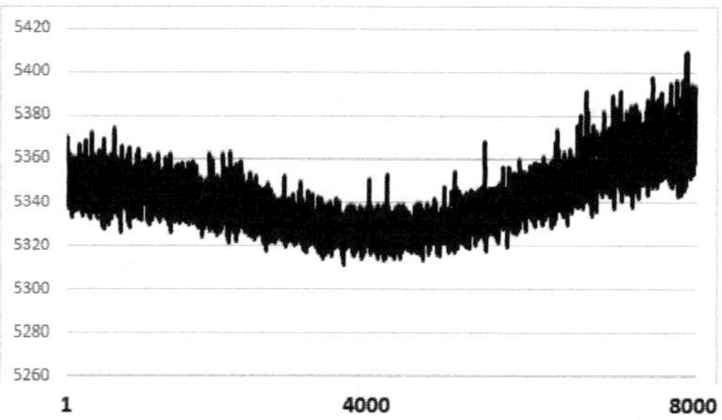

Figure 8. Uncorrected green DN vs detector index in dark test.

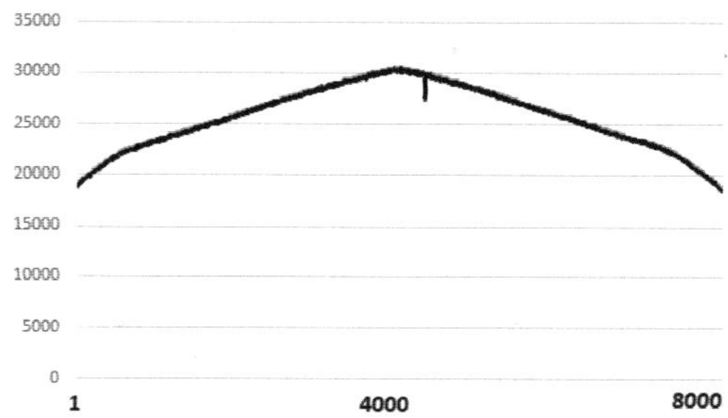

Figure 9. Uncorrected green DN vs detector index in flat test.

3.4. Radiometric model

Radiometric model formulated by mathematic model of hardware level of multispectral imager and convert from DN to voltage after subtract dark DN. Multiple voltage to radiance each detector index based on figure 6 and resampling use average gain on table 3 last column. Detail formula describes in formula 1. Example DN subtract from image 9 and 8 show in figure 10. DN after correction describes in figure 11. Correction result from table 2 show in table 3. In this table PRNU can reduce significant.

$$DNnew_i = Int\left(\frac{E.R_i.(((DN_i - DNdark_i).4/2^{16}/G)) - Offset)}{Radiance}\right) \quad (1)$$

$DNnew_i$: DN with Uniform radiance representation
Int : Float to Integer

IOP Conf. Series: Earth and Environmental Science **149** (2018) 012068 doi:10.1088/1755-1315/149/1/012068

E : Exposure Time
i : number of pixel sensor in Line imager
R_i : Responsivity each pixel i
DN_i : Digital number of raw images
$DNdark_i$: Dark Digital Number
G : Gain PGA of ADC
Offset : Offset of ADC
Radiance : Average DN to Radiance value

DN images will convert to CCD voltage output depend of imager setting and multiple by imager responsivity each pixel to find radiance pixel value. New DN is find by radiance image divide with common radiance value. These noise levels are converted from DN to radiance units using the band averaged absolute gains [8].

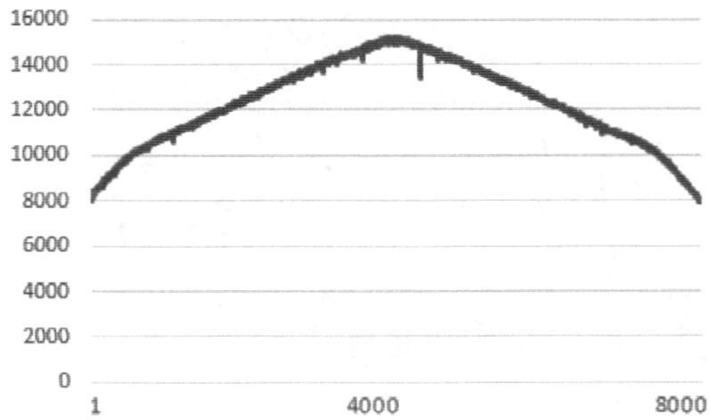

Figure 10. Green DN vs Detector Index after subtract with DN dark.

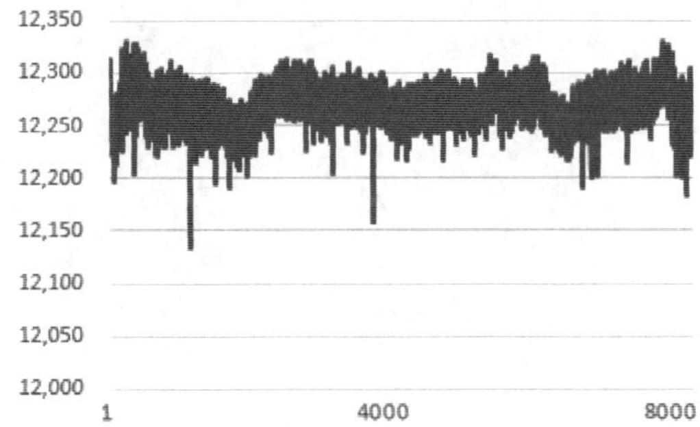

Figure 11. Green DN Corrected vs Detector Index.

Based on figure 10 relative radiometric correction can be done by normalize each pixel like figure 11 with error smaller than raw data. Relative radiometric correction is necessary to account for the detector-to-detector non-uniformity seen in raw imagery [9]. Image Before and after correction show in figure 12. On this image have little different between before and after except uniformity of sky.

However, the standard for designating the reference image needed for these methods is not unified [10]. For comparison with another image in different time need unified factor for time series analysis purpose. Two levels of radiometric correction, absolute and relative, have been developed for remote sense imagery [11]. Table 3 show DN to Radiance value get from input integrating sphere compare with DN value and parameter setting of camera and FWHM each channel. In absolute radiometric correction, atmospheric radiative-transfer codes are used to obtain the reflectance at the Earth's surface from the measured spectral radiances [12].

Figure 12. Before and after correction.

Table 3. Result of PRNU calculation after correction and gain DN to radiance (mW/cm2-sr-μm) with offset 0 radiance.

PRNU (%)	1	2	3	4	5	6	7	8	9	10	DN to Rad
Blue	1.82	1.98	1.77	1.91	1.02	0.99	1.00	1.60	1.47	2.32	0.001505
Green	0.79	0.77	0.88	0.73	0.46	0.40	0.40	0.68	0.60	0.89	0.000599
Red	0.51	0.52	0.50	0.49	0.31	0.35	0.32	0.40	0.38	0.81	0.000738
NIR	0.47	0.47	0.39	0.42	0.55	0.57	0.57	0.47	0.44	1.21	0.000851

4. Conclusion

Radiometric model can be formulation by hardware level process to radiance conversion and normalization. With this simple model can reduce response non uniformity less 30% to less than 2%. With statistic data experiment this formula more accurate prediction. This model can be implement to radiometric correction for multispectral imager in LAPAN-IPB Satellite on Orbit.

Acknowledgement

With the completion of this scientific paper the authors say many thanks to friend's fellow researchers and engineers at the Satellite Technology Centre for advice, encouragement, motivation and discussion that build. The chairman of Satellite Technology Centre LAPAN and Bogor Agriculture University (IPB) who provide tasks and support funds and personnel and cooperation in conducting research activities.

References

[1] Leroy M, Henry P, Guenther B and McLean J 1990 Comparison of CNES spherical and NASA hemisphere large aperture integration sources *Remote Sens. Environ.* **31** 97-104

[2] Misgaiski-Hass M. and Hieronymus J 2014 Radiometric Calibration of dual Sensor Camera System, a Comparison of classical and low cost Calibration *Int. Arch. Photogramm. Remote Sens. Spatial Inf. Sci.* **XL-5** 421-424, https://doi.org/10.5194/isprsarchives-XL-5-421-2014

[3] Mansouri A, Marzani F and Gouton P 2005 Development of a protocol for CCD calibration: application to a multispectral imaging system *International Journal of Robotics and Automation* **20**(2) 94–100

[4] Hamey L 2008 Radiometric Camera Calibration *Wiley Encyclopedia of Computer Science and Engineering* 2329-2338 DOI: 10.1002/9780470050118.ecse590

[5] Shoji T 1999 Spectral imaging by a multichannel camera *Journal of Electronic Imaging* **8**(4) http://dx.doi.org/10.1117/1.482702

[6] Thorne K, Markharn B, Barker I, Slater P, and Biggar S 1997 Radiometric Calibration of Landsat *Photogrammetric Engineering & Remote Sensing* **63**(7) 853-858

[7] Goldman D B 2010 Vignette and exposure calibration and compensation *IEEE transactions on pattern analysis and machine intelligence* **32**(12) 2276–88

[8] Morfitt R, Barsi J, Levy R, Markham B, Micijevic E, Ong L, Scaramuzza P, Vanderwerff K 2015 Landsat-8 Operational Land Imager (OLI) Radiometric Performance On-Orbit *Remote Sens.* **7** 2208-2237

[9] Bradley G. Henderson, Keith S. Krause 2004 Relative radiometric correction of QuickBird imagery using the side-slither technique on orbit *Proc. SPIE* **5542** Earth Observing Systems IX ; doi: 10.1117/12.559910

[10] Yong X, Jie H, Xingfa G, Qiyue L 2017 On-Orbit Radiometric Calibration for a Space-Borne Multi-Camera Mosaic Imaging Sensor *Remote Sensing* **9** 1248 ISSN 2072-4292

[11] Hu Y, Liu L, Liu L, Jiao Q 2011 Comparison of absolute and relative radiometric normalization use Landsat time series images *Proceedings of SPIE - The International Society for Optical Engineering* 8006. 41-. 10.1117/12.902076

[12] Júnior DC, Osmar & Guimarães R, Silva N, Alan RG, Roberto G, Cristiano S, Carvalho PVD 2013 Radiometric Normalization of Temporal Images Combining Automatic Detection of Pseudo-Invariant Features from the Distance and Similarity Spectral Measures, Density Scatterplot Analysis, and Robust Regression *Remote Sensing* **5** 2763-2794. 10.3390/rs5062763

IOP Conf. Series: Earth and Environmental Science **149** (2018) 012069 doi:10.1088/1755-1315/149/1/012069

Barium strontium titanate thin film growth with variation of lanthanum dopant compatibility as sensor prototype in the satellite technology

Mulyadi[1], Rika Wahyuni[1], Hendradi Hardhienata[2] and Irzaman[2]

[1]Department of Electrical Engineering, Borneo University, Indonesia
[2]Department of Physics, Bogor Agriculture University, Indonesia

Email: irzaman@apps.ipb.ac.id

Abstract. Electrical properties of barium strontium titanate thin films were investigated. Three layers of barium strontium titanate thin films have been prepared by chemical solution deposition method and spin coating technique at 8000 rpm rotational speed for 30 seconds and temperature of annealing at 850°C for eight hours with temperature increment of 1.67°C/minute. Materials produced by the process of lanthanum dopant with doping variations of 2%, 4% and 6% above type-p silicon (100) substrates. Film obtained was then carried out the characterization using USB 2000 VIS-NIR and tauc plot method. As a result, the barium strontium titanate thin film has the value of band gap energy of 1.58 eV, 1.92 eV and 2.24 eV respectively. The characterization of electrical properties shows that the band gap value of barium strontium titanate thin film with lanthanum dopant was in the range of semiconductor value. Barium strontium titanate thin films with lanthanum dopant are sensitive to temperature changes, so it potentially to be applied to temperature monitoring on satellite technology.

1. Introduction

Ferroelectric materials such as barium strontium titanate (BST) get special attention from many researchers because of various technical advantages such as chemical stability, high permittivity, high dielectric constant, good thermal stability and high tunability [1-7]. In some studies it has been reported that the temperature range in ferroelectric materials such as barium strontium titanate can be controlled through the barium to strontium ratio of BST thin films. This class of materials possesses wide range of useful physical properties due to their unique combination of piezoelectric and dielectric characteristics [8]. Dielectric properties of BST films depend on the substrate, crystal structure, composition, grain size, orientation and deposition parameters even thickness of BST thin films [9-10]. The strontium-substituted ferroelectric BST thin films investigated for microwave device applications because of its high permittivity and low loss at microwave frequencies and its Curie temperature lies below room temperature. The addition of proper dopant can make materials suitable for applications in medical device technology and other electronic devices [11-16]. Therefore, it is required to investigate the characteristics of BST films deposited at different deposition conditions for optimization as well as for investigating the correlations between optical and electrical properties. In this paper the $Ba_{0.5}Sr_{0.5}TiO_3$ deposited on the Silicon substrate by CSD method and an investigation of their electrical properties at lanthanum dopant with doping variations are reported.

IOP Conf. Series: Earth and Environmental Science **149** (2018) 012069 doi:10.1088/1755-1315/149/1/012069

2. Experimental method

In the preparation of the $Ba_{0.5}Sr_{0.5}TiO_3$ powder, we used 2.5 ml of 2-methoxy ethanol as a solvent. Precursor materials to produce $Ba_{0.5}Sr_{0.5}TiO_3$ doped Lanthanum by chemical solution deposition method involved [$Ba(CH_3COO)_2$; Aldrich-243671 99%], [$Sr(CH_3COO)_2$; Aldrich - 41283 with normal purity of 99%] and [$Ti(C_{12}H_{28}O_4)$; Aldrich-21642200 with normal purity of 99%] to obtain 1 M. The solutions were stirred with an ultrasonic stirrer for 90 minutes. Thin films were prepared by spin coating technique at 8000 rpm for 30 seconds which a pre-washed p-type silicon substrate is mounted on a spin coating plate with double-sided tape at its centre. Then 1/3 surface of the p-type silicon substrate had been cut with cubic area 1 cm^2 and affixed to the surface of the spin coating plate covered with glue seal tape. Seal tape adhesion aims to avoid all p-type silicon substrate surface coated by BST solution. This procedure was repeated for 2%, 4% and 6% lanthanum oxide dopant. The BST thin film was annealing at 850°C for 8 hours with temperature increment of 1.67°C/min using Nabertherm furnace [17]. The thin films thicknesses were determined by the volumetric method after weighed mass of the sample.
The chemical formula of BST thin film as follow:

$$_{0,5}Ba(Ch_3COO)_2 + {}_{0,5}Sr(CH_3COO)_2 + Ti(C_{12}H_{28}O_4) + {}_{22}O_2 \rightarrow Ba_{0,5}Sr_{0,5}TiO_3 + {}_{17}H_2O + {}_{16}CO_2$$

Optical properties of barium titanate thin films have been studied like UV-visible spectrophotometer. The tool used in this test was a UV-Vis Spectroscopy Ocean Optics USB 2000. Energy gap is obtained by creating plotting the relationship between

$$\left[\ln\left(\frac{R_{max}-R_{min}}{R-R_{min}}\right)\right]^2 \quad to \quad h\nu \tag{1}$$

Extrapolation performed on the curve that has the highest gradient and cutting $h\nu$ axis. Values were cut $h\nu$ axis represents the value of energy gap.

3. Result and discussion

Changes in surface structure and grain size of thin films resulting from the addition of lanthanum oxide dopants are also experienced by $Ba_{0.5}Sr_{0.5}TiO_3$ thin films while undergoing annealing. In the process once again the structure, morphology and thickness of the crystal changes as planned [18]. The energy gap of $Ba_{0.5}Sr_{0.5}TiO_3$ thin films which are heated at temperature of 850°C with the increasing rate of 1.67°C/minute for 8 hours. Figure 1 shows the energy gap of undoped $Ba_{0.5}Sr_{0.5}TiO_3$ thin films was 3.07 eV. According to Irzaman, this energy amount of electron excitation from the valence to the conduction band [19]. Figure 2 shows the addition of lanthanum dopant affects the energy gap in barium strontium titanate thin films, addition 2% lanthanum dopant decreases the energy gap to 1.58 eV than undoped $Ba_{0.5}Sr_{0.5}TiO_3$ thin films while figure 3 and figure 4 shows that 4% and 6% lanthanum dopant increases the energy gap to 1.92 eV and 2.24 eV respectively. We suspect that the addition of lanthanum oxide dopant in 2% $Ba_{0.5}Sr_{0.5}TiO_3$ thin film has enhanced the BST conduction properties as indicated by the experimental results. The energy gap was obtained using [20].

$$\left(\frac{\alpha hc}{\lambda}\right)^2 = Cons.\left(\frac{hc}{\lambda} - \mathrm{Eg}\right) \tag{2}$$

Where α is the absorption coefficient, *hc* is the incident photon energy, and Eg is the energy gap. The energy gap obtained via linear plotting $\left(\frac{\alpha hc}{\lambda}\right)^2$ versus Eg.

IOP Conf. Series: Earth and Environmental Science **149** (2018) 012069 doi:10.1088/1755-1315/149/1/012069

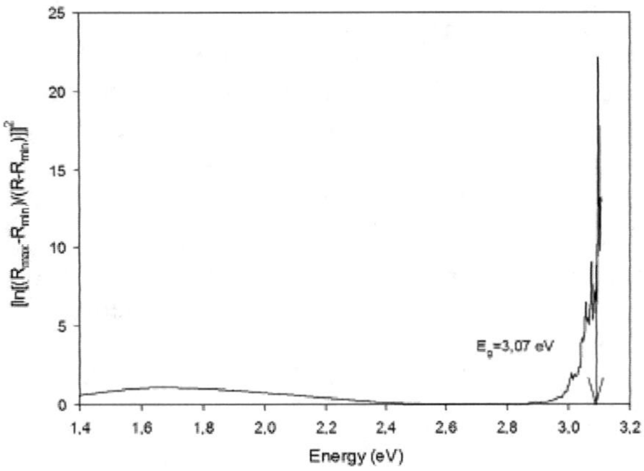

Figure 1. Energy gap of undoped BST thin films

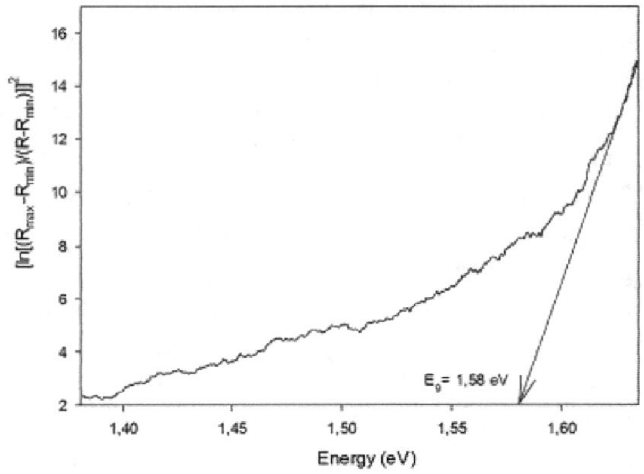

Figure 2. Energy gap of 2% La-doped BST thin films

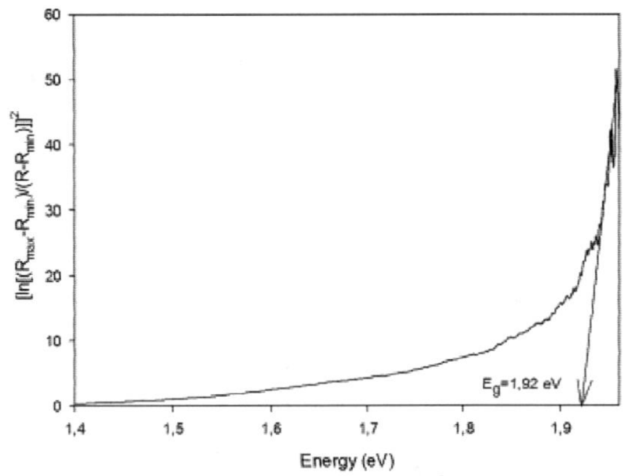

Figure 3. Energy gap of 4% La-doped BST thin films

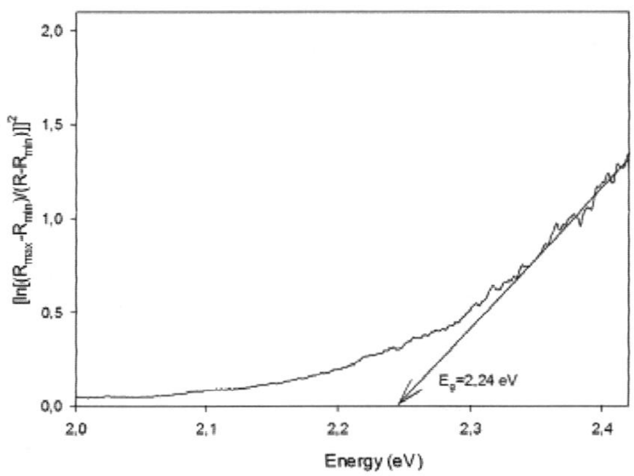

Figure 4. Energy gap of 6% La-doped BST thin films

We expect the lanthanum oxide density on BST thin films might increase the energy gap as expected from our previous work. Figure 2 shows that an energy gap on a BST thin film doped with 2% lanthanum oxide decreased to 1.58 eV compared to 3.07 eV on undoped BST energy gap. We think this is because electrons are more difficult to move from the valence band to the conduction band [8]. The energy gap decreases to 1.58 eV caused by the addition of donor atoms to the $Ba_{0.5}Sr_{0.5}TiO_3$ thin films, the energy level reach the conduction band. The existence of the conduction band decreases the energy gap of the $Ba_{0.5}Sr_{0.5}TiO_3$ thin films with the three-valences lanthanum dopant. The energy gap decreases to 1.58 eV caused by the addition of donor atoms to the $Ba_{0.5}Sr_{0.5}TiO_3$ thin films, the energy level reach the conduction band. The existence of the conduction band decreases the energy gap of the $Ba_{0.5}Sr_{0.5}TiO_3$ thin films with the three-valences lanthanum dopant. We think it's happening in BST doped lanthanum oxide because the electrons unable to absorb photons with the amount of energy is less than the amount of energy in the energy gap, the undoped BST does not happen similar condition.

4. Conclusions

The addition of lanthanum to the thin film of $Ba_{0.5}Sr_{0.5}TiO_3$ has a significant effect on the electrical and optical properties of the material. The addition of lanthanum oxide to BST thin film resulted in the value of energy gap that tends to decreased when compared with BST films that are not given additional lanthanum oxide, from the experimental results it was found that for BST thin films were given additional lanthanum oxide the resulting energy gap tends to increase in proportion to the addition of lanthanum oxide concentration. The tendency of decreasing energy gaps value on the lanthanum doped BST thin films might be caused by the effect of physical change of crystal and increased grain size due to addition of lanthanum concentration. Based on experimental results it's concluded that the energy gap in the BST thin film doped with lanthanum oxide is influenced by the concentration of the dopant, adding dopant concentrations will increase the value of energy gaps in thin films. $Ba_{0,5}Sr_{0,5}TiO_3$ with lanthanum dopant has an energy gap in the semiconductor value, thus BST thin film doped with lanthanum oxide can be utilized for sensor materials.

Acknowledgments
This research was supported by penelitian kerjasama antar perguruan tinggi project (Grant no.062/ADD/SP2H/LT/DRPM/VIII/2017) Kementerian Riset, Teknologi dan Pendidikan Tinggi Republik Indonesia.

References

[1] Liu M, Ma C, Collins G, Liu J, Chen C, Alemayehu A D, Subramanyam G, Ding Y, Chen J and Dai C 2013 *Nanoscale Res. Lett.* **8** 1-6
[2] Li S, Yao Y, Yuzyuk Y I, Jia Y, Wang X, Xie S and Jing Y 2014 *Appl. Phys.Lett.* **105** (2014) 092901
[3] Jin K X, Yang B, Zhang Y, Luo B C, Chen L Y and Chen C L 2016 *Scr. Mater.* **112** 62-6
[4] Yan D, Luo L, Zhang Z, Peng Z, Liu H, Xiao D, Liu T, Lai X and Zhu J 2015 *Cerm. Int.* **41** S520-525
[5] Moya X, Narayan S K and Mathur N D 2014 *Nat. Mater.* **13** 439M
[6] Xu Z, Yan D, Xiao D, Yu P, Zhu J. 2013 Dielectric enhancement of $BaSrTi_{1.1}O_3/BaSrTi_{1.05}O_3/BaSrTiO_3$ multilayer thin films prepared by RF magnetron sputtering *Ceram Int* **39** 1639–43
[7] Moore R J 2011 *Lanthanum: Compounds, Production and Applications, Chemistry Research and Applications* (New York: Nova Science Publisher) p 349-363
[8] Mulyadi, W Rika, Sulidah, Irzaman and Hendradi Hardhienata 2017 Barium strontium titanate thin film growth with rotational speed variation as a satellite temperature sensor prototype *IOP Conf. Series: Earth and Environmental Science* **54** (2017) 012094
[9] Zimmermann F, Voigts M, Weil C, Jakoby R, Wang P, Menesklou W and E Ivers-Tiffee 2001 Investigation of barium strontium titanate thick films for tunable phase shifters, *Journal of the European Ceramic Society* **21** 2019–23
[10] Jong Y H, Ji W C, Chong Y K, Karmanenko S F, Doo J C, Seok J Y and Hyun J K 2006 Investigation of barium strontium titanate thick films for tunable phase shifters, *J. Electroceram* **17** 141–144
[11] Mazon T, Zaghete M A, Varela J A and Longo E 2007 Barium strontium titanate nanocrystalline thin films prepared by soft chemical method *Journal of the European Ceramic Society* **27** 3799–3802
[12] Falcao E A, Eiras J A, Garcia D, Santos A I, Medina A N, Baesso M L, Catunda T and Balla A S 2016 Effects of lanthanum content on the thermo-optical properties of $(Pb,La)(Zr,Ti)O_3$ *Ferroelectrics.* **494** (1) 33–42
[13] Tran V D N, Ullah A and Dinh T H 2016 Effect of lanthanum doping on ferroelectric and strain properties of $_{0.96}Bi$ $1/2(Na_{0.84}K_{0.16})$ $1/2TiO_3$-$_{0.04}SrTiO_3$ lead-free ceramics *Journal of the Electronic Materials* **45** 2639–43
[14] Mamoun S, Merad A S and Guilbert L 2013 Electronic and optical properties of lithium niobate from density functional theory *Computational Materials Science.* **79** 125–131

IOP Conf. Series: Earth and Environmental Science **149** (2018) 012069 doi:10.1088/1755-1315/149/1/012069

[15] Bae J, Kimb S S, Choib E K, Songc T K, Kimb W J and Lee Y I 2005 Ferroelectric properties of lanthanum-doped bismuth titanate thin films grown by a sol–gel method *Thin Solid Films.* **472** (1-2) 90–94

[16] Xiangrong W, Yong Z, Jia Z, Xiaozhen S, Ruifang L, Jianan L and Meimei Z I 2014 Lanthanum doping effect in barium strontium titanate glass ceramics *Ceramics International.* **40** 16557–62

[17] Irzaman, Darmasetiawan H, Hardhienata H, Hikam M, Arifin P, Jusoh, Taking S N, Jamal Z and Idris M A 2009 Electrical properties of photodiode Ba0.25Sr0.75TiO3 (BST) thin film doped with ferric oxide on p-type Si (100) substrate using chemical solution deposition method *Atom Indonesia.* **35** (1) 133–138

[18] Irzaman, Pebriyanto Y, Apipah E R, Noor I and Alkadri A 2015 Characterization of optical and structural of lanthanum doped $LiTaO_3$ thin films *Integrated Ferroelectrics.* **167** (1) 137–145

[19] Irzaman, *et al* 2013 The effect of Ba/Sr ratio on electrical and optical properties of Ba_xSr $Ba_xSr_{(1-x)}TiO_3$ (x D 0.25; 0.35; 0.45; 0.55) thin film semiconductor *Ferroelectrics.* **445** 4–17

[20] Itskovsky M A 1999 *Japanese Journal of Applied Physics.* **38** (8) 4812-17.

Contribution of BeiDou satellite system for long baseline GNSS measurement in Indonesia

I Gumilar*, B Bramanto, W Kuntjoro, H Z Abidin, N F Trihantoro
Geodesy Research Group, Institut Teknologi Bandung LABTEK IX-C, Jalan Ganesha 10 Bandung, Indonesia

E-mail: igumilar@gd.itb.ac.id

Abstract. The demand for more precise positioning method using GNSS (Global Navigation Satellite System) in Indonesia continue to rise. The accuracy of GNSS positioning depends on the length of baseline and the distribution of observed satellites. BeiDou Navigation Satellite System (BDS) is a positioning system owned by China that operating in Asia-Pacific region, including Indonesia. This research aims to find out the contribution of BDS in increasing the accuracy of long baseline static positioning in Indonesia. The contributions are assessed by comparing the accuracy of measurement using only GPS (Global Positioning System) and measurement using the combination of GPS and BDS. The data used is 5 days of GPS and BDS measurement data for baseline with 120 km in length. The software used is open-source RTKLIB and commercial software Compass Solution. This research will explain in detail the contribution of BDS to the accuracy of position in long baseline static GNSS measurement.

1. Introduction

Global Navigation Satellite System (GNSS) has been widely used for positioning application, such as infrastructure, deformation monitoring and atmospheric monitoring. Currently, GNSS is consist more than one satellite constellations. There are three global satellite constellation (GPS, GLONASS and Galileo) and there is one regional satellite constellation which rapidly developing, BeiDou Navigation Satellite System (BDS). By combining GPS and BDS constellation, the number of observed satellites has increased in conjunction with the accuracy and precision of the estimated position for short baseline GNSS data processing [1, 2]. In General, with the addition of frequencies and GNSS constellation satellites, it is posibble to improve the reliable ambiguity resolution, reduce the various of error sources and imporve the positioning accuracy and precision [3, 4, 5, 6, 7].

In Indonesia, the use of GNSS is mainly used to establish the control point network. The control point network is divided into 5 classes, namely Orde-0 to Orde 4. Orde-0 and Orde 1 are maintained by Badan Infromasi Geospasial (BIG) while Orde-2 to Orde-4 are maintained by Badan Pertanahan Nasional (BPN). Those classes mainly divided by its baseline length. The baseline length for Orde-0, 1, 2, 3 and 4 are 200 to 1000 km, 100 to 200 km, 10 to 15 km, 1 to 2 km and up to 500 m respectively [8]. Those control point networks are represented by benchmark or monument which has coordinate as a reference point.

Control points network should be distributed evenly, however, in Indonesia, those control points network were not distribution evenly, especially for those outside the Java Island (figure 1). Therefore, the baseline length of reference point that used was so far away. Its time consuming when we use traverse method to established the control point, so that, GNSS method could overcome that problem.

IOP Conf. Series: Earth and Environmental Science **149** (2018) 012070 doi:10.1088/1755-1315/149/1/012070

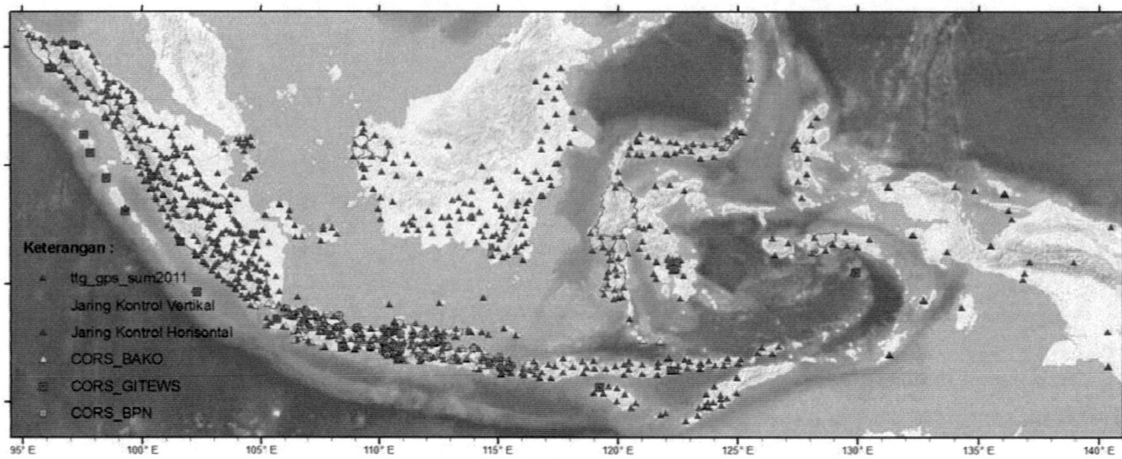

Figure 1. Distribution of horizontal control network of Indonesia [9]

GNSS data processing for longer baseline requires a specific processing strategy which can be done by using a scientific GNSS processing software. Orbital, ionosphere and troposphere error are can be resolved by using scientific GNSS processing software, meanwhile, GNSS data processing for longer baseline by using commercial GNSS processing software should be done carefully.

On the long baseline GNSS data processing, the un-differenced biases in orbital, ionosphere and troposphere error are the main factors of poor position solution [10]. This research aims to investigate the contribution of BDS in GNSS long baseline data processing in Indonesia.

2. Data and Method

An experimental observation points was carried out in Bandung and Jakarta (figure 3) with baseline length up to 120 km. ITB1 which is located in Bandung was used as a reference point, while A001 which is located in Jakarta was used as a rover. Both of ITB1 and A001 were Continuous Operating Reference System (CORS) that can observed BDS data. ITB1 used ComNav M300 GNSS receiver while A001 used CHC N72 (figure 2).

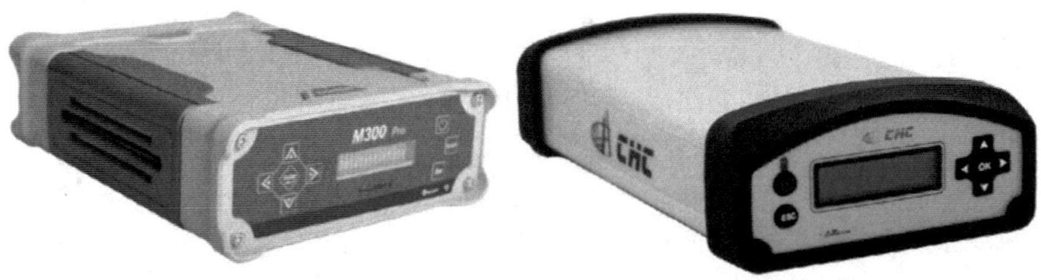

Figure 2. ComNav M300 (left) and CHC N72 (right)

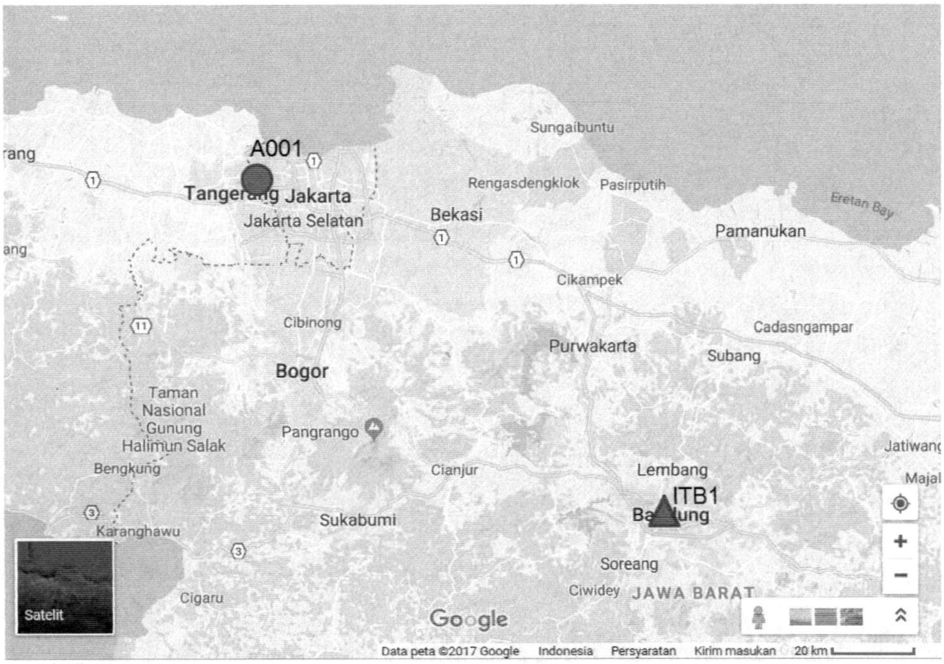

Figure 3. Experimental observation points

2.1. *Geometry of the Satellites*

Better satellite geometry leads a better position accuracy and precision. Satellite geometry is assessed by its satellite visibilities and the distribution of the satellites. The quality of the satellite geometry is quantified by the value of Dilution of Precision (DOP).

DOP is divided into several terms, namely vertical DOP (VDOP), horizontal DOP (HDOP), position DOP (PDOP) and time DOP (DOP). These terms can be concluded by using geometric DOP (GDOP) term. In general, GDOP is defined based on the user-equivalent range error (UERE), which is the standard deviation of the satellite's pseudorange error at the determined position. DOP can be defined as follows [11]

$$GDOP = \frac{\sqrt{\sigma_{x_u}^2 + \sigma_{y_u}^2 + \sigma_{z_u}^2 + \sigma_{ctb}^2}}{\sigma_{UERE}} \quad (1)$$

$$\sqrt{\sigma_{x_u}^2 + \sigma_{y_u}^2 + \sigma_{z_u}^2} = PDOP.\sigma_{UERE} \quad (2)$$

$$\sqrt{\sigma_{x_u}^2 + \sigma_{y_u}^2} = HDOP.\sigma_{UERE} \quad (3)$$

$$\sigma_{z_u} = VDOP.\sigma_{UERE} \quad (4)$$

$$\sigma_{ct_b} = TDOP.\sigma_{UERE} \quad (5)$$

where σ_x, σ_y, σ_z is the standard deviation of the three-dimensional position and σ_{ctb} is the standard deviation of the clock timing defined in distance at the specified location ($_u$). Better satellite geometry is represented by low DOP value which lead to a better accuracy and precision. Better and poor satellites-receiver geometry is represented in figure 4.

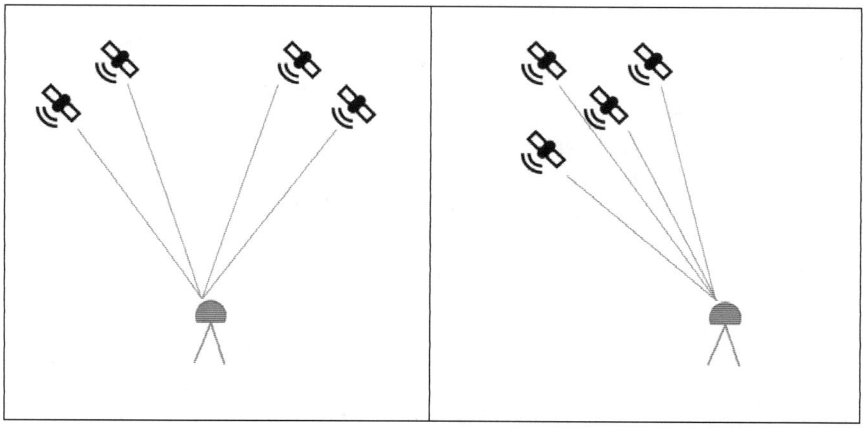

Figure 4. Illustration of better DOP (left) and poor DOP (right)

2.2. GNSS Positioning Algorithm

The GNSS positioning algorithm in relative positioning was used double difference (DD) positioning method. DD used carrier phase range that is constructed by differencing two single difference (SD) GNSS observation. Below is the brief equation for GNSS positioning algorithm [12]:

$$P_i = \rho + d\rho + d_{trop} + d_{ion} + c(dt - dT) + MP_i + \vartheta P_1 \quad (6)$$

$$L_i = \rho + d\rho + d_{trop} - d_{ion} + c(dt - dT) + ML_i + \lambda_i N_i + \vartheta L_i \quad (7)$$

where P_i and L_i are pseudorange and carrier phase range on selected frequency (i =1,2), ρ is related to geometrical range from receiver to satellite, $d\rho$ is the orbital error, d_{trop} and d_{ion} are troposphere and ionosphere biases, c is the speed of light (299,729,458 m/s), dt and dT are time error of receiver and satellites, MP_i and ML_i are related to multipath error of pseudorange and carrier phase range, λ_i and N_i are related to wavelength and ambiguity number, while ϑP_i and ϑL_i are related to noise error. SD can be described as follows:

$$\Delta L^j_{A,B} = \Delta\rho^j{}_{A,B} + d^j{}_{tropA,B} - d^j{}_{ionA,B} + c(dt - dT)^j{}_{A,B} + ML^j{}_{A,B} + \lambda N^j{}_{A,B} + \vartheta L^j{}_{A,B} \quad (8)$$

where Δ is the difference between receivers A and B. The superscript -j is the observed satellite. The satellite clock error is eliminated by taking single difference between receivers that observed the same satellite, while the atmospheric biases like tropospheric and ionospheric may be eliminated depending to the length of the baseline. Multipath would be considered as a noise error that could not be eliminated.

The remaining receiver clock error is eliminated by subtracting two SD observation. Mathematically, a DD is defined as follows:

$$\nabla\Delta L^{jk}_{A,B} = \Delta\rho^{jk}{}_{A,B} + +ML^{jk}{}_{A,B} + \lambda N^{jk}{}_{A,B} + \vartheta L^{jk}{}_{A,B} \quad (9)$$

where the two satellites are denote as he superscript-j and k. The atmospheric biases are negligible in Equation 8, however, for a longer baseline the atmospheric biases should be carefully handle.

This research used RTKLib and Tersus Geomatic Office (TGO) to process the kinematic and static position respectively. RTKLib [13] is an open source program package to handle satellite based positioning. RTKLib consists of a portable program library which has support to handle GPS,

GLONASS, Galileo, QZSS, BDS and BDS with various positioning mode for both real-time and post processing. TGO is an integrated GNSS processing software to manage baseline processing and network adjustment.

The processing strategy was designed as follow:

- The data were processed with cutoff angle of 15°.
- The orbit used was broadcast ephemeris.
- The troposphere model is Saastomoinen
- The ionosphere model is from broadcast ephimeris

3. Result and Discussion

3.1. Geometry of the Satellite

Figure 5 shows the number of observed satellites along with DOP values for GPS, BDS and Combined GPS-BDS respectively. It can be found that at least 10 satellites of BDS can be observed for all observation, while at least 7 satellites of GPS can be observed. The number of BDS and GPS observed satellites is vary from 7 to 11 and 10 to 14 satellites respectively. Although BDS gave a better number of observed satellites than GPS, the DOP value for BDS was worse than GPS. The geometry of BDS satellites give smaller volume compared with the geometry of GPS satellites. This is due to the 5 geocentric satellites of BDS that makes the geometry of BDS satellites clustered into horizontal axis (figure 6). The combination of GPS-BDS can be significantly increased the number of observed satellites up to 25 satellites. This circumstances led to a better GDOP compared with GPS only observation, the GDOP improved from 1.694 to 1.195.

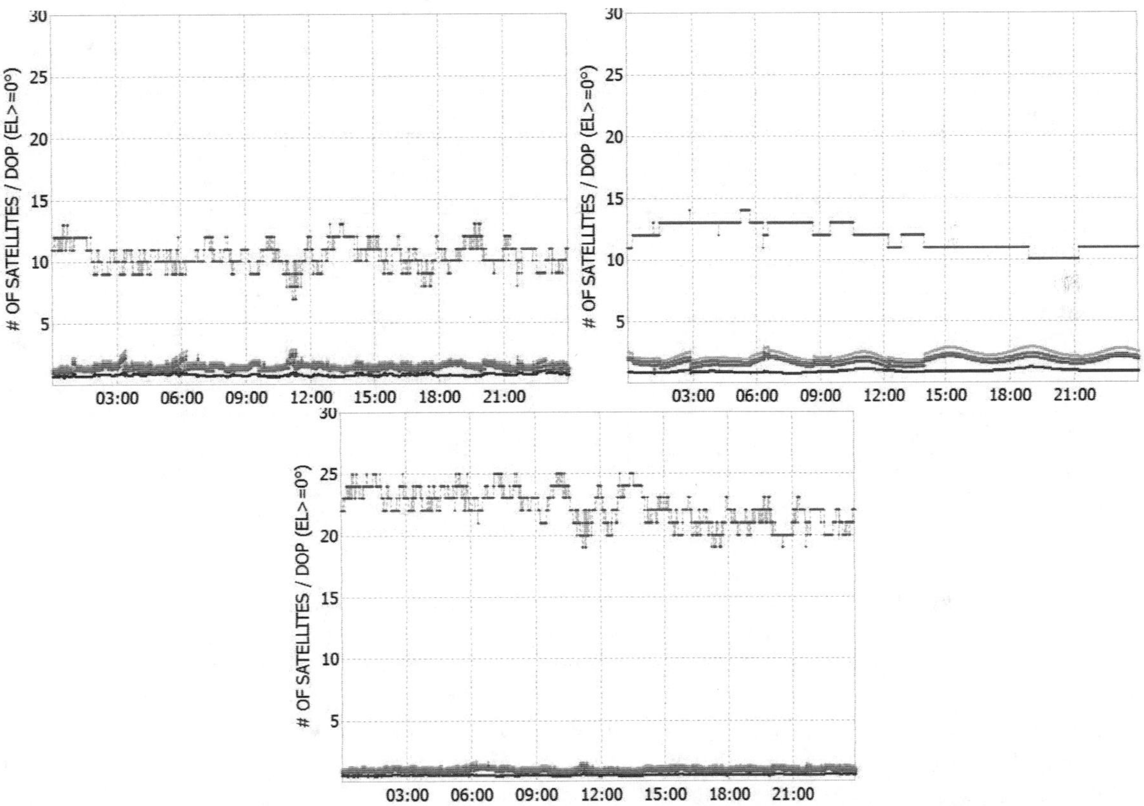

Figure 5. Number of observed satellites for GPS (upper left), BDS (upper right) and GPS-BDS (below)

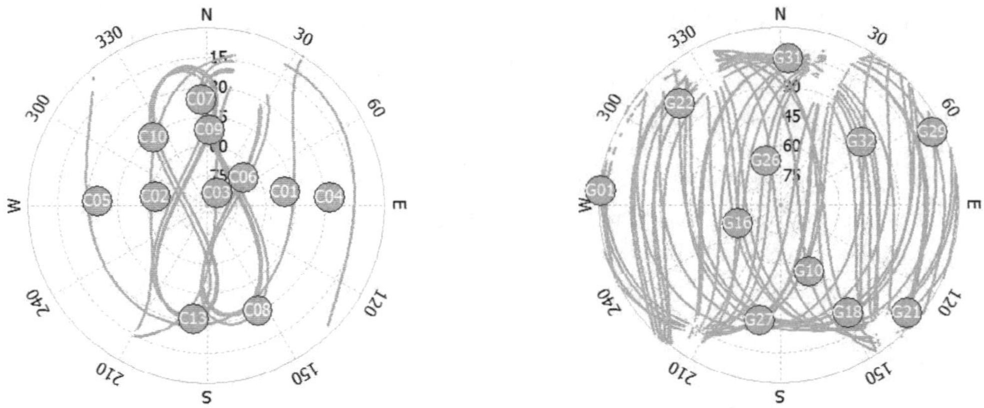

Figure 6. Satellite geometry of BDS (left) and GPS (right) at 2017-5-30 00:00:00 UTC

Table 1. Statistics of the number of satellite, GDOP, PDOP, HDOP and VDOP for each system

		GPS			
Statistics	**NSAT**	**GDOP**	**PDOP**	**HDOP**	**VDOP**
Average	10.421	1.694	1.531	0.804	1.298
SD	1.039	0.239	0.202	0.099	0.204
Min	7	1.3	1.2	0.6	0.9
Max	13	2.9	2.5	1.1	2.3
		BDS			
Statistics	**NSAT**	**GDOP**	**PDOP**	**HDOP**	**VDOP**
Average	11.805	2.303	1.94	0.967	1.68
SD	1.051	0.292	0.236	0.112	0.244
Min	10	1.6	1.4	0.8	1.2
Max	14	3	2.5	1.3	2.3
		GPS-BDS			
Statistic	**NSAT**	**GDOP**	**PDOP**	**HDOP**	**VDOP**
Average	22.226	1.195	1.045	0.587	0.867
SD	1.408	0.119	0.097	0.042	0.096
Min	19	1	0.9	0.5	0.7
Max	25	1.7	1.4	0.7	1.3

3.2. Combined GPS and BDS Estimated Position

Figure 7 shows the daily solution of 5 days GNSS observations. The precision of daily solution is improved by using combined GPS-BDS observations. The precision of GPS-BDS observations is improved to cm level compared with GPS only observations. Kinematic solution were also analysed in this research, figure 8 shows the kinematic epoch-wise solution from each system. Unlike the daily solution, the combined solution gives poor result at around 2017-6-1 21:00 UTC. This is likely due to propagation error from GPS observation, hence, further investigation is required. In general, BDS only solution gives better precision compared with GPS only solution.

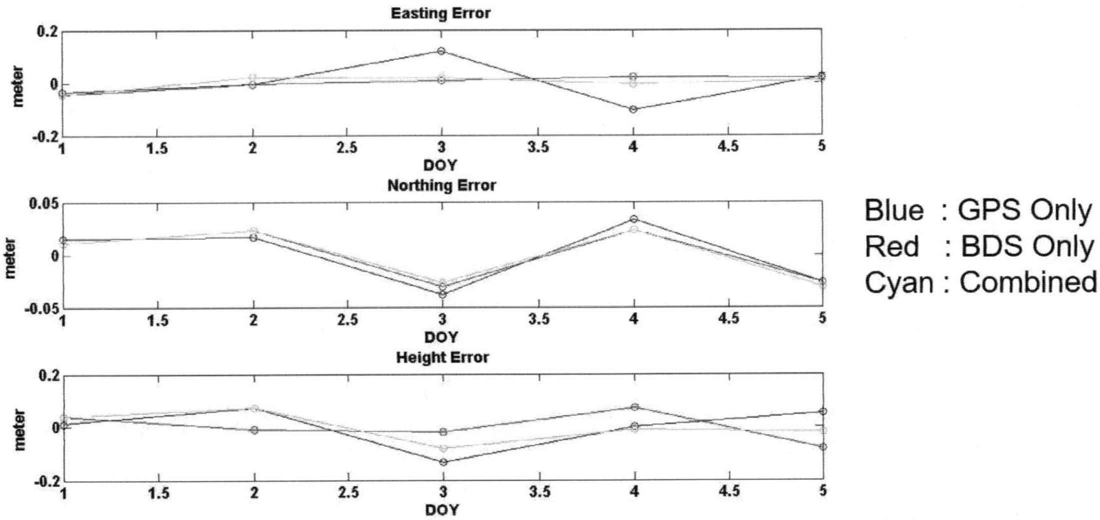

Figure 7. Daily position error solution for each constellation

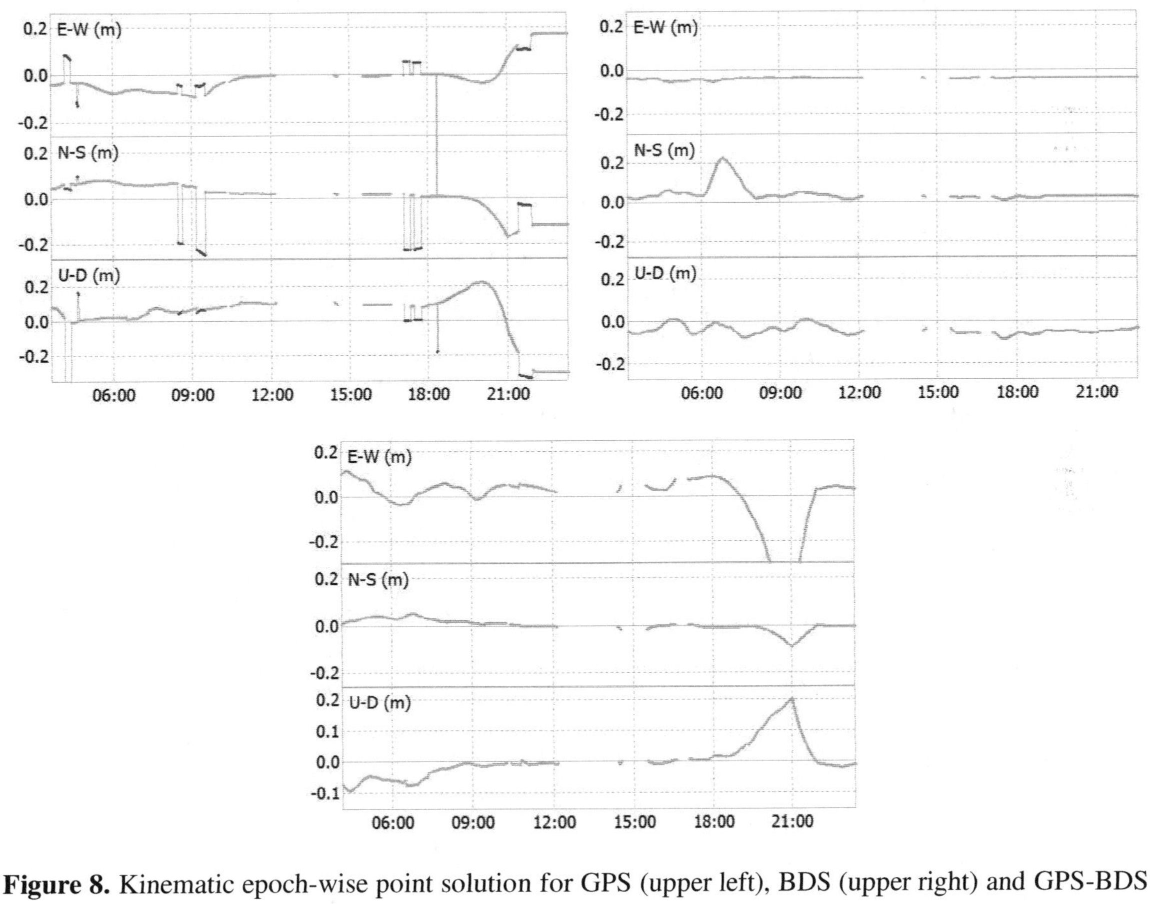

Figure 8. Kinematic epoch-wise point solution for GPS (upper left), BDS (upper right) and GPS-BDS (below)

IOP Conf. Series: Earth and Environmental Science **149** (2018) 012070 doi:10.1088/1755-1315/149/1/012070

4. Result and Discussion

The use of high point positioning accuracy and precision is not unneglectable in Indonesia. In accordance with the highly growth of positioning system, BDS will significantly help to achieve the high accuracy and precision positioning. The accuracy and precision for combined GPS-BDS daily solution is within 5 cm for horizontal position and 15 cm for vertical position, however, further researches to investigate the utilization of BDS in Indonesia is still needed.

References

[1] Gumilar I, Bramanto B, Pamungkas A I, Abidin H Z, Adi F S 2017 Contribution of BeiDou Positioning System for Accuracy Improvement: A Perspective from Bandung, Indonesia *Journal of Aeronautics, Astronautics and Aviation* **49** 171-184

[2] Brian B, Gumilar I, Abidin H Z, Prijatna K, Adi F S 2017 Assessment of the BeiDou Data Quality and the Positioning Performance: A Perspective from Bandung, Indonesia

[3] Hatch R, Jung J, Enge P, Pervan B 2000 Civilian GPS: The benefits of three frequencies *GPS Solutions*, **3** (4) 1-9

[4] Richert T, El-Sheimy N 2007 Optimal linear combinations of triple frequency carrier phase data from future global navigation satellite systems *GPS Solutions* **11** (1) 11-19

[5] Hernandez-Pajares M, Zomoza J M J, Subirana J S, Colombo O L 2003 Feasibility of wide-area subdecimeter navigation with GALILEO and modernized GPS *IEEE Transactions on Geoscience and Remote Sensing* **41** (9) 2128-2131

[6] Li B, Feng Y, Shen Y 2010 Three carrier ambiguity resolution: Distance-independent performance demonstrated using semi-generated triple frequency GPS signals *GPS Solutions* **14** (2) 177-184

[7] Feng Y, Li B 2010 Wide area real time kinematic decimetre positioning with multiple carrier GNSS signals *Science China: Earth Sciences* **53** (5) 731-740

[8] Badan Standarisasi Nasional 2002 Jaring Kontrol Horizontal SNI 19-6724-2002

[9] Badan Informasi Geospasial http://www.big.go.id/jaring-kontrol-geodesi/

[10] Li J, Yang Y, Xu J, He H, Guo H 2012 *Ionosphere-free combinations for triple-frequency GNSS with application in rapid ambiguity resolution over medium-long baselines* China Satellite Navigation Conference 2012 Proceedings, pp. 173-187

[11] Kaplan E D 1996 *Understanding GPS: Principle and Applications* (London: Artech House Boston)

[12] Guichang X 2010 *GPS: Theory, Algorithms and Applications* (Springer)

[13] Takasu 2009 *RTKLIB: Open Source Program Package for RTK-GPS, FOSS4G* (Japan)

IOP Conf. Series: Earth and Environmental Science **149** (2018) 012071 doi:10.1088/1755-1315/149/1/012071

Bias correction of daily satellite precipitation data using genetic algorithm

A W Pratama[1], A Buono[1], R Hidayat[2], H Harsa[1,3]

[1]Department of Computer Science, Faculty of Mathematics and Natural Science, Bogor Agricultural University, Bogor, Indonesia
[2]Department of Geophysics and Meteorology, Faculty of Mathematics and Natural Science, Bogor Agricultural University, Bogor, Indonesia
[3]Meteorological, Climatological, and Geophysics Agency of Indonesia (BMKG), Kemayoran, Jakarta, Indonesia

Email: angga_jisung13@apps.ipb.ac.id

Abstract. Climate Hazards Group InfraRed Precipitation with Stations (CHIRPS) was producted by blending Satellite-only Climate Hazards Group InfraRed Precipitation (CHIRP) with Stasion observations data. The blending process was aimed to reduce bias of CHIRP. However, Biases of CHIRPS on statistical moment and quantil values were high during wet season over Java Island. This paper presented a bias correction scheme to adjust statistical moment of CHIRP using observation precipitation data. The scheme combined Genetic Algorithm and Nonlinear Power Transformation, the results was evaluated based on different season and different elevation level. The experiment results revealed that the scheme robustly reduced bias on variance around 100% reduction and leaded to reduction of first, and second quantile biases. However, bias on third quantile only reduced during dry months. Based on different level of elevation, the performance of bias correction process is only significantly different on skewness indicators.

1. Introduction

In field of precipitation analysis, the problem which is commonly faced, is availability of observation data [1]. Rain gauge stations in Indonesia are unevenly distributed and the quantity is limited [2]. To address limitations of observation data, Satelite Precipitaion Products (SPPs) is frequently used as alternative [1]. Generally, near global coverage SPPs are only available on spatial resolution 0.25° or lower [3]. However, *Climate Hazard Group InfraRed Precipitation* (CHIRP) dan *Climate Hazard Group InfraRed Precipitation with Stations* (CHIRPS) provide 0.05° spatial resolution. CHIRPS was developed by U.S Geological Survey (USGS) and University of California (UCSB). CHIRPS was developed to support drought monitoring system *Famine Early Warning Systems Network* (FEWS NET) and focus over Africa area [4]. However, [5] and [3] revealed that CHIRPS still showed systematic bias. [6] performed preliminary study to find out the performance of CHIRPS over Java Island, and concluded that CHIRPS still has systematic bias on statistical value.

Bias correction is required to do before uses SPPs for further analysis, because the accuracy of SPPs is affected by coastal and topography aspecs [2, 7]. Bias correction process on several earlier studies evidently improved performance of SPPs data [1, 2, 7]. In this paper, Nonlinear Power Transformation

(NL) combined with Single Objective Genetic Algorithm (SOGA) is utilized to address systematic bias of CHIRPS over Java Island. The equation of GA used in this research is adopted from Pratama *et. al.* (2018), however we focus to investigate the impact of the equation and search space based on spatial area (separated by elevation).

2. Methodology

Flow charts of this research is showed by Figure 1. Precipitation data of fourteen observation stations are used (Table 1). January precipitation data of Pondok Betung station is used as training data in training process. Training process is aimed to search optimal genetic algorithm parameter, contain of probability crossover (pc), probability mutation (pm), population size, and maximum generation. Temporal resolution of data is daily-type data, from January 2005 to December 2016.

Table 1. List of station.

No	Location	Latitude	Longitude	Elevation (masl)
1	Pondok Betung	-6.2615	106.751	27
2	Kemayoran	-6.1556	106.84	4
3	Darmaga	-6.5	106.75	207
4	Bandung	-6.8836	107.597	791
5	Tegal	-6.8682	109.121	1
6	Cilacap	-7.7189	109.015	8
7	Semarang	-6.9847	110.381	6
8	Sangkapura	-5.85	112.63	0
9	Juanda	-7.3846	112.783	3
10	Karang Ploso	-7.9008	112.598	590
11	Karang Kates	-8.15	112.45	285
12	Kalianget	-7.05	113.97	0
13	Sawahan	-7.74	111.79	835
14	Banyuwangi	-8.215	114.355	52

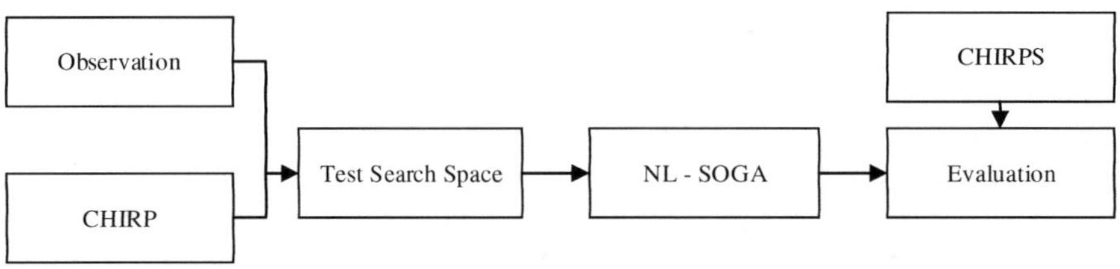

Figure 1. Research method.

2.1. Single Objective Genetic Algorithm

Search space and equation used are based on Pratama *et. al.* (2018). The equation and search space is shown in Equation 1. Parameter of genetic algorithm is explored to minimize time required in searching process of solution. Reference value of pc and pm is adopted from Arkeman *et. al.* (2012), pc is between 0.7 and 0.9 then explored value of pm is between 0.01 and 0.2. Population sizes explored are 10, 15, 20, 25 while number of maximal generation is 100, 200, 300, 400, 500.

The objectives on estimating parameter of NL method are to minimize the difference in both mean and CV of corrected data to observation data. In this paper, post process is added to mitigate possibility of more than one solution which have maximum fitness value. When the mentioned condition occurs, single solution is selected based on criteria of minimal root mean square error (RMSE) on monthly precipitation. Figure 2 shows the process of SOGA with additional post process.

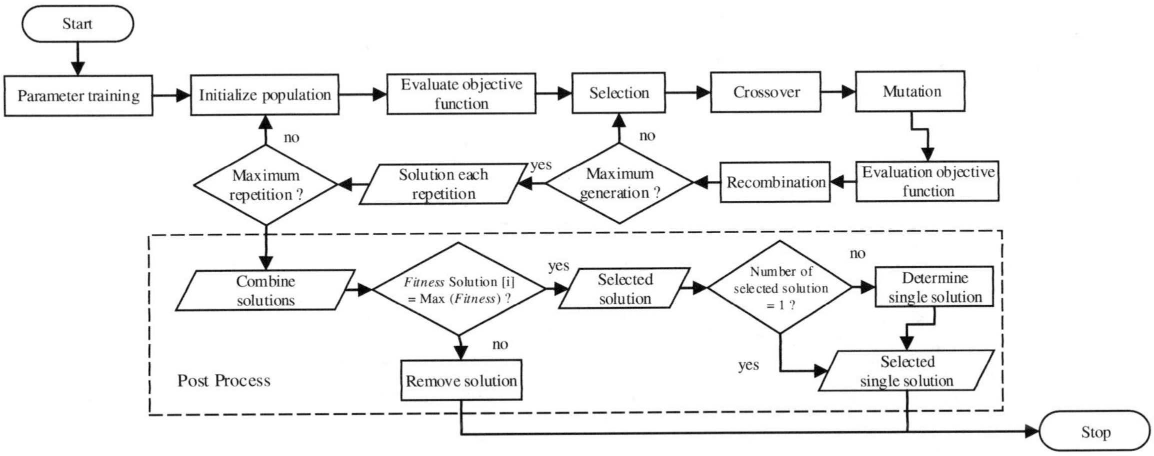

Figure 2. Process of SOGA with post process.

$$P^* = \begin{cases} a_1.P^{b_1}, \ P > \theta_1 \\ P \quad , \ otherwise \\ a_2.P^{b_2} \ , 2\,\theta_3 < P < \theta_2 \ or \ \theta_2 < P < 2\theta_3 \\ 0 \quad , \quad P < \theta_3 \end{cases} \tag{1}$$

Subject to,

$$\theta_1 > \theta_2 > \theta_3$$
$$\bar{x}_{CHIRP} + s_{CHIRP} - 2 < \theta_1 < \bar{x}_{CHIRP} + 2 * s_{CHIRP} + 2$$
$$(Q_{1_{CHIRP}} * 2) < \theta_2 < (Q_{2_{CHIRP}} + 3)$$
$$(Q_{1_{CHIRP}}/2) < \theta_3 < (Q_{1_{CHIRP}} + 1)$$
$$0.1 < a_2 < 1$$
$$0.1 < b_2 < 1$$
$$0.7 < a_1 < 4$$
$$0.7 < b_1 < 4$$

Objective function of SOGA in this study is a combination of the objectives. As a replacement of CV, we use variance (S^2) and combine it with mean ($\bar{x}$) to construct objective function. Objective function of SOGA is shown in Equation 2. c is constants which is weighted the absolute difference of mean value. In this paper, value of c selected is three. Equation 2 is invers of the objective. Maximum value of Equation 2 is one.

$$max\, f(obs, cor) = \frac{1}{1 + \frac{|S^2{}_{obs} - S^2{}_{cor}| + c * |\bar{x}_{obs} - \bar{x}_{cor}|}{2}} \tag{2}$$

2.2. Evaluation
Evaluation on this research is evaluating the obtained results based on statistical moment and quantile value. Furthermore, corrected CHIRP constructed is evaluated on different month and elevation. Elevation is grouped by 0 – 20 masl, 20 – 200 masl, 200 – 500 masl, and above 500 masl.

3. Result and Analysis

3.1. Single Objective Genetic Algorithm

Optimal parameters of SOGA obtained from training process were probability crossover (pc) 0.75, probability mutation (pm) 0.01, population size 15, and number of generations 500. Result of SOGA is shown in Table 2.

Tabel 2. Fitness Values of Stations.

Month	BMKG Stations													
	1	2	3	4	5	6	7	8	9	10	11	12	13	14
1	0.52	0.27	0.61	0.77	0.21	0.84	0.49	0.82	0.94	0.41	0.93	0.29	0.42	0.46
2	1.00	0.68	0.60	0.72	0.34	0.97	0.95	0.73	0.79	0.78	0.94	0.88	0.21	0.81
3	0.83	0.73	0.98	0.99	0.90	0.49	0.31	0.43	0.88	0.29	0.40	0.30	0.24	0.21
4	0.56	0.21	0.31	0.88	0.31	0.95	0.48	0.64	0.92	0.39	0.69	0.64	0.35	0.99
5	0.81	0.29	0.76	0.26	0.54	0.29	0.84	0.25	0.62	0.24	0.28	0.23	0.41	0.57
6	0.98	0.96	0.88	0.82	0.49	0.87	0.97	0.56	0.57	0.33	0.98	0.48	0.66	0.95
7	0.99	0.98	0.59	0.99	0.66	0.93	0.61	0.45	0.71	0.35	0.60	0.75	0.31	0.68
8	0.44	0.50	0.46	0.66	0.52	0.41	0.69	0.99	0.94	0.70	0.88	1.00	0.74	0.99
9	0.75	0.90	0.49	0.82	0.51	0.70	0.94	0.46	0.82	0.22	0.84	0.93	0.97	0.68
10	0.41	0.48	0.56	0.29	0.49	0.57	0.85	0.66	0.78	0.31	0.55	0.59	0.44	0.40
11	0.96	0.96	0.63	0.80	0.44	0.02	0.94	0.34	0.28	0.59	0.83	0.41	0.57	0.67
12	0.45	0.51	0.36	0.69	0.39	0.67	0.29	0.83	0.46	0.45	0.74	0.88	0.99	0.47

GA doesn't always product solution with high fitness value. The low fitness value represents the problem should not solve with single objective. However, high fitness value reveals we can adjust mean and variance of corrected CHIRP equal to observation data with some degree, higher fitness value means higher the similarity of mean and variance of corrected data to observation.

3.2. Evaluation

Both CHIRP, CHIRP, and corrected CHIRP (using SOGA) product bias against observation data. with calculating bias of fourteen station, we have mean absolute error (MAE) of those stations. Statistical test analysis of variance (ANOVA) is used to investigate the difference of MAE, Tuckey HSD test is applied when MAE is significantly different. P value of ANOVA is shown by Table 3.

Table 3. P value of ANOVA.

	DJF	MAM	JJA	SON
Q1	8.51E-25**	1.46E-18**	3.07E-11**	2.78E-05**
Q2	2.05E-33**	6.28E-17**	5.74E-14**	7.17E-06**
Q3	0.638	0.407	0.0005**	0.296
Mean	6.67E-16**	4.57E-09**	3.52E-08**	3.12E-11**
Variance	6.21E-38**	1.32E-23**	5.38E-15**	3.66E-16**
Skewness	1.89E-10**	1.09E-05**	0.003**	0.0097**
Kurtosis	0.325	0.552	0.714	0.683

** significant with level $\alpha = 0.01$

Kurtosis indicator of three data is not significantly different (Table 3), the bias correction can not reduce bias on this indicator. Bias on first and second quantile was significantly different, therefore bias on third quantile is different only during dry season (JJA).

Performance of CHIRPS during dry months is great. It is represented by quantile values which are nearly identic to observation and the bias on variance reduces relative to CHIRP (Table 5). This result corresponds to the background of CHIRPS establishment which is to support drought monitoring [4]. Comparing to our corrected CHIRP, MAE of Q_3 indicator between CHIRPS and corrected CHIRP is not significantly different (Table 4). Most of stations used in this paper have first quantile value equal to zero ($Q_1 = 0$). This condition occurs also during wet months, the number of non-rain days are around 25% from total wet months. Both CHIRP and CHIRPS cannot estimate number of non-rain days well during wet months, it is represented from first quantile value not equal to zero. After bias correction process, bias on first quantile reduces between 74 % and 96 % (Table 5).

Table 4. Mean Absolute Error of quantile indicator with tuckey HSD index.

Q1	DJF	MAM	JJA	SON
CHIRP	4.187a	2.817a	0.167a	0.901a
CHIRPS	2.876b	0.525b	0b	0.126b
NL-SOGA	0.994c	0.661b	0.008b	0.227b
Q2	**DJF**	**MAM**	**JJA**	**SON**
CHIRP	5.607a	4.83a	1.055a	2.51a
CHIRPS	5.94a	3.432b	0.005b	1.422b
NL-SOGA	1.912b	1.55c	0.352c	0.697b
Q3	**DJF**	**MAM**	**JJA**	**SON**
CHIRP	3.319a	4.122a	2.353a	2.738a
CHIRPS	3.889a	4.784a	1.104b	2.48a
NL-SOGA	3.319a	4.052a	1.315b	2.078a

Bias on third quantile is reduced only during dry months (Table 5). Proposed approach robustly reduces bias on first and second quantile. However Bias on kurtosis increases after bias correction process (shown with negative values).

Table 5. Mean of percentage of bias reduction relative to CHIRP (%).

Month	Q1	Q2	Q3	Mean	Var	Skew	Kurt
DJF	76.25	65.90	0.00	69.38	99.96	11.66	-43.22
MAM	76.55	67.91	1.71	46.83	99.82	35.13	5.43
JJA	95.37	66.63	44.13	53.84	99.92	48.92	19.77
SON	74.83	72.24	24.14	51.81	97.90	40.55	22.27

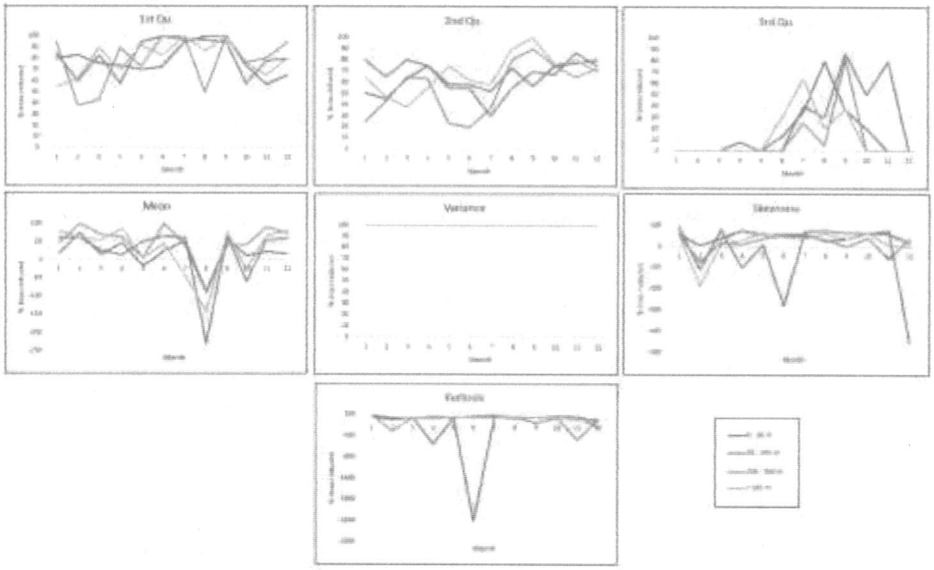

Figure 3. Percentage of bias reduced based on different station elevation.

For different spatial area, bias on mean indicator continually increases on august (Figure 3). It is impact of adjusting variance of corrected chirp identic to observation data thus bias on mean increases, because basically mean value of chirp approaches observation although the variance value is significantly different. Skewness and kurtosis represent shape and asymmetry of distribution [9]. Bias on skewness can be handled robustly only on 0 - 20 masl elevation station group. Value of skewness relates to extreme value. Second threshold in Equation 1 was added to transform high precipitation of CHIRP becoming extreme precipitation value because Pratama *et. al.* (2018) found that CHIRP cannot predict extreme precipitation value of observation data. However, when high precipitation values of CHIRP is equivalent to observation, the bias correction process will cause the high precipitation value getting further to observation which products higher bias. In other hand, bias reduction on third quantile during dry months does not happen on stations which have elevation between 200 and 500 masl, it means number of not-rain days during dry months on the stations are less than 75% of total data.

4. Conclusion

The brief result of this study is similar to Pratama *et. al.* (2018), the bias correction process cannot handle bias on mean, skewness, kurtosis robustly since the bias on several months increase after correction process. The performance of corrected CHIRP is not separated by different spatial area (elevation). However, performance of corrected CHIRP is separated by different season.

Selecting low value of c (in Equation 2) makes genetic algorithm less sensitive to difference of mean between corrected data and observation data. The impact is, model cannot handle bias on mean because the model only focus on address the difference on variance.

References

[1] Mamenun 2014 Validasi dan koreksi data satelit TRMM pada tiga pola hujan di Indonesia *Jurnal Meteorologi dan Geofisika*, **15** 13-23

[2] Vernimmen R. R. E., Hooijer A, Mamenun and Aldrian E 2012 Evaluation and bias correction of satellite rainfall data for drought monitoring in Indonesia *Hydrology and Earth System Sciences*, **16** 133–146

[3] Katsanos D, Retalis A and Michaelides S 2016 Validation of a high-resolution precipitation database (CHIRPS) over Cyprus for a 30-year period *Atmospheric Research*, **169** 459-464

[4] Funk CC, Peterson PJ, Landsfeld MF, Pedreros DH, Verdin JP, Rowland JD, et al. A quasi-global

precipitation time series for drought monitoring. US Geological Survey, 2014 2327-638X.

[5] Toté C, Patricio D, Boogaard H, Wijngaart R V D., Tarnavsky E and Funk C 2015. Evaluation of satellite rainfall estimates for drought and flood monitoring in Mozambique *Remote Sensing,* **7** 1758-1776

[6] Pratama AW, Buono A, Hidayat R, and Harsa H 2018 Estimating parameter of nonlinear bias correction method using NSGA-II in daily precipitation data *TELKOMNIKA Accepted,* **17**

[7] Valdés-Pineda R, Demaría E M C, Valdés J B, Wi S and Serrat-Capdevil A 2016 Bias correction of daily satellite-based rainfall estimates for hydrologic forecasting in the Upper Zambezi, Africa *Hydrology and Earth System Sciences Discuss*

[8] Arkeman Y, Gunawan H, and Seminar K B 2012 *Algoritma genetika teori dan aplikasinya untuk bisnis dan industri* (Bogor: IPB Press)

[9] Press W H, Teukolsky S A, Vetterling W T and Flannery B P 1996 *Numerical recipes in C* (Cambridge: Cambridge University press) pp 610-613

IOP Conf. Series: Earth and Environmental Science **149** (2018) 012072 doi:10.1088/1755-1315/149/1/012072

Development of Nanosatellite Technology with APRS Module for Disaster Mitigation

S Y Prahyang[1], M Z Dhiya'Ulhaq[1], O P Golim[1], R Gunawan[1], Suhandinata[1], E Jahja[2], E R G Nelwan[1], C Ananta[1], I M Chow[1] and N D F Mali[1]

[1]Engineering Physics Department, Surya University, Tangerang, Banten, 15810 Indonesia
[2]Computer Sciences Department, Surya University, Tangerang, Banten, 15810 Indonesia

Abstract. Development of nanosatellite technology has enabled satellites to be developed with multiple capabilities for a specific mission in a short time with a low cost. Satellite communications are proved to be more effective in delivering information due to its large coverage area. Surya Satellite-1 will become the first Indonesian nanosatellite developed by undergraduate students. It is designed with low-cost commercial payloads, including an APRS module for communication and operated on VHF and UHF amateur radio frequencies. The mission of the satellites focused on disaster mitigation through APRS communication network with remote stations located on disaster-prone areas.

1. Introduction

Indonesia is an archipelago country which has 16,056 islands. Referring to data from Central Bureau of Statistic in 2015, Indonesia has 3,544,743.9 km² area of sea and 1,913,578.68 km² land area, which means over 60% of Indonesia area was covered by water [1]. The location of Indonesia, above the assembly area of 3 major tectonic plates, makes many volcanoes had arisen on its land [2]. Indonesia also has average rainfall more than 1000 mm per year [3]. These geographical conditions make Indonesia must to stay alert when disaster like flood, earthquake, and volcanic eruption will happen.

National Board of Disaster Management (Badan Nasional Penanggulangan Bencana or BNPB) noted disasters in Indonesia have increased gradually from 2000 until now. BNPB reported there were 2,369 disasters happened in 2016 [4]. Therefore, a quick, low-cost and accurate disasters mitigation system is important and necessary, especially for small islands which are separated from central island. Disaster mitigation designed nanosatellite is an option for these conditions [5,6,7,8,9].

Nanosatellite needs an ability to forward or broadcast data collected from ground segment for disaster mitigation [10,11], APRS is one of suitable payload for this mission. Automatic Packet Reporting System (APRS) is a communication protocol founded by amateur radio member, WB4APR, Bob Bruninga [12]. It is applicable in many fields of study, based on many amateur radio projects that have been made [7,13]. It has open access status which allows anyone who has amateur radio license to communicate with APRS protocol, which is especially practical for emergency communication service when the disaster happened. [14]

There are some differences that APRS has compared to other radio packet. Its features include giving real-time weather data, maps and other data presentations. Since it is integrable with other communication platforms, such as internet, data updates can be disseminated to public in real time. Technically, it provides generic and intelligent digital repeating, a means to relay digital data through a data network, to avoid network flooding and more user-friendly with callsign for each user. While it also capable to do quick broadcasting of text information through full-duplex messaging using AX.25 UI-frame. [13,15]

Surya Satellite-1 (SS-1) is a nanosatellite which brings APRS payload to fulfil disaster mitigation mission. SS-1 will be the first nanosatellite which made by Indonesian students. Interfaces design of this nanosatellite is based on Interface Control Documents from JAXA.

For disaster mitigation purpose, SS-1 three mission goal stages of Minimum, Full and Extended missions have been set as the success indicator. Minimum mission achieved if SS-1 is able to beacon APRS packet data at interval time and ground station will be able to capture the data of that beacon. While, Full mission success achieved if SS-1 has the ability to provide short text messages communication services for disaster mitigation function, such as early fire hotspot warning in forest, or early flood warning, and Extended mission will be to improve education environment in Indonesia in satellite and communication technology with nanosatellite workshop and training for university and high school student.

2. Satellite Development

To fulfil the needs of low-cost satellites for communication and disaster mitigation function, the nanosatellite was designed as a low earth orbit satellite consisting of four common subsystems: Electrical Power System (EPS), Data Handling System (DHS), Payloads which are radio communication modules and Antenna Deployment System [16,17,18,19]. These subsystems are enclosed in aluminium case structure based on JAXA Interface Control Document (ICD) for satellite design.

Electrical Power System (EPS) is responsible for harvesting, storing and distributing power for satellite [20,21,22]. EPS includes current limiting and battery protection function for safety measures. The system consists of GaAs triple-junction solar cells as a power source, Maximum Power Point Tracking (MPPT) as power absorption optimizer, charge controller as battery protector, ideal diode as current limiter, battery as energy storage media, boost converter for regulated 5V power source. [23, 24]

Data Handling System (DHS) function as satellite monitoring and control system [21,25,26,27]. Monitoring includes battery voltage as battery state of charge (SOC) indicator, battery temperature for heater operation indicator, and charging current as power production indicator. Furthermore, additional temperature sensor is attached to measure satellite environment temperature for future studies. On the other side, controlling includes timer, antenna deployment, battery heater, and subsystem power switch control. DHS consists of On-Board Computer (OBC) as data processor and system decision maker, sensors for monitoring, signal converter for bi-directional data conversion between microcontroller and radio modem, and power switch to control power flow. [28,29,30]

SS-1 consist of two APRS payloads, one in UHF-band for uplink channel while the other is VHF-band for downlink channel. Both frequencies have been set to work in half-duplex mode, meaning they can receive and transmit signals at VHF-band and UHF-band, but not to do both at the same time. Each payload consists of APRS module, which converts radio wave data to digital data and vice versa, APRS Tracker as On-Board Data Handling (OBDH) unit, and monopole antenna since it does not have altitude control function [31]. The architecture of communication system is shown in figure 1 below [32], while the whole subsystem configurations are shown in figure 2.

IOP Conf. Series: Earth and Environmental Science **149** (2018) 012072 doi:10.1088/1755-1315/149/1/012072

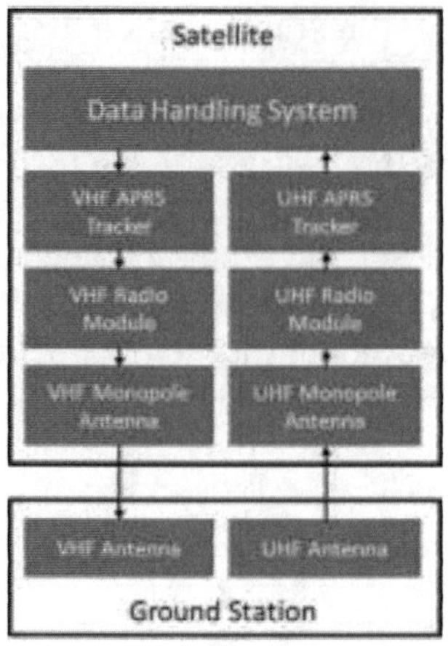

Figure 1. Communication Architecture.

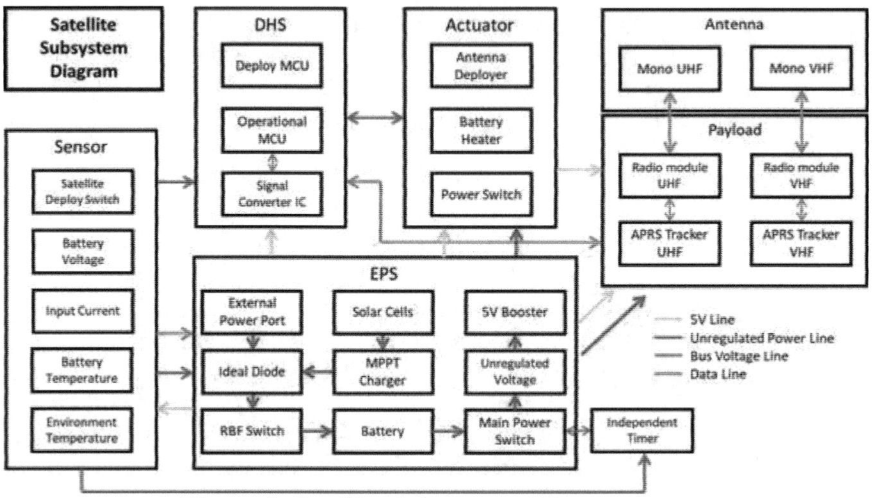

Figure 2. Satellite Subsystems Diagram.

Antenna deployment system is mechanism for folding the antenna before satellite launch and to deploy it at the right altitude [33,34,35,36]. It is responsible for deploying two monopole antennas. Initially, both antenna will be concealed inside the deploy module in nanosatellite by using nylon string to hold the Polytetrafluoroethylene (PTFE) gate cap. To deploy the antenna, the system injected nichrome wire with high current electricity from the battery to heat up nylon string until it breaks off. Once the nylon string is broken, PTFE gate will open, antenna will be deployed, and two limit switches at each antenna will give feedback to OBC to stop nichrome wire heating and indicate successful antenna deployment. Below are procedures needed to launch the nanosatellite via International Space Station (ISS) nanosatellite launcher module shown in figure 3.

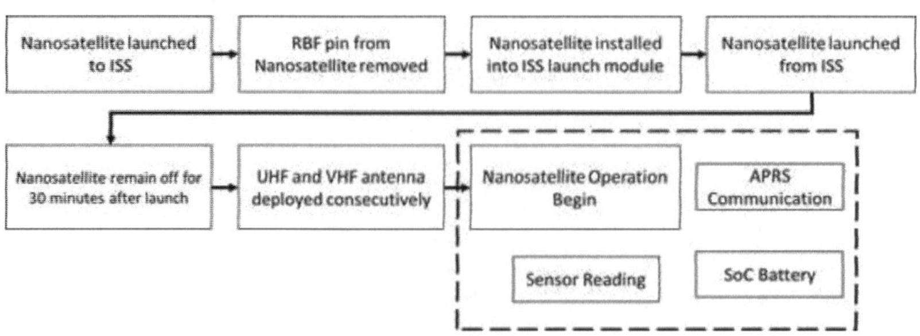

Figure 3. Nanosatellite launching procedures with ISS nanosatellite launcher module.

Produced nanosatellite model will be accepted if it passes vibration test, vacuum test, thermal test, and radiation test. Vibration test will simulate the vibration at the magnitude that satellite will experience during rocket launching. JAXA ICD dictates the magnitude of vibration Quasi-Static Acceleration to be above 8G with additional random vibration test on various frequency depends on launch vehicle used. Vacuum test simulates near 0 Pa depressurization in space environment, with depressurization rate depends on rocket type. Temperature test ranging from -15 °C to 60 °C in Low Earth Orbit environment. Lastly, RF radiation test which limits allowable Electric Field level and allowable Power Density for each VHF-band and UHF-band, including output power limit for exceptional cases. [37]

3. Mission Analysis

The mission of SS-1 will be reviewed through simulation using the AGI STK software [38,39,40,41] [42,43]. Simulated objectives of the CubeSat will be as follows:

1. Able to cover over Indonesian Region at least once a day
2. Able to receive or transmit to a remote station
3. Have at least a year lifetime

The CubeSat simulation parameter will be similar to the simulation of MYSAT nanosatellite, starting from its deployment from ISS to its designated orbit with the nanosatellite launcher module [44]. The simulation scenario was plotted for a year from 1 Jan 2018 05:00:00 UTC until 1 Jan 2019 04:59:59.999 UTC. The modelled nanosatellite can only be tracked 30 minutes after the deployment due to the operating conditions mentioned in the JAXA ICD. Initial conditions of the satellite will be shown in table 1.

Table 1. Initial Conditions of Surya Satellite – 1.

Orbital Elements	Value
Semi-Major Axis (Km)	6775.734821
Eccentricity	0.000406
Inclination (deg)	51.714
Right Ascending of Ascension Node (deg)	127.92
Argument of Perigee (deg)	76.008
True Anomaly (deg)	133.805
Apogee Altitude (Km)	400.34877
Perigee Altitude (Km)	394.84687

SS-1 lifetime was predicted with three atmospheric models used, 1976 Standard, Jacchia-Roberts, and NRLMSISE 2000. The 1976 standard model used an idealised version of atmospheric model, while Jacchia-Roberts model considered some environmental data such as solar irradiation. The CubeSat does not equipped with an altitude control, causing it to be dragged by earth gravitational pull until a certain altitude, and finally plunged into the earth atmosphere as shown in figure 4. The three atmospheric models used in the simulation affects the rate of decay of the satellite. The simulation results show that the developed satellite will decay within 0.7 until 1.8 years, which is common for nanosatellite without altitude control. [44]

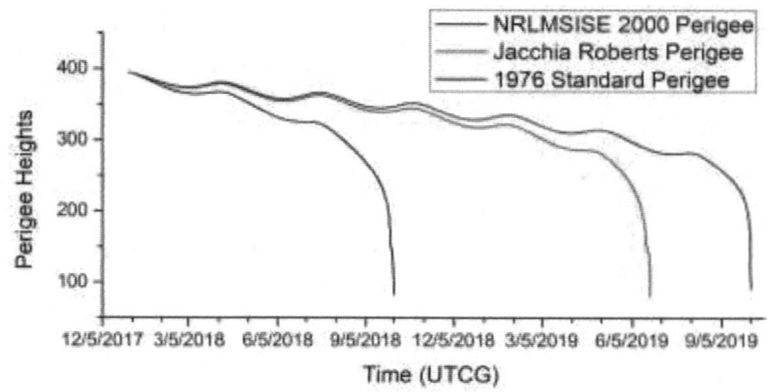

Figure 4. Lifetime Prediction of Surya Satellite - 1 with Different Atmosphere Model (AGI STK Simulation Result).

The CubeSat coverage over Indonesian region was simulated for one-week time range between 2 January 2018 until 9 January 2018 to represent the CubeSat cyclic movement as shown in figure 5. According to the simulation result, SS-1 will pass Indonesia eight times a day, with at least twice has a coverage area more than 70%. Therefore, it is very probable for the satellite to establish a communication with ground stations located in the covered area every day. However, due to the speed of satellite orbit, there would be only a short duration of communications [45,46]. The results of the simulation show that the satellite will obtain maximum coverage area for only 5-6 minutes at a time. Figure 6 shows an illustration of SS-1 radio coverage for more than 50% of Indonesia region.

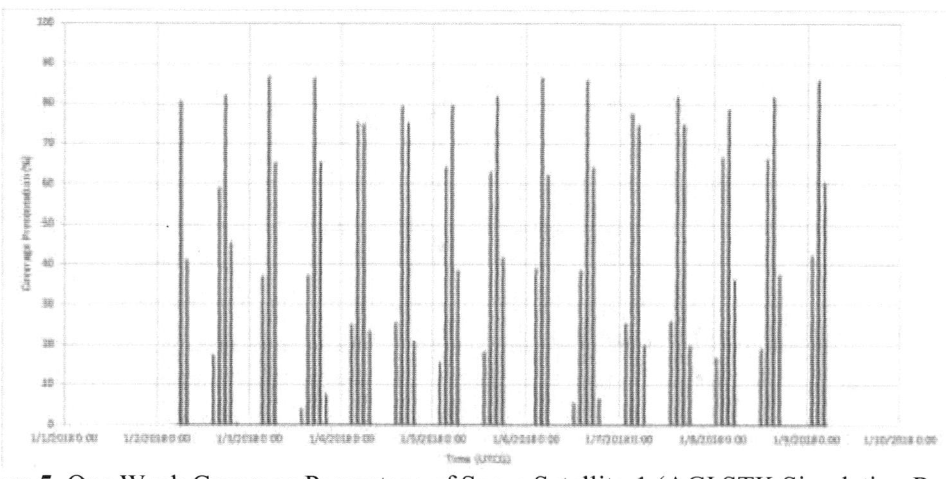

Figure 5. One Week Coverage Percentage of Surya Satellite-1 (AGI STK Simulation Result).

SS-1 ability to establish a communication link to a ground station, during the short time it is within the communication range is determined through the calculation of communication links [47,48,49,50]. The value of received power when communication was made from ground station to the satellite and vice versa could be obtained from equation 1 (Link Budget) and equation 2 (Free space path loss) [45,47,51,52]. Table 2 shows the value of the parameter used and the value of received power when uplink and downlink. Results show that uplink value on UHF was -98.3 dBm respectively, while the designed satellite radio component have a floor threshold of -122 dBm. Downlink power of -107.8 on VHF could be used as the threshold when designing ground station. Therefore, according to the calculation results, it is highly probable for the satellite to establish communication to the ground station.

IOP Conf. Series: Earth and Environmental Science **149** (2018) 012072 doi:10.1088/1755-1315/149/1/012072

Figure 6. Coverage Illustration of Surya Satellite-1 (AGI STK Simulation Result).

Link Budget

$$\text{Received Power} = \text{Power Output} + \text{Transmitter gain} - \text{Transmitter loss} - \text{Free space path loss} - \text{Misc. Loss} + \text{Receiver Gain} - \text{Receiver loss} \quad (1)$$

Free space path loss

$$FSPL = 20 \text{ Log (distance)} + 20 \text{ Log (frequency)} + 32.44 \quad (2)$$

Table 2. Link Budget Parameter and Calculations.

Parameters	Value
Satellite Antenna Gain (dBi)	0
Satellite Transmitted Power (W)	1
Satellite Antenna Feed Loss (dB)	0.58
Ground Station Antenna Gain (dBi)	13
Ground Station Transmitted Power (W)	5
Ground Station Antenna Feed Loss (dB)	1.5
Frequency (MHz)	435.825
Distance (Km)	1000
UHF Uplink Received Power (dBm)	-98.3
VHF Downlink Received Power (dBm)	-107.8

4. Conclusions

Surya Satellite - 1 has been designed based on accepted launcher standard with the necessary subsystems and APRS payload. The result of the simulation shows that nanosatellite launched from the International Space Station able to circulate around the earth 13 times a day with at least two circulations covered

70% of Indonesia region. Additionally, it has the necessary link budget to send and receive data from ground station. This concludes high probability of nanosatellite development for disaster mitigation.

Acknowledgements
The author would like to express gratitude to LAPAN and ORARI that provide technical and legal support in the satellite development project.

References

[1] Convention on the Law of Sea, Montego Bay, 10 December 1982, *United Nations Treaty Series*, vol. 1833, No. 31363, p. 3, available from https://treaties.un.org/doc/Publication/MTDSG/Volume%20II/Chapter%20XXI/XXI-6.en.pdf
[2] Hamilton W 1979 *Tectonics of the Indonesian Region* (Washington: United States Government Printing Office)
[3] Badan Pusat Statistik 2017 *Jumlah Curah Hujan dan Jumlah Hari Hujan di Stasiun Pengamatan BMKG, 2011-2015* (Jakarta: BPS)
[4] Pusdatinmas BNPB 2017 *Info Bencana Desember 2016* (Jakarta: BNPB)
[5] Anubhav T, Sarwesh P, Narayan V, Varma P A, Prasad R A, Loganathan M, Rao D N, Sriram S, and Venkatesh M 2013 *IEEE Aerospace Conf.* (Montana: USA)
[6] Kameche M, Benzeniar H, Benbouzid A B, Amri R and Bouanani N 2014 *J. Aerosp. Technol. Manag.* **6** p 93-100
[7] Kenny J 2012 All-In-One APRS Transmitter (San Luis Obispo: California Polytechnic State University) p 4
[8] Selva D and Krejci D 2012 *Acta Astronautica* **74** p 63
[9] Woellert K, Ehrenfreund P, Ricco A J, and Hertzfeld H 2011 *Advances in Space Research* **47** p 663
[10] Kramer H J and Cracknell A P 2008 *Int. J. of Remote Sensing* **29** p 4285
[11] Sandau R, Brie K, and D'Errico M 2010 Small Satellites for global coverage: Potential and limits *ISPRS J. of Photogrammetry and Remote Sensing* **65** p 492
[12] Fort S 2008 *ARRL's VHF Digital Handbook* (Connecticut: American Radio Relay League)
[13] The APRS Working group 2000 *APRS Protocol Reference: Protocol Version 1.0* (Arizona: Tucson Amateur Packet Radio Corp)
[14] Asri A K, Elyaa H, Duantari N, Suryaningsih E and Victoria L D D D 2016 *Procedia Social and Behavioral Sciences* **227** p 435
[15] Abdelkarim A A Y 2015 *Int. Conf. On Computing, Control, Networking, Electronics, and Embedded Systems Engineering* (Khartoum: Sudan)
[16] Addaim A, Kherras A and Zantou B 2007 *IEEE Aerospace Conf.* p 3-5
[17] Jaffer G and Schwingenschuh K 2011 A LEO Nano-Satellite Mission for the Detection of Lightning VHF Sterics *Adaptive Filtering Applications* ed L Garcia (Intech)
[18] Woellert K, Ehrenfreund P, Ricco A J and Hertzfeld H 2010 *Advances in Space Research* **47** p 665-671
[19] You Z 2017 Micro/Nano Satellite System Technology *Space Microsystems and Micro/Nano Satellites* (Elsevier) p 3-32
[20] Kim S Y, Castet J-F and Saleh J H 2011 Satellite Electrical Power Subsystem: Statistical Analysis of On-Orbit Anomalies and Failures *IEEE Aerospace Conf.* p 1
[21] Poghosyan A and Golkar A 2016 CubeSat evolution: Analyzing CubeSat capabilities for conducting science missions *Progress in Aerospace Sciences*
[22] Zoppi M, Tipaldi M and Cerbo A D 2018 Cross-model verification of the electrical power subsystem in space projects *Measurement*
[23] Chang Y K, Kang S J, Moon B Y, and Lee B H *Proc. 4th Responsive Space Conf.* (Los Angeles: USA)
[24] Gopakumar S, Eega S, Suresh S V S, Reddy M S S, Antony A, Ramachandran H and Koilpillai D 2015 *Proc. Int. Conf. On Space Science and Communication* (Langkawi: Malaysia)

[25] Jensen J D and Swenson C M 2016 Command and Data Handling Subsystem Design for the Ionosphere Observation Nanosatellite Formation *14th Annual AIAA/USU Conf. on Small Satellite*
[26] Thakar L, Dutta C, Sura P S and Udupa S 2016 Design & Realization of Multi Mission Data Handling System for Remote sensing satellite *IEEE India Conf.*
[27] Zhou L and An J 2013 Design of a Payload Data Handling System for Satellites *Third Int. Conf. on Instrumentation, Measurement, Computer, Communication and Control*
[28] Busch S and Schilling K 2013 *19th IFAC Symp. on Automatic Control in Aerospace* (Wurzburg: Germany) p 218
[29] Hanafi A, Karim M, Latachi I, Rachidi T, Dahbi S, and Zougarr *Proc. Int. Conf. on Advanced Technologies* (Fez: Morroco)
[30] Harris K, McGarvey M, Chang H Y, Ryle M, Rusciti T, Udrea B and Nayak M 2013 *27th Annual AIAA/USU Conf. On Small Satellites* (Logan: USA)
[31] Leao T F C, Mooney-Chopin V, Trueman C W and Gleason S 2013 Design and Implementation for a Diplexer and a Dual-Band VHF/UHF Antenna for Nanosatellites *IEEE Antennas and Wireless Propagation Letters* **12**
[32] Humad Y A I, TagElsir A and Daffalla M M 2017 Design and Implementation of Communication Subsystem for ISRASAT1 Cube Satellite *Int. Conf. on Communication, Control, Computing and Electronics Engineering* (Khartoum: Sudan)
[33] Constantine J, Tawk Y, Ernest A and Christodoulou 2011 Deployable Antennas for CubeSat and Space Communications *6th European Conference on Antennas and Propagation*
[34] Keny P, Menon A, Rao M, Gaitonde U and Sriharsha A 2013 Development of Antenna Deployment Circuit for Nano-Satellites *European Conference on Circuit Theory and Design*
[35] Lim L S, Bui T D V, Lau Z, Tissera M S C, Soon J J, Lew J M, Aung H, Ye C, Low K S, Goh S T, and Chen S S 2015 *Proc. Int. Conf. on Space Science and Communication* (Langkawi:Malaysia) p 158
[36] Prodonignrum R T, Wijanto H and Prasetyo A D 2015 Antenna Deployment For Automatic Packet Reporting System of Nanosatellite Using Global Positioning System As A Height Sensor *Int. Conf. on Quality in Research*
[37] Japan Aerospace Exploration Agency 2013 JEM Payload Accommodation Handbook: Small Satellite Deployment Interface Control Document **8** available from http://aerospacebiz.jaxa.jp/wp-content/uploads/2016/07/jem_handbook_eng.pdf
[38] Abbott J 2012 *IEEE Aerospace Conf.* (Montana: USA)
[39] Antonini K, Langer M, Farid A and Walter U 2017 SWEET CubeSat – Water detection and water quality monitoring for the 21st century *Acta Astronautica*
[40] Mahdi M C 2016 *J. of Science and Arts* **34** p 67
[41] Lazreg N, Bahri O B and Besbes K 2017 Analysis and design of CubeSat constellation for the Mediterranean south coastal monitoring against illegal immigration *Advances in Space Research*
[42] Navarathinam N, Lee R and Chesser H 2011 Characterization of Lithium-Polymer batteries for CubeSat applications *Acta Astronautica*
[43] Qiao L, Rizos C and Dempster G 2012 Analysis and Comparison of CubeSat Lifetime *Proc. Of the 12th Australian Space Conf.* (Melbourne: Australia)
[44] Thaheer A M S and Ismail N A 2017 Orbit Design and Lifetime Analysis of MYSat: A 1U CubeSat for Electron-density Measurement *NatGrad Conference*
[45] Popescu O, Harris J S, and Popescu D C 2016 *IEEE Southeastcon*
[46] Cakaj S, Kamo B, Lala A and Rakipi A 2014 *Int. J. of Advanced Computer Science and Applications* **5** p 6
[47] Berrezzoug S, Bendimerad F T and Boudjemai A 2015 Communication Satellite Link Budget Optimization Using Gravitational Search Algorithm *3rd Int. Conf. on Control, Engineering & Information Technology*
[48] Tomaello A, Bonato C, Deppo V D, Naletto G and Villoresi P 2010 Link budget and background noise for satellite quantum key distribution *Advances in Space Research* **47** p 802-10

IOP Conf. Series: Earth and Environmental Science **149** (2018) 012072 doi:10.1088/1755-1315/149/1/012072

[49] Popescu O 2017 Power Budgets for CubeSat Radios to Support Ground Communication and Inter-Satellite Links *IEEE Access* **5**
[50] Vassaki S, Pitsiladi G T, Kourogiorgas C, Poulakis M, Panagopoulos A D, Gardikis G and Costicoglou S 2014 Satellite-Based Sensor Networks: M2M Sensor Communications and Connectivity Analysis *Int. Conf. on Telecommunications and Multimedia*
[51] Dey S and Sahoo B B 2014 *Int. J. of Innovations in Engineering and Technology* **4** p 173 -9
[52] Kundu A K, Khan M T H, Sharmin W, Goni M O and Barket K A 2013 Designing a Mobile Satellite Communication Antenna and Link Budget Optimization *Int. Conf. on Electrical Information and Communication Technology* (Khulna:Bangladesh)

IOP Conf. Series: Earth and Environmental Science **149** (2018) 012073 doi:10.1088/1755-1315/149/1/012073

Sensing of the atmospheric variation using Low Cost GNSS Receiver

Brian Bramanto, Irwan Gumilar, Teguh P. Sidiq, Wedyanto Kuntjoro, Daniel A. Tampubolon
Geodesy Research Group, Institut Teknologi Bandung, LABTEK IX-C, Jalan Ganesha 10 Bandung, Indonesia

E-mail: brian.bramanto@s.itb.ac.id

Abstract. As the GNSS signals transmitted through the atmosphere, they are delayed by interference of TEC (Total Electron Content) in the ionosphere and water vapor in the troposphere. By using inverse-problem, name GNSS Meteorology, those parameters can be obtained precisely and several researches has approved and supported that method. However, the geodetic GNSS receivers are relatively high cost ($30,000 to $70,000 each) to be established on a regular and uniform network. This research aims to investigate the potential use of low cost GNSS receiver (less than $2,000) to observe the atmospheric dynamic both in ionosphere and troposphere. Results indicated that low cost GNSS receiver is a promising tools to sensing the atmospheric dynamic, however, further processing is needed to enhance the data quality. It is found that both of ionosphere and troposphere dynamic has diurnal periodic component.

1. Introduction

Atmosphere consists at least two layers, like ionosphere and troposphere. The ionosphere can be described as the upper layer (20 to 2,000 km) of atmosphere where the thickness of free elections and ions is sufficiently high to interfere the electromagnetic waves propagation [1]. The troposphere can be defined as the lower layer (up to 20 km) of the atmosphere where consists of water vapor and aerosol [2]. Nowadays, atmospheric study is important for the communication and the climate change issues.

Atmospheric can be monitor by using ionosonde (for ionosphere) and Automatic Weather Station (for troposphere), however, in accordance with the emergence and the development of GNSS (Global Navigation Satellite System) technology and methods, it leads to diverse application of GNSS, such as ionosphere and troposphere monitoring. As the GNSS signals transmitted through the atmosphere, the GNSS signals are interfered by the TEC (Total Electron Content) in the ionosphere, while water vapor in the troposphere.

In ionosphere and troposphere monitoring, the geodetic GNSS receiver (high cost $30.000 to $70.000 each) leads to a problem, especially in Indonesia. As seen on Figure 1, at about 1.9 million square kilometers, Indonesia needs a lot of atmosphere monitoring stations. Assumed that there will be at least one monitoring station every 20 kilometers, Indonesia demand at least 4,000 monitoring stations. Recently several researches on low cost GNSS receiver (less than $2,000) had been done by many researchers, such as precise point positioning [3], landslide monitoring [4-5] and UAV applications [6]. Those researches indicate the accuracy and the potential use of the low cost GNSS receiver for a wider applications. Furthermore, as mentioned before, by replacing those geodetic GNSS receivers with low cost GNSS receiver, it lead to a cost efficiency for the site monitoring establishment itself. Therefore, this research aims to investigate the potential use of low cost GNSS receiver to observe the atmospheric dynamic both in ionosphere and troposphere. The data and method are presented in detail in Section 2. In section 3, the ionosphere and troposphere parameter derived from low-cost GNSS will be evaluated.

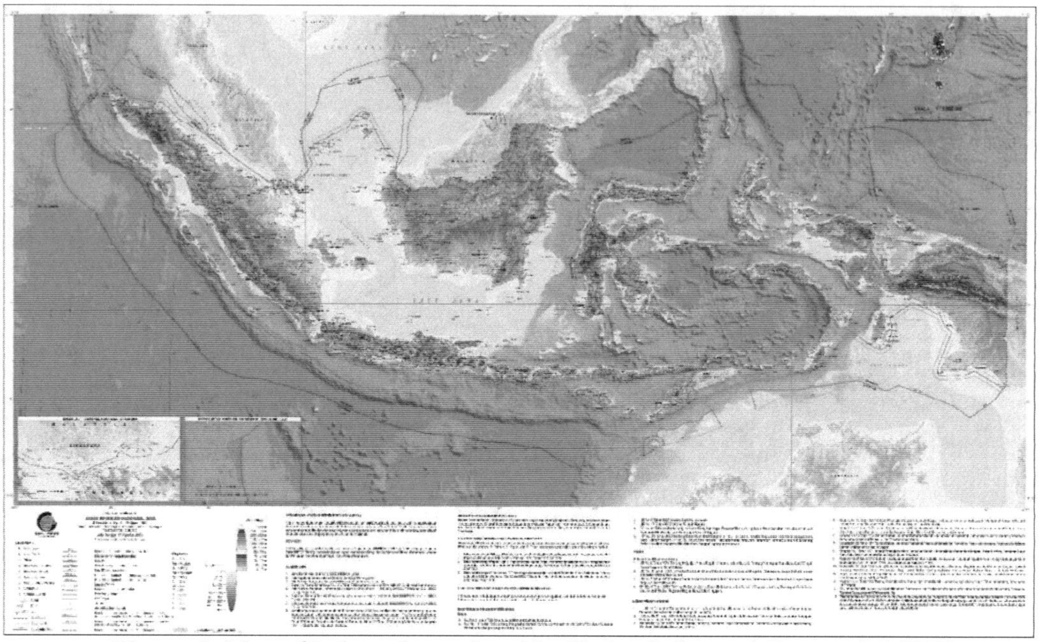

Figure 1. Map of Indonesia produced by Badan Informasi Geospasial (BIG).

2. Data and Method

An experimental observation points was established in Bandung, by using Astech ProFlex 500 as the reference geodetic GNSS receiver and Tersus BX305 as the low cost GNSS receiver (Figure 2). Figure 3 shows the basic flowchart used in this research. This research includes several steps, such as data observation and satellite navigation collection, data quality assessment, ionospheric parameter determination and tropospheric parameter determination.

Figure 2. Astech Proflex 500 (left) and Tersus BX305 (right).

IOP Conf. Series: Earth and Environmental Science **149** (2018) 012073 doi:10.1088/1755-1315/149/1/012073

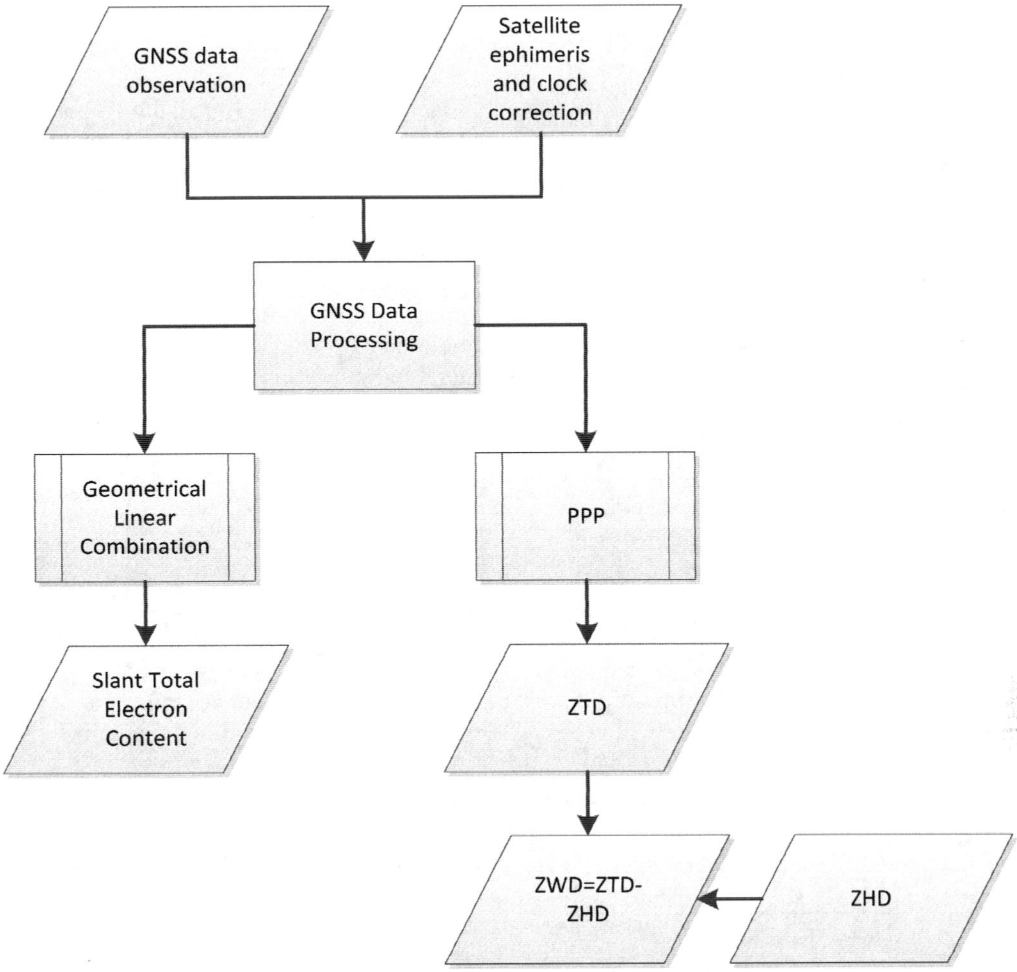

Figure 3. Basic flowchart used in this research.

2.1. Data Quality Assessment Method

GNSS data quality can be determined by evaluating the Signal to Noise (SNR) ratio an the multipath error of GNSS signal propagation. SNR indicates the degree of distorted GNSS signal propagation which is affected by surrounding environment. The high value of SNR (in dbHZ) indicates that noise is significantly low compared with the signal strength and vice versa. In simplified, SNR can be defined as follows:

$$SNR = \frac{Singal\ Strength}{Noise} \tag{1}$$

The multipath of GNSS signal can be investigate by applied the Multipath Combination (MPC) algorithm. The MPC can be expressed as [7]:

$$MP1 = P_1 - \frac{f_1^2+f_2^2}{f_1^2-f_2^2} L_1 + \frac{2f_2^2}{f_1^2-f_2^2} L_2 \tag{2}$$

IOP Conf. Series: Earth and Environmental Science **149** (2018) 012073 doi:10.1088/1755-1315/149/1/012073

$$MP2 = P_2 - \frac{f_2^2}{f_1^2 - f_2^2} L_1 + \frac{f_1^2 + f_2^2}{f_1^2 - f_2^2} L_2 \tag{3}$$

where P and L are the pseudorange and carrie phase range respectively and f is the frequency of the carrier phase. By using these combination, the first order of ionospheric delay and geometric range from satellite and receiver are eliminated, while noise are assumed to be negligible due to the accuracy of the pseudorange observation.

2.2. *GNSS Signal Propagation and Error*

GNSS observation data is consist of carrier phase and pseudorange code data observation. Error and other noises are also contained within the GNSS observation data. Mathematically, GNSS data observation, as well as the error and noise can be defined as follows [8]:

$$P_i = \rho + d\rho + d_{trop} + d_{ion} + c(dt - dT) + MP_i + \vartheta P_1 \tag{4}$$

$$L_i = \rho + d\rho + d_{trop} - d_{ion} + c(dt - dT) + ML_i + \lambda_i N_i + \vartheta L_i \tag{5}$$

where P_i and L_i are pseudorange and carrier phase range on selected frequency (i =1,2), ρ is related to geometrical range from receiver to satellite, $d\rho$ is the orbital error, d_{trop} and d_{ion} are troposphere and ionosphere biases, c is the speed of light (299,729,458 m/s), dt and dT are time error of receiver and satellites, MP_i and ML_i are related to multipath error of pseudorange and carrier phase range, λ_i and N_i are related to wavelength and ambiguity number, while ϑP_i and ϑL_i are related to noise error

Table 1. Error in GNSS signal propagation [9].

Error Source	Magnitude
Satellite orbital error	~ 100 m
Satellite clock error	~ 500 m
Ionospheric bias	~ 50 m
Tropospherc bias	~ 5 m

By understanding the GNSS signal propagation, TEC and water vapor can be estimated. Geometrical linear combination of GNSS signal is used to determine the TEC, while PPP (Precise Point Positioning) method is used to determine the tropospheric delay which lead to water vapor estimation.

2.3. *GNSS Ionospheric Algorithm*

The Slant TEC is defined as the line integral of the electron density from all GNSS-Satellites visible from each of these receiver above a user-specified elevation cut-off angle (usually 15^0). Mathematically, slant TEC can be defined as follows [10]:

$$STEC = \int Ne(r)dl = \frac{[P1 - P2] - [RCB - SCB]}{\frac{Cx}{2}\left(\frac{1}{f1^2} - \frac{1}{f2^2}\right)} \tag{6}$$

where Cx, is the constant 80.62 (m^3/s^2), RCB and SCB are receiver code bias/receiver clock bias and satellite code bias/satellite clock bias respectively.

2.4. *GNSS Tropospheric Algorithm*

Tropospheric parameter can be determined by Precise Point Positioning (PPP) algorithm. PPP uses both of pseudorange and carrier phase GNSS observation data from a single dual-frequency receiver. PPP

IOP Conf. Series: Earth and Environmental Science **149** (2018) 012073 doi:10.1088/1755-1315/149/1/012073

uses ionospheric-free linear combination in order to eliminate the ionosphere biases. Ionospheric-free linear combination can be describe as follows [11]:

$$P_{IF} = \frac{f_1^2 P_1 - f_2^2 P_2}{f_1^2 - f_2^2} = \rho + d\rho + c(dt - dT) + d_{trop} + MP_{IF} + \vartheta P_{IF} \quad (7)$$

$$L_{IF} = \frac{f_1^2 L_1 - f_2^2 L_2}{f_1^2 - f_2^2} = \rho + d\rho + c(dt - dT) + d_{trop} + \frac{cf_1 N_1 - cf_2 N_2}{f_1^2 - f_2^2} + ML_{IF} + \vartheta L_{IF} \quad (8)$$

where MP_{if} and ML_{IF} are multipath error due to linear combination, while ϑP_{IF} and ϑL_{IF} are noise error due to linear combination. To eliminate the satellite orbital error and satellite clock error, precise orbit and clock correction are needed. Multipath error can be reduced by using a proper antenna. As a result, PPP can be simplified as follows:

$$P_{IF} = \frac{f_1^2 P_1 - f_2^2 P_2}{f_1^2 - f_2^2} = \rho + cdt + mf.ztd + \vartheta P_{IF} \quad (9)$$

$$L_{IF} = \frac{f_1^2 L_1 - f_2^2 L_2}{f_1^2 - f_2^2} = \rho + cdt + mf.ztd + \frac{cf_1 N_1 - cf_2 N_2}{f_1^2 - f_2^2} + \vartheta L_{IF} \quad (10)$$

where mf is refer to tropospheric mapping function and ztd is refer to zenith troposphere delay. The PPP estimates four types of parameters, the receiver position (X, Y, Z), the receiver clock (dt), the troposphere delay (ztd) and the ambiguity (N_1 and N_2).

3. Result and Discussion

3.1. Data Quality Assessment

Figure 4 shows the SNR and multipath for both Astech Proflex 500 (geodetic GNSS) receiver and Tersus BX305 (low-cost GNSS) receiver. Both observed the same satellites in the same observation time. It could be seen that SNR for low-cost GNSS receiver is almost the same as the reference geodetic GNSS receiver which vary from 30 dbHZ to 50 dbHZ. On the other hand, Low-cost GNSS receiver gives worse performance in code multipath analysis. Low-cost GNSS receiver code's multipath is vary from -3 meter to 3 meter, while the multipath for geodetic GNSS receiver is vary from -1 meter to 1 meter. This condition leads to a worse code point positioning accuracy in low-cost GNSS receiver rather than geodetic GNSS receiver.

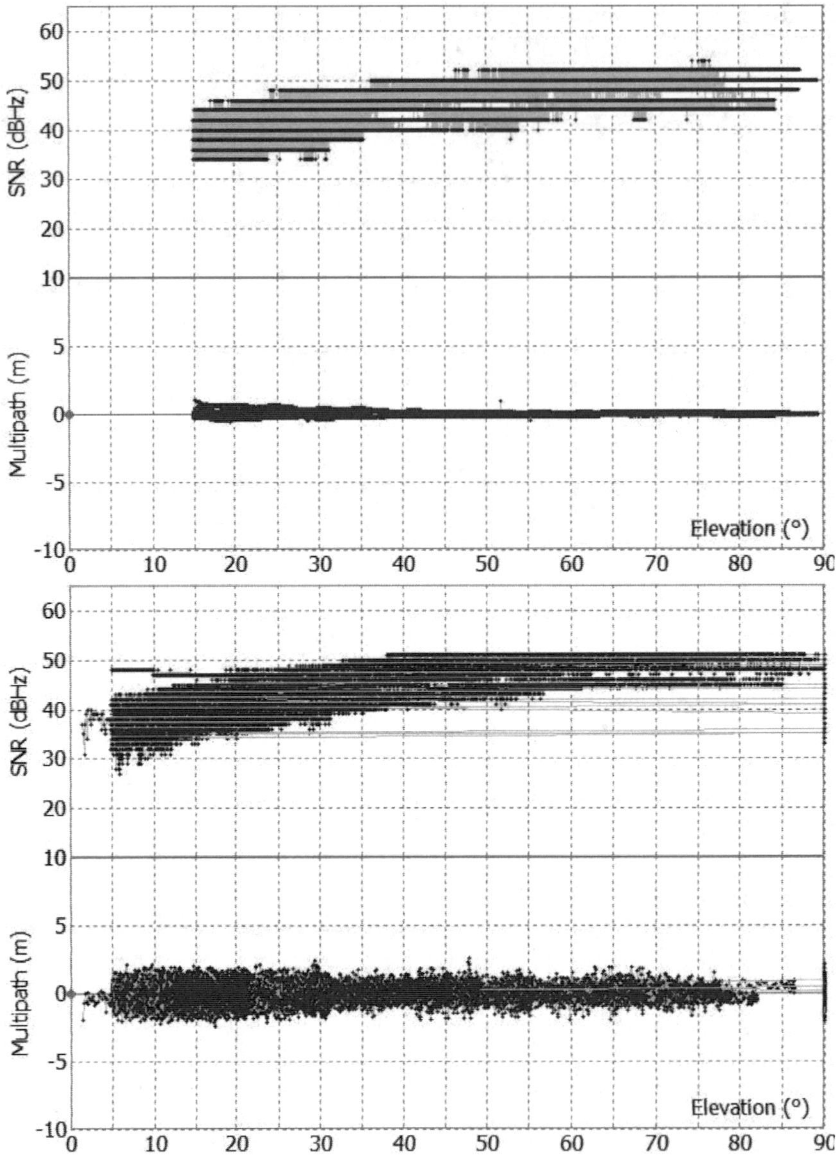

Figure 4. SNR and multipath over elevation for geodetic GNSS receiver (upper) and low-cost GNSS receiver (bellow).

3.2. Ionosphere Parameter Estimation

Figure 5 shows the estimated relative slant TEC from code pseudorange data for both of low-cost GNSS receiver and geodetic GNSS receiver for each observed satellite. Both of them show similar magnitude of slant TEC which is up to 120 TECU, however, the estimated slant TEC from low-cost GNSS receiver showed a significant noise compared with the estimated slant TEC from geodetic GNSS receiver. According to equation (4), possible caused that lead estimated slant TEC has significant noise is there receiver clock bias. Receiver clock bias is related with the effect of oscilator instability. That factor may also be the main factor of the high deviation of code mulipath error [12].

To improve the quality of the data observation, further data processing is needed. Data smoothing is applied in this research. Data smoothing is conducted in several steps [13]:

IOP Conf. Series: Earth and Environmental Science **149** (2018) 012073 doi:10.1088/1755-1315/149/1/012073

- Checking the receiver clock consistency
- Data screening based on Melbourne-Wübbena linear combination
- Data screening based on Geometry-Free linear combination
- Data screening based on Ionosphere-Free linear combination
- Code smoothing

Figure 6 shows the estimated slant TEC from smoothed low-cost GNSS data. It could be seen that the estimated slant TEC has small biases. Slant TEC has maximum magnitude around 3:00-7:00 UTC or 10:00-14:00 at local time and minimum at the 12:00-00:00 UTC or 19:00-7:00 at local time. These diurnal variation is related to the sun occurance [14,15]

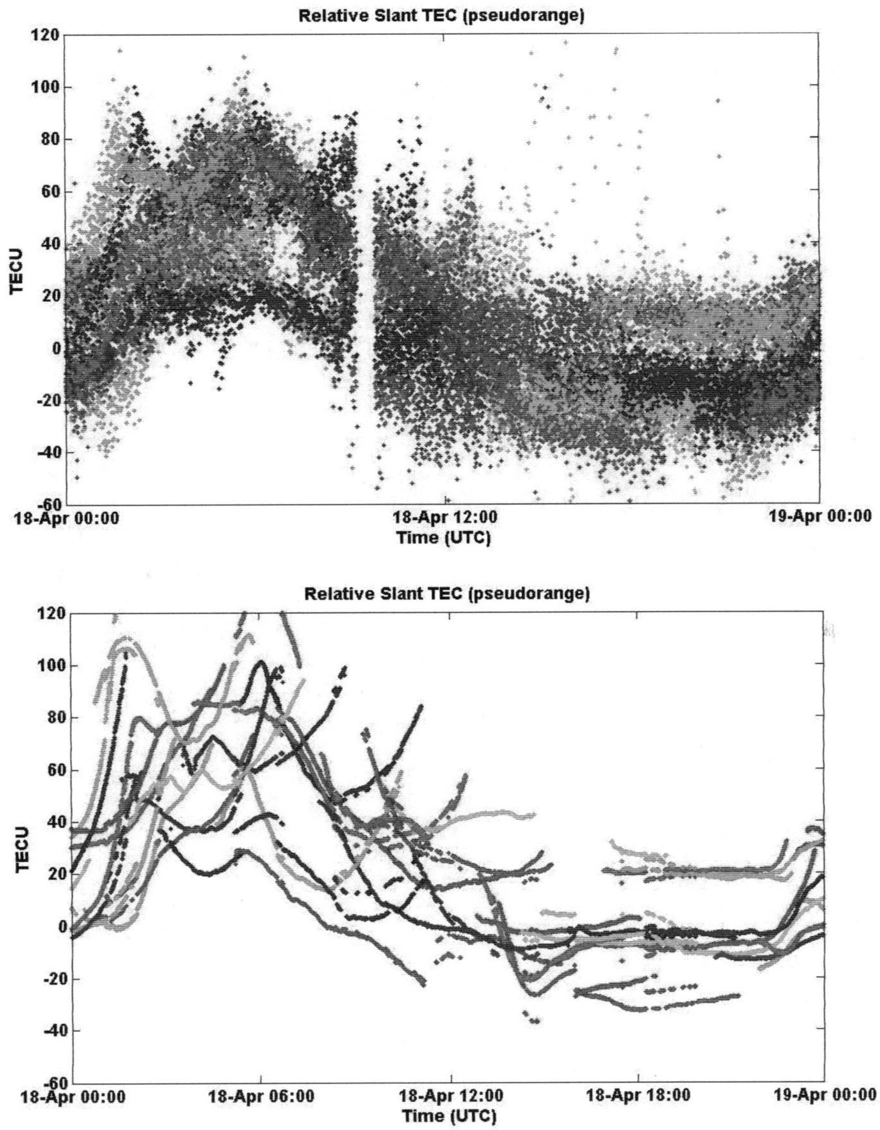

Figure 5. Relative slant TEC that is estimated from low-cost GNSS receiver (upper) and geodetic GNSS receiver (bellow).

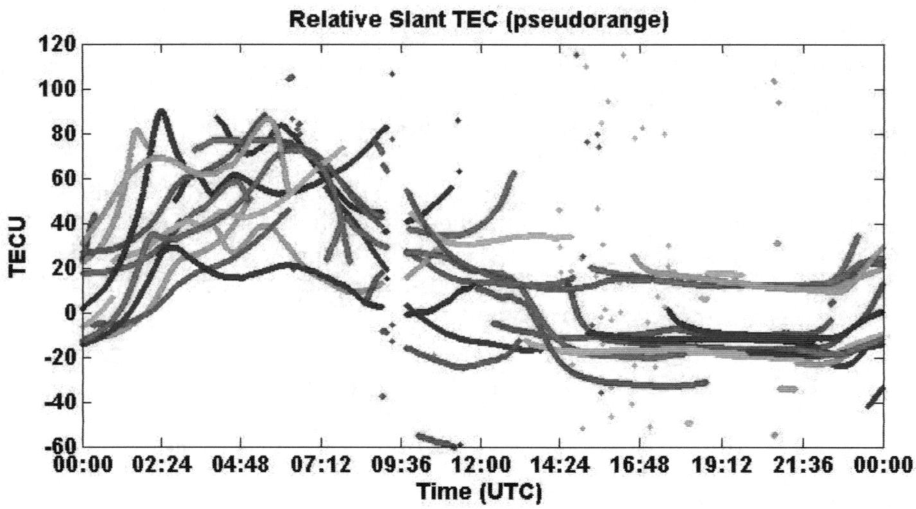

Figure 6. Relative slant TEC estimated from smoothed low-cost GNSS receiver data.

3.3. Troposphere Parameter Estimation

Figure 7 shows the zenith wet delay (ZWD) from three days observation by substracting the zenith tropospheric delay with zenith hidrostatic delay using Saastamoinen model. ZWD is vary from 0.22 meter to 0.34 meter. Similar with TEC, ZWD also indicates diurnal variation. Precitable water vapor (PWV) can be estimated from zenith wet delay as follows:

$$PWV = \Pi \, ZWD \tag{11}$$

where Π is 0.15 as constant. Figure 8 shows the PWV that is vary from 35 mm to 50 mm.

4. Conclusion

It is now possible to accurately observe variations of atmospheric parameters using a single ground-based low cost GNSS receiver, however, further processing is needed to enhance the data quality. As the implications, atmospheric monitoring using dense low-cost GNSS receiver can provide continuous regional/global monitoring of the ionosphere and troposphere parameters with various spatio-temporal scales (even in real-time).

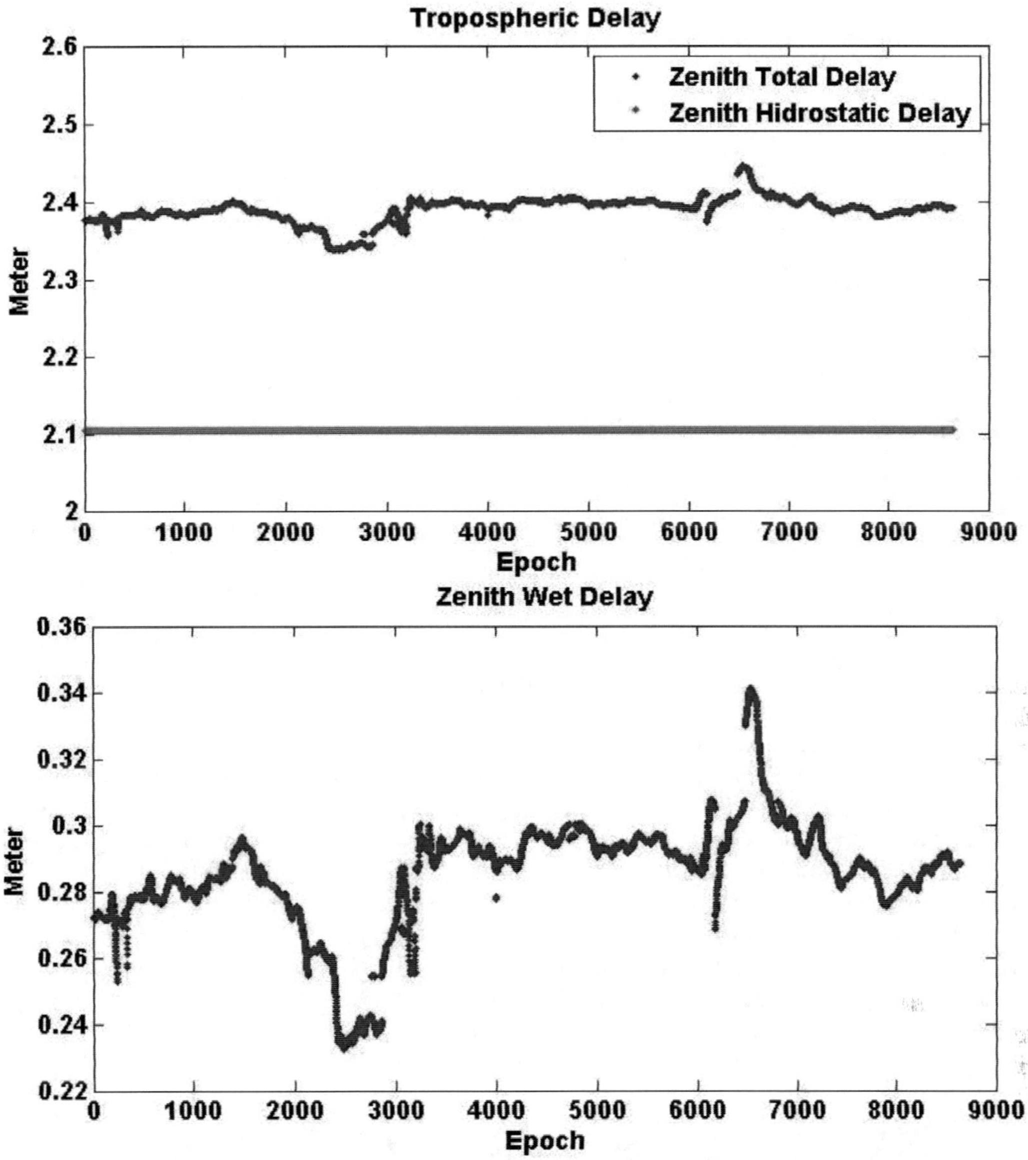

Figure 7. Tropospheric delay (upper) and zenith wet delay (bellow) from 17 April 2017 with 30 second of interval observation.

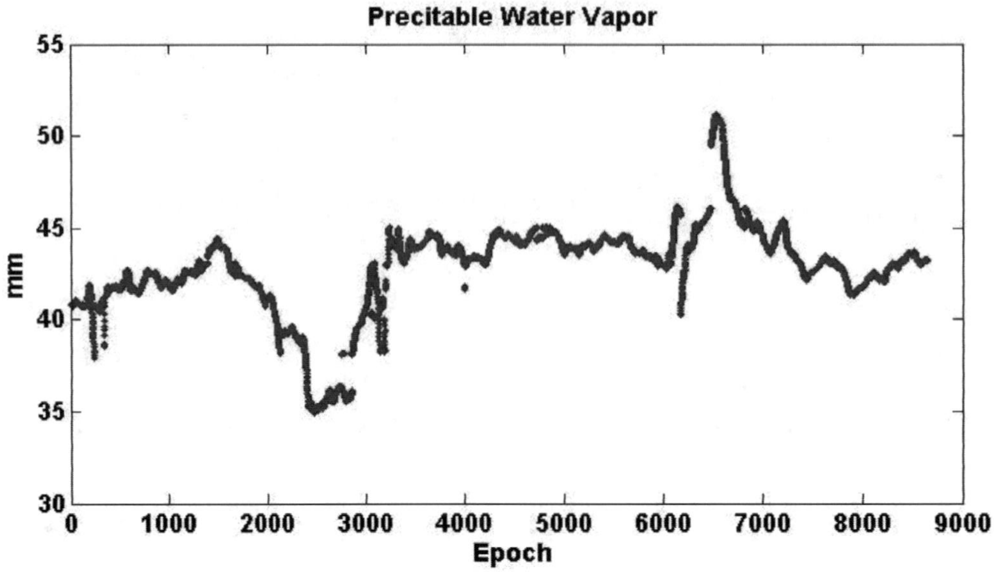

Figure 8. Estimated precitable water vapor from 17 April 2017 with 30 second of interval observation.

References

[1] Zolesi B and Cander L R 2004 *Ionospheric Prediction and Forecasting* (UK: Springer Geophysics)

[2] Bevis M, Businger S, Chiswell S, Herring T, Anthes R, Rocken C, Ware R 1994 GPS meteorology: Mapping zenith wet delays onto precipitable water *Journal of Applied Meteorology* **33** 379–386

[3] Bramanto B, Gumilar I, Sidiq T P, Abidin H Z, Hermawan M D A, Wijayanto B M 2016 On the Performance of a Single-Frequency Low-Cost GPS (Yogyakarta :FIT-ISI and CGISE 2016)

[4] Cina A, Piras M 2015 Performance of low-cost GNSS receiver for landslides monitoring: test and results *Journal of Geomatics Natural Hazards and Risk* **6** 497-514

[5] Biagi L, Grec F C. and Negretti M 2016 Low-cost GNSS receivers for local monitoring: experimental simulation, and analysis of displacement *Sensor* **16** 2140

[6] Stempfhuber W, Buchholz M 2011 (Zurich: *Proceeding of ISPRS volume XXXVIII-1/C22*)

[7] Estey L H, Meertens C M 1999 TEQC: the multi-purpose toolkit for GPS/GLONASS data *GPS Solutions* **3** 42-49

[8] Xu G 2010 *GPS Theory, Algorithms and Application* (Berlin: Springer)

[9] Seeber G 2003 *Satellte geodesy, 2nd edition* (Berlin: de Gruyter)

[10] Cabrera M A, Ezquer R G, and Radicella S M 2005 Predicted and measured slant ionospheric electron content *J. Atmos. Terr. Phys.* **67** 1566–1577

[11] Gao Y and Chen K 2004 Performance Analysis of Precise Point Positioning Using Real-Time Orbit and Clock Products *Journal of Global Positioning Systems* **3** 95-100

[12] Gaggero P O 2008 *Effect of oscillator instability on GNSS signal integration time* Master thesis (Neuchâtel: University of Neuchâtel)

[13] Dach R, Lutz S, Walser P, Fridez P 2015 Bernse GNSS Sofrware Version 5.2 (Bern: Astronomical Institute, University of Bern)

[14] Bramanto B, Rudin O and Wijaya D D 2013 Sensing of the Ionosphere Dynamics over the Indonesian Region using GNSS Observations (Bali: Asian Conference of Remote Sensing 2013)

[15] Pastrav A, Puschita E, Palade T 2017 Ionospheric propagation monitoring and TEC measurements using GPStation-6 *GNSS receiver European Conference on Antennas and Propagation 11th*

Institute of Physics
Dirac House, Temple Back
Bristol BS1 6BE UK

ISSN: 1755-1307
ISBN 978-1-5108-6310-1